POLYMERASE CHAIN REACTION

Edited by **Patricia Hernandez-Rodriguez**
and **Arlen Patricia Ramirez Gomez**

Polymerase Chain Reaction
Edited by Patricia Hernandez-Rodriguez and Arlen Patricia Ramirez Gomez

Published by InTech
Janeza Trdine 9, 51000 Rijeka, Croatia

Publishing Process Manager Bojan Rafaj
Technical Editor Teodora Smiljanic
Cover Designer InTech Design Team

Additional hard copies can be obtained from orders@intechopen.com

Polymerase Chain Reaction,
Edited by Patricia Hernandez-Rodriguez and Arlen Patricia Ramirez Gomez
p. cm.
ISBN 978-953-51-0612-8

For EU product safety concerns:
IN TECH d.o.o., Prolaz Marije Krucifikse Kozulić 3, 51000 Rijeka, Croatia,
info@intechopen.com or visit our website at intechopen.com.

Meet the editors

Dr Patricia Hernández-Rodríguez is the Director of Research Group BIOMIGEN (Molecular Biology and Immunogenetics) in the Biology Program, Department of Basic Science; she is also a member of Animal Medicine and Reproduction Research Center (CIMRA) of the Faculty of Agricultural Sciences, Universidad de La Salle (Colombia). She holds a Biology degree, a Specialization in Epidemiology, a Master of Science in Molecular Biology and she is a PhD student. She has served as a professor to undergraduate and graduate students. He has lectured at scientific events in Colombia and other countries such as Argentina, Cuba, Spain, Mexico and as guest lecturer at the National University of Asuncion in Paraguay. Fields of Research: Infectious Diseases, Genetics and Molecular Biology.

Dr Patricia Gómez is a Veterinarian and holds a PhD from the Universidad Nacional de Colombia. She is an associate lecturer in the Faculty of Agricultural Sciences, Universidad de La Salle and a member of Animal Medicine and Reproduction Research Center (CIMRA) of Universidad de La Salle. Her professional focus is on the Avian Medicine and Molecular Biology. She has a teaching experience on Histology, Physiology and Avian Medicine as well as wide training in research and laboratory.

Contents

Preface

This book is intended to present current concepts in molecular biology with the emphasis on the application to animal, plant and human pathology, in various aspects such as etiology, diagnosis, prognosis, treatment and prevention of diseases as well as the use of these methodologies in understanding the pathophysiology of various diseases that affect living beings.

It is known today that molecular biology has revolutionized the study and understanding of health and disease. Significant developments occurred after 1953, based on the impact generated in many disciplines, especially those life-related such as medicine. Furthermore, the advances in molecular biology have revolutionized industry, agriculture, pharmacology, and animal and plant production, among others. Technology based on Molecular Chain reaction Polymerase (PCR) is advancing rapidly since it is fundamental for improving the health of all living beings.

Importantly, the most of the research in biology and medicine requires a series of molecular strategies that allow the generation of new knowledge, in order to enable better understanding of the mechanisms of life and the cellular changes that affect all living things. Molecular biology has transformed the way we see and understand the physiological and pathological changes of cells, organs and systems. In this sense, this book presents the fundamentals, applications, advantages and disadvantages of various molecular techniques from the research process in biology, medicine, agriculture and environment in basic and applied science. Each chapter explains molecular techniques through various experiments offering new knowledge in different disciplines with applications trying to ultimately improve the conditions of life.

The book includes the participation of different authors and co-authors of various nationalities, all of them experts in the field. The book will be useful to professionals, students, teachers and researchers interested in expanding their knowledge in molecular biology, one of the most exciting areas of work today.

I am grateful for the possibility of editing this book and sending a message to all readers: perform with passion, responsibility and dedication your projects in life; in my case – it is the research.

Patricia Hernandez-Rodriguez
Universidad De La Salle, Bogota,
Colombia

1

Application of PCR in Diagnosis of Peste des Petits Ruminants Virus (PPRV)

Muhammad Abubakar*, Farida Mehmood,
Aeman Jeelani and Muhammad Javed Arshed
National Veterinary Laboratory (NVL), Park Road, Islamabad
Pakistan

1. Introduction

a. Global perspective of PPRV

A Peste des petits ruminant (PPR) is a viral disease of sheep, goats and wild ruminants. It is acute disease which is endemic in many countries of Africa, Arabian Peninsula, Middle east and India. [7, 12, 13]

It was first reported in Côte d'Ivoire in West Africa [14] and was named as Kata, psuedorinderpest, pneumoenteritits complex and stomatitis-pneumenteritis syndrome [15] . Then in 1972 a sort of disease in goats in Sudan was identified to be PPR [16] . In recent years either the presence of antibodies to the virus or viral nucleic acid has been confirmed from the countries like Burkina Faso (2008), Ghana (2010), Nigeria (2007) and Senegal (2010) [17] .

Recently detection of PPRV in East Africa countries is shown by the detection of Antibodies in Kenya (1999 and 2009) and Uganda (2005 and 2007) [18]. It has also been detected in North Africa (Egypt) in 1987 and 1990.

In Saudi Arabia, an outbreak of PPRV has been reported in April, 2002 in Sheeps and Goats [1]. In Pakistan PPRV has been reported since 1991 which was confirmed by PCR in 1994. [19] In India the was first reported in 1987 [11]. In Iran the disease was reported in 1995 [20] while in Iraq it was first detected in 2000 [21] .

b. Disease picture of PPRV

Peste des petits ruminants (PPR) represents one of the most economically important animal diseases in areas that rely on small ruminants. Outbreaks tend to be associated with contact of immuno-naïve animals with animals from endemic areas. In addition to occurring in extensive-migratory populations, PPR can occur in village and urban settings though the number of animals is usually too small to maintain the virus in these situations.

- Morbidity rate in susceptible populations can reach 90–100%
- Mortality rates vary among susceptible animals but can reach 50–100% in more severe instances

---

* Corresponding Author

- Both morbidity and mortality rates are lower in endemic areas and in adult animals when compared to young ones.

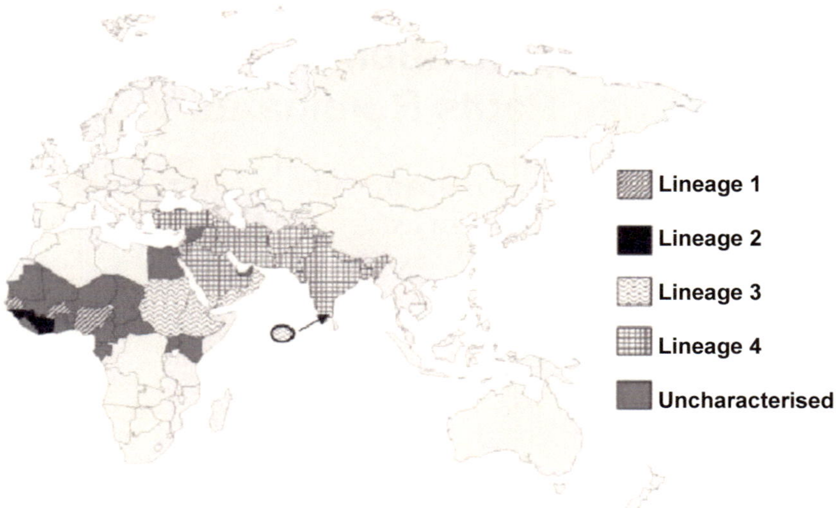

Fig. 1. Geographic distribution of PPRV lineages (Dhar *et al.*, 2002)

c. Hosts Range

- Goats (predominantly) and sheep
    - Breed-linked predisposition in goats

Fig. 2. Clinical Picture and Severity of the Disease

Wildlife host range not fully understood

- documented disease in captive wild ungulates: Dorcas gazelle (Gazelle dorcas), Thomson's gazelles (Gazella thomsoni), Nubian ibex (Capra ibex nubiana), Laristan sheep (Ovis gmelini laristanica) and gemsbok (Oryx gazella)
- Experimentally the American white-tailed deer (Odocoileus virginianus) is fully susceptible

- Cattle and pigs develop in-apparent infections and do not transmit disease
- May be associated with limited disease events in camels

2. Molecular epidemiology of PPRV

A close contact between the infected animals which is in the febrile stage and susceptible animals is a source of transmission of the disease [15]. During sneezing and coughing the virus spread from animal to animal [22]. Indirect transmission seems to be unlikely in view of the low resistance of the virus in the environment and its sensitivity to lipid solvent. [4]

Epidemiology pattern vary from area to area, for example in the humid Guinean zone where PPR occurs in an epizootic form can cause mortality between 50-80% while in arid and semi-arid regions, PPR is seldomly fatal but usually occurs as a subclinical or inapparent infection opening the door for other infections such as Pasteurellosis [4] . In Saudi Arabia a high morbidity of 90% was reported, [2] 3-8 months animal are more susceptible to disease than either of adults or unweaned animals [23].

a. Genome Organization of PPRV:

PPRV belong to *Morbillivirus* genus. For a long time it was thought to be a variant of RP that was adapted to sheeps and goats and had lost its virulence for cattles. [3] The causative agent of PPR is RNA virus which is single strand and non-segmented. It belongs to the family *Paramyxoviridae* and *genus Morbillivirus* which also includes measles virus, rinderpest virus (RPV), canine-distemper virus, phocinedistemper virus, and dolphin and porpoise morbilliviruses[24]. All the viruses belonging to the genus morbilli are serologically related. Phylogenetic analysis also shows that there is high degree of homology.

The genome contains six tandemly arranged transcription units which encodes six structural proteins i.e the surface glycoproteins F and H, the nucleocapsid (N), the matrix (M), the polymerase or large (L) and the polymerase-associated (P) proteins. The cistron directing the synthesis of this later protein is encoding the virus non-structural proteins C and V by the use of two other open reading frames (ORF) of the messengers. The gene order is 3'N-P-M-F-H-L5', as determined by transcriptional mapping. [25] The genome is flanked by extragenic sequences at the 3' ((52 nucleotides, leader) and 5' ends (37 nucleotides, trailer).

For viruses of the family *Paramyxoviridae*, the genome promoter (GP) contains 107 nucleotides comprising the leader sequence and the adjacent non-coding region of the N gene at the 3' end of the negative-strand. While antigenome promoter (AGP) contain 109 nucleotides that encompass the trailer sequence and the proximal untranslated region of the L gene. Both GP and the AGP contains the polymerase binding sites and the RNA encapsidation signals for the replication of the full genome while the production of messengers (m-RNA) is a function of the GP [26]. So GP and AGP have an impact on the virulence of virus.

Genes and promoters of *Morbillivirus;* the protein coding regions (N, P, V, C, M, F, H, and L), noncoding intergenic regions and the leader and trailer regions along with the specialized sequence motifs are shown. The genome promoter includes the leader sequence and the non coding regions N at the 3' end of the genomic RNA. The antigenome promoter includes the trailer sequence and the untranslated regions of the L gene at 5' end. Gene start (GS) and gene end (GE), enclosing the intergenic trinucleotide motifs are also shown.

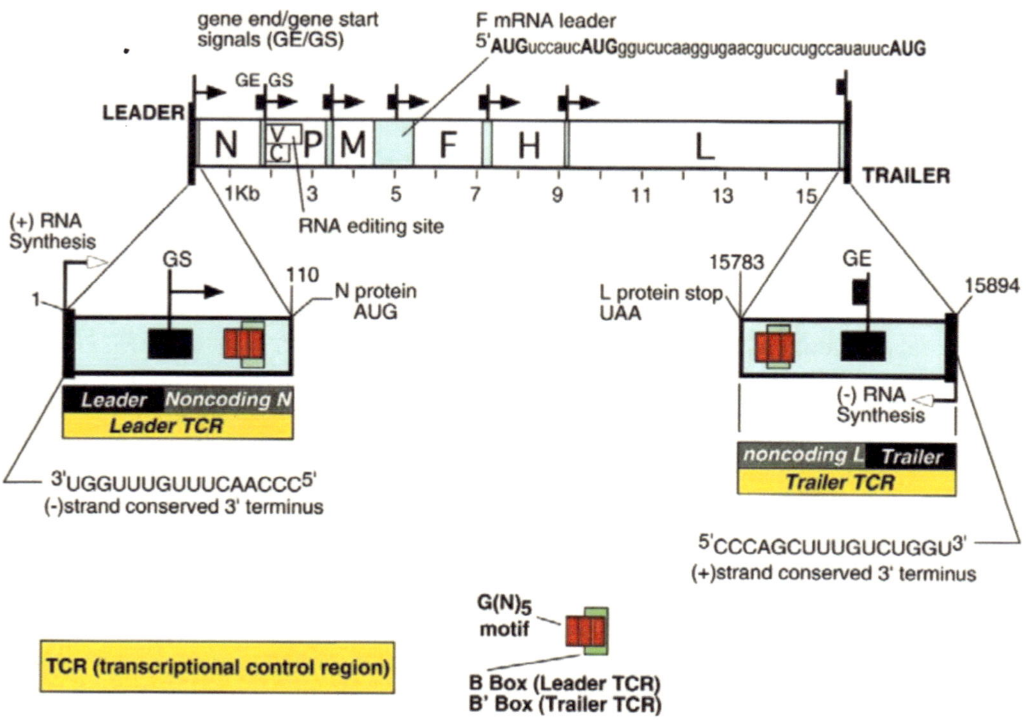

Fig. 3. Genome of PPR virus

b. Antigenic and Immunogenic Epitopes:

Surface glycoproteins hemagglutinin (H) and fusion protein (F) of morbilliviruses are highly immmunogenic and helps in providing the immunity. PPRV is closely related to rinderpest virus (RPV). Antibodies against PPRV are both cross neutralizing and Cross protective. A vaccinia virus double recombinant expressing H and F glycoproteins of RPV has been shown to protect goats against PPR disease though the animals developed virus-neutralizing antibodies only against the RPV and not against PPRV. Capripox recombinants expressing the H protein or the F protein of RPV or the F protein of PPRV conferred protection against PPR disease in goats, but without production of PPRV-neutralizing antibodies[27] or PPRV antibodies detectable by ELISA (Berhe *et al*, 2003). These results suggested that cell-mediated immune responses could play a crucial role in protection. Goats immunized with a recombinant baculovirus expressing the H glycoprotein generated both humoral and cell-mediated immune responses.[28] The responses generated against PPRV-H protein in the experimental goats are also RPV crossreactive suggesting that the H protein presented by the baculovirus recombinant 'resembles' the native protein present on PPRV.[28]

Lymphoproliferative responses were demonstrated in these animals against PPRV-H and RPV-H antigens [28]. N-terminal T cell determinant and a C-terminal domain harboring potential T cell determinant(s) in goats were mapped. Though the sub-set of T cells (CD4+ and CD8+ T cells) in PBMC that responded to the recombinant protein fragments and the synthetic peptide could not be determined, this could potentially be a CD4+ helper T cell epitope, which has been shown to harbor an immunodominant H restricted epitope in mice[28] .Identification of B- and T-cell epitopes on the protective antigens of PPRV would open up avenues to design novel epitope based vaccines against PPR.

Sheep and goats are unlikely to be infected more than once in their economic life[12]. Lambs or kids receiving colostrum from previously exposed or vaccinated with RP tissue culture vaccine were found to acquire a high level of maternal antibodies that persist for 3-4 months. The maternal antibodies were detectable up to 4 months using virus neutralization test compared to 3 month with competitive ELISA[29]. Measles vaccine did not protect against PPR, but a degree of cross protection existed between PPR and canine distemper. [30]

3. Specimen collection, processing and shipment

Before collecting or sending any samples from animals with a suspected foreign animal disease, the proper authorities should be contacted. Samples should only be sent under secure conditions and to authorized laboratories to prevent the spread of the disease. In live animals, swabs of ocular and nasal discharges, and debris from oral lesions should be collected; a spatula can be rubbed across the gum and inside the lips to collect samples from oral lesions. Whole, unclotted blood (in heparin or EDTA) should be taken for virus isolation and PCR. Biopsy samples of lymph nodes or spleen may also be useful. Samples for virus isolation should be collected during the acute stage of the disease, when clinical signs are present; whenever possible, these samples should be taken from animals with high fever and before the onset of diarrhea. At necropsy, samples can be collected from lymph nodes (particularly the mesenteric and mediastinal nodes), lungs, spleen, tonsils and affected sections of the intestinal tract (e.g. ileum and large intestine). These samples should be taken from euthanized or freshly dead animals. Samples for virus isolation should be transported chilled on ice. Similar samples should be collected in formalin for histopathology. Whenever possible, paired sera should be taken rather than single samples. However, in countries that are PPR-free, a single serum sample (taken at least a week after the onset of clinical signs) may be diagnostic.

4. Laboratory diagnosis of PPR

a. Conventional Methods of PPRV Diagnosis

Conventional techniques such as the Agar Gel Immuno Diffusion (AGID) test are not routinely used for standard diagnosis as they lack sensitivity when compared to other assays. However, Haemagglutination tests (HA) and Haemagglutination Inhibition tests (HI) tests can be used for routine screening purposes in control programmes as they display comparative sensitivity alongside being simple to perform and cheap to produce.

Virus isolation in cell culture can be attempted with several different cell lines where samples permit. Although Vero cells have been the choice for isolation and propagation of PPRV, it is reported that B95a, an adherent cell line derived from Epstein-Barr virus-transformed marmoset B-lymphoblastoid cells, is more sensitive and support better growth of PPRV lineage IV as compared to Vero cells. More recently, Vero cells expressing the SLAM receptor have been used as an effective alternative for isolation in cell culture. The fragility of morbillivirus virions generally renders techniques such as virus isolation redundant for routine diagnostic use, especially where sample quality is poor. Such techniques are also considered to be time-consuming and cumbersome. Virus isolation does, however, play an important role from a research perspective.

ELISA tests using monoclonal antibodies are often used for serological diagnosis and antigen detection for diagnostic and screening purposes. For PPR antibodies detection, the competitive ELISA is the most suitable choice as it is sensitive, specific, reliable, and has a high diagnostic specificity (99.8%) and sensitivity (90.5%). Immunocapture ELISA (ICE) is a rapid, sensitive and virus specific test for PPRV antigen detection and it can differentiate between RPV and PPRV and has been reported to be more sensitive than the AGID test.

For rapid diagnosis to enable a swift implementation of control measures, further development and validation of pen-side tests such as the chromatographic strip test and the dot ELISA that can be performed without the need for equipments or technical expertise are highly desirable.

Sr #	**Test Name**	**Acronym**	**Application (Lab or Field)**	**Feature Detected (Antigen or Antibody)**
1	Agar gel immuno-diffusion	AGID	Both	Both
2	Counter Immuno-electrophoresis	CIEP	Both	Both
3	Dot enzyme immunoassay	--	Lab	Antigen
4	Differential immuno-histo-chemical staining of tissue sections	IH staining	Lab	Antigen
5	Haemagglutination and Haemagglutination inhibition tests	HA and HI	Both	Both
6	Immuno-filtration	IF	Lab	Antigen
7	Latex agglutination tests	LA	Field	Antigen
8	Virus isolation	VI	Lab	Antigen
9	Competitive enzyme-linked Immuno-sorbent assay (c-ELISA)	cELISA	Lab	Antibody
10	Novel sandwich ELISA	sELISA	Lab	Antigen
11	Immuno-capture enzyme-linked immunosorbent assay	Ic-ELISA	Lab	Antigen

Table 1. Detail of conventional methods for the detection and confirmation of PPR

b. Molecular Methods for PPRV Diagnosis

Molecular techniques such as reverse transcription polymerase chain reaction (RT- PCR) and nucleic acid hybridization are generally used. These genome based techniques are largely used because of their high specificity and sensitivity. However, modern one step real-time RT-PCR assays specific for PPRV and loop-mediated isothermal amplification techniques are more sensitive techniques for PPRV detection but do not allow genetic typing of positive samples. RT-PCR coupled with ELISA have also been used to increase the analytical sensitivity of visualization of RT-PCR products and to overcome the drawbacks of electrophoresis-based detection such as use of ethidium bromide, exposure to UV light etc. The assay is reported to detect viral RNA in infected tissue culture fluid with a virus titre as low as 0.01 TCID50/100 μL and has been reported as being 100 and 10,000 times more sensitive than the sandwich ELISA and RT-PCR, respectively. [31]

5. Potential and application of PCR technique for future advances in diagnosis of PPR

Among the various techniques developed for the detection of PPRV, PCR technique has been the most popular and highly sensitive tool so far for diagnosis of PPR. The routine serological techniques and virus isolation are normally used to diagnose morbillivirus infection in samples submitted for laboratory diagnosis. However, such techniques are not suitable for use on decomposed tissue samples, the polymerase chain reaction (PCR), has proved invaluable for analysis of such poorly preserved field samples. The PCR test consists of repetitive cycles of DNA denaturation, primer annealing and extension by a DNA polymerase effectively doubling the target with each cycle leading, theoretically, to an exponential rise in DNA product. There placement of the polymerase now fragment by thermo-stable polymerase derived from Thermus aquaticus (Taq) has greatly improved the usefulness of PCR. These qualities have made the PCR one of the essential techniques in molecular biology today and it is starting to have a wide use in laboratory disease diagnosis. Since the genome of all Morbilliviruses consists of a single strand of RNA, it must be first copied into DNA, using reverse transcriptase, in a two-step reaction known as reverse transcription polymerase chain reaction (RT-PCR).Among the various techniques developed for the detection of PPRV, however, polymerase chain reaction (PCR) technique developed using F-gene primers has been the most popular tool so far, for diagnosis as well as molecular epidemiological studies. RT-PCR using phosphoprotein (P) universal primer and fusion (F) protein gene specific primer sets to detect and differentiate between PPR and RP are described by [8, 24, 32] developed a RT-PCR test, using phosphoprotein (P) gene and fusion protein(F) gene specific primer sets to detect and differentiate RPV and PPRV. They observed that RT-PCR was able to detect virus secretion in ocular swabs at four days post infection (PI) in experimentally infected goats, as compared to eight days PI by IcELISA. RT-PCR assay preclude the need for virus isolation and, because of the rapidity with which completely specific results could be obtained, the assay appeared to be the test of choice for PPRV detection. Relative specificity and sensitivity of F-gene based RT-PCR with sandwich-ELISA was 100 and 12.5 percent, respectively [31] .

6. Conclusion

The conventional techniques are largely replaced by genome-based detection techniques for the diagnosis and confirmation of PPR virus. Molecular-biological techniques such as RT-PCR and nucleic acid hybridization are now in use. These genome based techniques are largely used because of their high specificity and sensitivity. However one step real-time RT-PCR assays specific for PPRV and loop-mediated isothermal amplification techniques are more sensitive techniques for PPRV detection.

7. References

[1] Housawi, F., Abu Elzein, E., Mohamed, G., Gameel, A., Al-Afaleq, A., Hagazi, A. & Al-Bishr,B. (2004). Emergence of peste des petits ruminants virus in sheep and goats in Eastern Saudi Arabia. *Rev Elev Med Vet Pays Trop* 57, 31–34. http://osp.mans.edu.eg/elsawalhy/Inf-Dis/PPR.htm

[2] Abu Elzein, E.M.E., Hassanien, M.M., Alfaleg, A.I.A, Abd Elhadi, M.A., Housawi, F.M.T. (1990) Isolation of PPR virus from goats in Saudi Arabia. Vet. Rec., 127: 309-310.

[3] Laurent, A. (1968) Aspects biologiques de la multiplication du virus de la peste des petits ruminants sur les cultures cellulaire. Rev. Elev. Méd. Vét. Pays trop. 21: 297-308.

[4] Lefèvre, P.C. and Diallo, A. (1990) Peste des petites ruminants. Revue Scientifique Office of rinderpest virus. Vet. Microbiol. 41: 151-163.

[5] Radostits OM, CC Gay, DC Blood and KW Hinchcliff, 2000. Veterinary Medicine. 9th Ed, WB Saunders Company Ltd, London, UK, pp: 563-565.

[6] Dhar P, BP Sreenivasa, T Barrett, M Corteyn, RP Singh and SK Bandyopadhyay, 2002. Recent epidemiology of peste des petits ruminants virus (PPRV). Vet Microbiol, 88: 153-15

[7] Shaila, M.S., Shamaki, D., Morag, A.F., Diallo, A., Goatley, L., Kitching, R.P. and Barrett, T. (1996). Geographic distribution and epidemiology of peste des petits ruminants viruses Virus Res. 43: 149-153.

[8] Forsyth, M.A. and T. Barrett, 1995. Evaluation of polymerase chain reaction for the detection and characterization of rinderpest and peste des petits ruminants viruses for epidemiological studies. Virus Res. 39: 151–63

[9] Farooq, U., Q.M. khan and T. Barrett. (2008). Molecular Based Diagnosis of Rinderpest and Peste Des Petits Ruminants Virus in Pakistan.international journal of agriculture & biology. 10 (1): 93-96

[10] Albayrak, H and F.Alkan. (2009). PPR virus infection on sheep in blacksea region of Turkey: Epidemiology and diagnosis by RT-PCR and virus isolation. Veterinary research communications. 33 (3) 241-249.

[11] Shaila, M.S., V. Purushothaman, D. Bhavasar, K. Venugopal and R.A. Venkatesan, 1989. Peste des petits ruminants of sheep in India. Vet. Rec., 125: 602

[12] Taylor, W.P. (1984). The distribution and epidemiology of peste des petits ruminants. Prey. Vet. Med., 2: 157-166.

[13] Wamwayi, H.M, M. Fleming, T. Barrett. (1995). Characterisation of African isolates of rinderpest virus. Veterinary Microbiology 44 (2–4): 151–163.

[14] Gargadennec, L. and A. Lalanne, 1942. La peste des petits ruminants. Bulletin des Services Zoo Techniques et des Epizooties de I'Afrique Occidentale Francaise, 5: 16–21

[15] Braide, V.B. (1981) Peste des petits ruminantss. World anim. Review.39: 25-28.

[16] Diallo, A., Barrett, T., Barbron, M., Subbarao, S.M., Taylor, W.P., 1988. Differentiation of rinderpest and peste des petits ruminants viruses using specific cDNA clones. J. Virol. Methods 23, 127–136.

[17] El-Yuguda, A., Chabiri, L., Adamu, F. & Baba, S. (2010). Peste des petits ruminants virus Experimental PPR (goat plague) in Goats and sheep. Canadian J. Vet. Res. 52, 46-52.

[18] Saeed, I. K., Ali, Y. H., Khalafalla, A. I. & Rahman-Mahasin, E. A. (2010). Current situation of peste des petits ruminants (PPR) in the Sudan. *Trop Anim Health Prod* 42, 89–93.

[19] Amjad, H., Qamar ul, I., Forsyth, M., Barrett, T. & Rossiter, P. B. (1996). Peste des petits

[20] Bazarghani, T. T., Charkhkar, S., Doroudi, J. & Bani Hassan, E. (2006). A review on peste des petits ruminants (PPR) with special reference to PPR in Iran. *J Vet Med B Infect Dis Vet Public Health* 53 (Suppl. 1), 17–18. Medline

[21] Barhoom, S., Hassan, W. & Mohammed, T. (2000). Peste des petits ruminants in sheep in Iraq. *Iraqi J Vet Sci* 13, 381–385.

[22] Housawi, F., Abu Elzein, E., Mohamed, G., Gameel, A., Al-Afaleq, A., Hagazi, A. & Al-Bishr,B. (2004). Emergence of peste des petits ruminants virus in sheep and goats in Eastern Saudi Arabia. *Rev Elev Med Vet Pays Trop* 57, 31–34. http://osp.mans.edu.eg/elsawalhy/Inf-Dis/PPR.htm

[23] Taylor, W. P. Abusaidy, S., Barret, T. (1990) The epidemiology of PPR in the sultanate of Oman. Vet. Micro. 22: 341-352.Taylor, W.P. (1979a) Protection of goats against PPR with attenuated RP virus. Res. Vet. Sci. 27: 321-324.

[24] Barrett, T., C. Amarel-Doel, R.P. Kitching and A. Gusev, 1993a. Use of the polymerase chain reaction in differentiating rinderpest field virus and vaccine virus in the same animals. Rev. Sci. Tech. Off. Int. Epiz., 12: 865–72

[25] Dowling, P.C., Blumberg, B.M., Menonna, J., Adamus, J.E., Cook, P., Crowley, J.C., 1986. (PPRV) infection among small ruminants slaughtered at the central abattoir, Maiduguri, Nigeria.

[26] Walpita, P. (2004). An internal element of the measles virus antigenome promoter modulates replication efficiency. Virus Research 100: 199-211.

[27] Romero, C.H., Barrett, T., Kitching, R.P., Bostock, C., Black, D.N. (1995) Protection of goats against peste des petits ruminants with a recmbinant capripox viruses expressing the fusion and haemagglutinin protein genes of rinderpest virus. Vaccine 13 : 36-40 ruminants in goats in Pakistan. *Vet Rec* 139, 118–119.

[28] Sinnathamby, G., G.J. Renukaradhya, M. Rajasekhar, R. Nayak, M.S. Shaila (2001) Immune responses in goats to recombinant hemagglutinin-neuraminidase glycoprotein of peste des petits ruminants virus: identification of a T cell determinant. Vaccine 19: 4816-4823.

[29] Libeau, G., A. Diallo, F. Colas and L. Gaerre, 1994. Rapid differential diagnosis of rinderpest and peste des petits ruminants using an immunocapture ELISA. Vet. Rec., 134: 300–4

[30] Gibbs, P.J.E., Taylor, W.P. Lawman, M.P. and Bryant, J. (1979) Classification of the peste des petits ruminants virus as the fourth member of the genus Morbillivirus. Intervirology. 11: 268 - 274.

[31] Abubakar M, HA Khan, MJ Arshed, M Hussain M and Ali Q, 2011. Peste des petits ruminants (PPR): Disease appraisal with global and Pakistan perspective. Small Rum Res, 96: 1-10.

[32] Couacy-Hymann, E., Roger, F., Hurard, C., Guillou, J.P., Libeau, G., Diallo,A., 2002. Rapid andsensitive detection of peste despetitsruminants virus by a polymerase chain reaction assay. J. Virol.Methods 100, 17–25.

2

Application of PCR-Based Methods to Dairy Products and to Non-Dairy Probiotic Products

Christophe Monnet[1] and Bojana Bogovič Matijašić[2]
[1]*UMR782 Génie et Microbiol. des Procédés Alimentaires INRA, AgroParisTech, Thiverval-Grignon*
[2]*Institute of Dairy Science and Probiotics, Biotechnical Faculty, University of Ljubljana*
[1]*France*
[2]*Slovenia*

1. Introduction

Many types of cheeses and fermented dairy products are produced throughout the world. They contain various types of bacteria and fungi. In many cases, their exact microbiological composition is not well known because the deliberately added microorganisms are only part of the final microbiota. These microorganisms contribute to the manufacturing of the product (aroma compound production, acidification, impact on texture, colour etc.). Occasionally, dairy products may also be contaminated by spoilage microorganisms and pathogens. PCR-based methods have many interesting applications for dairy products. They can be used to detect, identify and quantify either unwanted or beneficial microorganisms. They can also provide culture-independent microbial fingerprints. Another application is the detection or the quantification of specific genes or groups of genes, such as those involved in the generation of the functional properties. In addition, the abundance of specific mRNA transcripts can be quantified by reverse transcription real-time PCR, which is very useful for a better understanding of the physiology and activity of the microorganisms present in dairy products.

Probiotics have been defined as "live microorganisms that, when administered in adequate amounts, confer a health benefit on the host" (FAO/WHO, 2002). The deficiencies of the quality of probiotic products in terms of too-low numbers or the absence of labelled species are commonly observed. The facts that probiotic functionality is a strain specific trait and that several probiotic strains have very similar phenotypic properties dictate the need for more powerful and rapid methods than conventional cultivation-based methods which have several disadvantages and very limited selectivity. The use of PCR based methods especially has greatly expanded during recent years.

Conventional PCR, combined with gel electrophoresis, has been successfully used for the genus-, species- or strain-specific determination of the presence of probiotic organisms in the products or in the biological samples (faeces). An important feature of probiotics, however, is the viability which is a prerequisite for the probiotic functionality. In this regard, a common DNA-based quantification by real-time PCR is not very useful for quantification purposes since the DNA released from dead or damaged cells also contributes to the results of analysis. One of the alternative approaches for selective detection of viable bacteria is the treatment of the samples with DNA-intercalating dyes such as ethidium monoazide (EMA) or propidium monoazide (PMA) that they can penetrate only into membrane-compromised bacterial cells or dead cells where they are by photo-activation covalently linked to DNA and prevent it from PCR amplification.

2. Application of PCR-based methods to dairy products

2.1 Nucleic acid extraction from dairy products

2.1.1 DNA extraction

Most of the DNA present in cheeses and other fermented dairy products is from the microorganisms that are present. This DNA has to be purified before performing PCR analyses. Dairy products are compositionally complex and there are several reports of dairy matrix-associated PCR inhibition (Niederhauser et al., 1992; Rossen et al., 1992; Herman and Deridder, 1993). One can distinguish two types of DNA extraction methods from dairy products: either direct extractions, or extractions after prior separation of the cells from the food matrix. In all cases, the DNA extraction protocols have to be adapted to the cheese under investigation.

Most methods described in the literature involve prior separation of the cells (Allmann et al., 1995; Herman et al., 1997; Serpe et al., 1999; Torriani et al., 1999; McKillip et al., 2000; Coppola et al., 2001; Ogier et al., 2002; Randazzo et al., 2002; Ercolini et al., 2003; Furet et al., 2004; Ogier et al., 2004; Baruzzi et al., 2005; Rudi et al., 2005; Rademaker et al., 2006; El-Baradei et al., 2007; Lopez-Enriquez et al., 2007; Parayre et al., 2007; Rossmanith et al., 2007; Trmcic et al., 2008; Van Hoorde et al., 2008; Alegría et al., 2009; Dolci et al., 2009; Zago et al., 2009; Le Dréan et al., 2010; Mounier et al., 2010). The recovery of cells from milks or fermented milks is easier to perform than from cheeses. In most cases, homogenisation of the samples and casein solubilisation is done in a sodium citrate solution, using a mechanical blender or glass beads, and the cells are recovered subsequently by centrifugation. Part of the fat is eliminated at this step because it forms a layer at the surface after centrifugation. Serpe et al. (1999) homogenised cheese samples in a Tris-HCl buffer containing the non-anionic detergent Tween 20 to emulsify the fat fraction of the sample. Depending on the type of cheese and the ripening stage, the cell pellet obtained after centrifugation may contain a large amount of caseins. These may be removed by washing the cell pellet with a buffer once or several times, and compounds such as Triton X-100 may be added for a better removal (Baruzzi et al., 2005). Caseins may also be eliminated by pronase digestion before recovery of the cells by centrifugation (Allmann et al., 1995; Furet et al., 2004; Ogier et al., 2004; Flórez and Mayo, 2006). It has been reported that the recovery of the bacterial cells may be improved by addition of polyethylene glycol during the homogenisation step (Stevens and Jaykus, 2004). A matrix lysis buffer containing urea and SDS combined with an homogenisation in a Stomacher laboratory blender has been used by Rossmanith et al. (2007) to recover Gram-positive cells from various food samples, including cheeses. In the procedure described by Herman et al. (1997) and Bonetta et al. (2008), bacterial cells are recovered from homogenised cheese by centrifugation after chemical extraction of fat and proteins. At the surface of some cheeses, for example smear-ripened cheeses, there is a high microbial density, and therefore, a simple surface scraping is sometimes sufficient to recover the microbial cells without need to eliminate the components from the cheese matrix (Rademaker et al., 2005). After their recovery, the cells are disrupted and DNA is purified from the lysed cells. Cell disruption may involve bead-beating, addition of lytic enzymes such as lysozyme, lyticase, mutanolysin or lysostaphine, addition chemical compounds, or a combination of these treatments. After cell lysis, purification of DNA may be performed by classical phenol/chloroform extraction. Phenol is a strong denaturant of proteins that leads to the partition of the proteins into the organic phase and at the interface of the organic and aqueous phases. Procedures avoiding the use of phenol, which is a toxic chemical, have been described. For example, Coppola et al. (2001), Rademaker et al. (2006), and Moschetti et al. (2001) used a commercial kit containing a synthetic resin which removes the cell lysis products that interfere with the PCR amplification. Baruzzi et al. (2005), Trmcic et al. (2008), and Furet et al. (2004) used a commercial kit in which proteins are eliminated by the use of a protein precipitation solution. Column-based or DNA-binding matrix purification methods have also been used (Rudi et al., 2005; Parayre et al., 2007; Zago et al., 2009; Le Dréan et al., 2010), sometimes as a final purification step after phenol/chloroform extraction (Stevens and Jaykus, 2004; Lopez-Enriquez et al., 2007). Separation of cells from the food matrix simplifies the subsequent steps of DNA extraction because most undesirable compounds such as matrix-associated reaction inhibitors are eliminated at the first step of extraction. In addition, large amounts of cheeses (for example more than 10 grams) can be processed in each extraction, which yields a large final amount of DNA. This is important in dairy products containing a low concentration of cells, for example at the initial steps of cheese-manufacturing, where direct DNA extraction is in most cases not possible. Furthermore, the separation of cells from the dairy food matrix eliminates in some cases the need for cultural enrichment prior to detection of pathogens. In contrast to RNA, it is unlikely that there is a large quantitative or qualitative change of the DNA present inside of the cells during the separation of the cells from the dairy food matrix. One of the drawbacks of the DNA extraction methods based on cell separation is that some DNA may be lost during the separation, due to cell lysis, especially for yeasts and Gram-negative strains.

In direct DNA extraction procedures (McKillip et al., 2000; Duthoit et al., 2003; Feurer et al., 2004a; Feurer et al., 2004b; Callon et al., 2006; Monnet et al., 2006; Delbes et al., 2007; Masoud et al., 2011), the cheese samples are first homogenised in a liquid solution by a method involving bead-beating, a mortar and pestle or other mechanical treatments. Efficient treatments of casein degradation and cell lysis, followed by phenol/chloroform extractions, are then needed to remove most contaminating compounds. Contaminating RNA can be removed by a treatment with RNase. Subsequent alcohol precipitation or column-based purification is then used to further purify and/to concentrate the DNA. Carraro et al. (2011) used a column-based purification method for direct extraction of DNA from cheese samples.

2.1.2 RNA extraction

Reverse transcription PCR analyses of RNA may be used in microbial diversity evaluation or for the detection or quantification of mRNA transcripts. Like for DNA, there are two types of extraction methods for RNA from dairy products, either direct extractions, or extractions after prior separation of the cells from the food matrix. The amount of RNA that can be recovered from dairy products is in general higher than for DNA. Indeed, the RNA content of microbial cells is higher than DNA. For example, in *Escherichia (E.) coli*, Bremer and Dennis (1996) reported a concentration varying from 7.6 to 18.3 μg of DNA per 10^9 cells, and from 20 to 211 µg of RNA per 10^9 cells, depending on the growth rates. Messenger RNA (mRNA) accounts for only 1-5% of the total cellular RNA. Compared to DNA, RNA is relatively unstable. This is largely due to the presence of ribonucleases (RNases), which break down RNA molecules. RNases are very stable enzymes and are difficult to inactivate. They can be present in the sample or introduced by contamination during RNA handling.

RNA extraction methods involving prior separation of the cells from cheeses and other dairy products have been used in several studies (Randazzo et al., 2002; Bleve et al., 2003; Sanchez et al., 2006; Bogovic Matijasic et al., 2007; Smeianov et al., 2007; Makhzami et al., 2008; Rantsiou et al., 2008a; Rantsiou et al., 2008b; Ulvé et al., 2008; Duquenne et al., 2010; Falentin et al., 2010; Cretenet et al., 2011; La Gioia et al., 2011; Masoud et al., 2011; Rossi et al., 2011; Taïbi et al., 2011). The recovery of microbial cells is done following similar protocols than for DNA extraction methods (see above). It is unlikely that the abundance of ribosomal RNA is modified during the cell separation procedure, but changes may occur with mRNA transcripts. Indeed, steady-state transcript levels are a result of both RNA synthesis and degradation. The mean half-life of *E. coli* mRNA measured by Selinger et al. (2003) was 6.8 min. It is likely that mRNA synthesis and degradation occurs also during the separation of the cells from the food matrix. This is why all treatments before the complete inactivation of cellular processes should be as short as possible. Ulvé et al. (2008) separated bacterial cells from cheeses by homogenisation in a citrate solution at a temperature of +4 °C, and extracted RNA using a column-based purification method after disruption of the cells by bead-beating. This method was compared to a direct RNA extraction, by measurement of the transcript abundance of 29 genes (Monnet et al., 2008). For most genes, there was no difference, but a higher level was measured for genes which expression is known to be modified by heat, acid, or osmotic stresses. Different methods of bacterial cell disruption were tested by Ablain et al. (2009) for the extraction of *Staphylococcus (S.) aureus* DNA and RNA. The best results were obtained with a combination of lysostaphin treatment and bead-beating. The cell pellets recovered from Camembert cheeses were treated with Chelex beads to remove contaminating compounds that may interfere in subsequent PCR analyses. *Propionibacterium (P.) freundereichii*, a species involved in Emmental cheese ripening, has a thick cell wall surrounded with capsular exopolysaccharides. For an efficient lysis of *P. freundereichii* cells recovered from cheeses, Falentin et al. (2010) used a combination of lysozyme treatment, bead-beating and phenol-chloroform extraction. Sanchez et al. (2006) recovered lactic acid bacteria cells from milk cultures after dispersion of caseins with EDTA, and extracted RNA using guanidinium thiocyanate-phenol-chloroform (commercial TRIzol reagent), a reagent that inactivates cellular processes and allows separation of RNA from DNA and proteins (Chomczynski and Sacchi, 1987). Duquenne et al. (2010) also used this type of extraction, after disruption of the cells by bead-beating. Bacterial cells may also be separated from cheese matrices using a Nycodenz gradient (Makhzami et al., 2008). In order to limit the changes in mRNA transcript composition inside of the cells during their separation from the dairy food matrix, Taïbi et al. (2011) added to the samples a stopping solution consisting of a mixture of phenol and ethanol. Smeianov et al. (2007) added the commercial reagent RNAprotect and rifampin, an antibiotic that suppresses the initiation of RNA synthesis, during the recovery of *Lactobacillus (Lb.) helveticus* cells from milk cultures.

So far, only a few studies have involved direct RNA extraction procedures from dairy products (Duthoit et al., 2005; Bonaiti et al., 2006; Monnet et al., 2008; Carraro et al., 2011;

Trmcic et al., 2011). In the method described by Monnet et al. (2008), the cellular processes are stopped at the very beginning of the procedure, by addition of a guanidinium thiocyanate-phenol-chloroform solution to the cheese sample, and bead-beating is immediately performed. The reagent also inactivates RNases that may be present. At this step, the samples can be kept several weeks at -80 °C without any decrease of RNA integrity, which is not possible when the cheese samples are frozen before the RNA extraction. It was found that the amount of cheese sample should not exceed 100 mg per ml of reagent, as a higher ratio affects the quality and quantity of the purified RNA. The fat, caseins and DNA are removed after recovery of the aqueous phase which is formed after addition of chloroform. Subsequent acidic phenol-chloroform extraction and column-based purification is then performed to get RNA extracts suitable for reverse transcription PCR analyses and which can be stored several months at -80 °C. Use of 7-ml bead-beating tubes allows the processing of 500 mg samples of cheese (Trmcic et al., 2011). In addition, several samples may be pooled and concentrated during the column-based purification step, which allows higher amounts of RNA to be recovered. With this procedure, sufficient amounts of RNA could be obtained for analysing gene expression of a *Lactococcus (L.) lactis* strain whose concentration was about 10^8 CFU per gram of cheese, with a corresponding RNA extraction yield of 4.9×10^{-6} ng RNA per CFU.

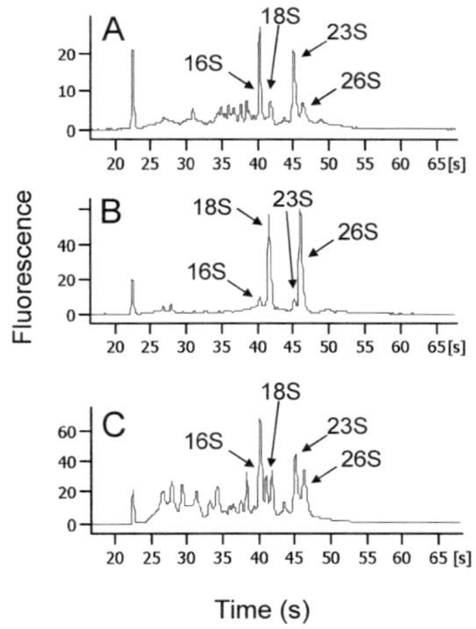

Fig. 1. RNA quality assessment with the Agilent Bioanalyzer: electrophoregrams of RNA preparations from various commercial smear-ripened cheeses using the method described by Monnet et al. (2008). 16S and 23S rRNA are from bacterial origin, and 18S and 26S rRNA are from fungi. Cheese B contains more RNA from fungi than cheeses A and C, and shows a higher overall RNA integrity.

The quality of the RNA samples has to be assessed. Absence of contaminating DNA can be checked by performing PCR amplifications with controls in which reverse transcription has not been performed. RNA concentration can be measured with a spectrophotometer at 260 nm or with a fluorometer after addition of fluorescent dyes. The RNA integrity is evaluated by gel electrophoresis or by automated capillary-based electrophoresis (e.g. 2100 Bioanalyzer equipment, Agilent). RNA is mostly constituted of ribosomal RNA (rRNA), and the sharpness of the small (16S or 18S) and large (23S or 26S) rRNA subunit bands is indicative of the global degree of RNA integrity. From the 2100 Bioanalyzer electrophoresis profile, a value, named RIN (RNA Integrity Number), is calculated. A RIN value of 10 corresponds to apparently intact material. RIN calculations can be done with either eukaryotic or prokaryotic RNA, but not when both types of RNA are present in the same sample, which would be the case for RNA samples from numerous types of cheeses. Examples of RNA electrophoregrams of RNA preparations from cheese samples are shown in Figure 1. During the ripening or storage of cheeses, some microbial populations may decline, for example by autolysis. This has a detrimental effect on RNA integrity and, in consequence, a poor RNA integrity level is not necessarily due to an inadequate sampling or RNA extraction procedure.

2.2 Amplification targets

All PCR analyses rely on amplification of DNA target sequences. Concerning PCR applications to dairy products, one can distinguish targets used for PCR-based microbial diversity evaluation, and targets for PCR analysis of specific microbial groups.

2.2.1 Amplification targets for microbial diversity evaluation methods

In methods of microbial diversity evaluation involving PCR, the amplification target is a sequence which has to be present in a large part of the bacterial or fungal population. The sequence variations allow the subsequent differentiation of the generated amplicons. In most cases, these techniques involve amplification of ribosomal RNA or housekeeping genes. In both prokaryotes and eukaryotes, rRNA genes usually show a high sequence homogeneity within a species (Liao, 1999), which explains why they are widely used in species identification and makes them a good target in molecular microbial diversity evaluation methods.

Bacterial 16S, 23S and 5S rRNA genes are organised into a co-transcribed operon. The typical length of theses genes is ~2900 bp (23S), ~1500 bp (16S) and ~120 bp (5S). There are multiple copies (generally <10) of the rRNA genes in most bacteria, and the rRNA operons are generally dispersed throughout the chromosome. 16S rRNA sequences are frequently used as amplification target. All 16S rRNA genes share nine hypervariable (polymorphic) regions (Neefs et al., 1993) and the sequences are easily available from public databases. The hypervariable regions are flanked by conserved sequences, which can serve for amplification with "universal" primers (Baker et al., 2003). The variable V1 (Cocolin et al., 2004; Bonetta et al., 2008), V3 (Coppola et al., 2001; Ercolini et al., 2001; Ogier et al., 2002; Duthoit et al., 2003; Ercolini et al., 2003; Mauriello et al., 2003; Andrighetto et al., 2004; Ercolini et al., 2004; Feurer et al., 2004a; Feurer et al., 2004b; Lafarge et al., 2004; Ogier et al., 2004; Duthoit et al., 2005; Flórez and Mayo, 2006; Delbes et al., 2007; El-Baradei et al., 2007; Parayre et al., 2007; Abriouel et al., 2008; Ercolini et al., 2008; Gala et al., 2008; Van Hoorde et al., 2008; Alegría et al., 2009; Casalta et al., 2009; Dolci et al., 2009; Giannino et al., 2009; Mounier et al., 2009; Serhan et al., 2009; Dolci et al., 2010; Fontana et al., 2010; Van Hoorde et al., 2010; Masoud et al., 2011), V2 (Duthoit et al., 2003; Delbes and Montel, 2005; Saubusse et al., 2007), V4-V5 (Ercolini et al., 2003), V1-V3 (Randazzo et al., 2002), V4-V8 (Randazzo et al., 2006), V5-V6 (Le Bourhis et al., 2005; Le Bourhis et al., 2007) and V6-V8 (Randazzo et al., 2002; Ercolini et al., 2008; Nikolic et al., 2008; Randazzo et al., 2010) regions of the 16S rRNA genes and the 16S-23S-spacer region (Coppola et al., 2001; Henri-Dubernet et al., 2004) are widely used in studies of the bacterial diversity of dairy products. Several distinct amplicons may be produced with some strains, due to differences in sequences of the rRNA copies.

In fungi, the internal transcribed spacer (ITS) is a region located between the 18S rRNA and 26S rRNA genes. It includes the 5.8S rRNA gene that splits the ITS into two parts: ITS1 and ITS2. The 18S, 5.8S, 26S and 5S rRNA sequences form up to hundreds of tandem repeats. The ITS region undergoes a faster rate of evolution than rRNA but its sequence remains homogenous within a species. The ITS2 region has been chosen as target for the study of the fungal biodiversity of smear-ripened cheeses (Mounier et al., 2010), and the ITS1 region for the study of the fungal diversity in cow, goat and ewe milk (Delavenne et al., 2011). Primers targeting regions of the 26S rRNA (Feurer et al., 2004b; Flórez and Mayo, 2006; Bonetta et al., 2008; Alegría et al., 2009; Dolci et al., 2009; Mounier et al., 2009) and the 18S rRNA (Callon et al., 2006; Arteau et al., 2010) were chosen to investigate the dominant yeast microflora of several types of cheeses.

Housekeeping genes are less used than rRNA in molecular studies of microbial diversity of dairy products. This is due to a lower availability in sequence databases. However, this may change in the near future, due to the rapid increase of the number of sequenced genomes. The *rpoB* gene, encoding the RNA polymerase beta subunit has been used as a target for PCR-DGGE analysis to follow lactic acid bacterial population dynamics in cheeses (Rantsiou et al., 2004).

2.2.2 Amplification targets for specific microbial groups

Defined groups of microorganisms may be studied by amplification of specific targets, either by PCR or by real-time PCR. In the latter case, quantitative data can be obtained. The primers have to be designed so that amplification occurs only from DNA of the group of interest. As for PCR-based methods of microbial diversity evaluation, rRNA sequences are frequently used as target and the specificity may be evaluated *in silico* by comparing the rRNA sequences of the group of interest to that of other microorganisms that are present in the same habitat. A high level of specificity is achieved when there is a large sequence difference with non-target microorganisms for one or both of the PCR primers. Presence of mismatches near the 3' of the primers ensures a better specificity than at the 5' end. In addition, absence, or presence of only one or two G or C residues in the last five nucleotides at the 3' end of primers, makes them less likely to hybridise transiently and to be available for non-specific extension by the DNA polymerase (Bustin, 2000). *Corynebacterium casei* cells could be quantified in cheeses by real-time PCR using a couple of primers targeting the V6 region of the 16S rRNA gene (Monnet et al., 2006). The assay was specific, as no amplification occurred with DNA from other *Corynebacterium* species present in cheeses. Primers targeting 16S rRNA genes were also used for the quantification of *Carnobacterium* cells in cheeses (Cailliez-Grimal et al., 2005), of *L. lactis* subsp. *cremoris* in fermented milks (Grattepanche et al., 2005), of *Streptococcus (Str.) thermophilus* and lactobacilli in fermented milks (Furet et al., 2004), of thermophilic bacilli in milk powder (Rueckert et al., 2005) and of bacterial species that can develop during the cold storage of milk (Rasolofo et al., 2010). Primers targeting the 16S-23S-spacer region were used for the specific detection of *Clostridium tyrobutyricum* in semi-soft and hard cheeses (Herman et al., 1997) and for the quantification of *Listeria (List.) monocytogenes* in foods, including fresh and ripened cheeses

(Rantsiou et al., 2008a). rRNA sequence primers were also advised for the quantification of fungi in cheeses by real-time PCR. The variable D1/D2 domain of the 26S rRNA and the ITS1 region of the rRNA genes were targeted for the study of yeasts (Larpin et al., 2006; Makino et al., 2010) and *Penicillium roqueforti* (Le Dréan et al., 2010).

Primers of specific protein-encoding genes have been designed for the detection or the quantification of various groups of cheese microorganisms. Proteolytic lactobacilli can be detected in stretched cheeses by amplification of cell envelope proteinase genes (Baruzzi et al., 2005). Successful detection of specific bacteriocin biosynthesis genes could be achieved in microbial DNA extracted directly from several types of cheeses (Moschetti et al., 2001; Bogovic Matijasic et al., 2007; Trmcic et al., 2008). Allman et al. (1995) used specific PCR amplifications for the detection of pathogenic bacteria in dairy products. The targets were the *List. monocytogenes* listeriolysin O (*hlyA*), the *E. coli* heat-labile enterotoxin type 1 (*elt*) and heat-stable toxin 1 (*est*), and the *Campylobacter jejuni* and *Campylobacter coli* flagellin proteins (*flaA/flaB*). *List. monocytogenes* has also been quantified in gouda-like cheeses by real-time PCR, through *hlyA* gene amplification (Rudi et al., 2005). Another pathogen, *Brucella* spp., can be detected in soft cheeses by amplification of a fragment from a characteristic membrane antigen, protein BCSP-31 (Serpe et al., 1999). Thermonuclease (*nuc*) gene amplification has been applied for the quantification of *S. aureus* cells in cheese and milk samples (Hein et al., 2001; Hein et al., 2005; Alarcon et al., 2006; Studer et al., 2008; Aprodu et al., 2011). Manuzon et al. (2007) monitored the pool of tetracyclin resistance genes in retail cheeses in order to estimate the amount of tetracyclin resistant bacteria, which may pose a potential risk to consumers. Coliforms are a broad class of bacteria, whose presence can be used to assess the hygienic quality of foods. A real-time PCR detection method of all coliform species in a single assay has been set up (Martin et al., 2010). It is based on the amplification of a fragment of the beta-galactosidase gene (*lacZ*). *Enterococcus (E.) gilvus*, which is found in some types of cheeses, was quantified by real-time PCR using the phenylalanyl-tRNA synthase gene (*pheS*) as target (Zago et al., 2009). The procedure was selective against the highly phylogenetically related species *E. malodoratus* and *E. raffinosus*, and the *pheS* gene seems able to differentiate enterococcal species better than 16S rRNA sequences. Histamine is a toxic biogenic amine that is sometimes involved in food poisoning. In order to quantify histamine-producing bacteria in cheeses by real-time PCR, Fernandez et al. (2006) designed consensual primers targeting the histidine decarboxylase (*hdcA*) gene of Gram-positive species. Another type of undesired bacteria, *Clostridium tyrobutyricum*, responsible for late-blowing in hard and semi-hard cheeses, can be quantified in milk samples by real-time PCR amplification of the flagellin (*fla*) gene (Lopez-Enriquez et al., 2007).

It is likely that in the future, the increased availability of genome sequences will facilitate the selection of amplification targets for specific microbial groups. A good example is the study of Chen el al. (2010), in which real-time PCR primers were designed for the detection of *Salmonella enterica* strains. In this study, specific targets were generated by using a genomic analysis workflow, which compared 17 *Salmonella enterica* genome sequences to 827 non-*Salmonella* bacterial genomes.

2.3 PCR-based methods for microbial diversity investigation

Dairy products, especially cheeses, have diverse microbial compositions, which may be analysed by culture-dependent or culture-independent methods. Culture-independent methods involving PCR amplification are based on the analysis of DNA or RNA extracted from the food product. Even if they have several potential biases, they are faster and potentially more exhaustive than culture-dependent methods.

Denaturing gradient gel electrophoresis (DGGE), temperature gradient gel electrophoresis (TGGE) and temporal temperature gradient gel electrophoresis (TTGE) are widely used to study cheese microbial communities (Coppola et al., 2001; Ercolini et al., 2001; Ogier et al., 2002; Randazzo et al., 2002; Ercolini et al., 2003; Mauriello et al., 2003; Andrighetto et al., 2004; Cocolin et al., 2004; Ercolini et al., 2004; Henri-Dubernet et al., 2004; Lafarge et al., 2004; Ogier et al., 2004; Rantsiou et al., 2004; Le Bourhis et al., 2005; Flórez and Mayo, 2006; Randazzo et al., 2006; Cocolin et al., 2007; El-Baradei et al., 2007; Le Bourhis et al., 2007; Parayre et al., 2007; Abriouel et al., 2008; Bonetta et al., 2008; Ercolini et al., 2008; Gala et al., 2008; Henri-Dubernet et al., 2008; Nikolic et al., 2008; Rantsiou et al., 2008b; Van Hoorde et al., 2008; Alegría et al., 2009; Casalta et al., 2009; Dolci et al., 2009; Giannino et al., 2009; Serhan et al., 2009; Dolci et al., 2010; Fontana et al., 2010; Fuka et al., 2010; Randazzo et al., 2010; Van Hoorde et al., 2010; Masoud et al., 2011). Target sequences from rRNA or housekeeping genes are amplified and separated by electrophoresis. Separation is based on decreased electrophoretic mobility of partially melted double-stranded DNA molecules in polyacrylamide gels with a thermal gradient (TGGE) or which contain a gradient of DNA denaturants (DGGE). In TTGE, the separation is based on a temporal temperature gradient that increases in a linear fashion over the length of the electrophoresis time. Even if the DNA molecules have the same size, they may be separated because of their melting temperature behaviour, which depends on the sequence. A GC-rich clamp of about 40 bases is added at the 5' end of one of the primers to stabilize the melting behaviour and to prevent the complete dissociation of the DNA fragments during electrophoresis. Assignment of the migration bands is done by comparison to a database containing the migration profiles of reference strains. DNA bands can be recovered from the gel and sequenced in order to confirm the assignments, or to find an assignment for bands which are not present in the database. DGGE, TGGE and TTGE profiles reveal a picture of the microbial diversity and can be used to compare different dairy products or to follow a given product at different fabrication stages. However, these methods are only semi-quantitative.

Single-strand conformation polymorphism-PCR (SSCP-PCR) is another PCR-based method for microbial diversity investigation that has been applied to dairy products (Duthoit et al., 2003; Feurer et al., 2004a; Feurer et al., 2004b; Delbes and Montel, 2005; Duthoit et al., 2005; Callon et al., 2006; Delbes et al., 2007; Saubusse et al., 2007; Mounier et al., 2009). This technique is based on the sequence-dependent differential intra-molecular folding of single strand DNA, which alters the migration speed of the molecules under non-denaturing conditions. Single strand DNA fragments having the same size may thus be separated, if their sequences generate different intramolecular interactions. After denaturation, the fluorescently labelled PCR products are separated using a capillary-based automated sequencer. In some cases, several stable conformations can be formed from one single strand DNA fragment, resulting in multiple bands. As for DGGE, TGGE and TTGE, SSCP provides community fingerprints that cannot be phylogenetically assigned directly. A database containing the migration profile of reference strains has to be created. One disadvantage of this technique is that the labelled single strand DNA fragments cannot be sequenced to confirm the assignations.

Another PCR-based technique that has been applied to dairy products is terminal restriction fragment length polymorphism (TRFLP) (Rademaker et al., 2005; Rademaker et al., 2006; Arteau et al., 2010; Cogan and John, 2011). In TRFLP analyses, marker genes are amplified using one or two fluorescently labelled primers. The amplicons are then cut with one or several restriction enzymes and separated using a capillary-based automated sequencer. Only the end-labelled fragments are detected by the laser detector and their size can be determined by comparison with DNA size standards. One advantage of this technique is that the size of the fragments of any known DNA sequence can be determined *in silico*. This is why 16S rRNA genes, whose sequences are easily available from public databases, are frequently used in TRFLP studies. As for SSCP, a drawback of capillary electrophoresis-based TRFLP is that bands remaining unknown cannot be extracted from the gel to be identified by DNA sequencing.

In denaturing high-performance liquid chromatography (DHPLC), PCR amplicons are partially denatured and separated on a liquid chromatography column which contains chemical agents that bind more strongly to double-stranded DNA molecules. Amplicons of the same size but with sequence differences resulting in modified melting behaviours will thus have different retention times. DHPLC analyses are rapid and the elution fraction corresponding to the different amplicons can be sequenced for confirmation or identification purposes. There are not many papers concerning DHPLC analyses of dairy products (Ercolini et al., 2008; Mounier et al., 2010; Delavenne et al., 2011), but this technique will probably be increasingly used in the future.

Bacterial diversity may also be assessed by sequencing clones libraries generated from 16S rRNA gene amplification of DNA extracted from dairy products (Feurer et al., 2004a; Feurer et al., 2004b; Delbes et al., 2007; Rasolofo et al., 2010; Carraro et al., 2011). The main advantage of this technique is that no dedicated database is needed, as the sequences are already available in public genomic databases. In addition, in most cases, the 16S rRNA gene sequences permit assignments at the species level. But this technique is expensive and time-consuming, which is why it is not widely used. Second-generation DNA sequencing is a promising alternative to clone library sequencing (Cardenas and Tiedje, 2008). Masoud et al. (2011) studied the bacterial populations in Danish raw milk cheeses by pyrosequencing of tagged amplicons of the V3 and V4 regions of the 16S rRNA gene. After amplification of the 16S rRNA targets, a second PCR is done by using, for each sample, a different bar-coded primer. The amplified fragments of the different samples are then mixed and sequenced together, and the sequences are assigned to bacterial taxa. A very good agreement was found with the results of PCR-DGGE analysis. In addition, minor bacterial populations that were not detected by PCR-DGGE, were found by pyrosequencing. Furthermore, pyrosequencing provides a more reliable estimate of the relative abundance of the individual bacteria. Second-generation DNA sequencing appears thus to be a powerful and promising method, which will allow a deeper investigation of the bacterial populations in dairy products.

PCR-based methods for microbial diversity investigation can also be applied to RNA samples, after reverse transcription. As the ribosomal RNA content inside of the cells increases with the growth rate (Bremer and Dennis, 1996), one can assume that higher amounts of rRNA targets will be detected in active growing cells. In addition, since RNA is less stable than DNA, it will degrade more quickly in dead cells. In a study of the bacterial community from an artisanal Sicilian cheese, Randazzo et al. (2002) compared the intensity of bands from DNA and RNA-derived DGGE profiles and concluded that some species of the samples were not very metabolically active. Other studies of RNA profiles involving either DGGE (Rantsiou et al., 2008b; Dolci et al., 2010; Masoud et al., 2011), TTGE (Le Bourhis et al., 2007), SSCP (Le Bourhis et al., 2005), T-RFLP (Sanchez et al., 2006), clone library sequencing (Carraro et al., 2011) or pyrosequencing (Masoud et al., 2011) have been published.

2.4 Real-time PCR methods

Real-time PCR (qPCR) uses fluorescent reporter dyes to combine the amplification and detection steps of the PCR reaction in a single tube format. The assay relies on measuring the increase in fluorescent signal, which is proportional to the amount of DNA produced during each PCR cycle. A quantification cycle (Cq) value is determined from the plot relating fluorescence against the cycle number. Cq corresponds to the number of cycles for which the fluorescence is higher than the background fluorescence. qPCR offers the possibility to quantify microbial populations through measurements of the abundance of a target sequence in DNA samples extracted from food products (Postollec et al., 2011). Combined with reverse transcription (RT), qPCR can also be used to estimate the amount of RNA transcripts.

Several applications of qPCR for the quantification of microbial populations in dairy products have been described (Table 1). In general, the experimental approach is the following: after extraction of DNA from the sample, qPCR is performed together with a standard curve, and the results are expressed as colony-forming-units (CFU), cell, or DNA target number per amount of dairy product. For an accurate quantification, several technical considerations have to be taken into account. First, the efficiency of recovery of the DNA from the dairy products should be constant and as high as possible. This may be verified in experiments where target cells are added to a control dairy matrix. Larpin et al. (2006) observed significant DNA losses during the extraction of DNA from cheese samples containing yeast species, and it appeared that cheese composition affected the extraction yields. DNA losses may occur during alcohol precipitation steps, especially in samples containing low amounts of DNA. A better recovery can be obtained by addition of co-precipitants such as exogenous DNA and glycogen. When column-based purification methods are used, it should be made sure that the amount of DNA loaded onto the columns does not exceed the column capacity. Another important technical consideration is that the amount of qPCR inhibitors in the DNA sample should be as limited as possible. One convenient way to evaluate the presence of inhibitors is to analyse by qPCR several dilutions of the DNA samples. The samples that need high dilution factors to reach the maximum PCR efficiency contain more inhibitors than those that need a lower dilution factor. The amount of PCR inhibitors has an impact on the detection level, as it determines the dilution factor that has to be applied in the qPCR assays. Absence of inhibitors can also be verified by inclusion of an internal amplification control (IAC). An IAC is a non-target DNA fragment that is co-amplified with the target sequence, ideally with the same primers used for the target. The forward and reverse target sequences are fused to both ends of a non-target fragment, to which a second fluorescent probe (the IAC probe) hybridises. The simultaneous use in a single reaction of two differently labelled fluorescent probes makes it possible to quantify the target and to assess PCR efficiency at the same time. If negative results are obtained for the target PCR, the absence of a positive IAC signal indicates that amplification has failed. Phenol extraction and repeated washing of alcohol-precipitated DNA pellets are efficient in reducing the impact of PCR inhibitors. In phenol-based purifications, the amount of PCR inhibitors may also be reduced by using a gel (Phase Lock Gel tubes) improving separation between the liquid and organic phases. For accurate qPCR quantification of microbial populations in dairy products, the level of cross-contaminations of DNA during DNA extraction and subsequent steps should be as limited as possible. This can be checked by adding several controls during the qPCR, such as water or DNA extracted from a dairy matrix that does not contain the target population. If complete absence of cross-contamination cannot be achieved, one may define a maximum Cq (quantification cycle) value, which is lower than the value obtained with the controls (e.g. five cycles lower), and over which the assay will not be considered. After qPCR amplification, melting curve analysis is carried out to confirm the absence of secondary amplification products. It is also possible to confirm amplification specificity by sequencing the resulting amplicon. Several types of standards may be used for calculating the concentration of targets in the dairy product. In the method used by Monnet et al. (2006), a standard curve is generated from different dilutions of a genomic DNA sample extracted from a pure culture of the target microorganism in liquid broth. The amount of target genomic DNA present in cheeses is then calculated and converted to colony-forming-units values, using a conversion factor determined from the pure culture DNA extract. Such calculation is valid only if the DNA recovery yield from cheeses is similar to that from cells grown in the liquid broth. Le Dréan et al. (2010) quantified *Penicillium camemberti* and *Penicillium roqueforti* mycelium in cheeses. To imitate cheese matrix effects, DNA was extracted from curd mixed with known amounts of fresh mycelium and was used as standard for further qPCR analyses. The mycelium concentration was then expressed as weight of mycelium per weight of cheese. Microbial cells may also be quantified using standard curves obtained with PCR-amplified targets. For example, Furet et al. (2004) determined the number of 16S rRNA gene targets in DNA samples prepared from dairy products and converted this value to cell numbers, taking into account the number of 16S rRNA gene copies in the chromosome of each species (http://rrndb.mmg.msu.edu, (Lee et al., 2009). Rasolofo et al. (2010) used a similar procedure for the quantification of *Staphyloccous aureus, Aerococcus viridans, Acinetobacter calcoaceticus, Corynebacterium variabile, Pseudomonas fluorescens* and *Str. uberis* in milk samples, except that standard curves were obtained from plasmids in which 16S rRNA gene sequences of the target species were inserted.

The quantification limit values for microbial cells in dairy products reported for qPCR methods are heterogeneous. They depend on factors such as the type of dairy product (cheese or fermented milk), the efficiency of DNA extraction, the target microbial population and the target DNA sequence. A value of 10^5 CFU/g has been reported for *Corynebacterium casei* (Monnet et al., 2006) and *Carnobacterium* species (Cailliez-Grimal et al., 2005), of 10^3-10^4 CFU/g for *List. monocytogenes* (Rantsiou et al., 2008a), of 10^4 CFU/g for *E. gilvus* (Zago et al., 2009), and of 10^3 cells/ml for lactic acid bacteria (Furet et al., 2004). In some cases, higher amounts of microorganisms are measured with qPCR analyses than with classical agar counts, which may be explained by the fact that DNA from dead cells can also be amplified. In order to lower the detection levels of pathogens, it is possible to perform culture enrichment of the food samples before qPCR (Rossmanith et al., 2006; Chiang et al., 2007; Karns et al., 2007; O'Grady et al., 2009; Omiccioli et al., 2009). However, in that case, the results can only be used for detection, and not quantification.

Target population	Target sequence	Food matrix	References
Str. thermophilus	*rim*M (16S rRNA processing protein)	Commercial yoghurt samples	(Ongol et al., 2009)
L. lactis subsp. *cremoris*	16S rRNA	Experimental fermented milks, mixed culture with *Lb. rhamnosus* and *L. lactis* subsp. *lactis* biovar. *diacetylactis*	(Grattepanche et al., 2005)
Str. thermophilus, Lb. delbrueckii, Lb. casei, Lb. paracasei, Lb. rhamnosus, Lb. acidophilus, Lb. johnsonii	16S rRNA	Commercial fermented milks	(Furet et al., 2004)
Carnobacterium sp.	16S rRNA	Artificially contaminated cheeses and commercial cheeses	(Cailliez-Grimal et al., 2005)
Corynebacterium casei	16S rRNA	Commercial smear-ripened cheese	(Monnet et al., 2006)
P. freudenreichii and *Lb. paracasei*	16S rRNA, *tuf* (elongation factor TU), *Gro*L (chaperonin GroEL)	Experimental Emmental cheese	(Falentin et al., 2010)
Str. thermophilus and *Lb. helveticus*	16S rRNA, *tuf* (elongation factor TU), *Gro*L (chaperonin GroEL)	Experimental Emmental cheese	(Falentin et al., 2012)
E. gilvus	*pheS* (phenylalanyl-tRNA synthase	Artisanal raw milk cheeses	(Zago et al., 2009)
E. faecium	Conserved *E. faecium* sequence	Lebanese raw goat's milk cheeses	(Serhan et al., 2009)
Clostridium tyrobutyricum	*fla* (flagellin)	Artificially contaminated milks	(Lopez-Enriquez et al., 2007)
Histamine-producing bacteria	*hdc*A (histidine decarboxylase)	Experimental cheeses and commercial cheeses	(Fernandez et al., 2006; Ladero et al., 2008; Ladero et al., 2009)
Tetracyclin resistant bacteria	*tet*S (tetracycline resistance protein)	Artificially contaminated cheeses and commercial cheeses	(Manuzon et al., 2007)
Thermophilic bacilli	16S rRNA	Artificially contaminated milk powder	(Rueckert et al., 2005)
Coliform species	*lac*Z (beta-galactosidase)	Artificially contaminated cheeses	(Martin et al., 2010)

Target population	Target sequence	Food matrix	References
E. coli O157:H7	*eae* (intimin adherence protein)	Market dairy food samples	(Singh et al., 2009)
E. coli O157:H7	virulence genes	Milk samples	(Karns et al., 2007)
S. aureus	*nuc* (thermonuclease)	Commercial food samples, including cheeses	(Omiccioli et al., 2009)
S. aureus	*nuc* (thermonuclease)	Artificially contaminated and naturally contaminated milk samples	(Studer et al., 2008; Aprodu et al., 2011)
S. aureus	*nuc* (thermonuclease)	Artificially contaminated cheeses, bovine and caprine milk samples	(Hein et al., 2001; Hein et al., 2005)
S. aureus	*egc* (enterotoxin gene cluster)	Artificially contaminated and naturally contaminated milk samples	(Fusco et al., 2011)
S. aureus genotype B	*sea* (enterotoxin A), *sed* (enterotoxin D), *luk*E (leucotoxin E)	Milk samples	(Boss et al., 2011)
Brucella spp.	*rnp*B (RNA component of ribonuclease P), bcsp31 (311 kDa cell surface protein)	Buffalo milk samples	(Marianelli et al., 2008; Amoroso et al., 2011)
List. monocytogenes	*prfA* (transcriptional activator)	Commercial food samples, including cheeses	(Omiccioli et al., 2009)
List. monocytogenes	16S-23S-spacer region	Various foods, including milk and soft cheese	(Rantsiou et al., 2008a)
List. monocytogenes	*hly*A (listeriolysin O)	Artificially contaminated cheeses and commercial gouda-like cheeses	(Rudi et al., 2005)
List. monocytogenes	*ssr*A (tmRNA)	Commercial dairy products	(O'Grady et al., 2009)
Mycobacterium avium subsp. *paratuberculosis*	MAP F57 sequence	Commercial raw milk cheeses	(Stephan et al., 2007)
Mycobacterium avium subsp. *paratuberculosis*	Insertion element IS*900*	Milk samples and commercial cheeses	(Rodríguez-Lázaro et al., 2005; Donaghy et al., 2008; Herthnek et al., 2008; Slana et al., 2008; Botsaris et al., 2010)

Target population	Target sequence	Food matrix	References
Mycoplasma bovis	*uvrC* (deoxyribodipyrimidine photolyase)	Bovine milk samples	(Rossetti et al., 2010)
S. aureus, Aerococcus viridans, Acinetobacter calcoaceticus, Corynebacterium variabile, Pseudomonas fluorescens and *Str. uberis*	16S rRNA	Milk during cold storage	(Rasolofo et al., 2010)
Salmonella spp., *List. monocytogenes* and *E. coli* O157	*Salmonella* spp: *ttr* cluster (tetrathionate reductase genes) *List. monocytogenes*: *hlyA* (listeriolysin O) *E. coli* O157: *rfbE* (perosamine synthetase homolog)	Artificially contaminated milk	(Omiccioli et al., 2009)
Debaryomyces hansenii, Geotrichum candidum, Kluyveromyces sp., *Yarrowia lipolytica*	*G. candidum*: *cgl* (cystathionine-gamma-lyase), *Kluyveromyces* sp.: *lac4* *Y. lipolytica*: topoisomerase II	Commercial Livarot cheeses	(Larpin et al., 2006)
Penicillium roqueforti and *Penicillium camemberti*	*P. roqueforti*: ITS1 region of rRNA *P. camemberti*: beta-tubulin gene	Model cheeses and commercial Camembert-type cheeses	(Le Dréan et al., 2010)
Candida albicans, Candida glabrata, Candida parapsilosis, Candida tropicalis, Clavispora lusitaniae, Filobasidiella neoformans, Issatchenkia orientalis, Trichosporon asahii, and Trichosporon jirovecii	D1/D2 domain of 26S rRNA	Artificially contaminated fermented milk	(Makino et al., 2010)
Lb. delbrueckii bacteriophages	bacteriophage lysin genes	Artificially contaminated milk samples	(Rossetti et al., 2010)

Table 1. Examples of applications of qPCR for the quantification or detection of microbial populations in dairy products.

The study of gene expression within natural environments such as dairy products is an emerging field in microbial ecology that is especially promising in the study of bacterial function even though only a few applications of reverse-transcription qPCR to dairy products have been described so far (Table 2). Reverse-transcription qPCR experiments involve the following steps: RNA extraction, evaluation of RNA integrity, DNase treatment, reverse-transcription and qPCR (Nolan et al., 2006; Bustin et al., 2009). Reverse transcriptions can be done with random hexamers, specific primers or oligo-dT primers (only for eukaryotic mRNA). Two types of quantification methods may be used: absolute quantification and relative quantification (Wong and Medrano, 2005; Nolan et al., 2006; Bustin et al., 2009; Cikos and Koppel, 2009). Absolute quantification is based on comparison of Cq values with a standard curve generated from the target sequence. The determination of a concentration of target RNA in the samples requires generating a standard curve with known amounts of RNA targets (and not DNA) that have been transcribed *in vitro*. This is necessary because the efficiencies of reverse transcription reactions are not known and vary from target to target. In addition, the reverse transcription step has been proposed as the source of most of the variability in reverse-transcription qPCR (Freeman et al., 1999), owing to the sensitivity of reverse transcriptase enzymes to inhibitors that may be present in the samples. As the production of *in vitro*-transcribed RNA standards is fastidious and time-consuming, and there is no guarantee that the reverse transcription efficiency with these standards will be similar to that with the biological RNA samples, there are not many reports of absolute quantification in reverse transcription qPCR involving RNA standards. Absolute quantification of RNA transcripts with DNA standards (e.g. with standards that have not been reverse transcribed) is sometimes used. In that case, the exact number of RNA targets in the biological samples cannot be determined and results are expressed as "DNA gene equivalent" (Nicolaisen et al., 2008) or "cDNA". If it is assumed that the reverse transcription efficiencies for a given target are constant whatever the sample, these results can be used to compare the abundance of the same RNA target in several samples. Smeianov et al. (2007) used absolute quantification to compare the expression of *Lb. helveticus* genes during growth in milk and in MRS medium. In these experiments, the amount of cDNA before qPCR was standardised. Ulvé et al. (2008) standardised the amount of RNA before reverse transcription and compared the Cq values obtained for genes of *L. lactis* in cheeses at different ripening times. Even if it is not possible by this method to quantitatively compare the abundance of different RNA targets in the same sample (which would need *in vitro*-transcribed RNA standards), large differences in abundance may be shown. Direct comparisons of Cq values with a standardised amount of RNA have also been used to investigate the effect of cell separation from the cheese matrix before RNA extraction (Monnet et al., 2008). Bleve et al. (2003) observed a correlation between standard plate counts of yeasts and moulds present in spoiled commercial food products and the Cq values obtained by reverse transcription qPCR analysis with primers targeting the fungal actin gene. To follow gene expression of *P. freudenreichii* and *Lb. paracasei* during cheese-making, Falentin et al. (2010) measured the amount of cDNA copies of the target sequence after reverse transcription, and divided this value by the corresponding number of cells, which was measured by qPCR analysis of DNA extracted from the cheese samples. From these analyses, it could be concluded that the metabolic activity of *Lb. paracasei* cells reached a maximum during the first part of ripening, whereas the maximum activity of *P. freudenreichii* was reached later. A similar approach was used for the study of the metabolic activity of *Lb. helveticus* and *Str. thermophilus* cells during the ripening of Emmental cheese (Falentin et al., 2012).

One disadvantage of all absolute quantification analyses is the significant reduction in the number of experimental samples that can be run on a single plate because a standard curve has to be included in each reaction run. In relative quantification methods, the amount of RNA targets in samples is expressed relative to the amount of the same target present in another sample, which is designated as the calibrator. This calibrator is chosen among the samples being compared (Cikos and Koppel, 2009). The advantage of this method is that standard curves don't have to be included in each run. However, this does not compensate for variations in reverse transcription efficiency and in RNA extraction efficiency from one sample to another. To compensate for this sample-to-sample variation, the quantity of RNA target is usually normalised to the quantity of one or several internal reference genes. These reference genes must be shown to be stable under the experimental conditions being examined, and are evaluated using software programmes such as geNorm or Bestkeeper. Two ideal reference genes are expected to have an identical expression ratio in all samples, whatever the experimental conditions. In the geNorm procedure (Vandesompele et al., 2002), the Cq values of each sample are transformed into relative quantities (Q) with a calibrator (cal) sample and using the gene-specific PCR efficiency (E), calculated as follows: $Q = E^{(calCq\ -\ sampleCq)}$. Normalisation is then applied by dividing the relative quantities of genes of interest by the geometric mean of the relative quantities of selected reference genes (normalisation factor). The 16S rRNA gene was used as reference gene to follow the expression of *L. lactis* nisin genes in a model cheese (Trmcic et al., 2011) . Several groups of genes could be distinguished based on expression profiles as a function of time, which contributed to a better knowledge of the regulation of nisin biosynthesis. For normalisation of gene transcripts from *Pseudomonas* spp., *Enterococcus* spp., *Pediococcus (P.) pentosaceus* and *Lb. casei* during the manufacturing of an experimental Montasio cheese, Carraro et al. (2011) used one couple of primers targeting the 16S rRNA of all bacteria present. The calculated fold-change does not reflect the specific gene expression of each population, but rather an expression taking into account the total amount of 16S rRNA. Cretenet et al. (2011) quantified the expression of several genes from *L. lactis* in model cheeses made from ultra-filtered milk, using *gyrB* (DNA gyrase subunit B) as reference gene. The histidine decarboxylase gene (*hdcA*) present in certain *Str. thermophilus* strains is involved in the synthesis histamine, a biogenic amine which may be accumulated in cheeses. The expression of *hdcA* was studied under conditions common to cheese-making, using the gene encoding the alpha subunit of the RNA polymerase (*rpoA*) as reference gene (Rossi et al., 2011). In this case, the stability of reference gene expression was assessed by absolute quantification of the transcripts obtained from fixed amounts of RNA. Up-regulation of *hdcA* occurred in the presence of free histidine and salt, and repression after thermisation. In bacteria, the gene encoding the elongation factor TU (*tuf*) is frequently used as reference gene in reverse transcription qPCR analyses. The expression of this gene by *L. lactis* was investigated in model cheeses by relative quantification using the total amount of RNA for normalisation, i.e. with reverse transcriptions performed with a fixed amount of RNA (Monnet et al., 2008). In this case, one has to check that potential biases, such as differences of reverse transcription efficiencies among the samples being studied, do not interfere. With this method, the calculated gene expression does not represent the expression relative to other mRNA transcripts, but rather the expression relative to the ribosomal RNA, which form most RNA. A large decrease of *tuf* expression, up to 100-fold, was observed after a few days. This decrease probably reflected the global decrease of mRNA transcription in the cheese matrix, after the end of growth of *L. lactis*. Duquenne et al. (2010) were able to quantify the expression of *Staphyloccus aureus* enterotoxins genes in model cheeses using a set of three stably expressed reference genes. A similar approach was applied for the study of the growth of *L. lactis* subsp. *cremoris* strains under conditions simulating cheddar cheese manufacture (Taïbi et al., 2011) and for the study of iron acquisition by *Arthrobacter arilaitensis* in experimental cheeses (Monnet et al., 2012).

Target population	Target sequence	Food matrix	References
L. lactis subsp. *lactis*	16S rRNA, 23S rRNA and 27 protein-encoding genes	Experimental cheeses	(Monnet et al., 2008)
L. lactis	11 genes involved in nisin biosynthesis	Experimental cheeses	(Trmcic et al., 2011)
L. lactis subsp. *lactis*	*tuf* (elongation factor Tu), *gapB* (glyceraldehyde 3-phosphate dehydrogenase), *purM* (phosphoribosyl-aminoimidazole synthetase), *cysK* (cysteine synthase), *ldh* (L-lactate dehydrogenase), *citD* (citrate lyase acyl-carrier protein), *gyrA* (DNA gyrase subunit A)	Experimental cheeses	(Ulvé et al., 2008)
L. lactis subsp. *lactis*	*bcaT, codY, serA, cysK, gltD, lacC, gapA, gapB, pdhB, aldB, butA, noxE, murF, dnaK, chiA, pepN, gyrB, pi139, pi302*	Experimental cheeses	(Cretenet et al., 2011)
L. lactis subsp. *cremoris*	*bcaT, clpE, dnaG, gapA, glyA, groEL, oppA, pepQ, purD, ldh, holin1, holin2*	Experimental cheeses	(Taïbi et al., 2011)
Lb. helveticus	*asnA, cysE, dapA, serA, L-ldh, clpP, oppA, oppC, pepO2, pepT2, prtH, prtH2, purA, pyrR*	Milk cultures	(Smeianov et al., 2007)
Str. thermophilus	*hdcA* (histidine decarboxylase)	Milk cultures	(Rossi et al., 2011)
Str. thermophilus	*tdcA* (tyrosine decarboxylase)	Milk cultures	(La Gioia et al., 2011)
P. freudenreichii and *Lb. paracasei*	16S rRNA, *tuf* (elongation factor TU), *Gro*L (chaperonin GroEL)	Experimental Emmental cheese	(Falentin et al., 2010)

Target population	Target sequence	Food matrix	References
Lb. helveticus and *Str. thermophilus*	16S rRNA, *tuf* (elongation factor Tu), *groL* (chaperonin GroEL)	Experimental cheeses	(Falentin et al., 2012)
Arthrobacter arilaitensis	16S rRNA, *gyrB* (DNA gyrase subunit B), *ftsZ* (cell division protein), *recA* (recombinase A), *rpoB* (RNA polymerase beta chain), *rpoA* (RNA polymerase alpha chain), *tuf* (elongation factor Tu), *dnaG* (DNA primase), and genes involved in iron acquisition	Experimental cheeses	(Monnet et al., 2012)
Str. thermophilus	two-component system genes	Milk cultures	(Thevenard et al., 2012)
Lb. casei, *P. pentosaceus*, *Str. thermophilus*, *Enterococcus* spp., *Pseudomonas* spp.	16S rRNA	Montasio cheese manufacturing	(Carraro et al., 2011)
Yeasts and moulds	actin gene	Commercial food products, including milk and yoghurt	(Bleve et al., 2003)
S. aureus	*gyrB* (DNA gyrase subunit B), *ftsZ* (cell division protein), *hu* (DNA-binding protein), *rplD* (50S ribosomal protein L4), *recA* (recombinase A), *sodA* (superoxide dismutase), *gap* (glyceraldehyde-3-phosphate dehydrogenase), *rpoB* (RNA polymerase beta chain), *pta* (phosphate acetyltransferase), *tpi* (triose phosphate isomerase), *sea* (enterotoxin A), *sed* (enterotoxin D)	Experimental cheeses	(Duquenne et al., 2010)
S. aureus	16S rRNA, *nuc* (thermonuclease)	Artificially contaminated Camembert cheeses	(Ablain et al., 2009) (Fumian et al., 2009)
Noroviruses	ORF1-ORF2 junction region	Artificially contaminated cheeses	(Fumian et al., 2009)

Table 2. Examples of applications of reverse-transcription qPCR to dairy products.

3. Application of PCR-based methods to non-dairy probiotic products

3.1 Nucleic acid extraction from non-dairy probiotic products

Probiotic products comprise probiotic dairy products and probiotic food supplements which appear in several forms, like powders, capsules, tablets, suspensions etc. containing the lyophilised, dried or microencapsulated bacterial cells. Since an overview of the nucleic acid extraction and PCR application in dairy products in general have already been addressed in this chapter, we focus here on the non-dairy probiotic products such as food supplements or pharmaceutical preparations. The protocols of DNA or RNA extraction from different probiotic products have to be properly adapted to the matrix in order to achieve satisfactory yield and efficient PCR amplification. It is important to evaluate whether the components of the product other than microbial cells influence the extraction and amplification steps. Probiotic formulations may contain polysaccharides, salts, oils (microencapsulated) or proteins (milk-based) which have been demonstrated to affect the extraction or inhibit amplification by direct interaction with DNA or by interference with the polymerases used in PCR. DNA isolation from the samples containing milk which is among the common ingredients of probiotic formulations, requires multiple steps such as centrifugation, heating or cation exchange to remove proteins, calcium ions and fats (Cressier and Bissonnette, 2011).

An increasing amount of non-dairy probiotic products contain microencapsulated probiotic cells. Depending on the microencapsulation technique (spray-drying, coacervation, co-crystallisation, molecular inclusion) and the matrix and coating materials used, the physico-chemical properties of microcapsules differ much. Microcapsules containing probiotic bacteria are often insoluble in water, in order to allow their controlled release in the intestine. In order to enable the release of bacterial cells and DNA to the medium, particular treatment and diluents different from the commonly used (Ringer solution, peptone saline solution, water) are needed, for example addition of emulsifiers (anionic, cationic) or non-ionic detergents such as Tween 80 (Champagne et al., 2010; Burgain et al., 2011).

When probiotics are microencapsulated in alginate beads, a calcium-binding solution such as phosphates or citrates is most often used to dissolve the particles. Another problem presents dried, fat-based spray-coated probiotic bacteria which can be found in different products in a form of powders, capsules, tablets, suspension in oil or for example in chocolate. One of the concerns could be that fat coating on the particles would prevent hydration, resulting in unsatisfactory recovery of viable bacteria and under-estimation of CFU counts.

The selection of rehydration method and solutions significantly influenced the results of CFU determination by plate counting in microencapsulated *Lb. rhamnosus* R0011 or *Bifidobacterium (B.) longum* ATCC 15708 cultures spray-coated with fat. Tween 80 did not result in direct improvement of the recovery of CFU, while the addition of fat improved it. The authors concluded that the methods appropriate for the analysis of free cells in dried probiotics may not be optimal for the analysis of spray-coated ME cultures (Champagne et al., 2010). The recovery of dried probiotic cultures is greatly dependent on the reconstitution conditions. Maximum recovery of *B.standardised longum* NCC3001 was achieved at 30-min reconstitution at pH 8, in the presence of 2% l-arabinose and with a ratio of 1:100 of powder to diluent, while *Lb. johnsonii* La1 showed highest recovery after reconstitution, when mixed with maltodextrin at pH 4 (Muller et al., 2010). The published data on the optimisation of DNA isolation from microencapsulated bacteria are scarce however since the first step of bacterial DNA isolation from the product is separation of the bacterial cells from the matrix, the conditions and procedures found suitable for viable count (CFU) determination in samples containing microencapsulated bacteria may be a good starting point also for DNA isolation.

Due to the specificities described above, there are no universal standard procedures and media/buffers for the rehydration of probiotic products and quantification of probiotics in such products, either by the assessment of viable counts or by PCR-based methods. Often the authors do not explain in detail the preparation of the samples of probiotic products but refer to the standards such as ISO 6887-1:2000 on the general rules for the preparation of the initial suspension and decimal dilutions of food and animal feeding stuffs, or ISO 6887-5:2010 including specific rules for the preparation of milk and milk products which are applicable also to dried milk products and milk-based infant foods. ISO 20838:2006 provides the overall framework for qualitative methods for the detection of food-borne pathogens in or isolated from food and feed matrices using the polymerase chain reaction (PCR), but can also be applied to other matrices, for example environmental samples, or to the detection of other microorganisms under investigation. However, the standards do not contain detailed protocols which have to be developed specifically considering the properties of the products.

Champagne et al. (2011) recently published recommendations for the viability assessment of probiotics as concentrated cultures and in food matrices by plate counting, but the recommendations relevant for the DNA isolation are not available.

Microbial analysis of probiotic food supplements and pharmaceutical preparations require standardised and accurate procedures for the reactivation of dehydrated cultures. Among the resuspension buffers, ¼ Ringer solution with or without cysteine (0,05 %), peptone physiological solution (0.1% wt/vol peptone, 0.85% wt/vol NaCl) or water are used most often (Temmerman et al., 2003; Masco et al., 2005; Masco et al., 2007; Kramer et al., 2009; Bogovic Matijasic et al., 2010). For the preparation of mesophilic cultures for qPCR analysis, which present similar medium as probiotic formulations, Friedrich and Lenke (2006) used PBS and sodium citrate (1% wt/vol).

Usually the probiotic cells are removed by centrifugation from the product matrix before being exposed to the cell lysis. Drisko et al. (2005) exceptionally resuspended the products directly in TE buffer (10 mM Tris-HCl with pH 8.0, 1 mM EDTA) and proceeded with SDS and proteinase K treatment. After the lysis of bacterial cells, phenol/chloroform extraction or different kits such as the QIAamp®DNA stool mini kit (Qiagen), the NucleoSpin® food kit (Macherey-Nagel), Wizard Genomic DNA Purification kit (Promega), Maxwell 16 Cell DNA Purification Kit (Promega) are most commonly used.

Lyophilised probiotic products can also be resuspended in water and the suspension added directly in PCR mixture, without previous isolation of bacterial DNA. This way Vitali et al. (2003) for instance carried out the real-time PCR quantification of three *Bifidobacterium* strains in a pharmaceutical product VSL-3 containing lyophilised bacteria and excipients.

Target population	Method	Target sequence	Form of product	References
B. bifidum, Bacillus coagulans, Lb. acidophilus, Lb. casei, Lb. delbrueckii subsp. *bulgaricus, Lb. delbrueckii* subsp. *lactis, Lb. helveticus, Lb. kefiri, Lb. paracasei, Lb. plantarum, Lb. reuteri, Lb. rhamnosus, Lb. salivarius, Lc. Lactis, P. freudenreichii subsp. freudenreichii, P. freudenreichii subsp. shermanii, Str. thermophilus*	PCR	16S rDNA, 16S-23S IS, *htrA, pepIP, rpoA*	capsules, tablets, powder sachets, chewable tablets, bottled products	(Aureli et al., 2010)
Lb. gasseri, E. faecium, B. infantis	real-time PCR	16S rDNA, 16S-23S IS	capsules	(Bogovic Matijasic et al., 2010)
Lb. acidophilus, Lc. lactis, E. faecium, B. bifidum, B. lactis, Lb. rhamnosus, Lb. helveticus, Bacillus cereus, Lb. delbrueckii subsp. bulgaricus, Str. thermophilus	PCR-DGGE	16S rDNA	capsules, tablets	(Temmerman et al., 2003)
Lb. delbrueckii subsp. *bulgaricus, Lb. salivarius, Lb. plantarum, Lb. rhamnosus, Lb. acidophilus, B. infantis, Lb. casei, Lb. brevis, B. lactis, Str. thermophilus, B. bifidum*	PCR	16S rDNA, 16S-23S IS, β-galactosidase gene	not stated	(Drisko et al., 2005)
Lb. acidophilus, B. animalis subsp. lactis	real-time PCR	16S rDNA	capsules	(Kramer et al., 2009)
B. animalis subsp. lactis, B. longum biotype *longum, B. bifidum, B. animalis* subsp. *lactis, B. bifidum, B. breve, B. longum* biotype *longum, B. longum* biotype *infantis*	nested PCR-DGGE	16S rDNA	not stated	(Masco et al., 2005)
B. animalis subsp. *lactis, B. breve, B. bifidum, B. longum* biotype *longum*	real-time PCR	16S rDNA, *recA* genes	capsules, powders, tablets	(Masco et al., 2007)
B.standardised infantis Y1, B.standardised breve Y8, B.standardised longum Y10	PCR, real-time PCR	16S rDNA, 16S-23S IS	powder sachets	(Vitali et al., 2003)
Lb. acidophilus, B.standardised infantis v. liberorum, Ent. faecium, B. bifidum, Lb. delbrueckii subsp. *bulgaricus, Str. thermophilus, B. longum, B. breve, Lb. rhamnosus, L. lactis*	PCR	16S rDNA, 16S-23S IS	Capsules, powder, pastilles	(Bogovic Matijasic and Rogelj, 2006)

Table 3. Examples of applications of PCR, qPCR or PCR-DGGE to probiotic food supplements or pharmaceutical products.

3.2 Detection or quantification of probiotics in non-dairy probiotic products by PCR

3.2.1 PCR detection of labelled probiotic bacteria in probiotic food supplements or pharmaceutical preparations

Probiotic food supplements and pharmaceutical preparations are widespread and commercially important. The most important parameters of their quality are appropriate labelling of probiotic bacteria and adequate number of them in the products. This is still not such an easy task since standardised methods are available for only avery limited number of probiotic bacteria in dairy products such as Lb. *acidophilus* (ISO 20128/IDF 192:2006) and *Bifidobacterium* (ISO 29981/IDF 220:2010). This speaks in favour of using molecular techniques which are rapid, sensitive and specific. Several PCR tests for detection of pathogens in foods have been validated, harmonised, and commercialised to make PCR a standard tool used by food microbiology laboratories (Maurer, 2011; Postollec et al., 2011). In the probiotic field there is still much to do in terms of the application of PCR-based methods for the control of probiotic products. Conventional PCR is very useful for the detection of labelled species or genera in the probiotic products. While several applications of this technique in food, including probiotic fermented dairy products, can be found in the literature (Table 1), the reports dealing with probiotic food supplements or pharmaceutical preparations are still few (Table 3). Among the targets which have been used in PCR analysis of probiotic products in the form of capsules, tablets or powders there are most often 16S rDNA or 16S-23S intergenic spacer (IS) regions which appear in the cells in multiple copies, contain several species or genus-specific regions and enable higher sensitivity than single copy genes. In addition to the ribosomal genes, several monocopy genes have also already been used for PCR or real-time PCR of probiotics such as *htrA*, *pepIP*, *rpoA*, β-galactosidase gene, or *recA* gene (Table 3). Primers for *htrA*-trypsin-like serine protease gene were used originally by Fortina et al. (2001), for *pepIP*-immunopeptidase proline gene pepIP by Torriani et al. (2007) and for *rpoA*-RNA polymerase alpha subunit gene by Naser et al. (2007). The main advantage of the application of genes that usually appear in one copy is that they enable accurate quantification by real-time PCR also in the mixed populations of bacteria belonging to different species, while the number of rRNA genes copies differs among the species.

3.2.2 Real-time PCR quantification of probiotic bacteria in non-dairy products

It is well known that many food ingredients, including fats, proteins, divalent cations, and phenolic compounds, can act as PCR inhibitors. Some of the ingredients may also hinder the adequate microbial cell separations from the sample matrix. Another common problem is non-heterogeneous distribution of target cells in the samples, the presence of microbial aggregates which are difficult to disrupt or high amounts of non-target microbiota (Brehm-Stecher et al., 2009). In the analysis of probiotic products in general the usual approach is to separate first the bacterial target cells from the matrix, which in the case of lyophilised or dried products is usually not such a difficult task and may be successfully performed by rehydration of the samples followed by centrifugation. This way most of the potential inhibitory compounds are removed. Inhibitors are further removed also during the nucleic acids purification steps which have been described above. However, as some of the inhibitors may still be present in the samples intended for quantitative PCR (qPCR) analysis, the examination of possible inhibition of PCR reaction is always required.

In order to exclude possible inhibition, Masco et al. (2007) prepared bacteria-free sample matrices of the food supplement, spiked them with known quantities of reference bifidobacteria and compared the standard curve slopes and efficiencies obtained during PCR amplification of pure cultures and spiked samples. The finding that amplification of pure cultures and spiked samples was equally efficient indicated that the product's matrix did not have a significant impact on DNA extraction and subsequent real-time PCR performance.

Similarly Kramer et al. (2009) prepared the standard curves from the mixture of bacterial cells of *Lb. acidophilus* or *B. animalis* ssp. *lactis* with a suspension of filler ingredients of probiotic capsules. The concentrations of Beneo synergy (0,73%), saccharose (0,11%), dextrose anhydrous (0,10%), microcrystalline cellulose (0,026 %), potato starch (0,026 %) and Mg-stearate (0,019 %) in the standard samples were the same as in the 1:100 diluted product. In addition, the negligible effect of the product ingredients on the PCR amplification efficiency was demonstrated also by the comparison of the standard curves prepared from the DNA derived from pure cultures of from the suspensions of cultures in the simulated filler.

In a further study of the same probiotic pharmaceutical preparation (Bogovič Matijašić, not published) the authors treated 1% (w/v) suspension of the product with heat (two times 120 °C/15 min). The total DNA in the suspension was mostly degraded as was demonstrated by real-time PCR amplification using *Lactobacillus* (LactoR'F/LBFR, (Songjinda et al., 2007)) or *Bifidobacterium* (Bif-F/Bif-R, (Rinttila et al., 2004). The two-times autoclaved suspension was spiked with either of the two strains isolated from the product, and after that DNA isolated from the spiked suspension was used for the generation of standard curves.

Bogovič Matijašić et al. (2010) prepared the simulated matrix with Mg stearate (0.22%), lactose (0.39%) and starch (0.39%) corresponding to the concentrations of these ingredients in a 100-fold sample of the product in capsules. DNA was isolated by different procedures from the standard samples containing simulated matrix with a known amount of added probiotic bacteria of *Lb. gasseri, B. infantis* or *Ec. faecium*. When DNA was isolated by heat treatment (100 °C/5 min) of the standard bacterial suspensions in 1% Triton X-100, the ingredients of the prepared suspension affected the real-time PCR result. Since the filler ingredients themselves did not show any fluorescence interaction when included directly in PCR reactions, the lower concentration of probiotic determined in real-time PCR was attributed to the less effective DNA extraction by heat-triton treatment due to the presence of Mg stearate, lactose and starch. Any effect was however observed when DNA was isolated by the Maxwell system (Promega) based on the use of MagneSil paramagnetic particles (Bogovic Matijasic et al., 2010).

In all studies presented in Table 3, the real- time PCR analyses were performed by SYBR® Green I chemistry. The species specificity of the PCR was ensured by using species-specific oligonucleotide primers and additionally validated by melting point analysis.

3.2.3 Viability determination of probiotics by PCR-based methods

The viability of probiotic bacteria is traditionally assessed by plate counting which has several limitations, such as unsatisfactory selectivity, too-low a recovery, long incubation time, underestimation of cells in aggregates or chains morphology etc. (Breeuwer and Abee,

2000). Real-time PCR has a potential to replace conventional enumeration of probiotic bacteria, used for routine monitoring of quality of a probiotic product and for stability studies. However, since probiotic bacteria have to be viable to exert their activity the contribution of DNA arising from non-viable cells to the result of quantification has to be excluded.

An approach using PMA or EMA treatment of the samples before the DNA isolation seems promising in this regard. Such DNA-intercalating dyes are able to bind upon exposure to bright visible light to DNA and, consequently, to inhibit PCR amplification of the DNA which is free or inside the bacterial cells with the damaged membrane. Although probiotic bacteria in the products are represented in different stages not only as viable or dead (Bunthof and Abee, 2002), the most important criterion for distinguishing between viable and irreversibly damaged cells is membrane integrity. The treatment of bacteria with EMA as a promising tool of DNA-based differentiation between viable and dead pathogenic bacteria was first proposed by Nogva et al. in 2003 (Nogva et al., 2003). In the following years several applications of this approach have been reported, where the method was optimised for different complex media such as faeces, fermented milk and environmental samples (Garcia-Cayuela et al., 2009; Fittipaldi et al., 2011; Fujimoto et al., 2011). Since ethidium monoazide has been suggested as being toxic to some viable cells, PMA has been proposed as a more appropriate alternative to EMA (Nocker et al., 2006; Fujimoto et al., 2011).

The PMA treatment in combination with real-time PCR was applied for determination of probiotic strains *Lb. acidophilus* LA-5 and *B. animalis* ssp. *lactis* BB-12 bacteria in a pharmaceutical formulation in the form of capsules (Kramer et al., 2009). The possible effects of the ingredients of the product on PMA treatment of the samples including the photo-activation step, as well as on the PCR reaction were evaluated in the study. The ability of PMA to inhibit amplification of DNA derived from damaged bacterial cells was confirmed on bacteria from pure cultures of *Lb. acidophilus* or *B. animalis* ssp. *lactis* in a 1% (w/v) suspension of ingredients which are otherwise present in the product and on probiotic product (1% w/w). Other examples of direct application of PMA-real time PCR on the lyophilised probiotic products have not been found in the literature. The efficient PMA treatment of fermented dairy products containing the same two strains, *Lb. acidophilus* LA-5 and *B. animalis* ssp. *lactis* BB-12, have also been described (Garcia-Cayuela et al., 2009). In order to eliminate the milk ingredients prior to the PMA treatment, the samples were adjusted to pH 6.5 with 1 M NaOH, then casein micelles were dispersed by theaddition of 1 M trisodium citrate, and bacterial cells were harvested by centrifugation. The obtained cells were resuspended in water, treated with PMA and used for DNA isolation. Fujimoto et al. (2010) evaluated strain-specific qPCR with PMA treatment for quantification of viable *B. breve* strain Yakult (BbrY) in human faeces. The quantification was carried out on faecal samples spiked with BbrY strain, on the BbrY culture and on the faecal samples collected from the healthy volunteers who ingested a commercially available fermented milk product containing BbrY, once daily for 10 days. They confirmed the use of a combination of qPCR with PMA treatment and BbrY-specific primers as a quick and accurate method for quantification of viable BbrY in faecal samples (Fujimoto et al., 2011).

Viable probiotics may be enumerated also by a qPCR-based method targeting mRNA of different housekeeping genes. The advantage of using mRNA targets over the use of DNA or rRNA is mainly in the instability of mRNA molecules which is degraded soon after the cell death. Reimann et al. (2010) demonstrated in *B. longum* NCC2705 a good correlation between measured mRNA levels of cysB and purB, two constitutively expressed housekeeping genes and plate counts. The 400-bp fragment of purB was degraded more quickly than the 57-bp fragments of cysB and purB, and is therefore a better marker of cell viability (Reimann et al., 2010).

With the availability of new highthroughput molecular technologies such as microarray technology and next-generation sequencing, new possibilities are now open to further development of the viability PCR approach also in the probiotic field, as has already been similarly demonstrated for selected pathogenic bacteria in environmental samples (Nocker et al., 2009; Nocker et al., 2010).

3.3 Strain-specific detection or quantification of probiotics

While species- or genus-specific primers are not so difficult to construct, the problem arises when we intend to confirm different strains of the same species in the product. A variety of PCR-based genotyping techniques such as random amplified polymorphic DNA analysis (RAPD), repetitive sequence-based PCR (rep-PCR), pulsed-field gel electrophoresis (PFGE), amplified fragment length polymorphism (AFLP) ribotyping etc., are successfully used everywhere to distinguish different strains also closely related among each other (Li et al., 2009). The genotyping methods, however, require the cultivation of pure cultures of examined strains and do not enable quantification. For PCR quantification of individual probiotic strains in the probiotic products or different environments (faeces, mucosa...) strain specific primers or probes are needed. So far it has been very difficult to find strain-specific genome sequences as a target for the construction of strain-specific primers or probes.

In the study of Vitali et al. (2003), the 16S rDNA and 16S-23S rDNA-targeted strain-specific primers were designed for the quantitative detection of *B. infantis* Y1, *B. breve* Y8 and *B. longum* Y10 used in a pharmaceutical probiotic product VSL-3. These were applied in PCR, and real-time PCR techniques with the selected primers were employed for the direct enumeration of the bifidobacteria in the probiotic preparation and for studying their kinetic characteristics in batch cultures (Vitali et al., 2003).

Maruo et al. (2006) generated a *L. lactis* subsp. *cremoris* FC-specific primer pair by using a specific 1164-bp long RAPD band sequence. The specificity of this primer pair has been proven with 23 *L. lactis* subsp. *cremoris* strains and 20 intestinal bacterial species, and real-time PCR determination of FC strain in the faeces was demonstrated to be successful. Marzotto et al. (2006) selected specific primers for the putative probiotic strain *Lb. paracasei* A LcA-Fw and LcA-Rv from the terminal regions of the 250-bp RAPD fragment sequence tested the selectivity with 20 different *Lactobacillus* species and 39 *Lb. paracasei* strains. The primers were successfully applied in PCR analysis of faecal samples (Marzotto et al., 2006).

Strain-specific PCR primers and probes for real-time PCR and for conventional PCR were designed based on the sequence of RAPD products, also for *Lb. rhamnosus* GG which is one of the most studied probiotic strains (Ahlroos and Tynkkynen, 2009). The strain specificity of the primers was verified in conventional PCR using a set of strains – six *Lb. rhamnosus*, one *Lb. casei* and one *Lb. zeae*, while the applicability of the GG strain-specific primer probe set was confirmed on the human faecal samples by LightCycler (Roche Diagnostics) real-time PCR.

A similar approach was applied to *B. breve* strain Yakult (BbrY) by Fujimoto et al. (2011). The specificity of the BbrY-specific primer set was confirmed by PCR using DNA from 112 bacterial strains belonging to *B. breve* species, of other *Bifidobacterium* species and representatives of 11 other genera. The BbrY-specific primers were used in a real-time PCR with PMA treatment to measure the number of BbrY in the faeces of subjects who drank a fermented milk product containing BbrY (Fujimoto et al., 2011).

The qPCR method based on the amplification of a strain-specific DNA fragment identified by suppressive subtractive hybridisation was developed recently for specific and sensitive monitoring of *P. acidipropionici* P169 in animal feed and rumen fluid by Peng et al. (2011). The specificity and amplification efficiency was assessed on 44 *Propionibacterium* strains and also in complex microbial communities containing *P. acidipropionici* P169 (Peng et al., 2011).

Certain strains have specific features that distinguish them from the other related strains, such as for example bacteriocin production. Treven et al. (submitted) evaluated the possibility of using bacteriocin-specific primers for the detection and quantification of *Lb. gasseri* K7 probiotic strain, a producer of at least two two-component bacteriocins (Zoric Peternel et al., 2010). Two pairs of primers, namely GasA_401/610F/R and GasB_2610-2807F/R showed specificity for total gene cluster of gassericin K7 A (Genbank EF392861) or gassericin K7 B (Genbank AY307382) respectively as established by PCR assays using DNA of 18 reference strains belonging to *Lb. acidophilus* group and 45 faecal samples of adult volunteers who have never consumed K7 strain. GasA_401/610F/R primers were also found to be especially useful also for real-time PCR quantification of gassericin K7 A gene cluster in faecal samples and also for *Lb. gasseri* K7-specific detection or quantification in the biological samples (Treven et al. submitted).

4. Conclusions

Microorganisms are very important components of fermented dairy products, including probiotic food, as well as of probiotic food supplements and pharmaceutical preparations. PCR-based methods have become indispensable in the microbiological analysis of these groups of products. In the field of fermented dairy products, several applications based on PCR have been developed with the aim to detect, identify and quantify either unwanted bacteria, which may negatively influence the sensory properties of food or may be pathogenic, or beneficial microorganisms which are added as starter cultures or probiotic cultures. Beside PCR analysis of DNA, reverse transcription real-time PCR analysis of mRNA transcript is particularly useful, especially in studies of the physiology and functionality of bacteria in the food environment. In the probiotic field, PCR is expected to be increasingly applied in quality control in terms of detection and quantification of labelled probiotic bacteria in probiotic food supplements or pharmaceutical preparations, and in viability analysis of probiotics in the products. In addition to the already well-established methods described in this chapter, ever easier access to the next generation sequencing may replace some PCR approaches as molecular fingerprint, metagenomic and metatranscriptomic analyses. The access to increasing number of complete bacterial genomes may also facilitate the strain-specific analysis of probiotics or other bacteria through identification of strain-specific sequences.

5. References

Ablain, W., Hallier Soulier, S., Causeur, D., Gautier, M., Baron, F., 2009. A simple and rapid method for the disruption of *Staphylococcus aureus*, optimized for quantitative reverse transcriptase applications: Application for the examination of Camembert cheese. Dairy Sci. Technol. 89, 69-81.

Ablain, W., Hallier Soulier, S., Causeur, D., Gautier, M., Baron, F., 2009. A simple and rapid method for the disruption of *Staphylococcus aureus*, optimized for quantitative reverse transcriptase applications: Application for the examination of Camembert cheese. Dairy Sci. Technol. 89, 69-81.

Abriouel, H., Martin-Platero, A., Maqueda, M., Valdivia, E., Martinez-Bueno, M., 2008. Biodiversity of the microbial community in a Spanish farmhouse cheese as revealed by culture-dependent and culture-independent methods. Int. J. Food Microbiol. 127, 200-208.

Ahlroos, T., Tynkkynen, S., 2009. Quantitative strain-specific detection of *Lactobacillus rhamnosus* GG in human faecal samples by real-time PCR. J. Appl. Microbiol. 106, 506-514.

Alarcon, B., Vicedo, B., Aznar, R., 2006. PCR-based procedures for detection and quantification of *Staphylococcus aureus* and their application in food. J. Appl. Microbiol. 100, 352-364.

Alegría, Á., Álvarez-Martín, P., Sacristán, N., Fernández, E., Delgado, S., Mayo, B., 2009. Diversity and evolution of the microbial populations during manufacture and ripening of Casín, a traditional Spanish, starter-free cheese made from cow's milk. Int. J. Food Microbiol. 136, 44-51.

Allmann, M., Hofelein, C., Koppel, E., Luthy, J., Meyer, R., Niederhauser, C., Wegmuller, B., Candrian, U., 1995. Polymerase chain reaction (PCR) for detection of pathogenic microorganisms in bacteriological monitoring of dairy products. Res. Microbiol. 146, 85-97.

Amoroso, M.G., Salzano, C., Cioffi, B., Napoletano, M., Garofalo, F., Guarino, A., Fusco, G., 2011. Validation of a Real-time PCR assay for fast and sensitive quantification of *Brucella* spp. in water buffalo milk. Food Control 22, 1466-1470.

Andrighetto, C., Marcazzan, G., Lombardi, A., 2004. Use of RAPD-PCR and TTGE for the evaluation of biodiversity of whey cultures for Grana Padano cheese. Lett. Appl. Microbiol. 38, 400-405.

Aprodu, I., Walcher, G., Schelin, J., Hein, I., Norling, B., Rådström, P., Nicolau, A., Wagner, M., 2011. Advanced sample preparation for the molecular quantification of *Staphylococcus aureus* in artificially and naturally contaminated milk. Int. J. Food Microbiol. 145, S61-S65.

Arteau, M., Labrie, S., Roy, D., 2010. Terminal-restriction fragment length polymorphism and automated ribosomal intergenic spacer analysis profiling of fungal communities in Camembert cheese. Int. Dairy J. 20, 545-554.

Aureli, P., Fiore, A., Scalfaro, C., Casale, M., Franciosa, G., 2010. National survey outcomes on commercial probiotic food supplements in Italy. Int. J. Food Microbiol. 137, 265-273.

Baker, G.C., Smith, J.J., Cowan, D.A., 2003. Review and re-analysis of domain-specific 16S primers. J. Microbiol. Meth. 55, 541-555.

Baruzzi, F., Matarante, A., Caputo, L., Morea, M., 2005. Development of a culture-independent polymerase chain reaction-based assay for the detection of lactobacilli in stretched cheese. J. Rapid Meth. Aut. Mic. 13, 177-192.

Bleve, G., Rizzotti, L., Dellaglio, F., Torriani, S., 2003. Development of reverse transcription (RT)-PCR and real-time RT-PCR assays for rapid detection and quantification of viable yeasts and molds contaminating yogurts and pasteurized food products. Appl. Environ. Microbiol. 69, 4116-4122.

Bogovic Matijasic, B., Koman Rajsp, M., Perko, B., Rogelj, I., 2007. Inhibition of *Clostridium tyrobutyricum* in cheese by *Lactobacillus gasseri*. Int. Dairy J. 17, 157-166.

Bogovic Matijasic, B.B., Obermajer, T., Rogelj, I., 2010. Quantification of *Lactobacillus gasseri, Enterococcus faecium* and *Bifidobacterium infantis* in a probiotic OTC drug by real-time PCR. Food Control 21, 419-425.

Bogovic Matijasic, B.B., Rogelj, I., 2006. Demonstration of suitability of probiotic products - An emphasis on survey of commercial products obtained on Slovenian market. Agro Food Ind. Hi-Tech 17, 38-40.

Bonaiti, C., Parayre, S., Irlinger, F., 2006. Novel extraction strategy of ribosomal RNA and genomic DNA from cheese for PCR-based investigations. Int. J. Food Microbiol. 107, 171-179.

Bonetta, S., Bonetta, S., Carraro, E., Rantsiou, K., Cocolin, L., 2008. Microbiological characterisation of Robiola di Roccaverano cheese using PCR-DGGE. Food Microbiol. 25, 786-792.

Boss, R., Naskova, J., Steiner, A., Graber, H.U., 2011. Mastitis diagnostics: Quantitative PCR for *Staphylococcus aureus* genotype B in bulk tank milk. J. Dairy Sci. 94, 128-137.

Botsaris, G., Slana, I., Liapi, M., Dodd, C., Economides, C., Rees, C., Pavlik, I., 2010. Rapid detection methods for viable *Mycobacterium avium* subspecies *paratuberculosis* in milk and cheese. Int. J. Food Microbiol. 141, S87-S90.

Breeuwer, P., Abee, T., 2000. Assessment of viability of microorganisms employing fluorescence techniques. Int. J. Food Microbiol. 55, 193-200.

Brehm-Stecher, B., Young, C., Jaykus, L.A., Tortorello, M.L., 2009. Sample Preparation: The Forgotten Beginning. J. Food Prot. 72, 1774-1789.

Bremer, H., Dennis, P.P., 1996. Modulation of chemical composition and other parameters of the cell by growth rate, in: Neidhart, F.C. (Ed.), Escherichia coli and Salmonella: Cellular and Molecular Biology, ASM Press, Washington DC, pp. 1553-1569.

Bunthof, C.J., Abee, T., 2002. Development of a flow cytometric method to analyze subpopulations of bacteria in probiotic products and dairy starters. Appl. Environ. Microbiol. 68, 2934-2942.

Burgain, J., Gaiani, C., Linder, M., Scher, J., 2011. Encapsulation of probiotic living cells: From laboratory scale to industrial applications. J. Food Eng. 104, 467-483.

Bustin, S.A., 2000. Absolute quantification of mRNA using real-time reverse transcription polymerase chain reaction assays. J. Mol. Endocrinol. 25, 169-193.

Bustin, S.A., Benes, V., Garson, J.A., Hellemans, J., Huggett, J., Kubista, M., Mueller, R., Nolan, T., Pfaffl, M.W., Shipley, G.L., Vandesompele, J., Wittwer, C.T., 2009. The MIQE Guidelines: Minimum Information for Publication of Quantitative Real-Time PCR Experiments. Clin. Chem. 55, 611-622.

Cailliez-Grimal, C., Miguindou-Mabiala, R., Leseine, M., Revol-Junelles, A.M., Milliere, J.B., 2005. Quantitative polymerase chain reaction used for the rapid detection of

Carnobacterium species from French soft cheeses. FEMS Microbiol. Lett. 250, 163-169.
Callon, C., Delbes, C., Duthoit, F., Montel, M.C., 2006. Application of SSCP-PCR fingerprinting to profile the yeast community in raw milk Salers cheeses. Syst. Appl. Microbiol. 29, 172-180.
Cardenas, E., Tiedje, J.M., 2008. New tools for discovering and characterizing microbial diversity. Curr. Opin. Biotechnol. 19, 544-549.
Carraro, L., Maifreni, M., Bartolomeoli, I., Martino, M.E., Novelli, E., Frigo, F., Marino, M., Cardazzo, B., 2011. Comparison of culture-dependent and -independent methods for bacterial community monitoring during Montasio cheese manufacturing. Res. Microbiol. 162, 231-239.
Casalta, E., Sorba, J.-M., Aigle, M., Ogier, J.-C., 2009. Diversity and dynamics of the microbial community during the manufacture of Calenzana, an artisanal Corsican cheese. Int. J. Food Microbiol. 133, 243-251.
Champagn, C.P., Ross, R.P., Saarela, M., Hansen, K.F., Charalampopoulos, D., 2011. Recommendations for the viability assessment of probiotics as concentrated cultures and in food matrices. Int. J. Food Microbiol. 149, 185-193.
Champagne, C.P., Raymond, Y., Tompkins, T.A., 2010. The determination of viable counts in probiotic cultures microencapsulated by spray-coating. Food Microbiol. 27, 1104-1111.
Chen, J., Zhang, L., Paoli, G.C., Shi, C., Tu, S.-I., Shi, X., 2010. A real-time PCR method for the detection of *Salmonella enterica* from food using a target sequence identified by comparative genomic analysis. Int. J. Food Microbiol. 137, 168-174.
Chiang, Y.-C., Fan, C.-M., Liao, W.-W., Lin, C.-K., Tsen, H.-Y., 2007. Real-Time PCR Detection of *Staphylococcus aureus* in Milk and Meat Using New Primers Designed from the Heat Shock Protein Gene htrA Sequence. J. Food Prot. 174; 70, 2855-2859.
Chomczynski, P., Sacchi, N., 1987. Single-step method of RNA isolation by acid guanidinium thiocyanate-phenol-chloroform extraction. Anal. Biochem. 162, 156-159.
Cikos, S., Koppel, J., 2009. Transformation of real-time PCR fluorescence data to target gene quantity. Anal. Biochem. 384, 1-10.
Cocolin, L., Diez, A., Urso, R., Rantsiou, K., Comi, G., Bergmaier, I., Beimfohr, C., 2007. Optimization of conditions for profiling bacterial populations in food by culture-independent methods. Int. J. Food Microbiol. 120, 100-109.
Cocolin, L., Innocente, N., Biasutti, M., Comi, G., 2004. The late blowing in cheese: a new molecular approach based on PCR and DGGE to study the microbial ecology of the alteration process. Int. J. Food Microbiol. 90, 83-91.
Cogan, T.M., John, W.F., 2011. Cheese - Microbiology of Cheese, Encyclopedia of Dairy Sciences, Academic Press, San Diego, pp. 625-631.
Coppola, S., Blaiotta, G., Ercolini, D., Moschetti, G., 2001. Molecular evaluation of microbial diversity occurring in different types of Mozzarella cheese. J. Appl. Microbiol. 90, 414-420.
Cressier, B., Bissonnette, N., 2011. Assessment of an extraction protocol to detect the major mastitis-causing pathogens in bovine milk. Journal of Dairy Science 94, 2171-2184.
Cretenet, M., Laroute, V., Ulve, V., Jeanson, S., Nouaille, S., Even, S., Piot, M., Girbal, L., Le Loir, Y., Loubiere, P., Lortal, S., Cocaign-Bousquet, M., 2011. Dynamic Analysis of

the *Lactococcus lactis* Transcriptome in Cheeses Made from Milk Concentrated by Ultrafiltration Reveals Multiple Strategies of Adaptation to Stresses. Appl. Environ. Microbiol. 77, 247-257.

Delavenne, E., Mounier, J., Asmani, K., Jany, J.-L., Barbier, G., Le Blay, G., 2011. Fungal diversity in cow, goat and ewe milk. Int. J. Food Microbiol. 151, 247-251.

Delbes, C., Ali-Mandjee, L., Montel, M.-C., 2007. Monitoring Bacterial Communities in Raw Milk and Cheese by Culture-Dependent and -Independent 16S rRNA Gene-Based Analyses. Appl. Environ. Microbiol. 73, 1882-1891.

Delbes, C., Montel, M.C., 2005. Design and application of a *Staphylococcus*-specific single strand conformation polymorphism-PCR analysis to monitor *Staphylococcus* populations diversity and dynamics during production of raw milk cheese. Lett. Appl. Microbiol. 41, 169-174.

Dolci, Barmaz, Zenato, Pramotton, Alessandria, Cocolin, Rantsiou, Ambrosoli, 2009. Maturing dynamics of surface microflora in Fontina PDO cheese studied by culture-dependent and -independent methods. J. Appl. Microbiol. 106, 278-287.

Dolci, P., Alessandria, V., Rantsiou, K., Bertolino, M., Cocolin, L., 2010. Microbial diversity, dynamics and activity throughout manufacturing and ripening of Castelmagno PDO cheese. Int. J. Food Microbiol. 143, 71-75.

Donaghy, J.A., Rowe, M.T., Rademaker, J.L.W., Hammer, P., Herman, L., De Jonghe, V., Blanchard, B., Duhem, K., Vindel, E., 2008. An inter-laboratory ring trial for the detection and isolation of *Mycobacterium avium* subsp. *paratuberculosis* from raw milk artificially contaminated with naturally infected faeces. Food Microbiol. 25, 128-135.

Drisko, J., Bischoff, B., Giles, C., Adelson, M., Rao, R.V.S., McCallum, R., 2005. Evaluation of five probiotic products for label claims by DNA extraction and polymerase chain reaction analysis. Dig. Dis. Sci. 50, 1113-1117.

Duquenne, M., Fleurot, I., Aigle, M., Darrigo, C., Borezee-Durant, E., Derzelle, S., Bouix, M., Deperrois-Lafarge, V., Delacroix-Buchet, A., 2010. Tool for Quantification of Staphylococcal Enterotoxin Gene Expression in Cheese. Appl. Environ. Microbiol. 76, 1367-1374.

Duthoit, F., Godon, J.-J., Montel, M.-C., 2003. Bacterial community dynamics during production of registered designation of origin salers cheese as evaluated by 16S rRNA gene single-strand conformation polymorphism analysis. Appl. Environ. Microbiol. 69, 3840-3848.

Duthoit, F., Tessier, L., Montel, M.C., 2005. Diversity, dynamics and activity of bacterial populations in 'Registered Designation of Origin' Salers cheese by single-strand conformation polymorphism analysis of 16S rRNA genes. J. Appl. Microbiol. 98, 1198-1208.

El-Baradei, G., Delacroix-Buchet, A., Ogier, J.-C., 2007. Biodiversity of Bacterial Ecosystems in Traditional Egyptian Domiati Cheese. Appl. Environ. Microbiol. 73, 1248-1255.

Ercolini, D., Frisso, G., Mauriello, G., Salvatore, F., Coppola, S., 2008. Microbial diversity in natural whey cultures used for the production of Caciocavallo Silano PDO cheese. Int. J. Food Microbiol. 124, 164-170.

Ercolini, D., Hill, P.J., Dodd, C.E.R., 2003. Bacterial community structure and location in Stilton cheese. Appl. Environ. Microbiol. 69, 3540-3548.

Ercolini, D., Mauriello, G., Blaiotta, G., Moschetti, G., Coppola, S., 2004. PCR-DGGE fingerprints of microbial succession during a manufacture of traditional water buffalo mozzarella cheese. J. Appl. Microbiol. 96, 263-270.

Ercolini, D., Moschetti, G., Blaiotta, G., Coppola, S., 2001. The potential of a polyphasic PCR-dGGE approach in evaluating microbial diversity of natural whey cultures for water-buffalo Mozzarella cheese production: bias of culture-dependent and culture-independent analyses. Syst. Appl. Microbiol. 24, 610-617.

Falentin, H., Henaff, N., Le Bivic, P., Deutsch, S.-M., Parayre, S., Richoux, R., Sohier, D., Thierry, A., Lortal, S., Postollec, F., 2012. Reverse transcription quantitative PCR revealed persistency of thermophilic lactic acid bacteria metabolic activity until the end of the ripening of Emmental cheese. Food Microbiol. 29, 132-140.

Falentin, H., Postollec, F., Parayre, S., Henaff, N., Le Bivic, P., Richoux, R., Thierry, A., Sohier, D., 2010. Specific metabolic activity of ripening bacteria quantified by real-time reverse transcription PCR throughout Emmental cheese manufacture. Int. J. Food Microbiol. 144, 10-19.

FAO/WHO, 2002. Guidelines for the Evaluation of Probiotics in Food. London, Ontario, Canada. April 30 and May 1, 2002.

Fernandez, M., del Rio, B., Linares, D.M., Martin, M.C., Alvarez, M.A., 2006. Real-time polymerase chain reaction for quantitative detection of histamine-producing bacteria: use in cheese production. J. Dairy Sci. 89, 3763-3769.

Feurer, C., Irlinger, F., Spinnler, H.E., Glaser, P., Vallaeys, T., 2004a. Assessment of the rind microbial diversity in a farm house-produced *vs* a pasteurized industrially produced soft red-smear cheese using both cultivation and rDNA-based methods. J. Appl. Microbiol. 97, 546-556.

Feurer, C., Vallaeys, T., Corrieu, G., Irlinger, F., 2004b. Does smearing inoculum reflect the bacterial composition of the smear at the end of the ripening of a french soft, red-smear cheese? J. Dairy Sci. 87, 3189-3197.

Fittipaldi, M., Codony, F., Adrados, B., Camper, A.K., Morato, J., 2011. Viable Real-Time PCR in Environmental Samples: Can All Data Be Interpreted Directly? Micr. Ecol. 61, 7-12.

Flórez, A.B., Mayo, B., 2006. Microbial diversity and succession during the manufacture and ripening of traditional, Spanish, blue-veined Cabrales cheese, as determined by PCR-DGGE. Int. J. Food Microbiol. 110, 165-171.

Fontana, C., Cappa, F., Rebecchi, A., Cocconcelli, P.S., 2010. Surface microbiota analysis of Taleggio, Gorgonzola, Casera, Scimudin and Formaggio di Fossa Italian cheeses. Int. J. Food Microbiol. 138, 205-211.

Fortina, M.G., Ricci, G., Mora, D., Parini, C., Manachini, P.L., 2001. Specific identification of *Lactobacillus helveticus* by PCR with pepC, pepN and htrA targeted primers. FEMS Microb. Lett. 198, 85-89.

Freeman, W.M., Walker, S.J., Vrana, K.E., 1999. Quantitative RT-PCR: pitfalls and potential. BioTechniques 26, 112-122, 124-125.

Friedrich, U., Lenke, J., 2006. Improved enumeration of lactic acid bacteria in mesophilic dairy starter cultures by using multiplex quantitative real-time PCR and flow cytometry-fluorescence in situ hybridization. Appl. Environ. Microbiol. 72, 4163-4171.

Fujimoto, J., Tanigawa, K., Kudo, Y., Makino, H., Watanabe, K., 2011. Identification and quantification of viable *Bifidobacterium breve* strain Yakult in human faeces by using strain-specific primers and propidium monoazide. J. Appl. Microbiol. 110, 209-217.

Fuka, M.M., Engel, M., Skelin, A., Redzepovic, S., Schloter, M., 2010. Bacterial communities associated with the production of artisanal Istrian cheese. Int. J. Food Microbiol. 142, 19-24.

Fumian, T.M., Leite, J.P.G., Marin, V.A., Miagostovich, M.P., 2009. A rapid procedure for detecting noroviruses from cheese and fresh lettuce. J. Virol. Methods 155, 39-43.

Furet, J.-P., Quenée, P., Tailliez, P., 2004. Molecular quantification of lactic acid bacteria in fermented milk products using real-time quantitative PCR. Int. J. Food Microbiol. 97, 197-207.

Fusco, V., Quero, G.M., Morea, M., Blaiotta, G., Visconti, A., 2011. Rapid and reliable identification of *Staphylococcus aureus* harbouring the enterotoxin gene cluster (egc) and quantitative detection in raw milk by real time PCR. Int. J. Food Microbiol. 144, 528-537.

Gala, E., Landi, S., Solieri, L., Nocetti, M., Pulvirenti, A., Giudici, P., 2008. Diversity of lactic acid bacteria population in ripened Parmigiano Reggiano cheese. Int. J. Food Microbiol. 125, 347-351.

Garcia-Cayuela, T., Tabasco, R., Pelaez, C., Requena, T., 2009. Simultaneous detection and enumeration of viable lactic acid bacteria and bifidobacteria in fermented milk by using propidium monoazide and real-time PCR. Int. Dairy J. 19, 405-409.

Giannino, M.L., Marzotto, M., Dellaglio, F., Feligini, M., 2009. Study of microbial diversity in raw milk and fresh curd used for Fontina cheese production by culture-independent methods. Int. J. Food Microbiol. 130, 188-195.

Grattepanche, F., Lacroix, C., Audet, P., Lapointe, G., 2005. Quantification by real-time PCR of *Lactococcus lactis* subsp. *cremoris* in milk fermented by a mixed culture. Appl. Microbiol. Biotechnol. 66, 414-421.

Hein, I., Jorgensen, H.J., Loncarevic, S., Wagner, M., 2005. Quantification of *Staphylococcus aureus* in unpasteurised bovine and caprine milk by real-time PCR. Res. Microbiol. 156, 554-563.

Hein, I., Lehner, A., Rieck, P., Klein, K., Brandl, E., Wagner, M., 2001. Comparison of different approaches to quantify *Staphylococcus aureus* cells by real-time quantitative PCR and application of this technique for examination of cheese. Appl. Environ. Microbiol. 67, 3122-3126.

Henri-Dubernet, S., Desmasures, N., Guéguen, M., 2004. Culture-dependent and culture-independent methods for molecular analysis of the diversity of lactobacilli in "Camembert de Normandie" cheese. Lait 84, 179-189.

Henri-Dubernet, S., Desmasures, N., Gueguen, M., 2008. Diversity and dynamics of lactobacilli populations during ripening of RDO Camembert cheese. Can. J. Microbiol. 54, 218-228.

Herman, L., Block, J.d., Renterghem, R.v., 1997. Isolation and detection of *Clostridium tyrobutyricum* cells in semi-soft and hard cheeses using the polymerase chain reaction. J. Dairy Res. 64, 311-314.

Herman, L., Deridder, H., 1993. Cheese Components Reduce the Sensitivity of Detection of *Listeria monocytogenes* by the Polymerase Chain-Reaction. Neth. Milk Dairy J. 47, 23-29.

Herthnek, D., Nielsen, S.S., Lindberg, A., Bölske, G., 2008. A robust method for bacterial lysis and DNA purification to be used with real-time PCR for detection of *Mycobacterium avium* subsp. *paratuberculosis* in milk. J. Microbiol. Meth. 75, 335-340.

Karns, J.S., Van Kessel, J.S., McClusky, B.J., Perdue, M.L., 2007. Incidence of *Escherichia coli* O157:H7 and *E. coli* Virulence Factors in US Bulk Tank Milk as Determined by Polymerase Chain Reaction. J. Dairy Sci. 90, 3212-3219.

Kramer, M., Obermajer, N., Bogovic Matijasic, B., Rogelj, I., Kmetec, V., 2009. Quantification of live and dead probiotic bacteria in lyophilised product by real-time PCR and by flow cytometry. Appl. Microbiol. Biotechnol. 84, 1137-1147.

La Gioia, F., Rizzotti, L., Rossi, F., Gardini, F., Tabanelli, G., Torriani, S., 2011. Identification of a Tyrosine Decarboxylase Gene (tdcA) in *Streptococcus thermophilus* 1TT45 and Analysis of Its Expression and Tyramine Production in Milk. Appl. Environ. Microbiol. 77, 1140-1144.

Ladero, V., Fernández, M., Alvarez, M.A., 2009. Effect of post-ripening processing on the histamine and histamine-producing bacteria contents of different cheeses. Int. Dairy J. 19, 759-762.

Ladero, V., Martínez, N., Martín, M.C., Fernández, M., Alvarez, M.A., 2008. qPCR for quantitative detection of tyramine-producing bacteria in dairy products. Food Res. Int. 43, 289-295.

Lafarge, V., Ogier, J.C., Girarda, V., Maladena, V., Leveau, J.Y., Delacroix-Buchet, A., 2004. Le potentiel de la TTGE pour l'étude bactérienne de quelques laits crus. Lait 84, 169-178.

Larpin, S., Mondoloni, C., Goerges, S., Vernoux, J.-P., Gueguen, M., Desmasures, N., 2006. *Geotrichum candidum* dominates in yeast population dynamics in Livarot, a French red-smear cheese. FEMS Yeast Res. 6, 1243-1253.

Le Bourhis, A.G., Dore, J., Carlier, J.P., Chamba, J.F., Popoff, M.R., Tholozan, J.L., 2007. Contribution of *C. beijerinckii* and *C. sporogenes* in association with *C. tyrobutyricum* to the butyric fermentation in Emmental type cheese. Int. J. Food Microbiol. 113, 154-163.

Le Bourhis, A.G., Saunier, K., Dore, J., Carlier, J.P., Chamba, J.F., Popoff, M.R., Tholozan, J.L., 2005. Development and validation of PCR primers to assess the diversity of *Clostridium* spp. in cheese by temporal temperature gradient gel electrophoresis. Appl. Environ. Microbiol. 71, 29-38.

Le Dréan, G., Mounier, J., Vasseur, V., Arzur, D., Habrylo, O., Barbier, G., 2010. Quantification of *Penicillium camemberti* and *P. roqueforti* mycelium by real-time PCR to assess their growth dynamics during ripening cheese. Int. J. Food Microbiol. 138, 100-107.

Lee, Z.M.-P., Bussema, C., Schmidt, T.M., 2009. rrnDB: documenting the number of rRNA and tRNA genes in bacteria and archaea. Nucleic Acids Res. 37, D489-D493.

Li, W.J., Raoult, D., Fournier, P.E., 2009. Bacterial strain typing in the genomic era. FEMS Microbiol. Rev. 33, 892-916.

Liao, D., 1999. Concerted evolution: molecular mechanism and biological implications. Am. J. Hum. Genet. 64, 24-30.

Lopez-Enriquez, L., Rodriguez-Lazaro, D., Hernandez, M., 2007. Quantitative Detection of *Clostridium tyrobutyricum* in Milk by Real-Time PCR. Appl. Environ. Microbiol. 73, 3747-3751.

Makhzami, S., Quenee, P., Akary, E., Bach, C., Aigle, M., Delacroix-Buchet, A., Ogier, J.C., Serror, P., 2008. In situ gene expression in cheese matrices: application to a set of enterococcal genes. J. Microbiol. Meth. 75, 485-490.

Makino, H., Fujimoto, J., Watanabe, K., 2010. Development and evaluation of a real-time quantitative PCR assay for detection and enumeration of yeasts of public health interest in dairy products. Int. J. Food Microbiol. 140, 76-83.

Manuzon, M.Y., Hanna, S.E., Luo, H., Yu, Z., Harper, W.J., Wang, H.H., 2007. Quantitative Assessment of the Tetracycline Resistance Gene Pool in Cheese Samples by Real-Time TaqMan PCR. Appl. Environ. Microbiol. 73, 1676-1677.

Marianelli, C., Martucciello, A., Tarantino, M., Vecchio, R., Iovane, G., Galiero, G., 2008. Evaluation of Molecular Methods for the Detection of *Brucella* Species in Water Buffalo Milk. J. Dairy Sci. 91, 3779-3786.

Martin, M.C., Martinez, N., Rio, B.d., Ladero, V., Fernandez, M., Alvarez, M.A., 2010. A novel real-time polymerase chain reaction-based method for the detection and quantification of lactose-fermenting *Enterobacteriaceae* in the dairy and other food industries. J. Dairy Sci. 93, 860-867.

Marzotto, M., Maffeis, C., Paternoster, T., Ferrario, R., Rizzotti, L., Pellegrino, M., Dellaglio, F., Torriani, S., 2006. *Lactobacillus paracasei* A survives gastrointestinal passage and affects the fecal microbiota of healthy infants. Res. Microbiol. 157, 857-866.

Masco, L., Huys, G., De Brandt, E., Temmerman, R., Swings, J., 2005. Culture-dependent and culture-independent qualitative analysis of probiotic products claimed to contain bifidobacteria. Int. J. Food Microbiol. 102, 221-230.

Masco, L., Vanhoutte, T., Temmerman, R., Swings, J., Huys, G., 2007. Evaluation of real-time PCR targeting the 16S rRNA and recA genes for the enumeration of bifidobacteria in probiotic products. Int. J. Food Microbiol. 113, 351-357.

Masoud, W., Takamiya, M., Vogensen, F.K., Lillevang, S., Al-Soud, W.A., Sørensen, S.J., Jakobsen, M., 2011. Characterization of bacterial populations in Danish raw milk cheeses made with different starter cultures by denaturating gradient gel electrophoresis (DGGE) and pyrosequencing. Int. Dairy J. 21, 142-148.

Maurer, J.J., 2011. Rapid Detection and Limitations of Molecular Techniques. Ann. Rev. Food Sci. Techn., Vol 2 2, 259-279.

Mauriello, G., Moio, L., Genovese, A., Ercolini, D., 2003. Relationships between flavoring capabilities, bacterial composition, and geographical origin of natural whey cultures used for traditional water-buffalo mozzarella cheese manufacture. J. Dairy Sci. 86, 486-497.

McKillip, J.L., Jaykus, L.A., Drake, M.A., 2000. A comparison of methods for the detection of *Escherichia coli* O157:H7 from artificially-contaminated dairy products using PCR. J. Appl. Microbiol. 89, 49-55.

Monnet, C., Back, A., Irlinger, F., 2012 (in press). Growth of Aerobic Ripening bacteria at the Cheese Surface is Limited by the Availability of Iron. Appl. Environ. Microbiol.

Monnet, C., Correia, K., Sarthou, A.-S., Irlinger, F., 2006. Quantitative detection of *Corynebacterium casei* in cheese by real-time PCR. Appl. Environ. Microbiol. 72, 6972-6979.

Monnet, C., Ulvé, V., Sarthou, A.-S., Irlinger, F., 2008. Extraction of RNA from cheese without prior separation of microbial cells. Appl. Environ. Microbiol. 74, 5724-5730.

Moschetti, G., Blaiotta, G., Villani, F., Coppola, S., 2001. Nisin-producing organisms during traditional 'Fior di latte' cheese-making monitored by multiplex-PCR and PFGE analyses. Int. J. Food Microbiol. 63, 109-116.

Mounier, J., Blay, G.L., Vasseur, V., Floch, G.L., Jany, J.L., Barbier, G., 2010. Application of denaturing high-performance liquid chromatography (DHPLC) for yeasts identification in red smear cheese surfaces. Lett. Appl. Microbiol. 51, 18-23.

Mounier, J., Monnet, C., Jacques, N., Antoinette, A., Irlinger, F., 2009. Assessment of the microbial diversity at the surface of Livarot cheese using culture-dependent and independent approaches. Int. J. Food Microbiol. 133, 31-37.

Muller, J.A., Stanton, C., Sybesma, W., Fitzgerald, G.F., Ross, R.P., 2010. Reconstitution conditions for dried probiotic powders represent a critical step in determining cell viability. J. Appl. Microbiol. 108, 1369-1379.

Neefs, J., Van de Peer, Y., De Rijk, P., Chapelle, S., De Wachter, R., 1993. Compilation of small ribosomal subunit RNA structures. Nucleic Acids Res. 21, 3025-3049.

Nicolaisen, M.H., Baelum, J., Jacobsen, C.S., Sorensen, J., 2008. Transcription dynamics of the functional tfdA gene during MCPA herbicide degradation by *Cupriavidus necator* AEO106 (pRO101) in agricultural soil. Environ. Microbiol. 10, 571-579.

Niederhauser, C., Candrian, U., Hofelein, C., Jermini, M., Buhler, H.P., Luthy, J., 1992. Use of Polymerase Chain-Reaction for Detection of *Listeria monocytogenes* in Food. Appl. Environ. Microbiol. 58, 1564-1568.

Nikolic, M., Terzic-Vidojevic, A., Jovcic, B., Begovic, J., Golic, N., Topisirovic, L., 2008. Characterization of lactic acid bacteria isolated from Bukuljac, a homemade goat's milk cheese. Int. J. Food Microbiol. 122, 162-170.

Nocker, A., Cheung, C.Y., Camper, A.K., 2006. Comparison of propidium monoazide with ethidium monoazide for differentiation of live vs. dead bacteria by selective removal of DNA from dead cells. J. Microbiol. Meth. 67, 310-320.

Nocker, A., Mazza, A., Masson, L., Camper, A.K., Brousseau, R., 2009. Selective detection of live bacteria combining propidium monoazide sample treatment with microarray technology. J. Microbiol. Meth. 76, 253-261.

Nocker, A., Richter-Heitmann, T., Montijn, R., Schuren, F., Kort, R., 2010. Discrimination between live and dead cells in bacterial communities from environmental water samples analyzed by 454 pyrosequencing. Int. Microbiol. 13, 59-65.

Nogva, H.K., Dromtorp, S.M., Nissen, H., Rudi, K., 2003. Ethidium monoazide for DNA-based differentiation of viable and dead bacteria by 5 '-nuclease PCR. Biotechniques 34, 804-813.

Nolan, T., Hands, R.E., Bustin, S.A., 2006. Quantification of mRNA using real-time RT-PCR. Nature Protocols 1, 1559-1582.

Ogier, J.-C., Lafarge, V., Girard, V., Rault, A., Maladen, V., Gruss, A., Leveau, J.-Y., Delacroix-Buchet, A., 2004. Molecular fingerprinting of dairy microbial ecosystems by use of Temporal Temperature and Denaturing Gradient Gel Electrophoresis. Appl. Environ. Microbiol. 70, 5628-5643.

Ogier, J.-C., Son, O., Gruss, A., Tailliez, P., Delacroix-Buchet, A., 2002. Identification of the bacterial microflora in dairy products by temporal temperature gradient gel electrophoresis. Appl. Environ. Microbiol. 68, 3691-3701.

O'Grady, J., Ruttledge, M., Sedano-Balbas, S., Smith, T.J., Barry, T., Maher, M., 2009. Rapid detection of *Listeria monocytogenes* in food using culture enrichment combined with real-time PCR. Food Microbiol. 26, 4-7.

Omiccioli, E., Amagliani, G., Brandi, G., Magnani, M., 2009. A new platform for Real-Time PCR detection of *Salmonella* spp., *Listeria monocytogenes* and *Escherichia coli* O157 in milk. Food Microbiol. 26, 615-622.

Ongol, M.P., Tanaka, M., Sone, T., Asano, K., 2009. A real-time PCR method targeting a gene sequence encoding 16S rRNA processing protein, rimM, for detection and enumeration of *Streptococcus thermophilus* in dairy products. Food Res. Int. 42, 893-898.

Parayre, S., Falentin, H., Madec, M.N., Sivieri, K., Le Dizes, A.S., Sohier, D., Lortal, S., 2007. Easy DNA extraction method and optimisation of PCR-Temporal Temperature Gel Electrophoresis to identify the predominant high and low GC-content bacteria from dairy products. J. Microbiol. Meth. 69, 431-441.

Peng, M., Smith, A.H., Rehberger, T.G., 2011. Quantification of *Propionibacterium acidipropionici* P169 Bacteria in Environmental Samples by Use of Strain-Specific Primers Derived by Suppressive Subtractive Hybridization. Appl. Environ. Microbiol. 77, 3898-3902.

Postollec, F., Falentin, H., Pavan, S., Combrisson, J., Sohier, D., 2011. Recent advances in quantitative PCR (qPCR) applications in food microbiology. Food Microbiol. 28, 848-861.

Rademaker, J.L.W., Hoolwerf, J.D., Wagendorp, A.A., te Giffel, M.C., 2006. Assessment of microbial population dynamics during yoghurt and hard cheese fermentation and ripening by DNA population fingerprinting. Int. Dairy J. 16, 457-466.

Rademaker, J.L.W., Peinhopf, M., Rijnen, L., Bockelmann, W., Noordman, W.H., 2005. The surface microflora dynamics of bacterial smear-ripened Tilsit cheese determined by T-RFLP DNA population fingerprint analysis. Int. Dairy J. 15, 785-794.

Randazzo, C.L., Pitino, I., Ribbera, A., Caggia, C., 2010. Pecorino Crotonese cheese: Study of bacterial population and flavour compounds. Food Microbiol. 27, 363-374.

Randazzo, C.L., Torriani, S., Akkermans, A.D.L., de Vos, W.M., Vaughan, E.E., 2002. Diversity, dynamics, and activity of bacterial communities during production of an artisanal sicilian cheese as evaluated by 16S rRNA analysis. Appl. Environ. Microbiol. 68, 1882-1785.

Randazzo, C.L., Vaughan, E.E., Caggia, C., 2006. Artisanal and experimental Pecorino Siciliano cheese: microbial dynamics during manufacture assessed by culturing and PCR-DGGE analyses. Int. J. Food Microbiol. 109, 1-8.

Rantsiou, K., Alessandria, V., Urso, R., Dolci, P., Cocolin, L., 2008a. Detection, quantification and vitality of *Listeria monocytogenes* in food as determined by quantitative PCR. Int. J. Food Microbiol. 121, 99-105.

Rantsiou, K., Comi, G., Cocolin, L., 2004. The rpoB gene as a target for PCR-DGGE analysis to follow lactic acid bacterial population dynamics during food fermentations. Food Microbiol. 21, 481-487.

Rantsiou, K., Urso, R., Dolci, P., Comi, G., Cocolin, L., 2008b. Microflora of Feta cheese from four Greek manufacturers. Int. J. Food Microbiol. 126, 36-42.

Rasolofo, E.A., St-Gelais, D., LaPointe, G., Roy, D., 2010. Molecular analysis of bacterial population structure and dynamics during cold storage of untreated and treated milk. Int. J. Food Microbiol. 138, 108-118.

Reimann, S., Grattepanche, F., Rezzonico, E., Lacroix, C., 2010. Development of a real-time RT-PCR method for enumeration of viable *Bifidobacterium longum* cells in different morphologies. Food Microbiol. 27, 236-242.

Rinttila, T., Kassinen, A., Malinen, E., Krogius, L., Palva, A., 2004. Development of an extensive set of 16S rDNA-targeted primers for quantification of pathogenic and indigenous bacteria in faecal samples by real-time PCR. J. Appl. Microbiol. 97, 1166-1177.

Rodríguez-Lázaro, D., D'Agostino, M., Herrewegh, A., Pla, M., Cook, N., Ikonomopoulos, J., 2005. Real-time PCR-based methods for detection of *Mycobacterium avium* subsp. *paratuberculosis* in water and milk. Int. J. Food Microbiol. 101, 93-104.

Rossen, L., Norskov, P., Holmstrom, K., Rasmussen, O.F., 1992. Inhibition of Pcr by Components of Food Samples, Microbial Diagnostic Assays and DNA-Extraction Solutions. Int. J. Food Microbiol. 17, 37-45.

Rossetti, B.C., Frey, J., Pilo, P., 2010. Direct detection of Mycoplasma bovis in milk and tissue samples by real-time PCR. Mol. Cell. Probes 24, 321-323.

Rossi, F., Gardini, F., Rizzotti, L., La Gioia, F., Tabanelli, G., Torriani, S., 2011. Quantitative Analysis of Histidine Decarboxylase Gene (hdcA) Transcription and Histamine Production by *Streptococcus thermophilus* PRI60 under Conditions Relevant to Cheese Making. Appl. Environ. Microbiol. 77, 2817-2822.

Rossmanith, P., Krassnig, M., Wagner, M., Hein, I., 2006. Detection of *Listeria monocytogenes* in food using a combined enrichment/real-time PCR method targeting the prfA gene. Res. Microbiol. 157, 763-771.

Rossmanith, P., Su, B., Wagner, M., Hein, I., 2007. Development of matrix lysis for concentration of gram positive bacteria from food and blood. J. Microbiol. Meth. 69, 504-511.

Rudi, K., Naterstad, K., Dromtorp, S.M., Holo, H., 2005. Detection of viable and dead *Listeria monocytogenes* on gouda-like cheeses by real-time PCR. Lett. Appl. Microbiol. 40, 301-306.

Rueckert, A., Ronimus, R.S., Morgan, H.W., 2005. Development of a rapid detection and enumeration method for thermophilic bacilli in milk powders. J. Microbiol. Meth. 60, 155-167.

Sanchez, J.I., Rossetti, L., Martinez, B., Rodriguez, A., Giraffa, G., 2006. Application of reverse transcriptase PCR-based T-RFLP to perform semi-quantitative analysis of metabolically active bacteria in dairy fermentations. J. Microbiol. Meth. 65, 268-277.

Saubusse, M., Millet, L., Delbes, C., Callon, C., Montel, M.C., 2007. Application of Single Strand Conformation Polymorphism --PCR method for distinguishing cheese bacterial communities that inhibit *Listeria monocytogenes*. Int. J. Food Microbiol. 116, 126-135.

Selinger, D.W., Saxena, R.M., Cheung, K.J., Church, G.M., Rosenow, C., 2003. Global RNA Half-Life Analysis in Escherichia coli Reveals Positional Patterns of Transcript Degradation. Genome Res. 13, 216-223.

Serhan, M., Cailliez-Grimal, C., Borges, F., Revol-Junelles, A.M., Hosri, C., Fanni, J., 2009. Bacterial diversity of Darfiyeh, a Lebanese artisanal raw goat's milk cheese. Food Microbiol. 26, 645-652.

Serpe, L., Gallo, P., Fidanza, N., Scaramuzzo, A., Fenizia, D., 1999. Single-step method for rapid detection of *Brucella* spp. in soft cheese by gene-specific polymerase chain reaction. J. Dairy Res. 66, 313-317.

Singh, J., Batish, V.K., Grover, S., 2009. A scorpion probe-based real-time PCR assay for detection of *E. coli* O157:H7 in dairy products. Foodborne Pathog. Dis. 6, 395-400.

Slana, I., Kralik, P., Kralova, A., Pavlik, I., 2008. On-farm spread of *Mycobacterium avium* subsp. *paratuberculosis* in raw milk studied by IS900 and F57 competitive real time quantitative PCR and culture examination. Int. J. Food Microbiol. 128, 250-257.

Smeianov, V.V., Wechter, P., Broadbent, J.R., Hughes, J.E., Rodriguez, B.T., Christensen, T.K., Ardo, Y., Steele, J.L., 2007. Comparative High-Density Microarray Analysis of Gene Expression during Growth of *Lactobacillus helveticus* in Milk versus Rich Culture Medium. Appl. Environ. Microbiol. 73, 2661-2672.

Songjinda, P., Nakayama, J., Tateyama, A., Tanaka, S., Tsubouchi, M., Kiyohara, C., Shirakawa, T., Sonomoto, K., 2007. Differences in developing intestinal microbiota between allergic and non-allergic infants: A pilot study in apan. Biosci. Biotechn. Biochem. 71, 2338-2342.

Stephan, R., Schumacher, S., Tasara, T., Grant, I.R., 2007. Prevalence of *Mycobacterium avium* Subspecies *paratuberculosis* in Swiss Raw Milk Cheeses Collected at the Retail Level. J. Dairy Sci. 90, 3590-3595.

Stevens, K.A., Jaykus, L.A., 2004. Direct detection of bacterial pathogens in representative dairy products using a combined bacterial concentration-PCR approach. J. Appl. Microbiol. 97, 1115-1122.

Studer, E., Schaeren, W., Naskova, J., Pfaeffli, H., Kaufmann, T., Kirchhofer, M., Steiner, A., Graber, H.U., 2008. A Longitudinal Field Study to Evaluate the Diagnostic Properties of a Quantitative Real-Time Polymerase Chain Reaction-Based Assay to Detect *Staphylococcus aureus* in Milk. J. Dairy Sci. 91, 1893-1902.

Taïbi, A., Dabour, N., Lamoureux, M., Roy, D., LaPointe, G., 2011. Comparative transcriptome analysis of *Lactococcus lactis* subsp. *cremoris* strains under conditions simulating Cheddar cheese manufacture. Int. J. Food Microbiol. 146, 263-275.

Temmerman, R., Scheirlinck, I., Huys, G., Swings, J., 2003. Culture-independent analysis of probiotic products by denaturing gradient gel electrophoresis. Appl. Environ. Microbiol. 69, 220-226.

Thevenard, B., Rasoava, N., Fourcassié, P., Monnet, V., Boyaval, P., Rul, F., 2011. Characterization of Streptococcus thermophilus two-component systems: In silico analysis, functional analysis and expression of response regulator genes in pure or mixed culture with its yogurt partner, *Lactobacillus delbrueckii* subsp. *bulgaricus*. Int. J. Food Microbiol. 151, 171-181.

Torriani, S., Zapparoli, G., Dellaglio, F., 1999. Use of PCR-based methods for rapid differentiation of *Lactobacillus delbrueckii* subsp. *bulgaricus* and *L. delbrueckii* subsp. *lactis*. Appl. Environ. Microbiol. 65, 4351-4356.

Treven, P., Trmcic, A., Obermajer, T., Rogelj, I., Bogovic Matijasic, B. Prevalence of gassericin K7 A and K7 B gene determinants in the *Lactobacillus acidophilus* group and in adult faecal microbiota, submitted.

Trmcic, A., Monnet, C., Rogelj, I., Bogovic Matijasic, B., 2011. Expression of nisin genes in cheese-A quantitative real-time polymerase chain reaction approach. J. Dairy Sci. 94, 77-85.

Trmcic, A., Obermajer, T., Rogelj, I., Bogovic Matijasic, B., 2008. Short Communication: Culture-Independent Detection of Lactic Acid Bacteria Bacteriocin Genes in Two Traditional Slovenian Raw Milk Cheeses and Their Microbial Consortia. J. Dairy Sci. 91, 4535-4541.

Ulvé, V.M., Monnet, C., Valence, F., Fauquant, J., Falentin, H., Lortal, S., 2008. RNA extraction from cheese for analysis of *in situ* gene expression of *Lactococcus lactis*. J. Appl. Microbiol. 105, 1327-1333.

Van Hoorde, K., Heyndrickx, M., Vandamme, P., Huys, G., 2010. Influence of pasteurization, brining conditions and production environment on the microbiota of artisan Gouda-type cheeses. Food Microbiol. 27, 425-433.

Van Hoorde, K., Verstraete, T., Vandamme, P., Huys, G., 2008. Diversity of lactic acid bacteria in two Flemish artisan raw milk Gouda-type cheeses. Food Microbiol. 25, 929-935.

Vandesompele, J., De Preter, K., Pattyn, F., Poppe, B., Van Roy, N., De Paepe, A., Speleman, F., 2002. Accurate normalization of real-time quantitative RT-PCR data by geometric averaging of multiple internal control genes. Genome Biology 3, research0034.0031 - research0034.0011.

Vitali, B., Candela, M., Matteuzzi, D., Brigidi, P., 2003. Quantitative detection of probiotic *Bifidobacterium* strains in bacterial mixtures by using real-time PCR. Syst. Appl. Microbiol. 26, 269-276.

Wong, M.L., Medrano, J.F., 2005. Real-time PCR for mRNA quantitation. BioTechniques 39, 75-85.

Zago, M., Bonvini, B., Carminati, D., Giraffa, G., 2009. Detection and quantification of *Enterococcus gilvus* in cheese by real-time PCR. Syst. Appl. Microbiol. 32, 514-521.

Zoric Peternel, M., Canzek Majhenic, Andreja, Holo, Helge, Nes, Ingolf, Salehian, Zhian, Berlec, Ales, Rogelj, Irena., 2010. Wide-inhibitory spectra bacteriocins produced by *Lactobacillus gasseri* K7, Probiotics & Antimicro. Prot. 2, 233-240.

3

Role of Polymerase Chain Reaction in Forensic Entomology

Tock Hing Chua[1] and Y. V. Chong[2]
[1]*Department of Parasitology and Medical Diagnostics, Universiti Malaysia Sabah, Jalan UMS, Kota Kinabalu, Sabah,*
[2]*Monash University, Jalan Lagoon Selatan, Bandar Sunway, Selangor Darul Ehsan Malaysia*

1. Introduction

The history of forensic entomology dates back to the 13th century when Song Ci (or Sung Tz'u) (1186–1249), an outstanding forensic scientist in the Southern Song Dynasty, documented the first forensic entomology case in his book "Collected Cases of Injustice Rectified" (*Xi Yuan Ji Lu* which means "Washing Away of the Wrongs"). In his investigation, Song Ci identified the murderer through forensic flies which flew to the sickle used in committing the crime. This sickle had bits of soft tissue, blood, bone and hair attached to it and thus attracted the flies. The owner of the sickle then admitted to his crime (translated by McKnight [1]).

Insects (mainly flies and beetles) are the main resources in forensic entomology. They could be found in every part of the world making them a useful forensic indicator by providing useful clues and evidence in death and criminal cases in forensic investigations.

With the advancement of biotechnology, forensic entomology has become a technically well developed field. Molecular biology tools are incorporated into this field where DNA based techniques are used to help solving complicated criminal or death cases. Often, police seek scientist's help to perform these genetic techniques mainly involving Polymerase Chain Reaction (PCR) analysis.

Today, the use of PCR-based methods in forensic entomology to help solve criminal and death investigations is continually increasing. In fact, it is a standard tool in most forensic laboratories, and police officers are trained in molecular technology.

In this chapter, we will look into the various PCR-based methods that have been developed elsewhere and adopted in forensic entomology, focusing on medicocriminal entomology. We will also look into how each method contributes to the field, as well as discussing its strengths and weaknesses.

2. Estimation of Post Mortem Interval (PMI)

Forensic entomologists estimate the post-mortem interval (PMI) or the minimum time of death by analyzing the correlation between the developmental stages of the collected insect specimens with the approximate weather data at the time when the crime or death occurs. Within minutes of the death of a person, forensic insects are able to locate the body through the sense of smell. The female fly deposits eggs (in the case of Calliphorid flies) or larvae (for Sarcophagid flies) on open wounds or natural openings of the corpse. These larvae hatch from eggs or born alive would then feed on the corpse. The larva undergoes three developmental stages and moults into a pupa. Metamorphosis occurs within the pupa, and an adult fly emerges in about a week.

Flies are usually the insects that arrive first at the decomposing corpse, starting with Calliphorids such as *Chrysomya megacephala* (Fabricius) and *Chrysomya rufifacies* (Macquart), and Sarcophagids. Following the flies, a succession of other arthropods and species of other phyla such as beetles, ants, moth, butterflies, earthworms and snails would arrive and join in decomposing the corpse. These include the beetles (Family Dermestidae and Silphidae), wasps (Family Vespidae), ants (Order Hymenoptera) and mites (gamasid and oribatid mites).

The arrival time of each individual species differ; some species such as blowfly and flesh fly arrive within five minutes prior to death while other species such as soldier fly and beetles arrive when the corpse is at the advanced decay stage [2]. The developmental rate of each species and each immature stage differs, and variation exists even among closely related species. Thus, correct species identification of the collected specimens and realistic values of the development rate of immature stages are very crucial for accurate PMI estimation.

3. Species identification using molecular methods

Insect species identification has been traditionally carried out using morphological characteristics. However, morphological characteristic keys for the immature stages of many forensically important species are either not constructed yet or not easily available or appear confusing to the non experts. To overcome this problem, forensic workers have started using Polymerase Chain Reaction (PCR) in insect species identification for forensic entomology since 1994.

The DNA of the insect specimen collected from the corpse or criminal scene is extracted usually using a commercial extraction kit and the extracted DNA is amplified using a specific primer designed for a certain gene. Then, the desired amplified fragment is purified for sequencing, and the sequence obtained is analysed further using Bioinformatics tools.

For species identification in forensic entomology, further investigation subsequent to the simple PCR analysis is commonly carried out by random fragment length polymorphism (RFLP), randomly amplified polymorphic DNA (RAPD), inter simple sequence repeat (ISSR) and sequence-characterized amplified region (SCAR) marker methods and real time PCR analysis.

3.1 Simple PCR

The DNA-based method as an alternative to using morphological keys for species identification was first proposed by Sperling et al. [3]. In his research, mitochondrial DNA (mtDNA), mitochondrial cytochrome oxidase I (COI), cytochrome oxidase II (COII) and tRNA leucine genes of blowflies (*Phormia regina* (Meigen), *Phaenicia sericata* (Meigen) and *Lucilia illustris* (Meigen)) were amplified using PCR and followed by direct sequencing. He found that there were nucleotide differences in the DNA sequences between these three species which could be used to differentiate their immature larval stages.

Subsequent to this research, mtDNA has been widely used for DNA analysis in forensic entomology, using COI and COII gene sequences analysis for distinguishing forensically important blow flies and flesh flies [4-14]. In purpose, such molecular work is similar to other non-entomological forensic methods in that it provides supplementary evidence in the form of PMI estimate to support the charge of a suspect to the crime.

In China, a 278 bp region of COI was used for the identification of nine forensic flies, namely *Ophyra capensis* (Wiedemann), *Chrysomya megacephala, Phaenicia sericata, Lucilia curpina,* and *Boettcherisca peregrina* (Robineau-desvoidy) [15]. They found the species could be easily separated by molecular means except for *Phaenicia sericata* and *Lucilia curpina* because of low sequence divergence between these two species.

In 2010, Mazzanti et al. [16] demonstrated that PCR could successfully amplify the mtDNA from empty puparial case and also fragments of the case. They could also correctly determine the eight Dipteran species (*Calliphora vicina* (Robineau-Desvoidy), *Sarcophagidae crassipalpis* (Macquart), *Phormia regina, Phaenicia sericata, Sarcophaga argyrostoma* (Robineau-Desvoidy), *Calliphora vomitoria* (Linnaeus), *Chrysomya megacephala, Synthesiomyia nudiseta* (Van Der Wulp)) through the amplified DNA from pupal case. This finding is particularly important as empty pupa cases left after adult emergence or fragments of it are commonly found on the corpse or in the area surrounding the corpse. The mtDNA has also been used for identification of beetle species found on corpses. In addition, COI and COII sequences have been used to study the phylogenetic relationships of carrion beetles species (Silphidae) (*Nicrophorus investigator* Zetterstedt, *Oiceoptoma novaboracense* (Förster), *Necrophilia americana* (Linnaeus) [17].

However molecular identification of species may not be accurate if it uses only the mtDNA gene [18]. More recent works also make use of internal transcribed spacer (ITS) which is a piece of non-functional RNA situated between structural ribosomal RNAs (rRNA) on a common precursor transcript. This transcript contains the 5' external transcribed sequence (5' ETS), 18S rRNA, ITS1, 5.8S rRNA, ITS2, 28S rRNA and finally the 3'ETS. The ITS region is widely used in taxonomy and molecular phylogeny because it is easy to amplify even from small quantities of DNA (due to the high copy number of rRNA genes) and has a high degree of variation even between closely related species.

For example, Song et al. [19] analyzed the nuclear ribosomal DNA especially internal transcribed spacer-II (ITS2) for species identification of some common necrophagous flies in southern China by phenetic approach. ITS2 gene was amplified from each individual specimen and sequences obtained were analyzed using ClustalX to construct a neighbour joining (NJ) tree. The results showed that species could be differentiated, and the identification was not affected by intra and interspecific variations. However, because of the high sequence homology between some congeneric species, more sequencing of specimens is required before such method can be used for forensic investigations.

A different section of the COI, a 250 base pair region of the gene for 16S rDNA has also being sequenced and tested [20]. They examined eight forensically important species from ten sites distributed at nine provinces in China. These were *Chrysomya megacephala, Chrysomya rufifacies, Calliphora vicina, Lucilia caesar* (Linnaeus), *Lucilia porphyrina, Phaenicia sericata* (Meigen), *Lucilia bazini* (Seguy), *Lucilia illustris* (Meigen). Their analysis of 16S rDNA sequences indicated abundant phylogenetically informative nucleotide substitutions which could identify most of the species tested except for specimens of *Lucilia caesar* and *Lucilia porphyrina.*

In a more recent study [21], species diagnosis of blowflies (*Chrysomya megacephala, Chrysomya pinguis* (Walker), *Phaenicia sericata, Lucillia porphyrina* (Walker), *Lucillia illustris* (Meigen), *Hemipyrellia ligurriens* (Wiedemann), *Aldrichina grahami* (Aldrich) and *Musca domestica* L.) from China and Pakistan was explored using phylogenetic analysis with five gene segments. They found that more accurate results were achieved through multi gene trees compared to single gene especially in resolving evolutionary relationship between species.

Although the mitochondrial cytochrome c oxidase gene is a favourite amongst forensic entomologists resulting in vast amount of DNA data being generated, there is little agreement as to which portion of the gene to be sequenced in forensic work, as different workers used different primers and obtained different sequence lengths from different regions. This can be seen from the above works quoted in this paper, and thus sequence analysis across species may be difficult. If agreement can be reached between various workers, a COI barcode identification system can be developed for use internationally. For example, such a system, using a 658-bp fragment of the COI, was found to be suitable for the identification of *Chrysomya* species from Australia [22]. This COI barcode region can facilitate the rapid generation of a barcode database and subsequent identification of specimens.

3.2 Restriction fragment length polymorphism (RFLP)

PCR-RFLP is the next method developed for species identification and separation. It is robust, easy and inexpensive. It detects the difference in homologous DNA sequences in the form of fragments of different lengths after digestion of the DNA samples in question with specific restriction endonucleases. Thus this technique is a combination of PCR amplification and RFLP analysis, in which the desired amplified product is digested with one or more restriction enzymes. Banding patterns that are specific for each species produced from the restriction digestion can be used for identification.

Schroeder et al. [23], for example, analyzed three forensically important species in Germany using the PCR-RFLP technique. They amplified specific fragments of the COI and COII region of the mitochondrial DNA (mtDNA) which were then digested with different restriction enzymes (either DraI or HinfI). The results revealed that a short sequence of 1.3 kb of COI and COII regions could differentiate the three species (*Phaenicia sericata, Calliphora vicina* and *Calliphora vomitoria*). Similarly the restriction enzyme SfcI was utilised on cytochrome oxidase I gene region to distinguish between *Calliphora vicina* and *Calliphora vomitoria*, two of the main UK blowfly species [12].

The utility of COI gene for identification of important forensic blow fly species found in Taiwan (*Chrysomya megacephala, Chrysomya rufifacies, Chrysomya pinguis, Hemipyrellia ligurriens* (Wiedemann), *Lucillia bazini* Seguy, *Lucilia cuprina, Lucillia hainanesis* Fan and *Lucilia prophyrina*) using different stages and different parts of the fly individual was also tested [11], and high support for congeneric grouping of species were obtained.

The PCR-RFLP techniques have also been employed elsewhere, using the internal transcribed spacer (ITS) in addition to COI. For example, three major blow fly species in Taiwan (*Chrysomya megacephala, Chrysomya pinguis* and *Chrysomya rufifacies*) could be successfully differentiated using COI and internal transcribed spacer I (ITS1) [24]. In Australia, the potential use of internal transcribed spacer II (ITS2) was investigated using PCR-RFLP analysis on all known *Chrysomya* species known from Australia [22]. All the species produced distinct restriction profiles except for the closely related species pairs, viz. between *Chrysomya latifrons* Malloch and *Chrysomya semimetallica* Malloch, and between *Chrysomya incisuralis* Macquart and *Chrysomya rufifacies.*

Recently, we carried out research on the PCR-RFLP assay for twelve Malaysian forensically important fly species. Our results (unpublished) indicate that the twelve species (*Chrysomya megacephala, Chrysomya rufifacies, Chrysomya pinguis, Chrysomya bezziana* (Villeneuve), *Chrysomya villenuevi, Chrysomya nigripes* (Aubertin), *Lucillia cuprina* (Wiedemann), *Ophyra spinigera* (Stein), *Sarcophaga ruficornis* (Liopygia), *Sarcophaga dux* (Thomson), *Sarcophaga peregrina* (Robineau-Desvoidy) and *Hermetia illucens* (Linnaeus)) in the study could be differentiated through COI gene digestion with three restriction enzymes (HpaII, SspI and HpyCH4V). We found that this method could be applied to immature stages and also incomplete specimens collected from the criminal scene.

3.3 Randomly amplified polymorphic DNA (RAPD)

RAPD is another commonly used method for species identification. This method uses non-specific primers for PCR amplification by which different regions of the DNA sample are amplified. The first RAPD typing of forensic insects was reported 1998 [4]. Eleven RAPD primers were tested to differentiate closely related species of flies and beetles found on corpse such as 'green bottle' blow flies, 'blue bottle' blow flies (Diptera: Calliphoridae) and beetles (Coleopyera: Silphidae). He found one particular primer (REP1R XIIIACGTCGICATCAGGC) was sufficient in resolving a practical forensic situation, but suggested for forensic purposes a set of at least six primers should be used to establish similarity coefficients. Nevertheless, he cautioned that in medico-legal matters, RAPD results may only be reported for so-called exclusions (where two specimens are definitely proven to be different) since an inclusion (where two specimens are shown to be similar or directly related) might induce the question of the likelihood of finding the same RAPD pattern by chance in any other animal.

3.4 Inter simple sequence repeat (ISSR) and sequence-characterized amplified region (SCAR) markers methods

ISSRs are DNA fragments of about 100-3000 bp located between adjacent, oppositely oriented microsatellite regions, and the variation in the regions between these microsatellites is used in ISSR PCR genotyping. The primers used are microsatellite core sequences with a few selective nucleotides as anchors into the non-repeat adjacent regions (16-18 bp). The advantage of ISSRs is that no sequence data for primer construction are needed.

The inter simple sequence repeat (ISSR) method was used to analyze the DNA polymorphism among the five forensic fly species in China, namely, *Phaenicia sericata, Aldrichina grahami, Chrysomya megacephala, Parasarcophaga crassipalpis* and *Musca domestica* using [25]. They found that nine ISSR primers could amplify 95 polymorphic bands which can be used to identify these species. They further converted these species-specific ISSR fragments into the sequence-characterized amplified region (SCAR) markers that can be used for the molecular diagnosis of these species.

Determination of specimens using ISSR is based on the similarity and difference in the electrophoresis result when compared with other individuals. For a high reliability of identification, such method requires a reference sample from the same species in a large database containing all species likely to be attracted to corpses in the same geographic region. On the other hand, SCAR is a genomic fragment localized in a single genetically defined locus that can be amplified by PCR using a pair of specific primers. SCARs are less sensitive to reaction conditions when compared to ISSR markers, thus allowing for a higher reliability and reproducibility among different laboratories which may use different brands of reagents and equipment. Therefore, SCARs are more appropriate diagnostic tool for practical applications.

3.5 Real time PCR assay

Further development in molecular identification of species was achieved in 2010 [26]. The investigators designed a species-specific real-time polymerase chain reaction (PCR) assay to target the ribosomal DNA internal transcribed spacer 1 (rDNA ITS1) of *Chrysomya bezziana*. It was very specific and can exclude other morphologically similar and related *Chrysomya* and *Cochliomyia* species. With this they were able to detect one *Chrysomya bezziana* in a sample of 1000 non-target species. Similar specific system can be developed to confirm the identity of other *Chrysomya* spp.

4. Determination of insect developmental rate

Immature stages of flies particularly the larvae and pupae, are often recovered from the death scene. Obtaining a better estimate of the time needed for the immature insect to develop to a certain stage will help to give a more accurate PMI. The larva and pupa stages occupy more than half of the immature development time. The developmental rate of larvae is determined by specific morphological changes and measurement of the specimen length [27-29]. For the egg and pupa which do not change size, physical measurement is not very useful. However, the changes of the pupal case and measurement of hormone level have been used for determining the pupa developmental rate of *Protophormia terraenovae* (Robineau-Desvoidy) [30].

Recently, molecular techniques involving PCR analysis have been developed for the developmental time of immature stages of the blowfly, *Phaenicia sericata*. Tarone et al. [31]

profiled the expression of three genes (bcd, sll, cs) throughout the maturation of blow fly eggs, and found the expression data could predict more precisely the blow fly age (within 2 h of true age). Later, they continued to work on the larvae and pupae [32]. Samples were collected from the carcass at different time intervals for gene expression evaluation. The RNA of the sample was extracted and the complementary DNA (cDNA) was synthesized from the RNA using specific gene primers. The desired developmentally regulated gene expression levels were assessed by quantitative PCR, and these levels were incorporated into traditional stage and size data. They tested on 86 immature *Phaenicia sericata,* and obtained a better precision in ageing blow flies, especially for postfeeding third instars and pupae.

Real time PCR and differentially expressed genes have also been used in the determination of pupal age in *Calliphora vicina* [33]. This research indicated that expression of Arylphorin and Gene G genes is possible to determine the age of the immature stages. Arylphorin gene is highly expressed at the early stage of pupae development (at 4500 accumulated degree 14 hours or ADH) whereas Gene G is highly expressed at the end of pupae stage (at 8640ADH). On the other hand, the changes in gene expression using differential display PCR has also been investigated [34]. The data showed that different genes are expressed at different levels during pupal development of *Phaenicia sericata*. However, they admitted that their method was not able to determine a pupa's age as yet.

Further research was carried out to improve estimation of the age of blow fly (*Phaenicia sericata*) with the aim of achieving a more accurate and precise PMI approximate, through gene expression where 20 genes were analyzed using RT-PCR [32]. Nine of these genes viz. resistance to organophosphate 1 (rop-1), acetylcholine esterase (ace), chitin synthase (cs), ecdysone receptor (ecr), heat shock protein 60 and 90 (hsp60, hsp90), slalom (sll), ultraspiracle (usp), and white (w) evaluated in this study were found to be useful in increasing the accuracy of PMI estimation for post feeding third instars and pupae.

5. Genetic variation of forensic species population for detecting postmortem relocation

Research on genetic variation of common forensically important species between populations is important for forensic studies. The genetic data of these species is likely to be different among different populations. If specimens from only one location were used in the research, the data collected might only be accurate for that particular location and it could not be applied to death investigations which occur at other locations. It might also be possible to detect post-mortem relocation of a corpse through the study of genetic variation among populations within a species. For study in a particular geographical area, PCR analysis is usually coupled with RAPD, amplified fragment length polymorphism (AFLP) and inter simple sequence repeat (ISSR). In these methods, non-specific primers are used for PCR amplification. RAPD fingerprinting could be a valuable tool for separating various populations [35].

The intraspecific genetic variation of *Phaenicia sericata* between two populations in southern England has been investigated using RAPD analysis [**36**]. The genetic homogeneity of *Phaenicia sericata* was determined, basing on the RAPD data which was analysed using a similarity coefficient method and a randomization test. They found that banding profiles (which were defined with ten random primers) from RAPD could differentiate among closely related individuals of the species. Such investigation can be used to elucidate relationships between even closely related populations of *Phaenicia sericata* and differentiate between populations, if more than a population is found on the corpse, thus helping to make a conclusion if a body had been relocated prior to its discovery.

Similarly, AFLP analysis has also been used for genetic population study of *Phormia regina* from sites spanning the contiguous United States [37]. They found there was only a very weak correlation between individual genetic and geographic distances. More interestingly, they found that adult *Phormia regina* that arrived together to the baits were closely related individuals compared to a random sample. They later applied the same method for investigating the population genetic structure of *Phaenicia sericata* from North America based on AFLP genotypes with 249 loci [38]. Although the study could not find any regional genetic variation, they nevertheless detected high local relatedness among the females in the samples. This led them to suggest that a pattern of local relatedness might support a genetic test for inferring the post-mortem relocation of a corpse.

We have conducted using similar methods a preliminary study of the population genetic variation among *Chrysomya megacephala* individuals in Malaysia. We tested the usefulness of COI gene for differentiating Malaysian *Chrysomya megacephala* individuals from four locations. Our results showed that the individuals could be put into two geographical groups based on a single nucleotide polymorphism (SNP) observed (unpublished results). It would appear possible then to infer if a corpse has been relocated from one location to another by comparing the SNP of larvae or pupae left behind at one place and those on the corpse which has been moved postmortem.

6. Recovery of human DNA from insects

Many studies found that human DNA can be recovered from insects found at the scene. The recovery of DNA provides useful information for forensic cases. For example, the identity of the suspect or the deceased could be identified from a fed mosquito, fly larvae or bed bugs [39-43]. The detection of insect gut content by PCR amplification is useful for forensic entomology. DNA is extracted from the collected insect, often from the insect gut contents. Then PCR amplification usually is conducted using either short tandem repeat (STR), human mtDNA hypervariable region (HVR) or insect mtDNA for profiling. STR and HVR typing are commonly used for human profiling.

Coulson et al. [44] demonstrated the possibility of human DNA extraction, amplification and fingerprinting from *Anopheles gambiae* mosquitoes stored at different storing conditions. The results showed that it is possible to use PCR for the amplification of human DNA extracted from mosquitoes. A very interesting casework has been demonstrated in 2006 [41], where only a fresh mosquito blood stain from a smashed mosquito was found in a room of the death scene. DNA was successfully extracted from the blood stain, and PCR amplification and STRs profiling at 15 human genetic loci was then performed on the extracted DNA, using AmpFLSTR Identifiler. This produced a complete genetic profile which aided the identification of the suspect.

A number of researches were conducted on the DNA extraction from the digestive tract of necrophagous larvae or the 'last meal' of these maggots. This is a useful study as the DNA profile of the host could be obtained from the extracted DNA and to determine whether the maggots used in the investigation are associated with the crime or death [40, 45]. Kondakci et al. [46] found that a complete human profile could be obtained using STR and SNP profiling of *Phaenicia sericata* third instar larvae. The STR and SNP profiles matched the identity of the host which showed that this analysis could be used to relate the maggots studied to the corpse in the investigation.

7. Conclusion

With the advent of molecular techniques, forensic entomology has certainly come a long way since the days of Song Ci. Molecular technology has changed the manner by which forensic entomological investigations are being carried out, making it a sophisticated a science. This has resulted in quicker, more accurate determination of the species, as well as the age of specimens recovered from the corpse, and consequently a more accurate of PMI.

From the initial use simple PCR in forensic entomology, it has progressed to RFLP, then RAPD, ISSR, SCAR and finally to RT- PCR Assay.

Initially the molecular techniques were used mainly for species identification [3, 23, 24]. However, later works extended to ageing the pupa [33], which is very useful as the size does not change during metamorphosis, and physical ageing is not possible. The accuracy of PMI estimates increases if ageing of the immature stages becomes more precise.

It would appear that future research is in the direction of RT-PCR assay, as this is a faster and more accurate method. Similarly studies on the intraspecific genetic variation on forensic insect populations will result in more accurate methods of population identification, and aid in deciding if a victim's body has been relocated by the criminals for burial to avoid suspicion, or to mislead criminal investigation.

Another area which has great potential use is identification of suspect from the gut content of a fed mosquito [44] or body louse [47] at the crime scene. In the case where the body has been moved, the carrion fly larvae or pupae may help identify the deceased indicating the relocation of the corpse. Although the blood meal may be partially digestion and makes DNA extraction difficult, future research will likely to yield better technology for genotyping profile with degraded or low-copy DNA template.

Although molecular methods have advanced forensic entomology, the validity and reliability of the methods, and have also questioned the statistical basis of the sampling size have been questioned [18], and suggestions to improve have been offered. Among many things, it was suggested (a) the DNA extraction procedure should include a negative control, and the genotyping procedures should include both positive and negative controls, (b) a portion of the original tissue should be saved so that it is available for independent testing, (c) there must be an extensive record of reproducibility under specified working conditions, both when performed by the same analyst and by different analysts, (e) the analyst should have considerable experience with the particular genotyping method, and publications based on the same kind of analysis, (f) the analyst should provide a description of all aspects of the laboratory protocol used (e.g., PCR primer sequences) in response to a reasonable request, and (g) a forensic insect species identification must include phylogenetic analysis of sequence data. They also asked (a) what research sample size is adequate for a species-diagnostic test to be used in court, (b) whether the DNA-Based species identification using BLAST search of the huge and easily queried GenBank /EMBL/DDBJ sequence database is critical enough, bearing in mind there are possible errors in some of these sequences, and (c) whether a taxonomic expert had confirmed the identification of the specimen the gene sequence of which was uploaded on the web.

These are important considerations as the analyst may need to testify in court about his findings, and above all, a forensic scientist must take great care to avoid a miscarriage of justice arising from careless interpretation of molecular data.

8. Acknowledgements

We thank Universiti Malaysia Sabah and Monash University Sunway for research facilities made available to us in the preparation of this paper.

9. References

[1] McKnight, B.E (1981). *The washing away of wrongs: Forensic medicine in thirteenth-century China* by Tz'u Sung. Translated by McKnight, B.E. University of Michigan, Ann Arbor,. 181 pp, ISSN 0892648007

[2] Gunn, A. (2006). *Essential Forensic Biology*. John Wiley & Sons, Ltd, ISBN -10: 0470012773

[3] Sperling, F.A.H.; Anderson, G.S. & Hickey, D.A. (1994). A DNA-based approach to the identification of insect species used for postmortem interval estimation. *Journal Forensic Science,* Vol. 39, No. , pp. 418–27. Erratum. 2000. *Journal of Forensic Sciences,* Vol. 45, pp. 1358–59, ISSN 1556-4029

[4] Benecke, M. (1998). Random amplified polymorphic DNA (RAPD) typing of necrophageous insects (Diptera, Coleoptera) in criminal forensic studies: validation and use in practice. *Forensic Science International,* Vol. 98, No. , pp. 157-168, ISSN 0379-07385. Wallman, J.F. & Donnellan, S.C. (2001). The utility of mitochondrial DNA sequences for the identification of forensically important blowflies (Diptera: Calliphoridae) in southeastern Australia. *Forensic Science International,* Vol. 120, pp.60-67, ISSN 0379-0738

[5] Wallman, J.F. & Donnellan, S.C. (2001). The utility of mitochondrial DNA sequences for the identification of forensically important blowflies (Diptera: Calliphoridae) in southeastern Australia. *Forensic Science International,* Vol. 120, pp.60-67, ISSN 0379-0738

[6] Wells, J.D. & Sperling, F.A.H. (2001). DNA-based identification of forensically important Chrysomyinae (Diptera: Calliphoridae). *Forensic Science International,* Vol. 120, pp. 110-115, ISSN 0379-0738

[7] Wells, J.D.; Pape, T., & Sperling, F.A. (2001). DNA-based identification and molecular systematic of forensically important Sarcophagidae (Diptera). *Journal of Forensic Sciences,* Vol. 46, pp. 1098-102, ISSN 1556-4029

[8] Harvey, M.L.; Dadour, I.R. & Gaudieri, S. (2003a). Mitochondrial DNA cytochrome oxidase I gene: potential for distinction between immature satges of some forensically important fly species (Diptera) in Western Australia. *Forensic Science International*, Vol. 131, pp. 134-139, ISSN 0379-073

[9] Harvey, M.L; Mansell, M.W.; Villet, M.H. & Dadour, I.R. (2003b). Molecular identification of some forensically important blowflies of southern Africa and Australia. *Medical and Veterinary Entomology*, Vol. 17, pp. 363-369, ISSN 0269-283X

[10] Zehner, R.;Amendt J.; Schutt S.; Sauer J.; Krettek, R. & Povolny, D. (2004). Genetic identification of forensically important flesh flies (Diptera: Sarcophagidae). *International Journal of Legal Medicine*, Vol. 118, pp. 245-247, ISSN 0937-9827

[11] Chen, W-Y.; Hung, T-H. & Shiao, S.F. (2004). Molecular Identification of Forensically Important Blow Fly Species (Diptera: Calliphoridae) in Taiwan. *Journal of Medical Entomology*, Vol. 41, No. 1, pp. 47-57, ISSN 0022-2585

[12] Ames, C.; Turner, B. & Daniel, B. (2006a). The use of mitochondrial cytochrome oxidase I gene (COI) to differentiate two UK blowfly species - *Calliphora vicina* and *Calliphora vomitoria*. *Forensic Science International*, Vol. 64, pp.179–182, ISSN 0379-0738

[13] Wells, J.D. & Williams, D.W. (2007). Validation of a DNA-based method for identifying Chrysomyinae (Diptera: Calliphoridae) used in death investigation. *International Journal of Legal Medicine,* Vol. 121, pp.1–8, ISSN 0937-982718. Wells, J. D. & Stevens, J. R. (2008). Application of DNA-Based Methods in Forensic Entomology. *Annual Review of Entomology*, Vol. 53, pp.103–20. ISSN 0066-4170

[14] Park, S.H.; Zhang, Y.; Piao, H.; Yu, D.H.; Jeong, H.J.; Yoo, G.Y.; Chung, U.; Jo, T-H. & Hwang, J-J. (2009). Use of Cytochrome c Oxidase Subunit I (COI) Nucleotide Sequences for Identification of Korean Luciliinae Fly Species (Diptera: Calliphoridae) in Forensic Investigations. *Journal of Korean Medical Science*, Vol. 24, No. , pp. 1058-1063, ISSN 1011-8934

[15] Cai, J-F.; Liu, M.; Ying, B-W.; Deng, R-L.;Dong, J-G.; Zhang, L.; Tao, T.; Pan, H-F; Yang, H-T. & Liao, Z-G. (2005). The availability of mitochondrial DNA cytochrome oxidase I gene for the distinction of forensically important flies in China. *Acta Entomologica Sinica*, Vol. 48, No. 3, pp. 380-385, ISSN 0454-6296

[16] Mazzanti, M.; Alessandrini, F.; Tagliabracci, A.; Wells, J.D. & Campobasso, C.P. (2010). DNA Degradation and genetic analysis of empty puparia: Genetic identification limits in forensic entomology. *Forensic Science International,* Vol. 195, pp. 99-102, ISSN 0379-0738

[17] Dobler, S. & Muller, J. K. (2000). Resolving phylogeny at the family level by mitochondrial cytochrome oxidase sequences: phylogeny of carrion beetles (Coleoptera: Silphidae). *Molecular Phylogenetics and Evolution,* Vol. 15, No. 3, pp. 390–402, ISSN

[18] Wells, J. D. & Stevens, J. R. (2008). Application of DNA-Based Methods in Forensic Entomology. *Annual Review of Entomology*, Vol. 53, pp.103–20. ISSN 0066-4170

[19] Song, Z-K.; Wang, X-Z. & Liang, G-Q. (2008). Species identification of some common necrophagous flies in Guangdong province, southern China based on the rDNA internal transcribed spacer 2 (ITS2). *Forensic Science International*, Vol. 175, pp. 17-22, ISSN 0379-0738

[20] Wang, X.; Cai, J.; Guo, Y.; Chang, Y.; Wu, K.; Wang, J., Yang, L.; Lan, L.; Zhong, M.; Wang, X. ; Liu, Q.; Cheng, Y. S.; Liu, Y.; Chen, Y.; Li, J.; Zhang, J. & Xin, P. (2010). The availability of 16SrDNA gene for identifying forensically important blowflies in China. *Romanian Society of Legal Medicine,* Vol.1, pp. 43 - 50, ISSN 1221-8618

[21] Zaidi, F., Wei, S-j., Shi, M. & Chen, X-x. (2011). Utility of multi-geneloci for forensic species diagnosis of blowflies. *Journal of Insect Sciences*, Vol. 11, pp. 59, ISSN 1536-2442

[22] Nelson, L.A.; Wallman, J.F. & Dowton, M. (2008). Identification of forensically important *Chrysomya* (Diptera: Calliphoridae) species using the second ribosomal internal transcribed spacer (ITS2). *Forensic Science International*, Vol. 177, No. , pp. 238-247, ISSN 0379-0738

[23] Schroeder, H., Klotzbach, H., Elias, S., Augustin C. & Pueschel, K. (2003). Use of PCR-RFLP for differentiation of calliphorid larvae (Diptera: Calliphoridae) on human corpses. *Forensic Science International,* Vol. 132, Pp.76-81, ISSN 0379-0738

[24] Chen, C-H. & Shih, C-J. (2003). Rapid identification of three species of blowflies (Diptera: Calliphoridae) by PCR-RFLP and DNA sequencing analysis. *Formosan Entomologist,* Vol. 23, No. , pp. 59–70, ISSN 1680-7650

[25] Lin, H.; Wang, S.B.; Miao, X.X.; Wu, H. & Huang, Y.P. (2007). Identification of necrophagous fly species using ISSR and SCAR markers. *Forensic Science International,* Vol. 168, No. 2-3, pp. 148–153,ISSN 0379-0738

[26] Jarrett, S.; Morgan, J.A.T.; Wlodek, B.M.; Brown, G.W.; Urech, R.; Green, P.E. & Lew-Tabor A.E. (2010). Specific detection of Old World screwworm fly, *Chrysomya bezziana,* in bulk fly trap catches using real-time PCR. *Medical and Veterinary Entomology*, Vol. 24, pp. 227-235, ISSN 0269-283X

[27] Donovan, S.E.; Hall, M.J.R.; Turner, B.D. & Moncrieff, C.B. (2006). Larval growth rates of the blowfly, *Calliphora vicina*, over a range of temperatures. *Medical and Veterinary Entomology*, Vol. 20, No.1, pp. 106-114, ISSN 0269-283X

[28] Anderson, G.S. (2000). Minimum and maximum development rates of some forensically important Calliphoridae (Diptera). *Journal of Forensic Sciences*, Vol. 45, No. 2, pp. 842-832, ISSN 1556-4029

[29] Clark, K.; Evans, L. & Wall, R. (2006). Growth rates of the blowfly, *Lucillia sericata*, on different body tissues. *Forensic Science International*, Vol. 156, No. 2-3, pp. 145-149, ISSN 0379-0738

[30] Gaudry, E.; Blais, C.; Maria, A. & Dauphin-Villemant, C. (2006). Study of steriodogenesis in pupae of the forensically important blow fly *Protophormia terraenovae* (Robineau-Desvoidy) (Diptera: Calliphoridae). *Forensic Science International,* Vol. 160, No. 1, pp. 27-34, ISSN 0379-0738

[31] Tarone, A.M.; Jennings, K.C. & Foran, D.R. (2007). Aging Blow Fly Eggs using Gene expression: A Feasibility Study. *Journal of Forensic Sciences*, Vol. 52, No.6, pp. 1350-1354, ISSN 1556-4029

[32] Tarone, A.M. & Foran, D.R. (2011). Gene expression during blow fly development: improving the precision of age estimates in forensic entomology. *Journal of Forensic Sciences*, Vol. 56, pp. S114-S122, ISSN 1556-4029

[33] Ames, C.; Turner, B. & Daniel, B. (2006b). Estimating the post-mortem interval (II): The use of differential temporal gene expression to determine the age of blowfly pupae. *International Congress Series* 1288, pp. 861-863, ISSN 0531-5131

[34] Zehner, R.; Mösch, S. & Amendt, J. (2006). Estimating the post-mortem interval by determining the age of fly pupae: Are there any molecular tools? *International Congress Series,* Vol. 1288, pp. 619-621, ISSN 0531-5131

[35] Hadrys, H., Balick, M. & Schierwater, B. (1992). Applications of random amplified polymorphic DNA (RAPD) in molecular ecology. *Molecular Ecology,* Vol. 1, pp. 55-63, ISSN 0962-1083

[36] Stevens, J. & Wall, R. (1995). The use of random amplified polymorphic DNA (RAPD) analysis for studies of genetic variation in populations of the blowfly *Lucilia sericata* (Diptera: Calliphoridae) in southern England. *Bulletin of Entomological Research,* Vol. 85, pp. 549–555, ISSN 0007-4853

[37] Picard, C.J. & Wells, J.D. (2009). Survey of the genetic diversity of *Phormia regina* (Diptera: Calliphoridae) using amplified fragment length polymorphisms. *Journal of Medical Entomology,* Vol. 46, No. 3, pp. 664-670, ISSN 0022-2585

[38] Picard, C.J. & Wells, J.D. (2010). The population genetic structure of North American *Lucillia sericata* (Diptera: Calliphoridae), and the utility of genetic assessment methods for reconstruction of post-mortem corpse relocation. *Forensic Science International,* Vol. 195, pp. 63-67, ISSN 0379-0738

[39] Kester, K.M.; Toothman, M.T.; Brown, B.L.; Street, W.S. & Cruz, T.D. (2010). Recovery of environmental human DNA by insects. *Journal of Forensic Sciences,* Vol. 55, No. 6, pp. 1543-1551, ISSN 1556-4029

[40] Li, K.; Ye, G-Y.; Zhu, J-Y. & Hu, C. (2007). Detection of food source by PCR analysis of the gut contents of *Aldrinchina graham* (Aldrich) (Diptera: Calliphoridae) during post-feeding period. *Insect Science,* Vol. 14, pp. 47-52, ISSN 1744-7917

[41] Spitaleri, S.; Romano, C.; Luise, E.D.; Ginestra, E. & Saravo, L. (2006). Genptyping of human DNA recovered from mosquitoes found on a crime scene. *International Congress Series,* Vol. 1288, pp. 574-576, ISSN 0531-5131

[42] Szalanski, A.L.; Austin, J.W.; Mckern, J.A.; McCoy T.; Steelman, C.D. & Miller, D.M. (2006). Time course analysis of bed bug, *Cimex lectularius* L., (Hemiptera: Cimicidae) blood meals with the use of polymerase chain reaction. *Journal of Agricultural Urban Entomology,* Vol. 23, No. 4, pp. 237-241, ISSN 1523-5475

[43] Mumcuoglu, K.Y.; Gallili, N.; Reshef, A.; Brauber, P. & Grant, H. (2004). Use of Human Lice in Forensic Entomology. *Journal of Medical Entomology,* Vol. 41, No. 4, pp. 803-806, ISSN 0022-2585

[44] Coulson, R.M.R.; Curtis, C.F.; Ready, P.D.; Hill, N. & Smith, D.F. (1990). Amplification and analysis of human DNA present in mosquito blood meals. *Medical and Veterinary Entomology,* Vol. 4, No. , pp. 357-366, ISSN 0269-283X

[45] Zehner, R.; Amendt, J. & Krettek, R. (2004). STR Typing of Human DNA from Fly Larvae Fed on Decomposing Bodies. *Journal of Forensic Sciences,* Vol. 49, No. 2, pp. 1-4, ISSN 1556-4029

[46] Kondakci, G.O.; Bulbul, O.; Shahzad, M.S.; Polat, E.; Cakan, H.; Altuncul, H. & Filoglu, G. (2009). STR and SNP analysis of human DNA from *Lucillia sericata* larvae's gut contents. *Forensic Science International, Genetic Supplement Series,* Vol. 2, pp. 178-179, 1875-1768, ISSN 1875-1768

[47] Lord, W.D.; DiZinno, J.A.; Wilson, M.R.; Budowle, B.; Taplin, D. & Meinking, T.L. (1998). Isolation, amplification, and sequencing of human mitochondrial DNA obtained from human crab louse, *Pthirus pubis* (L.), blood meals. *Journal of Forensic Sciences*, Vol. 43, No. 2, pp. 1097-1100, ISSN 1556-4029

4

PCR for Screening Potential Probiotic Lactobacilli for Piglets

Maurilia Rojas-Contreras[1],
María Esther Macías-Rodríguez[2] and José Alfredo Guevara Franco[1]
[1]*Universidad Autónoma de Baja California Sur, Área de Conocimientos Ciencias Agropecuarias, Food Science and Technology Laboratory, La Paz, Baja California Sur,*
[2]*Universidad de Guadalajara, Centro Universitario de Ciencias e Ingenierías, Department of Pharmacobiology, Sanitary Microbiology Laboratory, Guadalajara, Jalisco, México*

1. Introduction

1.1 Screening of potential probiotic lactobacilli

To continuously select probiotic bacteria, is needed to look for new strategies to make easy this task. In this chapter the characterization and identification by PCR of presumptive adhering lactobacilli to piglet gastrointestinal tract components is described and compared with previous reports. *Lactobacillus* is one of the major bacterial groups in the gastrointestinal tract of humans and animals (Smith, 1965; Dubos, 1965). Moreover, there is accumulating scientific evidence which strongly suggest that lactobacilli are associated with health (Bibel, 1988; Sanders, 2011). Consequently lactobacilli are frequently used as probiotics. This term refers to preparations of living microbes that can be added to the diet to improve health in humans and in farm animals (Fuller, 1989; Guilliland et al., 2001). The number of reports of health-promoting effects attributed to *Lactobacillus* strains has been increased in recent years where antagonistic activities against enteropathogens and modulation of immune system are well documented (Collado, 2006). The worldwide impact of advances in the scientific knowledge in this area is being enormous. For instance, diarrheal diseases affect millions of people throughout the world, having the greatest impact among children in developing countries (Guerrant et al., 1990; Guarino et al., 2011; Mondal et al., 2011). *Lactobacillus* have been shown to possess inhibitory activity toward the growth of pathogenic bacteria such as *Listeria monocytogenes* (Ashenafi 2005; Harris et al., 1989), *Escherichia coli*, *Salmonella* spp. (Chateau & Castellanos, 1993; Hudault et al., 1997), and others (Coconnier et al., 1997). When lactobacilli could be commonly used to prevent or alleviate some of the infections by enteropathogens, e. g. *E. coli*, *Salmonella*, *Shigella*, *Campylobacter*, etc. it could be an achievement for human beings. From an economical point of view, lactobacilli could reduce the risk for major economic losses due to decreased performance and health in the farm industry. For example, pig rising has become more industrialized and intestinal disturbances, e. g. diarrhea, affect significantly the piglet health and decrease intestinal performance (Goswami et al., 2011; Oostindjer et al., 2010).

Antibiotics have been used successfully against these infections, however there is an increasing concern consuming meat containing antibiotic residues as well as the potential hazards from spreading of resistance factors. Lactobacilli *Lactobacillus* is an alternative to maintain the health of growing pigs, mainly were environmental conditions are not controlled (Chiduwa et al., 2008). Under these conditions are a large number of pig farms worldwide. These conditions stress the animals, causing susceptibility to gastrointestinal diseases. It is well known that lactobacilli is a habitant of the intestinal tract of pigs and has been found as dominant microbiota. However confinement in small yards, large variations in temperatures, diet and other conditions, stress the animals, causing susceptibility to gastrointestinal diseases (Shimizu & Shimizu, 1978). Lactobacilli should retain special features to survive under these harsh conditions. At birth, piglets are exposed to a huge variety of microorganisms. Most of them come from the vagina, faeces, and skin of the mother as well as the environment (Jonsson & Conway, 1992). Composition of gut microbiota can be modulated by host, environmental, and bacterial factors (Thompson-Chagoyán et al., 2007). The colonization potential of lactobacilli has been investigated using small intestinal mucus extracts from 35 day old pigs. Numbers of lactobacilli in different portions of the small intestine of 35 days old pigs were enumerated. Mucus isolated from the small intestine of pigs was investigated for its capacity to support the growth of lactobacilli and results confirmed that *Lactobacillus* spp inhabit the mucus layer of the small intestine and can grow and adhere to ileal mucus (Rojas & Conway, 1996). The survivability and colonization of probiotics in the digestive tract are considered critical to ensure optimal functionality and expression of health promoting physiological functions. Muralidhara (Muralidhara, 1977) reported that viable counts of lactobacilli in tissue homogenates from the duodenum and upper jejunum of 3 weeks old pigs were 5.5-6.21 $\log_{10}$ per g mucosa. In addition, when segments of the small intestine of piglets, from the duodenum to the ileum were examined, it was found that lactobacilli increased from 6.4 to 8.2 $\log_{10}$ per g of mucosa (McAllister et al., 1979). From the total numbers of identified strict anaerobic organisms associated with the cecal mucosa, anaerobic lactobacilli were much lower (4.0-5.7 $\log_{10}$) per cm^2 than the numbers of obligated anaerobes. Although differences in the counts of the different groups of organisms have been quite large for the various reports, *Lactobacillus* appears to be dominant group in cecal and colonic content.

Screening for functional and probiotic attributes in lactobacilli new isolates is commonly performed, following these assays: Gram stain, acid and bile salt tolerance, cell surface hydrophobicity, adhesion to mucus and mucin, autoaggregation, Caco-2 cell-binding as well as antibacterial activity against *E. coli, L. monocytogenes, S. typhi,* etc. and antioxidative activities (Jacobsen et al., 1999; Macías-Rodríguez et al., 2008; Kaushik et al., 2009). Recently a screening of predominant *Lactobacillus* strains from healthy piglets has been performed in order to select specific probiotics for arid land piglets. Among the 164 isolates, 27 adhesive strains were identified using comparisons with 16S rDNA and intergenic 16-23S sequences. Results indicated that *L. fermentum* and *L. reuteri* were the most common species in faeces and mucus, respectively (Macías-Rodríguez et al., 2009). Likewise probiotics are increasingly used as nutraceuticals, functional foods or prophylactics and considering that probiotics strains have shown to be population-specific due to variation in gut microbiota, food habits and specific host-microbial interactions (Kaushik et al., 2009), screening of new indigenous probiotic strains in different region of the world is necessary.

1.2 Colonization by lactobacilli

Colonization studies of lactobacilli to the gastrointestinal tract first were concentrated on the attachment to the non secretory epithelium from the stomach. Cell morphology by electron microscopy, viable counts and biochemical test have been very important tools to identify lactobacilli attached to the keratinized squamous epithelium of the stomach of mice (N. Suegara et al., 1975; Moser & Savage, 2001; Savage, 1992; Tannock & Savage, 1974; Conway & Adams, 1989) and pig (Fuller et al., 1978; Pedersen & Tannock, 1989; Tannock et al., 1987; Henriksson et al., 1991). Later other reports on colonization by lactobacilli to other regions in the intestinal tract were found. Colonization of lactic acid bacteria isolated from rats and humans in the gastrointestinal tract of gnotobiotic rats has been studied by performing viable counts of the contents and tissue homogenates from the different regions of the intestinal tract. It was observed that lactobacilli seem to be retained, and to multiply on the mucosal surfaces along the intestinal tract (Kawai et al., 1982). In other report lactobacilli were ingested by human volunteers and samples of jejunal fluid at varying intervals were cultured for lactobacilli (Robins-Browne & Levine, 1981). It was shown that lactobacilli entered the small intestine and persisted there for 3-6 h after which time, levels returned to the base-line (Dixon, 1960). Studies on the possible interaction of lactobacilli with mammalian extracellular proteins have been performed. It was shown that specific collagen binding is common among lactobacilli of various origins (Aleljung et al., 1991).

Attention has been focused on interactions of lactobacilli with the mucosa of the intestinal tract. The gastrointestinal tract is covered by a protective mucus layer consisting of glycolipids and a complex mixture of large and highly glycosylated proteins called mucins as the main components. Mucus layer represents the first barrier of contact between bacteria contained in the lumen and the epithelial cell layer of the host (Tassell et al., 2011). Ability of commensal bacteria to adhere mucus is an important characteristic that is evaluated in probiotic bacteria (Ma et al., 2005). Adherence of lactobacilli to the intestinal epithelium and mucus is associated with stimulation of the immune system and inhibition of adhesion of pathogens (Herías et al., 1999). Caco-2 and HT-29 cells and a subpopulation of mucus secreting HT29-MTX cells have been used to study the adhesion of human isolated *L. acidophilus* BG2F04 strain. These studies showed scanning electron micrographs where mucus secreting HT29-MTX monolayer covered by the dense mucus gel produced by these typical goblet cells, bound to lactobacilli. In addition they proposed a model for the adherence of this *Lactobacillus* strain to human intestinal cells (Coconnier et al., 1997). Other workers used human colon mucosa in an *in vitro* assay, to test the capacity of five *Lactobacillus* strains to colonize; a dense population of lactobacilli was observed covering the whole mucosal surface of the colon tissue (Sarem-Daamerdji et al., 1995). Other contributions for understanding the interactions between gastrointestinal mucosa and lactobacilli have been reported. The diversity of *Lactobacillus spp* on healthy and diseased human intestinal mucosa biopsies has be studied (Molin, 1993). These workers assessed the potential of the *Lactobacillus* isolates for treating intestinal disorders, suggesting that there are no general differences in the type of dominating *Lactobacillus* microbiota between mucosa from different regions of the intestine. In another report, different *Lactobacillus* strains in fermented oatmeal soup were administered to healthy human volunteers. Biopsies were taken from both the upper jejunum region and the rectum before one and eleven days after administration. Results showed significantly increased counts of lactobacilli on the jejunum mucosa and high levels of all those strains that remained eleven days after administration (Johansson et al., 1993). Colonization experiments in mice, also showed that the number of lactobacilli detected in samples collected from various regions of the gastrointestinal tract, two weeks after inoculation, were not statistically significant different, no matter which strain had been used to colonize mice. In addition, it was concluded that bile salt hydrolase production was not an essential attribute for lactobacilli to colonize the murine gastrointestinal tract. Furthermore, the growth rate of mice that consumed a nutritionally balanced diet were not affected by the presence of bile salt hydrolase producing or not lactobacilli in the gastrointestinal tract (Bateup et al., 1995). The capacity of different lactobacillus strains to grow in and adhere to small intestinal mucus as well as the characteristics of binding was studied. It was shown that six *Lactobacillus* strains isolated from porcine small intestinal mucosa, one isolated from faeces, one isolated from stomach and one more isolated from human feces, all grew equally well in intestinal mucus extract. Growth was monitored by enumerating the colony forming units. During growth in mucus, a visible precipitation was developed because lactobacilli formed clusters surrounded by mucus. In this study it was observed that when lactobacilli were grown in mucus, the ability to adhere to mucus was reduced from 35% to 10% of the adhesion. This could occur because adhesin(s) on the surface of the bacteria were being blocked by receptors or receptors-like components in the mucus (Rojas & Conway 1996). Adhesion assays of *Lactobacillus fermentum* 104R (Actually identified as *L. reuteri* 104R) indicated that this strain adhered to mucus when it was grown in synthetic media. Adhesion data were analyzed by Scatchard plot and it was noted that the binding of lactobacilli to mucus is not mediated by a single adhesin-receptor interaction. The quantitative interpretation of the binding data for this system was not possible to perform because the complexity of the system. These results correlate with other report suggesting that lactobacillus species adhere to intestinal cells via mechanisms which involve different combination of factors on the bacterial cell surface (Greene & Klaenhammer, 1994). Adhesion promoting compound(s) from *L. reuteri* 104R were found in the spent culture medium on the late stationary phase of growth. The spent culture fluid was used to inhibit adhesion to mucus of whole *L. reuteri* 104R strain, revealing that proteinaceous compound(s) were involved in the binding (Rojas & Conway, 1996).

1.3 *Lactobacillus* adhesins

Bacteria can have many types of surfaces, including sheaths, S-layers, capsules and walls. In the laboratory certain surface types are usually expressed. For example, *E. coli* K12 contains only core polysaccharide plus lipid A in its lipopolisaccharide that was why this strain is restricted to a laboratory habitat since it cannot withstand the rigors of a natural environment. This strain possesses only an outer membrane as its surface component surfaces components, but a related strain, K-30, is enclosed in a capsule. Frequently, it is the natural environment and their intrinsic stress that elicit expression of the surface attributes of a bacterium (Costerton, 1988; Brown et al., 1988). A bacterium in its native habitat will often possess a wall overlaid by a multiplicity of superficial layers. After several subcultures in laboratory medium these layers are not longer required and are lost (Costerton, 1988). This surface character could makes difficult the correlation of laboratory studies on adhesins of the bacteria with the *In vivo* state. Intestinal mucus extract from the small intestine of pig was used for lactobacilli growth and for studying the production and expression of the mucus and mucin adhesion promoting proteins.

Cell wall of Gram positive bacteria is composed primarily of peptidoglycan, which often contains peptide interbridge and large amounts of teichoic acids (polymers of glycerol or ribitol joined by phosphate groups). Amino acids and sugars are attached to the glycerol and ribitol groups. These molecules are important for maintaining the structure of the wall. Capsules, slims S-layers, sheaths or even pili (fimbriae) can occur as superficial layers above the cell wall. They can occur singly or in combination. Distinction among them is based primarily on their structural attributes (Beveridge, 1989; Beveridge & Graham, 1991). The term adhesin has been used to denote functions that are involved in one or more of the three following activities: 1) they may promote attachment and then initiate colonization of surface habitats, 2) They may be responsible for the organization of microbial communities and assemblages, and 3) they may be instrumental in promoting cell to cell contact as a phase preceding the transfer of genetic information between cells. The term adhesion has been used to describe the relatively stable, irreversible attachment of bacteria to surfaces, and the term receptor has been used for both known and putative entities on surfaces to which adhesins bind to effect specific adhesion (Jones & Isaacson, 1984). While there is a considerable amount of information published about proteinaceous bacterial adhesins and their receptors on pathogenic bacteria (Jones & Isaacson, 1984; Klemm, 1994; Bonazzi & Cossart, 2011), there are fewer studies on the mechanisms of adhesion of lactobacilli to gastrointestinal mucosa. Adhesion of *L. acidophilus* to avian intestinal epithelial cells mediated by the crystalline bacterial cell surface layer protein (S-layer) protein was reported (Schneitz & Lounatma, 1993), and the adhesion to collagen by *L. reuteri* NCIB 11951 was shown to be mediated by a 29 KDa protein (Aleljung et al., 1994) and to *L. crrispatus* JCM 5810 was mediated by a 120 KDa S-layer protein (Toba et al., 1995). Another interesting finding was a 32 KDa protein, an aggregation promoting factor on *L. plantarum* strain 4B2 which increased the frequency of conjugation (Reniero et al., 1992; Reniero et al., 1993). The ability of probiotic bacteria to aggregate should be considered a desirable characteristic because they potentially inhibit adherence of pathogenic bacteria to intestinal mucosa either by direct coaggregation with the pathogens to facilitate clearance, by forming a barrier via self-aggregation or coaggregation with commensal organisms on the intestinal mucosa. Surface proteins from lactobacilli have been reported to be affected by freeze drying (Ray & Johnson, 1986; Brennan et al., 1986) and by the composition of the culture media (Pavlova et al., 1993; Cook et al., 1988).

Purification and characterization of proteins from lactobacilli which promote the adhesion to mucus have been well studied. The purification of a mucus and mucin adhesion promoting protein (MAPP) from the surface of *L. reuteri* 104R was performed by using LiCl (1M). A variety of different agents to extract proteins have been used. EDTA (0.1M), urea (8M) or $MgCl_2$ have been used to effectively release surface associated material from bacterial cells of various genera. Solutions of detergents such as sodium lauryl sarcosinate, triton X 100 (1% v/v final concentration), sonication and sodium dodecyl sulphate (SDS, 1%, w/v) have been shown to be effective in extracting proteins from *L. reuteri* strain 100-23 (Boot et al., 1993; Chagnaud et al., 1992). Guanidine hydrochloride (4M, GHCl) was used to extract regular arrays from the cell walls of different strains (Masuda & Kawata, 1983) and an S-layer protein from *L. acidophilus* ATCC 4356 (Boot et al., 1993). GHCl (2M) was used to extract a collagen bindig S-layer protein from *L. crispatus* JCM 5810 (Toba et al., 1995) while LiCl (1M) for 20 h at 20°C after treatment with lysozyme (2 mg per ml) for 1 h, was used to extract another collagen binding protein from *L. reuteri* NCIB 11951 (Aleljung et al., 1994).

The MAPP protein from *L. reuteri* 104R was extracted from the surface, by treating the cells after 14-16 h growth in a semi-defined medium (LDM), with LiCl (1M) for 1 h with gently mixing at 4°C. However, when other lactobacilli strains isolated also from intestinal tract which presented binding to mucus and mucin were treated as above did not show the characteristic band of the MAPP adhesin as it was visualized by western blot with the horse radish peroxidase labeled mucin (Rojas et al., 2002). The adhesion of *L. reuteri* JCM1081 to HT-29 cells mediated by a cell surface protein was reported. Results showed a 29-kDa surface protein which displays significant peptide sequence similarity to the Lr0793 protein from *L. reuteri* ATCC55730 (71.1% identity), whereas the protein Lr0793 is homologous to the ABC transporter component CnBP, which previously has been described as a collagen binding protein. The 29-kDa surface protein of *L. reuteri* JCM1081 probably is classified as a member of the ABC transporter family, as well as CnBP from *L. reuteri* NCIB11951 and MapA from *L. reuteri* 104R (Wang et al., 2008). The mucus-binding properties of a large collection of *L. reuteri* strains isolated from a range of vertebrate hosts and the correlation of the adherence of a subset of strains to the presence and expression of MUB was performed by immunodetection, microscopic immunolocalization of MUB on the bacteria, characterization of cell-surface extracts and spent media by gel electrophoresis, Western blotting and mass spectrometry, quantification of *mub* gene expression by qRT-PCR, cell aggregation and cell-surface MUB quantification. Results revealed that the particular MUB investigated is highly specific to a very small set of closely related strains of *L. reuteri.* This was observed despite the fact that 17 proteins with a putative MucBP domain were found in the available genomes of *L. reuteri.* strains 100-23, DSM 20016T, MM2-3, MM4-1, ATCC 55730 and CF48-3A, nine of which were present in the rodent isolate 100-23 (Mackenzie et al., 2010).

1.4 Adhering probiotic *Lactobacillus*

Two requirements have been identified as desirable properties for *Lactobacillus* to be considered as an effective probiotic microorganism, these include the ability to adhere (Reid, 1999), and then to consequently colonize mucous surfaces. Mucus layer is the first physical barrier to host-cell stimulation by bacteria in the gut. Adhesion to mucus is therefore the first step required for probiotic organisms to interact with host cells and elicit any particular response. Adherence to intestinal mucus has been associated to competitive exclusion of pathogens (Gueimonde et al., 2006; Lee et al., 2008) considering it as a critical event for colonization not only for lactobacilli but also for pathogenic bacteria (Beachey, 1981; Soto & Hultgren, 1999). In the gastrointestinal tract, mucus is the outermost luminal layer, and is the first intestinal component of surface that microorganisms are likely to contact before they reach epithelial cells. Mackie (Mackie et al., 1999) suggested that during a colonization event, bacterial population remains stable in size, with no need of periodic reintroduction of bacteria by oral doses. This implies that colonizing bacteria multiply in a given intestinal niche at a rate that equals or exceeds their rate of washout or elimination from the intestinal site. However, in practical terms it is well known that external factors can arise such as antibiotic treatments or a change in the nutritional regime that can disrupt the equilibrium of the normal bacterial population (Jernberg et al., 2005). In these cases, it is necessary to supplement the feed with probiotics to restore the balance. Therefore, the ability to replicate in mucus represents an important parameter to evaluate in potential probiotic strains.

Additionally, it is recognized that resistance of potential probiotic to bile salts is a testable and is a necessary property (Moser & Savage, 2001).

The mechanisms used by lactobacilli to recognize and adhere to gastrointestinal components, until now is not completely understood. Protein and carbohydrate play an important role in mediating the adhesion to mucosal and or epithelial host surfaces. Some cell-surface biomolecules as exopolysaccharides and proteins have been recognized by their ability to bind gastrointestinal components (Vélez et al., 2007; Rojas et al., 2002; Sun et al., 2007). The best characterized are proteins present in the surface of lactobacilli that can be attached covalently or not to the cell wall (Vélez et al., 2007). Recently, proteins that adhere to mucus or mucins have been described and characterized. Adhering protein molecules characterized from *Lactobacillus* are Mucus-binding protein (Mub) of *L. reuteri* 1063 (Roos & Jonsson, 2002), the lectine-like mannose-specific adhesin (Msa) of *L. plantarum* WCFS1 (Pretzer et al., 2005), the mucus adhesion promoting protein (MAPP or MapA) from *L. reuteri* 104R reported by its ability to bind porcine mucus and mucin (Rojas et al., 2002) and Caco-2 cells (Miyoshi et al., 2006) and the Mub of *L. acidophilus* NCFM (Buck et al., 2005). Moreover, two proteins EF-Tu (Elongation Factor-Tu) and GroEL (a class of heat shock protein) of *L. johnsonii* La1 NCC533 showed abilities to adhere to mucins at specific conditions of pH (Granato et al., 2004; Bergonzelli et al., 2006). Recently a piglet mucus adhesion protein was completely characterized from the potential probiotic *L. fermentum* strain BCS87 (Macías-Rodríguez et al., 2009).

1.5 Genes codifying for *Lactobacillus* adhesins

Genetic research on *Lactobacillus* is underway in many laboratories around the world. Research has centered on 1) characterization and construction of vectors based on endogenos *Lactobacillus* plasmids which are capables to replicateof replicate and express molecules in specific lactobacilli strains, 2) molecular cloning of genes and operons from lactobacilli encoding important metabolic pathways, proteinases and adhesins 3) methods for introduction of genes *In vivo* and *In vitro* through conjugation, transfection and transformation (Chassy, 1987), and more recently 4) the global analysis of proteins and genes using the new tools of proteomic and genomic and the data base information of diferent species of *Lactobacillus* which are in public data bases. The development of cloning systems of *Lactobacillus* have increased in the last years. Methods for the introduction and stable maintenance of DNA into *Lactobacillus* are routine now and can be applied to almost any *Lactobacillus* species. Both broad host-range and narrow host range multi-copy plasmid vectors based on a variety of replicons have become available for the introduction and expression of homologous and heterologous genes (Pouwels & Leer, 1993). The sequenced genomes of lactobacilli are increasing and their availability might lead to the identification of the adhesin domain containing proteins in other species of *Lactobacillus* and in the specific functions of this surface proteins. Genes codifying for above adhesins are well known. The cloning and sequencing of the *L. reuteri* 104R gene encoding the adhesion promoting protein (MAPP) that binds to porcine gastrointestinal mucus was also studied. The sequence revealed one open reading frame consisting of 744 nucleotides corresponding to 244 aminoacids with deduced pI of 10.57, net charge at pH 7 of 16.23 and a molecular mass of 26.4 KDa. No putative promoter was found, however a start codon (ATG) appeared 6 bases downstream from the beginning of the sequence. The open reading frame ended with stop codons in all three reading frames (TGA A TAA T TAA). Computer search of the nucleotide and aminoacid sequences, showed that this adhesin is related to proteins encoding adherence factors from several pathogenic bacteria, as well as amino acid transporter

binding protein precursors (Rojas, 1996; E. Satoh et al., 2000). Expression by real time PCR of the genes *Mub* and *MapA*, adhesion-like factor *EF-Tu* and bacteriocin gene *plaA* by *L. plantarum* 423 grown in the presence of bile, pancreatin and at low pH, was reported. It was found that under normal physiological concentration of bile and pancreatin, expression of the *Mub* gene was affected, the *MapA* gene was over expressed and the *EF-Tu* gene remained stable, suggesting that whilst the expression of certain mucus genes may be affected by bile and pancreatin, others mucus genes are switched on, enabling the strain to adapt to physiological conditions and adhere to the gastrointestinal tract (Ramiah et al., 2007). To confirm the *MapA* results will be interesting to search in *L. plantarum* genome the compete sequence of this gene and find the adhering function in the reported specie or strain. By searching bacterial genome sequences and the UniProt protein data base for potential mucus binding proteins based on the sequence of the Mub domains of *L. reuteri* and *L. plantarum*. Boaekhorst et al, 2006. found that MUB domain is variable in size and sequence, making it difficult to determine precise domain boundaries. However the high variability in the number of MUB domain in putative mucus-binding proteins suggested that the MUB domain is often duplicated or deleted in evolution and appears to be only present in lactic acid bacteria, with the highest abundance in lactobacilli of the gastrointestinal tract, fulfilling an important function in host-microbe interactions (Boekhorst et al., 2006). Characterization of 32 Mmubp and *32-mmubp* gene from the potential probiotic strain previously isolated from piglet *L. fermentum* BCS87 was reported (Macías-Rodríguez et al., 2008). In the adhesion of this wild type strain to mucus and mucin, two proteins were identified, one of them, the 32Mmubp was characterized and the gene that codes for it was reported. Results indicate that the gene encoding this adhesin is conserved for *L. fermentum.* Other results suggested that 32Mmubp is released to the medium, but it could be anchored to cell wall by electrostatic interactions with acidic groups. It was indicated that Mmubp protein is a member of an ABC transporter system and is part of the OpuAC family. Based on homology and sequence domain search and in a phylogenetic tree with sequences of a seed group of the OpuAC family were shown conserved sequences between prokaryotic proteins of substrate-binding region on ABC type glycine/betaine transport systems. Some members of the corresponding taxa having similar ecological niches to those occupied by lactobacilli (gastrointestinal and respiratory tracts), i.e. *Helicobacter pylori* and *Mycobacterium tuberculosis,* did not group together suggesting that adhesion mechanisms is not a phylogenetic associated trait (Macías-Rodríguez et al., 2009). Recently was discovered only in the genome of the probiotic *Lactobacillus rhamnosus* GG, two different pilus fiber in the *spaCBA* and *spaFED* gene clusters. Moreover the expression and localization of intact SpaCBA pili on the cell surface of this strain were confirmed by immunoblotting and immunogold-labeled electron microscopy using antiserum specific for the Spa pilin. SpaCBA pilus-mediated binding of *L. rhamnosus* GG cells to human intestinal mucus was revealed (Kankainen et al., 2009). More recently pilin subunits SpaA, SpaB, SpaD, SpaE and SpaF encoded by genes in the *spaCBA* and *spaFED* genes clusters were cloned in *E. coli*. Recombinant, overproduced proteins were purified and assessment of the adherence to human intestinal mucus was performed. Results suggested that SpaC and

SpaB may be involved in SpaCBA pilus-mediated adherence to intestinal mucus. It was established that the SpaF minor pilin is the only mucus binding component in the putative SpaFED pilus fiber (von Ossowski et al., 2010). Aggregation promoting factors (Apf) are secreted proteins that have been associated with a diverse number of functional roles in lactobacilli, including self aggregation, coaggregation with other commensal or pathogenic bacteria, maintenance of cell shape and the bridging of conjugal pairs. Genes encoding Apf´s have been characterized for several *Lactobacillus* species, including *L. crispatus, L. johnsonii, L. gasseri, L. paracasei* and *L. coryniformis*. Investigation of the functional role of the putative *apf* gene (LBA0493) in *L. acidophilus* NCFM by mutational analysis was performed. It was observed that survival rates mutant strain NCK2033 decreased when stationary phase cells were exposed to simulated small intestinal and gastric juices. Furthermore, NCK2033 in the stationary phase showed a reduction of *In vitro* adherence to Caco-2 intestinal epithelial cells, mucin glycoproteins and fibronectin. It was suggested that the Apf-like proteins may contributes to the survival of *L. acidophilus* during transit through the digestive tract and, potentially, participate in the interactions with the host intestinal mucosa (Goh & Klaenhammer, 2010). The ability to tolerate the toxic levels of bile salts accumulated therein is the essential requirement to survive in the gut and it is generally included among the criteria used for selection of the potential probiotic strains and their application as functional ingredients in foods and nutraceuticals. Expression of bile salt hydrolase and surface proteins were targeted to look at their expression profile in two putative probiotic *L. plantarum* Lp9 and Lp91, (compared with standard strain CSCC5276) by quantitative real time PCR (RT - qPCR). Expression ratio for *bsh, mub, mapA* and *EF-Tu* genes under *In vitro* simulated gut conditions was tested for significance by qBase-Plus software. Amongst the three probiotic strains used in that study, Lp91 showed the highest level of *bsh* gene expression when the medium was supplemented with 0.01% mucin along with 1% of both bile and pancreatin in all the three strains. Results suggested that the expression of *mub* is a characteristic of not only the specie but could also be strain specific. The highest level of expression of *mapA* gene was recorded when normal gut conditions (Mucin, 0.01% and 0.3% each of bile and pancreatin, 0.3% supplemented in MRS at pH 6.5) were used. The relative expression of EF-Tu gene was significantly up-regulated in Lp9 in presence of mucin along at 0.01 and 0.05%, respectively at pH 7.0. It was concluded that the efficacy of both Lp9 and Lp91 with regards to expression of *mub, mapA* and *EF-Tu* was found to be either superior or comparable to that of standard probiotic strain (Duary et al., 2011). To confirm the *MapA* results in this last report it is important to find if the *L. plantarum* genome contains this gene to probe then its functionality.

1.6 Methods for screening mucus or mucin adhering bacteria

Mucus provides protective functions in the gastrointestinal tract and plays an important role in the adhesion of microorganisms to host surfaces. Mucin glycoprotein forms a framework to which microbial population can adhere, including probiotic *Lactobacillus* strains. Numerous factors have been shown to influence binding of lactobacilli to mucus *in vitro*. Experimental methods should be reviewed and compared to get a better understanding of the bacteria-mucosa interaction. The mechanism of this interaction could help to determine the degree of probiotic functionality imparted by adhesion (Tassell et al., 2011). Different methods to measure adhesion to mucus have been reported. Mucus contains about 80% of carbohydrates which occur as oligosaccharides and most of the glycans are present in clusters flanked by naked regions of the protein core (Clamp & Sheehan, 1978). Since mucins from different sources could be substituted with different oligosaccharides, properties such as the linear charge density could vary considerably. Porcine and rat mucin differ markedly in glycosylation and charge density (Malmsten et al., 1992). This characteristic of mucin, need to be considered when performing experiment to test the interaction between bacteria and mucus or mucin. A common method used to test *E. coli* adhesion to mucus extract prepared from the large and small intestine of mice involved immobilizing the mucus extracts on polystyrene. Radioactively labeled bacterial suspensions were added to the immobilized mucus compound and after a short inoculation time, the unbound cells were removed and adhesive cells were enumerated by measuring the amount of radioactivity (Laux et al., 1984; Laux, 1986). This method was adapted for studying *E. coli* adhesion to ileal mucus extracts from pigs (Conway et al., 1990; Blomberg & Conway, 1989). It has also been used to study the adhesion of *L. reuteri* 104R to small intestinal mucus extracts from pig (Rojas & Conway, 1996). This method still is used with some modifications (Mackenzie et al., 2010), however, it was not suitable for studying adhesion to mucin since it bound poorly to the polystyrene. In a control experiment where horse radish peroxidase labeled mucus and mucin were used, it was shown that mucin adhered to polystyrene a less extent than mucus. These results are consistent with other finding where rat and pig mucin layers on hydrophobic surfaces were studied. It was found by ellipsometry and surface force measurements, by using mica and silica surfaces, that the adsorption equilibrium of rat gastric mucin was reached after 5 hours, however for pig gastric mucin equilibrium it was not reached. It was demonstrated that for such layers, as the repulsive forces become weaker the slower the surfaces are brought together (Malmsten et al., 1992). Dot blot assay, a qualitative *In vitro* assay to detect the binding of bacterial cell surface components to mucus extracts was developed whereby extracts containing bacterial components and fractionated proteins were immobilized in a solid phase matrix and then blotted with enzymatically labelled mucus (Rojas and Conway, 2001). Results were compared to those obtained using the inhibition assay. In addition, whole cells of *Lactobacillus* and *E. coli* were tested in the dot blot assay and results compared with a modification of the method of Laux and coworkers (Conway et al., 1990). The results obtained using the dot blot assay provided further information about the binding of *Lactobacillus* and *E. coli* to gastrointestinal mucus, not only because adhesion promoting compounds could be detected in fractionated extracts but also because porcine gastric mucin as well as small intestinal mucus could be used for blotting (Rojas & Conway, 2001). Other methods have used to study adhesion to mucosa. Cultured cells have been suggested to be the best available models to study intestinal attachment of bacteria and viruses (Coconnier et al., 1997). Particularly, mucus secreting cells could be the best to study *Lactobacillus*-mucus and mucin interactions. Unfortunately this method has the same limitations as the mucus immobilization method of Laux et al. (Conway et al., 1990) for studing adhesins in soluble extracts.

1.7 Genetic tools to study the expression of genes encoding adhesins

The number of genetic tools that have been developed has increased tremendously during the last 20 years. Genetic analysis is made possible for several lactobacilli strains of known probiotic action, such as *L. plantarum* WCFS1, *L. acidophilus* NCFM, *L. johonsonii* NCC533, *L. salivarius* UCC118, *L. reuteri* ATCC 55730 and *L. rhamnosus* GG. Mutant studies are of the utmost importance in the unraveling of modes of action of lactobacilli as they can often directly relate genotype to phenotype. Nevertheless the number of currently identified genetic loci hypothesized to encode features supporting probiotic action confirmed by mutant analysis is still limited (Lebeer et al., 2008). Although the availability of genome sequences will certainly advance the field, they need to be complemented with functional studies. Methods that start to be applied for differential gene expression analysis of lactobacilli under relevant conditions are genome-wide comparisons of RNA profiles using microarrays, comparison of protein profiles with two dimensional (2D) difference gel electrophoresis, *In vivo* expression technology (IVET) using a promoter probe library and differential-display PCR (DD-PCR) (Lebeer et al., 2008).

2. Materials and methods for screening probiotic potential lactobacilli

2.1 Animals

Newborn piglets *(Landrace-Duroc)* from a pig farm were maintained with their mothers in maternity cages with grid floors during 23 days before weaned. Piglets received an intramuscular Fe injection (100 mg Fe, VITALECHON DEXTRAN) the second day after birth. Mother's milk fed piglets were given free access to commercial starter feed (17.5% crude protein, 2.5% crude fat, 5% crude fiber, 12% moisture, salts, vitamins, and minerals) and water (<900 ppm) 2-5 days before weaning. Maternity cages were maintained at room temperature and warmed up with lamps during the night when needed. To avoid excessive stress caused by high temperatures, piglets were bathed every day at midday.

2.2 Sampling

Faecal samples of healthy 23-day-old preweaned piglets from different cages with weights of 10 to 12 Kg were collected in sterile falcon tubes just at the time of defecating and transported to the laboratory at 4 °C. Piglets randomly selected, were sacrificed by a humanitarian method in the laboratory and immediately the small intestine and cecum were removed and sectioned with a sterile dissection kit. These pieces were opened and rinsed with sterile ice-cold phosphate-buffer saline (PBS) (145 mM NaCl, 2.87 mM KH_2PO_4, and 6.95 mM K_2HPO_4, pH 7.2) in order to remove loosely associated intestinal material. Mucus was then released by gently scraping the small intestine and cecum with a spatula and used to isolate lactic acid bacteria.

2.3 Isolation of bacteria

Isolation and characterization of bacteria was previously performed as reported before (Rojas & Conway, 1996; Macías-Rodríguez et al., 2008) . Briefly, lactic acid bacteria from faeces and from associated small intestine and cecum mucus of healthy preweaned piglets were isolated. Both faecal and mucosal samples were diluted in PBS and serial dilutions were plated on Rogosa SL agar (Difco). Plates were incubated at 37 °C for 24 h in an anaerobic jar with a Gaspack system. Counts of colony forming units (CFU) per gram and for cm^2 were reported. Colonies from each faecal or mucosal piglet sample were randomly selected from the last dilutions, purified on Rogosa SL plates and grown in MRS broth (Mann, Rogosa and Sharpe, Difco). Aliquots of each strain were kept in 1.5 ml tubes with 50% of glycerol at -85° C. Fresh cultures were used to perform the adhesion assay.

Source	Strain	Accession numbers (16-23S/ 16Sr DNA)	% identity
			Based on 16S rDNA sequence
Faeces	BCS9	EF113967/ EF113958	*99% to Lactobacillus fermentum*
Faeces	BCS10	EF113968/ EF113959	*99% to Lactobacillus fermentum*
Faeces	BCS12	EF113969/ EF113960	*99% to Lactobacillus fermentum*
Faeces	BCS13	EF113970/ EF113961	*99% to Lactobacillus fermentum*
Faeces	BCS14	EF113971/ EF113962	*99% to Lactobacillus fermentum*
Faeces	BCS21	EU547278/ EU547296	*99% to Lactobacillus fermentum*
Faeces	BCS24	E113972/ EF113963	*99 % to Lactobacillus fermentum*
Faeces	BCS25	EU547279/ EU547297	*99% to Lactobacillus fermentum*
Faeces	BCS27	EU547280/ EU547298	*99% to Lactobacillus fermentum*
Faeces	BCS30	EU547281/ EU547299	*99% to Lactobacillus fermentum*
Faeces	BCS36	EU547282/ EU547300	*100 % Lactobacillus fermentum*
Faeces	BCS41	EU547283/ EU547301	*100% Lactobacillus johnsonii;*
Faeces	BCS46	EF113973/ EF113964	99% to *Lactobacillus fermentum*
Faeces	BCS68	EU547284/ EU547302	99% to *Lactobacillus vaginalis*
Faeces	BCS75	EF113974/ EF113965	99% to *Lactobacillus fermentum*
Faeces	BCS80	EU547285/ EU547303	99% to *Lactobacillus fermentum*
Faeces	BCS81	EU547286/ EU547304	99% to *Lactobacillus fermentum*
Faeces	BCS82	EU547287/ EU547305	99% to *Lactobacillus fermentum*
Faeces	BCS87	EF113975/ EF113966	99% to *Lactobacillus fermentum*
SI mucus*	BCS113	EU547288/ EU547306	92 % to *Lactobacillus delbrueckii subsp. bulgaricus*
SI mucus*	BCS125	EU547289/ EU547307	99% to *Lactobacillus crispatus*
SI mucus*	BCS127	EU547290/ EU547308	99% to *Lactobacillus reuteri*
SI mucus*	BCS154	EU547294/ EU547312	99% to *Lactobacillus vaginalis*
C mucus**	BCS134	EU547291/ EU547309	99% to *Lactobacillus reuteri*
C mucus**	BCS136	EU547292/ EU547310	99% to *Lactobacillus reuteri*
C mucus**	BCS142	EU547293/ EU547311	99% to *Lactobacillus reuteri*
C mucus**	BCS159	EU547294/ EU547313	99% to *Lactobacillus reuteri*

*Intestinal tract mucus, ** Cecum mucus

Table 1. Strains isolated from faeces and mucus of healthy piglets used in this study (Macías-Rodríguez et al., 2008).

2.4 Oligonucleotide design and synthesis

Oligonucleotides used for PCR amplifications were designed with the Primer Select tool of the Laser gene software (Version 5) and synthesized at the Instituto de Biotecnología, UNAM (Mexico). All are listed in Table 2.

Oligonucleotide name	Orientation	Sequence
MAP1F	Forward	5' ATGCCTGCAGGAATCACAA 3′
MAP1R	Reverse	5' AGTAATATCTGCACCGAAGTA 3'
MEF7	Forward	5´ ATTTACGCCCTGGCCCTGGAAAAG-3´
MER9	Reverse	5´ AGAGGGTGTATTTGTTGCCATTGG-3´
MAP2F	Forward	5′ TCTTATGCGACCCACAGTTTG 3′
MAP2R	Reverse	5′ CTAAGAGCCCCGTCGTTC 3′

Table 2. Oligonucleotides used for PCR amplifications

2.5 PCR amplification of the *32-Mmubp* gene

Amplification of the *32-Mmubp* gene of *L. fermentum* previously reported by (Macías-Rodríguez et al., 2009) was performed using as template the chromosomal DNA of *Lactobacillus* strains previously characterized as potential probiotic by traditional methods (Table 1). A combination of gene specific oligonucleotides for an internal fragment MEF7 and MER9 was used to perform the amplification. The PCR solution contained a final concentration of 1× Taq polymerase buffer, 3mmol l^{-1} $MgCl_2$, 0.4mmol l^{-1} for each dNTP, 120 pmol of each primer, 250 ng chromosomal DNA and 1 U of Taq DNA polymerase in a total volume of 25 µl. Amplification reaction was performed in a thermocycler (Perkin-Elmer mod. GeneAmp 2400) with the following temperature program: 1 cycle at 94°C for 5 min; 30 cycles consisted in a denaturation step at 94° C for 1 min, an annealing step at 55°C for 1 min and an extension step at 72° C for 1 min. A final extension was performed at 72° C for 5 min. PCR products were then analyzed in a 1.5% agarose gel.

2.6 PCR amplification of the *mapp* or *mapA* gene

Amplification of the gene *mapp* or *mapA* (Genebank accession number AJ293860) previously described (Rojas 1996, Satoh et al., 2000 and Miyoshi et al., 2006) was performed using as template the same chromosomal DNA of *Lactobacillus* strains used for amplification of the 32*Mmubp* gene. A combination of gene specific oligonucleotides for an internal fragment of the open reading frame MAP1F and MAP1R (Table 2) was used. The PCR solution contained a final concentration of 1× Taq polymerase buffer, 3mmol $MgCl_2$, 0.4mmol for each dNTP, 60 pmol of each primer, 300 ng chromosomal DNA and 1 U of Taq DNA polymerase in a total volume of 25 µl. Amplification reaction was performed in a thermocycler (Perkin-Elmer mod. GeneAmp 2400) with the following temperature program: 1 cycle at 94°C for 5 min; 28 cycles consisted in a denaturation step at 94° C for 1 min, an annealing step at 49°C for 1 min and an extension step at 72° C for 2 min. A final extension was performed at 72° C for 5 min. PCR products were analyzed in a 1.5% agarose gel.

2.7 PCR amplification of the operon containing the *MapA* gene

Polymerase Chain Reactions was performed with primers MAP2F and MAP2R (Table 2) for an internal fragment of the operon containing the *mapA* gene (Genebank LOCUS AJ293860) using as template the chromosomal DNA of *Lactobacillus* strains *map*A positive. The PCR solution contained a final concentration of 1× Taq polymerase buffer, 2mmol $MgSO_4$, 0.4mmol for each dNTP, 100 pmol of each primer, 300 ng chromosomal DNA and 2 U of Platinum *Taq* DNA Polymerase (Invitrogene), in a total volume of 25 µl. Amplification reaction was performed in a thermocycler (Perkin-Elmer mod. GeneAmp 2400) with the following temperature program: 1 cycle at 94°C for 5 min; 30 cycles consisted in a denaturation step at 94° C for 1 min, an annealing step at 62°C for 1 min and an extension step at 68° C for 2 min. A final extension was performed at 68° C for 5 min. PCR products were analyzed in a 1.0% agarose gel.

3. Results and discussion

The association of lactobacilli with the epithelial and mucosal surfaces and their presence in faeces in pigs has been well studied (Rojas & Conway 1996; Macías-Rodríguez et al., 2008). It was shown that *Lactobacillus* population in faeces ranged between 10^7 and 10^9 CFU gr^{-1}. Likewise, in intestinal mucosa, counts of $3.8x10^6$ and $3.2 x10^6$ CFU per cm^2 of small intestine and cecum respectively were reported. Cultivable *Lactobacillus* strains has been found in similar amounts in faeces and intestinal mucus of pigs that inhabit different environmental conditions, cool countries (Rojas and Conway, 1996) and warm arid coasts (Macías-Rodríguez et al., 2008). It was found too that *L. fermentum* and *L. reuteri* are the major strains which colonize the gastrointestinal tract of pigs. Therefore the screening of *Lactobacillus* with probiotic potential for piglets with the ability to interact with the host should be addressed to this species. It has been reported that species of *Lactobacillus* which colonize humans, differ in number and specie from one region to other in the world. Likewise *In vivo* trials have been shown that probiotic effect of one strain in one region of the world could produce confused results in other. This finding supports the idea to look for a new generation of specific probiotics for animals and humans inhabiting specific region in the world.

Traditionally the screening of *Lactobacillus* with probiotic potential involve the isolation and purification of many colonies of lactic acid bacteria, confirmation that correspond to presumptive lactobacilli (grown in selective medium, Gram stain, catalasa reaction, etc), selection according to adhesion profile, growth in mucus, bile salt resistance, growth in broad range of temperature and salt concentration, bacteriocin production, growth and adhesion inhibition of enteropathogens, molecular identification, etc. Previously, more than 150 strains were isolated from mucus and feaces of piglets. Results showed that 64% of presumptive *Lactobacillus* presented abilities to grow in the presence of 680 mM of NaCl. Additionally 75% of the isolates were able to grow at 50 °C. These abilities are important considering that probiotic bacteria are exposed to high temperatures and presence of NaCl during their technological preparation as pelleted or dried feed for pigs. The adhesion assay of the 164 isolates to porcine mucus and mucin allowed visualize strains that bind mucus or gastric mucin in a qualitative manner. Results indicated that 88 isolates representing 53.7% of the 164 strains, presented adhesion to both mucus and gastric mucin similar to the positive control *L. reuteri* 104R, (Rojas et al., 2002). From the total of faecal strains 45% showed binding ability, whereas from intestinal and cecal mucus strains, 64 and 78%

presented adhesion ability respectively. These results showed the highest percentage of adhesive strains in the cecum and intestine compared with faeces. Adhesive strains isolated from faeces could be released to the lumen during the renewal of mucus. Different adhesive abilities between faecal and mucosal strains could be also explained if is considered that microbiota in the intestine differs from that in faeces (Marteau, 2002). Moreover, adhesive properties are strain-dependent and differences exist even if strains were isolated from the same source (Kinoshita et al., 2007).

For molecular identification the most common amplified sequences by PCR are the 16-23S intergenic region and 16S rDNA gene. The 27 strains used in this work were identified by these methods. Analysis of 16S rDNA gene sequences showed that 17 strains belong to *L. fermentum* specie (between 98 to 100% identity), one strain *to L. johnsonii*, 2 strains to *L. vaginalis*, one strain to *L. crispatus* and 5 strains *to L. reuteri* species (Table 1). Except strain BCS113 that showed 92% identity to 16S rDNA of *L. delbrueckii* subsp. *bulgaricus*. These results showed that *L. fermentum* was predominant in faecal adhesive isolates whereas *L. reuteri* was the principal in mucus of cecum. In small intestinal mucus there was not predominant specie. These observations agree with previously reported by Lin *et al.* (Lin et al., 2007) and (De Angelis et al., 2006) who found both species in faeces and mucus of pigs. This result confirmed the relevance of these species in the intestinal tract of pigs. Moreover, *L. fermentum* and *L. reuteri* species have been reported as good candidates as probiotics (De Angelis et al., 2007; Zoumpopoulou et al., 2008). Another species identified as *L. johnsonii, L. delbrueckii* subsp. *bulgaricus, L. vaginalis* and *L. crispatus* have been reported by their probiotic potential in humans and animals (Chen et al., 2007; Matijasic et al., 2006; Ohashi et al., 2007).

To understand the relevance of surface proteins in the adhesion of *Lactobacillus* to mucus and mucin, the purification and characterization of the adhesins should be performed. In previous reports proteins have been obtained by treatment with chaotropic agents as LiCl. From the spent, centrifuged growth medium and from soluble cytoplasmic extracts. A western blot assay using labelled mucus and mucin has been usually performed to show the protein bands with their relative molecular weight (MW) and in order to characterize them, N-terminal and internal peptide sequences has been determined. The MAPP adhesin of *L. reuteri* and the *Mmubp of L. fermentum* have been characterized in that manner (Rojas et al., 2002; Macías-Rodríguez et al., 2008). Recently the mucus-binding proteins (MUBs) have been revealed as one of the effectors molecules involved in mechanisms of the adherence of lactobacilli to the host; *mub*, or *mub*-like, genes were found in all of the six genomes of *L. reuteri* that are available but the MUB was only detectable on the cell surface of two highly related isolates when using antibodies that were raised against the protein (Mackenzie et al., 2010).

The complete process to get new strains of probiotic potential lactobacilli has been long and complex. Above a review of the different methods and results used was exposed and the results of a proposal are described.

The strains listed in Table 1 were selected because they were the predominant cultivable lactic acid bacteria in a selective medium (Rogosa agar, DIFCO); attached strongly to mucus and mucin when tested by the Dot Blot adhesion assay; grew in mucus, in presence of bile salt and in a broad range of temperatures. Likewise the molecular identification confirmed that *L. fermentum* and *L. reuteri* were the main isolates with probiotic potential for piglets (Macías-Rodríguez et al., 2008). In addition the genes *mapp* or *mapA* of *L. reuteri* and *mmub* of

L. fermentum, which codified for mucus adhesins have been well characterized. Here these genes are described and the results of this proposal are discussed.

3.1 Amplification and sequencing of the *32-Mmubp* encoding gene (*32-mmub*)

Primers MEF7 and MER9 were previously deduced from the complete nucleotide sequence of *32-mmub* gene. The gene presented an ORF (open reading frame) of 903 bp encoding a predicted primary protein of 300 amino acids. This protein presented a signal peptide of 28 amino acids. Cleavage site between residues 28 and 29 were detected with the Signal P 3.0 prediction software. The prediction of transmembrane helices showed that the first 1 to 7 amino acids are predicted to be inside of the cell whereas residues 7 to 29 could be in the membrane and finally the region encompassing amino acids 30 to 300 could be outside. The mature protein consists of 272 residues with a molecular mass of 29,974 Da, an isoelectric point of 9.78 and a positive net charge of 21.22 at pH 7.0. This adhesin protein showed high identity only to *L. fermentum* (BAG27284). A search of homology (BLAST) with the genome of *L. fermentum* IFO 3956 recently published (Morita et al., 2008) showed that 32-Mmubp in *L. fermentum* BCS87 is part of an ABC transporter system and belongs to the PBPb superfamily. It showed to be conserved between prokaryotic protein sequences of substrate binding domains on the ABC-type glycine/betaine transport systems of the OpuAc familiy (PF04069). This family is part of a high-affinity multicomponent binding proteins-dependent transport system involved in bacterial osmoregulation and members of this family are often integral membrane proteins or predicted to be attached to the membrane by a lipid anchor. Some members of the corresponding taxa having similar ecological niches to those occupied by lactobacilli (gastrointestinal and respiratory tracts), i.e. *Helicobacter pylori* and *Mycobacterium tuberculosis*, do not group together suggesting that adhesion mechanisms is not a phylogenetic associated trait.

To confirm that 32-Mmubp of *L. fermentum* BCS87 is specific for this especie, a PCR using the MEF7 and MER9 oligonucleotides was performed. Chromosomal DNA of the 26 adhering strains of Table 1 was used as template to amplify an internal product of *32-mmub* gene. PCR products of the same size (550 bp) were observed in *L. fermentum* strain BCS87 and in all strains which belong to the same specie (Figure 1). Moreover a weak band was also observed in species *L. johnsonii* BCS41, *L. vaginalis* strains BCS68 and BCS154, *L. delbrueckii* subsp. *bulgaricus* BCS113, *L. crispatus* BCS125 and *L. reuteri* strains BCS127, BCS134, BCS136, BCS142 and BCS159 (Figure 1) suggesting *32-mmub* gene is conserved in piglets adhesive *L. fermentum*.

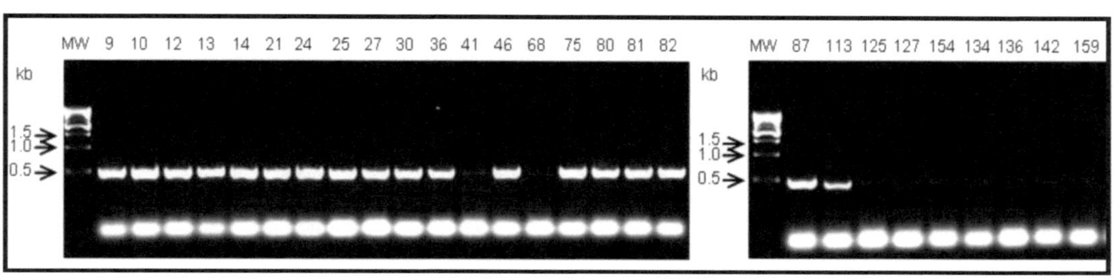

Fig. 1. Amplification of internal fragment of the *32-Mmubp* gene in adhesive strains of *L. fermentum* isolated from piglets intestinal tract. Lane MW, Molecular weight. Numbers 9 to 159 represents the identification code for each *Lactobacillus* strains from Table 1.

3.2 Amplification and sequencing of the *mapp* or *mapA* gene

A mucus adhesion promoting protein (MAPP) from *L. reuteri* 104R was reported (Rojas et al., 2002; Rojas, 1996). The gene encoding this MAPP adhesin (*mapp* gene) was found by using a PCR strategy were peptide derived oligonucletides were carefully devised and PCR reactions performed using chromosomal DNA of *L. reuteri* 104R as template. A PCR product was cloning and sequencing. Southern blotting of digested chromosomal DNA with selected enzyme mixtures was performed by using a 189 bp PCR product as a probe. Then a subgenomic DNA library of the hybridized fragment approximately of 4600 bp was running out. DNA fragments in this region were ligated in the pGEM3 vector and cloned in *E. coli*. Hybridization with the same probe showed a 4500 bp fragment containing the *mapp* gene. A subcloning and sequencing strategy (Figure 2) was used to determine the nucleotide sequence of the *mapp* gene. Nucleotide sequence analysis and search of the nucleotide and deduced aminoacid sequenced were searched in different data bases (NCBI). The complete gene *mapp* was sequenced. The sequence revealed one open reading frame which consists of 744 nucleotides corresponding to a protein of 244 aminoacids with a deduced pI of 10.57

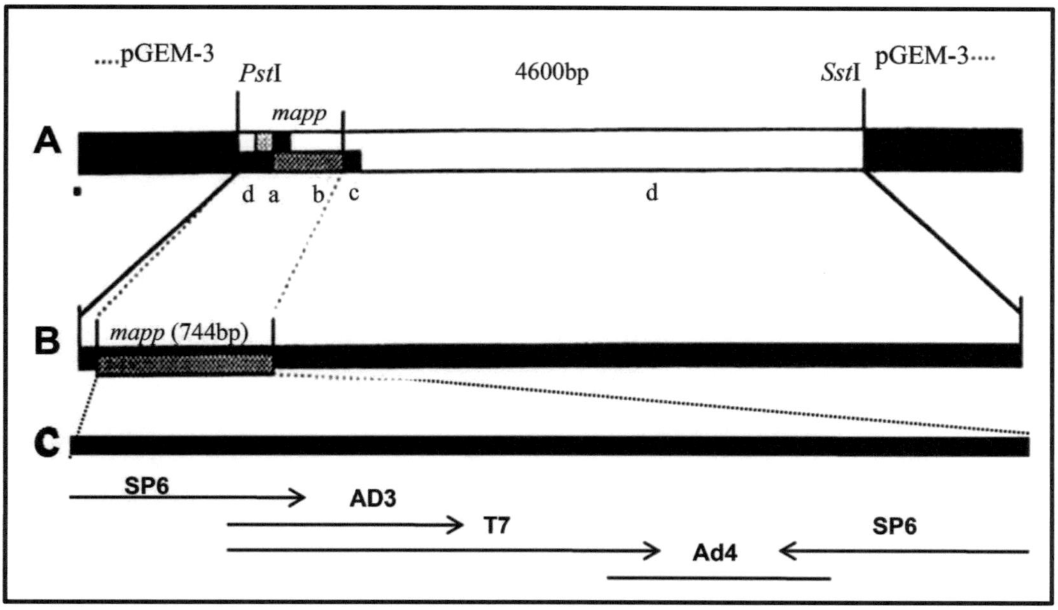

Fig. 2. Schematic drawing of the subcloning and sequencing strategy to determine the nucleotide sequence of the *mapp* gene. A) The stippled box represents the pGEM-3 vector used to clone the chromosomal DNA fragment (4500 bp) from *L. reuteri* 104R and to subclone the fragments a, b, c and d. (inside boxes): a)189 bp, PCR fragment b)610 *Bgl*II-*Bgl*II fragment, c)146 bp *Bgl* II-*Bgl* II fragment and d) vector plus fragment without the two *Bgl* II fragments. B) The largest box represent the 4500bp fragment and the inside box represent the *mapp* gene. Universal primers are indicated, arrows indicate nucleotides determined and the heads of the arrow indicate the transcription direction. C) The box represents the 744bp open reading frame of the *mapp* gene. Universal and sequence specific primers are indicates, arrows indicate nucleotides determined and the heads of the arrow indicate the transcription direction.

and a molecular mass of 26380.90 Da. No putative promoter was found, however, a start codon (ATG) was noted 6 bases downstream of the beginning of the sequence and 30 bases upstream of the first N terminal aminoacid derived codon. The open reading frame ends with stop codons in all three reading frames (TGA A TAA T TAA) (Rojas, 1996).

The *mapp* gene described in Rojas, 2006, was later reported in Gene Bank as *MapA* and as part of one operon whose expression is controlled by a mechanism of transcription attenuation involved cysteine, with accession number AJ 293860 (Satoh et al., 2000). The relation between MapA and adhesion of *L. reuteri* to human intestinal (Caco 2) cells was reported. Quantitative analysis of adhesion of *L. reuteri* strains to Caco 2 cells showed that various strains bind also intestinal epithelial cells. In addition purified MapA bound to Caco 2 cells and this binding inhibited the adhesion of *L. reuteri* in a concentration dependent manner. Additionally it was concluded that multiple receptor-like molecules are involved in the MapA binding to Caco 2 cells (Miyoshi et al., 2006).

To confirm that *MapA* gene is specific for adhesive *L. reuteri* strains, a PCR using the MAPF1 and MAPR1 oligonucleotides was performed. Chromosomal DNA of the 26 adhering strains of Table 1 was also used as template to amplify the *MapA* gene. PCR products of the same size were observed only in the *L. reuteri* strains tested (Figure 3) but not in other species. This result strongly suggests that *MapA* gene is conserved in piglet adhesive *L. reutri* strains.

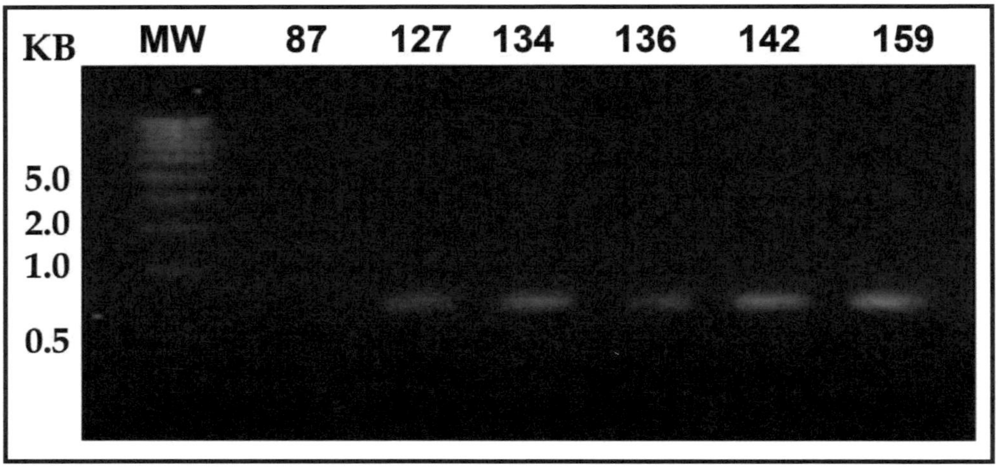

Fig. 3. Amplification of the gene *MapA* in adhesive *L. reuteri* strains isolated from piglets intestinal tract. KB Kilobases. MW; Molecular weight. Names on the lanes represent the identification code for each *Lactobacillus* strains from Table 1.

Expression of the mucus adhesion genes *Mub* and *MapA*, adhesion-like factor *EF-Tu* and bacteriocin gene *plaA* by *L. plantarum* 423 was reported. Growth in the presence of bile, pancreatin and at low pH, was studied by real-time PCR. It was found that *Mub, MapA* and *EF-Tu* were up-regulated in the presence of mucus, proportional to increasing concentrations. Expression of *Mub* and *MapA* remained unchanged at pH 4.0, whilst expression of *EF-Tu* and *plaA* were up-regulated. Expression of *MapA* was down-regulated in the presence of 1.0 g/l l-cysteine HCl, confirming that the gene is regulated by transcription attenuation that involves cysteine (Ramiah et al., 2007). However the gene and operon *MapA* were not found in *L. plantarum* by a nucleotide data base search in blastn suite (NCBI). However results in this work suggested that functional MapA gene is specific for at least adhesive *L. reuteri* strains.

Mucus-binding proteins (MUBs) are molecules involved in mechanisms of the adherence of lactobacilli to the host (Roos & Jonsson, 2002). It was suggested that MUB domain is an LAB -specific functional unit that performs its task in various domain contexts and could fulfils an important role in host-microbe interactions in the gastrointestinal tract (Boekhorst et al., 2006). Recently was reported that in spite that *mub*, or *mub*-like, genes are found in all of the six genomes of *L. reuteri* and further demonstrated that MUB and MUB-like proteins are present in many *L. reuteri* isolates, MUB was only detectable on the cell surface of two highly related isolates when using antibodies that were raised against the protein. There was considerable variation in quantitative mucus adhesion *in vitro* among *L. reuteri* strains, showing a high genetic heterogeneity among strains (Mackenzie et al., 2010). Different results were observed for the *MapA* gene which was present in all the adhesive *L. reuteri* strains used to amplify this gene.

Recently was reported a well-defined degradation product with antimicrobial activity obtained from the mucus adhesion-promoting protein (MapA) termed AP48-MapA from *L. reuteri* strain. The peptide was purified and characterized. This finding gave a new perspective on how some probiotic bacteria may successfully compete in this environment and thereby contribute to a healthy microbiota (Bøhle et al., 2010). This finding correlate with a report where trypsin digestion of the MapA protein resulted in peptides that bound to mucin suggesting that MapA protein could be involved in colonization of the intestinal mucosa of piglet, since the adhesive capacity could be retained in the intestinal mieleu (Rojas et al., 2002).

To find if *L. reuteri* strains which contain the *MapA* gene present the same operon as strain 104R, amplification was run out (Figure 4).

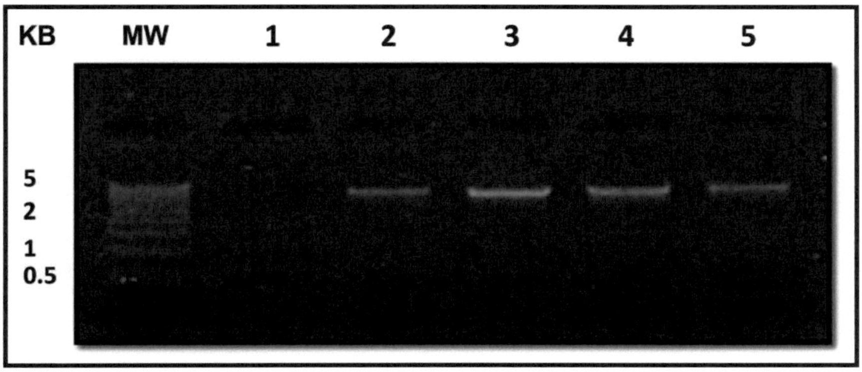

Fig. 4. Amplification of the *MapA* operon (3.9Kb) from different adhesive *L. reuteri* strains isolated from piglets intestinal tract. Lane MW) 500-5000 bp ladder lane 1) Control strain, *L. fermentum* BCS87 lane 2) *L. reuteri* BCS136, lane 3) *L. reuteri* BCS127, lane 4) *L. reuteri* BCS159 and lane 5) *L. reuteri* BCS142

These results together with the review of adhesins from *L. fermentum* and *L. reuteri* and their genes indicate that *Mmubp* and *MapA* genes are conserved in these species, at least in adhesive strains isolated from intestinal tract of piglets. In Addition these strains are considered the main *Lactobacillus* species which colonize the intestinal tract of piglets. Therefore the traditional methods for screening new probiotic strains for piglets could be reduced as described.

Take faeces and intestinal tract mucus samples from healthy piglets and make a viable count in a selective medium (Rogosa Agar, DIFCO). Incubate at 36°C in anaerobic conditions for 24-48 h and select colonies from the plates with the more diluted samples to grow and purify the DNA. Perform a PCR reaction using the specific primers for the *Mmub* and *MapA* genes. Strains which amplify a fragment with the size mentioned above should be *L. fermentum* for the *Mmub* gene and *L. reuteri* for the *MapA* gene.

4. Conclusion

Bacteria cultivated in the laboratory for long time could mutate and lost probiotic attributes, therefore it is important to look for an easy strategy to routinely screening for probiotics. Screening for new probiotic *Lactobacillus fermentum* and *Lactobacillus reuteri*, which are the dominant microbiota in healthy piglets and present the ability to adhere the intestinal tract mucus is described in this chapter. The main advantage of this method is the expend time.

5. Acknowledgment

This study was supported by Universidad Autónoma de Baja California Sur, México and Conacyt, Project No. 29410-B

6. References

Aleljung, P. et al., 1991. Collagen Binding by Lactobacilli. *Current Microbiology*, 23, pp.33-38.

Aleljung, P. et al., 1994. Purification of collagen-binding proteins of *Lactobacillus reuteri* NCIB 11951. *Current Microbiology*, 28, pp.231-236.

Ashenafi, M., 2005. Growth of *Listeria monocytogenes* in fermenting tempeh made of various beans and its inhibition by *Lactobacillus plantarum*. *Food Microbiology*, 8(4), pp.1991-1991.

Bateup, J.M. et al., 1995. Comparison of *Lactobacillus* strains with respect to bile salt hydrolase activity, colonization of the gastrointestinal tract, and growth rate of the murine host. *Applied and Environmental Microbiology*, 61(3), pp.1147-11499.

Beachey, E.H., 1981. Bacterial adherence: Adhesin-receptor interactions mediating the attachment of bacteria to mucosal surface. *Journal of Infectious Diseases.*, 143, pp.325-345.

Bergonzelli, G.E. et al., 2006. GroEL of *Lactobacillus johnsonii* La1 (NCC 533) is cell surface associated: potential role in interactions with the host and the gastric pathogen *Helicobacter pylori*. *Infection and immunity*, 74(1), pp.425-434.

Beveridge, T.J., 1989. Role of cellular design in bacterial metal accumulation and mineralization. *Annual Review of Microbiology*, 43, pp.147-171.

Beveridge, T.J. & Graham, L.L., 1991. Surface layers of bacteria. *Microbiological Reviews*, 55(4), pp.684-705.

Bibel, D.J., 1988. Elie Metchnikoff's Bacillus of long life. *ASM News*, 54(12), pp.661-665.

Blomberg, L. & Conway, P L, 1989. An *In vitro* study of ileal colonisation resistance to *Escherichia coli* K88 to piglet ileal mucus by *Lactobacillus* spp. *Microbial Ecology in Health and Disease*, 2, pp.285-291.

Boekhorst, J. et al., 2006. Comparative analysis of proteins with a mucus-binding domain found exclusively in lactic acid bacteria. *Microbiology (Reading, England)*, 152(Pt 1), pp.273-80.

Bonazzi, M. & Cossart, P., 2011. Host-pathogen interactions: Impenetrable barriers or entry portals? The role of cell-cell adhesion during infection. *The Journal of Cell Biology*, 195(3), pp.349-358.

Boot, H.J. et al., 1993. S-Layer protein of *Lactobacillus acidophilus* ATCC 4356: purification, expression in *Escherichia coli*, and nucleotide sequence of the corresponding gene. *Journal of bacteriology*, 175(19), pp.6089-6096.

Brennan, M. et al., 1986. Cellular damage in dried *Lactobacillus acidophilus*. *Journal of Food Protection*, 49, pp.47-53.

Brown, M.R.W., Anwar, H. & Casterton, J.W., 1988. Surface antigens *In vivo*: a mirror for vaccin development. *Canadian Journal of Microbiology*, 34, pp.494-498.

Buck, B.L. et al., 2005. Functional Analysis of Putative Adhesion Factors in *Lactobacillus acidophilus* NCFM. *Applied and Environmental Microbiology*, 71(12), pp.8344-8351.

Bøhle, L.A. et al., 2010. Specific degradation of the mucus adhesion-promoting protein (MapA) of *Lactobacillus reuteri* to an antimicrobial peptide. *Applied and Environmental Microbiology*, 76(21), pp.7306-9.

Chagnaud, P., Jenkinnson, H.F. & Tannock, G. W., 1992. Cell surface associated proteins of gastrointestinal strains of lactobacilli. *Microbial Ecology in Health and Disease*. 5(3), pp.121-131.

Chassy, B.M., 1987. Pospects for the genetic manipulation of lactobacilli. *FEMS Microbiol Rev*, 46, pp.297-312.

Chateau N, Castellanos I, D.A., 1993. Distribution of pathogen inhibition in the *Lactobacillus* isolates of a commercial probiotic consortium. *Journal of Applied Bacteriology*. 74(1), pp.36-40.

Chen, X. et al., 2007. The S-layer proteins of *Lactobacillus crispatus* strain ZJ001 is responsible for competitive exclusion against *Escherichia coli* O157:H7 and *Salmonella typhimurium*. *International Journal of Food Microbiology*. 115, pp.307-312.

Chiduwa, G. et al., 2008. Herd dynamics and contribution of indigenous pigs to the livelihoods of rural farmers in a semi-arid area of Zimbabwe. *Tropical Animal Health and Production*, 40(2), pp.125-36.

Clamp, J.R. & Sheehan, J.K., 1978. Chemical aspects of mucus. *British Medical Bulletin*, 34(1), pp.25-41.

Coconnier, M.H., Liévin, V. & Hudault, S, 1997. Antibacterial effect of the adhering human *Lactobacillus acidophilus* strain LB. *Microbiology*, 41(5), pp.1046-1052.

Collado, M.C. et al., 2006. Protection mechanism of probiotic combination against human pathogens: in vitro adhesion to human intestinal mucus. *Asia Pacific Journal of Clinical Nutrition*. 15(4), pp.570-575.

Conway, P L, Welin, A. & Cohen, P.S., 1990. Presence of K88-specific receptors in porcine ileal mucus is age dependent. *Infection and Immunity*, 58(10), pp.3178-3182.

Conway, P L & Adams, R.F., 1989. Role of erythrosine in the inhibition of adhesion of *Lactobacillus fermentum* strain 737 to mouse stomach tissue. *Journal of General*

Microbiology, 135(5), pp.1167-73. Available at: http://www.ncbi.nlm.nih.gov/pubmed/2559943.

Cook, R.L., Harris, R.J. & Reid, G, 1988. Effect of culture media and growth phase on the morphology of lactobacilli and on their ability to adhere to epithelial cells. *Current Microbiology*, 17, pp.159-166.

Costerton, J.W., 1988. Structure and plasticity at various organization levels in the bacterial cell. *Canadian Journal of Microbiology*, 34, pp.513-521.

De Angelis, M., et. al., 2006. Selection of potential probiotic lactobacilli from pig feces to be used as additives in pelleted feeding. *Research in Microbiology*, 157, pp.792-801.

Dixon, J.M.S., 1960. The fate of bacteria in the small intestine. *Journal of Pathology &Bacterology*, 79, pp.131-141.

Duary, R.K., Batish, V.K. & Grover, S., 2011. Relative gene expression of bile salt hydrolase and surface proteins in two putative indigenous *Lactobacillus plantarum* strains under *In vitro* gut conditions. *Molecular Biology Reports*, DOI: 10.1007/s11033-011-1006-9.

Dubos, R., et al., 1965. Indigenous, normal and autochthonous flora of the Gastrointestinal Tract. *Journal of Experimental Medicine*, 122, pp.67-76.

Fuller, R., 1989. Probiotics in man and animals. *Journal of Applied Bacteriology*, 66, pp.365-378.

Fuller, R., Barrow, P.A. & Brooker, M.E., 1978. Bacteria associated with the gastric epithelium of neonatal pigs. *Applied and Environmental Microbiology*, 35(3), pp.582-591.

Goh, Y.J. & Klaenhammer, T.R, 2010. Functional roles of aggregation-promoting-like factor in stress tolerance and adherence of *Lactobacillus acidophilus* NCFM. *Applied and Environmental Microbiology*, 76(15), pp.5005-12.

Goswami, P.S. et al., 2011. Preliminary investigations of the distribution of *Escherichia coli* O149 in sows, piglets, and their environment. *Canadian Journal of Veterinary Research*, 75(1), pp.57-60.

Granato, D. et al., 2004. Cell surface-associated elongation factor Tu mediates the attachment of *Lactobacillus johnsonii* NCC533 (La1) to human intestinal cells and mucins. *Infection and Immunity*, 72(4), pp.2160-2169.

Greene, J.D. & Klaenhammer, T R, 1994. Factors involved in adherence of lactobacilli to human Caco-2 cells. *Applied and Environmental Microbiology*, 60(12), pp.4487-94.

Guarino, A. et al., 2011. The management of acute diarrhea in children in developed and developing areas: from evidence base to clinical practice. *Expert Opinion on Pharmacotherpy*, pp.22106840-22106840.

Gueimonde M, Sakata S, Kalliomaki M, Isolauri E, Benno Y, S.S., 2006. Effect of maternal consumption of *Lactobacillus* GG on transfer and establishment of fecal bifidobacterial microbiota in neonates. *Journal of Pediatric Gastroenterology and Nutrition*, 42(2), pp.266-270.

Guerrant, R. L., Hughes, J. M., Lima, N. L., Crane, J., 1990. Diarrhea in developed and developing countries: magnitude, special settings, and etiologies. *Reviews of Infectious Diseases*, 12(1), pp.41-50.

Guilliland, S. E., Morelli, L., R.G., 2001. Health and nutritional properties of probiotics in food including powder milk with live lactic acid bacteria. . In *Joint FAO/WHO expert consultation, Cordova Argentina.*

H, Kinoshita et al., 2007. Quantitative evaluation of adhesion of lactobacilli isolated from human intestinal tissues to human colonic mucin using surface plasmon resonance (BIACORE assay). *Journal of Applied Microbiology*, 102, pp.116-123.

Harris, L. J., et al., 1989. Antimicrobial activity of lactic acid bacteria against *Listeria monocytogenes*. *Journal of Food Protection*, 52, pp.384-387.

Henriksson, A.R., Szewzyk, R. & Conway, P. L., 1991. Characteristics of the adhesive determinants of *Lactobacillus fermentum* 104. *Applied and Environmental Microbiology*, 57(2), pp.499-502.

Herías M.V. et al., 1999. Immunomodulatory effects of *Lactobacillus plantarum* colonizing the intestine of gnotobiotic rats. *Clinical & Experimental Immunology*, 116, pp.283-290.

Hudault, S et al., 1997. Antagonistic activity exerted *In vitro* and *In vivo* by *Lactobacillus casei* (strain GG) against *Salmonella typhimurium* C5 infection. *Microbiology*, 63(2), pp.513-518.

Jacobsen, C.N. et al., 1999. Screening of probiotic activities of forty-seven strains of *Lactobacillus* spp. by *In vitro* techniques and evaluation of the colonization ability of five selected strains in humans. *Applied and Environmental Microbiology*, 65(11), pp.4949-56.

Jernberg, C. et al., 2005. Monitoring of antibiotic-induced alterations in the human intestinal microflora and detection of probiotic strains by se of Terminal Restriction Fragment Length Polymorphism. *Applied and Environmental Microbiology*, 71(1), pp.501-506.

Johansson, M.L. et al., 1993. Administration of different *Lactobacillus* strains in fermented oatmeal soup: *In vivo* colonization of human intestinal mucosa and effect on the indigenous flora. *Applied and Environmental Microbiology*, 59(1), pp.15-20.

Jones, G.W. & Isaacson, R.E., 1984. Proteinaceous bacterial adhesins and their receptors. *Critical Reviews in Microbiology*, 10(3), pp.229-260.

Jonsson, E. & Conway, P.L., 1992. Probiotics for pigs. In R. Fuler, ed. *Probiotics, the scientific basis*. London: Chapman Press, pp. 260-314.

Kankainen, M. et al., 2009. Comparative genomic analysis of *Lactobacillus rhamnosus* GG reveals pili containing a human- mucus binding protein. *Proceedings of the National Academy of Sciences of the United States of America*, 106(40), pp.17193-8.

Kaushik, J.K. et al., 2009. Functional and probiotic attributes of an indigenous isolate of *Lactobacillus plantarum*. *PloS one*, 4(12), pp.1-11.

Kawai, Y., Suegara, Y.N. & Shimohashi, H., 1982. Colonization of lactic acid bacteria isolated from rats and humans in the gastrointestinal tract of rats. *Microbiology and Immunology*, 26(5), pp.363-373.

Klemm, P., 1994. *Fibriae: adhesion, genetics, biogenesis and vaccines* Per Klemm, ed., London: CRC Press.

Laux, D.C., 1986. Identification and characterization of mouse small intestine mucosal receptors for *Escherichia coli* K-12(K88ab). *Infection and Immunity*, 52(1), pp.18-25.

Laux, D.C., McSweegan, E.F. & Cohen, P.S., 1984. Adhesion of enterotoxigenic *Escherichia coli* to immobilized intestinal mucosal preparations: a model for adhesion to mucosal surface components. *Journal of Microbiological Methods*, 2, pp.27-39.

Lebeer, S., Vanderleyden, J. & De Keersmaecker, S.C.J., 2008. Genes and molecules of lactobacilli supporting probiotic action. *Microbiology and Molecular Biology Reviews*, 72(4), pp.728-64.

Lee, N.K. et al., 2008. Screening of Lactobacilli derived from chicken feces and partial characterization of *Lactobacillus acidophilus* A12 as an animal probiotics. *Journal of Microbiology and Biotechnology*, 18(2), pp.338-342.

Lin, W.H. et al., 2007. Different probiotic properties for *Lactobacillus fermentum* strains isolated from swine and poultry. *Anaerobe*, 13, pp.107-113.

Ma, Y.L. et al., 2005. Effect of *Lactobacillus* isolates on the adhesion of pathogens to chicken intestinal mucus *in vitro*. *Letters in Applied Microbiology*, 42, pp.369-374.

Mackenzie, D. et al., 2010. Strain-specific diversity of mucus-binding proteins in the adhesion and aggregation properties of *Lactobacillus reuteri*. *Microbiology (Reading, England)*, 156(Pt 11), pp.3368-78.

Mackie, R.I., Sghir, a & Gaskins, H.R., 1999. Developmental microbial ecology of the neonatal gastrointestinal tract. *The American Journal of Clinical Nutrition*, 69(5), p.1035S-1045S.

Macías-Rodríguez, M.E. et al., 2009. *Lactobacillus fermentum* BCS87 expresses mucus- and mucin-binding proteins on the cell surface. *Journal of Applied Microbiology*, 107(6), pp.1866-74.

Macías-Rodríguez, M.E. et al., 2008. Potential probiotic *Lactobacillus* strains for piglets from an arid coast. *Annals of Microbiology*, 58(4), pp.641-648.

Malmsten, M. et al., 1992. Mucin layers on hydrophobic surfaces studied with ellipsometry and surface force measurements. *Journal of Colloid and Interface*, 151, pp.579-590.

Marteau, P.R., 2002. Probiotics in clinical conditions. *Clinical reviews in Allergy & Immunology*, 22(3), pp.255-73. Available at: http://www.ncbi.nlm.nih.gov/pubmed/12043384.

Masuda, K. & Kawata, T., 1983. Distribution and chemical characterization of regular arrays in the cell walls of strains of the genus *Lactobacillus*. *FEMS Microbiology Letters*, 20, pp.145-150.

Matijasic, B., Stojkovic, S. & Rogelj, I., 2006. Survival and *In vivo* adhesion of human isolates *Lactobacillus gasseri* LF221 and K7 in weaned piglets and their effects on coliforms, clostridia and lactobacilli viable counts in faeces and mucosa. *Journal of Dairy Research*, 73, pp.417-422.

McAllister, J.S., Kurtz, H.J. & Short, E.C., 1979. Changes in the intestinal flora of young pigs with postweaning diarrhea or edema disease. *Journal of Animal Science*, 49, pp.868-679.

Miyoshi, Y. et al., 2006. A Mucus Adhesion Promoting Protein, MapA, Mediates the Adhesion of *Lactobacillus reuteri* to Caco-2 Human Intestinal Epithelial Cells. *Bioscience, Biotechnology, and Biochemistry*, 70(7), pp.1622-1628.

Molin, G., 1993. Numerical taxonomy of *Lactobacillus* spp. associated with healthy and diseased mucosa of the human intestines. *Journal of Applied Bacteriology*, 74, pp.314-323.

Mondal, D. et al., 2011. Contribution of enteric infection, altered intestinal barrier function, and maternal malnutrition to infant malnutrition in Bangladesh. *Clinical Infectious Diseases*, doi:10.1093/cid/cir807

Moser, S.A. & Savage, D.C., 2001. Bile salt hydrolase activity and resistance to toxicity of conjugated bile salts are unrelated properties in Lactobacilli. *Applied and Environmental Microbiology*, 67(8), pp.3476-3480.

Muralidhara, K.S., 1977. Effect of feeding lactobacilli on the coliform and lactobacillus flora of intestinal tissue and feces from piglets. *Journal of Food Protection*, 40(5), pp.288-295.

Ohashi, Y. et al., 2007. Stimulation of indigenous lactobacilli by fermented milk prepared with probiotic bacterium, *Lactobacillus delbrueckii subsp. bulgaricus* strain 2038, in the pigs. *Journal of Nutritional Science and Vitaminology (Tokyo) 53:82-86*, 53(82-86).

Oostindjer, M. et al., 2010. Effects of environmental enrichment and loose housing of lactating sows on piglet performance before and after weaning. *Journal of Animal Science*, 88(11), pp.3554-62.

von Ossowski, I. et al., 2010. Mucosal adhesion properties of the probiotic *Lactobacillus rhamnosus* GG SpaCBA and SpaFED pilin subunits. *Applied and Environmental Microbiology*, 76(7), pp.2049-57.

Pavlova, S.I. et al., 1993. Effect of medium composition on the ultrastructure of *Lactobacillus* strains. *Archives of Microbiology*, 160, pp.132-136.

Pedersen, K. & Tannock, G W, 1989. Colonization of the porcine gastrointestinal tract by lactobacilli. *Applied and Environmental Microbiology*, 55(2), pp.279-83.

Pouwels, P H & Leer, R J, 1993. Genetics of lactobacilli: Plasmids and gene expression. *Antonie van Leeuwenhoek*, 64(2), pp.85-107.

Pretzer, G. et al., 2005. Biodiversity-based identification and functional characterization of the mannose-specific adhesin of *Lactobacillus plantarum*. *Journal of Bacteriology*, 187(17), pp.6128-6136.

Ramiah, K., van Reenen, C. a & Dicks, L.M.T., 2007. Expression of the mucus adhesion genes Mub and MapA, adhesion-like factor EF-Tu and bacteriocin gene plaA of *Lactobacillus plantarum* 423, monitored with real-time PCR. *International Journal of Food Microbiology*, 116(3), pp.405-409.

Ray, B. & Johnson, M.C., 1986. Freeze drying injury of surface layer protein and its protection in *Lactobacillus acidophilus*. *CryoLetters*, 7, pp.210-217.

Reid, G., 1999. The scientific basis for probiotic strains of *Lactobacillus*. *Applied and Environmental Microbiology*, 65(9), pp. 3763-3766.

Reniero, R. et al., 1992. High frequency of conjugation in *Lactobacillus* mediated by an aggregation-promoting factor. *Journal of General Microbiology*, 138(4), pp.763-768.

Reniero, R. et al., 1993. Purification of *Lactobacillus* secreted proteins. *Biotechnology Techniques*, 7(7), pp.401-406.

Robins-Browne, R.M. & Levine, M.M., 1981. The fate of ingested lactobacilli in the proximal small intestine. *American Journal of Clinical Nutrition*, 34, pp.514-519.

Rojas, M & Conway, P L, 1996. Colonization by lactobacilli of piglet small intestinal mucus. *The Journal of Applied Bacteriology*, 81(5), pp.474-80.

Rojas, M. and Conway, P., 2001. A dot blot assay for adhesive components relative to probiotics. In R. J. Doyl, ed. *Methods of enziyology Vol. 336. Microbial Growth and Biofilms. Part A. Developmental and Molecular Biological Aspects*. San Diego, California, U.S.A.: Academic Press, pp. 389-402.

Rojas, M., 1996. Studies on an adhesion promoting protein from *Lactobacillus* and its role in the colonisation of the gastrointestinal tract. *PhD thesis Goteborg University, Goteborg, Sweden*. I. Maurilia Rojas, ed., Goteborg: Goteborg University.

Rojas, M., Ascencio, F. & Conway, P.L., 2002. Purification and characterization of a surface protein from *Lactobacillus fermentum* 104R that binds to porcine small intestinal mucus and gastric mucin. *Applied and Environmental Microbiology*, 68(5), pp.2330-2336.

Roos, S. & Jonsson, H., 2002. A high-molecular-mass cell-surface protein from *Lactobacillus reuteri* 1063 adheres to mucus components. *Microbiology (Reading, England)*, 148(Pt 2), pp.433-42.

Sanders, M.E., 2011. Impact of probiotics on colonizing microbiota of the gut. *Journal of Clinical Gastroenterology*, 45(5), pp.115-119.

Sarem-Daamerdji, L. et al., 1995. *In vitro* colonization ability of human colon mucosa by exogenous *Lactobacillus* strains. *FEMS Microbiology Letters*, 131(2), pp.133-137.

Satoh, E. et al., 2000. The gene encoding the adhesion promoting protein MapA from *Lactobacillus reuteri* 104R is part of one operon whose expression is controlled by a mechanism of transcription attenuation, involving cysteine. Gene Bank Accession Number AJ 293860.

Savage, D C, 1992. Growth phase, cellular hydrophobicity, and adhesion in vitro of lactobacilli colonizing the keratinizing gastric epithelium in the mouse. *Applied and Environmental Microbiology*, 58(6), pp.1992-5.

Schneitz, C.L. & Lounatma, K., 1993. Adhesion of *Lactobacillus acidophilus* to avian intestinal epithelial cells mediated by the crystalline bacterial cell surface layer (S-layer). *Journal of Applied Bacteriology*, 74, pp.290-294.

Shimizu M, Shimizu Y, K.Y., 1978. Effects of ambient temperatures on induction of transmissible gastroenteritis in feeder pigs. *Infection and Immunity*, 21, pp.747-752.

Smith, H.W., 1965. Observations on the flora of the alimentary tract of animals and factors affecting its composition. *Journal of Pathology Bacteriology*, 89, pp.95-122.

Soto, G.E. & Hultgren, S.J., 1999. Bacterial adhesins: common themes and variations in architecture and assembly. *Journal of Bacteriology*, 181(4), pp.1059-1071.

Suegara, N. et al., 1975. Behavior of microflora in the rat stomach: adhesion of lactobacilli to the keratinized epithelial cells of the rat stomach *In vitro*. *Infection and Immunity*, 12(1), pp.173-179.

Sun, J. et al., 2007. Factors involved in binding of *Lactobacillus plantarum* Lp6 to rat small intestinal mucus. *Letters in Applied Microbiology*, 44(1), pp.79-85.

Tannock, G. W., Blumershine, R. & Archibald, R., 1987. Demonstration of epithelium-associated microbes in the oesophagus of pigs, cattle, rats and deer. *FEMS Microbiology Ecology*, 45, pp.199-203.

Tannock, G. W. & Savage, D. C., 1974. Influences of dietary and environmental stress on microbial populations in the murine gastrointestinal tract. *Infection and Immunity*, 9(3), pp.591-598.

Van Tassell, M.L. & Miller, M.J., 2011. *Lactobacillus* adhesion to mucus. *Nutrients*, 3(5), pp.613-636.

Thompson-Chagoyán, O.C., Maldonado, J. & Gil, A., 2007. Colonization and impact of disease and other factors on intestinal microbiota. *Digestive Diseases and Sciences*, 52(9), pp.2069-2077.

Toba, T. et al., 1995. A Collagen-binding S-Layer protein in *Lactobacillus crispatus*. *Applied and Environmental Microbiology*, 61(7), pp.2467-2471.

Vélez, M.P., De Keersmaecker, S.C.J. & Vanderleyden, J., 2007. Adherence factors of *Lactobacillus* in the human gastrointestinal tract. *FEMS Microbiology Letters*, 276(2), pp.140-148.

Wang, B. et al., 2008. Identification of a surface protein from *Lactobacillus reuteri* JCM1081 that adheres to porcine gastric mucin and human enterocyte-like HT-29 cells. *Current Microbiology*, 57(1), pp.33-38.

5

Polymerase Chain Reaction for Phytoplasmas Detection

Duška Delić
University of Banjaluka, Faculty of Agriculture
Bosnia and Herzegovina

1. Introduction

This chapter inspired treat caused by phytoplasmas diseases in food production, and increased need for sensitive and accurate detection of these microorganisms. Early and sensitive detection and diagnosis of phytoplasmas is of paramount importance for effective prevention strategies and it is prerequisite for study of the diseases epidemiology and devising of pathogen management.

Phytoplasmas are prokaryotes lacking cell walls that are currently classified in the class *Mollicutes* (2). To the class *Mollicutes* (cell wall-less prokaryotes) belonging both pathogenic groups: mycoplasma-like organisms (MLOs) and mycoplasmas. However, in contrast to mycoplasmas, which cause an array of disorders in animals and humans, the phytopathogenic MLOs resisted all attempts to culture them *in vitro* in cell free media (89). Following the application of molecular technologies the enigmatic status of MLOs amongst the prokaryotes was resolved and led to the new trivial name of "phytoplasma", and eventually to the designation of a new taxon named 'Candidatus phytoplasma' (73).

Diseases associated with phytoplasma presence occur worldwide in many crops, although individual phytoplasmas may be limited in their host range or distribution. There are more than 300 distinct plant diseases attributed to phytoplasmas, affecting hundreds of plant genera (70). Many of the economically important diseases are those of woody plants, including coconut lethal yellowing, peach X-disease, grapevine yellows, and apple proliferation. Following their discovery, phytoplasmas have been difficult to detect due to their low concentration especially in woody hosts and their erratic distribution in the sieve tubes of the infected plants (15). First detection technique which indicated presence of some intercellular disorder was based on graft transmission of the pathogen to healthy indicator plants. The establishment of electron microscopy (EM) based techniques represents an alternative approach to the traditional indexing procedure for phytoplasmas. EM observation (17, 33) and less frequently scanning EM (59) were the only diagnostic techniques until staining with DNA-specific dyes such as DAPI (148) was developed. Lately, protocols for the production of enriched phytoplasma-specific antigens have been developed, thus introducing serological-based detection techniques for the study of these pathogens in plants or insect vectors (65).

Phytoplasma detection is now routinely done by different nucleic acid techniques based on polymerase chain reaction (PCR) (144, 12, 52, 165). The procedures developed in the last 20 years are now used routinely and are adequate for detecting phytoplasma infection in plant propagation material and identifying insect vectors, thus helping in preventing the spread of the diseases and their economical impact.

Therefore, aim of this chapter is to provide an overview of the PCR-based techniques for detection, identification and characterisation of this plant-pathogenic *Mollicutes* (cell wall-less prokaryotes).

1.1 Relevant features of phytoplasmas

Phytoplasmas, previosly known as 'Mycoplasma-like organisms' or MLOs, are wall-less bacteria obligate parasites of plant phloem tissue, and of several insect species (Fig. 1). Phytoplasma-type diseases of plants for long time were believed to be caused by viruses considering their infective spreading, symptomatology, and transmission by insects (84, 85, 86, 119, 90). Etiology of these pathogens was explored accidentally by group of Japanese sciences (45). They demonstrated that the causes agent of the yellows-type diseases are wall-less prokaryotes related to bacteria, pleomorphic incredibly resembling to mycoplasmas.

Phytoplasmas have diverged from gram-positive bacteria, and belong to the '*Candidatus* Phytoplasma' genus within the Class *Mollicutes* (73). Through evolution the genomes of phytoplasmas became greatly reduced in size and they also lack several biosynthetic pathways for the synthesis of compounds necessary for their survival, and they must obtain those substances from plants and insects in which they are parasites (11) thus they can't be cultured *in vitro* in cell-free media.

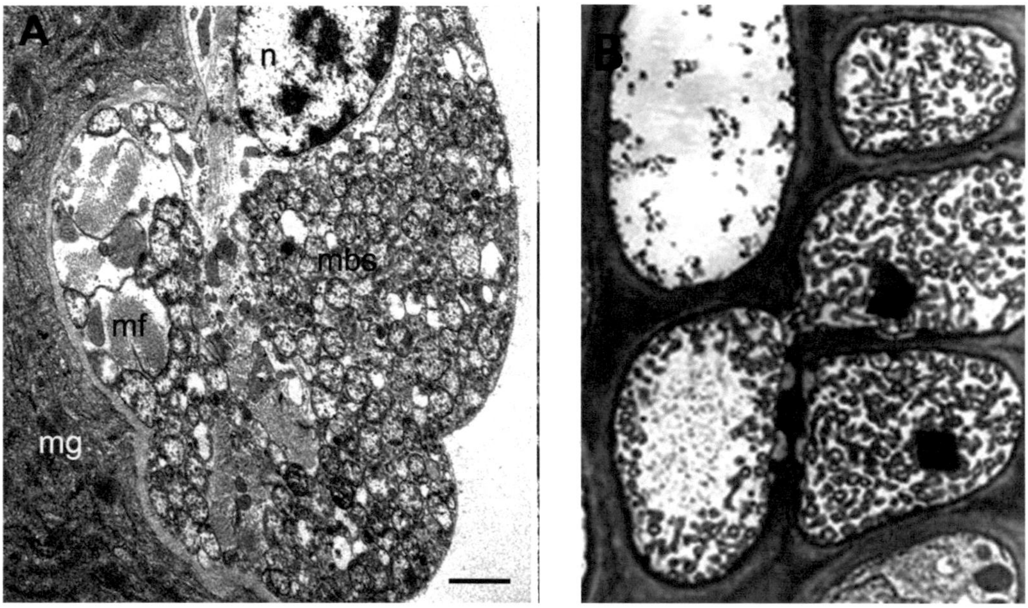

Fig. 1. Electron microscopy: of cross sections: A) of the vector leafhopper muscle cells around the midgut; B) sieve tubes of phytoplasmas infecting plants. http://www.jic.ac.uk/staff/saskia-hogenhout/insect.htm

Not all plant species infected with phytoplasmas have disease symptoms, but infected plants normally show symptoms such as virescence, phyllody, yellowing, witches' broom, leaf rool and generalized decline (19). The most common symptoms of the infected plants are yellowing caused by the breakdown of chlorophyll and carotenoids, whose biosynthesis is also inhibited (21). Induced expression of sucrose synthase and alcohol dehydrogenase I genes in phytoplasma-infected grapevine plants grown in the field was also recently demonstrated (72).

Phytoplasmas are mainly spread by insects of the families *Cicadellidae* (leafhoppers), *Fulgoridae* (planthoppers), and *Psyllidae*, which feed on the phloem tissues of infected plants acquiring the phytoplasmas and transmitting them to the next plant they feed on (136, 2). They enter the insect's body through the stylet and then move through the intestine and been absorbed into the haemolymph. From here they proceeded to colonize the salivary glands, a process that can take up to some weeks (5, 80). Another pathway of phytoplasma survival and transmission is vegetative propagating plant material. As it mentioned phytoplasma invading phloem tissue and it is mostly find that in woody plants they disappear from aerial parts of trees during the winter and survive in the root system to re-colonize the stem and branches in spring (149, 150, 58).

1.2 Laboratory diagnostic of phytoplasmas

In time when phytoplasmas were discovered as plant pathogens diagnostic was difficult since detection was based on symptoms observation insect or dodder/graft transmission to host plant and electron microscopy of ultra-thin sections of the phloem tissue. Serological diagnostic techniques for the detection of phytoplasma began to emerge in the 1980's with ELISA based methods. However, serological methods weren't always sensitive enough to detect various phytoplasmas (13, 47). Finally, in the early 1990's PCR coupled with RFLP analysis allowed the accurate identification of different strains and species of phytoplasma (127, 91, 145). Nowadays, diagnosis of phytoplasmas is routinely done by PCR and can be divided into three phases: total DNA extraction from symptomatic tissue or insects; PCR amplification of phytoplasma-specific DNA; characterization of the amplified DNA by sequencing, RFLP analysis or nested PCR with group-specific primers (117).

For the DNA extraction of known phytoplasma, several protocols for isolation from infected plant material and insects have been developed. Control samples are drowning from plants commonly infected by phytoplasmas. Reference phytoplasma strain collections are maintained in experimentally infected periwinkle (*Catharanthus roseus*) which is available for research and classification purposes (18, 26).

In the second stage of the testing, DNA extracted from plants or insects is amplifying by using the polymerase chain reaction or PCR. PCR is a standardised technique in gene analysis to provide sufficient genetic material for detection (153). It works through the use of short lengths of DNA called primers that have a known sequence. Double stranded DNA is melting in a heating step exposing two single strands to which the primer can anneal. For the final stage, study of genetic variability is performing in order to differentiate between gene sequences from different phytoplasma.

In adition to sequencing, there are several strategies which allow study of genetic variability in PCR products: Restriction fragment length polymorphism (RFLP) (93, 162); Terminal restriction fragment length polymorphism (T-RFLP) (66); Heteroduplex Mobility Assays (HMAs) (160); Single Strand Conformation Polymorphisms (SSCP) (126).

Alternative diagnostic methods have been established such as real-time PCR (12, 71, 161) and recently developed method for rapid detection of several phytoplasma species called loop-mediated isothermal amplification (LAMP) (155, 68).

2. Sampling procedure

Quality of DNA is of key importance in molecular diagnostics, since it can affect the final result. On other hand, for preparations of good quality and enriched in phytoplasma DNA, sampling material is of essential importance. Nevertheless, the quality of DNA depends on which plant tissue is examined.

2.1 Sampling of plants

It is generally more accurate sampling in the growing season, and although it can be used in the dormant season, this is not appropriate for the plant health inspections under the certification scheme. Due to the seasonal variation the optimal time for the diagnosis of phytoplasmas is from June to late autumn (30). Phytoplasmas could be detected using the polymerase chain reaction (PCR) from leaf midribs or phloem shaves from shoots, cordons, trunks and roots (117). Phytoplasmas were not always detected in samples from the same sampling area, from one sampling period to the next, firstly due to the uneven distribution, seasonal movement. Having this in mind, when collecting samples the best is to take leaves from different part of plant if it is possible symptomatic one, total amount should be around 20 g. If symptoms are absent phytoplasma detection by PCR can be improved by sampling from shoots, cordons and trunks, especially during October or early spring. In this case the best is to sample roots near to the plant bases though small feeding roots are the best tissue for extraction. Sampling of dry and rotted plant parts is not recommended since phytoplasmas are obligatory parasites. Palmano (2001) (134) demonstrated importance of proper identification of plant parts sampling; in this case the leaves have to show obvious symptoms but without being necrotic or completely yellow. In addition, variance in phytoplasma titters between infected plants of the same species has been observed by Berges et al. (2000) (15) and may be caused by different stages of development and age of plants.

It is recommended to record sampling area and plants by GPS device taking the coordinates and keep samples on cold (4 °C) till laboratory delivery.

2.2 Sampling of insects

Collection of the insect vectors for phytoplasma PCR analyses should be done in period where insects carry phytoplasma, furthermore knowledge about insects host plants and habitats are crucial things for successful collection.

Different traps and sampling techniques can be applied to collect and monitor phytoplasma vectors according to the objective of the study. The most common trapping techniques are sticky chromotropic traps, emergence traps, sweep net and vacuum insect collectors (107, 40). Collected insects should be place in ethanol and/or frozen.

3. Preparation of DNA templates

3.1 Samples preparation for homogenization

Prior to start extraction from collected plant samples, leaf midribs and/or phloem shaves are preparing for homogenization. Homogenization in liquid nitrogen with mortar and pestles is the most used method although some automatic homogenizers such as Fast Prep (MP Biomedicals, USA) (137) and Homex 6 (Bioreba, Switzerland) (52, 131) are available as faster alternative for the standard method.

3.2 DNA extraction

Accuracy of molecular analysis for pathogen detection in plant material requires efficient and reproducible methods to access nucleic acids. The preparation of samples is critical and target DNA should be made as available as possible for applying the different molecular techniques. However the suitability of most of the molecular methods depends closely on the amount of phytoplasma cells or nucleic acid in the extract. Approximately, 1% of phytoplasma DNA is extracted from tissue of total DNA (20). Since the concentration of this phloem-inhabiting pathogens is subjected to significant variations according to season (151), and is very low especially in woody hosts (79, 88), the importance of obtaining phytoplasma DNA at a concentration and purity high enough for precise analysis is aparent.

There are a great many published methods for preparing the plant tissues or other type of samples before molecular detection of phytoplasmas; however, they all pursue access the nucleic acid, avoiding the presence of inhibitory compounds that compromise the detection systems. Target sequences are usually purified or treated to remove DNA polymerase inhibitors, such as polysaccharides, phenolic compounds or humic substances from plants (121, 63, 164, 122).

Depending on the material to be analyzed the extraction methods can be quite simple or more complex. Generally there are three main approaches for obtaining of DNA template: protocols including a phytoplasma enrichment step, CTAB (cetyltrimethylammonium bromide) buffer-extraction protocols and DNA extraction using commercial kits.

Phytoplasma enrichment extraction protocols (1, 138, 108) including preparation of plant extract in the phytoplasma enrichment buffer (PGB), after one or two centrifugations the obtained pellet is dissolving in the CTAB buffer following chloroform and/or phenol extraction and precipitation in isopropanol.

Simple laboratory protocols based on preparation of plant extract in CTAB-buffer have also been published by several authors (35, 46, 6, 106, 165, 120, 152) with few steps and minimal handling, reducing the risk of cross contamination, cost and time, with similar results to those of longer and more expensive protocols.

CTAB based-protocols were also adopted for extraction of phytoplasmas DNA from hemipterian vectors (107, 46, 116, 50, 51).

The use of commercial kits, either general or specifically designed for plant material or for insect individuals, in some cases with magnetic separation has gained acceptance for extraction, given the ease of use and avoidance of toxic reagents during the purification process. Among those: DNeasy Plant kits, Qiagen (52, 42); Genomic DNA Purification kit,

Fermentas (143, 77); High Pure PCR Template Preparation kit, Roche (132); Wizard Genomic DNA Purification kit, Promega (104); NucleoSpin PlantII kit, Macherey-Nagel (135); FastDNA spin kit MP, Biomedicals (10); while InviMag Plant DNA Mini kit, Invitek; and QuickPick Plant DNA kit, Bio Nobile are optimized for extraction with a King Fisher mL Thermo Science workstation (137, 24, 99, 41).

Recently a new method (LFD) (37, 155) has been developed for rapid DNA extraction which processing DNA in loop-mediated isothermal amplification (LAMP) procedure for the detection of phytoplasmas from infected plant material. LFD method allows DNA extraction from leaf and wood material just in two minutes. Plant extract prepared in commercial buffer supplied with the LFD (Forsite Diagnostics Ltd) commercial kit is placing onto LFD membranes of lateral flow devices, and small sections of these membranes are then adding directly into the LAMP reaction mixture and incubating for 45 min at 65 °C. Moreover, Hodgetts et al. (2011) (68) obtained also satisfied results with LAMP using DNA prepared with an alkaline polyethylene glycol (PEG). This DNA extraction method (31) involves gently maceration of a small amount of plant tissue in the PEG buffer and then transfer of the macerate to the LAMP reaction.

Nevertheless, the choice of one or another system for nucleic acid extraction relies in practice on the phytoplasma to be detected and the nature of the sample, the experience of the personnel, the number of analyses to be performed per day, and the type of technique. As there are no universally validated nucleic-acid extraction protocols for all kinds of material and phytoplasma pathogens, those available should be compared before selecting one method for routine.

4. Nucleic acid amplification method

Detection and identification of phytoplasmas is necessary for accurate disease diagnosis. Sensitive methods need to be implemented in order to monitor the presence and spread of phytoplasma infections. Hence, it is necessary to devise a rapid, effective and efficient mechanism for detecting and identifying these microorganisms. Molecular diagnostic techniques for the detection of phytoplasma introduced during the last two decades have proven to be more accurate and reliable than biological criteria long used for phytoplasma identification (95). Polymerase Chain Reaction (PCR) is the most versatile tool for detecting phytoplasmas in their plant and insect hosts (153). One of the most utilized protocols for phytoplasma detection and characterization encompasses nested-PCR and RFLP analyses.

4.1 Nested PCR

Nested-PCR assay, designed to increase both sensitivity and specificity, is the leading method for the amplification of phytoplasmas from samples in which unusually low titer, or inhibitors are present that may interfere the PCR efficacy (56). The use of nested-PCR has been reported for diagnostic purposes particularly in plants when phytoplasmas occur in low titer in the phloem vessels of their host-plants and their concentration may be subjected to seasonal fluctuation (57, 75, 100, 117).

DNA consists of long sequences of paired bases called genes which code for a particular trait. Some of these gene sequences are consistent across bacteria but vary in their detailed sequence. These differences can be compared and used as a diagnostic test for a particular phytoplasma. Phytoplasma diagnostics has been routinely based on phytoplasma-specific universal (generic) (Table 1) or phytoplasma group specific (Table 2) Polymerase Chain Reaction (PCR) primers designed on the basis of the highly conserved 16S ribosomal RNA (rRNA) gene sequences (1, 38, 44, 61, 77, 144, 153). Nevertheless, to detect phytoplasmas in DNA samples universal phytoplasma primers designed on sequences of the 16S-23S rRNA spacer region (SR) (153) are generally using.

Nested-PCR is performing by preliminary amplification using a universal primers pair followed by second amplification using a second universal primer pair. By using a universal primer pair followed by PCR using a group specific primer pair, nested-PCR is capable of detection of dual or multiple phytoplasmas present in the infected tissues in case of mixed infection (92). Until the reliability of universal primers detecting phytoplasmas is determined, it is advisable to use at least 2 different primer pairs to test a sample (eg P1/P7 (44) and R16F2/R16R2 (91); 6F/7R (146) and fU5/rU3 (102). Unfortunately, some of the primers can induce dimers or unspecific bands. They also have sequence homology in the 16S-spacer region to chloroplasts and plastids increasing the risk of false positives (64). Therefore, more specific universal phytoplasma primers are currently being developed (66, 112) and it may be that these will be more suitable for diagnostics from samples.

Primer set	Location	PCR product length	Reaction	References
P1 P7	16S/23SR	1800 bp	Direct PCR	(44) (153)
R16F2 R16R2	16S/IS	1245 bp	Nested PCR	(91)
R16F2n R6R2	16S/IS	1240 bp	Nested PCR	(55)
F1 B6	16S	1050 bp	semi-nested PCR	(38) (133)
6F 7R	16S/23	1700 bp	Direct PCR	(146)
fU3 fU5	16S	880 bp	Nested PCR	(102)
SecAfor 1 SecArev 3	secA gene	840 bp	Direct PCR	(67)
SecAfor 2 SecArev 3	sec A gene	480 bp	semi-nested PCR	(67)

Table 1. PCR universal primers commonly used for the detection of phytoplasma

Phytoplasma group-specific primers have also been designed on ribosomal protein gene, SecA, SecY genes (coding for the translocase protein) (28, 98), *vmp*1 gene (stolbur phytoplasma membrane protein) (28), *imp* gene (coding immunodominant membrane protein (112, 36), non-ribosomal gene *aceF* (115) and *tuf* gen (encoding the translation elongation factor Tu) (Table 2) (56, 67, 109, 147, 87,).

Primer set	Specificity	Location	Expected size of PCR product	References
fTufAy rTufAy	16SrI	tuf gene	940 bp	(147)
AysecYF1 AysecYR1	16SrI	secY gene	1400 bp	(98)
rp(I)F1A rp(I)R1A	16SrI	Ribosomal protein	1200 bp	(96)
rp(II)F1 rp(II)R1	16SrII	Ribosomal protein	1200 bp	(112)
rp(III)F1 rp(III)R1	16SrIII	Ribosomal protein	1200 bp	(112)
LY 16Sf LY16Sr	16SrIV	16S	1400 bp	(62)
LYC24F LYC24R	16SrIV	nonribosomal	1000 bp	(60)
rp(V)F1A rp(V)R1A	16SrV	Ribosomal protein	1200 bp	(97)
rp(VI)F2 rp(VI)R2	16SrVI	Ribosomal protein	1000 bp	(112)
rp(VIII)F2 rp(VIII)R2	16SrVII, 16SrVIII	Ribosomal protein	1000 bp	(112)
rp(IX)F2 rp(IX)R2	16SrIX	Ribosomal protein	800 bp	(112)
rpStolIF rpStolIR	16SrXII-A	Ribosomal protein	1372 bp	(112)
rpAP15f rp/AP15r	16SrX-A	Ribosomal protein	1000 bp	(114)
AP13/AP10 AP14/AP15	16SrX-A	nonribosomal	776 bp	(27)
f01 r01	16SrX	16S	1100 bp	(102)
AceFf1/AceFr1 AceFf2/AceFr2	16SrX	aceF	500 bp	(115)
FD9R FD9F	16SrV	secY	1300	(35)
FD9F3b FD9R2	16SrV	secY	1300 bp	(29) (6)
STOL11R1 STOL11F2	16SrXII	secY	990 bp	(35)
STOL11R2 STOL11F3	16SrXII	secY	720 bp	(29)
fStol rStol	16SrXII-A	16S/SR	570 bp	(106)
fAY rEY	16S	16SrV	300 bp	(1)

Table 2. Several group specific primers used for phytoplasma detection

The search for phytoplasma-specific primers has led to evaluation of primers based on these regions appears to offer more variation than that of the 16S gene. Nevertheless, design of primers based on various conserved sequences such as 16S rRNA gene, ribosomal protein gene operon, *tuf* and *SecY* genes was the major breakthrough in detection, identification, and classification of phytoplasmas (57, 147, 109, 161, 111, 112).

Primers previously designed for specific amplification of DNA from stolbur phytoplasma were recently found to prime amplification of DNA from other phytoplasmas (39, 77); therefore, it may be advisable to supplement use of phytoplasma-specific primers with RFLP analysis of amplified DNA sequences.

The choice of primer sets for phytoplasma diagnosis by nested PCR mostly depends on the phytoplasma we are looking for. Nested-PCR with a combination of different universal primers (Table 1) can improve the diagnosis of unknown phytoplasmas present with low titter in the symptomatic host. Universal ribosomal primers followed with nested with group-specific primers (Table 2) are extremely useful when the phytoplasma to be diagnosed belongs to a well-defined taxonomic group (117).

PCR products are usually visualised on 1% agarose gel prepared in 1xTAE buffer, stained with ethidium bromide (40).

The efficiency of nested-PCR has shown that it can reamplify the direct PCR product in dilution of 1: 60 000 (81). However, the system has not yet been devised to identify all the taxonomic groups, and this approach requires more than one PCR step, increasing the chances of contamination between samples, and does not provide the rapid and simple diagnostic tool required.

4.1.1 Restriction fragment length polymorphism (RFLP)

For identification of all detected phytoplasmas as well as for molecular characterisation of certain phytoplasmas strains Restriction Fragment Length Polymorphism, or RFLP is commonly used. RFLP is a technique that exploits variations in homologous DNA sequences. It refers to a difference between samples of homologous DNA molecules that come from differing locations of restriction enzyme sites, and to a related laboratory technique by which these segments can be illustrated.

Phytoplasma amplified PCR products are cutting into fragments at specific sites using enzymes. More specific detection methods involve using phytoplasma-specific primers or differentiation on the basis of phylogenetic RFLP analysis of PCR amplified sequences (91, 145). RFLP analysis of PCR amplified DNA sequences using a number of endonuclease restriction enzymes (93). The pattern of cut DNA is viewing using 5% polyacrilamid gel (95) or 2,5% to 3% agarose gel electrophoresis. Analysis of a known genomic sequence can show what size of fragments to expect depending upon the enzymes chosen for the cuts e.g., providing that 6 or more frequently cutting restriction enzymes are used in the RFLP analysis, specific identification of the phytoplasma may be obtained.

Moreover this analysis is very useful for identification of new phytoplasmas, or phytoplasmas from a poorly studied region or crop. Because the RFLP patterns characteristics of each phytoplasmas are conserved, unknown phytoplasmas can be identified by comparing the patterns of the unknown with the available RFLP patterns for known phytoplasmas without co-analyses of all reference representative phytoplasmas (94, 162, 163, 25). In this case it is preferable to use bigger number of enzymes to achieve identification (38). Enzymes found valuable for these analyses include AluI, BamHI, BfaI, DraI, HaeIII, HhaI, HinfI, HpaI, HpaII, KpnI, MseI, RsaI, Sau3AI, TaqI and ThaI.

Phytoplasma has not been cultured in cell-free medium, thus cannot be differentiated and classified by the traditional methods which are applied to culturable prokaryotes. The highly conserved 16S rRNA gene sequence has been widely used as the very useful primary molecular tool for preliminary classification of phytoplasmas. A total of 19 distinct groups, termed 16S rRNA groups (16Sr groups), based on actual RFLP analysis of PCR-amplified 16S rDNA sequences or 29 groups based on RFLP with new computer-simulated RFLP *in silico* analysis have been identified (93, 162).

4.1.2 Terminal restriction fragment length polymorphism (T-RFLP)

A protocol based on the Terminal Restriction Fragment Length Polymorphism (T-RLFP) analysis of 23S rDNA sequence using a DNA sequence analysis system has been developed to provide the simultaneous detection and taxonomic grouping of phytoplasmas (66). Terminal-restriction fragment length polymorphism (T-RFLP) analysis is a direct DNA-profiling method that usually targets rRNA (82). This genetic fingerprinting method uses a fluorescently labelled oligonucleotide primer for PCR amplification and the digestion of the PCR products with one or more restriction enzymes. This generates labelled terminal restriction fragments (TRFs) of various lengths depending on the DNA sequence of the bacteria present and the enzyme used to cut the sequence. The results of T-RFLP are obtaining through TRF separation by high-resolution gel electrophoresis on automated DNA sequencers. The laser scanning system of the DNA sequencer detects the labelled primer (141) and from this signal the sequencer can record corresponding fragment sizes and relative abundances. Resulting data is very easy to analyse, being presented as figures for statistical analysis and graphically for rapid visual interpretation.

The method was also designed to allow simple and easy testing of phytoplasmas and at the same time gave indication of their taxonomic group (9, 66). Comparing with the conventional nested-PCR/RFLP, method is less time-consuming and the approach is less expensive than sequencing.

4.1.3 Single Strand Conformation Polymorphisms (SSCP)

Single-strand conformation polymorphism (SSCP) analysis is a broadly used technique for detection of polymorphism in PCR-amplified fragments. SSCP was also assessed for the application in detection of the molecular variability phytoplasmas (125, 126). Amplified phytoplasma regions (16S rDNA, *tuf* gene, and *dnaB* gene), respectively are mixing with denaturing buffer after incubation, results of the SSCP are visualising on a non-denaturing polyacrylamide gel, optimized for each fragment length. SSCP revealed the presence of polymorphism undetected by routine RFLP analyses in all analyzed phytoplasma regions. Advantages of the SSCP in comparison with RFLP are sensitivity, time and cost consumption as well as suitability when large number of samples are screening for molecular variability.

4.1.4 Heteroduplex Mobility Assay (HMA)

Heteroduplex mobility assay (HMA) has been recently developed as fast and inexpensive method for determining relatedness between phytoplasmas DNA sequences. Initially, it was developed by Delwart et al. (1993) (43) to evaluate viral heterogeneity and for genetic typing of human immunodeficiency virus (HIV).

So far, HMA was used in studies for differentiation of phytoplasmas in the aster yellows group and clover proliferation group (159) determination of genetic variability among isolates of Australian grapevine phytoplasmas (32); study of the genetic diversity of 62 phytoplasma isolates from North America, Europe and Asia (160); for phylogenetic relationships among flavescence dorée strains and related phytoplasmas belonging to the elm yellows group (7); and to determine genomic diversity among African isolates of coconut lethal yellowing phytoplasmas causing Cape St. Paul wilt disease (CSPD, Ghana), lethal disease (LD, Tanzania), and lethal yellowing (LYM, Mozambique) (110).

Amplified PCR products from positive phytoplasma strains are combining with the amplified products of reference strain mixing with annealing buffer and submitting to HMA analyses (110, 160) following visualization of HMA products on polyacrylamide gel. Heteroduplexes migrate more slowly than a homoduplex in polyacrylamide gel electrophoresis. The extent of the retardation has been shown to be proportional to the degree of divergence between the two DNA sequences. It was noticed, that presence of an unpaired base influence the mobility of a heteroduplex more than a mismatched nucleotide (158, 157). Performing HMA, Marihno et al (2008) (110) succeeded to identified three groups of phytoplasmas associated with various coconut lethal yellowing diseases. Moreover, this grouping was consistent with the genetic diversity described in the coconut yellowing-associated phytoplasmas detected after cloning, sequencing, and phylogenetic analyses.

Further optimisations of this approach could facilitate phylogenetic study and diagnosis of many other phytoplasmas and development of a comprehensive PCR-based classification system. Considering simplicity and rapidness of the method, HMA could be used for initial screening among a large number of isolates and rapid identification of phytoplasmas as well as other organisms.

4.2 Immuno-capture PCR

Immuno-capture PCR assay, in which the phytoplasma of interest is first selectively captured by specific antibody adsorbed on microtiter plates, and then the phytoplasma DNA is released and amplified using specific or universal primers, can be an alternative method to increase detection sensitivity (139, 64). This method is aimed at avoiding the lengthy extraction procedures to prepare target DNAs. Nonetheless, this method is not suitable for detection of fruit tree and grapevine phytoplasmas.

4.3 Real-time PCR

Since the most universal as well as specific diagnostic protocols rely on nested PCR which, although extremely sensitive, is also time-consuming and posses risk in terms of carry-over contamination between the two rounds of amplification, real-time PCR has recently replaced the traditional PCR in efforts to increase the speed and sensitivity of detection for mass screening.

The main principle of real-time PCR is based on fluorescent chemistries for labelling of the amplicons. During a real-time PCR run, accumulation of newly generated amplicons is monitored by each cycle by fluorescent detection methods, and so there is no need for post-PCR manipulation such as electrophoresis, which is required at the end of regular PCR. Moreover, the amount of fluorescent, monitored at each cycle is proportional to the log of concentration of the PCR target, and for this reason real-time PCR is also powerful technique for quantification of specific DNA. There are several labelling techniques, most of which specially bind to a target sequence on the amplicon, while others aspecifically stain double-stranded (ds) DNA amplicons. In addition, numbers of protocols have been developed for real-time PCR universal and specific detection phytoplasma.

For preliminary screening, 16S rDNA gene were adapted for the universal diagnosis of phytoplasmas using direct real-time PCR amplification (30, 48, 71) (Table 3) and all of them exploited a TaqMan probe for detection. TaqMan probes are labelled at the 5′end with reporter dye and at the 3′ end with a quenching molecule; during each PCR cycle in the presence of the specific target DNA, the TaqMan probe, bound to its target sequence, which is then degraded by the 5′-3′ exonuclease activity of the Taq polymerase as it extend the primer. The fluorescence moiety of the probe is therefore freed from its quencher-labelled portion and the fluorescence is detected by the optical system of the apparatus. The sensitivity of the 16S rDNA-based primer/probe system can be used to detect phytoplasmas belonging to several ribosomal subgroups and they showed sensitivity similar to that of conventional nested-PCR.

Group specific phytoplasma primers and probes for real-time PCR system have been designed to overcome problem with the time-consuming methods for phytoplasma strains identification and to further enhance the specificity of detection. Several laboratories have proposed rapid, specific and sensitive diagnostic protocols for detection of quarantine and economically important phytoplasmas of fruit trees and grapevine such as flavescance dorée (FD) and bois noir (BN) phytoplasmas infecting grapevine (22, 48, 8, 53, 71, 14); '*Ca.* Phytoplasma mali' (apple proliferation, AP), '*Ca.* Phytoplasma pyri' (pear decline, PD), '*Ca.* Phytoplasma pruni' (European stone fruit yellows, ESFY) important pathogens of fruit trees (12, 76, 48, 156, 3,4, 113, 23, 128, 41). Most of the primer/probe systems are targeting 16S rDNA gene though some others genes or even randomly cloned DNA fragments to which no specific function is assigned have been used (Table 3). For fluorescent detection SYBR Green I has been applied for the diagnosis of AP, PD, ESFY and FD, all quarantine phytoplasmas of fruit trees and grapevine in Europe. Real-time PCR assays were also developed using TaqMan minor groove binding (MGB) probe to detect AP in plant material (12, 3) as well as for FD, BN and other phytoplasmas less frequently infecting grapevines (71, 128). MGB (minor groove binding) probe has an MGB ligand and non-fluorescent quencher conjugated to the 3′ end, plus a fluorescent reporter dye at the 5′ end. The MGB ligand allows the use of shorter and more specific probes by increasing the stability of the probe-target bond. This property allows the use of shorter probes, with higher specificity than conventional TaqMan ones and the discrimination of even single nucleotide mismatched (83, 128). Furthermore, applying the same protocols, phytoplasmas DNA could be also detected in insect samples (113, 76, 48, 71) what is also decisive in the search for other potential vectors.

Specificity	Target gene	References
Universal	16S rDNA	(30)
Universal	16S rDNA	(48)
Universal	16S rDNA	(71)
FD	16S rDNA	(48)
FD	16S rDNA	(8)
FD	Sec Y	(71)
FD	16S rDNA	(22)
BN	Genomic fragment	(48)
BN	16S rDNA	(8)
BN	Genomic fragment	(71)
AP	Nitro reductase	(48)
AP	Genomic fragment	(76)
AP	16S rDNA	(12)
AP	16S rDNA	(4)
AP	16S rDNA	(23)
AP	16S–23S rRNA	(128)
PD	16S–23S rRNA	(128)
ESFY	16S–23S rRNA	(128)
ESFY	Ribosomal protein	(113)
'Ca. P. asteris'(onion yellows)	tuf	(161)
'Ca. P. asteris'(aster yellows)	16S rDNA	(8)
'Ca. P. asteris'(aster yellows)	16S rDNA	(69)
Beet leafhopper transmitted virescence virus	16S rDNA	(34)

Table 3. Oligonucleotide primers and probes used for phytoplasma detection by real-time PCR

A well-optimized reaction is essential for accurate results, which must be further analysed. As it is mentioned before, diagnosis of the pathogens in woody plants is often hampered by the presence of PCR inhibitors such as polyphenolics, polysaccharides and other molecules that may produce false negative results even from heavily infected samples. Additional problem may be also caused by amplification of other bacteria with universal phytoplasma primers/probe which could be present on the surface of some plants (49). Therefore, to avoid false positives specific probe can be included. So far, several sequence-specific detection tools are available: the chloroplast chaperonin 21 gene (8); cytochrome oxidase gene (71); the chloroplast gene for tRNA leucine (12); and the 18S rDNA gene (30, 118, 113,

128) addressed as targets to control the quality of total DNA extracted. SYBR Green I is one of the cheapest chemistry for real-time PCR detection, but the specificity of the reaction is extremely low, and needs to be checked. SYBR Green I dye chemistry will detect all double-stranded DNA, including non-specific reaction products. Therefore, amplification of non specific DNA may occur and analyses of melting curve is usually indispensible (48, 156).

One of the biggest advantages of real-time PCR is suitability of the method for quantification of nucleic acids of many plant pathogens, including phytoplasmas. In past competitive PCR was applied to monitor multiplication of '*Candidatus* Phytoplasma asteris' in vector *Macrosteles quadrinlineatus* (101). Quantification was achieved following co-amplification of phytoplasma DNA and several dilutions of an appropriate internal standard. This approach was complex, several steps, such as electrophoresis, image analysis of gel, compensating for differences in intensity due to the different sizes of the product from the pathogen target and the internal standard, were required before the band intensities could be plotted for linear regression analysis. However, nowadays absolute quantification of phytoplasma DNA was achieved per gram of extracted tissue (161, 23) or per insect vector (76). Possibility of the method to quantify amount of phytoplasma DNA in plant tissue and insect vectors gave opportunity to better understand biology and epidemiology of the pathogens, to allow examination of different multiplication rates and to calculate the concentration in their plant and vector host (161, 142, 23) as well as to study interactions of different phytoplasma species or strains present in mixed infection (100, 19). These results will find application in development of resistant plant varieties, a hot topic for economically important woody crops such as palms, fruits and grapevines.

4.4 Loop-mediated isothermal amplification assay (LAMP)

Methods described above require relatively expensive equipment for amplification of the phytoplasma DNA and/or analysis of the results. In addition, standard methods for DNA extraction involve buffers, such as a CTAB buffer combined with phenol / chloroform extraction and isopropanol precipitation (46, 165), which are time-consuming and cannot be performed in the field. Whilst leaf tissue is usually used as the source of DNA for detection of many phytoplasmas, in other cases, such as coconuts, trunk borings or roots are often used, and DNA is then extracted from this woody tissue either by grinding in liquid nitrogen, or when this is unavailable, the sawdust is left in the CTAB extraction buffer for 48 h before the subsequent phenol chloroform extraction and alcohol precipitation (129). For that reason there is increase need for development of the method for a more rapid diagnostic assay for phytoplasmas that can be used to produce a diagnosis within an hour of sampling in the field or on site in case of imported material in quarantine stations.

Several attempts to produce field-based systems, e.g. using phytoplasma-specific antibodies and ELISA-based or lateral flow devices (LFD)-based systems, fall down because of a lack of sensitivity, and whilst a phytoplasma IgG antibody based system is commercially available for few phytoplasmas (103). Recently, Fera (Food and Environment Research Agency) developed isothermal amplification assays, such as the Loop-Mediated Isothermal Amplification (LAMP) procedure for detection of several human and plant pathogens including phytoplasmas (130, 140, 37, 154). In the method the cycling accumulates stem-loop

DNAs with several inverted repeats of the target and cauliflower-like structures with multiple loops, and produces up to 10^9 copies of the target in less than 1 h at 65°C. Further, LAMP products can be detected by conventional agarose gel electrophoresis; using spectrophotometric equipment to measure turbidity (124); in real-time using intercalating fluorescent dyes (105); or by visual inspection of turbidity or colour changes (123, 74).

For the routine diagnosis colorimetric assay that uses hydroxyl napthol blue to detect the magnesium pyrophosphate by-product in successful LAMP amplification (54) showed the best suitability. The hydroxyl napthol blue can be incorporated into the LAMP reaction and the colour change visualized immediately after amplification has been completed, and amplification can subsequently be confirmed by agarose gel electrophoresis when necessary.

Two methods for extraction of nucleic acid from plant material were adopted for LAMP application: LFD (37, 155) and an alkaline polyethylene glycol (PEG) DNA extraction method (31, 68).

Primers for the LAMP assays were designed as described in Tomlison et al. (2010) (155) and Bekele et al (2011) (16) based on the 16S-23S intergenic spacer region. In addition *cox* gene primers were used to confirm that all DNA extractions supported LAMP (16). Primers for LAMP assays were designed against range of ribosomal group (16SrI, 16SrII, 16SrIII, 16SrIV, 16SrV, 16SrXI, 16SrXII, 16SrXXII) (68).

Developed protocol for LAMP-based diagnostic for a range of phytoplasmas can be conducted in the field and used to provide diagnosis within 1-hour of DNA extraction (68). According to the same author, PEG extraction method showed several advantages such is rapidness and requires less equipment than the LFD-based method, reducing the likelihood of sample contamination though the disadvantage of this method is that the DNA cannot be stored reliable long-term. Further efforts are doing to develop a hand held device capable of performing extraction, set-up and real-time detection for grapevine phytoplasmas. The device will make a single step homogeneous system from sampling to result, further reducing the risk of sample-to-sample contamination and enabling testing by non-specialists in the field (68).

5. Conclussions

In this review, molecular approaches for phytoplasma detection, identification and characterisation have been discussed. Before molecular techniques were developed, the diagnosis of phytoplasma diseases was difficult because they could not be cultured. Thus classical diagnostic techniques, such as observation of symptoms, were used. Ultrathin sections were also examined for the presence of phytoplasmas in the phloem tissue of suspected infected plants. Treating infected plants with antibiotics such as tetracycline to see if this cured the plant was another diagnostic technique employed. Diagnostic techniques such as ELISA test which allowed the specific detection of the phytoplasma began to emerge in the 1980s. In the early 1990s, PCR-based methods were developed that were far more sensitive than those that used ELISA, and RFLP analysis allowed the accurate identification of different strains and species of phytoplasma. Restriction fragments length polymorphism (RFLP) analysis together with the sequencing of 16Sr phytoplasma genes was the first step on this way enabling the construction of phylogenetic trees. Nowadays, polymerase chain reaction with primers from sequencing of randomly cloned phytoplasma DNA, from 16S rRNA, from ribosomal protein gene sequences, from SecY and Tuf genes, and from membrane associated protein genes opened new paths for research on phytoplasma identification and classification.

Nested PCR has been applied to overcome problems related to sensitivity of phytoplasma detection, although this approach is more time consuming and subject to template. Unfortunately, nested-PCR also meets some difficulties: unspecific bands, false positives or negatives caused by DNA and contamination of single or nested PCR. Therefore, confirmation of PCR results by using different primer pairs combinations (generic and group-specific) with subsequent RFLP and/or sequencing of PCR amplicons seems to be the way for correct phytoplasma identification in the examined samples.

More recently, real-time PCR has replaced the traditional PCR in efforts to increase the speed and sensitivity of detection and improve techniques for mass screening as well as to bypass post-PCR manipulations. Moreover, the techniques as quantitative real-time PCR (QPCR) have been developed to allow assessment of the level of infection in plants and vectors.

T-RFLP, SSCP and HMA analyses provide simultaneous detection and group characterisation of phytoplasmas.

Isothermal amplification of nucleic acid has recently been described as an alternative to PCR and applied for specific detection of several phytoplasmas. This method has potential for testing in field or in under equipped laboratories.

Despite the developments of all protocols which overcome most of the difficulties of phytoplasma diagnosis, the detection of these pathogens is still quite laborious. Therefore, future work is needed to develop quicker procedures to extract phytoplasma-enriched nucleic acids, giving accent on automation which involving silica or magnetic beads. Furthermore, developments for phytoplasma detection should be stressed on improvements of methods which enable simultaneous detection and taxonomic grouping of phytoplasmas. Use of high-throughput, sensitive, rapid and quantitative techniques will help to understand how phytoplasmas exploit their unique ecological niches.

6. References

[1] Ahrens U., Seemüller E. 1992. Detection of plant pathogenic mycoplasmalike organisms by a polymerase chain reaction that amplify a sequence of the 16S rRNA gene. Phytopathology, 82: 828-832.

[2] Agrios G. N. 1997. Plant diseases caused by *Mollicutes*: phytoplasmas and spiroplasmas. *In* Plant Pathology, 4th, pp. 457-470. Edited by G. N. Agrios. New York: Academic Press.

[3] Aldaghi M., Massert S., Roussel S., Jijaki M.H. 2007. Development of a new probe for specific and sensitive detection of *Candidatus* Phytoplasma mali in inoculated apple trees. Annals of Applied Biology 151: 251-258.

[4] Aldaghi M., Massert S., Roussel S., Dutrecq O., Jijaki, M.H. 2008. Adaptation of real-time PCR assay for specific detection of apple proliferation phytoplasma. Acta Horticulturae 781: 387-393.

[5] Alma, A., Bosco, D., Danielli, A., Bertaccini, A., Vibrio, M., and Arzone, A. 1997. Identification of phytoplasmas in eggs, nymphs, and adults of Scaphoideus titanus Ball reared on healthy plants. *Insects Molecular Biology,* 6: 115-121.
[6] Angelini E., Clair D., Borgo M., Bertaccini A., Boudon-Padieu E. 2001. Flavescence dorée in France and Italy - occurrence of closely related phytoplasma isolates and their near 101 relationships to palatinate grapevine yellows and an alder yellows phytoplasma. Vitis, 40: 79-86.
[7] Angelini E., Negrisolo E., Clair D., Borgo M., Boudon-Padieu E. 2003. Phylogenetic relationships among Flavescence dorée strains and related phytoplasmas determined by heteroduplex mobility assay and sequence of ribosomal and nonribosomal DNA. Plant Pathology, 52:663-672.
[8] Angelini E., Bianchi G. L., Filippin L., Morassutti C.,Borgo M. 2007. A new TaqMan method for the identificationof phytoplasmas associated with grapevine yellows byreal-time PCR assay.- Journal of Microbiological Methods,68: 613-622.
[9] Anthony RM, Brown TJ, French GL, 2000. Rapid diagnosis of bacteremia by universal amplification of 23S ribosomal DNA followed by hybridisation to an oligonucleotide array. Journal of Clinical Microbiology 38, 781–8.
[10] Arocha-Rosete Y., Kent P., Agrawal V., Hunt D., Hamilton A., Bertaccini A., Scott J., Crosby W., Michelutti R. 2011. Preliminary investigations on Graminella nigrifrons as a potential vector for phytoplasmas identified at the Canadian Clonal Genebank. Bulletin of Insectology, 64 (supplement), pp. 133-134.
[11] Bai X, Zhang J, Ewing A, Miller S.A., Radek A.J., Shevchenko D.V., Tsukerman K., Walunas T., Lapidus A., Campbell J.W., and Hogenhout S.A. 2006. Living with genome instability: the adaptation of phytoplasmas to diverse environments of their insect and plant hosts. Journal of Bacteriology, 188: 3682–3696.
[12] Baric S., Dalla Via J. 2004. A new approach to apple proliferation detection: a highly sensitive real-time PCR assay. Journal of Microbiological Methods, 57: 135-145.
[13] Batlle A., Laviña A., García-Chapa, M., Sabaté J., Folch C., Asin L. 2004. Comparative results between different detection methods of virus and phytoplasmas for pear and apple certification program. Acta Horticulturae (ISHS), 657:71-77.
[14] Berger J., SchweilgkoFfler W., Kerschbamer C., Roschatt C., Dalls Via J., Baric S. 2009. Occurrence of Stolbur phytoplasma in the vector *Hyalesthes obsoletus,* herbaceous host plants and grapevine in South Tyrol (Northern Italy). Vitis, 48 (4), 185–192.
[15] Berges R., Rott M., Semüller E. 2000. Range of phytoplasmas concentrations in various plant hosts as determined by competitive polymerase chain reaction. Phytopathology, 90: 1145-1152.
[16] Bekele B., Hodgetts J., Tomlinson J., Boonham N.,Nikolic P., Swarbrick P., Dickinson M. 2011. Use of a real-time LAMP isothermal assay for detecting 16SrII and XII phytoplasmas in fruit and weeds of the Ethiopian Rift Valley. Plant Pathology, 60: 345-355.
[17] Bertaccini A., Marani F. 1982. Electron microscopy of two viruses and mycoplasma-like organisms in lilies with deformed flowers. Phytopathologia mediterranea, 21: 8-14.
[18] Bertaccini A., Davis R.E., Lee I.-M. 1992. In vitro micropropagation for maintenance of mycoplasmalike organisms in infected plant tissues. Horticultural Science, 27(9): 1041- 1043.

[19] Bertaccini A., Franova J., Botti S., Tabanelli D. 2005. Molecular characterization of phytoplasmas in lilies with fasciations in the Czech Republic. Fems Microbiology Letters, 249: 79-85.

[20] Bertaccini A. 2007. Phytoplasmas: diversity, taxonomy, and epidemiology. Frontieres in Bioscience, 12: 673-689.

[21] Bertamini M., Nedunchezhian N. 2001. Effect of phytoplasma, stolbur-subgroup (Bois noir- BN)] of photosynthetic pigments, saccarides, ribulose-1,5-bisphosphate carboxylase, nitrate and nitrite reductases and photosynthetic activities in field-grow grapevine (*Vitis vinifera* L. cv Chardonnay) leaves. Photosynthetica, 39(1): 119-122.

[22] Bianco P.A., Casati P., Marziliano, N. 2004. Detection of phytoplasmas associated with grapevine flavescance dorée disease using real-time PCR. Journal of Plant Pathology, 86:257-261.

[23] Bisognin C., Schneider B., Salm H., Grando M.S., Jarausch W., Moll E., Seemüller E. 2008. Apple proliferation resistance in apomictic rootstocks and its relationship to phytoplasma concentration and simple sequence repeat genotypes. Phytopathology 98, 153-158.

[24] Boben J., Mehle N., Ravnikar M. 2007. Optimization of extraction procedure can improve phytoplasma diagnostic. Bulletin of Insectology, 60 (2), pp. 249-250

[25] Cai H., Wei W., Davis R.E., Chen H., Zhao Y. 2008. Genetic diversity among phytoplasmas infecting *Opuntia* species: virtual RFLP analysis identifies new subgroups in the peanut witches'-broom phytoplasma group. International Journal of Systematic and Evolutionary Microbiology, 58(6): 1448-1457.

[26] Carraro., Rugero O., Refatti E., Poggi P. 1988. Transmission of the possible agent of apple proliferation to *Vinca rosea* by dodder. Rivista di Patologia Vegetale, 24: 43-52.

[27] Casati P., Quaglino F., Tedeschi R., Spiga F.M., Alma A., Spadone P., Bianco P.A. 2010. Identification and molecular characterisation of 'Candidatus Phytoplasma mali' isolates in North-western Italy. Journal of Phytopathology, 158, 81-87.

[28] Cimerman A., Pacifico D., Salar P., Marzachì C., Foissac X. 2009. The striking diversity of *Vmp*1, a gene encoding a variable putative membrane protein of the Stolbur phytoplasma. Applied and Environmental Microbiology, 75: 2951-2957

[29] Clair, D., Larrue, J., Aubert, G., Gillet, J., Cloquemin, G., Boudon-Padieu, E. 2003. A multiplex nested-PCR assay for sensitive and simultaneous detection and direct identification of phytoplasma in the Elm yellows group and stolbur group and its use in survey of grapevine yellows in France. Vitis, 42, 151–157.

[30] Christensen N.M., Nicolaisen M., Hansen M., Schultz A. 2004. Distribution of phytoplasmas in infected plants as revealed by real time PCR and bioimaging. Molecular Plant–Microbe Interactions, 17: 1175-1184.

[31] Chomczynski P. and Rymaszewski M. 2006. Alkaline polyethylene glycol-based method for direct PCR from bacteria eukaryotic tissue samples and whole blood. BioTechniques, 40: 454-458.

[32] Constable FE., Symons RH. 2004. Genetic variability amongst isolates of Australian grapevine phytoplasmas. Australasian Plant Pathology 33:115-119.

[33] Cousin M.T., Sharma A.K., Isra S. 1986. Correlation between light and electron microscopic observations and identification of mycoplasmalikeorganisms using consecutive 350 nm think sections. Journal of Phytopathology, 115: 368-374.
[34] Crosslin J. M., Vandemark G. J., Munyaneza J. E. 2006. Development of a real-time, quantitative PCR for detection of the Columbia Basin potato purple top phytoplasma in plants and beet leafhoppers. Plant Diseases 90: 663–667.
[35] Daire X., Clair D., Larrue J., Boudon-Padieu E. 1997. Survey for grapevine yellows phytoplasmas in diverse European countries and Israel. Vitis, 36: 53-54.
[36] Da Rold G., Filippin L., Malembic-Maher S., Forte V., Borgo M., Foissac X., Angelini E. 2010. The imp gene in flavescence dorée and related phytoplasmas from grapevine, Clematis sp. and alder. In: 18th Congress of the International Organization for Mycoplasmology, Chianciano Terme (SI, Italy), 11th-16th July, p. 101.
[37] Danks C. and Boonham N. 2007. PurificationMethod and Kits. Patent WO/ 2007 / 104962.
[38] Davis R.E., Lee I.-M. 1993. Cluster-specific polymerase chain reaction amplification of 16SrDNA sequences for detection and identification of mycoplasmalike organisms. Phytopathology, 83: 1008-1011.
[39] Davis R.E., Dally E.L., Gundersen D.E., Lee I-M., Habili N. 1997. '*Candidatus* Phytoplasma australiense', a new phytoplasma taxon associated with Australian grapevine yellows. International Journal of Systematic Bacteriology, 47: 262-269.
[40] Delić, D., Seljak, G., Martini, M., Ermacora, P., Carraro, L., Myrta, A. Đurić, G. 2007. Surveys for grapevine yellows phytoplasmas in Bosnia and Herzegovina. *Bulletin of Insectology*, 60 (2), pp. 369-370
[41] Delić D., Mehle N., Lolić B., Ravnikar M. Đurić G. 2010. Current status of European stone fruit yellows in Bosnia and Herzegovina: Julius-Kühn-Archiv , Procedeeng of *21st International Conference on Virus and other Graft Transmissible Diseases of Fruit Crops,* July 5-10 2009, Neustadt, Germany, Neustadt: Julius Kühn-Institut, 427: 415-417.
[42] Delić D., Contaldo N., Paltrinieri S., Lolić B., Đurić Z., Hrnčić S., Bertaccini A. 2011. Grapevine yellows in Bosnia and Herzegovina: surveys to identify phytoplasmas in grapevine, weeds and insect vectors. Bulletin of Insectology, 64 (supplement), pp. 245-246.
[43] Delwart EL. and Gordon CJ. 1997. Tracking changes in HIV-1 envelope quasispecies using DNA heteroduplex analysis. Methods, 12:348-354.
[44] Deng S., Hiruki C. 1991. Amplification of 16S rRNA genes from culturable and nonculturable Mollicutes. Journal of Microbiology Methods, 14: 53-61.
[45] Doi Y., Teranaka M., Yora K., Asuyama H. 1967. Mycoplasma or PLT grouplike microrganisms found in the phloem elements of plants infected with mulberry dwarf, potato witches' broom, aster yellows or pawlonia witches'broom. Annals of Phytopathological Society Japan, 33: 259-266.
[46] Doyle, J.J., J.L. Doyle. 1990. Isolation of plant DNA from fresh tissue. Focus, 12: 13-15.
[47] Fos A., Danet J.L., Zreik J., Garnier M., Bové J.M. 1992. Use of a monoclonal antibody to detect the stolbur mycoplasma -like organism in plants and insects and to identify a vector in France. Plant Disease, 76: 1092-1096.
[48] Galetto L., Bosco D., Marzachi C. 2005. Universal and group-specific real-time PCR diagnosis of flavescance dorée (16Sr-V), bois noir (16Sr-XII) and apple proliferation

(16Sr-X) phytoplasmas from field-collected plant hosts and insect vectors. Annals of Applied Biology 147: 191-201.

[49] Galetto L. Marzachi C. 2010. Real-time PCR Diagnosis and Quantification of Phytoplasmas. In: Phytoplasmas Genomes, Plant Hosts and Vectors, Weintraub, P.G. and P. Jones (Eds.). CABI Publishers, USA., pp: 1-19.

[50] Garcia-Chapa M., Sabate J., Lavina A., Battle A.2005. Role of Cacopsylla pyri in the epidemiology of pear decline in Spain. European Journal of Plant Pathology, 111: 9-17.

[51] Gatineau F., Larruae J., Clair D, Lortoin F., Richard-Molard M., Boudon-Padieu E. 2001: A new natural planthopper vector of stolbur phytoplasma in the genus *Pentastiridius* (Hemiptera: Cixiidae). European. Journal of Plant Pathology, 107: 263-271.

[52] Green M. J., Thompson D. A., MacKenzie D. J. 1999. Easy and efficient DNA extraction from woody plants for the detection of phytoplasmas by polymerase chain reaction. Plant Dis 83, 482–485.

[53] Gori M., Monnanni R, Buiatti M., Goti E., Carnevale S., Da Prato L., Bertaccina A., Biricolti S. 2007. Establishing a real-time PCR detection procedure of "flavescence dorée" and "bois noir" phytoplasmas for mass screening. Bulletin of Insectology 60 (2): 255-256.

[54] Goto M., Honda E., Ogura A., Nomoto A., Hanaki KI. 2009. Colorimetric detection of loop-mediated isothermal amplification reaction by using hydroxyl napthol blue. BioTechniques 46, 167–72.

[55] Gundersen D.E., Lee I.-M. 1996. Ultrasensitive detection of phytoplasmas by nested-PCR assays using two universal primer pairs. Phytopathologia Mediterranea, 35: 144-151.

[56] Gundersen D.E, Lee I.-M., Rehner S.A., Davis R.E, Kingsbury D.T. 1994. Phylogeny of Mycoplasmalike organisms (Phytoplasmas): a basis for their classification. Journal of Bacteriology, 176: 5244-5254

[57] Gundersen D.E., Lee I.-M., Schaff D.A., Harrison N.A., Chang C.J., Davis R.E., Kinsbury D.T. 1996. Genomic diversity among phytoplasma strains in 16S rRNA Group I (Aster Yellows and related phytoplasmas) and III (X-Disease and related phytoplasmas). International Journal of Systematic Bacteriology, 46: 64-75.

[58] Guthrie J.N., White D.T., Walsh K.B., Scott P.T. 1998. Epidemiology of phytoplasma associated papaya diseases in Queensland, Australia. Plant Disease, 82: 1107-1111.

[59] Haggis G.H., Sinha R.C. 1978. Scanning electron microscopy of mycoplasmalike organisms after freeze fracture of plant tissues affected with clover phyllody and aster yellows. Phytopathology, 68: 677-680.

[60] Harrison N.A., Richardson P.A., Kramer J.B., Tsai J.H. 1994. Detection of the mycoplasmalike organism associated with lethal yellowing disease of palms in Florida by polymerase chain reaction. Plant Pathology, 43: 998-1008.

[61] Harrison N.A., Richardson, P.A., Tsai J.H., Ebbert M.A., Kramer J.B. 1996. PCR assay for detection of the phytoplasma associated with maize bushy stunt disease. Plant Disease, 80: 263-269.

[62] Harrison N.A., Womack M., Carpio M.L. 2002. Detection and characterization of a lethal yellowing (16SrIV) group phytoplasma in Canary Island date palms affected by lethal decline in Texas. Plant Disease, 86: 676-681.

[63] Hartung J.S., Pruvost, O.P., Villemot I., Alvarez A. 1996. Rapid and sensitive colorimetric detection of *Xanthomonas ax- - onopodis* pv. *citri* by immunocapture and a nested-polymerase chain reaction assay. Phytopathology 86, 95–101.

[64] Heinrich M., Botti SCaprara., L., Arthofer W., Strommer S. 2001. Improved detection methods for fruit tree phytoplasmas. Plant Molecular Biology Reporter, 19: 169-179.

[65] Hobbs H.A., Reddy D.V.R., Reddy A.S. 1987. Detection of a mycoplasma-lke organism in peanut plants with witches' broom using indirect enzyme-linked immunosorben assay (ELISA). Plant Pathology, 36: 164-167.

[66] Hodgetts J., Ball T., Boonham N., Mumford R , Dickinson M. 2007. Use of terminal restriction fragment length polymorphism (T-RFLP) for identification of phytoplasmas in plants. Plant Pathology, 56: 357-365.

[67] Hodgetts J., Boonham N., Mumford R., Harrison N., Dickinson M. 2008. Phytoplasma phylogenetics based on analysis of secA and 23S rRNA gene sequences for improved resolution of candidate species of *Candidatus* Phytoplasma. International Journal of Systematic and Evolutionary Microbiology. 58: 1826-1837.

[68] Hodgetts J., Tomlison J., Boonham N., Gonzales-Martin I., Nikolić P., Swarbrick P., Yankey E.N., Dickinson M. 2011. Development of rapid in-field loop mediated isothermal amplification (LAMP) assays for phytoplasmas. Bulletin of Insectology, 64:41-42.

[69] Hollingsworth C.R., Atkinson L.M., Samac D.A., Larsen L.E., Motteberg C.D., Abrahamson M.D., Glogoza P., MacRae I.V. 2008. Region and field level distributions of aster yellows phytoplasma in small grain crops. Plant Disease, 92: 623-630.

[70] Hoshi A., Ishii Y., Kakizawa S., Oshima K., Namba S. 2007. Host-parasite interaction of hosts as determined by competitive polymerase chain reaction. Phytopathology, 90: 1145-1152.

[71] Hren M., Boben J., Rotter A., Kralj P., Gruden K., Ravnikar M. 2007. Real-time PCR detection systems for Flavescence doree and Bois noir phytoplasmas in grapevine: comparison with conventional PCR detection and application in diagnostics. Plant Pathology, 56: 785-796.

[72] Hren I.M., Ravnikar M., Brzin J., Ermacora P., Carraro L., Bianco P.A., Casati P., Borgo M., Angelini E., Rotter A., Gruden K. 2009. Induced expression of sucrose synthase and alcohol dehydrogenase I genes in phytoplasma-infected grapevine plants grown in the field. Plant Pathology, 58(1): 170-180.

[73] IRPCM. 2004. '*Candidatus* Phytoplasma', a taxon for the wall-less, non-helical prokaryotesthat colonise plant phloem and insects. International Journal of Systematic and Evolutionary Microbiology, 54: 1243-1255.

[74] Iwamoto T., Sonobe T., Hayashi K. 2003. Loop-mediated isothermal amplification for direct detection of Mycobacterium tuberculosis complex, M. avium, and M. intracellulare in sputum samples. Journal of Clinical Microbiology 41: 2616–22.

[75] Jacobs K.A., Lee I.M., Griffiths H.M., Miller F.D., Bottner K.D. 2003. A new member of the clover proliferation phytoplasma group (16SrVI) associated with elm yellows in Illinois. Plant Disease, 87: 241-246.

[76] Jarausch W., Peccerella T., Schwind N., Jarausch B., Krczal G. 2004. Establishment of a quantitative real-time PCR assayfor the quantification of apple proliferation phytoplasmas in plants and insects. Acta Horticulturae 657, 415–20.

[77] Jomantiene R., Davis R.E., Maas J., Dally E.L. 1998. Classification of new phytoplasmas associated with diseases of strawberry in Florida, based on analysis of 16S rRNA and ribosomal protein gene operon sequences. International Journal of Systematic Bacteriology, 48: 269-277.

[78] Jomantiene R., Valiunas D., Ivanauskas A., Urbanaviciene L., Staniulis J., Davis R.E. 2011. Larch is a new host for a group 16SrI, subgroup B, phytoplasma in Ukraine. Bulletin of Insectology, 64 (supplement), pp. 101-102.

[79] Kartte S., Seemüller E. 1991. Susceptibility of grafted *Malus* taxa and hybrids to apple proliferation disease. Journal of Phytopathology, 131: 137-148.

[80] Kawakita, H., Saiki, T., Wei, W., Mitsuhashi, W., Watanabe, K. and Sato, M. 2000. Identification of mulberry dwarf phytoplasmas in genital organs and eggs of the leafhopper Hishimonoides sellatiformis . *Phytopathology,* 90: 909-914.

[81] Khan JA, Srivastava P, Singh SK. 2004. Efficacy of nested-PCR for the detection of phytoplasma causing spike disease of sandal. Current Science 86: 1530–1533.

[82] Klamer M, Roberts MS, Levine LH, Drake BG, Garland JL. 2002. Influence of elevated CO_2 on the fungal community in a coastal scrub oak forest soil investigated with terminal-restriction fragment length polymorphism analysis. Applied and Environmental Microbiology 68: 4370–6.

[83] Kostina E.V., Ryabinin V.A., Maksakova G.A., Sinyakov A.N. 2007. TaqMan probes based on oligonucleotide-hairpin minor groove binder conjugates. Russian Journal of Bioorganic Chemistry 33, 614-616.

[84] Kunkel L.O. 1926. Studies on aster yellows. American Journal of Botany, 13: 646-705.

[85] Kunkel L.O. 1931. Celery yellows of California not identical with the aster yellows of New York. Boyce Thompson Institute, 4: 405-414.

[86] Kunkel L.O. 1955. Cross protection between strains of aster yellow-type viruses. Advances in Virus Research, 3: 251-273.

[87] Langer M., Maixner M. 2004. Molecular characterization of grapevine yellows associated phytoplasmas of the Stolbur-group based on RFLP-analysis of non-ribosomal DNA. Vitis, 43: 191-199.

[88] Lederer W., Seemüller E. 1991. Occurrence of mycoplasmalikeorganisms in diseased and non-symptomatic alder trees (*Alnus* spp.).- European Journal of Forest Pathology, 21: 90-96.

[89] Lee I.M. and Davis R.E. 1986. Prospects for *in vitro* culture of plant-pathogenic mycoplasmalike organisms. Annual Review of Phytopathology, 24: 339-354.

[90] Lee G.T.N., Golino D.A., Hackett K.J., Kirkptrick B.C., Marwitz R., Petzold H., Shina R.H., Sugiura M., Whitcomb R.F., Yang I.L., Zhu B.M., Seemüller E. 1989. Plant Diseases Associated with Mycoplasmalike Organisms. In: The Mycoplasmas, vol. 5. (eds. Whitcomb R.F. e Tully J.G.). Academic Press, New York: 545-640.

[91] Lee I.-M., Hammond R.W., Davis R.E., Gundersen D.E. 1993. Universal amplification and analysis of pathogen 16S rDNA for classification and identification of mycoplasmalike organisms. Phytopathology, 83: 834-842.

[92] Lee I.-M., Gundersen D.E., Hammond R.W., Davis R.E. 1994. Use of mycoplasmalike organism (MLO) group-specific oligonucleotide primers for nested-PCR assays to detect mixed-MLO infections in a single host plant. Phytopathology, 84: 559-566.

[93] Lee I.M., Gundersen-Rindal D.E., Davis R.E., Bartoszyk I.M. 1998. Revised classification scheme of phytoplasma based on RFLP analyses of 16S rRNA and ribosomal protein gene sequences International Journal of Systematic Bacteriology, 48: 1153-1169.

[94] Lee I.-M., Gundersen-Rindal D.E., Bertaccini A. 1998. Phytoplasma: ecology and genomic diversity. Phytopathology, 88: 1359-1366.

[95] Lee I.M., Davis R.E., Gundersen-Rindal D.E. 2000. Phytoplasma: Phytopathogenic mollicutes. Annual Review of Microbiology, 54: 221-555.

[96] Lee I.-M., Gundersen-Rindal D., Davis R.E., Bottner K.D., Marcone C., Seemueller E. 2004.'*Candidatus* Phytoplasma asteris', a novel taxon associated with aster yellows and related diseases. International Journal of Systematic Bacteriology, 54: 1037-1048.

[97] Lee I.-M., Martini M., Marcone C., Zhu S.F. 2004. Classification of phytoplasma strains in the elm yellows group (16SrV) and proposal of '*Candidatus* Phytoplasma ulmi' for the phytoplasma associated with elm yellows. International Journal of Systematic and Evolutionary Microbiology, 54(2): 337-347.

[98] Lee I.-M., Zhao Y., Bottner K.D. 2006. SecY gene sequence analysis for finer differentiation of diverse strains in the aster yellows phytoplasma group. Molecular and Cellular Probes, 20(2): 87-91.

[99] Leifting L., Veerakone S., Gerrard R.G., Clover L., Ward L.I. 2011. An update on phytoplasma diseases in New Zealand. Bulletin of Insectology, 64 (supplement), pp. 93-94.

[100] Leyva-Lopez N.E., Ochoa-Sanchez J.C., Leal-Klevezas D.S., Martinez-Soriano J.P. 2002. Multiple phytoplasmas associated with potato diseases in Mexico. Canadian Journal of Microbiology, 48: 1062-1068.

[101] Liu H.W., Goodwin P.H., Kuske C.R. 1994. Quantificationof DNA from the aster yellows mycoplasma like organism in aster leafhoppers (*Macrosteles fascifrons* Stål) by a competitive polymerase chain reaction. Systematic and Applied Microbiology, 17: 274-280.

[102] Lorenz K.H., Schneider B., Ahrens U., Seemüller E. 1995. Detection of the apple proliferation and pear decline phytoplasmas by PCR amplification of ribosomal and nonribosomal DNA. Phytopathology, 85: 771-776.

[103] Loi N., Ermacora P., Carraro L., Osler R., Tsen An C. 2001. Production of monoclonal antibodies against apple proliferation phytoplasma and their use in serological detection. European Journal of Plant Pathology, 108: 81-86.

[104] Ludvikova H., Lauterer P., Sucha J., Franova J. 2011. Monitoring of psyllid species (Hemiptera, Psylloidea) in apple and pear orchards in East Bohemia. Bulletin of Insectology, 64 (supplement), pp. 121-122.

[105] Maeda H., Kokeguchi S., Fujimoto C. 2005. Detection of periodontal pathogen Porphyromonas gingivalis by loopmediated isothermal amplification method. FEMS Immunology and Medical Microbiology, 43: 233–239.

[106] Maixner M., Ahrens U., Seemüller E. 1995. Detection of the German grapevine yellows (Vergilbungskrankheit) MLO in grapevine, alternative hosts and a vector by a specific PCR procedure. European Journal of Plant Pathology, 101: 241-250.
[107] Maixner M. 2010. Grapevine yellows vector sampling and monitoring training school.- *COST Action FA0807, Working group 2-* Grapevine Yellows Vector Sampling and Monitoring Training School, *Bernkastel-Kues Germany,* 5th to 9th of July, 2010. From: http://costphytoplasma.eu/PDF%20files/Proceedings_Vector_TS.pdf
[108] Malisano G., Firrao G., Locci R. 1996. 16S rDNA-derived oligonucleotide probes for the differential diagnosis of plum leptonecrosis and apple proliferation phytoplasmas. EPPO Bulletin 26, 421–428.
[109] Marcone C., Lee I.-M., Davis R.E., Ragozzino A., Seemüller E. 2000. Classification of aster yellows-group phytoplasmas based on combined analyses of rRNA and tuf gene sequences. International Journal of Systematic and Evolutionary Microbiology, 50(5): 1703-1713.
[110] Marinho V L A., Sandrine F., Michel D. 2008. Genetic variability among isolates of *Coconut lethal yellowing phytoplasmas* determined by Heteroduplex Mobility Assay (HMA). Tropical Plant Pathology, 33 (5): 377-380.
[111] Martini M., Botti S., Marcone C., Marzachì C., Casati P., Bianco P.A., Benedetti R., Bertaccini A. 2002. Genetic variability among Flavescence dorée phytoplasmas from different origins in Italy and France. Molecular and Cellular probes, 16(3): 197-208.
[112] Martini M., Lee I.-M. Bottner K.D., Zhao Y., Botti S., Bertaccini A., Harrison N.A., Carraro L., Marcone C., Khan J., Osler R. 2007. Ribosomal protein gene-based filogeny for finer differentiation and classification of phytoplasmas. International Journal of Systematic and Evolutionary Microbiology, 57: 2037-2051.
[113] Martini M., Loi N., Ermacora P., Carraro L., Pastore M. 2007. A real-time PCR method for detection and quantification of '*Candidatus* Phytoplasma prunorum' in its natural host. Bulletin of Insectology 69(2): 251-252.
[114] Martini M., Ermacora P., Falginella L., Loi N., Carraro L. 2008. Molecular differentiation of '*Candidatus* Phytoplasma mali' and its spreading in Friuli Venezia Giulia region (North- East Italy). Acta Horticulturae, 781: 395-402.
[115] Martini M., Ferrini F., Danet J.L., Ermacora P., Gülşen S., Delić D., Nazia L.,Xavier F., Carraro L. 2010: PCR/RFLP based method for molecular characterization of "*Candidatus* Phytoplasma prunirum" strains using *aceF* gene: Julius-Kühn-Archiv , Procedeeng of *21st International Conference on Virus and other Graft Transmissible Diseases of Fruit Crops,* July 5-10 2009, Neustadt, Germany, Neustadt: Julius Kühn-Institut, 427: 386-391.
[116] Marzachi C., Verrati F., Bosco D. 1998. Direct PCR detection of phytoplasmas in experimentally infected insects. Annals of Applied Biology, 133: 45-54.
[117] Marzachi C. 2004. Molecular diagnosis of phytoplasmas. Phytopathology Mediterranean, 43: 228-231.
[118] Marzachi C. and Bosco D. 2005. Relative quantification of chrysanthemum yellows (16SrI) phytoplasma in its plantand insect host using real-time polymerase chain reaction. Molecular Biotechnology, 30: 117-127.
[119] Mc Coy R.E., Caudwell A., Chang C.J., Chen T.A., Chiykowskyi L.N., Cousin M.T., Dale de Leeuw G.T.N., Golino D.A., Hackett K.J., Kirkptrick B.C., Marwitz R., Petzold H., Shina R.H., Sugiura M., Whitcomb R.F., Yang I.L., Zhu B.M., Seemüller

E. 1989. Plant Diseases Associated with Mycoplasmalike Organisms. In: The Mycoplasmas, vol. 5. (eds. Whitcomb R.F. e Tully J.G.). Academic Press, New York: 545-640.

[120] Mikec I., Križanac I., Budinščak Ž., Šeruga Musić M., Krajačić M., Škorić D. 2006. Phytoplasmas and their potential vectors in vineyards of indigenous Croatian varieties. *Extended Abstracts 15th Meeting of the ICVG, Stellenbosch, South Africa*: 255-257.

[121] Minsavage G.V., Thompson C.M., Hopkins D.L., Leite RMVBC, Stall RE. 1994. Development of a polymerase chain-reaction protocol for detection of Xylella-fastidiosa in plant-tissue. Phytopathology, 84:456-461

[122] Munford R.A., Boonham N., Tomlinson J., Barker I. 2006. Advances in molecular phytodiagnostics - new solutions for old problems. European Journal of Plant Pathology, 116: 1-19.

[123] Mori Y., Nagamine K., Tomita N., Notomi T. 2001. Detection of loop-mediated isothermal amplification reaction by turbidity derived from magnesium pyrophosphate formation. Biochemical and Biophysical Research Communications, 289:150–154.

[124] Mori Y., Kitao M., Tomita N., Notomi T. 2004. Real-time turbidimetry of LAMP reaction for quantifying template DNA. Journal of Biochemistry and Biophysical Methods, 59: 145–157.

[125] Music M.S., Krajačić M., Škorić D. 2007. Evaluation of SSCP analysis as a tool for detection of phytoplasma molecular variability. Bulletin of Insectology, 60: 245-246.

[126] Music M.S., Krajačić M., Škorić D. 2008. The use of SSCP analysis in the assessment of phytoplasma gene variability. Journal of Microbiological Methods, 73: 69-72.

[127] Namba S., Kato S., Iwanami S., Oyaizu H., Shiozawa H., Tsuchizaki T. 1993. Detection and differentiation of plant-pathogenic mycoplasmalike organisms using polymerase chain reaction. Phytopathology, 83: 786-791.

[128] Nikolić P., Mehle N., Gruden K., Ravnikar M., Dermastia M. 2010. A panel of real-time PCR assays for specific detection of three phytoplasmas from the apple proliferation group. Molecular and Cellular Probes, 24, 303-309.

[129] Nipah J.O., Jones P., Dickinson M.J. 2007. Detection of lethal yellowing phytoplasma in embryos from coconut palms infected with Cape St Paul wilt disease in Ghana. Plant Pathology, 56: 777-784.

[130] Notomi T, Okayama H, Masubuchi H. 2000. Loop-mediated isothermal amplification of DNA. Nucleic Acids Research 28: 63.

[131] Oberhänsli T., Altenbach D., Bitterlin W. 2011. Development of a duplex TaqMan real-time PCR for the general detection of phytoplasmas and 18S rRNA host genes in fruit trees and other plants. Bulletin of Insectology, 64:37-38.

[132] Orsagova H., Brezikova M., Schlesingerova G. 2011. Presence of phytoplasmas in hemipterans in Czech vineyards. Bulletin of Insectology, 64 (supplement), pp. 119-120.

[133] Padovan A.C., Gibb K.S., Bertaccini A., Vibio M., Bonfiglioli R.E., Magarey P.A., Sears B.B. 1995. Molecular detection of the Australian grapevine yellows phytoplasma and comparison with a grapevine yellows phytoplasma from Emilia-Romagna in Italy. Australian Journal of Grape and Wine Research, 1: 25 31.

[134] Palmano S. 2001. A comparison of different phytoplasma DNA extraction methods using competitive PCR. Phytopathology Mediterranean, 40:99-107.

[135] Petrzik K., Sarkisova T., Čurnova L. 2011. Universal primers for plasmid detection and method for their relative quantification in phytoplasma-infected plants. Bulletin of Insectology, 64 (supplement), pp. 25-26.

[136] Ploaie, P. G. 1981. Mycoplasma-like organisms and plant diseases in Europe. Pages 61-104 *In* Plant Diseases and Vectors: Ecology and Epidemiology. Maramorosch, K., and Harris, K. F., eds. Academic Press, New York.

[137] Prezelj N., Mehle N., Nikolić P., Ravnikar M., Dermastia M. 2010. Rapid diagnostic for economically important phytoplasmas in grapevine and fruit trees. Knjiga povzetkov = Book of abstracts / [5. slovenski simpozij o rastlinski biologiji z mednarodno udeležbo, 6.-9. september 2010, Ljubljana = 5th Slovenian Symposium on Plant Biology with International Participation, September 6-9 2010, Ljubljana, Slovenia]; [organizator] Slovensko društvo za biologijo rastlin = [organized by] Slovenian Society of Plant Biology; p, 94.

[138] Prince J.P., Davies. R.E., Wolf T.K., Lee I.-M., Mogen B.D., Dally E.L., Bertaccini A., Credi R., Barba M. 1993. Molecular detection of diverse mycoplasmalike organisms (MLOs) associated with grapevine yellows and their classification with aster yellows, X-disease and elm yellows MLOs. Phytopathology, 83: 1130-1137.

[139] Rajan J., Clark M.F. 1995. Detection of apple proliferation and other MLOs by immunocapture PCR (IC-PCR). Acta Horticulturae, 386: 511-514.

[140] Saito R., Misawa Y., Moriya K., Koike K., Ubukata K., Okamura N. 2005. Development and evaluation of a loop-mediatedisothermal amplification assay for rapid detection of Mycoplasma pneumoniae. Journal of Medical Microbiology, 54:1037–41.

[141] Sakai M, Matsuka A, Komura T, Kanazawa S. 2004. Application of a new PCR primer for terminal restriction fragment length polymorphism analysis of the bacterial communities in plant roots. Journal of Microbiological Methods, 59: 81–89.

[142] Saracco P., Bosco D., Veratti F., Marzachi C. 2006. Quantification over time of chrysanthemum yellows phytoplasma (16Sr-I) in leaves and roots of the host plant Chrysanthemum carinatum (Schousboe) following inoculation with its insect vector. Physiological and Molecular Plant Pathology, 67: 212-219.

[143] Samuitiene M. and Navalinskiene M. 2006. Molecular detection and characterization of phytoplasma infecting *Celosia argentea* L. plants in Lithuania. Agronomy-Research, 4: 345-348.

[144] Schaff D.A., Lee I.-M., Davis R.E. 1992. Sensitive detection and identification of mycoplasmalike organisms by polymerase chain reactions. Biochemistry Biophysics Research Communications, 186: 1503-1509.

[145] Schneider B., Ahrens U., Kirkpatrick B.C., Seemüller E. 1993. Classification of plant pathogenic mycoplasma-like organisms using restriction-site analysis of PCR-amplified 16S rDNA. Journal of General Microbiology, 139: 519-527.

[146] Schneider B., Seemuller E., Smart C.D., Kirkpatrick B.C. 1995. Phylogenetic Classification of Plant Pathogenic Mycoplasmalike Organisms or Phytoplasmas. In: Molecular and Diagnostic Procedures in Mycoplasmology, Razin, S. and J.G. Tully (Eds.). Academic Press, San Diego, pp: 369-380.

[147] Schneider B., Gibb K.S., Seeümller E. 1997. Sequence and RFLP analysis of the elongation factor Tu gene used in differentiation and classification of phytoplasmas. Microbiology, 143: 3381-3389.
[148] Seemüller E. 1976. Investigation to demostrate mycoplasmalike organism in diseases plants by fluorescence microscopy. Acta Horticulturae, 67: 109-112.
[149] Seemüller E. Kunze L., Schaper U. 1984. Colonization behaviour of MLO and symptom expression of proliferation in diseased apple trees and decline-diseased pear trees over a period of several years. Journal of Plant Disease Protection, 91: 525-532.
[150] Seemüller E., Schaper U., Zimbelmann F. 1984. Seasonal variations in the colonization patterns of mycoplasma-like organisms associated with apple proliferation and pear decline. Z. Pflanzenkrankeiten Pflanzenschutz, 91: 371-382.
[151] Seemüller, E., Stolz, E., Kison, H. 1998. Persistence of European stone fruit yellows phytoplasma in aerial parts of *Prunus* taxa during the dormant season. Journal of Phytopathology, 146: 407–410.
[152] Seemüller E., Moll E., Schneider B. 2007. *Malus sieboldii*-based rootstocks mediate apple proliferation resistance to grafted trees. Bulletin of Insectology, 60(2): 301-302.
[153] Smart, C.D., B. Schneider, C.L. Blomquist, L.J. Guerra and N.A. Harrison *et al.*, 1996. Phytoplasma-specific PCR primers based on sequences of the 16S-23S rRNA spacer region. Phytopathology, 62: 2988-2993.
[154] Tomlinson JA., Barker I., Boonham N. 2007. Faster, simpler, more-specific methods for improved molecular detection of Phytophthoraramorum in the field. Applied and Environmental Microbiology 73: 4040–4047.
[155] Tomlinson J. A., Boonham N., Dickinson M. 2010. Development and evaluation of a one-hour DNA extraction and loop-mediated isothermal amplification assay for rapid detection of phytoplasmas. Plant Pathology, 59, 465–471.
[156] Torres E., Bertolini E., Cambra M., Montón C., Martín MP. 2005. Real-time PCR for simultaneous and quantitative detection of quarantine phytoplasmas from apple proliferation (16SrX) group. Molecular and Cellular Probes 19, 334–40.
[157] Upchurch DA, Shankarappa R, Mullins JI. 2000. Position and degree of mismatches and the mobility of DNA heteroduplexes. Nucleic Acids Research, 28 (12), 69-69.
[158] Wang YH., Griffith J. 1991. Effects of bulge composition and flanking sequence on the kinking of DNA by bulged bases. Biochemistry, 30:1358-1363.
[159] Wang K., Hiruki C. 2001. Use of heteroduplex mobility assay for identification and differentiation of phytoplasmas in the aster yellows group and the clover proliferation group. Phytopathology 91:546-552.
[160] Wang K., Hiruki C. 2005. Distinctions between phytoplasmas at the subgroup level detected by heteroduplex mobility assay. Plant Pathology, 54: 625-633.
[161] Wei W., Kakizawa S., Suzuki S., Jung H.Y., H. Nishigawa H., Miyata S., Oshima K., Ugaki M., Hibi T., Namba S. 2004. *In planta* dynamic analysis of onion yellows phytoplasma using localized inoculation by insect transmission. Phytopathology, 94(3): 244-250.
[162] Wei W., Davis R.E., Lee I.M., Zhao Y. 2007. Computer-simulated RFLP analysis of 16S rRNA genes: Identification of ten new phytoplasma groups. International Journal of Systematic and Evolutionary Microbiology, 57: 1855-1867.
[163] Wei W., Lee I.-M., Davis R.E., Suo X., Zhao Y. 2008. Automated RFLP pattern comparison and similarity coefficient calculation for rapid delineation of new and

distinct phytoplasma 16Sr subgroup lineages. International Journal of Systematic and Evolutionary Microbiology, 58(10): 2368-2377.

[164] Wilson IG. 1997. Inhibition and facilitation of nucleic acid amplification. Applied and Environmental Microbiology, 63: 3741–3751.

[165] Zhang Y., Uyemoto J.K., Kirkpatrick B.C. 1998. A small-scale procedure for extracting nucleic acids from woody plants infected with various phytopathogens for PCR assay. Journal of Virological Methods, 71: 45-50.

6

Molecular Diagnostics of Mycoplasmas: Perspectives from the Microbiology Standpoint

Saúl Flores-Medina[1,2,*], Diana Mercedes Soriano-Becerril[1]
and Francisco Javier Díaz-García[3]
[1]*Departamento de Infectología, Instituto Nacional de Perinatología, D.F.,*
[2]*CECyT No. 15 "DAE", IPN, D.F.,*
[3]*Departamento de Salud Pública, Facultad de Medicina, UNAM, D.F.,*
México

1. Introduction

Some of the smallest self-replicating bacteria, the wall-less mycoplasmas belonging to Class Mollicutes, are pathogenic for mammals and humans, showing tissue and host-specificity. In humans, the pathogenic species of the *Mycoplasma* or *Ureaplasma genus* cause covert infections that tend to chronic diseases. At present, 7 species of *Mycoplasma*, 2 species of *Ureaplasma* and 1 of *Acholeplasma* have been consistently isolated/detected from several specimens from diseased subjects, specially through the use of molecular detection techniques [Mendoza *et al.*, 2011; Waites & Talkington, 2005; Waites, 2006].

Current laboratory diagnosis of these infections relies on cultural methods, however this is complicated and emission of results may delay up to 5 weeks. Thus development and application of molecular methods, such as polymerase chain reaction (PCR), have allowed direct detection in clinical specimens and shortened the time to get the final results. Nevertheless some pitfalls still hampers the widespread use of these technologies, mainly due to technical difficulties in collecting representative specimens and optimizing sample preparation. There are countless reports on new nucleic acid-based tests (NATs) for mycoplasma detection, however there is a great variation between methods from study to study, including variability of target gene sequences, assay format and technologic platform [Waites *et al.*, 2000; Waites , 2006;].

The processing of the clinical samples is crucial for the improvement of PCR assays as part of routine diagnostic approaches. In general, for the strength of performance of any diagnostic PCR, the overall setting-up of the assay should consider the following four basic steps: 1) sampling, 2) sample preparation, 3) nucleic acid amplification, and 4) detection of PCR products [Rådström *et al.*, 2004].

As occurred with much of the emerging or reemerging pathogens, the molecular detection plays a key role in the discovery, identification and association or such pathogens with human disease [Relman & Persing, 1996]. Nevertheless, routine clinical microbiology

*Corresponding Author

laboratories still lack of skilled personnel in molecular detection techniques, and consequently in the appropriate sample preparation procedures [Cassell *et al,* 1994a; Talkington & Waites, 2009]. Unlike other fast-growing pathogens, the pathogenic Mollicutes species exhibit unique features that make them the last link in the diagnostic chain, only sought after failure in other diagnostic approaches [Cassell *et al.,* 1994a].

2. Relevant features of mycoplasmas

The term mycoplasmas will be used to refer to any member of the Class Mollicutes. The mycoplasmas are the smallest microorganisms (0.3 - 0.8 µm diameter) capable of self-replication, which lack a rigid cell wall. These bacteria also incorporate exogenous cholesterol into their own plasma membrane and use the UGA codon to encode tryptophan. Due to their reduced cell dimensions, they possess small genome sizes (0.58-2.20 Mb) and exhibit restricted metabolic alternatives for replication and survival. As a result of the above mentioned, the mycoplasmas show a strict dependence to their hosts for acquisition of biosynthetic precursors (aminoacids, nucleotides, lipids and sterols), in a host- and tissue-restricted manner, reflecting their nutritional demands and parasitic lifestyle [Baseman & Tully, 1997; Razin, 1992; Razin *et al.,* 1998].

Mycoplasmas infecting humans mainly colonize the mucosal surfaces of the respiratory and genitourinary tracts [Cassell *et al.,* 1994, Patel & Nyirjesy, 2010; Taylor-Robinson, 1996]. The mycoplasma species commonly isolated from humans and their attributes are listed in Table 1. Of the pathogenic species, *Mycoplasma pneumoniae* is found principally in the respiratory tract, whereas *M. genitalium, Ureaplasma parvum, U. urealyticum., M. hominis, M. fermentans*

Species	*Primary colonization sites*		*Main metabolic substrates*			*Pathogenicity*
	Respiratory tract	*Urogenital tract*	*Glucose*	*Arginine*	*Urea*	
Mycoplasma. salivarium	+	-	-	+	-	-
M. orale	+	-	-	+	-	-
M. buccale	+	-	-	+	-	-
M. faucium	+	-	-	+	-	-
M. lipophilum	+	-	-	+	-	-
M. pneumoniae	+	-	+	-	-	+
M. hominis	+	+	-	+	-	+
M. genitalium	+	+	+	-	-	+
M. fermentans	+	+	+	+	-	+[a]
M. primatum	-	+	-	+	-	-
M. spermatophilum	-	+	-	+	-	-
M. pirum	¿?	¿?	+	+	-	-
M. penetrans	-	+	+	+	-	+[b,c]
Ureaplasma urealyticum	+	+	-	-	+	+
U. parvum	+	+	-	-	+	+
Acholeplasma laidlawii	+	-	+	-	-	-
A. oculi	¿?	-	+	-	-	-

[a]Lo *et al*,1993; [b]Lo *et al*, 1992; [c]Yáñez *et al.*, 1999.

Table 1. Mycoplasmas which infect humans.
Adapted from: Taylor-Robinson, 1996.

and *M. penetrans* are primarily urogenital residents, but exceptionally they can be isolated from other unusual tissues and organs, especially in immunocompromised patients or in patients undergoing solid organ transplantation [Cassel *et al.*, 1993; Waites & Talkington, 2004; Waites *et al*, 2005; Waites *et al.*, 2008].

Most of mycoplasmal diseases are underdiagnosed because the specific laboratory diagnostic strategies are quite different than those for fast-growing bacteria. It is noteworthy that mycoplasmal etiology of diseases in humans is considered only after failure of diagnosis of other common bacterial etiologies. In addition, outside their hosts, the mycoplasmas are highly labile to environmental factors, such as changing osmotic pressure and temperature, desiccation and/or alkaline or acidic conditions [Cassell *et al.*, 1994a; Waites *et al.*, 2000]. Noteworthy, there are few specialized or reference laboratories for diagnosis of mycoplasmal diseases and therefore, limited skilled laboratory personnel [Cassell *et al.*, 1994a].

3. Routine laboratory diagnostic approaches

Several different detection techniques of mycoplasmal infections have been developed, each one of which has its advantages and limitations with respect to cost, time, reliability, specificity, and sensitivity. According to the laboratory's infrastructure, the most common methods include: a) culture-based isolation/detection/identification/antimicrobial susceptibility profile; b) antigen detection, c) mycoplasmal-specific serologic responses; and, d) PCR and other NATs. [Razin *et al.*, 1998; Talkington & Waites, 2009; Waites *et al.*, 2000; Yoshida *et al.*, 2002].

3.1 Culture

Relationship between mycoplasmas as etiologic agents and diseases in humans remains doubtful due to unsuccessful isolation/detection of these microorganisms in specimens from affected persons, as compared with healthy carriers [Taylor-Robinson, 1996]. Demonstration of growth of mycoplasma, by means of *in vitro* culture from clinical specimens, is still required to link the pathogen with the disease; thus culture is considered as the Gold Standard. However, current culture methods for detection of mycoplasmas in clinical specimens are arduous and emission of results may delay up to 5 weeks, which even then may be inconclusive or inaccurate [Cassell *et al.*, 1994a; Waites *et al.*, 2000].

In this context, detection or isolation of mycoplasmas from clinical specimens requires careful consideration of the type of specimen available and the organism (species) sought [Cassell *et al.*, 1994a]. The adequate specimens for culture include: a) normally sterile body fluids (sinovial, amniotic, cerebrospinal, urine, peritoneal, pleural, etc., b) secretions exudates or swabs from sites with associated flora (from nasopharinx, pharinx, cervix, vagina, urethra, surgical wounds, prostate, sputum, etc.), and c) cell-rich fluids (including blood and semen) or tissue biopsies. Overall, specimen collection should reflect the site of infection and/or the disease process. [Atkinson *et al.*, 2008; Waites & Talkington, 2004].

Liquid specimens or tissues do not require special transport media if culture can be performed within 1 hour, otherwise specimens should be placed in transport media, such as SP-4, 10B or 2SP broths. When swabbing is required, aluminum- or plastic-shafted calcium alginate or dacron swabs should be used, taking care to obtain as many cells as possible [Atkinson *et al.*, 2008; Cassell *et al.*, 1994a; Waites, 2006].

There is no ideal formulation of culture media for all pathogenic species, mainly due to their different substrate and pH requirements [Waites *et al.,* 2000]. Modified SP-4 media (broth and agar) [Lo *et al.*, 1993a], containing both glucose and arginine, can support the growth of all human pathogenic *Mycoplasma* species, including the fastidious *M. pneumoniae* and *M. genitalium*. A set of Shepard´s 10B broth and A8 agar can be used for cultivation of *Ureaplasma* species and *M. hominis*. For cultivation, specimens in transport media should be thoroughly mixed, and then should be 10-fold serially diluted in broth (usually up to 10^{-6}) in order to allow semiquantitative estimation of mycoplasmal load, but subcultures in agar media should also be performed [Cassel *et al.*, 1994a]. Inoculated media should be incubated under microaerophilic atmosphere at 37 °C.

Detection of *M. pneumoniae* in broth culture is based on its ability to ferment glucose, causing an acidic shift after 4 or more weeks, readily visualized by the presence of the phenol red pH indicator. Broths with any color change, and subsequent blind broth passages, should be subcultured to SP4 agar, incubated, and examined under the low-power objective of the light microscope in order to look for development of typical "fried egg"-like colonies of up to 100 µm in diameter. Examination of agar plates must be done on a daily basis during the first week, and thereafter every 3 to 4 days until completing 5 weeks or until growth is observed [Waites *et al.*, 2000; Waites & Talkington, 2004]. *M. genitalium, M. fermentans* and *M. penetrans* are also glucose-fermenting and form colonies morphologically indistinguishable from those of *M. pneumoniae,* thus serologic-based definitive identification can be done by growth inhibition, metabolic inhibition, and mycoplamacidal tests [Atkinson *et al.*, 2008].

Hidrolysis of urea by Ureaplasma and hidrolysis of arginine by *M. hominis* cause an alkaline shift, turning the colour of 10B broth from yellow to pink. Tiny brown or black irregular colonies of *Ureaplasma* species develop between 1-5 days on A8 agar plates, due to urease activity in the presence of manganese sulfate. Typical fried egg colonies are produced by *M. hominis* in this medium [Cassell *et al*, 1994a; Waites *et al.*, 2000].

3.2 Molecular assays

The nucleic acid-based techniques have several advantages over culture-based methods, including rapid results, low detection limits (theoretically a single copy of target sequence), and specific organism detection. This is critical in a hospital setting, since rapid pathogen detection is important for faster and improved patient treatment and consequently for shortening hospitalization time [Mothershed & Whitney, 2006].

In particular, for PCR-based detection tests, selection of the appropriate target sequences for amplification appears to be of major concern. Mycoplasmal sequences to be amplified can be chosen from published gene sequences or from a mycoplasma-specific cloned DNA fragments [Kovacic *et al.*, 1996; Razin, 2002]. The accelerated rate of genomic sequencing has led to an abundance of completely sequenced genomes. Annotation of the open reading frames (ORFs) (i.e., gene prediction) in these genomes is an important task and is most often performed computationally based on features in the nucleic acid sequence [Jaffe *et al.*, 2004; Razin 2002]. Besides complete or almost complete sequences of the 16S rRNA genes for almost all the established mycoplasma species, the published full genome sequences of the human pathogenic mycoplasma species [Fraser *et al.*, 1995; Glass *et al.*, 2000; Himmelreich *et al.*, 1996] will accelerate the process of identification of novel target sequences for PCR diagnostics. Selection of a variety of target sequences, starting with highly conserved regions of the genes, allowed design of primers of wide specificity ("universal primers") (Table 2) for detection of mycoplasmal infections in anatomic sites where at least 2 or 3 species are frequently found. The use of a single Mollicutes universal primer set in cases of life-threatening infections has the advantage of allowing a rapid positive or negative report to clinicians, and in turn to establish as soon as possible the appropriate treatment [Razin, 1994]. The approach of using *Mollicutes*-specific and *Ureaplasma* spp-specific universal primers allowed better discrimination between organisms of the *Mycoplasma* and *Ureaplasma* genus, and subsequent identification by species-specific primers in urine specimens from HIV-infected patients [Díaz-García *et al.*, 2004].

Targets	Applications	References
16S rRNA gene sequence		
Conserved regions of mycoplasmal 16S rRNA genes.	Screening for any mycoplasma species in clinical specimens and cell cultures.	van Kuppeveld *et al.*, 1992; van Kuppeveld *et al.*, 1994; Yoshida *et al.*, 2001.
Variable regions of mycoplasmal 16S rRNA genes:	Species-specific detection.	Blanchard *et al.*, 1993a; Grau *et al.*, 1994; van Kuppeveld *et al.*, 1992.
The 16S-23S intergenic regions	Detection of cell culture contamination.	Harasawa *et al.*, 1993.
Mycoplasmal protein genes		
P1 adhesin gene: *M. pneumoniae*	Selective detection and typing	Bernet *et al.*, 1989
MgPa adhesin gene: *M. genitalium*	Selective detection	Palmer *et al.*, 1991; Jensen *et al.*, 1991.
Elongation factor *tuf* gene of *M. pneumoniae*	Selective detection	Lüneberg *et al.*, 1993
Ureasa genes: *Ureaplasma spp.*	Genus-specific detection	Blanchard *et al.*, 1993b
tet M gene (tetracycline-resistance determinant)	Identification of tetracycline-resistant strains	Blanchard *et al.*, 1992
Mba gene	Species-specific detection and typing.	Kong *et al.*, 1996
Repetitive genomic sequences		
Is-like elements: *M. fermentans*	Selective detection	Wang *et al.*, 1992
Rep elements of P1: *M. pneumoniae*	Selective detection	Ursi *et al.*, 1992

Table 2. Nucleic acid sequences suitable for PCR-based mycoplasma testing
Adapted from: Razin, 1994.

When differentiation of the mycoplasmas is required, a multiplex PCR system consisting of a universal set of primers along with primer sets specific for the mycoplasma species commonly involved in a given disease process can be successfully applied [Razin, 2002, Choppa *et al.*, 1998]. Moreover, both conserved and variable regions within the mycoplasmal 16S rRNA genes can also be selected for detection at cluster-, genus- species-, subspecies-, biovar- or serovar-specific levels [Kong *et al.*, 2000; Razin, 2002].

For diagnostic purposes in mycoplasmology, the nucleic acid tests are more sensitive than culture, and showing a fair to good correlation with serology. PCR testing for species-specific mycoplasmal infection are suitable for both urogenital and respiratory samples [Povlsen *et al.*, 2001, 2002]. Interestingly, sample processing prior amplification must be optimized depending of the type of specimen to overcome the presence of undefined inhibitory substances for DNA polymerases, avoiding false negative results. For example, nasopharyngeal samples have a higher rate of PCR inhibition than throat swabs. In general, results obtained by means of NATs will be as good as the quality of the nucleic acid used for the test [Mothersehed & Whitney, 2006; Maeda *et al.*, 2004].

Early in the past decade, Loens *et al.*, 2003b, stated that the development and application of new nucleic acid tests (NATs) in diagnostic mycoplasmology required proper validation and standardization, and performance of different NATs must be compared with each other in order to define the most sensitive and specific tests. The NATs have demonstrated their potential to produce rapid, sensitive and specific results, and are now considered the methods of choice for direct detection of *M. pneumoniae, M. genitalium,* and *M. fermentans* [Cassell *et al.*, 1994a; Loens *et al.*, 2003b]. There is a great variation in methods used from study to study, including variability of target gene sequences (P1, 16S RNA, ATPase, *tuf*), assay format (single, multiplex) or technologies (end-point PCR, Real-time PCR, NASBA) [Loens *et al.*, 2003a, 2003b; 2010]. Also, target DNA has been obtained from different specimens, such as sputum, nasopharyngeal or pharyngeal swabs, brochoalveolar lavages or pleural fluid, and then comparisons of performance between these assays are difficult. For comprehensive understanding of the use of NATs for the detection of *M. pneumoniae,* genital mycoplasmas and other respiratory pathogens in clinical specimens, see the reviews done by Ieven, 2007; ; Lo & Kam, 2006; Loens *et al.*, 2003b, 2010.

As with any other diagnostic test, PCR assays designed for mycoplasma detection in the clinical setting offer several advantages over other non-molecular tests, but still have several drawbacks to take into account (Table 3). Notwithstanding, there are several primer sets that have been successfully applied for diagnosis of mycoplasmal diseases in humans (Table 4).

4. Importance of the specimen collection and processing

Clinical specimens must be collected with use of strict aseptic techniques from anatomic sites likely to yield pathogenic microorganisms [Taylor, 1998; Wilson, 1996]. In the case of mycoplasmal infections, these are clinically silent or covert, thus it is important to differentiate between asymptomatic carriage and disease. In this context, sampling of representative diseased body sites is critical for successful diagnosis.

The usefulness of a PCR assay for diagnostic purposes is rather limited; this is partially explained by the presence of inhibitory substances in complex biological samples, which then provoke a significant reduction or even blockage of the amplification activity of DNA polymerases in comparison with that obtained with the use of pure solutions of nucleic acids. This in consequence affects the performance and the analytical sensitivity of the PCR assays [Lo & Kam, 2006; Vaneechoute & Van Eldere, 1997].

Advantages	*Disadvantages*
• Overcomes the need for mycoplasma cultivation	• Presence of undefined inhibitors of DNA polymerases may yield false-negative results.
• Emission of results is faster than culture (less than 24 h vs. up to 5 weeks)	• Upon detection, PCR poorly discriminates between disease or carriage
• Allows detection of antibiotic-inhibited or uncultivable species in clinical specimens	• Risk of false-positive results due to carryover contamination with amplicons from previous reactions.
• Selective detection. Presence of nucleic acids from the host or from other microorganisms usually do not affect PCR results	• Setting-up quantitative determination of bacteria in clinical specimens may be a very complicated task.
• Higher sensitivity than other non-molecular diagnostic assays (culture, serology).	• Performance of PCR assays for routine diagnostic purposes in microbiology laboratories is still complex and expensive.
• Use as an epidemiological tool since it allows detection of asymptomatic carriers.	• Skilled personnel are required to carry out tests and analysis of results.
• Allows detection of mycoplasmas at the level of Family, Genus, Species, Subspecies and/or Type.	• Depending on the target sequences used, cross reactivity with closely related bacteria may occur.

Table 3. Considerations for using PCR assays in diagnosis of mycoplasma infections. Adapted from: Razin 1994.

Due to the above mentioned, improvement of PCR assays for routine diagnostic purposes clearly should begin with optimal processing of clinical specimens prior to amplification reaction, thus successful amplification of the target DNA sequence can be obtained in the context of trace amounts of sample-associated inhibitory substances [Horz *et al*, 2010; Lo & Kam, 2006; Rådström *et al.*, 2004; Vaneechoute & Van Eldere, 1997].

Group or species	Primer sets.		Sequence (5′→3′)	Target	Amplicon size (bp)	Refs.
Mollicutes-specific	Sen: Antisen:	GPO-1 MGSO	ACT CCT ACG GGA GGC AGC AGT A TGC ACC ATC TGT CAC TCT GTT AAC CTC	16S rDNA	715	van Kuppeveld *et al.*, 1992
	Sen: Antisen:	My-Ins MGSO	GTAATACATAGGTCGCAAGCGTTATC TGC ACC ATC TGT CAC TCT GTT AAC CTC	16S rDNA	520	Yoshida *et al.*, 2001
M. fermentans	Sen: Antisen: IP:	RW005 RW004 RW006	GGT TAT TCG ATT TCT AAA TCG CCT GGA CTA TTG TCT AAA CAA TTT CCC GCT GTG GCC ATT CTC TTC TAC GTT	Insertion sequence-like element	206	Wang *et al.*, 1992
	Sen: Antisen: IP:	Mf-1 Mf-2 GPO-1	GAA GCC TTT CTT CGC TGG AG ACA AAA TCA TTT CCT ATT CTG TC ACT CCT ACG GGA GGC AGC AGT A	rDNA 16s	272	van Kuppeveld *et al.*, 1992
M. genitalium	Sen: Antisen: IP:	MGS-1 MGS-2 MGS-I	GAG CCT TTC TAA CCG CTG C GTG GGG TTG AAG GAT GAT TG AAG CAA CGT AGT AGC GTG AGC	MgPa Adhesin gene	673	de Barbeyrac *et al.*, 1993
	Sen: Antisen: IP:	MGS-1 MGS-4 MGS-I	GAG CCT TTC TAA CCG CTG C GTT GTT ATC ATA CCT TCT GAT AAG CAA CGT AGT AGC GTG AGC	MgPa Adhesin gene	371	de Barbeyrac *et al.*, 1993
M. hominis	Sen: Antisen: IP:	MYCHOMP MYCHOMN MYCHOMS	ATA CAT GCA TGT CGA GCG AG CAT CTT TTA GTG GCG CCT TAC CGC ATG GAA CCG CAT GGT TCC GTT G	16s rDNA	170	Grau *et al.*, 1994
	Sen: Antisen: IP:	Mh-1 Mh-2 GPO-1	TGA AAG GCG CTG TAA GGC GC GTC TGC AAT CAT TTC CTA TTG CAA A ACT CCT ACG GGA GGC AGC AGT A	16s rDNA	281	van Kuppeveld *et al.*, 1992
	Sen: Antisen:	RNAH1 RNAH2	CAATGGCTAATGGCCGGATACGC GGTACCGTCAGTCTGCAAT	16S rDNA	334	Blanchard *et al.*, 1993b
M. penetrans	Sen: Antisen: IP:	MYCPENETP MYCPENETN MYCPENETS	CAT GCA AGT CGG ACG AAG CA AGC ATT TCC TCT TCT TAC AA CAT GAG AAA ATG TTT AAA GTC TGT TTG	16s rDNA	407	Grau *et al.*, 1994

Group or species	Primer sets.		Sequence (5′→3′)	Target	Amplicon size (bp)	Refs.
M. pneumoniae	Sen: Antisen: IP:	MP5-1 MP5-2 MP5-4	GAA GCT TAT GGT ACA GGT TGG ATT ACC ATC CTT GTT GTA AGG CGT AAG CTA TCA GCT ACA TGG AGG	Unknown gene	144	Bernet *et al.*, 1989
	Sen: Antisen: IP:	P1-F P1-R PI-P	GCC ACC CTC GGG GGC AGT CAG- GAG TCG GGA TTC CCC GCG GAG G CTG AAC GGG GGC GGG GTG AAG G-	P1 adhesin gene	209	Ieven *et al.*, 1996
	Sen: Antisen: IP:	16S-F 16S-R 16S-P	AAG GAC CTG CAA GGG TTC GT CTC TAG CCA TTA CCT GCT AA ACT CCT ACG GGA GGC AGC AGT A	16S rDNA	277	Ieven *et al.*, 1996
	Sen: Antisen: IP:	MP-P11 MP-P12 MP-I	TGC CAT CAA CCC GCG CTT AAC CCT TTG CAA CTG CTC ATA GTA CAA ACC GGG CAG ATC ACC TTT	P1 Adhesin gene	466	De Barbeyrac *et al.*, 1993
Ureaplasma spp.	Sen: Antisen: IP:	U5 U4 U9	CAA TCT GCT CGT GAA GTA TTA C ACG ACG TCC ATA AGC AAC T GAG ATA ATG ATT ATA TGT CAG GAT CA	Urease locus	429	Blanchard *et al.*, 1993a
	Sen: Antisen: IP:	Uu-1 Uu-2 UUSO	TAA ATG TCG GCT CGA ACG AG GCA GTA TCG CTA GAA AAG CAA C CAT CTA TTG CGA CGC TA	16s rDNA	311	van Kuppeveld *et al.*, 1992
U. parvum *U. urealyticum*	Sen: Antisen:	UMS-125 UMA-226-	GTA TTT GCA ATC TTT ATA TGT TTT CG CAG CTG ATG TAA GTG CAG CAT TAA ATT C	*mba* gene	402,403 (Up) 443 (Uu)	Kong *et al.*, 2000

bp, Base pairs; Sen, sense or downstream; Antisen, antisense or upstream; IP, internal probe.

Table 4. Primer sets used for end-point PCR detection of mycoplasmas in clinical specimens

Under certain conditions PCR detection/identification/confirmation of mycoplasmas could be attempted from culture broths used for primary isolation, whether or not it have bacterial growth. According to broth turbidity boiling of small aliquots can be sufficient to release the DNA, but presence of precipitated material may inhibit the amplification assay.

In our experience, an alkaline shift around pH 8 frequently results in bacterial lysis, mainly of ureaplasmas, therefore concentration of insoluble material by ultracentifugation prior to DNA extraction is unproductive. This is due to spontaneous release of mycoplasmal DNA that easily dissolves in the aqueous phase and cannot be sedimentated by centrifugation. In such cases, one can take advantage of the alkaline condition to precipitate the dissolved DNA by adding one tenth of 1M NaCl and twice the volume of cold 100% ethanol, and proceed with conventional DNA extraction protocols (unpublished data).

4.1 Exudates and secretions

These types of specimens are fluids closely associated with mucosal surfaces, in low quantities, so collection should be done with the aid of swabs, cytological brush or small syringes. Secretions and exudates can be taken from upper respiratory airways and from lower genital tract, and exceptionally from surgical wounds [Waites, 2006].

4.1.1 Respiratory tract

Respiratory *M. pneumoniae* infection can be assessed by culture and PCR in nasopharyngeal and oropharyngeal secretions, sputa, bronchoalveolar lavage and lung tissue obtained by biopsy. There are reports that nasopharyngeal and oropharyngeal specimens are equally effective for detection of *M. pneumoniae* by PCR, although it is desirable that both sites are screened in parallel for better diagnostic yield [Waites *et al.*, 2008].

When neonatal mycoplasmal infections are suspected, endotracheal, nasopharyngeal and throat secretions are appropriate to evaluate respiratory infection., though specimens for culture should be transported quickly to laboratory since they are likely to contain at least a few contaminating microorganisms [Waites *et al.*, 2005].

Presence of mucous material in this kind of specimens frequently hampers appropriate processing for culture or PCR. Use of aggressive mucolytic agents (NaOH, n-acetyl-cisteine) can damage as well the mycoplasma cells, thus thorough homogenization by wide-bore pippeting is required prior to culture attempt. For nucleic acid extraction, addition of starch has been of help to enhance recovery of total genomic DNA from sputum samples [Harasawa *et al.*, 1993]. In other study, dithiotreitol was used as the mucolytic agent without any apparent detrimental effect on mycoplasmal DNA integrity [Raty *et al.*, 2005].

It this worthy to note that differential sample preparation from the same specimen may be necessary when testing separate single-species PCRs on BAL, as described by [de Barbeyrac *et al.*, 1993]. In that report, freeze-thawing cycles were applied for sample preparation for *M. genitalium* detection, while standard DNA extraction was needed for *M. pneumoniae* detection. During a study of Finnish patients with radiologically confirmed pneumonia, [Raty *et al.*, 2005], evidence further supported the notion that selection of the appropriate specimen is crucial for diagnosis of *M. pneumoniae* infection. By means of a *M.*

pneumoniae-specific 16S rDNA PCR, they obtained positive amplification frequencies of 69%, 50% and 37.5% for sputum, nasopharyngeal aspirate and throat swab specimens, respectively.

4.1.2 Urogenital tract

Since genital tract mycoplasmas are closely associated to live epithelial cells, collection of exudates must be avoided and vigorous scraping of epithelia must be done to obtain as many cells as possible. In this case, a higher associated flora is frequently present in the samples; therefore use of transport liquid media (for culture) of buffered solutions (for DNA extraction) is required immediately after sampling.

4.2 Sterile body fluids

Collection of normally sterile body fluids is made through invasive procedures, usually performed by physicians under aseptic conditions [Wilson, 1996]. When specimens are going to be collected through puncture, careful disinfection of the skin spot must be done, this is crucial to both avoid contamination of the specimen with the skin's associated flora and to minimize the risk of introduction of bacteria into patient's body. Clinically, access of mycoplasmas to sterile body sites may be associated with an underlying immune compromise, and probably the bacteria spread from pulmonary or genital infectious foci [Cassell *et al.*, 1994b; Waites & Talkington, 2004]. Ureaplasmas and mycoplasmas should always be sought from synovial fluid when hypogammaglobulinemic patients develop acute arthritis [Waites *et al.*, 2000].

Since mycoplasma-containing body fluids rarely became turbid; these specimens should be concentrated 10-fold by high-speed centrifugation (aprox. 12,000 x g) and immediately resuspended in one tenth of the original supernatant if culture will be performed. Prior to DNA extraction, the resulting pellet can be washed 1-2 times with Hank's balanced salt solution or PBS, pH 7.4.

4.3 Cell-rich fluids and tissues

Unlike normally sterile body fluids, blood and semen are cell-rich fluid specimens, thus processing for culture or PCR is quite different. It is important to note that mycoplasmas have the ability to invade several cell types, including leukocytes and spermatozoa [Andreev *et al.*, 1995; Baseman *et al.*, 1995; Díaz-García *et al.*, 2006; Girón *et al.*, 1996; Jensen *et al.*, 1994; Lo *et al.*, 1993b, Rottem, 2003; Taylor-Robinson *et al.*, 1991; Yavlovich *et al.*, 2004], consequently a high input of cells into culture media may result in a higher probability of detection.

In contrast, when DNA extraction must be performed for PCR assays, depuration of the sample must be done, (i.e. erythrocyte lysis and selective enrichment for leukocytes in blood; density gradient-based purification of spermatozoa). Noteworthy, the average content of leukocyte DNA per milliliter of blood ranges from 32 to 76 μg, therefore surpasses considerably the amount of bacterial DNA in a specimen from an infected subject. [Greenfield & White, 1993], so a high amount of sample DNA should be added to the PCR reaction mixture to raise the chances to detect bacterial target sequences.

In the case of solid tissues, mechanical homogenization is required to release single cells, either for culture or DNA extraction. A challenge for DNA extraction is when tissues have been formalin-fixed and/or paraffin-embedded since there is high risk of DNA damage [Shi *et al.*, 2004].

A summary of the processing of different specimen types for intended mycoplasma detection is depicted in Figure 1.

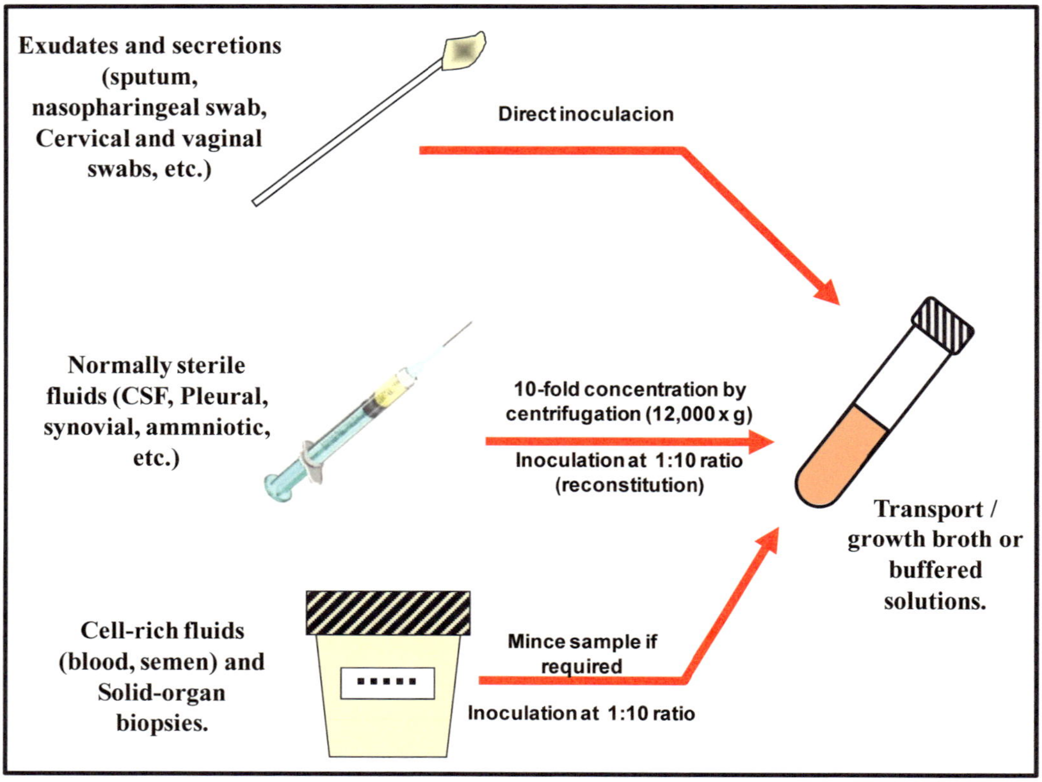

Fig. 1. Differential specimen processing for mycoplasma detection.

5. Culture vs. nucleic acid amplification methods

5.1 The gold standard for mycoplasmal infections

Some genital mycoplasmas, *Ureaplasma spp.* and *M. hominis*, are the fastest growing species among the *Mollicutes*, and due to this, culture-based detection is still the first-line diagnostic approach. However, extrapolating this particular feature to all pathogenic human mycoplasmas is inaccurate. PCR amplification has become essential if fastidious, slow-growing, mycoplasma species are sought in certain clinical conditions, especially in patients with high risk of invasive infections (neonates) or when invasive methods of sampling are required [Waites *et al.*, 2005]. It is well recognized that culture techniques are of poor or null value for detection of some mycoplasma species (i.e. *M. genitalium*) [Razin *et al.*, 1998].

Since culture rely on viability of mycoplasmas to give a positive result, analytical comparisons between culture and PCR invariably will regard the second as less sensitive and less specific, which by any means is wrong. The PCR assay ideally detect target DNA sequences present in the sample, whether it comes from live, dead or uncultivable bacteria [Persing, 1993].

Performance of non culture-based detection tests is frequently evaluated against culture, as the gold standard. Indirect assays measuring serologic responses as correlates of mycoplasma infections, have shown poor sensitivity and required at least 2 serum samples taken several days apart, to be informative [Waites, 2000].

5.2 Culture-enhanced PCR approach

When very few mycoplasma cells are present in a given specimen there is a high probability of obtaining false-negative results, even when the sensitivity of the specific PCR assay is high. To overcome this, several authors have developed culture-based pre-enrichment protocols for mycoplasmas, mycobacteria and *Actinobacillus* [Abele-Horn *et al.*, 1998; Díaz-García *et al.*, 2004; Flemmig *et al.*, 1995; Noussair *et al.*, 2009]. The effects of this procedure are, on one hand the dilution of potential undefined inhibitors, and on the other hand the promotion of short-term bacterial multiplication. This experimental approach has been termed as Culture-enhanced PCR (CE-PCR) [Abele-Horn *et al.*, 1998]. The genomic DNA content in overnight enriched mycoplasma cultures are extracted by standard or commercial techniques, and then subjected to broad-range or species-specific PCR assays. Under this approach, improved detection of *M. pneumoniae* has been achieved in respiratory specimens [Abele-Horn *et al.*, 1998], and of genital mycoplasmas in urine specimens [Díaz-García *et al.*, 2004].

Another culture-based enrichment approach for improvement of PCR detection of mycoplasmas is the cocultivation of these bacteria with permissive immortalized mammalian and/or insect cell lines [Kong *et al.*, 2007; Volokhov *et al.*, 2008]. Although this approach has been design for intentional screening of cell-derived biological and pharmaceutical products, including vaccines and cell culture substrates, it is a potential tool for biological enrichment of normally-sterile clinical specimens such as CFS, sera, synovial fluid, etc.

Interestingly, strains of mycoplasma-free *Trichomonas vaginalis* are readily infected *in vitro* by *M. hominis* isolates, but not by other urogenital mycoplasmas. The infection can be detected by a *M. hominis*-specific PCR assay after long-term incubation, since the mycoplasma can be transmitted between the protozoan cells [Dessi *et al.*, 2006; Rapelli *et al.*, 2001]. The symbiotic interplay between *M. hominis* and *T. vaginalis* has been well established, as well a significant correlation between detection of both microorganisms in vaginal specimens from infected women [Dessi *et al.*, 2006]. Thus it is likely to take advantage of such symbiosis and employ mycoplasma-free *T. vaginalis* cultures for specific enrichment of *M. hominis*-containing clinical specimens prior to PCR detection tests.

6. Commercial molecular diagnostic kits

Unlike the in-house PCR assays for diagnostic purposes, developed by several researchers, the commercial PCR kits are well standardized in terms of sensitivity and specificity, allowing their global use in clinical microbiology laboratories. Thus inter-laboratory performance comparisons of such kits are suitable, including testing of several specimen types. Indeed, according to the *In Vitro Diagnostic Medical Devices Directive* 98/79/EC, all commercial diagnostic kits used in European countries must have the CE (*Conformité Européene*) label [Dosà *et al.*, 1999].

Among commercially available real-time PCR kits are intended for *M. pneumoniae* detection, mainly targeting the P1 cytadhesin gene, including Nanogen Mycoplasma pn Q-PCR Alert kit (Nanogen Advanced Diagnostics); the Simplexa *Mycoplasma pneumoniae* kit (Focus Diagnostics, California); the Diagenode detection kit for *Mycoplasma pneumoniae*/*Chlamydophila pneumoniae* (Diagenode SA, Liège, Belgium); the Cepheid *Mycoplasma pneumoniae* ASR kit (Cepheid, Paris, France), and the Venor Mp-Qp PCR detection kit (Minerva Biolabs GmbH). It has been shown that these commercial kits had acceptable analytical sensitivity and performance with clinical specimens [Touati *et al.*, 2009].

Interestingly, many commercially available extraction kits incorporate a buffer to lyses the bacteria and a silica matrix membrane (typically in column format) to trap the DNA or RNA. Several wash steps are required to remove protein and other macromolecules, and the purified DNA and RNA is then eluted from the membrane. Many of the manual extraction methods require several centrifugation steps. To reduce hands-on time, operator error, and sample contamination, semi-automated DNA or RNA extraction kits and equipment have been designed and are commercially available.

7. PCR, sequencing, phylogeny and molecular epidemiology

The mycoplasmas may have evolved through regressive evolution from closely related Gram positive bacteria with low content of guanine plus cytosine (G+C), probably the Clostridia or Erysipelothrix [Bove, 1993; Brown *et al.*, 2007; Razin *et al.*, 1998]. The massive gene losses (i.e. genes involved in cell wall and aminoacid biosynthesis) had left mycoplasmas with a coding repertoire of 500 to 2000 genes [Sirand-Pugnet *et al.*, 2007]. The G+C content in DNA of mycoplasmas varies from 23 to 40 mol%, while genome size range is 580–2200 Kbp, much smaller than those of most walled bacteria [Razin *et al.*, 1998].

After PCR amplification and sequencing of the conserved 16S rDNA gene sequences from representative members of the Mollicutes, the resulting phylogenetic tree was shown to be monophyletic, arising from a single branch of the Clostridium ramosum branch [International Committee on Systematics of Prokaryotes- Subcommittee on the taxonomy of Mollicutes (ICSP-STM), 2010]. The Mollicutes split into two major branches: the AAP branch, containing the *Acholeplasma, Anaeroplasma* and *Asteroleplasma* genera, and the Candidatus *Phytoplasma* phyla; the other is the SEM branch that includes the *Spiroplasma, Entomoplasma, Mesoplasma, Ureaplasma* and *Mycoplasma* genera [Johansson *et al.*, 1998; Maniloff, 1992; Razin *et al.*, 1998]. Interestingly, the genus Mycoplasma is polyphyletic, with species clustering within the Spiroplasma, Pneumoniae and Hominis phylogenetic groups [Behbahani *et al.*, 1993; Johansson *et al.*, 1998; Maniloff, 1992]. Nevertheless, additional phylogenetic markers such as the elongation factor EF-Tu (tuf) gene, ribosomal protein genes, the 16S-23S rRNA intergenic sequences, etc, have been already used as complementary comparative data, thus there is no unique phylogenetic tree for Mollicutes [Razin *et al.*, 1998].

There are several in-house species-specific end-point or real-time PCR assays developed to detect mycoplasmas in diverse respiratory and urogenital tract infections [Blanchard *et al.*, 1993b; Loens *et al.*, 2003b; Sung *et al.*, 2006; van Kuppeveld *et al.*, 1992; Wang *et al*, 1992]. Of those clinical entities, more than one mycoplasma species are commonly associated as etiologic agents, i.e. urethritis, infertility, pelvic inflammatory disease, etc. [Cassell *et al*, 1994b; Taylor-Robinson, 1996]. Thus, simultaneous testing of several species-specific or multiplex PCRs to determine all possible pathogenic mycoplasmas associated with a particular clinical entity would be very complicated. Combination of PCR amplification of a given highly conserved target genome sequence with determination of its nucleotide sequences and phylogenetic analysis has been successfully applied for diagnosis and identification of mycoplasmal etiologies in male urethritis cases [Hashimoto *et al.*, 2006; Yoshida *et al.*, 2002].

Due to their fastidious growth conditions and frequent cross-reactive antigenic profile, identification and typing of human mycoplasmas is a very difficult task. Other approaches termed "Random Amplified Polymorphic DNA" (RAPD) or "Arbitrarily Primed PCR" (AP-PCR), and "Amplified-Fragment Length Polimorphism" (AFLP), are PCR-based typing methods used for intra- and inter-species differentiation of mycoplasma isolates. The RAPD / AP-PCR method involves PCR amplification with a single arbitrary primer at low stringency, while AFLP method selectively amplifies restriction fragments from whole genome. These PCR-based genotyping techniques have allowed faster and reproducible typing of mycoplasmas for epidemiologic studies [Cousin-Allery *et al.*, 2000; Geary & Forsyth, 1996; Grattard *et al.*, 1995; Iverson-Cabral, *et al.*, 2006; Kokotovic *et al.*, 1999; Rawadi, 1998; Schwartz *et al.*, 2009].

8. Conclusion

In today's clinical microbiology laboratory, introduction of PCR and other NATs has the potential to increase the speed and accuracy of bacterial detection/identification, especially of those fastidious microorganisms such as mycoplasmas. However, those molecular assays still have serious drawbacks that arise from inadequate acquisition, handling and processing of representative clinical specimens. False negative results ultimately can have a significant impact on patient management.

It is widely accepted that molecular methods are more sensitive and specific than culture- and serology-based diagnostic approaches but, what does a "positive" test result mean clinically?. This issue is a matter of controversy for genital mycoplasmas since the duality of their relationship with their host: Is it a commensal or is it a pathogen?. The answer depends of an integral clinical evaluation of patients, where a "signs and symptoms"-focused sampling will improve laboratory diagnosis.

In the clinical setting, when negative results after mycoplasma-specific PCR assays are reported, the type and quality of the specimen, history of antibiotic treatment of the patient, and how representative was the specimen used for the assay, should be taken into account.

Therefore, any set of diagnostic results must be reviewed and critically interpreted before diagnosis and intervention measures are made.

9. Acknowledgements

The authors want to give a very special thanks to the Instituto Nacional de Perinatología-SS, Grant; 2122250-077261 and the Facultad de Medicina, UNAM, for their support.

SFM is a Doctoral Student. Postgraduate Program; Molecular Biomedicine and Biotechnology at the Escuela Nacional de Ciencias Biológicas, IPN, México, D.F.; FJDG has been granted with the National Research Fellowship I of the CONACyT, México, D.F.

Corresponding author: M. Sc. Saúl Flores-Medina. Laboratorio de Biología Molecular, Departamento de Infectología, INPer. Montes Urales # 800, Col. Lomas de Virreyes. C.P. 11,000, México, D.F., México. Phone: +52(55)55209900 ext. 520. Email: s.flores@inper.mx

10. References

Abele-Horn, M., Busch, U., Nitschko, H., Jacobs, E., Bax, R., Pfaff, F., Schaffer, B., & Heesemann, J. (1998). Molecular Approaches to Diagnosis of Pulmonary Diseases Due to *Mycoplasma pneumoniae*. *Journal of Clinical Microbiology*, Vol. 36, No. 2 (February 1998), pp. 548-551, ISSN 0095-1137

Andreev, J., Borovsky, Z., Rosenshine, I., & Rottem, S. (1995). Invasion of HeLa Cells by *Mycoplasma penetrans* and the Induction of Tyrosine Phosphorylation of a 145 kDa Host Cell Protein. *FEMS Microbiology Letters*, Vol. 132, No. 3, (October 1995), pp. 189-194, ISSN 0378-1097

Atkinson, T.P., Balish, M.F., & Waites, K.B. (2008). Epidemiology, Clinical Manifestations, Pathogenesis and Laboratory Detection of *Mycoplasma pneumoniae* Infections. *FEMS Microbiology Reviews*, Vol. 32, No. 6, (November 2008), pp. 956–973, ISSN 0168-6445

Baseman, J.B., & Tully, J.G. (1997). Mycoplasmas: Sophisticated Reemerging and Burdened by Their Notoriety. *Emerging Infectious Diseases*, Vol. 3, No. 1, (January-March 1997), pp. 21-32, ISSN 1080-6040

Baseman, J.B., Lange, M., Criscimagna, N.L., Girón, J.A., & Thomas, C.A. (1995). Interplay Between *Mycoplasma*s and Host Target Cells. *Microbial Pathogenesis*, Vol. 19, No. 2, (August 1995), pp. 105-116, ISSN 0882-4010

Behbahani, N., Blanchard, A., Cassell, G.H., & Montagnier, L. (1993). Phylogenetic Analysis of *Mycoplasma penetrans*, Isolated From HIV-Infected Patients. *FEMS Microbiology Letters*, Vol. 109, No. 1, (May 1993), pp. 63-6, ISSN 0378-1097

Bernet, C., Garret, M., de Barbeyrac, B., Bebear, C., & Bonnet, J. (1989). Detection of *Mycoplasma pneumoniae* by Using Polymerase Chain Reaction. *Journal of Clinical Microbiology*, Vol. 27, No. 11, (November 1989), pp. 2492-2496, ISSN 0095-1137

Blanchard A, Crabb DM, Dybvig K, Duffy LB, Cassell GH. (1992). Rapid Detection of tetM in *Mycoplasma hominis* and *Ureaplasma urealyticum* by PCR: *tetM* Confers Resistance to Tetracycline But not Necessarily to Doxycycline. *FEMS Microbiology Letters*, Vol. 74, No. 2-3 (August 1992): pp. 277-281, ISSN 0378-1097

Blanchard A, Hentschel J, Duffy L, Baldus K, Cassell GH. (1993a). Detection of *Ureaplasma urealyticum* by Polymerase Chain Reaction in the Urogenital Tract of Adults, in Amniotic Fluid, and in the Respiratory Tract of Newborns. *Clinical Infectious Diseases,* Vol. 17, Suppl. 1 (August 1993), pp. S148-S153, ISSN 1058-4838

Blanchard, A., Yáñez, A., Dybvig, K., Watson, H.L., Griffiths, G., & Cassell, G.H. (1993b). Evaluation of Intraspecies Genetic Variation Within The 16S rRNA Gene of *Mycoplasma hominis* and Detection by Polymerase Chain Reaction. *Journal of Clinical Microbiology,* Vol. 31, No. 5, (May 1993), pp. 1358-1361, ISSN 0095-1137

Bové, J.M. (1993). Molecular Features of *Mollicutes. Clinical Infectious Diseases,* Vol. 17, Suppl 1, (August 1993), pp. S10-S31, ISSN 1058-4838

Brown, D.R., Whitcomb, R.F., & Bradbury, J.M. (2007). Revised Minimal Standards for Description of New Species of the Class *Mollicutes* (Division Tenericutes). *International Journal of Systematic and Evolutionary Microbiology,* Vol. 57, No. 11, (November 2007), pp. 2703–2719, ISSN 1466-5026

Cassell, G.H., Blanchard, A., Duffy, L., Crabb, D., & Waites, K.B. (1994a). *Mycoplasmas*, In: *Clinical and Pathogenic Microbiology,* B.J. Howard, J.F., Keiser, A.S. Weissfeld, T.F. Smith, & R.C. Tilton (Eds.), 491-502, Mosby, ISBN 978-0801664267, Boston, USA.

Cassell, G.H., Waites, K.B., Watson, H.L., Crouse, D.T., & Harasawa, R. (1993). *Ureaplasma urealyticum* Intrauterine Infection: Role in Prematurity and Disease in Newborns. *Clinical Microbiology Reviews,* Vol. 6, No. 1, (January-March 1993), pp. 69-87, ISSN 0893-8512

Cassell, G.H., Yáñez, A., Duffy, L. B., Moyer, J., Cedillo, L., Hammerschlag, M.R., Rank, R.G., & Glass, J.I. (1994b). Detection of *Mycoplasma fermentans* in the Respiratory Tract of Children with Pneumonia. 10th International Congress of the International Organization for Mycoplasmology (IOM), Bordeaux, France, July 1994. In *IOM Letters* Vol, 3, p. 456, ISSN 1023-1226

Choppa, P.C., Vojdani, A., Tagle, C., Andrin, R., & Magtoto, L. (1998). Multiplex PCR for the Detection of *Mycoplasma fermentans, M. hominis* and *M. penetrans* in Cell Cultures and Blood Samples of Patients With Chronic Fatigue Syndrome. *Molecular and Cellular Probes,* Vol. 12, No. 5, (October 1998), pp. 301-308, ISSN 0890-8508

Cousin-Allery, A., Charron, A., de Barbeyrac, B., Fremy, G., Jensen, J.S., Renaudin, H., & Bebear, C. (2000). Molecular Typing of *Mycoplasma pneumoniae* Strains by PCR-Based Methods and Pulse-Field Gel Electrophoresis. Application to French and Danish Isolates. *Epidemiology and Infection,* Vol. 124, No. 1, (February 2000), pp. 103-111, ISSN 0950-2688

de Barbeyrac, B., Bernet-Poggi, C., Febrer, F., Renaudin, H., Dupon, M., & Bebear, C. (1993). Detection of *Mycoplasma pneumoniae* and *Mycoplasma genitalium* in Clinical Samples by Polymerase Chain Reaction. *Clinical Infectious Diseases,* Vol. 17, Supp 1 (August 1993), pp. S83-S89, ISSN 1058-4838

Dessì, D., Delogu, G., Emonte, E., Catania, M.R., Fiori, P.L., & Rappelli, P. (2005). Long-Term Survival and Intracellular Replication of *Mycoplasma hominis* in *Trichomonas vaginalis* Cells: Potential Role of the Protozoon in Transmitting Bacterial Infection. *Infection and Immunity,* Vol. 73, No. 2 (February 2005), pp. 1180–1186, ISSN 0019-9567

Dessi, D., Rappelli, P., Diaz, N., Cappuccinelli, P., & Fiori, P.L. (2006). *Mycoplasma hominis* and *Trichomonas vaginalis*: A Unique Case of Symbiotic Relationship Between Two Obligate Human Parasites. *Frontiers in Bioscience*, Vol. 11, (September 2006), pp. 2028-2034, ISSN 1093-9946

Díaz-García, F.J., Giono-Cerezo, S., Tapia, J.L., Flores-Medina, S., López-Hurtado, M., & Guerra-Infante, F.M. (2004). Overnight Enrichment Culture Improves PCR-Based Detection of Genital Mycoplasmas in Urine Samples. 11th. ICID Abstracts. *International Journal of Infectious Diseases*, Vol. 8, Suppl. 1, (March 2004), pp. S130, ISSN 1201-9712

Díaz-García, F.J., Herrera-Mendoza, A.P., Giono-Cerezo, S., & Guerra-Infante, F. (2006). *Mycoplasma hominis* Attaches to and Locates Intracellularly On Human Spermatozoa. *Human Reproduction,* Vol. 21, No. 6, (June 2006), pp. 1591-1598, ISSN 0268-1161

Dosá, E., Nagy, E., Falk, W., Szöke, I., & Ballies, U. (1999). Evaluation of the Etest for susceptibility testing of *Mycoplasma hominis* and *Ureaplasma urealyticum*. *The Journal of Antimicrobial Chemotherapy,* Vol. 43, No. 4, (April 1999), pp. 575-578, ISSN 0305-7453

Flemmig, T.F., Rüdiger, S., Hofmann, U., Schmidt, H., Plaschke, B., Strätz, A., Klaiber, B., & Karch, H. (1995). Identification of *Actinobacillus actinomycetemcomitans* in Subgingival Plaque by PCR. *Journal of Clinical Microbiology,* Vol. 33, No. 12 (December1995), pp. 3102-3105, ISSN 0095-1137

Fraser, C.M., Gocayne, J.D., White, O., Adams, M.D., Clayton, R.A., Fleischmann, R.D.,*et al.* (1995). The minimal gene complement of *Mycoplasma genitalium*. *Science,* Vol. 270, No. 5235, (October 1995), pp. 397-404, ISSN 0036-8075

Geary, S.J., & M.H. Forsyth. (1996). PCR: Random Amplified Polymorphic DNA Fingerprinting. In: *Molecular and Diagnostic Procedures in Mycoplasmology,* J.G. Tully & S. Razin (Eds.), 81-85, Diagnostic procedures. Academic Press, ISBN 0125838069, San Diego, Ca.

Girón, J.A., Lange, M., & Baseman, J.B. (1996). Adherence, Fibronectin Binding, and Induction of Cytoskeleton Reorganization in Cultured Human Cells by *Mycoplasma penetrans*. *Infection and Immunity,* Vol. 64, No. 1, (January 1996), pp. 197-208, ISSN 0019-9567

Glass, J.I., Lefkowitz, E.J., Glass, J.S., Heiner, C.R., Chen, E.Y., & Cassell, G.H. (2000). The Complete Sequence of the Mucosal Pathogen *Ureaplasma urealyticum*. *Nature,* Vol. 407, No. 6805, (October 2000), pp. 757-62, ISSN 0028-0836

Grattard, F., Pozzetto, B., de Barbeyrac, B., Renaudin, H., Clerc, M., Gaudin, O.G., & Bébéar C. (1995). Arbitrarily-Primed PCR Confirms the Differentiation of Strains of *Ureaplasma urealyticum* Into Two Biovars. *Molecular and Cellular Probes,* Vol. 9, No. 6, (December 1995), pp. 383-389, ISSN 0890-8508

Grau, R., Kovacic, R., Griffais, R., Launay, V., & Montagnier, L. (1994). Development of PCR-Based Assays for the Detection of Two Human Mollicute species, *Mycoplasma penetrans and M. hominis*. *Molecular and Cellular Probes,* Vol. 8, No. 2 (April 1994), pp. 139–148. ISSN 0890-8508

Greenfield, L., & White, T.J. (1993). Sample Preparation Methods. In: *Diagnostic Molecular Microbiology, Principles and Applications,* D. H. Persing, T. F. Smith; F. C. Tenover, &

T. White (Eds.), pp. 122-137. American Society for Microbiology, ISBN 1-55581-056-X, U.S.A.

Harasawa R., Misuazawa H., & Nakagawa.T. (1993). Detection and Tentative Identification of Dominant Mycoplasma Species in Cell Cultures by Restriction Analysis of the 16S-23S rRNA Intergenic Spacer Regions. *Research in Microbiology*, Vol. 144, No. 6, (July-August 1993), pp. 489-493, ISSN 0923-2508

Hashimoto, O., Yoshida, T., Ishiko, H., Ido, M., & Deguchi, T. (2006). Quantitative Detection and Phylogeny-Based Identification of Mycoplasmas and Ureaplasmas from Human Immunodeficiency VirusType 1-Positive Patients. *Journal of Infection and Chemotherapy*, Vol. 12, No. 1, (February 2006), pp. 25-30, ISSN 1341-321X

Himmelreich R., Hilbert, H., Plagens, H., Pirkl, E., Li, B.C., Herrmann, R. (1996). Complete Sequence Analysis of the Genome of the Bacterium *Mycoplasma pneumoniae. Nucleic Acids Research*, Vol. 24, No. 22, (November 1996), pp. 4420-4449. ISSN: 0305-1048

Horz, H-P., Scheer, S., Vianna, M.E., & Conrads G. (2010). New Methods for Selective Isolation of Bacterial DNA from Human Clinical. *Anaerobe*, Vol. 16, No. 1, (February 2010), pp. 47-53, ISSN 2009.04.009

Ieven, M. (2007). Currently Used Nucleic Acid Amplification Tests for the Detection of Viruses and Atypicals in Acute Respiratory Infections. *Journal of Clinical Virology*, Vol.40, No.4, (December 2007), pp. 259-276, ISSN 1386-6532

Ieven, M., Ursi, D., Van Bever, H., Quint, W., Niesters, H.G., & Goossens, H. (1996). Detection of *Mycoplasma pneumoniae* by Two Polymerase Chain Reactions and Role of *M. pneumoniae* in Acute Respiratory Tract Infections in Pediatric Patients. *Journal of Infectious Diseases*, Vol. 173, No. 6 (June 1996), pp. 1445-1452, ISSN 0022-1899

International Committee on Systematics of Prokaryotes- Subcommittee on the taxonomy of Mollicutes (ICSP-STM). (2011). *International Journal of Systematic and Evolutionary Microbiology*, Vol. 61, No. 3 (March 2011), pp. 695–697, ISSN 1466-5026

Iverson-Cabral, S.L., Astete, S.G., Cohen, C.R., Rocha, E.P., & Totten, P.A. (2006). Intrastrain Heterogeneity of the mgpB Gene in *Mycoplasma genitalium* is Extensive *in vitro* and *in vivo* and Suggests that Variation is Generated via Recombination With Repetitive Chromosomal Sequences. *Infection and Immunity*, Vol. 74, No. 7, (July 2006), pp. 3715-26, ISSN 0019-9567

Jaffe, J.D., Berg, H.C., & Church, G.M. (2004). Proteogenomic Mapping as a Complementary Method to Perform Genome Annotation. *Proteomics*, Vol. 4, No. 1, (January 2004), pp. 59-77, ISSN 1615-9853

Jensen, J.S., Blom, J., & Lind, K. (1994). Intracellular Location of *Mycoplasma genitalium* in Cultured Vero Cells as Demonstrated by Electron Microscopy. *International Journal of Experimental Pathology*, Vol. 75, No. 2, (April 1994), pp. 91-98, ISSN 0959-9673

Jensen, J.S., Uldum, S.A., Søndergård-Andersen, J., Vuust, J., & Lind, K. (1991). Polymerase Chain Reaction for Detection of *Mycoplasma genitalium* in Clinical Samples. *Journal of Clinical Microbiology*, Vol. 29, No. 1, (January 1991), pp. 46-50. ISSN 0095-1137

Johansson, K.E., Heldtander, M.U.K., & Petterson, B. (1998). Characterization of Mycoplasmas by PCR and Sequence Analysis with Universal 16S rDNA Primers, In: *Mycoplasma protocols*, R. Miles, & R. Nicholas (Eds.), 145-165, Humana Press, ISBN 0-89603-525-5, Totowa, NJ

Kokotovic, B., Friis, N.F., Jensen, J.S., & Ahrens, P. (1999). Amplified-Fragment Length Polymorphism Fingerprinting of *Mycoplasma* Species. J. Clin. Microbiol, Vol. 37, No. 10, (October 1999), pp. 3300-3307, ISSN 0095-1137

Kong, F., Ma, Z., James, G., Gordon, S., & Gilbert, G.L. (2000). Species Identification and Subtyping of *Ureaplasma parvum* and *Ureaplasma urealyticum* Using PCR-Based Assays. *Journal of Clinical Microbiology*, Vol. 38, No. 3 (March 2000), pp. 1175–1179. ISSN 0095-1137

Kong, H., Volokhov, D.V., George, J., Ikonomi, P., Chandler, D., Anderson, C., & Chizhikov, V. (2007). Application of Cell Culture Enrichment for Improving the Sensitivity of Mycoplasma Detection Methods Based on Nucleic Acid Amplification Technology (NAT). Applied Microbiology and Biotechnology, Vol. 77, No. 1 (November 2007), pp. 223-232, ISSN 0175-7598

Kovacic, R., Grau, O., & Blanchard, A. (1996). PCR: Selection of Target Sequences. In: *Molecular and Diagnostic Procedures in Mycoplasmology*, J.G. Tully & S. Razin (Eds.), 53-60, Diagnostic procedures. Academic Press, ISBN 012-583806-9, San Diego, Cal

Lo, A.C.T., & Kam, K.M. (2006). Review of Molecular Techniques for Sexually Transmitted Diseases Diagnosis, In: *Advanced Techniques in Diagnostic Microbiology*, Y-W. Tang, & C.W. Stratton (Ed.), 353-386, ISBN 0387-32892

Lo, S.C., Hayes, M.M., Tully, J.G., Wang, R.Y., Kotani, H., Pierce, P.F., Rose, D.L., & Shih, J.W. (1992). *Mycoplasma penetrans* sp. nov., From the Urogenital Tract of Patients With AIDS. *International Journal of Systematic Bacteriology*, Vol. 42, No. 3 (July 1992), pp. 357-364, ISSN 0020-7713

Lo, S.C., Hayes, M.M., & Kotani, H. Pierce PF, Wear DJ, Newton PB 3rd, Tully JG, Shih JW. (1993a). Adhesion Onto and Invasion Into Mammalian Cells by *Mycoplasma penetrans* - A Newly Isolated Mycoplasma From Patients with AIDS. *Modern Pathology*, Vol. 6, No. 3, (May 1993), pp. 276-280, ISSN 0893-39520

Lo, S.C., Wear DJ, Green SL, Jones PG, Legier JF. (1993b). Adult Respiratory Distress Syndrome With or Without Systemic Disease Associated With Infections Due to *Mycoplasma fermentans*. *Clinical Infectious Diseases*, Vol. 17, Suppl. 1 (August 1993), pp. S259-S263, ISSN 1058-4838

Loens, K., Goossens, H., & Ieven. M. (2010). Acute Respiratory Infection Due to *Mycoplasma pneumoniae*: Current Status of Diagnostic Methods. *European Journal of Clinical Microbiology and Infectious Diseases*, Vol. 29, No. 9, (September 2010), pp. 1055–1069, ISSN 0934-9723

Loens, K., Ieven, M., Ursi, D., Beck, T., Overdijk, M., Sillekens, P., & Goossens, H. (2003a). Detection of *Mycoplasma pneumoniae* by Real-Time Nucleic Acid Sequence-Based Amplification. *Journal of Clinical Microbiology*, Vol. 41, No. 9, (September 2003), pp. 4448-4450, ISSN 0095-1137

Loens, K., Ursi, D., Goossens, H., & Ieven, M. (2003b). Molecular Diagnosis of *Mycoplasma pneumoniae* Respiratory Tract Infections. *Journal of Clinical Microbiology*, Vol. 41, No. 11, (November 2003), pp. 4915-4923, ISSN 0095-1137

Lüneberg, E., Jensen, J.S., & Frosch, M. (1993). Detection of *Mycoplasma pneumoniae* by Polymerase Chain Reaction and Nonradioactive Hybridization in Microtiter Plates.

Journal of Clinical Microbiology, Vol. 31, No. 5, (May 1993), pp. 1088-1094. ISSN: 0095-1137

Maeda, S., Deguchi, T., Ishiko, H., Matsumoto, T., Naito, S., Kumon, H., Tsukamoto, T., Onodera, S., & Kamidono, S. (2004). Detection of *Mycoplasma genitalium, Mycoplasma hominis, Ureaplasma parvum* (Biovar 1) and *Ureaplasma urealyticum* (Biovar 2) in Patients with Non-gonococcal Urethritis Using Polymerase Chain Reaction-Microtiter Plate Hybridization. *International Journal of Urology,* Vol. 11, No. 9, (September 2004), pp. 750-754. ISSN: 0919-8172

Maniloff, J. (1992). Phylogeny of Mycoplasmas. In: *Mycoplasmas: Molecular biology and Pathogenesis,* J. Maniloff, R.N. McElhaney, L.R. Finch , & J.B. Baseman (Eds.), pp. 549-559, American Society for Microbiology, ISBN 1-55581-050-0, U.S.A.

Mendoza N., Ravanfar, P., Shetty, A.K., Pellicane, B.L., Creed, R., Goel, S., & Tyring, S.K. (2011). Genital Mycoplasma Infection, In: *Sexually Transmitted Infections and Sexually Transmitted Diseases,* G. Gross, & S.K. Tyring (Eds.), 197-201, Springer-Verlag, ISBN 978-3-642-14663-3, Heidelberg, Berlin

Mothershed, E.A., & Whitney, A.M. (2006). Nucleic Acid-Based Methods for the Detection of Bacterial Pathogens: Present and Future Considerations for the Clinical Laboratory. *Clinica Chimica Acta,* Vol. 363, No. 1-2, (January 2006), pp. 206–220, ISSN 0009-8981

Noussair L, Bert F, Leflon-Guibout V, Gayet N, Nicolas-Chanoine MH. (2009) Early Diagnosis of Extrapulmonary Tuberculosis by a New Procedure Combining Broth Culture and PCR. *Journal of Clinical Microbiology,* Vol. 47, No. 5 (May 2009), pp. 1452-1457, ISSN 0095-1137

Palmer, H.M., Gilroy, C.B., Furr, P.M., & Taylor-Robinson, D. (1991). Development and Evaluation of the Polymerase Chain Reaction to Detect *Mycoplasma genitalium. FEMS Microbiology Letters,* Vol. 77, No. 2-3, (January 1991), pp. 199-204. ISSN 0378-1097

Patel, M.A., & Nyirjesy, P. (2010). Role of *Mycoplasma* and *Ureaplasma* Species in Female Lower Genital Tract Infections. *Current Infectious Diseases Report,* Vol. 12, No. 6, (November 2010), pp. 417–422, ISSN 1523-3847

Povlsen, K,. Thorsen, P., & Lind, I. (2001). Relationship of *Ureaplasma urealyticum* Biovars to the Presence or Absence of Bacterial Vaginosis in Pregnant Women and to the Time of Delivery. *European Journal of Clinical Microbiology and Infectious Diseases,* Vol. 20, No. 1, (January 2001), pp. 65-67, ISSN 0934-9723

Povlsen, K., Bjørnelius, E., Lidbrink, P., & Lind, I. (2002). Relationship of *Ureaplasma urealyticum* Biovar 2 to Nongonococcal Urethritis. *European Journal of Clinical Microbiology and Infectious Diseases,* Vol. 21, No. 2, (February) pp. 97-101, ISSN: 0934-9723

Rådström, P., Knutsson, R., Wolffs, P., Lövenklev, M. & Löfström, C. (2004). Pre-PCR Processing Strategies to Generate PCR-Compatible Samples. *Molecular Biotechnology,* Vol. 26, No. 2, (February 2004), pp. 133-146, ISSN 1073–6085

Rappelli, P.,· Carta, F., Delogu, G., Addis, M.F.,· Dessì, D., Cappuccinelli, P., & Fiori, P. (2001). *Mycoplasma hominis* and *Trichomonas vaginalis* Symbiosis: Multiplicity of Infection and Transmissibility of *M. hominis* to Human Cells. Archives of Microbiology (2001) 175 :70–74.

Raty, R., Ronkko, E. & Kleemola, M. (2005). Sample Type is Crucial to the Diagnosis of *Mycoplasma pneumoniae* Pneumonia by PCR. *Journal of Medical Microbiology*, Vol. 54, No. (Pt 3), pp. 287–291, ISSN 0022-2615

Rawadi, G.A. (1998). Characterization of Mycoplasmas by RAPD Fingerprinting. In: *Mycoplasma protocols*, R. Miles, & R. Nicholas (Eds.), 179-187, Humana Press, ISBN 0-89603-525-5, Totowa, NJ

Razin, S. (1992). Peculiar Properties of Mycoplasmas: The Smallest Self-Replicating Prokariotes. *FEMS Microbiology Letters*, Vol. 79, No. 1-3, (December 1992) pp. 423-432, ISSN 0378-1097

Razin, S. (1994). DNA Probes and PCR in diagnosis of Mycoplasma Infections. *Molecular and Cellular Probes*, Vol. 8, No. 6, (December 1994), pp. 497-511, ISSN 0890-8508

Razin, S. (2002). Diagnosis of Mycoplasmal Infections. In: *Molecular Biology and Pathogenicity of Mycoplasmas*, S. Razin & R. Herrmann (Eds.), 531-544, Kluwer Academic/Plenum Publishers, ISBN 0-306-47287-2, New York, NY

Razin, S., Yoguev, D., & Naot, Y. (1998). Molecular Biology and Pathogenicity of Mycoplasmas. *Microbiology and Molecular Biology Reviews*, Vol. 62, No. 4, (December 1998), pp. 1094-1156, ISSN 1092-2172

Relman, D.A., & Persing, D.H. (1996). Genotypic Methods for Microbial Identification. In: *PCR Protocols for Emerging Infectious Diseases*, Persing, D.H. (Ed), 3-31, ASM Press, ISBN 1-55581-108-6, Washington, D.C.

Rottem, S. (2003). Interaction of Mycoplasmas with Host Cells. *Physiology Reviews*, Vol. 83, No. 2 (Apr 2003), pp. 417-432, ISSN 0031-9333

Schwartz, S.B., Thurman, K.A., Mitchell, S.L., Wolff, B.J., &Winchell, J.M. (2009). Genotyping of *Mycoplasma pneumoniae* Isolates Using Real-Time PCR and High-Resolution Melt Analysis. *Clinical Microbiology Infection*. Vol. 15, No. 8, (August 2009), pp. 756-62, ISSN 1198-743X

Shi, S-R., Datar, R., Liu, C., Wu, L., Zhang, Z., Cote, R.J., & Taylor, C.R. (2004). DNA Extraction from Archival Formalin-Fixed, Paraffin-Embedded Tissues: Heat-Induced Retrieval in Alkaline Solution. *Histochemistry and Cellular Biology*, Vol. 122, No. 3, (September 2004), pp. 211–218, ISSN 0948-6143

Sirand-Pugnet, P., Citti, C., Barré, A., & Blanchard, A. (2007), Evolution of Mollicutes: Down a Bumpy Road with Twists and Turns. *Research in Microbiology*, Vol. 158, No. 10, (December 2007), pp. 754-766, ISSN 0923-2508

Sung, H., Kang, S.H., Bae, Y.J., Hong, J.T., Chung, Y.B., Lee, C.-K., & Song, S. (2006). PCR-Based Detection of Mycoplasma Species. *The Journal of Microbiology*, Vol. 44, No. 1, (February 2006), pp. 42-49, ISSN 1225-8873

Talkington, D.F., & Waites, K.B. (2009). *Mycoplasma pneumoniae* and Other Human Mycoplasmas, In: *Bacterial Infections of Humans*, A.S. Evans, & P.S., Brachman (Eds.), 519-541, Springer-Science+Business, ISBN 978-0-387-09843-2, New York, NY.

Taylor, P. (1998). Recovery of Human Mycoplasmas, In: *Mycoplasma protocols*, R. Miles, & R. Nicholas (Eds.), 25-35, Humana Press, ISBN 0-89603-525-5, Totowa, NJ

Taylor-Robinson, D. (1996). Infections Due to Species of *Mycoplasma* and *Ureaplasma*: an Update. *Clinical Infectious Diseases*, Vol. 23, No. 4, (October 1996), pp. 671-684, ISSN 1058-4838

Taylor-Robinson, D., Davies, H. A., Sarathchandra, P., & Furr, P. M. (1991). Intracellular Location of *Mycoplasmas* in Cultured Cells Demonstrated by Immunocytochemistry and Electron Microscopy. *International Journal of Experimental Pathology,* Vol. 72, No. 6, (December 1991), pp. 705-714, ISSN 0959-9673

Touati, A., Benard, A., Hassen, A.B., Bébéar, C.M., & Pereyre, S. (2009). Evaluation of Five Commercial Real-Time PCR Assays for Detection of *Mycoplasma pneumoniae* in Respiratory Tract Specimens. *Journal of Clinical Microbiology,* Vol. 47, No. 7, (July 2009), pp. 2269-2271, ISSN 0095-1137

Ursi, J.P., Ursi, D., Ieven, M., & Pattyn, S.R. (1992). Utility of an Internal Control for the Polymerase Chain Reaction. Application to Detection of *Mycoplasma pneumoniae* in Clinical Specimens. *Acta Pathologica Microbiologica et Immunologica Scandinavica,* Vol. 100, No. 7 (July 1992), pp. 635-639, ISSN 0903-4641

van Kuppeveld FJ, Johansson KE, Galama JM, Kissing J, Bölske G, van der Logt JT, Melchers WJ. Detection of Mycoplasma Contamination in Cell Cultures by a Mycoplasma Group-Specific PCR. *Applied and Environmental Microbiology,* Vol. 60, No. 1 (January 1994), pp. 149-152, ISSN 0099-2240

van Kuppeveld, F.J.M., van der Logt, J.T.M., Angulo, A.F., van Zoest, M.J., Quint, W.G., Niesters, H.G., Galama, J.M., & Melchers, W.J. (1992). Genus-and Species-Specific Identification of Mycoplasmas by 16S rRNA Amplification. *Applied and Environmental Microbiology,* Vol. 58, No. 8, (August 1992), pp. 2606-2615, ISSN 0099-2240

Vaneechoutte, M., & Van eldere, J. (1997). The Possibilities and Limitations of Nucleic Acid Amplification Technology in Diagnostic Microbiology. *Journal of Medical Microbiology,* Vol. 46, No. 3, (March 1997), pp. 188-194, ISSN 0022-2615

Volokhov, D.V., Kong, H., George, J., Anderson, C., & Chizhikov, V.E. (2008). Biological Enrichment of *Mycoplasma* Agents by Cocultivation with Permissive Cell Cultures. *Applied and Environmental Microbiology,* Vol. 74, No. 17 (September 2008), pp. 5383-5391, ISSN 0099-2240

Waites, K.B. (2006). Mycoplasma and Ureaplasma, In: *Congenital and Perinatal Infections: A Concise Guide to Diagnosis,* C. Hutto (Ed.), 271-288, Humana Press Inc., ISBN 1-58829-297-5, Totowa, NJ.

Waites, K.B., & Talkington, D.F. (2004). *Mycoplasma pneumoniae* and Its Role as a Human Pathogen. *Clinical Microbiology Reviews,* Vol. 17, No. 4, (October 2004), pp. 697–728, ISSN 0893-8512

Waites, K.B., & Talkington, D.F. (2005). New Developments in Human Diseases Due to Mycoplasmas. In: *Mycoplasmas: pathogenesis, molecular biology, and emerging strategies for control,* A. Blanchard, & G. Browning (Eds.), 289-354, Horizon Scientific Press, ISBN 0849398614, Norwich, U.K

Waites, K.B., Balish M.F., & Atkinson, T. P. (2008). New Insights Into the Pathogenesis and Detection of *Mycoplasma pneumoniae* Infections. *Future Microbiology,* Vol. 3, No. 6, pp. 635–648, ISSN 1746-0913

Waites, K.B., Bebear, C.M., Robertson, J.A., Talkington, D.F., & Kenny, G.E. (2000). Laboratory Diagnosis of Mycoplasmal Infections. Cumitech 34. *Coordinating ed. FS Nolte.* Washington: American Society for Microbiology.

Waites, K.B., Katz, B., & Schelonka, R. (2005). Mycoplasmas and Ureaplasmas as Neonatal Pathogens. *Clinical Microbiology Reviews,* Vol. 18, No. 4, (October 2005), pp. 757–789, ISSN 0893-8512

Wang, R.Y., Hu, W.S., Dawson, M.S., Shih, J.W., & Lo, S.C. (1992). Selective Detection of *Mycoplasma fermentans* by Polymerase Chain Reaction and By Using a Nucleotide Sequence Within the Insertion Sequence-Like Element. *Journal of Clinical Microbiology*, Vol. 30, No. 1 (January 1992), pp. 245-248, ISSN 0095-1137

Wang, R.Y.-H., Hu, W.S., Dawson, M.S., Shih, J.W.-K., Lo, S.-C. (1992) Selective Detection of *Mycoplasma fermentans* by Polymerase Chain Reaction and by Using a Nucleotide Sequence Within the Insertion Sequence-Like Element. *Journal of Clinical Microbiology*, Vol. 30, No. 1 (January 1992), pp. 245-248, ISSN 0095-1137

Wilson, M.L. (1996). General Principles of Specimen Collection and Transport. *Clinical Infectious Diseases*, Vol. 22, No. 5, (May 1996), pp. 766-77, ISSN 1058-4838

Yáñez A, Cedillo L, Neyrolles O, Alonso E, Prévost MC, Rojas R, Watson HL, Blanchard A, Cassell GH. 1999. *Mycoplasma penetrans* Bacteremia and Primary Antiphospholipid Syndrome. *Emerging Infectious Diseases,* Vol. 5, No. 1 (January 1999), pp. 164-167, ISSN 1080-6040

Yavlovich, A., Tarshis, M., & Rottem, S. (2004). Internalization and Intracellular Survival of *Mycoplasma pneumoniae* by Non-Phagocytic Cells. *FEMS Microbiology Letters,* Vol. 233, No. 2, (April 2004), pp. 241-246, ISSN: 1574-6968

Yoshida, T., Maeda, S., Deguchi, T., & Ishiko, H. (2002). Phylogeny-Based Rapid Identification of Mycoplasmas and Ureaplasmas from Urethritis Patients. *Journal of Clinical Microbiology,* Vol. 40, No. 1, (January 2002), pp. 105-10, ISSN 0095-1137

BRAF V600E Mutation Detection Using High Resolution Probe Melting Analysis

Jennifer E. Hardingham[1,2], Ann Chua[1],
Joseph W. Wrin[1], Aravind Shivasami[1], Irene Kanter[1],
Niall C. Tebbutt[3] and Timothy J. Price[1,2]
[1]*The Queen Elizabeth Hospital, Adelaide, SA, 5011*
[2]*University of Adelaide, SA, 5005*
[3]*Ludwig Institute for Cancer Research, Austin Health, Melbourne, VIC, 3084*
Australia

1. Introduction

Activation of oncogenic proteins is an important mechanism in carcinogenesis. The BRAF gene, located on chromosome 7q34, encodes a serine-threonine kinase that acts downstream of RAS in the RAS/RAF/MEK/ERK signaling pathway involved in regulating cell proliferation and survival. On activation of RAS, the BRAF kinase is activated and sequentially phosphorylates and activates MEK and ERK. A mutation in BRAF leads to constitutive hyperactivation of this pathway through evasion of the inhibitory feedback loop resulting in increased ERK signaling output which drives proliferative and anti-apoptotic signaling (Pratilas et al. 2009). Mutations in BRAF have been reported to occur at high frequency (66%) in melanoma with lower frequencies in colon and other tumours (Davies et al. 2002); BRAF is thus considered to be an important therapeutic target in melanoma (Bollag et al. 2010; Flaherty et al. 2010; Paraiso et al. 2011). Although over 30 single site missense mutations have been identified, 90% occur at nucleotide 1799 resulting in a T-A transition and an amino acid substitution at residue 600 (V600E) in the activation segment (Wan et al. 2004).

In colorectal cancer (CRC) mutations in BRAF have been found in about 9-12% of tumours overall (Di Nicolantonio et al. 2008); (Deng et al. 2004; Jensen et al. 2008). However there is a distinct difference in frequency of BRAF mutations between mismatch repair (MMR) deficient (the microsatellite unstable (MSI-H) tumours) and the mismatch repair intact, microsatellite stable (MSS) tumours (Jensen et al. 2008). This is important clinically as tumours that are MSI-H have a better prognosis (Popat, Hubner, and Houlston 2005). BRAF is mutated in almost all sporadic CRCs with MSI-H (Jensen et al. 2008) but not in tumours arising in patients with an inherited form of MMR deficiency, hereditary nonpolyposis colon cancer (HNPCC), known as Lynch syndrome. Thus a major indication for BRAF mutation testing is for a differential diagnosis of Lynch Syndrome in a CRC that is MSI-H. If BRAF is mutated, the tumour is more likely to be sporadic, rather than the heritable type (Sharma and Gulley 2010).

Mutated BRAF has also been associated with non response to anti-EGFR monoclonal antibody therapy (cetuximab or panitumamab) in metastatic CRC (mCRC) patients (Cappuzzo et al. 2008). In a larger study it was reported that 0/11 patients with a BRAF mutation responded to cetuximab or panitumumab, conversely none of the responders carried BRAF mutations (Di Nicolantonio et al. 2008). BRAF mutation has also been found to be a prognostic factor for poorer outcome in mCRC (Di Nicolantonio et al. 2008); (Price et al. 2011); (Samowitz et al. 2005); (Saridaki et al. 2010); (Souglakos et al. 2009; Tol, Nagtegaal, and Punt 2009); (Van Cutsem et al. 2011).

Although PCR-sequencing to detect BRAF mutations has been the gold standard technique, the improvement in instrumentation for high resolution analysis of PCR amplicon melt curves has opened up the way for the detection of single-base changes in short (approximately 100-200 bp) amplicons (Wittwer et al. 2003). Subsequently an improved method was developed, using melt curve analysis of an oligo-probe, annealing across the region of the mutation (Zhou et al. 2004). As the BRAF mutation is a class IV (T-A) change, we opted for this improved method using commercially available primer and probe sequences. Here we describe the optimisation and validation of this technique for the detection of the BRAF V600E mutation in formalin-fixed paraffin-embedded (FFPE) colorectal tumour tissue and, using the Kaplan-Meier method, the impact of this mutation on survival in the study cohort.

2. Materials and methods

2.1 Tumour collection and processing

Patient samples were obtained from the MAX phase III clinical trial colorectal tumour cohort, described in Price et al. (Price et al. 2011). The MAX study design and eligibility criteria have been reported previously (Tebbutt et al. 2010). Eligible patients were enrolled in this trial between July 2005 and June 2007. After enrollment, patients were randomly assigned to receive capecitabine (C), capecitabine and bevacizumab (CB), and capecitabine, bevacizumab and mitomycin C (CBM). Patient demographic and clinical characteristics are shown in Table 1. Patients in these three groups were evaluated for tumour response or progression every 6 weeks by means of radiologic imaging. Treatment was continued until the disease progressed or until the patient could not tolerate the toxic effects. Samples of tumour tissue from archived FFPE specimens collected at the time of diagnosis were retrieved from storage at participating hospital pathology departments. All patients participating in biomarker studies provided written informed consent at the time of study enrolment. Ethics approval was obtained centrally (Ethics Committee, Cancer Institute of NSW, Australia).

2.2 DNA extraction

DNA was extracted from 1-2 FFPE tissue sections (10 µm) mounted on plain glass slides, with an adjacent section stained with haematoxylin and eosin for reference. In cases that were deemed to have <50% presence of malignant crypts in the section (reviewed by a histopathologist), the tissue was manually dissected to ensure a high proportion of tumour cells. We used a single 10 µm section unless the size of the tissue section was <1 cm, in which case 2 10 µm sections were used. Paraffin was removed by xylene and DNA extracted using the QIAamp DNA FFPE tissue kit (Qiagen, Valencia, CA, USA), according to the manufacturer's protocol. DNA was quantified using the Nanodrop (Thermo Scientific, Wilmington, DE, USA), ensuring the ratio 260/280 was >1.7.

Baseline characteristic	All patients (%) (n=471)	BRAF MUT (%) (n=33)	BRAF WT (%) (n=280)	P
Age (years)				
Median	67	71	68	0.27
Range	32-86	36-85	32-86	
Sex Male	63	58	64	0.47
ECOG performance status				
0-1	94	88	94	0.11
2	6	12	6	
Capecitabine dosage				
2000mg/m^2/day	67	60	68	0.38
Disease-free interval > 12 months	27	18	30	0.17
Prior adjuvant chemotherapy	22	9	23	0.06
Prior Radiotherapy	13	6	10	0.47
Primary site of cancer				
Caecum	10	21	9	0.02
Ascending colon	10	24	11	0.04
Transverse colon	6	15	5	0.02
Descending colon	3	6	4	0.48
Sigmoid colon	30	18	32	0.11
Recto-sigmoid colon	11	3	13	0.1
Rectum	23	6	22	0.03
Primary tumour resected	79	91	86	0.47
Any metastases resected	10	3	9	0.23
Extent of disease at baseline				
Local disease (colon or rectum)	36	15	33	0.03
Liver metastases	75	62	75	0.19
Lymph node metastases	47	59	45	0.09
Lung metastases	39	21	41	0.03
Bone metastases	4	0	4	0.23
Peritoneal metastases	18	21	16	0.49
Other metastases	10	24	10	0.01

Table 1. Patient demographic and clinical characteristics
(Reproduced with permission from the Journal of Clinical Oncology).

2.3 Mutation analyses

Mutation status of BRAF was determined using high resolution melting analysis (HRM) PCR on the Rotorgene 6000 real-time instrument (Qiagen). BRAF HRM PCR (119 bp amplicon) was performed on 10 ng DNA in triplicate reactions using SsoFast™ EvaGreen® Supermix (Bio-Rad Laboratories Inc., Hercules, USA) and a primer/probe combination (RaZor® probe HRM assay, PrimerDesign, Southampton, UK). The sequences were 5'ATGAAGACCTCACAGTAAAAATAGG (sense), CTCAATTCTTACCATCCACAAAATG (antisense) and 5'GTGAAATCTGGATGGAGTGGGTCCCATCA (probe). Appropriate mutant and wild type (WT) controls were included. A 'touch-down' PCR cycling protocol was used for the first 9 cycles to avoid primer mis-priming events and, due to the asymmetric design, 50 cycles were performed according to the manufacturer's protocol. The sensitivity of detection of mutant sequences was determined by assaying dilutions (100%, 50%, 25%, 12.5%, 6.25%) of a tumour DNA sample, with known homozygous BRAF mutation status, in BRAF WT cell line DNA. Using the Rotor Gene 6000 (Qiagen) software analysis features for HRM, patient samples (n=315) were classified as having mutated (MUT) or WT BRAF respectively. Direct PCR sequencing was used to validate all mutant BRAF results and an additional 106 randomly chosen samples (45% of samples in total). The primers for BRAF sequencing reactions were designed in-house and obtained commercially (Geneworks, Thebarton, SA, Australia): 5'AATGCTTGCTCTGATAGGAAAA (sense) and 5'AGTAACTCAGCAGCATCTCAGG (antisense). PCR products were purified using ExoSAP-IT (GE Healthcare, Buckinghamshire, UK) to remove unwanted deoxynucleotides and primers according to the manufacturer's protocol. Sequencing was performed by Flinders Sequencing Facility (Flinders Medical Centre, Bedford Park, SA, Australia) using BigDye Terminator v3.1 chemistry and the Applied Biosystems 3130xl Genetic Analyser (Life Technologies, Carlsbad, CA, USA).

2.4 Statistical analyses

All randomly assigned patients for whom data on BRAF mutation status were available were included in the analysis (n=313). PFS, the primary endpoint, was defined as the time from randomisation until documented evidence of disease progression, the occurrence of new disease or death from any cause. The secondary endpoint was overall survival (OS), defined as the time from randomisation until death from any cause. The PFS and OS of patients according to BRAF status were summarised with the use of Kaplan–Meier curves, and the difference between these groups was compared with the use of the log-rank test. All reported P values were two-sided.

3. Results and discussion

Although significantly less DNA was isolated from the microdissected sections (P=0.0001), the range of values obtained overall, 60 ng -31.3 μg, meant that all samples were well within the amount required for the PCR (30 ng) (Figure 1).

In interpreting the HRM results, the first criterion of robust PCR amplification must be met (Figure 2A), so that the duplicates must show close Ct values (standard deviation <0.5) otherwise samples must be excluded from the HRM analysis and the PCR repeated.

Samples that show poor amplification with late Ct values may give erroneous results on HRM as shown in Figure 2B. The samples in the boxed area need to be excluded from the analysis to avoid misinterpretation of the difference plot as mutant calls. The poor amplification of a DNA sample may be due to the presence of inhibitors, and we have found that subsequent isolation of DNA from microdissected sections gave much better, more reproducible amplification results. This also suggests that minimising the amount of paraffin in the DNA preparation may be contributing to the improvement in PCR performance.

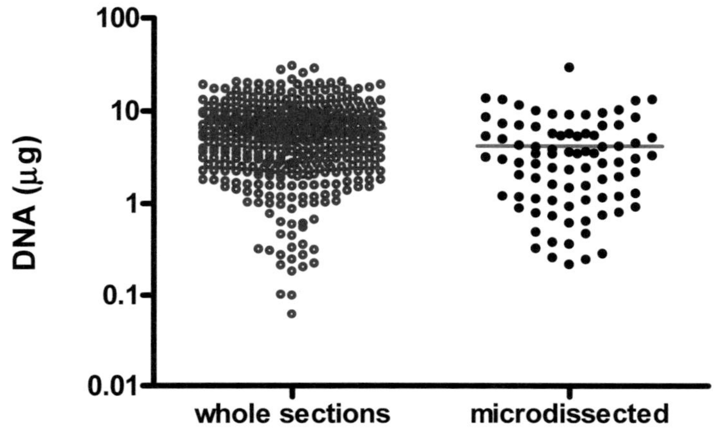

Fig. 1. Dot plot of DNA yields. The average amount of DNA obtained from whole sections was 6.5±0.25 μg and from manually microdissected sections 4.2±0.48 μg.

The positioning of the normalisation regions 1 and 2 in the first HRM analysis window is also a very important parameter in the correct calling of genotypes. This is user-defined and performed separately for HRM analysis of the probe region or the amplicon region. The correct positioning may be determined by monitoring the normalised graph to show the best separation of mutant versus WT curves.

To determine the level of sensitivity of detection, serial doubling dilutions of a tumour sample carrying a homozygous BRAF V600E mutation were tested. The difference graph, normalised to the WT control, shows that the mutation could be detected down to a dilution of 6.25% mutant DNA in WT DNA (Figure 3A). Although there is a distinct difference between the WT control used for normalisation and the 6.25% and 12.5% dilutions, in practice the software cannot call these with any confidence. From the normalised graph and the melt curves graph (Figure 3B and 3C), 25% mutant DNA appears to be the lower limit of detection. However to increase the probability of correctly assigning a genotype we aimed for at least 50% epithelial tumour cells, hence all of the tumour tissue in the cohort was reviewed to ensure at least 50% epithelial tumour cells were present. Manual microdissection was performed in 1/5 of the cohort to ensure >50% enrichment of tumour cells, relative to muscularis mucosa and other cell types such as lymphoid aggregates, in the sample.

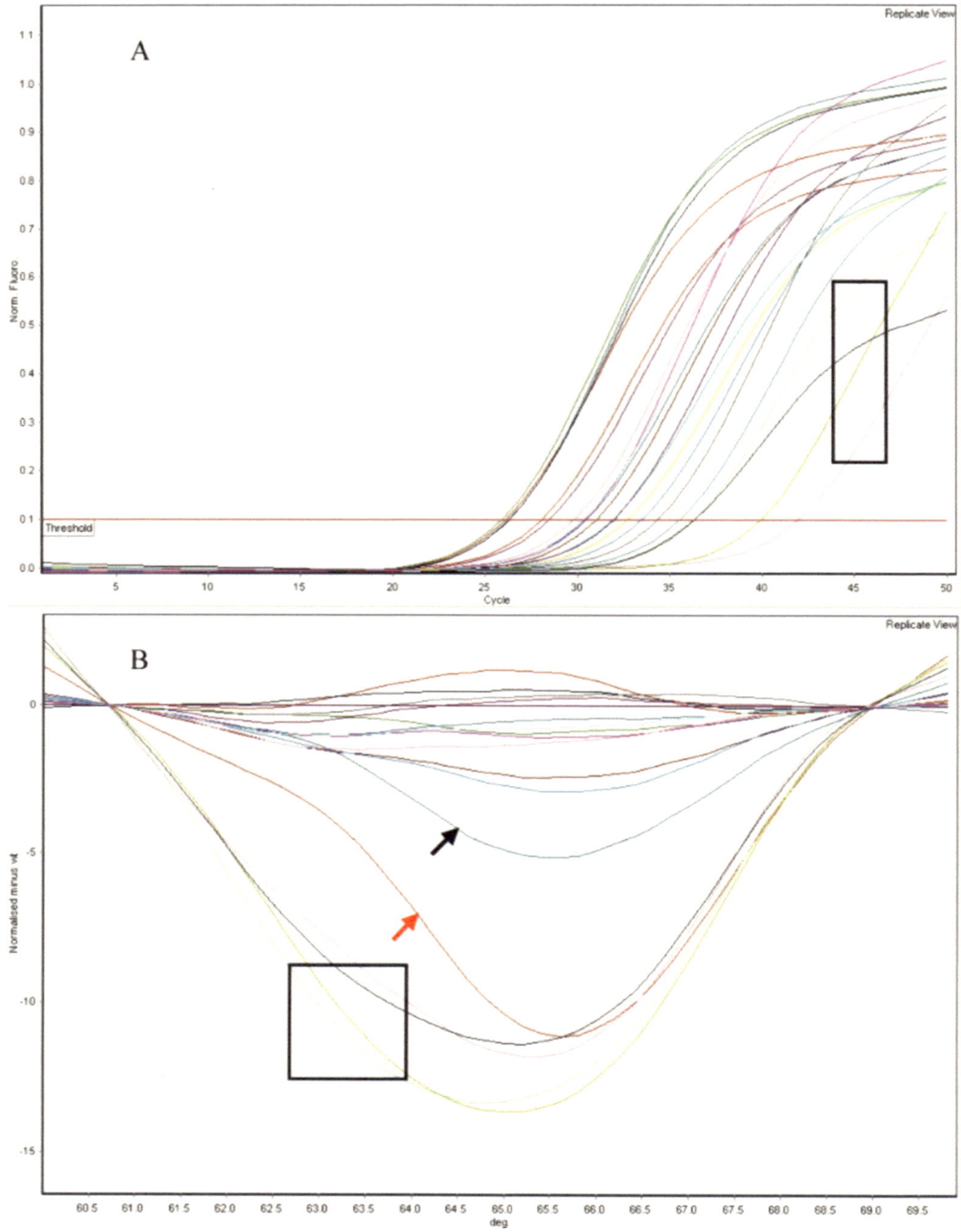

Fig. 2. A: amplification curves; B: difference plot normalised to WT. The boxed curves in A show samples with aberrant late amplification. The same samples boxed in B show the abnormal difference plots that could be incorrectly interpreted as mutant. Black arrow in B points to the heterozygous mutant control, red arrow shows the homozygous mutant control.

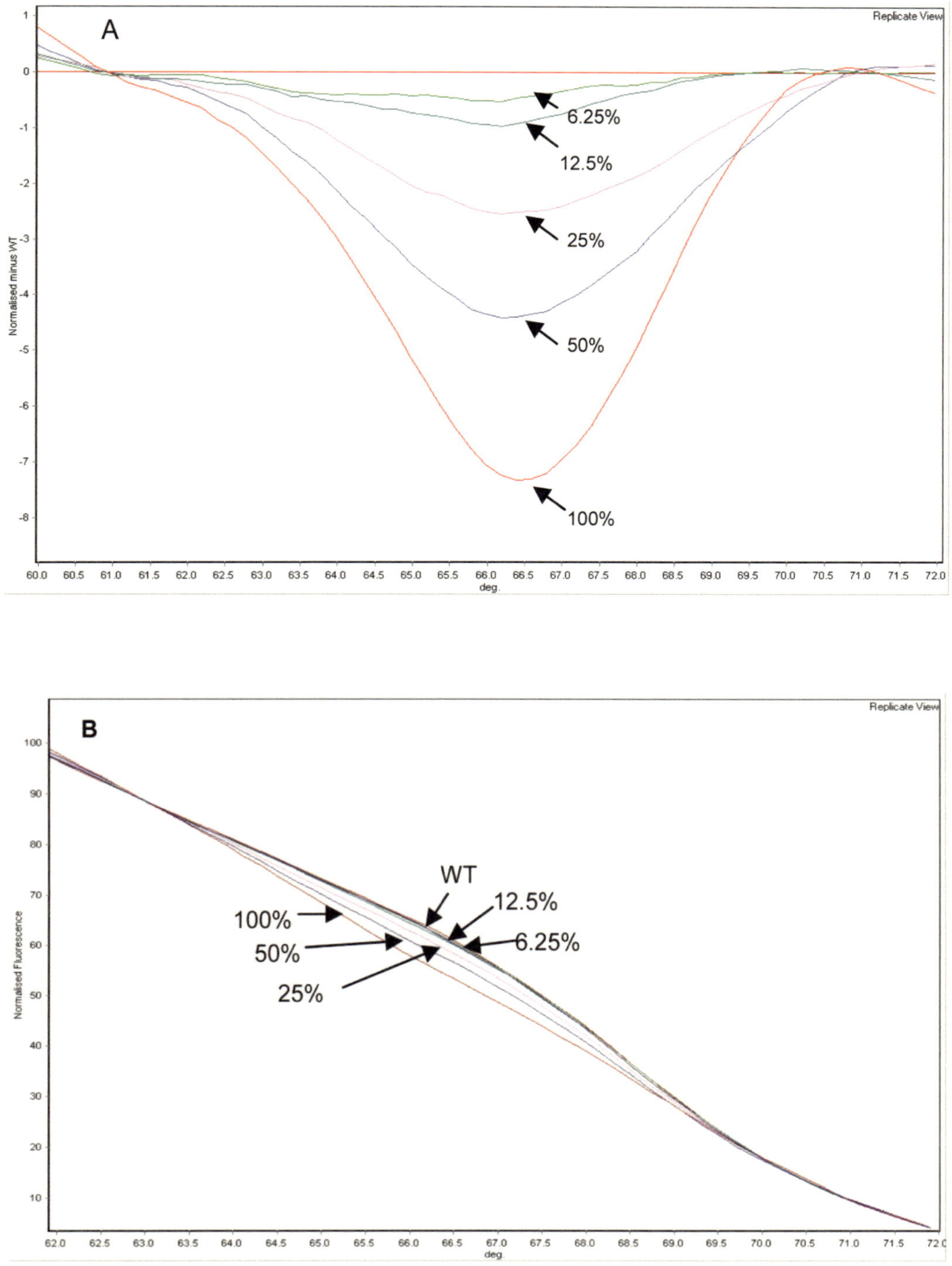
A
Replicate View
Normalised minus WT
6.25%
12.5%
25%
50%
100%
deg.
B
Replicate View
Normalised Fluorescence
WT
12.5%
100%
6.25%
50%
25%
deg.

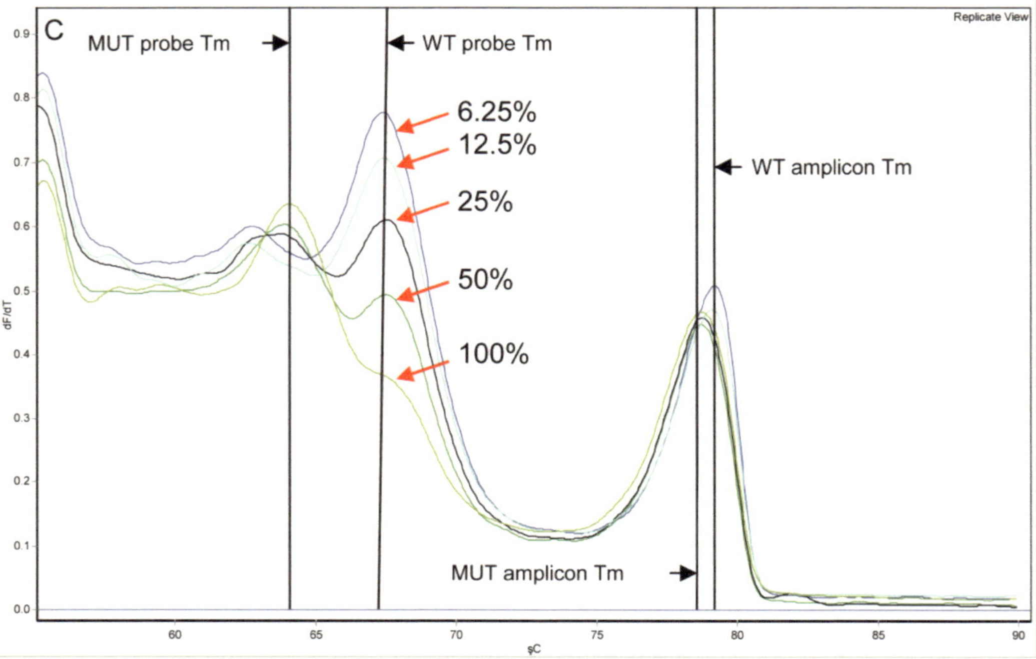

Fig. 3. A: Difference plot normalised to WT, with dilutions of homozygous MUT control DNA in WT DNA shown in replicate view (average of 3 for each dilution). Arrows indicate the plots for the dilutions of MUT control DNA in WT DNA from 100% MUT to 6.25%MUT; B: Normalised melting curves of the probe region. From this view it was not possible to distinguish the 12.5% or 6.25% dilutions of mutant sequence from WT; C: Melt curve showing Tm's for both the probe region and amplicon. The probe region HRM analysis was much easier to interpret than the amplicon HRM, however the 12.5% and 6.25%dilutions were indistinguishable from WT pattern.

We have found that it was of critical importance to select the control genotypes (WT or mutant) for the normalisation carefully. The DNA of these controls needed to be extracted from a similar tissue (i.e. colonic tissue FFPE), and be processed in exactly the same way as the test samples. Using cell line derived DNA as the controls resulted in too many mutation calls with low confidence (false positives), however when we used tumour samples of known *BRAF* status as the controls, the confidence of the software calls of the test samples reached >99%. Often we found it was more informative to look at the shape of the curves in the difference plot, even if a curve deflected away from the horizontal normalised line, the angle of deflection was much greater for mutant genotypes and shallower for WT (Figure 4). This visual interpretation usually correlated with the software calls and was a useful adjunct in interpretation where the confidence of the software calls was low.

Sequencing was used to validate the results and correlated with the HRM results. In some cases though sequencing showed a very small A peak which could be overlooked whereas HRM showed a very convincing shift and was called as a mutation with 99% confidence. An example is shown in Figure 5.

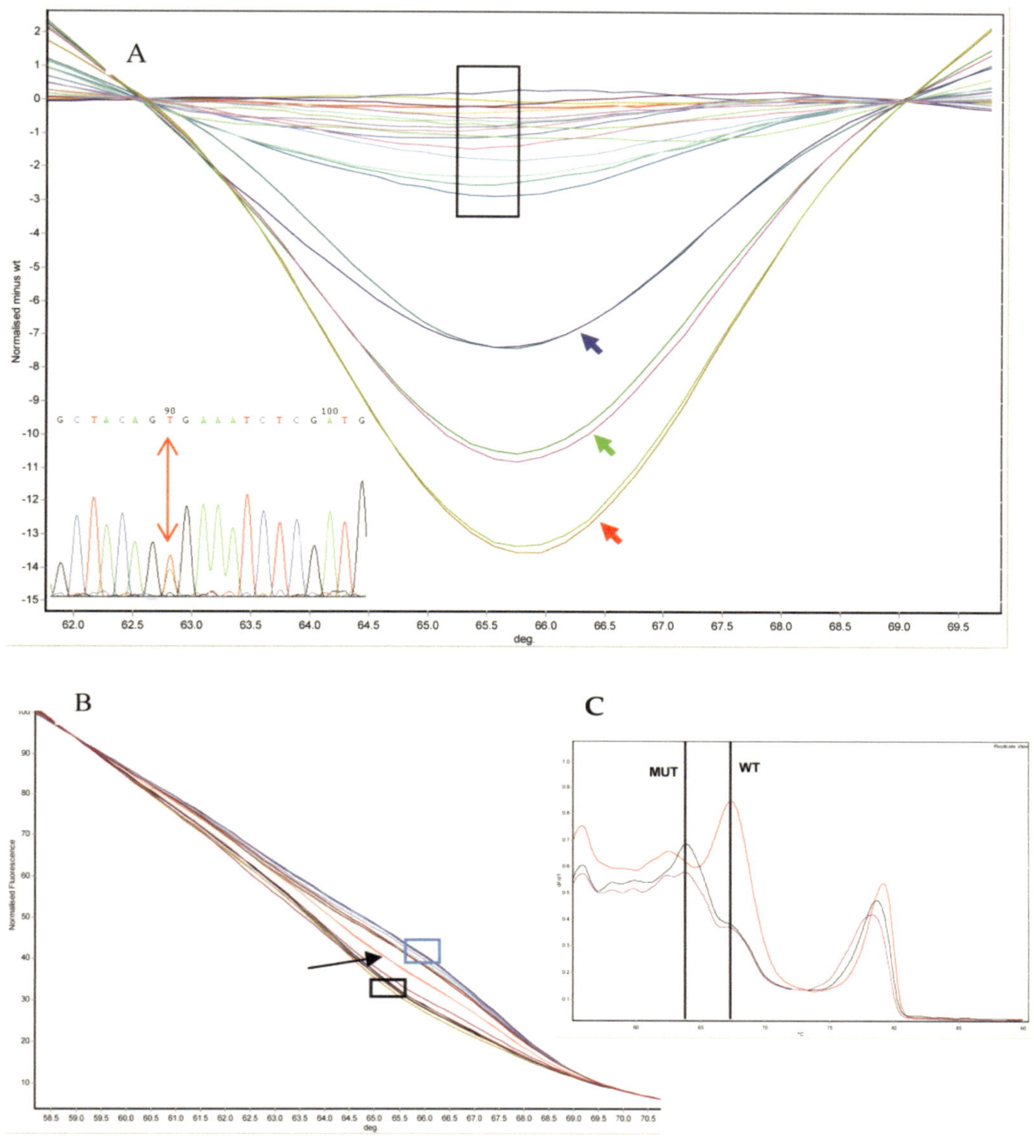

Fig. 4. Sequencing result and corresponding HRM analysis. A: Difference graph normalized to WT control (duplicates), blue arrow shows heterozygous control, green arrow pt 109, red arrow homozygous MUT control. Boxed area shows WT samples. Inset is the sequencing trace (Chromas Lite software) of patient (pt) 109, red arrow showing mutant (A) amongst WT (T) sequence. B: Normalised melt curve of probe region; black boxed area shows homozygous mutant control and 2 samples including pt 109, arrow points to the heterozygote mutant control, blue box shows WT control and WT samples. C: Melt curve analysis, red trace WT, black trace homozygous MUT control, purple trace pt 109.

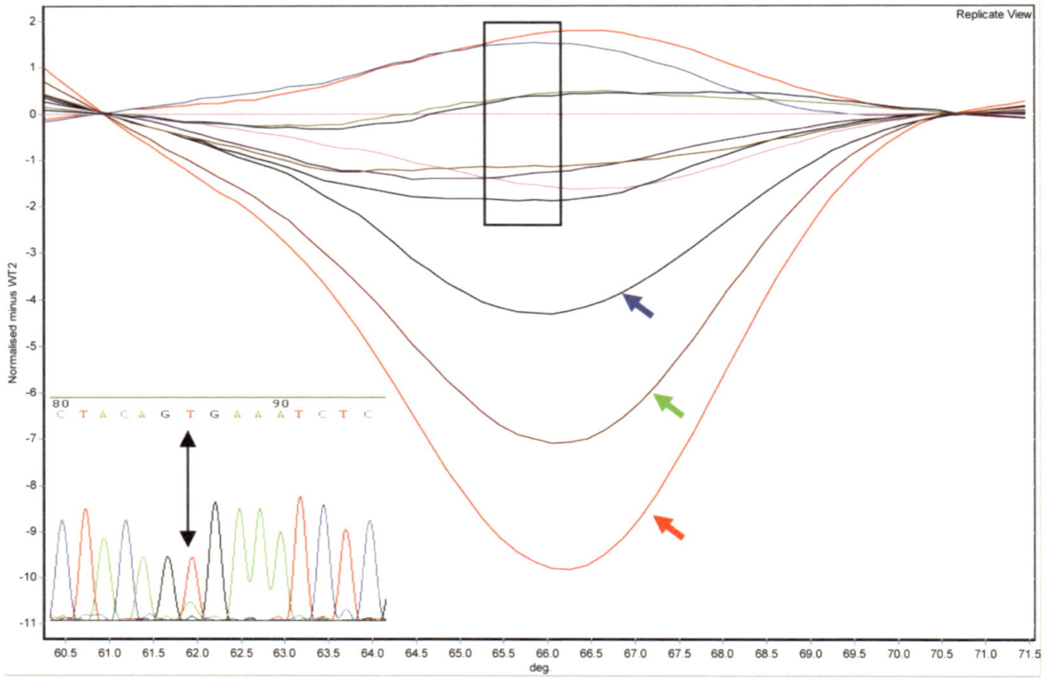

Fig. 5. An example of a sequencing result (pt 269) called WT (T) by the sequencing software that did in fact show a small A peak. The difference plot of the HRM analysis (normalized to WT control) showed a definite downward shifted curve (green arrow) between the homozygous BRAF MUT control (red arrow) and the heterozygous control (blue arrow). The boxed curves show the WT samples.

Of 471 patients who underwent random assignment, a total of 315 tumour specimens (n=103 from the capecitabine group, n=111 from the CB group, and n=101 from the CBM group, accounting for 66.9% of the total study population) were examined for *BRAF* mutation status by HRM. *BRAF* V600E mutations were detected in 10.5% of 313 tumours (2 samples were not evaluable). A proportion of samples were also genotyped using sequencing and showed 100% correlation with the HRM result.

A total of 313 patients were included in the survival analysis with a median follow-up time of 26.5 months (range, 0.4 to 37.6 months). There was no significant difference in PFS between patients with WT tumours and those with mutated tumours. The median PFS was 4.5 months among the patients with V600E tumours as compared with 8.2 months among those with WT tumours (HR: BRAF WT *vs* MUT, 0.80; 95% CI, 0.54 to 1.18; P=0.26). In contrast, there was a significant difference in OS between patients with WT tumours and those with V600E tumours. The median OS was 8.6 months among the patients with mutated BRAF tumours as compared with 20.8 months among those with WT tumours (HR: BRAF WT *vs* MUT, 0.49; 95% CI, 0.33 to 0.73; P=0.001) (Figure 6). BRAF status remained prognostically significant after adjustment of pre-defined baseline prognostic factors including age, sex, ECOG performance status, inoperable local disease, and prior chemotherapy (HR: BRAF WT *vs* MUT, 0.45; 95% CI, 0.30 to 0.68; P<0.0001).

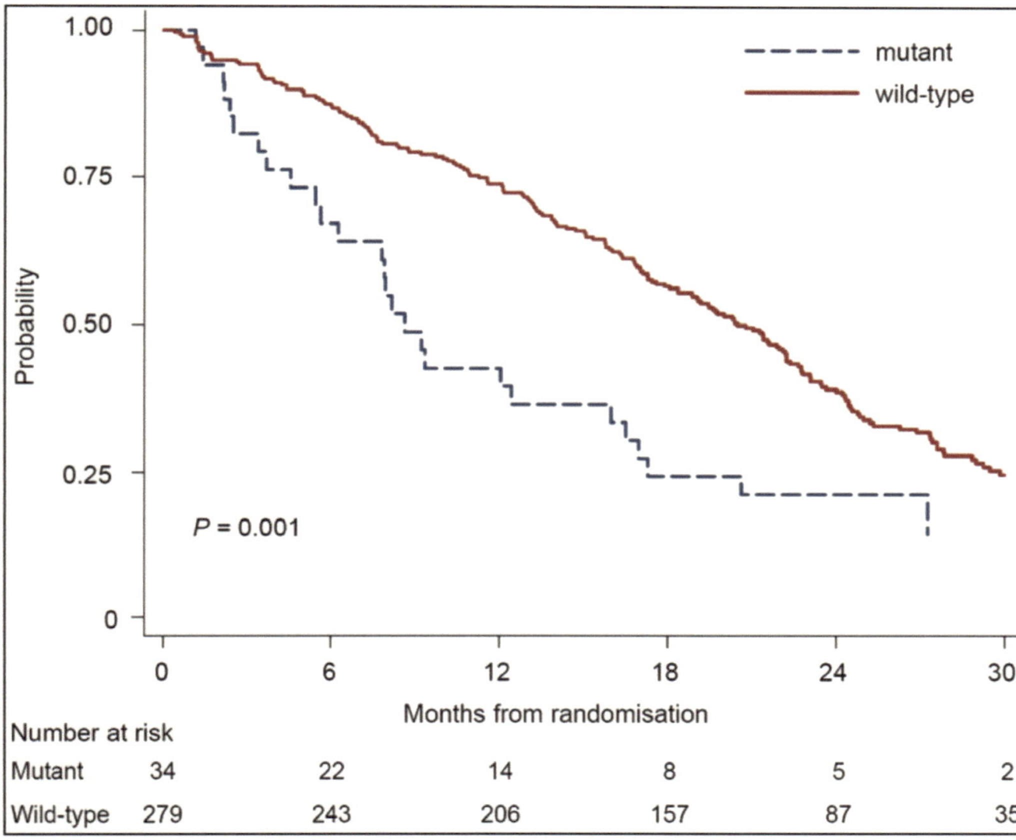

Fig. 6. Kaplan-Meier analysis for overall survival comparing patients WT or MUT for *BRAF* . The curves are significantly different (P=0.001, log-rank test). Reproduced with permission from the Journal of Clinical Oncology.

4. Conclusion

HRM analysis is a useful fast technique to determine BRAF mutations using the platform of real-time PCR. It is both reproducible and reliable provided the preceding guidelines are followed and rigorous attention is given to the PCR performance as well as to the use of the software analysis package. Here we have described how the technique can be applied to the analysis of DNA extracted from archived FFPE tissue sections, which in many cases is the only source of tumour tissue available for retrospective analyses. The survival analysis showed that metastatic CRC patients with tumours carrying the V600E mutation had significantly poorer overall survival outcomes compared to those without the mutation. This HRM analysis could equally be applied to the assessment of tumours from patients diagnosed with other diseases known to have a significant BRAF mutation rate.

5. References

Bollag, Gideon, Peter Hirth, James Tsai, Jiazhong Zhang, Prabha N. Ibrahim, Hanna Cho, Wayne Spevak, Chao Zhang, Ying Zhang, Gaston Habets, Elizabeth A. Burton, Bernice Wong, Garson Tsang, Brian L. West, Ben Powell, Rafe Shellooe, Adhirai Marimuthu, Hoa Nguyen, Kam Y. J. Zhang, Dean R. Artis, Joseph Schlessinger, Fei Su, Brian Higgins, Raman Iyer, Kurt D/'Andrea, Astrid Koehler, Michael Stumm, Paul S. Lin, Richard J. Lee, Joseph Grippo, Igor Puzanov, Kevin B. Kim, Antoni Ribas, Grant A. McArthur, Jeffrey A. Sosman, Paul B. Chapman, Keith T. Flaherty, Xiaowei Xu, Katherine L. Nathanson, and Keith Nolop. 2010. Clinical efficacy of a RAF inhibitor needs broad target blockade in BRAF-mutant melanoma. *Nature* 467 (7315):596-599.

Cappuzzo, F., M. Varella-Garcia, G. Finocchiaro, M. Skokan, S. Gajapathy, C. Carnaghi, L. Rimassa, E. Rossi, C. Ligorio, L. Di Tommaso, A. J. Holmes, L. Toschi, G. Tallini, A. Destro, M. Roncalli, A. Santoro, and P. A. Janne. 2008. Primary resistance to cetuximab therapy in EGFR FISH-positive colorectal cancer patients. *Br J Cancer* 99 (1):83-89.

Davies, Helen, Graham R. Bignell, Charles Cox, Philip Stephens, Sarah Edkins, Sheila Clegg, Jon Teague, Hayley Woffendin, Mathew J. Garnett, William Bottomley, Neil Davis, Ed Dicks, Rebecca Ewing, Yvonne Floyd, Kristian Gray, Sarah Hall, Rachel Hawes, Jaime Hughes, and Vivian Kosmidou. 2002. Mutations of the BRAF gene in human cancer. *Nature* 417 (6892):949.

Deng, Guoren, Ian Bell, Suzanne Crawley, James Gum, Jonathan P. Terdiman, Brian A. Allen, Brindusa Truta, Marvin H. Sleisenger, and Young S. Kim. 2004. BRAF Mutation Is Frequently Present in Sporadic Colorectal Cancer with Methylated hMLH1, But Not in Hereditary Nonpolyposis Colorectal Cancer. *Clin Cancer Res* 10 (1):191-195.

Di Nicolantonio, F., M. Martini, F. Molinari, A. Sartore-Bianchi, S. Arena, P. Saletti, S. De Dosso, L. Mazzucchelli, M. Frattini, S. Siena, and A. Bardelli. 2008. Wild-type BRAF is required for response to panitumumab or cetuximab in metastatic colorectal cancer. *J Clin Oncol* 26 (35):5705-12.

Flaherty, Keith T., Igor Puzanov, Kevin B. Kim, Antoni Ribas, Grant A. McArthur, Jeffrey A. Sosman, Peter J. O'Dwyer, Richard J. Lee, Joseph F. Grippo, Keith Nolop, and Paul B. Chapman. 2010. Inhibition of Mutated, Activated BRAF in Metastatic Melanoma. *New England Journal of Medicine* 363 (9):809-819.

Jensen, L. H., J. Lindebjerg, L. Byriel, S. Kolvraa, and D. G. Cruger. 2008. Strategy in clinical practice for classification of unselected colorectal tumours based on mismatch repair deficiency. *Colorectal Disease* 10 (5):490-497.

Paraiso, Kim H. T., Yun Xiang, Vito W. Rebecca, Ethan V. Abel, Y. Ann Chen, A. Cecilia Munko, Elizabeth Wood, Inna V. Fedorenko, Vernon K. Sondak, Alexander R. A. Anderson, Antoni Ribas, Maurizia Dalla Palma, Katherine L. Nathanson, John M. Koomen, Jane L. Messina, and Keiran S. M. Smalley. 2011. PTEN Loss Confers BRAF Inhibitor Resistance to Melanoma Cells through the Suppression of BIM Expression. *Cancer Research* 71 (7):2750-2760.

Popat, S., R. Hubner, and R. S. Houlston. 2005. Systematic review of microsatellite instability and colorectal cancer prognosis. *J Clin Oncol* 23 (3):609-18.

Pratilas, Christine A., Barry S. Taylor, Qing Ye, Agnes Viale, Chris Sander, David B. Solit, and Neal Rosen. 2009. V600EBRAF is associated with disabled feedback inhibition of RAF-MEK signaling and elevated transcriptional output of the pathway. *Proceedings of the National Academy of Sciences* 106 (11):4519-4524.

Price, T. J., J. E. Hardingham, C. K. Lee, A. Weickhardt, A. R. Townsend, J. W. Wrin, A. Chua, A. Shivasami, M. M. Cummins, C. Murone, and N. C. Tebbutt. 2011. Impact of KRAS and BRAF Gene Mutation Status on Outcomes From the Phase III AGITG MAX Trial of Capecitabine Alone or in Combination With Bevacizumab and Mitomycin in Advanced Colorectal Cancer. *J Clin Oncol* 29 (19):2675-82.

Samowitz, W. S., C. Sweeney, J. Herrick, H. Albertsen, T. R. Levin, M. A. Murtaugh, R. K. Wolff, and M. L. Slattery. 2005. Poor survival associated with the BRAF V600E mutation in microsatellite-stable colon cancers. *Cancer Res* 65 (14):6063-9.

Saridaki, Z., D. Papadatos-Pastos, M. Tzardi, D. Mavroudis, E. Bairaktari, H. Arvanity, E. Stathopoulos, V. Georgoulias, and J. Souglakos. 2010. BRAF mutations, microsatellite instability status and cyclin D1 expression predict metastatic colorectal patients' outcome. *Br J Cancer* 102 (12):1762-8.

Sharma, Shree G., and Margaret L. Gulley. 2010. BRAF Mutation Testing in Colorectal Cancer. *Archives of Pathology & Laboratory Medicine* 134 (8):1225-1228.

Souglakos, J., J. Philips, R. Wang, S. Marwah, M. Silver, M. Tzardi, J. Silver, S. Ogino, S. Hooshmand, E. Kwak, E. Freed, J. A. Meyerhardt, Z. Saridaki, V. Georgoulias, D. Finkelstein, C. S. Fuchs, M. H. Kulke, and R. A. Shivdasani. 2009. Prognostic and predictive value of common mutations for treatment response and survival in patients with metastatic colorectal cancer. *Br J Cancer* 101 (3):465-72.

Tebbutt, N. C., K. Wilson, V. J. Gebski, M. M. Cummins, D. Zannino, G. A. van Hazel, B. Robinson, A. Broad, V. Ganju, S. P. Ackland, G. Forgeson, D. Cunningham, M. P. Saunders, M. R. Stockler, Y. Chua, J. R. Zalcberg, R. J. Simes, and T. J. Price. 2010. Capecitabine, bevacizumab, and mitomycin in first-line treatment of metastatic colorectal cancer: results of the Australasian Gastrointestinal Trials Group Randomized Phase III MAX Study. *J Clin Oncol* 28 (19):3191-8.

Tol, J., I. D. Nagtegaal, and C. J. Punt. 2009. BRAF mutation in metastatic colorectal cancer. *N Engl J Med* 361 (1):98-9.

Van Cutsem, E., C. H. Kohne, I. Lang, G. Folprecht, M. P. Nowacki, S. Cascinu, I. Shchepotin, J. Maurel, D. Cunningham, S. Tejpar, M. Schlichting, A. Zubel, I. Celik, P. Rougier, and F. Ciardiello. 2011. Cetuximab Plus Irinotecan, Fluorouracil, and Leucovorin As First-Line Treatment for Metastatic Colorectal Cancer: Updated Analysis of Overall Survival According to Tumor KRAS and BRAF Mutation Status. *J Clin Oncol* 29 (15):2011-9.

Wan, Paul T. C., Mathew J. Garnett, S. Mark Roe, Sharlene Lee, Dan Niculescu-Duvaz, Valerie M. Good, Cancer Genome Project, C. Michael Jones, Christopher J. Marshall, Caroline J. Springer, David Barford, and Richard Marais. 2004. Mechanism of Activation of the RAF-ERK Signaling Pathway by Oncogenic Mutations of B-RAF. *Cell* 116 (6):855-867.

Wittwer, Carl T., Gudrun H. Reed, Cameron N. Gundry, Joshua G. Vandersteen, and Robert J. Pryor. 2003. High-Resolution Genotyping by Amplicon Melting Analysis Using LCGreen. *Clin Chem* 49 (6):853-860.

Zhou, Luming, Alexander N. Myers, Joshua G. Vandersteen, Lesi Wang, and Carl T. Wittwer. 2004. Closed-Tube Genotyping with Unlabeled Oligonucleotide Probes and a Saturating DNA Dye. *Clin Chem* 50 (8):1328-1335.

8

Polymerase Chain Reaction: Types, Utilities and Limitations

Patricia Hernández-Rodríguez[1] and Arlen Gomez Ramirez[2]
[1]*Molecular Biology and Immunogenetics Research Group (BIOMIGEN), Animal Medicine and Reproduction Research Center (CIMRA), Department of Basic Sciences, Biology Program, Universidad de La Salle, Bogotá*
[2]*Faculty of Agricultural Sciences, Veterinary Medicine Program, Animal Medicine and Reproduction Research Center (CIMRA), Universidad de La Salle, Bogotá*
Colombia

1. Introduction

1.1 Types, utilities and limitations of PCR

Nowadays, advances and applications of research in biochemistry and genetic play an important role in the field of health sciences. This has become necessary a molecular approach of the disease for a better interpretation of processes and as horizon in the development of new diagnostic and therapeutic strategies. Therefore, techniques in molecular biology have modified diagnosis, prevention and control of diseases in living beings. Molecular technology has become a crucial tool for identifying new genes with importance in medicine, agriculture, animal production and health, environment and the industry related to these areas. Among the applications of molecular techniques is important to highlight the use of the Polymerase Chain Reaction (PCR) in the identification and characterization of viral, bacterial, parasitic and fungal agents. This technique was developed by Kary Mullis in the mid 80's [1, 2, 3, 4] and since then it has been considered as an essential tool in molecular biology which allows amplification of nucleic acid sequences (DNA and RNA) through repetitive cycles *in vitro*. The mechanisms involved in this methodology are similar to those occurring *in vivo* during DNA replication. Each cycle had three temperature patterns carried out by a thermocycler. The first pattern of temperature is 94 °C (denaturation), the second one is 45 - 55 °C (alignment of the specific primers) and the third one is 72 °C (final extension). The amplification of specific nucleic acid sequences, even in the presence of millions of other DNA molecules, is achieved by thermostable DNA polymerase enzyme (as the name of this technique suggests: "polymerase chain reaction") and specific primers. Primers are short sequences of DNA or RNA (oligonucleotides) that initiate DNA synthesis. These are complementary to the template strand of DNA. The total duration of PCR reaction is around two hours; this depends on the specific conditions of the reaction. Therefore, the DNA polymerase enzyme is capable of producing a complementary strand of a template DNA. In summary, the requirements of PCR are as follows: i. Template DNA; ii. Four deoxyribonucleotides (dNTPs: dATP, dTTP, dGTP and dCTP) which are the base material to make the new strand from template DNA; iii. Two primers or oligonucleotides; iv. Mg^{2+} which joins to nucleotides to be recognized by the polymerase enzyme; and, v. Thermostable DNA polymerase enzyme. The synthesized product in each cycle can serve as a template in the next issue of copies of DNA, creating a chain reaction that can amplify a specific fragment of DNA. Requirements and purpose of PCR are showed in figure 1.

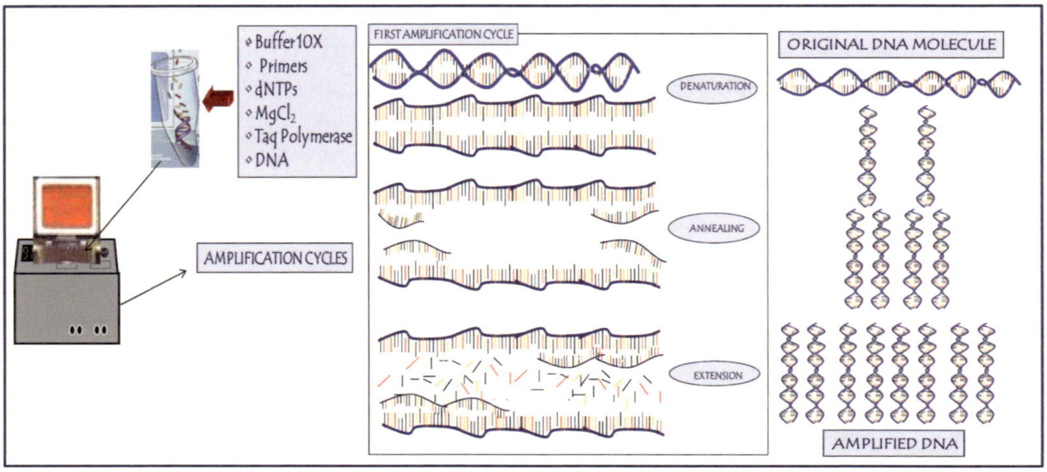

Fig. 1. Requirements and purpose of amplification cycles (denaturation, annealing and extension) in a polymerase chain reaction (PCR).

PCR is a relatively simple technique that can detect a nucleic acid fragment and amplify this sequence. In addition, this technique has other advantages that are described below. This technique offers *sensitivity* because from small amounts of genetic material can be detected target sequences in a sample. Also this offers *specificity* due to a specific sequence of DNA is amplified through strict conditions. It is considered a fast technique compared with other methods to detect microorganisms such as bacteria, fungus or virus, which require isolation and culture using culture media or cell lines. Finally we can mention that offers *versatility* due to the genetic sequences from various microorganisms can be identified with the same reaction conditions for diagnosis of different pathologies [4, 5, 6, 7].

In recent years, modifications or variants have been developed from the basic PCR method to improve performance and specificity, and to achieve the amplification of other molecules of interest in research as RNA. Some of these variants are: i. Multiplex PCR which simultaneously amplified several DNA sequences (usually exonic sequences); ii. Nested PCR increases the specificity of the amplified product for a second PCR with new primers that hybridize within the amplified fragment in the first PCR; iii. Semiquantitative PCR which allows an approximation to the relative amount of nucleic acids present in a sample; iv. RT-PCR which generates amplification of RNA by synthesis of cDNA (DNA complementary to RNA) that is then amplified by PCR; and, v. Real time PCR which performs absolute or relative quantification of nucleic acid copies obtained by PCR. The principles of each of the above techniques are described following.

1.2 Multiplex PCR

Multiplex PCR is an adaptation of PCR which allows simultaneous amplification of many sequences. This technique is used for diagnosis of different diseases in the same sample [8, 9]. Multiplex PCR can detect different pathogens in a single sample [10, 11, 12]. Also it can be used to identify exonic and intronic sequences in specific genes [13] (figure 2) and determination of gene dosage (figure 2, 3 and 4). This is achieved when in a single tube

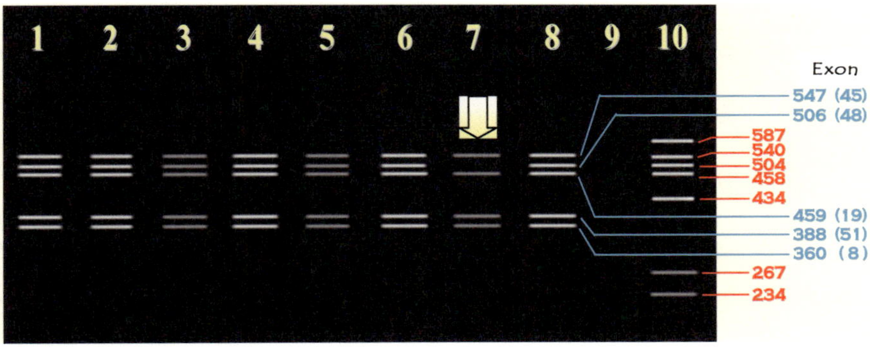

Fig. 2. Results of a multiplex PCR in a patient with Duchenne Muscular Dystrophy. Dystrophy gene has different mutations in exons; this is the cause of disease. In lane 7 is shown the absence of a band corresponding to exon 48 (506 bp) of the dystrophy gene (Hernández-Rodríguez et al., 2000; Hernández-Rodríguez & Restrepo, 2002).

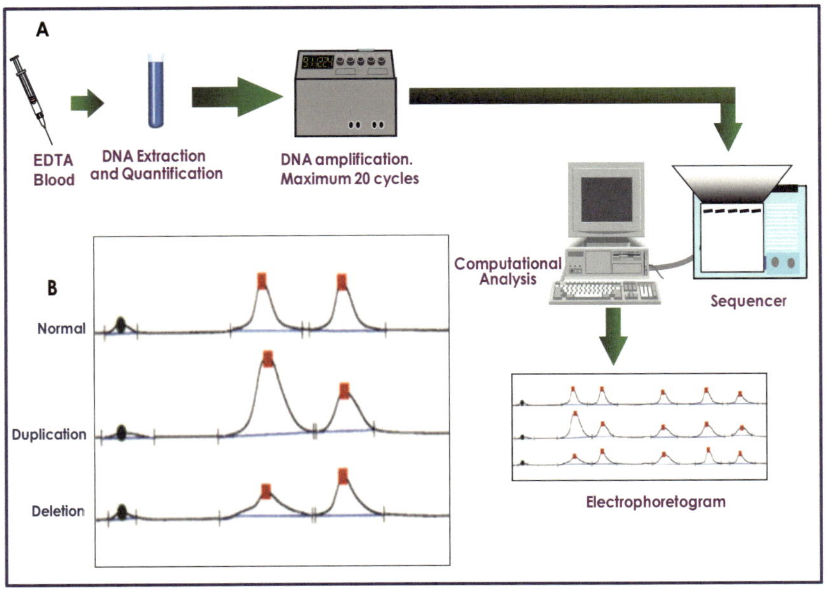

Fig. 3. A: Requirements for multiplex PCR. This molecular method is useful for identification of deletion and duplication mutations. B: Electrophoretogram showing duplication (area under the curve amplified compared to normal) and deletion (area under the curve reduced compared to normal) obtained by analysis of gene dosage. Results are accompanied by a statistical analysis, established by software, which determines areas under curve obtained by a sequencer.

include sets of specific primers for different targets. In this PCR is important the design of primers because they must be characterized by adherence to specific DNA sequences at similar temperatures. However, it may require several trials to achieve the standardization of the procedure [8, 9].

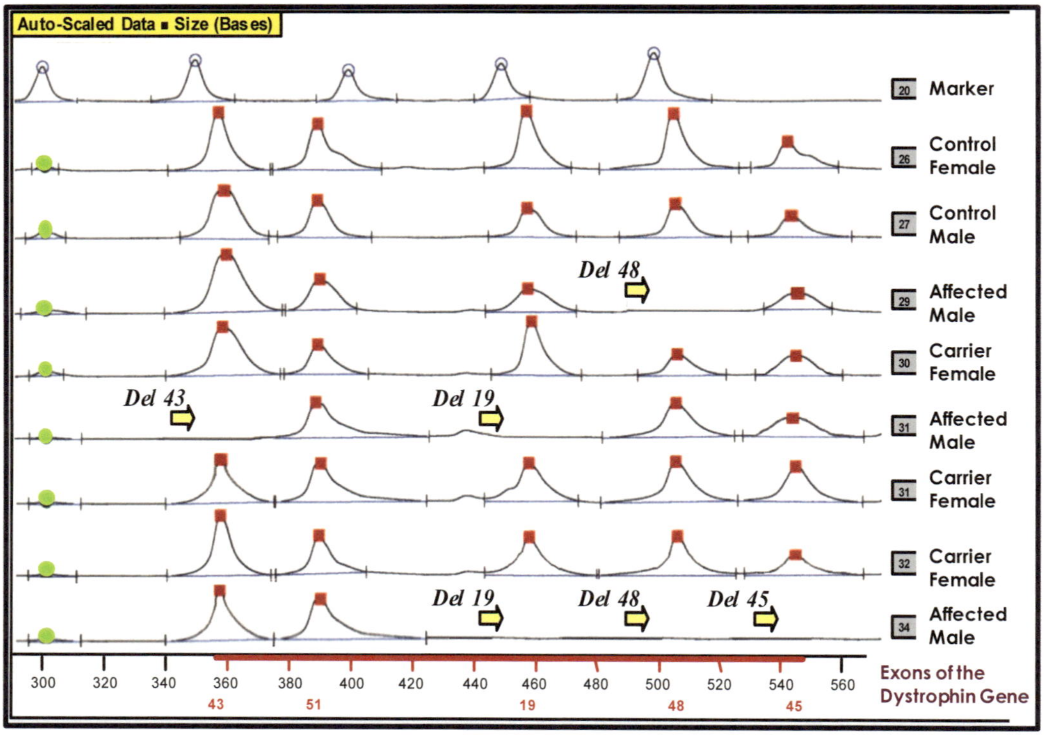

Fig. 4. Electrophoretogram which shows deletions associated with Duchenne Muscular Dystrophy (DMD). In this figure is noted the absence of peaks in men with deletions. Area under the curve in women with DMD is reduced compared to normal control [13, 14].

1.3 Nested PCR

This PCR increases the sensitivity due to small amounts of the target are detected by using two sets of primers, involving a double process of amplification [15, 16]. The first set of primers allows a first amplification. The product of this PCR is subjected to a second PCR using the second set of primers. These primers used in the second PCR are specific to an internal amplified sequence in the first PCR. Therefore, specificity of the first PCR product is verified with the second one. The disadvantage of this technique is the probability of contamination during transfer from the first amplified product into the tube in which the second amplification will be performed. Contamination can be controlled using primers designed to anneal at different temperatures. Contamination can also be controlled by adding ultra-pure oil to make a physical separation of two mixtures of amplification [15, 17, 18].

1.4 Reverse Transcriptase PCR (RT-PCR)

This PCR was designed to amplify RNA sequences (especially mRNA) through synthesis of cDNA by reverse transcriptase (RT). Subsequently, this cDNA is amplified using PCR. This type of PCR has been useful for diagnosis of RNA viruses, as well as for evaluation of antimicrobial therapy [18, 19, 20, 21]. It has also been used to study gene expression *in vitro*, due to the obtained cDNA retains the original RNA sequence. The main challenge of using this technique is the sample of mRNA, because this is considered difficult to handle by low level and concentration of mRNA of interest and low stability at room temperature together with sensitivity to action of ribonucleases and pH change [20, 21, 22].

1.5 Semiquantitative PCR

This technique allows an approximation to the relative amount of nucleic acids present in a sample, as mentioned above. cDNA is obtained by RT-PCR when sample is RNA. Then, internal controls (that are used as markers) are amplified. The markers commonly used are Apo A1 and B actin. Amplification product is separated by electrophoresis. Agarose gel is photographed after ethidium bromide staining, and optical density is calculated by a densitometer. The disadvantage of the technique is possibility of nonspecific hybridizations, generating unsatisfactory results. Control of specificity is performed using highly specific probes for hybridization [23, 24] (figure 5).

1.6 Real time PCR

Real time PCR or quantitative PCR (qPCR) is other adaptation of the PCR method to quantify the number of copies of nucleic acids during PCR. Thus, qPCR is used to quantify DNA o cDNA, determining gene or transcript numbers present within different samples [25, 26, 27]. qPCR offers advantages such as speed in the result, the reduced risk of contamination and the ease in handling technology [28, 29]. This PCR uses fluorescence detection systems which are generally of two types: intercalating agents and labeled probes with fluorophores.

Intercalating agents such as SYBR Green are fluorochromes that dramatically increase the fluorescence by binding to a double-stranded DNA [30, 31, 32]. Thus, the increase of DNA in each cycle reflects a proportional increase in the emitted fluorescence. However, it is considered that intercalating agents offer a low specificity because they can be bind to nonspecific products or primer dimers. Several studies have shown that careful selection of primers and using of optimal PCR conditions may minimize this nonspecificity [28, 32, 33]. The use of a high temperature to start the synthesis reaction (hot-start PCR) decreases the risk of nonspecific amplification. Another detection system used in real time PCR are specific hybridization probes labeled with two types of fluorochromes, a donor and an acceptor. The most commonly used probes are hydrolysis or TaqMan probes, molecular beacons probes, and FRET (fluorescent resonance energy transfer) [32, 33, 34]. The increase of DNA in each cycle is proportional to hybridization of probes, which in turn is proportional to the increase in the emitted fluorescence. The use of probes allows identifying polymorphisms and mutations; however, these are more complex and expensive than intercalating agents [35, 36, 37].

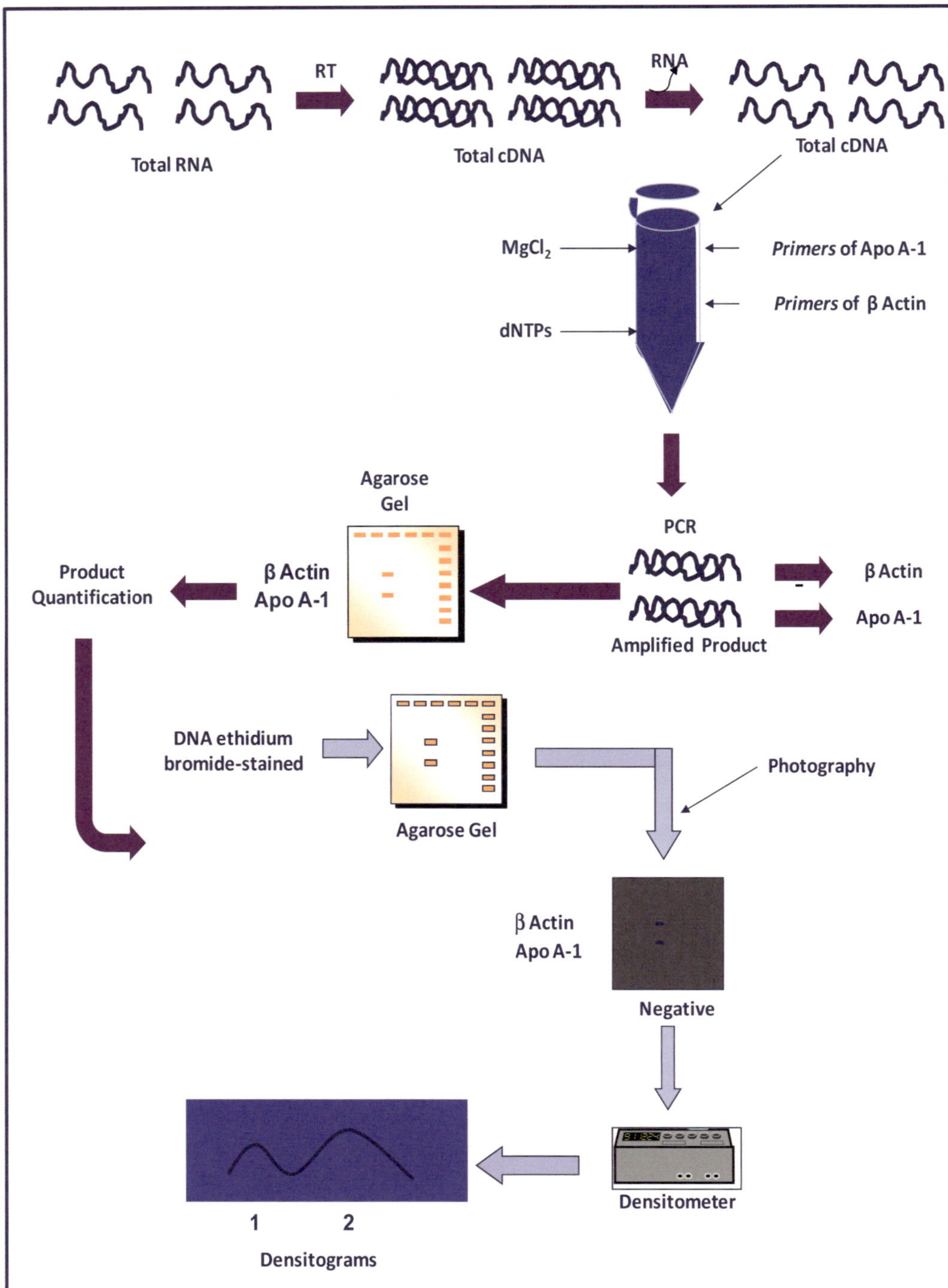

Fig. 5. Semiquantitative PCR procedure. This technique is useful for identifying small amounts of nucleic acids.

2. Applications of PCR and impact on science

During the past 30 years molecular techniques have been under development, however these have had a rapid and tremendous progress in recent year [38]. Among molecular techniques, PCR and its different variations are highlighted as the most commonly used in laboratories and research institutes. Thus, these have contributed to identification and characterization of several organisms and understanding of physiopathology of diverse diseases in human, animal and plant [39, 40]. Also these have provided clues for future research directions in specific topics with impact in public health such as genetics and biochemistry of antimicrobial resistance [41, 42]. The following describes some applications of PCR and its variants in studies in human medicine, forensic sciences, and agricultural science and environment.

2.1 Medicine

Molecular biology techniques, particularly PCR, have had a major impact on medicine. The versatility of molecular techniques has allowed advances and changes in all fields of medicine. The following is an overview of the main impacts generated for molecular biology in medical sciences.

Clinical microbiology has been transformed with the use of molecular technology because it has generated a benefit to the patient affected by infectious diseases. Molecular biology has allowed the development of clinical microbiology because it has been possible to identify microorganisms that are difficult to culture, that have many requirements of laboratory or dangerous for laboratory personnel. These problems have been reduced with the implementation of molecular diagnosis that provides high sensitivity, specificity, precision and speed with one small sample. These applications are transforming and complementing the work of biochemists, immunologists, microbiologists and other health professionals who see in the molecular tools new alternatives for a rapid diagnosis of microorganisms as well as for the determination of multiple factors associated with antibiotic resistance thus expanding the knowledge of microbial epidemiology and surveillance at the genetic level [43, 44, 45].

The usefulness of PCR in identification of microorganisms has led to the selection and quality assurance of blood that blood banks are using for patients with different pathologies [46]. The incorporation of molecular techniques has been of great importance in the identification and characterization of many viruses, including influenza, which through a rapid, sensitive, and effective molecular diagnosis has allowed inclusion of early treatment to benefit patients and control of a high impact infection [47, 48].

The implementation of molecular tools has allowed a transformation of pathological studies and has changed the clinical practice. This is how the diagnosis and treatment of complex diseases that require a multidisciplinary clinical team currently has a base of molecular biology due to histopathological evaluation of tissues, which is an important part in the morphological assessment, is insufficient by itself. Thus, the ability to define molecular alterations associated with the disease is increasingly required to clarify the diagnosis and therapeutic guidance [48]. At pathological level, molecular biology has allowed the identification of mutations and carriers of diseases as in diabetes, obesity, neurological, muscular, cardiac, metabolic, and congenital diseases and pathologies associated with sensory organs. At the ocular level, implementation of molecular biology has generated enormous advances in knowledge, diagnosis and treatment of ophthalmic diseases [49, 50]. The usefulness of this technique in the identification of mutations associated with ocular diseases has been widely used for the study of families at risk. Several reports show how PCR has allowed expanding the knowledge of certain diseases; thus, Woloschak et al. 1994 [51] showed the loss of heterozygosity in the retinoblastoma gene in pituitary human tumors. It was possible to demonstrate genetic heterogeneity in congenital fibrosis of extraocular muscles. With the advent of molecular technology, it has been possible to understand certain aspects of diseases as in retinitis pigmentosa, microftalmia, retinoblastoma, open-angle glaucoma, ocular diseases due to alterations in mitochondrial DNA and various types of corneal dystrophy, among others [50, 52, 53]. Also, genes that cause ocular diseases have been cloned at the anterior and posterior segment. In anterior segment basically aniridia and Peter's anomaly, autosomal dominant diseases in which have been identified candidate genes [54]. In posterior segment, the number of cloned genes has been higher; these are associated with different pathologies described as following. Retinitis pigmentosa with autosomal dominant inheritance pattern in most cases; however, it can also be found in a recessive or digenic form [55]. Congenital Stationary Night Blindness, a disease whose pattern of inheritance is autosomal dominant. Retinoblastoma which the genetic defect affects the retinoblastoma protein (Rb) whose gene rb has been cloned [56]. Cones degeneration inherited pattern linked to X, this means that the disease is transmitted by a carrier mother, where 50% of boys are likely to get the disease and 50% of their daughters are likely to pass it. Its alteration affects synthesis of red opsin [57, 58]. Leber hereditary optic neuropathy associated with alteration of mitochondrial DNA whose defect involves activity of mitochondrial enzymes [57]. These findings have strong implications for the understanding of physiopathology of these genetic entities and generate a new concept of ocular clinical practice due to advances in molecular biology not only can classify better the pathology but the diagnosis becomes specific and safe. On the other hand, in those ocular diseases attributed to mutations in genes located on chromosome X, it is possible to identify mothers or women on the mother line and to generate secondary prevention measures when inform the carrier or not carrier status of them [49].

Molecular tools have also allowed to perform preimplantation genetic diagnosis (PGD) being used for genetic analysis of embryos before transfer into the uterus. It was first developed in England in 1990, as part of the advances in reproductive medicine, genetics and molecular biology. PGD offers couples at risk of having an affected child the opportunity to have normal child by assisted fertilization. The molecular genetic analysis is performed on one or two blastomeres, and only unaffected embryos are transferred into the uterus. It is important to note that in many countries the using of this reproductive procedure has caused controversy. However, this technique provides an opportunity for couples whose children have shown earlier genetic abnormalities [59, 60, 61].

2.2 Forensic science

In forensic pathology, classic morphology remains as a basic procedure to investigate deaths, but recent advances in molecular biology have provided a very useful tool to research systemic changes involved in the pathophysiological process of death that cannot be detected by morphology. In addition, genetic basis of diseases with sudden death can also be investigated with molecular methods. Practical application of RNA analysis has not been accepted for post-mortem research, due to rapid decomposition after death. However, recent studies using variants of conventional PCR (qPCR and RT-PCR) have suggested that relative quantification of RNA transcripts can be applied in molecular pathology to research deaths ("molecular autopsy"). In a broad sense, forensic molecular pathology involves application of molecular biology in medical science to investigate the genetic basis of pathophysiology of diseases that lead to death. Therefore, molecular tools support and reinforce the morphological and physiological evidence in research of unexplained death [62].

Molecular methods are used in forensic science to establish the filiations of a person (paternity testing) or to obtain evidence from minimal samples of saliva, semen or other tissue debris [63]. Genetic profile of the alleles identified in different regions of DNA is performed in paternity tests using a genetic marker STR (Short Tandem Repeat). Each region has an allele contributed by mother and one from father. This profile is virtually unique to each individual, offering a high power molecular evidence of genetic discrimination [64, 65].

2.3 Agricultural sciences and environment

Applications of molecular techniques in research in agricultural sciences and environment have been very numerous and varied. It is possible that one of the most important contributions of the applications of some molecular techniques such as PCR has been the identification and characterization of multiple infectious agents that have great impact on human and animal health. Some applications in agricultural science and environment research are described below.

Currently, the genome of most domestic animals and major infectious agents that affect animals is known through the use of molecular tools, facilitating the study of mutations associated with disease (http://www.ncbi.nlm.nih.gov/genomes/leuks.cgi). Some of the most recent reports are listed below: i. the identification of polymorphisms in ABCB1 gene in phenobarbital responsive and resistant idiopathic epileptic Border Collies [66]; ii. A mutation of EDA gene associated with anhidrotic ectodermal dysplasia in Holstein cattle [67]; iii. The deletion of Meq gene which significantly decreases immunosuppression in chickens caused by Marek's disease virus [68]; iv. The MTM1 mutation associated with X-linked myotubular myopathy in Labrador Retrievers [69]; v. An insertion mutation in ABCB4 associated with gallbladder mucocele formation in dogs [70]; among others.

Molecular techniques such as conventional PCR or qPCR have also facilitated research in detection of pathogens in plants, animals, and the environment; understanding of their epidemiology; and, development of new diagnostic tests, treatments or vaccines. Conventional PCR or PCR based methods are being applied to identification and characterization of specific pathogens of animals, e.g., infectious bursal disease virus in avian samples [71]; bovine respiratory syncytial virus [72]; *Actinobacillus pleuropneumoniae* from samples of pigs [73]; canine parvovirus type 2 (CPV 2) in faecal samples of dogs [74]; feline leukemia virus (FeLV) and feline immunodeficiency virus (FIV) [75]; among others. Nucleic acid based detection methods are also important to identification of foodborne pathogens, such as *Listeria monocytogenes* [76]; *Campylobacter* spp., *Salmonella enterica,* and *Escherichia coli* O157:H7 [77].

Despite these important applications of molecular methods, one of the purposes with the greatest impact is the detection and characterization of agents with zoonotic potential, such as pandemic (H1N1) influenza [78]; leptospirosis [39]; Canine visceral leishmaniasis [79]; among others.

In summary, PCR has advantages as a diagnostic tool in conventional microbiology, particularly in the detection of slow growing or difficult to cultivate microorganisms, or under special situations in which conventional methods are expensive or hazardous. Due to the stability of DNA, nucleic acid based detection methods can be also used when inhibitory substances, such as antimicrobials or formalin, are present [80]. Therefore, through the use of molecular techniques has been able to identify different pathogens, to elucidate its epidemiology, to achieve standardization of diagnostic methods, and to establish strategies of prevention and control of diseases, advancing in sanitary regulations in different countries.

3. Conclusions

New knowledge has been generated in different fields of science with invention of PCR 25 years ago. The applications of molecular biology have transformed diagnosis, prognosis and treatment of many diseases. Likewise, molecular methodologies to measure and evaluate gene expression have become the key techniques of the post-genomic era. This correlates with the increasing number of reports of molecular technologies to identify and characterize multiple infectious agents and diseases affecting humans, plants, and animals. The above mentioned justifies the establishment of clear regulations and statistical models for evaluation and adoption of these protocols in laboratories of diagnosis [81]. Despite the continuing evolution of molecular biology, future efforts should continue to increase understanding of advantages and disadvantages of molecular methods in diagnosis, and its interpretation within the clinical context. In addition, it is necessary to increase research for the development of guideline for standardization, validation and comparison new molecular diagnostic methods with existing techniques regarding to sample type, sample preparation, PCR amplification, and reporting of results [80]. In conclusion, the development of molecular biology techniques such as PCR and its variants has led to advances in medicine, agriculture, animal science, forensic science and environment, among others; transforming the society and economy, and influencing the quality of life of people and the development of science and countries.

4. References

[1] Saiki, RK; Scharf S; Faloona F; Mullis, KB; Horn, GT; Erlich, HA; and Arnheim, N. (1985). Enzymatic amplification of β-globin genomic sequences and restriction siteanalysis for diagnosis of sickle cell anemia. *Science* 230: 1350-1354.

[2] Mullis, K; Faloona, F; Scharf S; Saiki, RK; Horn, GT; Erlich, HA. (1986). Specific enzymatic amplification of DNA *in vitro*: the polymerase chain reaction. Cold Spring Harbor Symp. *Quant. Biol.* 51: 263-273.

[3] Mullis, K and Faloona, F. (1987). Specific synthesis of DNA *in vitro* via a polymerase-catalyzed chain reaction. *Methods Enzymol.* 155:338-350.

[4] Louie, M; Louie, L; Simor, AE. (2000). The role of DNA amplification of technology in the diagnosis of infection diseases. *CMAJ.* 163(3): 301-9.

[5] Fredriscks, DN and Relman, DA. (1999). Application polymerase chain reactions to the diagnosis of infectious diseases. *Clin. Infect. Dis.* 29: 475-88.

[6] Erlich, HA; Gelfand, D; Snisky, JJ. (1991). Recent advances in the polymerase chain reaction. *Science.* 252: 1643-51.

[7] Tang, YM; Procop, GW; Persong, DH. (1997). Molecular diagnostics of infectious diseases. *Clin. Chem.* 43: 2021-38.

[8] Jackson, CR; Fedorka-Cray, PJ; Barret, JB. (2004). Use of a Genus and Species Specific Multiplex PCR for amplification of Enterococci. *J. Clin Microbiol.* 42: 3558-3565.

[9] Toma, C; Lu, Y; Higa, N; Nakasome, N; Chinen, I; Baschkier, A; Rivas, M; Iwanaga, M. (2003). Multiplex PCR Assay for identification of Human Diarrheagenic Eschericha coli. *J. Clin. Microbiol.* 41: 2669-2671.

[10] Pehler, K; Khanna, M; Water, CR; Henrickson, KJ. (2004). Detection and amplification of human adenovirus species by adenoplex, a multiplex PCR enzyme hybridization assay. *J. Clin. Microbiol.* 42:4070-4076.

[11] Echeverria, JE; Erdman, DD; Swierkosz, EM; Holloway, BP; Anderson, LJ. (1998). Simultaneous detection and identification of human parainfluenza viruses 1,2 and 3 from clinical samples by multiplex PCR. *J. Clin. Microbiol.* 36: 1388-91.

[12] Templeton, KE; Scheltinga, SA; Sillekens, P; Crielaard, JW; van Dam, AP; Goenssens, H; Claas, EC. (2003). Development and clinical evaluation of an internally controlled, single tube multiplex Real Time PCR assay for detection of Legionella pneumophila and other Legionella species. *J. Clin. Microbiol.* 41: 4016-4021.

[13] Hernández-Rodríguez, P; Gómez, Y and Restrepo, CM. (2000). Identification of carries Duchenne and Becker Muscular Dystrophy through gene dosage and DNA polymorphisms. *Biomedica.* 20 (3): 228-237.

[14] Hernández-Rodríguez, P and Restrepo, CM. (2002). Identification of Deletions in Affected of Duchenne and Becker Muscular Dystrophy (DMD/DMB) and Diagnostic of carrier for Molecular Methodologies. *Universitas Scientiarum.* 7 (1): 31-42.

[15] Jann-Yuan, W; Li-Na, N; Chin-Sheng, C; Chung-Yi, H; Shu-Kuan, W; Hsin-Chih, L; Po-Ren, H; Kwen-Tay, L. (2004). Performance assessment of a Nested-PCR assay (the RAPID BAP-MTB) and the BD ProbeTec ET system for detection of Micobacterium tuberculosis in clinical specimens. *J. Clin. Microbiol.* 42: 4599-4603.

[16] Zeaiter, Z; Fournier, PE; Greub, G; Raoult, D. (2003). Diagnosis of Bordertella pertussisand Bordertella parapertussis infections. *J. Clin. Microbiol.* 41: 919-25.

[17] Kitagawa, Y; Ueda, M; Ando, N; Endo, M; Ishibiki, K; Kobayashi, Y. (1996). Rapid diagnosis of methicillin-resistan Staphylococcus aureus bacteremia by Nested Polymerase Chain Reaction. *Ann. Surg.* 224: 665-71.

[18] Jou, NT; Yoshimori, RB; Mason, GR; Louei, JS; Liebling, MR. (2003). Single tube, nested, reverse transcriptase PCR for detection of viable Micobacterium tuberculosis. *J. Clin. Microbiol.* 35: 1161-1165.

[19] Salomon, RN. (1995). Introduction to quantitative reverse transcription polymerase chain reaction. *Diag. Mol. Pathol.* 4:82-84.

[20] Moon, SH; Lee, YJ; Park, SY; Song, KY; Kong, MH; Kim, JH. (2011). The Combined Effects of Ginkgo Biloba Extracts and Aspirin on Viability of SK-N-MC, *Neuroblastoma Cell Line in Hypoxia and Reperfusion Condition*. 49(1):13-19.

[21] Li, J; Huang, X; Xie, X; Wang, J; Duan, M. (2011). Human telomerase reverse transcriptase regulates cyclin D1 and G1/S phase transition in laryngeal squamous carcinoma. *Acta Otolaryngol.* 131(5):546-551.

[22] Puustinen, L; Blazevic, V; Huhti, L; Szakal, ED; Halkosalo, A; Salminen, M; Vesikari, T. (2011). Norovirus genotypes in endemic acute gastroenteritis of infants and children in Finland between 1994 and 2007. . *Epidemiol. Infect.* 14:1-8.

[23] Wang, J; Zhao, ZH; Luo, SJ; Fan, YB. Expression of osteoclast differentiation factor and intercellular adhesion molecule-1 of bone marrow mesenchymal stem cells enhanced with osteogenic differentiation]. Hua Xi Kou Qiang Yi Xue Za Zhi. 23(3):240-3.

[24] Panitsas, FP and Mouzaki, A (2004). Effect of splenectomy on type-1/type-2 cytokine gene expression in a patient with adult idiopathic thrombocytopenic purpura (ITP). *BMC Blood Disord.* 4(1):4.

[25] Higuchi, R; Fokler, C; Dollinger, G; Watson, R. (1993). Kinetic PCR analysis Real Time monitoring of DNA amplification reactions. *BioTechnology.* 11: 1026-30.

[26] Marty, A; Greiner, O; Day, PJR; Gunziger, S; Muhlemann, K; Nadal, D. (2004). Detection Haemophylus influenza Type b by real Time PCR. *J. Clin. Microbiol.* 42: 3813-3815.

[27] Lobert, S; Hiser, L; Correia, JJ. (2010). Expression profiling of tubulin isotypes and microtubule-interacting proteins using real-time polymerase chain reaction. *Methods Cell Biol.* 95:47-58.

[28] Mackay, IM; Arden, KE; Nitsche, A. (2002). Real Time PCR in virology. *Nucleic Acids Res.* 30: 1292-305.

[29] Maibach, RC and Altwegg, M. (2003). Cloning and sequencing an unknown gene of Tropheryma whipplei and deveploment of two LightCycler PCR assay. *Diagn. Microbiol. Infect Dis.* 46:181-7.

[30] Ke, D; Menard, C; Picard, FJ; Boissinot, M; Ouellette, M; Roy, PH. (2000). Development of conventional and Real Time PCR assays for the rapid detection of group B streptococci. *Clin. Chem.* 46: 324-31.

[31] Vlková, B; Szemes, T; Minárik, G; Turna, J; Celec P. (2010). Advances in the research of fetal DNA in maternal plasma for noninvasive prenatal diagnostics. *Med. Sci. Monit.* 16(4):RA85-91.

[32] Moretti, T; Koons, B; Budowle, B. (1998). Enhancement of PCR amplification yield and specificity using AmpliTaq Gold DNA polymerase. *Biotechniques.* 25: 716-22.

[33] Kellogg, DE; Rybalkin, I; Chen, S; Mukhamedova, N; Vlasic, T. (1994). TaqStart antibody: "hot start" PCR facilitated by a neutralizing monoclonal antibody directed against Taq DNA polymerase. *Biotechniques.* 16: 1134-7.

[34] Demeke, T and Jenkins GR (2010). Influence of DNA extraction methods, PCR inhibitors and quantification methods on real-time PCR assay of biotechnology-derived traits. *Anal. Bioanal. Chem.* 396(6):1977-90.

[35] Chagovetz, A and Blair, S. (2009). Real-time DNA microarrays: reality check. *Biochem. Soc. Trans.* 37(Pt 2):471-5.

[36] Smith, CJ and Osborn, AM. (2009). Advantages and limitations of quantitative PCR (Q-PCR)-based approaches in microbial ecology. *FEMS Microbiol. Ecol.* 67(1):6-20.

[37] Giasuddin, AS; Jhuma, KA; Haq AM. (2008). Applications of free circulating nucleic acids in clinical medicine: recent advances. *Bangladesh Med. Res. Counc. Bull.* 34(1):26-32.

[38] Fluit, AC; Visser, MR; Schmitz, FJ. (2001). Molecular detection of antimicrobial resistance. *Clin. Microbiol.* 14(4):836-71.

[39] Hernández-Rodríguez, P; Díaz, C; Dalmau, E; Quintero, G. (2011). A comparison between Polymerase Chain Reaction (PCR) and traditional techniques for the diagnosis of leptospirosis in bovines. *Journal of Microbiological Methods.* 2011. 84: 1-7.

[40] Gomez, AP; Moreno, MJ; Baldrich, RM; Hernández, A. (2008). Endothelin-1 messenger [corrected] ribonucleic acid expression in pulmonary hypertensive and nonhypertensive chickens. Poult Sci. 87(7):1395-401. *Erratum in: Poult Sci.* 2008 Aug;87(8):1689.

[41] Sundsfjord, A; Simonsen, GS; Haldorsen, BC; Haaheim, H; Hjelmevoll, SO; Littauer, P; Dahl, KH. (2004). Genetic methods for detection of antimicrobial resistance. *APMIS.* 112(11-12):815-37.

[42] Courvalin, P and Trieu-Cuot P. Minimizing potential resistance: the molecular view. *Clin. Infect. Dis.* 2001 Sep 15;33 Suppl 3:S138-46.

[43] Weile, J and Knabbe, C. (2009). Current applications and future trends of molecular diagnostics in clinical bacteriology. *Anal. Bioanal. Chem.* 394(3):731-42.

[44] Dreier, J; Störmer, M; Kleesiek K. (2007). Real-time polymerase chain reaction in transfusion medicine: applications for detection of bacterial contamination in blood products. *Transfus. Med. Rev.* 21(3):237-54.

[45] Hutchins, GG and Grabsch, HI. (2009). Molecular pathology the future?. *Surgeon.* 7(6):366-77.

[46] Dreier J, Störmer M, Kleesiek K. (2007). Real-time polymerase chain reaction in transfusion medicine: applications for detection of bacterial contamination in blood products. *Transfus. Med. Rev.* 21(3):237-54.

[47] Dale, SE. (2010). The role of rapid antigen testing for influenza in the era of molecular diagnostics. *Mol. Diagn. Ther.* 2010 Aug 1;14(4):205-14.

[48] Wyczałkowska-Tomasik A, Zegarska J. (2009). Real-time polymerase chain reaction applications in research and clinical molecular diagnostics. *Przegl. Lek.* 66(4):209-12. 46.

[49] Hernández-Rodríguez, P. (2003). Technical molecular: an advance in the diagnosis and knowledge of ocular pathologies. *Journal Science and Technology for Vision and Eye Health.* 1(1): 113-122. 47.

[50] Mahdy MA.(2010a). Gene therapy in glaucoma-part 2: Genetic etiology and gene mapping. *Oman J. Ophthalmol.* 3(2):51-9. 48.

[51] Woloschak, M; Roberts, J; and Kalmon, D. (1994). Loos of heterozygosity at the retinoblastoma locus in human pituitary tumors. *Cancer.* 74(2): 693-96. 49.

[52] Mahdy MA. (2010b). Gene therapy in glaucoma-3: Therapeutic approaches. *Oman J. Ophthalmol.* 3(3):109-16. 50.

[53] Traboulsi, E; Bjorn, A; Lee, AB;, Khamis, AR and Engle, E. (2000). Evidence of genetic heterogeneity in autosomal recessive congenital fibrosis of the extraocular muscles. *American journal of Ophthalmology.* 129(5): 658-662. 51.

[54] Jordan, T; Hanson, Y and Zaletayev, D. (1992). The human PAX6 gene is mutated in two patients with aniridia. *Nat. Genet.* 1: 328-32.52.

[55] Farrar, GJ; Kenna, PF and Humphries, P. (2002). On the genetics of retinitis pigmentosa and on mutation independent approaches to therapeutic intervention. *EMBO J.* 21[5]: 857-64. 53.

[56] Bogdanici, C; Miron, I and Gherghel, D. (2000). Actualities in retinoblastoma´s treatment. *Ophthalmology.* 50(4): 48-54. 54.

[57] Serratrice, J; Desnuelle, C ; Granel, B ; De Roux-Serratrice, C and Weiller, P. (2001). Mitochondrial diseases in adults. *Rev. Med. Interne.* 22 suppl 3 : 356-66 55.

[58] McLaughing, ME; Lin, D; Berson, EL and Dryja, TP. (1993). Recessive mutations in the gene encoding the beta subunit of rod phosphodiesterase in patients with retinitis pigmentosa. *Nat. Genet.* 4: 130-34. 56.

[59] Giasuddin, AS; Jhuma, KA; Haq, AM. (2008). Applications of free circulating nucleic acids in clinical medicine: recent advances. *Bangladesh Med. Res. Counc. Bull.* 34(1):26-32. 57.

[60] Basille, C; Frydman, R; El Aly, A; Hesters, L; Fanchin, R; Tachdjian, G; Steffann, J; LeLorc'h, M; Achour-Frydman, N. (2009). Preimplantation genetic diagnosis: state of the art. *Eur. J. Obstet. Gynecol. Reprod. Biol.* 145(1):9-13. 58.

[61] Vlková, B; Szemes, T; Minárik, G; Turna, J; Celec, P. (2010). Advances in the research of fetal DNA in maternal plasma for noninvasive prenatal diagnostics. *Med. Sci. Monit.* 1;16(4):RA85-91. 59.

[62] Maeda, H; Zhu, BL; Ishikawa, T; Michiue, T. (2010). Forensic molecular pathology of violent deaths. *Forensic. Sci. Int.* 203(1-3):83-92. 60.

[63] Butler, JM. Forensic DNA Typing: *Biology, Technology, and Genetics of STR.* 1 Edition. NY. Academic Press (2005). 61.

[64] Liu P, Li X, Greenspoon SA, Scherer JR, Mathies RA.. (2011). Integrated DNA purification, PCR, sample cleanup, and capillary electrophoresis microchip for forensic human identification. *Lab. Chip.* 21;11(6):1041-8. 62.

[65] Hurth C, Smith SD, Nordquist AR, Lenigk R, Duane B, Nguyen D, Surve A, Hopwood AJ, Estes MD, Yang J, Cai Z, Chen X, Lee-Edghill JG, Moran N, Elliott K, Tully G, Zenhausern F. (2010). An automated instrument for human STR identification: design, characterization, and experimental validation. *Electrophoresis.* 31(21):3510-3517. 63.

[66] Alves, L; Hülsmeyer, V; Jaggy, A; Fischer, A; Leeb, T; Drögemüller M. (2011). Polymorphisms in the ABCB1 Gene in Phenobarbital Responsive and Resistant Idiopathic Epileptic Border Collies. *J. Vet. Intern. Med.* doi: 10.1111/j.1939-1676.2011.0718.x. [Epub ahead of print]. 64.

[67] Ogino, A; Kohama, N; Ishikawa, S; Tomita, K; Nonaka, S; Shimizu, K; Tanabe, Y; Okawa, H; Morita, M. (2011). A novel mutation of the bovine EDA gene associated with anhidrotic ectodermal dysplasia in Holstein cattle. *Hereditas.* 148(1):46-9. 65.

[68] Li, Y; Sun, A; Su, S; Zhao, P; Cui, Z; Zhu, H. (2011). Deletion of the Meq gene significantly decreases immunosuppression in chickens caused by pathogenic Marek's disease virus. *Virol. J.* 8:2. 66.

[69] Beggs, AH; Böhm, J; Snead, E; Kozlowski, M; Maurer, M; Minor, K; Childers, MK; Taylor, SM; Hitte, C; Mickelson, JR; Guo, LT; Mizisin, AP; Buj-Bello, A; Tiret, L; Laporte, J; Shelton, GD. (2010). MTM1 mutation associated with X-linked myotubular myopathy in Labrador Retrievers. *Proc. Natl. Acad. Sci.* 107(33):14697-702. 67.

[70] Mealey, KL; Minch, JD; White, SN; Snekvik, KR; Mattoon, JS. (2010). An insertion mutation in ABCB4 is associated with gallbladder mucocele formation in dogs. *Comp. Hepatol.* 9:6. 68.

[71] Cardoso, TC; Rosa, AC; Astolphi, RD; Vincente, RM; Novais, JB; Hirata, KY; Luvizotto, MC. (2008). Direct detection of infectious bursal disease virus from clinical samples by in situ reverse transcriptase-linked polymerase chain reaction. *Avian. Pathol.* 37(4):457-61. 69.

[72] Hakhverdyan, M; Hägglund, S; Larsen, LE; Belák, S. (2005). Evaluation of a single-tube fluorogenic RT-PCR assay for detection of bovine respiratory syncytial virus in clinical samples. *J. Virol. Methods.* 123(2):195-202.70.

[73] Savoye, C; Jobert, JL; Berthelot-Hérault, F; Keribin, AM; Cariolet, R; Morvan, H; Madec, F; Kobisch, M. (2000). A PCR assay used to study aerosol transmission of Actinobacillus pleuropneumoniae from samples of live pigs under experimental conditions. *Vet. Microbiol.* 73(4):337-47. 71.

[74] Kumar, M and Nandi, S. (2010). Development of a SYBR Green based real-time PCR assay for detection and quantitation of canine parvovirus in faecal samples. *J. Virol. Methods*. 169(1):198-201. 72.

[75] Arjona, A; Barquero, N; Doménech, A; Tejerizo, G; Collado, VM; Toural, C; Martín, D; Gomez-Lucia, E. (2006). Evaluation of a novel nested PCR for the routine diagnosis of feline leukemia virus (FeLV) and feline immunodeficiency virus (FIV). *J. Feline Med. Surg.* 9(1):14-22. 73

[76] Leclercq, A; Chenal-Francisque, V; Dieye, H; Cantinelli, T; Drali, R; Brisse, S; Lecuit, M. (2011). Characterization of the novel Listeria monocytogenes PCR serogrouping profile IVb-v1. *Int. J. Food Microbiol.* Mar 21. [Epub ahead of print]. 74.

[77] Jokinen, C; Edge, TA; Ho, S; Koning, W; Laing, C; Mauro, W; Medeiros, D; Miller, J; Robertson, W; Taboada, E; Thomas, JE; Topp, E;, Ziebell, K; Gannon, VP. (2011). Molecular subtypes of Campylobacter spp., Salmonella enterica, and Escherichia coli O157:H7 isolated from faecal and surface water samples in the Oldman River watershed, Alberta, Canada. *Water Res.* 45(3):1247-57. 75.

[78] Slomka, MJ; Densham, AL; Coward, VJ; Essen, S; Brookes, SM; Irvine, RM; Spackman, E; Ridgeon, J; Gardner, R; Hanna, A; Suarez, DL; Brown, IH. (2009). Real time reverse transcription (RRT)-polymerase chain reaction (PCR) methods for detection of pandemic (H1N1) 2009 influenza virus and European swine influenza A virus infections in pigs. *Influenza Other Respi Viruses.* 4(5):277-93. 76.

[79] Travi, BL; Tabares, CJ; Cadena, H; Ferro, C; Osorio, Y. (2001). Canine visceral leishmaniasis in Colombia: relationship between clinical and parasitologic status and infectivity for sand flies. *Am. J. Trop. Med. Hyg.* 64(3-4):119-24. 77.

[80] Pusterla, N; Madigan, JE; Leutenegger, CM. (2006). Real-time polymerase chain reaction: a novel molecular diagnostic tool for equine infectious diseases. *J. Vet. Intern. Med.* 20(1):3-12. 78.

[81] Niedrig, M; Schmitz, H; Becker, S; Günter, S; Meulen, J; Meter, H; Ellerbrok, H; Nitsche, A; Gelderblom, HR; Drosten, C. (2004). First international quality assurance study on the rapid detection of viral agents of bioterrorism. *J. Clin. Microbiol.* 42: 1753–5. 79.

9

The Application of PCR-Based Methods in Food Control Agencies – A Review

Azuka Iwobi, Ingrid Huber and Ulrich Busch
Bavarian Health and Food Safety Authority, Oberschleissheim
Germany

1. Introduction

In food control laboratories the world over, molecular biological techniques play an increasingly central role in the analysis of food and food ingredients. Although the classical methods employing cultural, biochemical, cytological and immunological procedures are still being commonly practiced, molecular biological tools employing polymerase chain reaction (PCR) have become an increasingly popular alternative in many food control agencies in recent years. Factors responsible for the popularity of PCR-based detection assays include rapidity, specificity and enhanced sensitivity of the assays. With regard to the latter, often highly denatured food samples and ingredients can still be processed for PCR detection assays because the DNA may still be reliably amplified, as opposed to loss of processing material in detection methods relying on protein analytical tools.

Microbial source tracking (MST) which involves the ability to trace microbes, particularly food-borne pathogens, poses unique challenges to the food industry and food regulatory agencies (Santo Domingo and Sadowsky, 2007). Such information would assist regulatory agencies in localizing food producers or vendors responsible for supplying foods involved in human infections. Additionally, such knowledge would afford public health investigators the opportunity to track food-borne disease outbreaks to their point of origin, thereby preventing future occurrences. In providing such crucial information reliably and within the shortest possible time frame, MST employs a number of PCR-based detection assays. The recent outbreak of EHEC infections arising from verocytotoxin-producing *Escherichia coli* EHEC O104:H4, predominantly in Germany furnishes a good example of the importance of a rapid screening tool for the prompt identification of an infectious agent and surveillance monitoring. More than 16 countries in Europe and North America reported a total of 4,075 cases and 50 deaths as of July 21 2011, two months after the first reported case at the beginning of May 2011 (WHO International Health Regulations, Outbreaks of E. coli O104:H4 infection, Update 30) .

In this and other similar cases, PCR-based molecular biological methods are usually employed in the rapid and initial screening of samples, while complementing this approach with the classical cultural technique for reliable end-identification of the isolate. While not replacing the classical methodologies that have stood the test of time, PCR-based molecular approaches are rapidly becoming the initial screening tools in diverse food analytical processes. Commonly the molecular biological methods are supplemented with classical diagnostic tools to reach a definitive consensus before prosecution for negligent practice or falsified declaration by food producers and processors is effected by food control agencies. This review looks at the plethora of PCR-based approaches in food control laboratories, from pathogen detection and control, food allergen and GMO detection and quantitative determination, to animal species verification.

2. Molecular biology tools for detection of foodborne pathogens

In many food control agencies worldwide, continuous effort is devoted to risk monitoring assessments and evolvement of novel strategies for more rapid and reliable detection of the medically relevant enteropathogens. Although the ultimate goal is a zero-reduction of the pathogens in food, especially meat products and fresh produce, the quantitative microbiological risk-assessment has become an increasingly important parameter in predicting the infectious potential of a given food matrix (FAO/WHO, 2002). The medically relevant species are usually bacterial in origin, and include among others thermophilic *Campylobacter* spp., *Salmonella* spp. , enterohaemorrhagic *Escherichia coli* (EHEC), *Listeria monocytogenes, Bacillus cereus, Clostridium* spp. and *Shigellla* spp. Typical clinical symptoms include diarrhea, which could be self-limiting, invasive or bloody, and vomiting. In Europe, salmonellosis and campylobacteriosis account for the most cases of notified bacterial infections, while listeriosis, although less commonly reported accounts for the most mortalities. In the USA, bacterial pathogens like *Salmonella* and *Campylobacter* are also prevalent, but surveillance of food borne illness is complicated by underreporting (European Food Safety Authority, EFSA 2009, Mead *et al.*, 1999).

The traditional culture-based enumeration of the bacteria is often laborious and time-consuming. A typical detection assay for *Campylobacter* for example, requires up to 5 days, with enrichment. Additionally, the bacterial strain of interest can be frequently overlooked when only culture-based enumeration techniques are employed, due to a strong background of microflora that obscure the accurate detection and quantitative estimation of the pathogen. PCR-based detection of pathogens has therefore become increasingly popular in recent times. Effective PCR-detection assays have been successfully designed and implemented for a broad range of these bacterial food- borne pathogens such as *Salmonella, Campylobacter, Bacillus cereus*, pathogenic *Escherichia coli* (EHEC) and others (Anderson *et al*, 2010, Lehmann et al., 2010, Josefsen *et al.*, 2010, Fratamico *et al.*, 2011, Wang *et al.*, 2011).

2.1 PCR-based food - borne pathogen (bacteria) detection

On a global scale, the food sector remains a major player in the lives and well being of the general human population, and considerable trust and confidence is invested in it by consumers. When food-borne related illnesses or epidemics hit the headlines, the public is understandably disturbed and clamour for tighter regulations and more effective surveillance of food products. The food distribution chain is however a very complex one and tracing the origin of a food outbreak can be very difficult to achieve. In an attempt to address the challenges facing the food sector as regards protecting consumer trust and confidence, the Federation of Veterinarians of Europe (FVE) introduced the "stable to table approach" of food safety (FVE Food safety report). The concept involves a holistic approach embracing all elements, which may have an impact on the safety of food, at every level of the food chain from the stable to the table. Accordingly, the phrase is used to encompass not only the production of all foods of animal origin (including meat, milk, eggs, fish and other products from aquaculture), but fruits and vegetables as well. Applying this approach means that food safety is not solely a matter of inspection at the slaughterhouse or processing plants as has traditionally been the case. On the contrary, this system emphasises the need for interaction between all participants in the entire food chain, from the animal feed manufacturer down to the individual consumer.

In Europe, a Rapid Alert System for Food and Feed (RASFF) was implemented in 1979, to provide food and feed control authorities an effective tool to exchange information about measures taken in responding to serious risks detected in relation to food or feed. This exchange of information helps Member States to act more rapidly and in a coordinated manner in response to a health threat caused by food or feed. In 2010, more than 3,358 notifications were transmitted through the RASFF, with cases of food poisoning accounting for 60 of such reports (Rapid Alert Systems for Food and Feed (RASFF) Annual Report 2010).

A major advantage in the application of PCR-based methodologies lies in the fact that such assays are generally more specific, sensitive, and faster than conventional microbiological assays. However the inherent complexities and composition of food matrices hampers the direct application of PCR detection assays, requiring a pre-enrichment step, thus increasing the processing time for the analysis of the food sample. Nevertheless the simplicity and time saving feature of the PCR reaction has made it increasingly applicable for detection of bacterial pathogens in food. For reliable detection of possible contaminants in the PCR reaction, it is essential to include appropriate negative controls, both during DNA extraction procedures (extraction control) and during the PCR reaction (master mix control). Additionally, it is essential to monitor or detect possible inhibitors that could hamper the efficiency of the PCR reaction. There are a number of possibilities to detect such PCR inhibitors, the commonest of which is to include in each PCR run, an inhibition control, or an internal amplifications control (IAC). The requirement for inclusion of an appropriate IAC for each PCR run is non-negotiable and is in fact jointly stipulated by the International Standard Organization (ISO) and the European Standardization Committee (CEN) in a general guiding policy for PCR reactions in food analytical procedures (EN ISO22174). The choice of the IAC may vary from an artificial DNA molecule which is co-amplified with the same primers for the target DNA (competitive IAC), to a foreign DNA molecule which is coamplified in the PCR reaction with a different primer set (non-competitive) (Hoorfar *et al.*, 2004).

An example of a typical real-time PCR based approach for the detection of *Salmonella*, against the backdrop of the traditional cultural enumeration is outlined below. For the routine or traditional culture-based enumeration, an appropriate amount of the probe is inoculated in buffered peptone water. The culture is incubated at 37 °C for 18 - 24 h, followed by subculture in parallel, on a semi-solid MSRV plate (Rappaport-Vassilidis-Medium) and in Rappaport-Bouillon for 18-24 h at 43±1°C. On day 3, *Salmonella* suspects are then subcultured in parallel on XLD and Rambach agar, according to standard procedures. Presumptive *Salmonella* colonies are then confirmed by serotyping.

With the traditional culture enumeration, outlined above, up to 5 days must be allowed for a definite identification of the bacteria. Sometimes, *Salmonella* positive probes can be completely missed with the conventional cultural enumeration due to strong growth of accompanying flora as mentioned previously. In contrast, a real-time PCR assay for *Salmonella* detection can be completed in less than 2 days, with an initial and shortened pre-enrichment step. In a comprehensive study by Anderson *et al.*, 2010, such a real-time PCR assay was described for the qualitative detection of *Salmonella* in several food samples. More than 1,900 natural food samples were analyzed in this study and the method was found to be robust and resulted in reliable identification of the bacteria in as little as 28 hr, in contrast to 4 or 5 days with conventional *Salmonella* diagnostics. An internal amplification control, which is co-amplified in a duplex PCR reaction, was included in the assay.

As mentioned previously, a number of real-time PCR assays have been published for several important food pathogens. Fricker *et al.*, (2007) reported on the successful application of real-time PCR in the detection of *B. cereus,* which together with the closely associated *S. aureus* are the two most important bacteria responsible for food-associated intoxications. The traditional detection of the emetic toxin associated with these bacteria is often difficult, time consuming and expensive. With the described real-time PCR assay, a first diagnosis can be achieved within 30 hours, greatly accelerating the potential for rapidly implementing risk assessment studies for different food products or matrices. In another study, the successful implementation of multiplex real-time PCR assays in the detection of neurotoxin producing *Clostridium botulinum* in clinical, food and environmental samples was described (De Medici *et al.*, 2009, Messelhäusser *et al.*, 2011a and b).

A more recent approach is the quantitative real-time PCR assay. Various possibilities exist for quantification strategies, one of which is the employment of a CFU-based standard curve for quantification. Briefly, the bacteria of interest are grown or cultivated according to standard procedures and a serial dilution of the bacteria, spanning a representative colony concentration (say 10^1 to 10^8 cells) is plotted as a standard curve. With this curve, the unknown concentration of bacteria in a food sample can be calculated. A second possibility is the employment of a serial dilution of bacterial DNA for the generation of a standard curve for quantification (see fig. 3). In a recent study by Josefsen et al., 2010, a CFU-based standard curve was utilized in the quantitative determination of *Campylobacter* in chicken rinse (Josefsen *et al.*, 2010). In this work, the quantification method was compared with culture-based enumeration on 50 naturally infected chickens. The cell contents correlated with cycle threshold (C_T)* values with a quantification range of 1×10^2 to 1×10^7 CFU/ml). In a previous study, Yang *et al.*, (2003) also successfully applied a real-time PCR assay for quantitative detection of *C. jejuni* in poultry, milk and environmental water. Such quantification strategies are increasingly in demand and a number of commercial products are now available for such purposes.

Although the PCR method has evolved as a very powerful analytical tool indeed, a limitation of such methods is that the DNA analysis will generate results of all the bacteria present in the food sample or probe, irrespective of the status of the cells - whether the cells are live and viable or dead. Thus data for dead or inactivated bacteria which might not be significant from an epidemiological viewpoint are invariably included in such quantitative assays. An improvement in such analysis is the use of an appropriate DNA intercalating dye to distinguish dead from viable and viable, but non-culturable (VBNC) bacteria. Propidium monoazide (PMA) is one such chemical which selectively penetrates only into 'dead' bacterial cells with compromised membrane integrity but not into live cells with intact cell membranes (Nocker *et al.*, 2006, 2009, Pan and Breidt, 2007). PMA possesses an azide group whch permits cross-linking of the dye to DNA after exposure to strong visible light. When the PMA-treated cells are subjected to DNA extraction procedures and subsequently PCR for detection of the bacteria of interest, a reduction in the number of detectable bacteria is often observed with PMA-treated cells (Josefsen *et al.*, 2010). The PMA approach is currently being developed and validated in our laboratory for the reliable identification and quantification of viable and live bacterial pathogens in various food matrices.

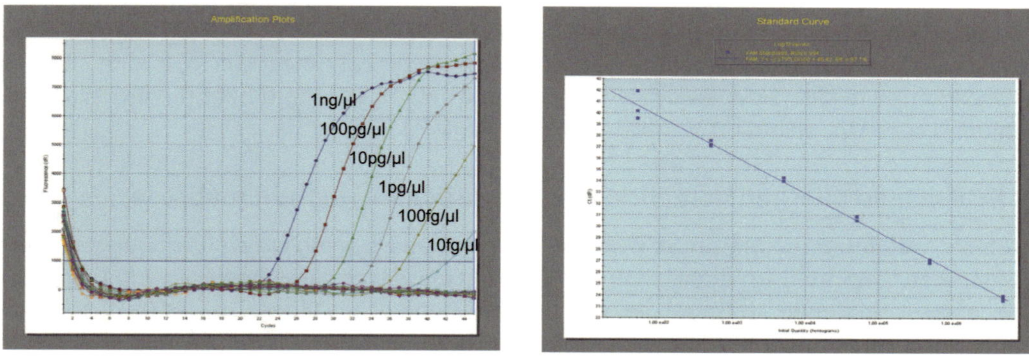

Fig. 1. Principle behind the quantitative PCR approach. A serial dilution of bacterial genomic DNA (fig. 1a) or DNA extracted from a dilution series of appropriate bacterial CFUs (fig. 1b) forms the basis for the calculation of a standard curve for quantification.*

2.2 PCR detection of food-borne viruses

A number of viruses associated with food infections are increasingly becoming important in recent years. The most relevant species are the norovirus, hepatitis-A virus, sapovirus, adenovirus, rotavirus, enterovirus and others. One category of implicated foods is those that are minimally processed, such as fresh produce and vegetables and bivalve molluscs. These are typically contaminated with viruses in the primary production environment. In addition, many of the documented outbreaks of foodborne viral illness have been linked to contamination of prepared, ready-to-eat food by an infected food handler. While in many countries viruses are now considered to be an extremely common cause of foodborne illness, they are rarely diagnosed as the analytical and diagnostic tools for such viruses are not widely available (Microbiological risk assessment series 13, 2008, WHO). Attempts have been made to implement PCR approaches in detection of food-bone viruses. While the overwhelming majority of food-associated viruses are RNA viruses, the RT-PCR (reverse transcription-PCR in which a reverse transcription step converting the viral RNA to template DNA precedes the PCR reaction) is the gold standard for analysis (Höhne and Schreier, 2004). Transferring the traditional and established methods for medical viral diagnosis to a food analytical setting is not readily implementable. While the viral particle load in human and animal tissues or organs is considerably great, the viral load in food samples is usually quite low - in some cases only 10-100 virions may be present in a food probe. Visualization of such a very low viral presence with electron microscopic means and detection of the viral protein through ELISA or latex tests would be impossible where the detection limits of such methods lie within the 10^5 to 10^6 virus particle range pro gram food. The PCR approach is in this regard the most promising of all techniques because the detection limit with RT-PCR lies in the 10^1 to 10^3 virus particle/ g food range (Koopmans und Duizer, 2004).

(* a threshold for detection of DNA-based fluorescence is set slightly above background. The number of cycles at which the fluorescence exceeds the threshold is called the cycle threshold).

Adequate care has to be however taken while subjecting the sample to extraction procedures for optimal yield of high quality nucleic acid (Croci *et al.*, 2003, De Husman *et al.*, 2007). Examples of successful application of the RT-PCR technique include the detection of norovirus in raspberries associated with a gastroenteritis outbreak, and the detection of the virus in oysters from China and Japan (Phan *et al.*, 2007). Other PCR-based methods that have been developed include a nested RT-PCR approach, real-time RT-PCR, and the limited application of nucleic acid sequence-based amplification, among others (Jean *et al.*, 2001, Kojima *et al.*, 2002, Nishida *et al.*, 2003, Beuret *et al.*, 2004).

3. PCR-based allergen detection and quantification in food matrices

Globally, millions of people suffer from allergic reactions to food, which fortunately in most cases range from mild to minor symptoms. In some extreme cases however, food allergies can trigger moderate to more severe life threatening reactions. In contrast to food intolerance, which is also a common form of an adverse reaction to food arising for example from an enzymatic deficiency, such as lactose intolerance, food allergies are immune-mediated. Usually a protein in the food is mistakenly recognized as harmful, triggering the recruitment of IgE antibody with a subsequent allergic reaction (Bush and Hefle, 1996). Symptoms may vary from dermatitis, gastrointestinal and respiratory distress to life-threatening anaphylactic shock. The most common food substances, accounting for almost 90 % of all allergic food reactions are milk, egg, peanut, tree nuts, fish, shellfish, soy, and wheat.

In order to protect consumer safety and health, the EU Labelling Directive (Directive 2000/13/EC) and its later amendments specifically mandate the labelling of allergenic foods. The Labelling Directive requires food manufacturers to declare all ingredients present in pre-packaged foods sold in the EU allowing very few exceptions. In order to respond to our rapidly changing times, this directive has been amended a number of times with regard to allergens. The two most important amendments are: Directive 2003/89/EC introduced Annex IIIa, which is a list of allergenic foods that must always be labelled when present as ingredient in a product, and Directive 2007/68/EC which contains the most recent amendment of Annex IIIa. The latter lists all the allergenic foods that must be labelled as well as a few products derived from those foods for which allergen labelling is not required (European Commission, 2000, 2003, and 2006).

Food allergies are present in about 1-3 % of the global adult population, while in children, a slightly higher incidence (4-6 %) has been documented (Bock et al., 2001). While some of these allergies may be shed when children approach adolescence and adulthood, a few of them are present for life, such as peanut and shellfish allergies. A need for careful labelling of food and food ingredients is strongly underscored by the fact that in some cases, even very minute amounts of an allergen can trigger such life-threatening anaphylactic responses like biphasic anaphylaxis and vasodilation, requiring immediate emergency intervention. Threshold doses for peanut allergic reactions have been found to range from as low as 100 μg up to 1g of peanut protein (Hourihane *et al.*, 1997, Poms *et al.*, 2007).

A variety of techniques have evolved over the years for the detection and possible quantification of the most common food allergens. Protein-based methods that have been employed include the RAST (radio-allergosorbent test, Holgate *et al.*, 2001), RIE (rocket immuno-electrophoresis, Malmheden, *et al.*, 1994) and the ELISA (enzyme-linked immunosorbent assay, Hefle *et al.*, 2001 and Hlywka *et al.*, 2000). The ELISA method is by far the most common and is routinely employed in various food analysis labs due to its high precision, simple handling and good potential for standardization. Additionally, quantitative data are possible with the ELISA technique (Shim and Wanasundara, 2008). However results generated with the ELISA method must be sometimes taken with caution as substantial differences in the detectable protein from the standard on which the test is based, resulting for example from variations in the processing of the food matrix, might lead to false results. Recently, PCR-based detection of allergens has become increasingly popular. A major advantage in the employment of PCR-based methods lies in the high specificity of the reaction. Additionally, proteins in foods that have been harshly processed, might not be detectable in the classical ELISA based approach for example, while the target DNA might be nevertheless efficiently extracted under such denaturing conditions. Another advantage that the PCR holds out against the classical protein-based analytical methods is its stability against the backdrop of geographical and seasonal variations in fruits and nuts for example, with accompanying variance in protein composition (Poms *et al.*, 2007).

Hupfer and colleagues have developed and validated a number of molecular-biology based methods for the detection of a number of allergens, notably celery, lupine and cashew nut (Demmel *et al.*, 2008, Hupfer *et al.*, 2006, Ehlert *et al.*, 2008). A typical scheme for the development and validation of an allergen, with celery as an example is described below (Fig. 2). Other studies have successfully identified and quantified allergens in various food matrices such as the work by Hirao and colleagues who developed a PCR method for quantification of buckwheat by using a unique internal standard. Food-labelling regulations in Japan require that buckwheat must be declared on the food label if its protein is present at concentrations higher than a few micrograms per gram, thus the relevance of this study (Hirao et al., 2006). More recently, Mujico and colleagues developed a highly sensitive real-time PCR for quantification of wheat contamination in gluten-free food for celiac patients. The method compared well with the ELISA in efficiency, with a quantification limit of 20 pg DNA/mg food sample (Mujico *et al.*, 2011). In addition to the conventional singleplex PCR or real-time PCR reactions for allergenic qualitative detection, attempts have also been made to detect simultaneously more than one allergenic event in a food matrix. This multiplex approach was recently demonstrated by Köppel and colleagues and allows the parallel detection of peanuts, hazelnuts, celery and soya in one multiplex reaction, and the quantitative detection of egg, milk, almond and sesame in another multiplex reaction. The tests exhibited good specificity and sensitivity in the 0.01 % range. Due to comparatively lower DNA content in milk and eggs, the authors reported lower sensitivities for these allergens. Initial comparisons of the generated results with conventional ELISA suggested a qualitative accordance, with low correlation of quantitative data (Köppel *et al.*, 2010a).

Another PCR-based approach partly developed and validated by our laboratory is the simultaneous detection of DNA from various food allergens by ligation-dependent probe amplification (LPA). Ligation dependent PCR is a technique originally used for detection of nucleic acids (Hsuih *et al.*, 1996). Briefly this method employs the ligation of bipartite hybridization probes that bind to a target DNA derived from the foodmatrix under investigation. The target DNA is first denatured according to standard protocols, and then incubated with the LPA probes, allowing binding of the LPA probes to the DNA strand, following which the two probes are ligated in a simple ligation reaction. The resulting oligonucleotide is turn subjected to PCR amplification. The arising PCR amplicon is then subjected to capillary electrophoresis and visualized with laser-induced fluorescence. With this method, the simultaneous detection of DNA from 10 allergens, notably peanuts, cashews, pecans, pistachios, hazelnuts, sesame seeds, macadamia nuts, almonds, walnuts and brazil nuts was possible (Ehlert *et al.*, 2009). Fig. 3 below outlines the principle of the LPA methodology.

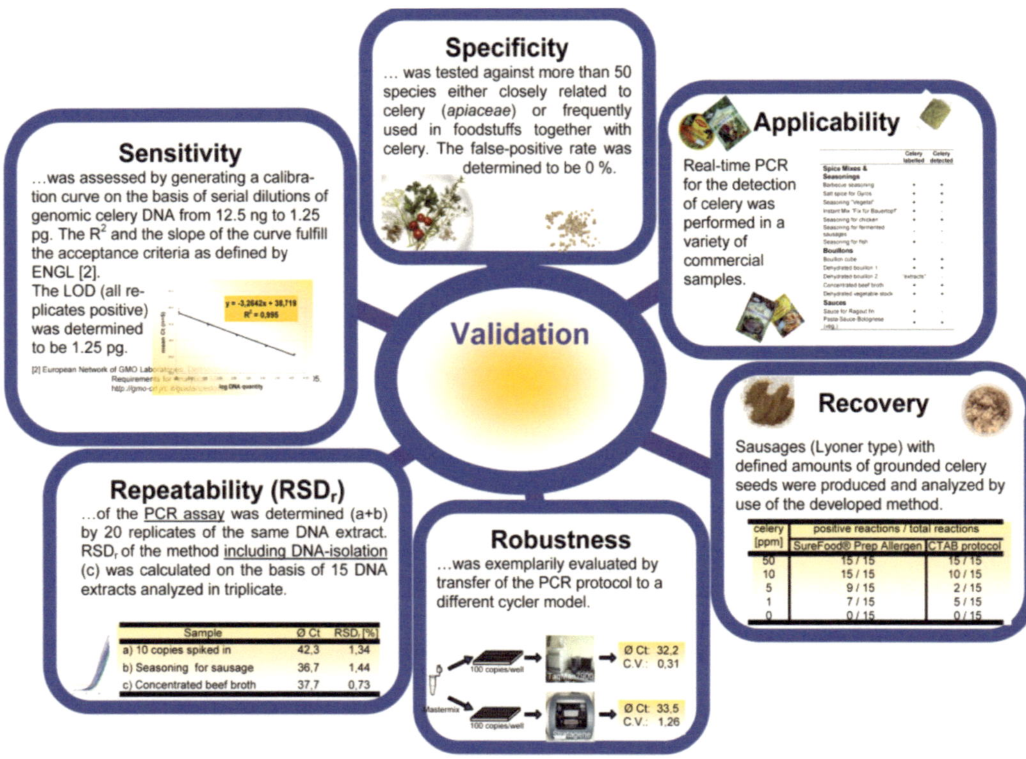

Fig. 2. Development and Validation of a Real-time PCR Detection Method for Celery in Food (Hupfer *et al.*, 2006)

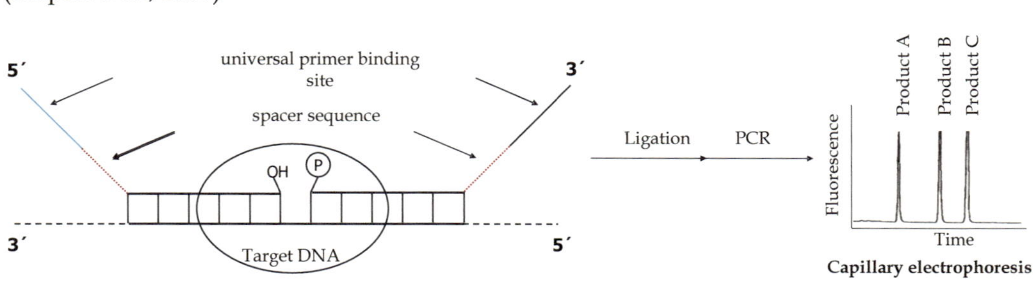

(Demmel *et al.*, 2011, Personal communication)

Fig. 3. Diagrammatic representation of the ligation dependent probe amplification (LPA) approach

4. Application of PCR in animal species detection and differentiation in meat products

A major challenge for food control agencies worldwide is the accurate determination of declared meat components for food and feed ingredients. For protection of consumer trust and confidence and to ensure the quality of meat produce, the verification of declared animal species is important for the following reasons: a) ethical considerations of some might reject the consumption of certain meat products, b) the underlying health condition of some might preclude consuming certain meat products, and c) possible economic loss from the fraudulent substitution of expensive meat components with inferior products (Commission Directive 2002/86/EC, Commission Recommendation 2004/787/EC).

A rapid and dependable detection system is therefore indispensable in a food control agency for protection of consumer trust. In the past, the traditional method for determination of animal species in food relied heavily on immunochemical and electrophoretic analysis of proteins. Although these protein-based analytical methods are still important tools in the food analytical industry, a major drawback in such applications is that in the case of highly processed food, the resulting protein denaturation affects the sensitivity of the procedure. Additionally, such methods may not enable the fine discrimination between closely related animal species like chicken and turkey, or sheep and goat. DNA-based detection systems have thus become increasingly popular in recent times. The distinct advantage of DNA-based detection lies in (1) the increased specificity (generally unambiguous identification of target sequences) and (2) relative stability of the DNA molecule, allowing detection of animal species even in food that have been seriously compromised by excessive processing.

In the early stages, molecular biological methods in species identification were largely based on the use of hybridization of homologous sequences, employing genomic DNA as a species-specific probe, hybridized to DNA extracted from meat samples (Lenstra *et al.*, 2001). Later improvements saw the development of probes derived from species-specific satellite repetitive DNA sequences, making detection of admixtures that account for less than 5 % of the product possible. These methods are however time consuming and quite laborious, with reduced sensitivity in some cases. PCR-based methods have thus become increasingly important in recent times, allowing enhanced sensitivity and specificity of the assays. In most PCR-based approaches, species-specific primers are employed that bind to sequences unique to the species under investigation. Another approach is the employment of universal primers that bind to consensus sequences in all the animal species present in the meat sample. Following amplification, the resulting DNA fragments are subjected to differing analytical procedures for accurate determination of the present species. A popular approach is the use of restriction fragment length polymorphism (RFLP, Fig. 4), which commonly employs restriction digestion assays to generate fragments that are unique to the different animal species present in the sample. Each species is then recognised by its unique restriction fragment pattern (Ong *et al.*, 2007, Girish *et al.*, 2005, Gupta *et al.*, 2008, Meyer *et al.*, 1995). In order to achieve a high level of sensitivity in these assays, especially when universal primers are employed for simultaneous amplification of all present meat species, genes present in multiple copies are usually employed as targets. Prime candidate genes are usually mitochondrial rRNA (12S or 18S) or the phylogenetically robust and highly conserved *cyt b* gene (Kocher *et al.*, 1989, Jain *et al.*, 2007).

In an attempt to simultaneously detect several meat species present in a food sample, several multiplex real-time PCR assays for species differentiation have been described in recent

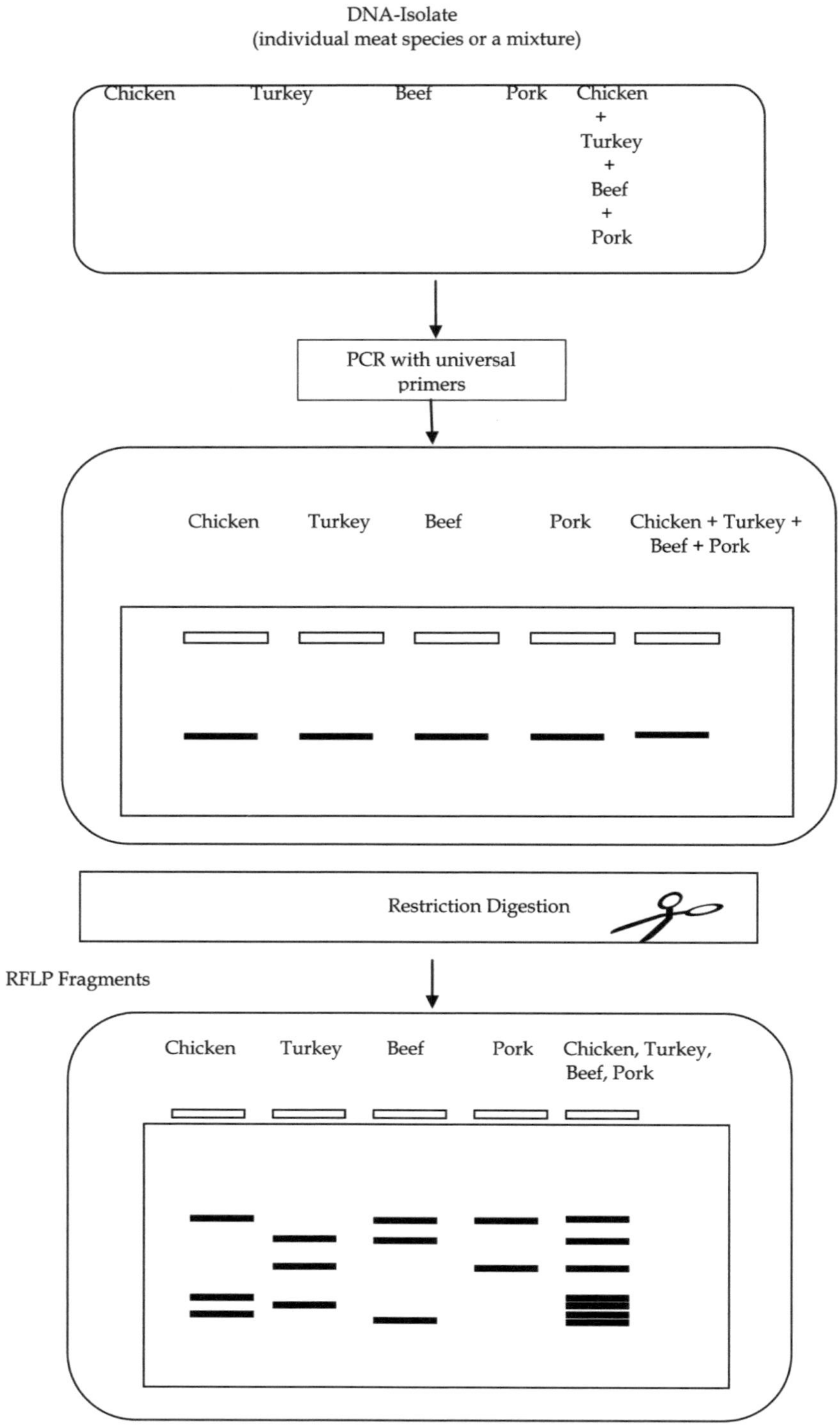

Fig. 4. PCR-Restriction Fragment Length-Polymorphism (PCR-RFLP)

times. Köppel et al. (2009) have for example described the implementation of a heptaplex Real-time PCR assay for the simultaneous identification and quantification of DNA from beef, pork, chicken, turkey, horse meat, sheep (mutton) and goat. Although such multiplex approaches will greatly accelerate meat species identification, results generated must be taken with caution as several meat products are produced with widely varying fat and tissue composition, thus the DNA extractable from similar meat products might vary greatly (Laube *et al.*, 2003).

As regards the accurate differentiation of fish species, several PCR assays have been developed. The majority of these assays rely on the application of universal primers for the generation of consensus sequences among various fish species and the subsequent use of restriction digestion to identify restriction fragments or patterns unique to various fish species. Here, as with meat species differentiation, molecular fish identification methods aim at ensuring that consumers get their money's worth when more expensive fish varieties are bought - substitution of expensive fish with much cheaper varieties can be unravelled by such techniques. Additionally, certain individuals are allergic to certain fish proteins and accurate identification of such potential fish allergens is another argument in favour of a robust fish differentiation method.

4.1 DNA Chip Technology in meat species differentiation

The 20th century saw an explosion of computer technology on all fronts. During the 1990s, molecular biology techniques met with computer electronics to see the birth of a DNA Microarray or DNA chip. One of the earliest attempts at microarray technology for global gene expression was reported by Shena et al., 1995, who designed a quantitative high-capacity system for monitoring of gene expression patterns with a complementary DNA microarray for *Arabidopsis*. Today microarray analyses are widely implemented in molecular biology laboratories, offering the unique advantage of simultaneous analysis of a variety of genetic events in an organism. In food control agencies, the biochip system has also come of age, enabling the quick and efficient analysis of meat products for answers as to their origin and composition.

The first commercial DNA-Chip for the detection of animal constituents in food products is the CarnoCheck Chip (Greiner Biosciences, http://www.greinerbioone.com). The chip allows the simultaneous identification of up to 8 different animal species in processed food and meat products with complex composition. The eight animal species detected by the CarnoCheck Test kit are pig, cattle, sheep, turkey, chicken, horse, donkey, and goat. Following sample homogenization and DNA extraction, a 389-bp fragment of the *cyt b* gene of all the animal species present in the food sample is amplified through polymerase chain reaction. By coupling the fluorophore Cy5 onto one of the primers, the amplified fragments are subsequently labelled in the applied PCR reaction. The labelled fragments are then hybridized to complementary oligonucleotide probes fixed as targets on the bottom of the biochip. The target probes themselves are coupled with the Cy3 fluorophore. Due to the use of fluorophore-labeled PCR primers (Cy5) and fluorophore-labeled target probes for the on-chip control system (Cy3), the analysis of the biochips is performed by microarray scanners using wavelengths of ~532 nm (Cy3) and ~635 nm (Cy5).

Another Biochip test system for species differentiation is the LCD-Array from Chipron. The LCD Array (Chipon Germany, http://chipron.com/index.html) allows the simultaneous detection of up to 14 animal species in food: cattle, buffalo, pig, sheep, goat, horse, donkey,

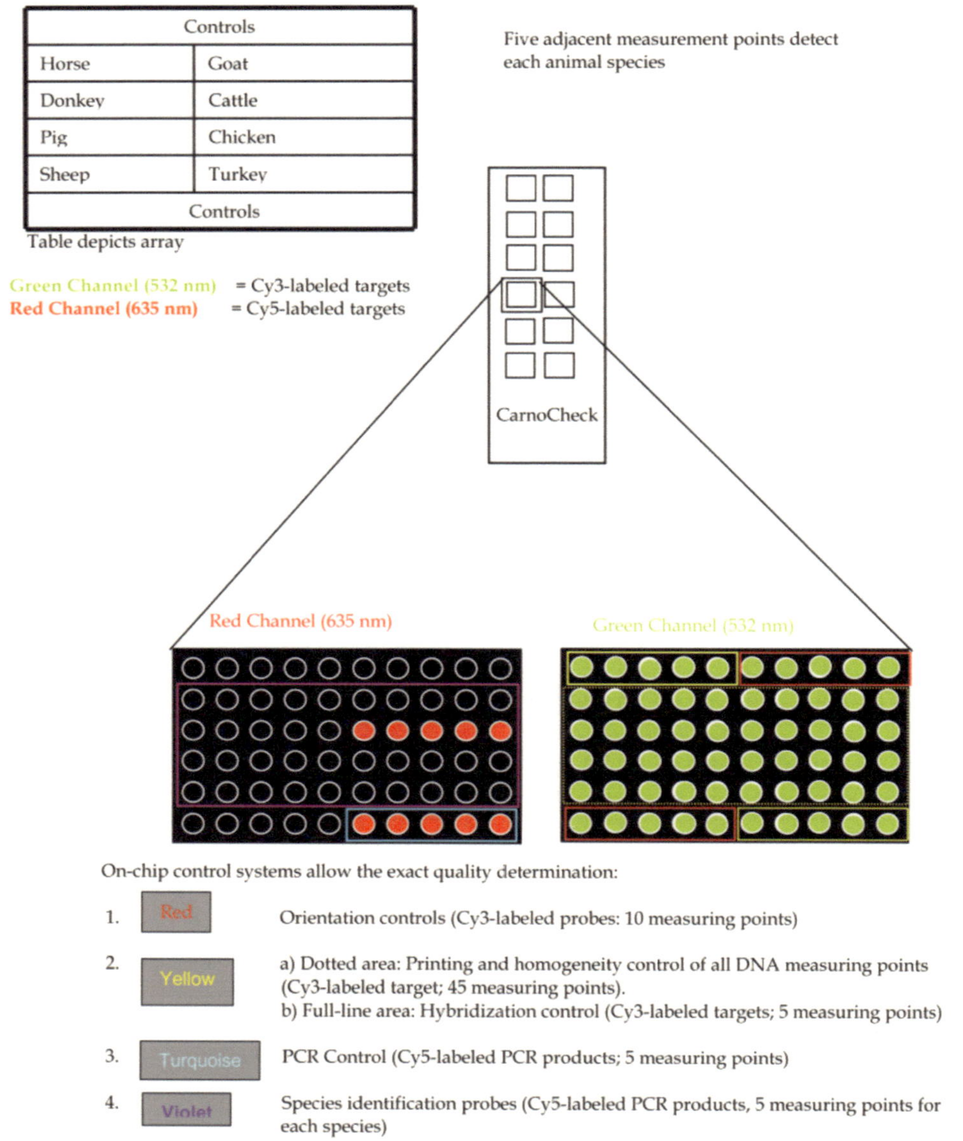

Fig. 5a. CarnoCheck Test kit for the detection of animal species in food. The small table above shows the order of the measurement points for the animal species while the figure below depicts the on-chip control systems for exact quality determination (orientation controls in red, printing controls in green). (CarnoCheck Handbook, Manual version: BQ-020-00, Greiner Bio-one).

rabbit, hare, chicken, turkey, goose, and two duck varieties. The test system here relies on the detection of specific sites within the 16S rRNA mitochondrial locus of all the meat species present in the tested food sample. Included in the test system is a consensus primer pair that amplifies the desired region of the animal species in a PCR. The pre-labeled PCR primer mix provided with the test kit generates biotinylated amplicons of the animal mtDNA present in the food sample. The labelled PCR fragments are then hybridized to the corresponding capture sequences on the individual array fields. The strong affinity between Biotin and streptavidin is exploited by the LCD Array test principle, and positive samples can be visually identified or by employing the scanner and software provided by the kit manufacturer. Figure 5 provides a schematic representation of the two test systems.

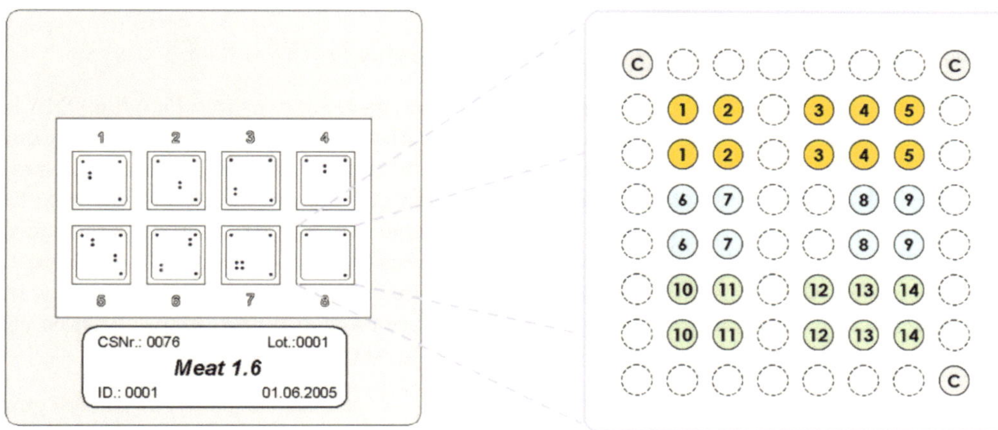

Capture probes

No	Probe	Specificity	No	Probe	Specificity
01	Beef	*Bos taurus*	08	Rabbit	*Oryctolagus cuniculus*
02	Buffalo	*Bubalus bubalis*	09	Hare	*Lepus europaeus*
03	Pork	*Sus scrofa*	10	Chicken	*Gallus gallus*
04	Sheep	*Ovis aries*	11	Turkey	*Meleagris gallopavo*
05	Goat	*Capra hircus*	12	Goose	*Ansa albifrons*
06	Horse	*Equus caballus 1)*	13	Mall. Duck	*Anas platyrhyncos*
07	Donkey	*Equus asinus 1)*	14	Musc. Duck	*Cairina moschata*
			C	Hyb-Contr.	Functional controls (Hybridization + stain)

Fig. 5b. LCD Array Meat 1.6 Test System for meat species identification.
The figure shows the spotting pattern of the array while the table lists the capture probes immobilized on each array (Data Sheet MeatSpecies 1.6, V-I-08, Chipron)

In a recent study, these two biochip test systems were thoroughly validated and approved for routine use in the meat labour of a food control agency (Iwobi *et al.*, 2011). In this study, the two animal species differentiation biochip methods compared well in efficiency and detection limits were found to be in the range of 0.1% to 0.5% in meat admixtures, with good reproducibility of results. More than 70 commercially available meat samples were analyzed in this work, with the results validated against traditional PCR methodology. Although such a simultaneous PCR approach will lead to accelerated analysis of meat species origin in food, while concomitantly revealing possible sources of deliberate adulteration or contamination, the efficiency of the approach is greatly influenced by the overall proficiency of the PCR reaction. In cases where very small amounts of a meat species is present in the food matrix, the amplification of such sequences might be hampered by the presence of other meat species present in more abundance in the sample, leading to possible false negative results. Bai *et al.* (2009) cited the inherent complexity, low amplification efficiency, and unequal amplification efficiency on different templates as major drawbacks of currently described multiplex PCR reactions, thus precluding their commercial application. The biochips here described nevertheless hold great promise in the parallel identification of meat species in food products or samples.

5. GM Food and Feed detection using PCR methods

Genetically modified organisms (GMOs) can be defined as organisms in which the DNA has been altered in a way that does not occur naturally. The technology used is often through recombinant DNA procedures and mainly involves the transfer of genetic material, usually from a microbe as donor to another host, in the context of this review, a plant. The resulting GM plants are then used to grow GM food crops. Generally, all GM crops available on the international market today have been designed to confer one of three basic traits to the plant: resistance to insect damage, resistance to viral infections and tolerance towards certain herbicides. Less common are genetic modifications resulting in plant varieties with altered nutritional values, or longer shelf lives (Holst-Jensen, 2007).

Although the DNA elements of interest mostly derive from microbes, such as the *cry* genes from *Bacillus thuringiensis*, which confer resistance to insects and the *cp4 EPSPS* gene encoded by *Agrobacterium* sp., other eukaryotic hosts may play a role, such as the plant *Petunia hybrida*, which is the source of a chloroplast transit peptide (CTP4). Transformation of the recipient plant cell might be characterized by one or more events or genetic rearrangements. Because current plant transformation procedures do not target specific locations in the recipient's genome, a second transformation event will be directed to a different location within the plant cell, thus making complex, detection of the genetic modification (Holst-Jensen *et al.*, 2006).

From its relatively small beginnings, GM plants have seen a recent explosion in recent times. Worldwide, more than 70 % of all soybeans cultivated are genetically modified, with genetically modified maize accounting for more than a quarter of global outputs. In 2009, for example, genetically modified corn was cultivated in approximately 91 % of all corn fields in the USA. In the most recent report on the Global Status of Commercialized Biotech/GM Crops in 2010, a total of 15.4 million farmers planted biotech crops on an estimated 148 million hectares in 29 countries (James, 2010). Detection and appropriate monitoring strategies are therefore indispensable in many food control agencies.

5.1 Regulation of GMOs

Worldwide, more than 100 genetically modified organisms (GMO) have received authorization for commercial use as food or feed.

Generally, GMOs are regulated by diverse legislation, aimed at protection of consumer safety and health. In the USA, the authorization process is simple and there is no requirement for traceability or labelling of de-regulated (approved) GMOs. In the EU, the GM legislation covering regulatory issues in the approval, detection and monitoring of GMOs is more complex. The authorization and use of genetically modified food and feed is covered by the provisions of regulation EC no. 1829/2003 and EC No. 1830/2003 (EC 2003a and b). In the EU appropriate thresholds have been set for both unintentional presence of GMOs in non-GMO food backgrounds (0.9 % per ingredient), and zero tolerance for non-approved varieties.

5.2 PCR-based detection and quantification of GMOs

Detection of GMOs usually relies on the identification of the altered genotypic locus or the detection of the novel trait or phenotype arising from the genetic modification event. The genetic modification event will usually result in a new phenotypic trait, arising from the production of a new protein of the modified organism. In the context of plants, which account for the greatest number of GM events, such traits could include resistance to herbicides or pests. For detection of the altered phenotypic traits, a number of immunological assays, typically ELISA tests have been developed and even marketed commercially (Anklam *et al.*, 2002, Stave, 2002). For detection of the genotypic trait, the PCR reaction is the most important approach in use. In this context, real-time PCR detection is the preferred method of choice because of its high specificity, its closed amplification system, resulting in fewer contamination incidents, and its potential for quantification of GMO events.

For a reliable PCR, good quality sample DNA is a prerequisite. Adequate care must be taken to ensure that the sample to be tested is truly representative of the matrix and that it has been adequately homogenized. Failure in extraction of adequate amounts of DNA for the PCR can be most readily overcome by increasing the volume of the sampling pool. Care however has to be taken in this regard as increasing the sample pool will also lead to an increased concentration of contaminants or inhibitors that could negatively hamper the PCR (Holst-Jensen, 2007, Anklam *et al.*, 2002).

In the event of a genetic transformation in an organism, not only the gene encoding the novel and desired trait is transferred, but also other important genetic control elements such as for example the strong 35S - Promoter from cauliflower mosaic virus (CaMV), which promotes high-level expression of the encoded trait, and *Agrobacterium tumefaciens nos* terminator (*nos3'*). Additionally, for easier identification of the transformed plant cells, reporter genes are included in the design of the transformation event. Because the above-mentioned markers are commonly found in many GMOs, they are readily employed for the routine screening of GMO events in food. However, the detection of these GMO markers is only an indication that the analyzed sample contains DNA from a GM plant, but does not provide unequivocal information on the specific trait that has been transformed in the plant. To achieve this, target sequences carrying the gene of interest that are characteristic for the transgenic organism must be reliably determined at their junctures with appropriate regulatory sequences (construct-specific detection). However this complete gene construct may have been transformed into different crops. To provide unambiguous verification of the transformation event in the particular plant under study, PCR reactions targeting the junction at the integration site between the plant genome and the inserted DNA or transgene provide the highest level of specificity (event-specific detection). An example of the principle behind the PCR-based detection of genetically modified plant is depicted below (Fig 6).

Several real-time PCR reactions for the detection of GMOs in food have been published in recent times (Gaudron *et al.*, 2009, Kluga *et al.*, 2011, Pansiot *et al.*, 2011). Reiting *et al.*, (2010) for example recently published a testing cascade for the real-time PCR detection of the genetically modified rice Kefeng6 which is unauthorized in Europe. While this work was based on the construct-specific detection of this rice line, our lab recently published and validated an event-specific detection of this rice line, allowing greater specificity in its identification (Guertler *et al.*, 2011, in Press). Additionally, we currently developed a modular approach allowing the simultaneous and parallel detection of several GMOs in a food matrix. With this approach, the detection systems for 15 transgenic maize events were combined in one setup, with additional detection of maize and soybean reference genes (see Fig. 7). The reactions are based on validated single detection systems and are run in parallel with identical temperature profiles, thereby allowing the simultaneous detection of all relevant transgenic events together with corresponding controls for DNA quality, reaction setup and contamination (Gerdes *et al.*, 2011, in Press).

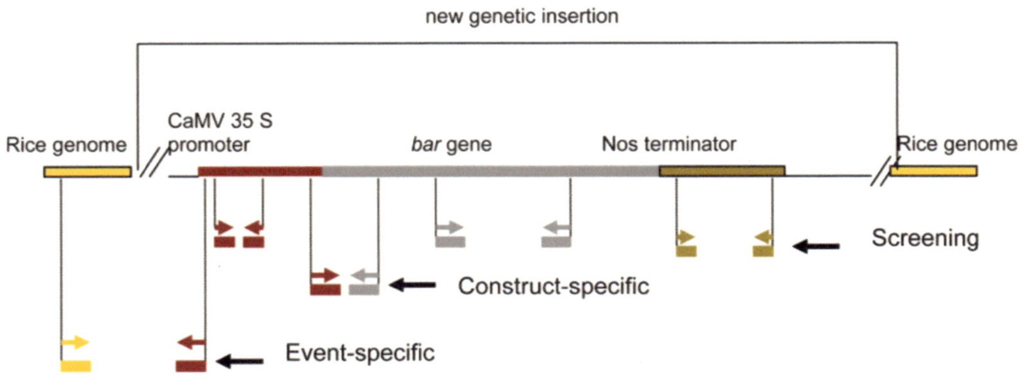

Fig. 6. Principle behind the molecular biological PCR-based detection of a genetic modification event in rice LL601. The commonly employed genetic elements CaMV 35S promoter, the bar gene (encoding herbicide resistance) and *nos* terminator are here depicted for rapid detection of a genetic modification event. The point of integration of the newly inserted genetic element is the basis for the event-specific detection (adapted from Waiblinger, 2010).

Presently, a major challenge in PCR approaches is the development of multiplex assays for the simultaneous quantification of several targets in the same sample. Multiplexing offers the advantage of lower costs and expenditure, and higher throughput compared to single-target assays. Kalogianni *et al.*, (2007) recently reported on a multiplex quantitative PCR based on a multianalyte hybridization assay performed on spectrally encoded microspheres. While these endpoint PCR approaches hold great promises, one major drawback is the requirement of separate steps for DNA amplification and detection of the products. Quantitative real-time PCR which allows continuous monitoring of the amplification products by a homogeneous fluorometric assay account therefore for the most widely used approach in GMO testing (Su *et al.*, 2011, Xu et al., 2011,). In this regard, Köppel and colleagues reported on the development of a multiplex real-time PCR assay for the simultaneous detection and quantification of DNA from three transgenic rice species and construction and application of an artificial oligonucleotide as reference material. Their test exhibited good specificity and sensitivity for the transgenes was in the range of 0.01-1% (Köppel *et al.*, 2010b). In summary, real-time PCR assays remain the gold standard in the analysis of GMO events in food. Because of the trend toward multiple detection events, multiplexing, with microarray-based methods will most likely continue to see greater applications in the future.

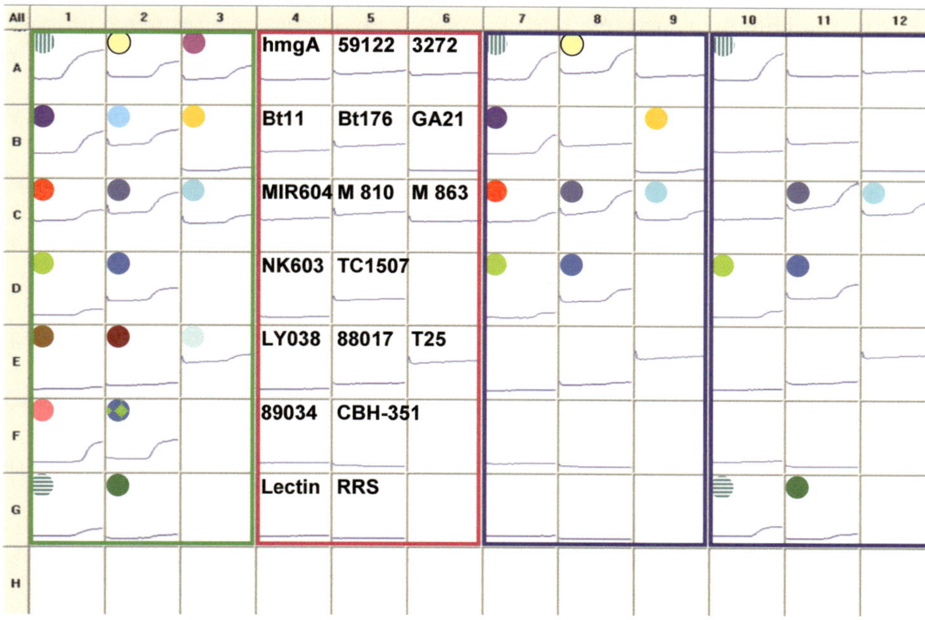

Fig. 7. Analysis of samples with the maize module
Two chocolate bar samples were analysed with the maize module on the Mx3005P. An overview of the recorded FAM fluorescence (R) of all 96 wells is shown. Positive control reactions are enclosed by green, negative control reactions by red, and samples by blue boundaries, respectively. Positive reactions were marked with a coloured dot in the upper left corner. All samples reacted positive for hmgA thus confirming that amplifiable DNA was present. One sample tested positive for eight maize events, the other was positive for four maize events, and RoundupReady soy (RRS).

6. Conclusion

PCR-based applications in food control agencies have seen a tremendous boost in recent years. The simplicity, specificity and rapidity inherent in molecular-based approaches continue to make them increasingly attractive in a wide spectrum of food analytical procedures. Multiplexing applications will continue to see an increase in the near future as the demand for simultaneous detection and quantification of various events in food matrices grows. Additionally, it is expected that increased instrumental development will push the drive toward automation of various analytical procedures commonly employed in food diagnostics.

7. References

Anderson, A., Pietsch, K., Zucker, R., Mayr, A., Müller-Hohe, E., Messelhäusser, U., Sing, A., Busch, U., and Huber, I. Validation of a Duplex Real-Time PCR for the Detection of *Salmonella* spp. in Different Food Products. Food Anal Meth. 2010. 4: 259-267

Anklam, E., Gadani, F., Heinze, P., Pijnenburg, H., Van den Eede, G. 2002. Analytical methods for detection and determination of genetically modified organisms in agricultural crops and plant-derived food products. Eur Food Res Technol 214:3-26

Bai, W., Xu, W., Huang, Y., Cao S., Luo, Y. 2009. A novel common primer multiplex PCR (CP-M-PCR) method for the simultaneous detection of meat species. Food Control 20: 366-370

Beuret, C. 2004. Simultaneous detection of enteric viruses by multiplex real-time RT-PCR. J Virol Methods. 115:1-8

Bock, S.A., Muñoz-Furlong, A., Sampson, H.A.2001. Fatalities due to anaphylactic reactions to foods. J Allergy Clin Immunol 107: 191-193

Bush, R.K., and Hefle, S. 1996. Food allergens. Critical reviews in food science and nutrition. 36: 119-163

Croci, L., De Medici, D., Ciccozzi, M., Di Pasquale, S., Suffredini, E., Toti, L. 2003. Contamination of mussels by hepatitis A virus: a public health problem in southern Italy. Food Control 14: 559-563

Commission Directive 2002/86/EC. L 305/19. 2002. Official Journal of the European Communities

Commission Recommendation 2004/787/EC. L 348/18. 2004. Official Journal of the European Union

De Husman, A.M., Lodder-Verschoor, F., van den Merg, H.H., Le Guyader, F. S., van Pelt, H., van der Poel, W.H., Rutjes, S.A., 2007. Rapid virus detection procedure for molecular tracing of shellfish associated with disease outbreaks. J. Food Prot. 70:967-974

Demmel, A., Hupfer, C., Ilg Hampe, E., Busch, U., and Engel, K.H. 2008. Development of a real-time PCR for the detection of lupine DNA (*Lupinus* species) in foods. J Agric Food Chem 56:4328-4332

De Medici, D., Anniballi, F., Wyatt, G., Lindström, M., Messelhäußer, U., Aldus, C., Delibato, E., Korkeala, H., Peck, M.W., Fenicia, L. 2009. Multiplex PCR for detection of botulinum neurotoxin-producing clostridia in clinical, food and environmental samples. Appl Environ Microbiol 75: 6457-6461

Ehlert, A., Hupfer, C., Demmel, A., Engel, K-H., and Busch, U 2008. Detection of cashew nut in foods by a specific real-time PCR method. Food Anal Methods. 1: 136-143

Ehlert, A., Demmel, A., Hupfer, C., Busch, U., and Engel, K-H. 2009. Simultaneous detection of DNA from 10 food allergens by ligation-dependent probe amplification. Food Additives and Contaminants. 26: 409-418

European Food Safety Authority (EFSA) (2009) Trend and sources of zoonoses and zoonotic agents in the European Union in 2007. EFSA J. 223

European Commission. Regulation (EC) No 1829/2003 of the European Parliament and of the Council of 22 September 2003 on genetically modified food and feed. Off. J. Eur. Union L 268 (2003a) 1-23

European Commission. Regulation (EC) No 1830/2003(b) of the European Parliament and of the Council of 22 September 2003 concerning the traceability and labeling of genetically modified organisms and the traceability of food and feed products produced from genetically modified organisms and amending Directive 2001/18/EC. Official Journal of the European Union L 268/24

FAO/WHO. 2002. Risk assessment of *Salmonella* in eggs and broiler chickens. Microbiological Risk Assessment series no. 2. Switzerland (6446): 566–8.

FVE Food Safety Report. http://www.fve.org/news/publications/pdf/stabletotable.pdf

Fratamico, P.M., Bagi, L.K., Cray Jr.,W.C., Narang, N., Yan, X., Medina, M., and Liu, Y. 2011. Detection by Multiplex Real-Time Polymerase Chain Reaction Assays and Isolation of Shiga Toxin–Producing *Escherichia coli* Serogroups O26, O45, O103, O111, O121, and O145 in Ground Beef. Foodborne Pathogens and Disease. 8: 601-607

Fricker, M., Messelhäußer, U., Busch, U., Scherer, S., and Ehling-Schulz, M. 2007. Diagnostic Real-time PCR assays for the detection of emetic *Bacillus cereus* strains in foods and recent food-borne outbreaks. Appl Environ Microbiol. 73: 1892-1898

Gaudron, T., Peters, C., Boland, E., Steinmetz, A. and Moris, G.2009. Development of a quadruplex-real-time PCR for screening food for genetically modified organisms. Eur Food Res Technol 229:295-305

Gerdes, L., Busch,U. and Pecoraro,S. 2011. Parallelised real-time PCR for identification of maize GMO events. Eur. Food Res. Technol. In Press. DOI: http://10.0.3.239/s00217-011-1634-2

Girish, P., Anjaneyulu, A., Viswas, K., Shivakumar,B., Anand, M., Patel, M., Sharma, B. 2005. Meat species identification by polymerase chain reaction-restriction fragment length polymorphism (PCR-RFLP) of mitochondrial 12S rRNA gene. Meat Science 70: 107-112

Gupta, A. R., Patra, R.C., Das, D.K., Gupta, P.K., Swarup, D., Saini, M. 2008. Sequence characterization and polymerase chain reaction-restriction fragment length polymorphism of the mitochondrial DNA 12S rRNA gene provides a method for species differentiation of deer. In Mitochondrial DNA

Gürtler, P., Huber, I., Pecoraro, S., and Busch, U. 2011. Development of an event-specific detection method for genetically modified rice (Kefeng 6) by means of quantitative real-time PCR. Journal of consumer protection and food safety. In Press.

Hefle, S.L., Jeanniton, E., and Taylor, S.L. 2001. Development of a sandwich enzyme-linked immunosorbent assay for the detection of egg residues in processed foods. Journal of Food Protection 64: 1812-1816

Hirao, T., Hiramoto, M., Imai, S., and Kato, H. 2006. A novel PCR method for quantification of buckwheat by using a unique internal standard material. Journal of Food Protection 69: 2478-2486

Hlywka, J.J., Hefle, S.L., and Taylor, S.L. 2000. A sandwich enzyme-linked immunosorbent assay for the detection of almonds in foods. Journal of Food Protection 63: 252-257

Höhne, M., Schreier, E. 2004. Detection and characterization of norovirus outbreaks in Germany: application of a one-tube RT-PCR using a fluorogenic real-time detection system. J Med Virol 72:312-319

Holgate, S.T., Church, M.K., and Lichtenstein, L.M. 2001. Allergy. 2nd edn. (St Louis, Mosby)

Holst-Jensen, A. Sampling, detection, identification and quantification of genetically modified organisms (GMOs). In: Pico, Y. (ed.) 2007. Food toxicants analysis. Techniques, Strategies and Developments. Elsevier, Amsterdam, Netherlands. ISBN-13:978-0-444-52843-8. Chapter 8, pp. 231-268

Holst-Jensen, A., De Loose, M. and Van den Eede, G. 2006. Coherence between legal requirements and approaches for detection of genetically modified organisms (GMOs) and their derived products. J. Agric. Food Chem. 54: 2799-2809

Hoorfar, J., Malorny, B., Abdulmawjood, A., Cook, N., Wagner, M., Fach, P. 2004. Practical considerations in design of internal amplification control for diagnostic PCR. J.Clin Microbiol 42:1863-1868

Hourihane, J. O'B., Kilburn, S.A., Nordlee, J.A:, Hefle, S.L., Taylor, S.L., and Warner, J. O. 1997. An evaluation of the sensitivity of subjects with peanut allergy to very low doses of peanut protein: a randomised, double-blind, placebo-controlled food challenge study. Journal of Allergy and Clinical Immunology. 100: 596-600

Hsuih, T.C., Park, Y.N., Zaretsky, C., Wu, F., Tyagi, S., Kramer, F.R., Sperling, R., Zhang, D.Y. 1996. Novel ligation-dependent PCR assay for detection of hepatitis C in serum. J Clin Microbiol 34: 501-507

Hupfer, C., Waiblinger, H.U., Busch, U. 2006. Development and validation of a real-time PCR detection method for celery in food. Eur Food Res Technol. 225: 329-335

Kalogianni DP, Elenis DS, Christopoulos TK, Ioannou PC (2007) Multiplex Quantitative Competitive Polymerase Chain Reaction Based on a Multianalyte Hybridization Assay Performed on Spectrally Encoded Microspheres. *Anal Chem* 79:6655–6661

Iwobi, A.N., Huber, H., Hauner, G., Miller, A., and Busch, U. 2011. Biochip technology for the detection of animal species in meat products. Food Analytical Methods. 4: 389-398

Jain, S., Brahmbhait, M.N., Rank, D.N., Joshi, C.G. and Solank, J.V. 2007. Use of *cytochrome b* gene variability in detecting meat species by multiplex PCR assay. Indian Journal of Animal Sciences 77:880-881

James C. 2010. Global Status of Commercialized Biotech/GM Crops. 2010. ISAAA Brief No. 42: ISAAA: Ithaca, NY.

Jean, J., Blais, B., Darveau, A., Fliss, I. 2001. Detection of hepatitis A virus by the nucleic acid sequence-based amplification technique and comparison with reverse transcription-PCR. Appl. Environ Microbiol 67: 5593-5600

Josefsen, M.H., Löfström, C., Hansen, T.B., Christensen, L.S., Olsen, J.E., and Hoorfar, J. 2010. Rapid quantification of viable *Campylobacter* bacteria on chicken carcasses, using real-time PCR and propidium monoazide treatment, as a tool for quantitative risk assessment. Appl. Environ. Microbiol.76: 5097-5104

Kocher, T.D:, Thomas, W.K., Meyer, A., Edwards, S.V., Paabo, S., Villablanca, F.X., Wilson, A.C. 1989. Dynamics of mitochondrial DNA evolution in animals: Amplification and sequencing with conserved primers. Proc. Natl. Acad. Sci. USA 86: 6196-6200

Köppel, R., Dvorak, V., Zimmerli, F., Breitenmoser, A., Eugster, A., Waiblinger, H.-U. 2010a. Two tetraplex real-time PCR for the detection and quantification of DNA from eight allergens in food. Eur Food Res Tech. 230: 367-374

Köppel, R., Zimmerli, F., Breitenmoser, A. 2010b. Multiplex real-time PCR for the simultaneous detection and quantification of DNA from three transgenic rice species and construction and application of an artificial oligonucleotide as reference molecule. Eur Food Res Technol 230:731-736

Köppel, R., Zimmerli, F., Breitenmoser, A. 2009. Heptaplex real-time PCR for the identification and quantification of DNA from beef, pork, chicken, turkey, horse meat, sheep (mutton) and goat. Eur Food Res Technol 230: 125-133

Kluga, L., Folloni, S., Van den Bulcke, M., Van den Eede, G., Querci, M. Applicability of the "Real-Time PCR-Based Ready-to-Use Multi-Target Analytical System for GMO Detection in processed maize matrices. 2011. Eur Food Res Technol. DOI 10.1007/s00217-011-1615-5

Kojima, S., Kageyama, T., Fukushi, S., Hoshino, F., Katayama, K.2002. Genogroup-specific PCR primers for the detection of Norwalk-like viruses. J Virol Methods 100:107-114

Koopmans, M., Duizer, E. 2004. Foodborne viruses: an emerging problem. Inter J Food Microbiol 90: 23-41

Laube, I., Spiegelberg, A., Butschke, A., Zagon, J., Schauzu, M., Kroh, L., Broll, H. 2003. Methods for the detection of beef and pork in foods using real-time polymerase chain reaction. Int J Food Sci Technol. 38: 111-118

Lehmann, L.E., Hunfeld, K.P., Steinbrucker, M., Brade, V., Book, M., Seifert, H., Bingold, T., Hoeft, A., Wissing, H., and Stüber, F. 2010. Improved detection of blood stream pathogens by real-time PCR in severe sepsis. Intensive care medicine 36: 49-56

Lenstra, J.A., Buntjer, J.B., and Janssen, F.W. 2001. On the origin of meat - DNA techniques for species identification in meat products. Veterinary Sciences Tomorrow -15 May 2001

Malmheden, Y.I., Eriksson, A., Everitt, G., Yman, L., and Karlsson, T. 1994. Analysis of food proteins for verification of contamination or mislabelling. Food and Agricultural Immunology 6: 167-172

Mead, P.S., Slutsker, L., Dietz, V., McCaig, L. F., Bresee, J, S., Shapiro, J., Griffin, P.M., and Tauxe R. V. 1999. Food-related illness and death in the United States. Emerg Infect Dis 5: 607-625

Messelhäusser, U., Kämpf, P., Hörmansdorfer, S., Wagner, B., Schalch, B., Busch, U., Höller, C., Wallner, P., Barth, G., Rampp, A. 2011a. Cultural and molecular method for detection of *Mycobacterium tuberculosis* complex und *Mycobacterium avium* ssp. paratuberculosis in milk and dairy products. Appl Environ Microbiol Nov. 4. In Press

Messelhäusser, U., Kämpf, P., Colditz, J., Bauer, H., Schreiner, H., Höller, C., and Busch, U. 2011b. Qualitative and quantitative detection of human pathogenic *Yersinia enterocolitica* in different food matrices at retail level in Bavaria. Foodborne Pathog Dis 1: 39-44.

Meyer, R., Hofelein, C., Lüthy, J., and Candrian, U. 1995. Polymerase chain reaction-restriction fragment length polymorphism analysis: a simple method for species identification in food. J Assoc Off Anal Chem Int 78: 1542-1551

Microbiological Risk assessments series 13: Viruses in food: Scientific advice to support risk management:
http://www.who.int/foodsafety/publications/micro/Viruses_in_food_MRA.pdf

Mujico, J.R., Lombardía, Mena, M., C., Méndez, E., Albar, J.P. 2011. A highly sensitive real-time PCR system for quantification of wheat contamination in gluten-free food for celiac patients. Food Chemistry DOI: 10.1016/j.foodchem.2011.03.061

Nishida, T., Kimura, H., Saitoh, M., Shinohara, M., Kato, M., Fukuda, S., Munemura, T., Mikami, T., Kawamoto, A., Akijama, M., Kato, Y., Nishi, K., Kozawa, K., Nishio, O. 2003. Detection, quantification and phylogenetic analysis of noroviruses in Japanese oysters. Appl Environ Microbiol 69:5782-5786

Nocker, A., Ceung, C.-Y., Camper, A.K. 2006. Comparison of propidium monoazide with ethidium monoazide for differentiation of live vs. dead bacteria by selective removal of DNA from dead cells. J Microbiol Meth 67:310-320.

Nocker, A., and Camper, A.K. 2009. Novel approaches toward preferential detection of viable cells using nucleic acid amplification techniques. FEMS Microbiol Lett 291:137-142

Ong, S.B., Zuraini, M.I., Jurin, W.G., Cheah, Y.K., Tunung, R., Chai, L.C., Haryani, Y., Ghazali, F.M., and Son, R. 2007. Meat molecular detection: sensitivity of polymerase chain reaction-restriction fragment length polymorphism in species differentiation of meat from animal origin. ASEAN Food Journal 14: 51-59

Pan, Y., and Breidt, Jr. 2007. Enumeration of viable *Listeria monocytogenes* cells by real-time PCR with propidium monoazide and ethidium monoazide in the presence of dead cells. Appl Environ Microbiol 73: 8028-8031

Pansiot, J., Chaouachi, M., Cavellini, L., Romaniuk, M., Ayadi, M., Bertheau, Y., Laval, V. 2011. Development of two screening duplex PCR assays for genetically modified organism quantification using multiplex real-time PCR master mixes. 2011. Eur Food Res Technol 232:327–334

Phan, T.G., Khamrin, P., Akiyama, M., Yagyu, F., Okitsu, S., Maneekarn, N., Nishio, O., Ushijima, H. 2007. Detection and characterization of norovirus in oysters from China and Japan. Clin. Lab. 53:405-412

Poms, R.E., Klein, C.L., and Anklam, E. 2007. Methods for allergen analysis in food. Food additives and contaminants 21:1-31

Rapid Alert Systems for Food and Feed (RASFF) Annual Report 2010. http://ec.europa.eu/food/food/rapidalert/docs/rasff_annual_report_2010_en.pdf

Reiting, R., Grohmann, L., Mäde. 2010. A testing cascade for the detection of genetically modified rice by real-time PCR in food and its application for detection of an authorised rice line similar to KeFeng6. Journal of consumer protection and food safety. 5: 185-188

Santo Domingo, J.W., Sadowsky, M.J., 2007. Microbial source tracking in: Emerging issues in food safety. pp. 65-91

Shena, M., Shalon, D., Davis, R.W., Brown, P.O. 1995. Quantitative monitoring of gene expression pattern with a complementary DNA microarray. Science 270: 467-470

Shim, Y-Y. and Wanasundara, J.P.D. 2008. Quantitative detection of allergenic protein *Sin a* 1 from yellow mustard (*Sinapsis alba* L.) seeds using enzyme-linked immunosorbent assay. J. Agric. Food Chem. 56: 1184-1192

Stave, J, W. 2002. Protein immunoassay methods for detection of biotech crops. Applications, limitations, and practical considerations. J AOAC Int. 85:780-786

Su, C., Sun, Y., Xie, J., Peng, Y. 2011. A construct-specific qualitative and quantitative PCR detection method of transgenic maize BVLA430101. Eur Food Res Technol 233:117–122

Waiblinger, H-U. 2010. Die Untersuchung auf gentechnische Veränderungen ("GVO – Analytik) In Molekularbiologische Methoden in der Lebensmittelanalytik (German). Springer Verlag Berlin Heideilberg. pp. 147-148

Wang, X., Mair, R., Hatcher, C., Theodore, M.J., Edmond, K., Wu, H.M., Harcourt, B.H, S. Carvalho, M., Pimenta, F., Nymadawa, P., Altantsetseg, D., Kirsch, M., Satola, S.W., Cohn, A., Messonnier, N.E., Mayer, L.W. 2011. Detection of bacterial pathogens in Mongolia meningitis surveillance with a new real-time PCR assay to detect *Haemophilus influenzae*. International Journal of Medical Microbiology 301: 303-309

WHO International Health Regulations, Outbreaks of *E. coli* O104:H4 infection, Update 30. http://www.euro.who.int/en/what-we-do/health-topics/emergencies/international-health-regulations/news/news/2011/07/outbreaks-of-e.-coli-o104h4-infection-update-30

Xu, W.T., Zhang, N., Luo, Y.B., Zhai, Z.F:, Shang, Y., Yan, X.H., Zheng, J.J., Huang, K.L. 2011. Establishment and evaluation of event-specific qualitative and quantitative PCR method for genetically modified soybean DP-356043-5. Eur Food Res Technol 233:685–695

Yang, C., Jiang, Y., Huang, K., Zhu, C., and Yin, Y. 2003. Application of real-time PCR for quantitative detection of *Campylobacter jejuni* in poultry, milk and environmental water. FEMS Immunol Med Microbiol. 38:265-271

10

PCR in Food Analysis

Anja Klančnik[1], Minka Kovač[2],
Nataša Toplak[2], Saša Piskernik[1] and Barbara Jeršek[1]
[1]*Dept. of Food Science and Technology, Biotechnical Faculty, University of Ljubljana,*
[2]*Omega d.o.o., Ljubljana*
Slovenia

1. Introduction

The aim of this chapter is to briefly present polymerase chain reaction (PCR)-based technologies for use in the detection and quantification of different microorganisms in foods, with an emphasis on sample preparation and evaluation of results. Furthermore, we indicate the PCR-based methods that are most commonly used for the typing of bacteria, and in the final section we provide examples of PCR application in the detection of unwanted components in foods.

2. PCR in the analysis of foods

The microbiological safety of food production is a significant concern of regulatory agencies and the food industry. The most important aspect is to avoid potential negative consequences to human health and economic losses, as well as the loss of consumer confidence.

2.1 The basics of PCR

What is PCR? PCR is a technique that is used to amplify a single or a few copies of a piece of nucleic acid, to generate thousands to millions of copies of a particular nucleic acid. It allows much easier characterisation and comparisons of genetic material from different individuals and organisms. Simply stated, it is a "copying machine for DNA molecules". PCR represented a revolution in biological techniques when it was first developed in 1983 by Kary Mullis (Saiki et al., 1985). Mullis won the Nobel Prize for Chemistry in 1993 for his work on the use and development of PCR. PCR allows the biochemist to mimic the natural DNA replication process of a cell in the test-tube.

DNA replication is a biological process in living cells that starts with one double-stranded DNA (dsDNA) molecule and produces two identical (double-stranded) copies of the original dsDNA. Each strand of the original dsDNA serves as a template for the production of the complementary strand. PCR is thus simply the *in-vitro* replication of dsDNA.

PCR is now a common, simple and inexpensive tool that is used in many different areas, from medical and biological research, to veterinary medicine, hospital analyses, forensic sciences, and paternity testing, and in the food and beverage, biotechnology and pharmaceutical industries, among others. PCR is used for different applications, like DNA-based phylogeny, DNA cloning for sequencing, functional analysis of genes, diagnosis of genetic and infectious diseases, human DNA identification, and identification and detection of bacteria and viruses. The principal of PCR is based on thermal cycling, which exploits the thermodynamics of nucleic-acid interactions. The vast majority of PCR machines now use thermal cycling, i.e., alternately heating and cooling of the PCR samples following a defined series of temperature steps. These thermal cycling steps are necessary first to physically separate the two strands in a dsDNA double helix, in the high-temperature process known as DNA melting. At lower temperatures, each strand is then used as a template in dsDNA synthesis, aided by the enzyme DNA polymerase, for the synthesis of the new, complementary, DNA strands.

Each cycle of PCR comprises three different temperature-step processes: denaturation, annealing and elongation. The thermal cycler consists of a metal thermal block with holes for the tubes holding the PCR reaction mixtures. The thermal cycler then raises and lowers the temperature of the block in preprogrammed steps. The first step in thermal cycling, the DNA melting, results in the denaturation of the dsDNA, as it unwinds and separates into single strands (ssDNA) through the breaking of the hydrogen bonding between the base pairs. This step is usually short, at between 10 s and 30 s at 92 °C to 96 °C. The second step is the annealing of the DNA primers to form the complementary sequences to the ssDNA through the formation of hydrogen bonds, which results in two new dsDNAs. These primers are short fragments of DNA that match up to the forming ends of the new DNA sequence of interest. The final step in temperature cycling is the elongation or enzymatic replication of the DNA. In this step, in combination with a positive cation as a catalyst and the required amounts of the complementary deoxynucleotides (dNTPs), the DNA polymerase enzyme is used to start DNA replication at the primer location. Then, to continue the cycling, the dsDNA is heated to separate the strands again, as the whole PCR process begins again. With each PCR cycle, the amount of the DNA segment of interest in a sample thus increases according to the exponent 2. This exponential increase means that one copy becomes two, which then becomes four, which then becomes eight, and so on with each PCR cycle, assuming 100% efficiency of template replication. As the PCR cycling progresses, with the DNA generated itself used as a template for further replication, this sets in motion a chain reaction in which the original DNA template is exponentially amplified. Generally speaking, 35 to 40 cycles are needed to provide sufficient DNA in a sample for further analysis.

The essential components in PCR reactions are the polymerase enzyme, primers, dNTPs, buffer and cations. Every PCR reaction contains a thermostable polymerase, as Taq polymerase or DNA polymerase. DNA polymerase was originally isolated from the bacterium *Thermus aquaticus*, by Thomas Brock in 1965 (Brock & Boylen, 1973; Chien et al., 1976). *T. aquaticus* is a bacterium that lives in hot springs and hydrothermal vents, and it has a DNA polymerase enzyme that can withstand the protein-denaturing conditions that are required during PCR (Chien et al., 1976; Saiki et al., 1988). Therefore, this replaced the DNA polymerase from *Escherichia coli* that was originally used in PCR (Saiki et al., 1985). The optimum temperature for the activity of DNA polymerase is 75 °C to 80 °C, and it has a half-life of 40 min at 95 °C and 9 min at 97.5 °C, although it can replicate a 1.000-base-pair strand of DNA in less than 10 s at 72 °C (Lawyer et al., 1993). Some thermostable DNA polymerases have been isolated from other thermophilic bacteria and archaea, such as Pfu DNA polymerase, which has a 'proofreading' activity, and which is being used instead of (or in combination with) Taq polymerase for high-fidelity DNA amplification. The use of a thermostable DNA polymerase eliminates the need for addition of new polymerase enzyme to the PCR reaction during the thermocycling process, and this represents the key to successful PCR.

The DNA polymerase requires a catalyst in the form of divalent cations, as either the magnesium (Mg^{2+}) or manganese (Mn^{2+}) cations. These cations also serve as a co-factor to help stabilises the two ssDNA strands. The usual concentration of Mg^{2+} in the PCR reaction is approximately 2.5 μM. The right concentration of cations is critical, because at higher concentrations they can promote greater promiscuity of the Taq polymerase.

The primers are oligonucleotides (in PCR, primer pairs are used) that are added as short synthesised DNA fragments that contain sequences that are complementary to the target region of the target DNA molecule. The primers anneal to terminal part of the target sequence that is to be amplified. These primers are key components for the selective and repeated amplification of the target DNA fragments from a pool of DNA, and they are typically 20-25 bases long, and usually not more than 30 bases long (Stock et al., 2009). A given set of primers is used for the amplification of one PCR product. One of the primers anneals to the forward strand and the other to the reverse strand of the DNA molecules during the annealing step. The primers themselves are most commonly synthesised from individual nucleoside phosphoramidites, in a sequence-specific manner. These primers thus readily bind to their respective complementary DNA or RNA strands in a sequence-specific manner, to form duplexes or, less often, hybrids of a higher order. As such, the primers are required for initiation of DNA synthesis, and they thus allow the DNA polymerase to extend the oligonucleotides and replicate the complementary strand. The DNA polymerase, starts replication at the 3'-end of the primer, and complements the opposite strand. A primer with an annealing temperature significantly higher than the reaction annealing temperature can miss-hybridise and extend the DNA at an incorrect location along the DNA sequence, while at a significantly lower temperature than the annealing temperature, the DNA can fail to anneal and extend at all. Primer sequences also need to be chosen to uniquely select for a region of DNA, and to avoid the possibility of miss-hybridisation to a similar sequence nearby. These primers are thus designed using specific tools, such as the Primer Express software (Life Technologies, Carlsbad, USA), or others. A commonly used method in primer design involves a BLAST search, which is a search tool in the GenBank database, whereby all of the possible regions to which a primer can bind are seen. Also, mononucleotide repeats should be avoided in primers, as loop formation can occur, which can contribute to miss-hybridisation. Primers should also not easily anneal with other primers in the PCR mixture, as this can lead to the production of 'primer dimer' products that can contaminate the PCR mixture. Primers should also not anneal to themselves, as internal hairpins and loops can also hinder annealing with the template DNA. Sometimes degenerate primers are used. These are actually mixtures of similar, but not identical, primers. These can be convenient to use if the same gene is to be amplified from different organisms, as the genes themselves are probably similar, but not necessarily identical. The use of such degenerate primers greatly reduces the specificity of the PCR process. Degenerate primers are widely used and have proven to be extremely useful in the field of microbial ecology. They allow for the amplification of genes from microorganisms that have not been cultivated previously, and they allow the recovery of genes from organisms where the genomic information is not available.

In the PCR reaction, the DNA polymerase enzymatically assembles a new dsDNA strand from the DNA building-blocks, the dNTPs, using the ssDNA as a template. These dNTPs are the molecules that when joined together, make up the structural units of RNA and DNA (Bartlett & Stirling, 2003).

Gel electrophoresis is usually performed following PCR, for analytical purposes. This is a method to separate a mixed population of DNA, RNA or PCR products according to their lengths. These nucleic acid molecules are separated by applying an electric field to move the negatively charged molecules through a particular matrix (agarose, polyacrylamide). After the electrophoresis is complete, the molecules in the gel can be stained to make them visible. DNA can be visualised using dyes that intercalate along dsDNA molecules, whereby the bound dyes fluoresce under ultraviolet light (e.g. ethidium bromide, SYBR Green I). The size of a PCR product is determined with the use of a DNA 'ladder'. This is a solution of DNA molecules of different known lengths that are also used in the gel electrophoresis, and these act as known references to estimate the sizes of the unknown DNA molecules (Robyt & White, 1990; Sambrook & Russel, 2001).

2.2 Principles of quantitative PCR

Over the last few years, the development of novel agents and instrumentation platforms that enable the detection of PCR products on a real-time basis has led to the widespread adoption of quantitative real-time PCR (qPCR; also known as Q-PCR/qrt-PCR). qPCR is one of the most powerful technologies in molecular biology. Using qPCR, specific sequences within a DNA or cDNA template can be copied, or 'amplified', many thousand-fold, or up to a million-fold. In conventional PCR, detection and quantification of the amplified sequence are performed at the end of the PCR, after the final PCR cycle, and this involves post-PCR analysis (gel electrophoresis and image analysis; as above). In qPCR, the amount of the PCR product is measured at each cycle. This ability to monitor the reaction during its exponential phase enables the user to determine the initial amount of the target with great precision (Holland et al., 1991; Higuchi et al., 1992; 1993). The quantity can be either an absolute number of copies or a relative amount when normalised to the DNA input or to additional normalising genes. The essential components in qPCR are the same as in standard PCR, the only differences are in the detection (fluorescence dye) of the amplified target, and the requirement for a specific instrumentation platform.

The qPCR instrumentation consists of a thermal cycler, a computer, optics for fluorescence excitation and emission collection (a fluorimeter), and the data acquisition and analysis software. The first qPCR machine was described in 1993 by Higuchi et al. qPCR monitors the actual progress of the PCR and the nature of the amplified products through the measurement of fluorescence. The benefit of qPCR is the use of a PCR 'master mix' (a mix containing all of the essential components for the qPCR). qPCR reactions are usually successfully carried out under the same reaction conditions, or under universal conditions. Also, the use of passive reference dyes is recommended (usually the ROX™ dye), to normalise for non-PCR-related fluctuations in the fluorescence signals.

The amplified target can be detected in two different ways: first, with non-specific fluorescent dyes that intercalate with any dsDNA; and second, with sequence-specific DNA probes that consist of oligonucleotides that are labelled with a fluorescent reporter dye that allows binding to, and thus detection of, only the target DNA that contains the probe sequence. With the use of non-specific fluorescent dyes an increase in the qPCR product during the qPCR leads to an increase in the fluorescence intensity that is measured at each cycle, thus allowing the dsDNA concentrations to be quantified. However, if the specify of the qPCR is limited as these dyes will bind to all of the dsDNA produced within the qPCR. Thus non-specific fluorescent dyes will measure not just the desired qPCR products, but also non-specific PCR products (e.g. including primer dimers). This can potentially interfere with, or prevent, the accurate quantification of the intended target sequence.

More specific detection is possible with the use of sequence-specific DNA probes, which detect only the target DNA sequence. The use of these probes significantly increases the detection specificity and the sensitivity of the method, and it also allows quantification even in the presence of non-specific DNA amplification. A variety of different probes are now used (Molecular Beacon, Scorpion probe, and others), although those most commonly used are hydrolysis probes (TaqMan probes). A hydrolysis probe is labelled with a fluorescent reporter at its 5′-end and with a molecule known as the 'quencher' at its opposite end. When the probe is intact, the close proximity of the reporter and the quencher prevents the detection of the reporter fluorescence. This quenching of the reporter fluorescence by the quencher occurs through the process of fluorescence resonance energy transfer. As the PCR reaction proceeds, during the annealing stage, the primers and the probe are hybridised to the complementary ssDNA strand and the reporter fluorescence remains quenched. Following initiation of polymerisation of the new DNA strand from the primers, the DNA polymerase then reaches the probe, and its 5'-3'-exonuclease activity degrades the probe, which physically breaks the reporter and quencher proximity (Lyamichev et al., 1993). The released emission of the separate fluorescent reporter can then be detected after excitation with an appropriate source of light, which results in an increase in fluorescence. Of note, probes with different fluorescence dye labels can be used in multiplex assays for the detection of several target nucleic acids in a single qPCR reaction.

To understand the benefits of qPCR, an overview of the fundamentals of PCR is necessary. At the start of a PCR reaction, the reagents are in excess, the template and product are at low enough concentrations that the product renaturation does not compete with the primer binding, and the amplification proceeds at a constant, exponential, rate. The point at which the reaction rate ceases to be exponential and enters a linear phase of amplification is extremely variable, even between replicate samples. Then, at a later cycle, the amplification rate drops to near zero (reaches a plateau), and little more PCR product is made. For the sake of accuracy and precision, it is necessary to collect quantitative data at a point in which every sample is in the exponential phase of amplification. Analysis of the reactions during the exponential phase at a given cycle number should theoretically provide several orders of magnitude of dynamic range, which would normally be from 5 to 9 orders of magnitude of quantification. The fluorescence of the PCR products for each sample in every cycle is detected and measured in the qPCR machine, and its geometric increase that corresponds to the exponential increase of the product is used to determine the threshold cycle in each reaction. This collected fluorescence for a positive qPCR reaction is actually seen as a sigmoidal amplification plot, where the fluorescence is plotted against the number of cycles. The different parts of the amplification curves are important. The baseline represents the noise level in the early cycles, and this is subtracted from the fluorescence obtained from the PCR products. The threshold is a level that is adjusted to a value above the baseline that must be located in the exponential phase of the amplification plot, and the threshold cycle (C_T) is the cycle at which the amplification plot crosses the threshold (Bustin & Nolan, 2004; Bustin, 2004; Logan et al., 2009; Raymaekers et al., 2009). A standard curve can be derived from the serial dilutions of positive sample. The slopes of the standard curves (S) and correlation coefficients (R^2) are used to estimate the qPCR efficiency (E) and to assess the linear range of detection and reliability of the qPCR assays used (Bustin, 2004; Rutledge & Côté, 2003).

There are numerous applications for qPCR. It is commonly used for both basic and diagnostic research. Diagnostic qPCR is used to rapidly detect nucleic acids that are diagnostic of, for example, infectious diseases, cancers or genetic abnormalities.

2.3 Food preparation and PCR-based detection of food-borne bacteria

Conventional methods for the detection of pathogens and other microorganisms are based on culture methods, but these are time consuming and laborious, and are no longer compatible with the needs of quality control and diagnostic laboratories to provide rapid results (Perry et al., 2007). In contrast, PCR is a specific and sensitive alternative that can provide accurate results in about 24 h, and this thus opens a lot of possibilities for the direct detection of microorganisms in a food product. The targets in the foods are DNA or RNA of pathogens, as spoilage microorganisms; DNA of moulds that can produce mycotoxins; DNA of bacteria that can produce toxins; and DNA associated with trace components (e.g. allergens, like nuts) or unwanted components for food authenticity (e.g. cows' milk in goats' milk cheese). However, when PCR is applied for detection of pathogens in food products, some problems can be encountered, although many of these can be solved by the use of suitable sample preparation methods (Lantz et al., 1994; Hill, 1996).

2.3.1 Sample preparation

Sample preparation is an important factor for PCR analysis and PCR sensitivity, especially in the direct implementation of PCR to complex foods. Sample treatment prior to PCR is also a complex issue. This mainly arises because of the need to concentrate the target DNA or RNA into the very small volumes used, which are usually 1 µl to 10 µl for PCR samples, and the presence of any PCR inhibitory substances in the samples (Rådström et al., 2004). Preparation of the sample is divided into the collection of the food sample, separation and concentration of any cells in the sample, treatment of these cells (lysis, for cell-wall decomposition), and isolation and purification of DNA (Lantz et al., 1994). The stomacher is the most widely used treatment technique for the recovery of microorganisms in food samples (Jay & Margitic, 1979). Compared to mechanical methods, hand massaging is a milder homogenisation technique (Kanki et al., 2009).

The objectives of sample preparation are to exclude PCR-inhibitory substances that can reduce the amplification capacity and efficiency, to increase the concentration of the target organism/DNA according to the PCR detection limit or quantification range, and to reduce the amount of the heterogeneous bulk sample for the production of a homogeneous sample for amplification, to insure reproducibility and repeatability (Rådström et al., 2004). Many sample-preparation methods are laborious, expensive, and time consuming, or they do not provide the desired template quality. Since sample preparation is a complex step in diagnostic PCR, a large variety of methods have been developed, and all of these methods can affect the PCR analysis differently, in terms of the specificity and sensitivity (Lantz et al., 2000; Germini et al., 2009).

2.3.1.1 Target-cell separation and sample concentration

The first challenge is to chose optimal sample collection and preparation protocols, and to know whether the pathogen contaminates the foods at high levels, or whether it will be necessary to amplify the bacteria with an enrichment culture, or to use other techniques.

The basic processes of the separation and concentration of the cells are centrifugation (physical separation of suspended particles from a liquid medium) and filtration (including ultrafiltration; physical separation of suspended particles by retention on the filtration medium). The homogeneity of a sample can also differ according to the kind of biological matrix from where it originates. Many sample preparation methods use multiple combinations of these basic processes, which can significantly reduce the presence of inhibitors while increasing the PCR sensitivity and specificity. Further modifications to these physical methods have been used, such as aqueous two-phase systems, buoyant-density centrifugation, differential centrifugation, filtration and dilution (Lantz et al., 1996; Lindqvist et al., 1997; Rådström et al., 2004; McKillip et al., 2000; Uyttendaele et al., 1999). Density media, such as Percoll (Pharmacia, Uppsala, Sweden) (Lindqvist et al., 1997) and BactXtractor (Quintessence Research AB, Bålsta, Sweden) (Thisted Lambertz et al., 2000), have been used to concentrate the target organism and to remove PCR-inhibitory substances of different densities. After this treatment, whole cells can be obtained, which can then be used directly as PCR samples. However, if components of the sample matrix have the same density as the cells, these can remain to inhibit the DNA amplification. An advantage of density centrifugation is that the target organism is concentrated, which allows a more rapid detection response. Furthermore, these methods are relatively user friendly (Rådström et al., 2004).

Alternatively, many sample treatment methods have been developed specifically for one type of organism and/or for a particular matrix, and studies have indicated that individual methods can work better for one organism than another. The flotation method, which is based on traditional buoyant density centrifugation, can concentrate the target cells and simultaneously separate them from PCR-inhibitory substances, the background flora and particles from the sample matrix, and it can reduce false-positive PCR results due to DNA from dead cells (Wolffs et al., 2004; 2007). More recent developments here include the concept of matrix solubilisation and the use of bacteriophage-derived capture molecules that are immobilised on beads (Mayrl et al., 2009; Aprodu et al., 2011).

However, pre-PCR processing methods without culture enrichment, such as flotation immunomagnetic separation and filtration, have a quantification limit of approximately 10^2–10^3 CFU/mL or g of sample, due to the loss of target material during sample preparation and the small volumes analysed (Wolfs et al., 2004; 2007; Löfström et al., 2010; Warren et al., 2007). This loss is usually still too high, as most samples in the food production chain are contaminated with something like 10^2 microorganisms/g (Krämer et al., 2011). Therefore, the optimal enrichment should inhibit the growth of background flora, while simultaneously recovering and multiplying the sublethally damaged cells in a standardised manner. An enrichment culture can amplify bacterial cells and PCR can detect the bacteria by sample collection of bacteria from an enrichment broth, extraction of DNA from the bacterial cells, and then PCR (Knutsson et al., 2002).

Most PCR-based assays currently applied to food samples include a pre-enrichment step, which can be 18 h or more, to increase the cell numbers while diluting any potential PCR inhibitors in the food matrix being sampled. There are also numerous reports of the successful application of PCR-based assays to samples enriched for 6 h. Recently, Krämer et al. (2011) presented a novel strategy to enumerate low numbers of *Salmonella* in cork borer samples taken from pig carcasses as a first concept and proof-of-principle for a new sensitive and rapid quantification method based on combined enrichment and qPCR. The novelty of this approach is in the short pre-enrichment step, where for most bacteria, growth is in the log phase. A number of commercial PCR-based kits are also available; e.g., the BAX system developed by Qualicon recommends short culture-based enrichment of the food sample and PCR amplification with gel-based detection of the PCR products (Stewart & Gendel, 1998). Increasingly, alternative methods have been suggested, such as immunomagnetic separation by magnetic beads coated with antibodies (Lantz et al., 1994; Hallier-Soulier & Guillot, 1999).

2.3.1.2 Treatment of cells and DNA extraction

DNA or RNA extraction is the first step in the analysis process, and the sample quality is probably the most important component to ensure the reproducibility of the analysis and to preserve the biological meaning (Bustin & Nolan, 2004; Postollec et al., 2011). Preparation of the template from cells requires lysis (rupture) of the cells (or viruses), to release the DNA or RNA (Lee & Fairchild, 2006). The DNA molecules inside the cell nucleus need be released from the cell by digestion of the cell walls (cell lysis) (Brock, 2000). The appropriate method for cell lysis is usually chosen according to the PCR detection limit, and the rapidity, preparation simplicity, and demand (Klančnik et al., 2003). The effectiveness of this nuclear extraction depends on several features of the bacterial cell wall, and the treatment that is used can be thermal, chemical, detergents, solvents, mechanical, osmotic shock or the action of enzymes.

Nowadays, it is relatively easy to isolate DNA at very high qualitative and quantitative yields. Most procedures use commercial extraction kits, and depending on the food matrix, these can provide satisfactory results as supplied, or after some modifications. Different commercial kits are also available for biochemical DNA extraction, such as Dr. Food™ (Dr. Chip Biotech Inc., Miao-Li, Taiwan), PrepMan (Life Technologies, Carlsbad, USA) (Dahlenborg et al., 2001), Purugene (Gentra Systems Inc., Minneapolis, MN, USA) (Fahle &

Fisher, 2000), QIAamp® (Qiagen, Valencia, CA, USA) (Freise et al., 2001), AccuProbe (Gne-Probe, San Diego, CA, USA), Gene-Trak (Gene-Trak Systems Corp., Hopkinton, MA, USA), BAX (Quallcon Inc., Wilmington, DE) (Bailey, 1998), Probelia (Sanofi-Daignostics Pasteur, Marnes-la-Coquette, France), and TaqMan (Life Technologies, Carlsbad, USA). A broad reactive TaqMan assay has also been reported for the detection of rotavirus serotypes in clinical and environmental samples (Jothikumar et al., 2009). In contrast to DNA, intact RNA extraction is more laborious, especially when from complex or fatty food matrices. Some extraction methods that are compatible with subsequent reverse transcription qPCR have been developed for various foods (de Wet et al., 2008; Ulve et al., 2008). Due to fast degradation, RNA has to be analysed rapidly.

The final stage of sample preparation (isolation and purification of the DNA) can be used with a combination of ultracentrifugation and purification by chromatography, extraction with phenol-chloroform, precipitation with ethanol, and treatment with enzymes (e.g. lizocim). The most useful method of removing the remains of other admixtures while also concentrating the sample is extraction with organic solvents and ethanol precipitation of DNA (Steffan & Atlas, 1991).

2.3.2 PCR-based detection of food-borne bacteria

There are numerous PCR-based methods for the detection of microorganisms cited in the scientific literature. There are also a number of commercially available PCR-based assays that have the convenience of providing most of the reagents and controls that are needed to perform the assay, and which appear to have high sensitivity for detecting microorganism contamination. Some examples are given in Table 1.

2.3.3 PCR inhibition

The use of conventional and qPCR can be restricted by inhibitors of PCR. This is particularly so when the techniques are applied directly to complex biological samples for the detection of microorganisms, such as clinical, environmental and food samples. PCR inhibitors can originate from the sample itself, or as a result of the method used to collect or to prepare the sample. Either way, inhibitors can dramatically reduce the sensitivity and amplification efficiency of PCR (Rådström et al., 2008). Inhibition of qPCR presents additional concerns, as slight variations in amplification efficiencies between samples can drastically affect the accuracy of template quantification (Ramakers et al., 2003).

2.3.3.1 Types of PCR inhibitors

Food samples produce some of the major problems associated with the use of PCR assays due to various PCR inhibitors that can be found in them. Furthermore, it is imperative to provide a method that has a flexible protocol that can be applied to numerous matrix types to efficiently remove these inhibitory substances that interfere with PCR amplification of the intended target. These PCR inhibitors can originate from the original sample or from sample preparation prior to PCR (Table 2).

PCR can be inhibited by inactivation of the thermostable DNA polymerase, degradation or capture of the nucleic acids, and interference with cell lysis (Rossen et al., 1992; Wilson, 1997;

Food matrix	Target bacteria	Sample preparation / DNA isolation	Detection limit	Reference
Milk, raw minced beef, cold smoked sausage, carrots, raw fish	*Yersinia enterocolitica*	Enrichment: tryptone soy broth with added 0.6% yeast extract. Incubation: 18-20 h at 25 °C/ DNeasy ® Blood and Tissue Kit (Qiagen)	0.5 CFU/10 g milk; 5.5 CFU/10 g cold-smoked sausage; 55 CFU/10 g raw minced beef, carrots and fish	Thisted-Lambertz et al., 2008
Chicken breast skin	*Salmonella Agona*, *Salmonella* Enteritidis	Enrichment: phosphate-buffered peptone water. Incubation: 24 h at 37 °C/ boiling with Triton-X	Approx. 1 CFU/10 g	Silva et al., 2011
Pasteurized liquid egg	*S. enterica*, *E. coli* O157:H7, *L. monocytogenes* Scott A	Enrichment: tryptone soy broth Incubation: 15 h at 37 °C/ Chelex100 Resin (Sigma) and Wizard® DNA Clean-Up system (Promega)	10 CFU/25 g whole liquid egg	Germini et al., 2009
Meat, smoked fish, and dairy products, dressing, crème.	*S. aureus*	Selective enrichment: Modified Giolitti and Cantoni broth. Incubation: 18 h at 37 °C/ boiling with Triton X-100	10^0 CFU/10 g sample	Trnčikova et al., 2009
Ground or minced beef, beef burgers, steak tartare, brunch beef, chicken juice.	*E. coli* O157:H7, *Salmonella* Enteritidis, *L. monocytogenes*	Enrichment: Universal enrichment broth. Incubation: 6 h or 20 h at 37 °C/ PrepMan Ultra sample preparation reagent (Life Technology)	2.1–12 CFU/10 g sample after 6 h of enrichment 1.6 CFU/10 g after 20 h of enrichment	Piskernik et al., 2010
Chicken skin rinse	*Campylobacter jejuni*, *Salmonella* Enteritidis	Flotation method for cell separation/ MagNa Pure system automated DNA extraction (Roche)	3×10^3 CFU/mL	Wolffs et al., 2007
Liquid eggs, infant formula.	*B. cereus*	No enrichment, direct extraction of DNA/ DNeasy Tissue kit (Qiagen)	40–80 CFU/mL food	Martinez-Blanch et al., 2009
Lettuce	*E. coli* O157:H7	Activated charcoal coated with bentonite/ Wizard® DNA Clean-Up system (Promega)	5.0 CFU/g	Lee & Levin, 2011

Table 1. Examples of PCR-based methods for detection of different bacteria, with details of food matrix, sample preparation, DNA isolation and detection limits. CFU, colony forming units.

Kainz, 2000; Opel et al., 2010). False-negative results can also occur because of degradation of the target nucleic acid sequences in the sample. The problem can increase with the isolation of bacteria and/or the bacterial DNA directly from a food matrix, with no single sample preparation protocol known to work for every application. When the target of the PCR is microorganisms, an enrichment step can be included if they are present in very low numbers, although most enrichment broths and selective agars contain substances that inhibit the PCR. The important step is to wash the cells collected from an enrichment or agar plate by pelleting them using centrifugation, removing the supernatant, and resuspending the cells in saline or water for the DNA extraction (Lee & Fairchild, 2006). A good sample preparation protocol will focus on the collection of the bacteria, the removal of potential inhibitors in the foodstuff or culture medium, and the concentration of the extracted DNA. Of note, PCR inhibitors are found in all food types, including meat, milk, cheese and spices (Wilson, 1997).

Sample	PCR inhibitor	Example references
DAIRY Milk (raw, skimmed, pasteurised, dry), cheese (dry, soft).	Fat, protein, calcium, chelators, dead cells	Kim et al., 2001; McKillip et al., 2000; Rådström et al., 2004
MEAT Chicken (meat, carcass rinse, skin homogenates, whole leg, sausage, muscle); turkey (leg, muscle, skin, internal organs); beef (ground, mince, roast); pork (ham, minced, raw whole leg, ground, sausage, meat rolls).	Fat, protein, collagen (blood)	Uyttendaele et al., 1999; De Medici et al., 2003; Hudson et al., 2001; Whitehouse & Hottel, 2007; Silva et al., 2011
SEAFOOD Fish (cakes, pudding, marinated, sliced); salmon (smoked), shrimps, shellfish (muscles, oysters).	Phenolic, cresol, aldehyde, protein, fat	Agersborg et al., 1997

Table 2. PCR inhibitors in dairy, meat and seafood samples.

Many potential inhibitors of PCR have not been identified, although some are indeed known. For example, milk contains high levels of cations (Ca^{2+}), proteases, nucleases, fatty acids, and DNA (Bickley et al., 1996). Studies have shown that high levels of oil, salt, carbohydrate, and amino acids have no inhibitory effects; while casein hydrolysate, Ca^{2+}, and certain components of some enrichment broths are inhibitory for PCR. In addition, haem, bile salts, fatty acids, antibodies, and collagen are PCR inhibitors that can be found in meat and liver samples (Lantz et al., 1997) (Table 2). These inhibitors all have variable effects on the PCR reaction, although in general they will make it more difficult to detect low numbers of bacteria or viruses (Lee & Fairchild, 2006).

Another important source of inhibitors of PCR is the materials and reagents that come into contact with the samples during their processing or the DNA purification. These include excess KCl, NaCl and other salts, ionic detergents like sodium deoxycholate, sarkosyl and sodium dodecyl sulphate, ethanol, isopropanol, phenol, xylene, cyanol, and bromophenol blue, among others (Weyant et al., 1990; Beutler et al., 1990; Hoppe et al., 1992).

2.3.3.2 Approaches to overcome inhibition

When PCR inhibition is suspected, the simplest course of action is to dilute the template (and thus also any inhibitors), and to take advantage of the sensitivity of PCR. Inhibition is problematic in many applications of PCR, particularly those involving degraded or low amounts of template DNA, when simply diluting the extract is not desirable. In standard PCR experiments, negative results or unexpectedly low product yields can be indicative of inhibition, provided that the template is known to be present; alternatively, a known amount of non-endogenous DNA can be added to a sample and amplified as an internal positive control. These controls can be used in qPCR, providing quantitative assessments of their performance. Based on modelling individual reaction kinetics and/or on the calculation of amplification efficiency, qPCR also allows inhibited samples to be identified without additional internal positive-control amplifications (Wilson, 1997; King et al., 2009).

The use of a DNA polymerase that is less susceptible to the effects of inhibitory substances is a possible solution to some PCR problems. For example, a number of the newer polymerases, such as Tfl and rTth, are more reliable than Taq polymerase when using PCR templates prepared from meat or cheese samples (Al-Soud & Rådström, 2000). Moreover, the activity of the DNA polymerases in the presence of inhibitors can be improved with the use of some facilitators, such as bovine serum albumin, dimethyl sulfoxide, Tween 20, Triton-X and betaine (Kreader, 1996; Pomp & Medrano, 1991; Al-Soud & Rådström, 2000; Rådström et al., 2004; Wilson, 1997).

2.4 PCR-based typing methods

Characterisation of microbial isolates below the species level generally involves the determination of the strains. Typing methods that describe the intraspecies variability of an organism can be important for many reasons: searching for the origin of an infectious disease outbreak (i.e. the contaminated food); relating individual cases to an outbreak; studying differences in pathogenicity, virulence and biocide resistance; seeking ways for food contamination or microbial source tracking; and selecting starter cultures. Over the last 25 years, the development of different molecular techniques for the study of microbial genomes has led to a large increase in the methods for typing microorganisms. The most ideal method is DNA sequencing, which allows the precise differentiation of strains. However, as this is still technically demanding and relatively expensive, many other DNA-based typing methods are used. Some of these can be used with PCR analyses, as follows.

2.4.1 Amplification profiling

Across any single microbial species, different genes can be particularly variable, and hence they can be used to determine the strain within the microbial species. Multiplex PCR (mPCR) offers one of the possibilities for the screening of different genes. mPCR provides simultaneous analysis of different genes that can be associated with virulence, toxins, antimicrobial resistance, or other properties of different strains. The presence/absence of different genetic factors can be screened for by mPCR, which can then provide the differentiation of strains. The analysis of mPCR amplicons can be performed with gel electrophoresis or directly by qPCR.

Akiba et al. (2011) applied mPCR to the identification of the seven major serovars of *Salmonella*; i.e., Typhimurium, Choleraesuis, Infantis, Hadar, Enteritidis, Dublin and Gallinarum. For this mPCR, they included the *Salmonella*-specific primers from the *invA* gene and serovar-specific primers. Using the primers that target six virulence genes (*fliC, stx1, stx2, eae, rfbE, hlyA*) in the mPCR, this allowed differentiation of the *E. coli* O157:H7 strains from the O25, O26, O55, O78, O103, O111, O127 and O145 *E. coli* serotypes (Bai et al., 2010). *Yersinia enterocolitica* strains have also been differentiated using mPCR, according to the presence or absence of genes that encode virulence-associated properties, by targeting the *ystA, ail, myfA* and *virF* genes (Estrada et al., 2011). mPCR was developed in a study by Kérouanton et al. (2010) as a rapid alternative method to *Listeria monocytogenes* serotyping. *Staphylococcus aureus* strains were typed with a system that used three mPCRs based on the nucleotide sequences of the *coa* genes (Sakai et al., 2008). This system allowed discrimination between eight main staphylocoagulase types (I–VIII) and three sub-types (VIa–VIc), and this represents a rapid method that can be used as an epidemiological tool for *S. aureus* infection. *S. aureus* strains isolated from different food samples were characterised according to the presence of genes encoding four enterotoxins (SEA, SEB, SEC and SED) (Trnčíková et al, 2010). Differentiation of enterotoxinogenic *Bacillus cereus* isolates has been achieved using three mPCRs that targeted first *hbl, nhe, ces* and *cytK1*, and the the *Hbl* (*hblC, hblD, hblA*) and *Nhe* genes (*nheA, nheB, nheC*) (Wehrle et al., 2009).

2.4.2 Amplified fragment length polymorphism

Amplified fragment length polymorphism (AFLP) consists of five steps (Savelkoul et al., 1999). First, microbial DNA is digested with restriction enzymes (for example EcoRI and MseI), to produce many restriction fragments. Next, there is the ligation of adaptors to correspond to the free ends of the restriction fragments. These adaptors contain sequences that are complementary to the restriction enzyme sites and these sequences are used as targets for PCR primer binding and the subsequent amplification of the restriction fragments. The PCR then uses selective primers that usually have 1-3 additional nucleotides on their 3`-ends. Each nucleotide added to the primer reduces the number of PCR products. Polyacrylamide gel electrophoresis usually yields a pattern of 40 to 200 bands. Improvements in AFLP was also obtained by using fluorescently labelled primers (e.g. FAM™, ROX™, JOE™, TAMRA™) for the detection of fragments in an automatic sequencer with a genetic analysis system and size standards, which can automatically analyse these fragments. This provides standardisation of the fragment sizes and facilitates identification of the polymorphic bands.

Hahm et al. (2003) analysed a total of 54 strains of *E. coli* that were isolated from food, clinical and faecal samples. Here they indicated that AFLP was not as good as pulsed-field gel electrophoresis to determine outbreak origins. In contrast, Leung et al. (2004) showed that AFLP analysis can provide discrimination of *E. coli* isolates from bovine, human and pig faecal samples although they used the same restriction enzymes (MseI and EcorI) as Hahm et al. (2003). In another study, Lomonaco et al. (2011) applied AFLP for the typing of 103 *L. monocytogenes* strains isolated from environmental and food samples. They used two sets of restriction enzymes (BamHI/EcoRI for AFLP I, and HindIII/HhaI for AFLP II), indicating that only with the second set of restriction enzymes and the corresponding adaptors and primers could all of the strains be typed and differentiated. AFLP has also been used for taxonomic studies, and Jaimes et al. (2006) suggested that according to AFLP fingerprinting of *Clostridium* spp. strains, two new species could be defined in this genus. Kure et al. (2003) used AFLP for typing *Penicillium commune* and *Penicillium palitans* strains isolated from different cheese factories (air, equipment, plastic film, brine, milk) and samples of semi-hard cheese, through which they demonstrated that the most critical point of unwanted contamination of the cheese was the air in the wrapping room.

2.4.3 Random amplified polymorphic DNA PCR

Random amplified polymorphic DNA (RAPD)-PCR involves PCR amplification of 'random' fragments of DNA with arbitrarily chosen primers that are selected without the knowledge of the sequence of the genome to be typed (Williams et al., 1990). These primers are generally 6–10 base pairs long and the amplification is usually run under low-stringency conditions. The primer can be expected to anneal to many sites in the DNA, and when two correctly oriented primers are close enough, the intervening sequence is amplified. The result is a fingerprint that consists of different amplicons when separated on agarose gels. The major drawback of RAPD-PCR is its reproducibility. The use of a combination of primers in a single PCR and a selection of primer sequences, primer lengths and primer concentrations represent the parameters that can improve its reproducibility (Tyler et al., 1997).

Abufera et al. (2009) characterised *Salmonella* isolates according to the fingerprints obtained with RAPD-PCR, and their results showed correlation between these RAPD profiles and the serogroups. There were close similarities among human isolates, and also among animal isolates. McKnight et al. (2010) use RAPD-PCR for the analysis of *Alicyclobacillus* strains isolated from passion fruit juice, and they showed that *Alicyclobacillus acidoterrestris* was the prevalent strain in these fruit juices, irrespective of the different batches. *A. acidoterrestris* is the main *Alicyclobacillus* species associated with fruit-juice spoilage. As an indicator of ochratoxin A formation isolated from wheat flours, *Penicillium verrucosum* strains were grouped into separate groups according to their RAPD-PCR fingerprints, as were *Penicillium nordicum* reference strains, which suggests a direct application of this method (Cabanas et al., 2008). RAPD-PCR has also been used for differentiation of *Penicillium expansum* strains, as patulin-producing fungi (Elhariry et al., 2011). These strains were isolated from healthy appearing and rot-spotted apples, but genomic fingerprints showed that although strains were clustered into two separate groups, all of strains of *P. expansum* represented potential hazards. A modification of conventional RAPD-PCR is also seen with the application of melting-curve analysis to RAPD-generated DNA fragments (McRAPD) (Deschaght et al., 2010).

2.4.4 Repetitive-element PCR

Repetitive-element (rep)-PCR is based on interspersed repetitive DNA elements of the repetitive extragenic palindrome and enterobacterial repetitive intergenic consensus, which are conserved throughout the eubacterial kingdom (Versalovic et al., 1991). The distribution and frequency of such repetitive DNA elements can be studied with PCR using outwardly directed primers that are specific for the repeat elements. If repeat elements are close enough to each other, amplification of the DNA sequences between them occurs (Versalovic et al., 1991). Rep-PCR products are then separated with agarose gel electrophoresis, and the fingerprints obtained are strain specific and can be used for typing. BOX elements represent another repetitive DNA element, which were introduced by van Belkum et al. (1996); these were also successfully used in rep-PCR. Automated rep-PCR technology is available as a commercial assay through the DiversyLab System®, which does not require gel electrophoresis (Healy et al., 2005). The amplicons are separated using the microfluidics LabChip device, and they are detected using a bioanalyser. The resulting data are automatically collected and analysed using the DiversiLab software.

Although repetitive DNA elements were discovered in the genomes of *E. coli* and the *Salmonella enterica* serovar Typhimurium, these elements were subsequently found in several diverse Gram-negative and Gram-positive bacteria. All species of *Listeria* show repetitive elements of repetitive extragenic palindromes and enterobacterial repetitive intergenic consensus (Jeršek et al., 1996). Jeršek et al. (1999) showed that rep-PCR allowed the grouping of strains of *L. monocytogenes* according to their origins of isolation (clinical, animal and food origins). Indeed, according to rep-PCR fingerprints, Blatter et al. (2010) identified potential *L. monocytogenes* contamination sources in a sandwich-production plant. Finally, rep-PCR was also used for typing *Aspergillus* strains (Healey et al., 2004) for the determination of strain relatedness.

2.4.5 Variable number of tandem repeat assay and its multiple-locus assay

Variable number of tandem repeat (VNTR) assays use the variation in the number of tandem DNA repeats at a specific locus to distinguish between isolates (Keim et al., 2000). Short nucleotide sequences that are repeated several times often vary in the copy number that can be detected with PCR using flanking primers, thus creating length polymorphisms that can be strain specific. To increase the discrimination, Keim et al. (2000) developed the multiple-locus VNTR assay (MLVA). The VNTR and MLVA assays require knowledge of specific DNA sequences and the appropriate design of the primers to amplify the tandem DNA repeats. The PCR products can be separated and detected on agarose gels, and the fingerprints thus produced are analysed. The other possibility is to use fluorescently labelled primers that allow the PCR products to be electrophoretically analysed with an automated capillary DNA sequencer (Keim et al., 2000). Recently, there have been a number of studies that have used VNTR and MLVA for the genotyping of different strains. Keim et al. (2000) developed an MLVA assay for typing *Bacillus anthracis* strains, where they used eight genetic loci that allowed the typing of 426 isolates, which were divided into 89 MLVA genotypes. Cluster analysis of the fingerprints identified six genetically distinct groups, with some of these types showing a worldwide distribution, and others restricted to particular

Food component	Target	PCR	Genetic marker (specific amplicon)	Detection limit	Reference
Allergen	Hazelnut (*Corylus* spp.)	qPCR	*Cor a 1* hazelnut gene (82 bp)	0.1 ng	Arlorio et al., 2007
	Celery (*Apium graveolens*)	qPCR	Mannose-6-phosphate reductase mRNA (77 bp)	10 pg, LOD 0.005% celery	Fuchs et al., 2012
	Lupin (*Lupinus*), soya (*Glycine max*)	Duplex PCR	Mitochondrial tRNA-MET gene (168 bp, 175 bp)	0.001 ng; 2.5 mg per kg food matrix	Galan et al., 2011
Mycotoxin	Aflatoxigenic *Aspergillus* spp.	mPCR	Structural genes *omtB* (1333 bp), *omtA* (1032 bp), *ver-1* (895 bp), and regulatory gene *aflR* (797 bp)	125 pg/µL, 10^5 spores/g meju	Kim et al., 2011
	Patulin producing *Aspergillus* and *Penicillim* spp.	PCR	Isoepoxydon dehydrogenase (*idh*) gene (496 bp)	0.5 ng DNA, 1.8×10^2 to 2.7×10^3 conidia/g in foods	Luque et al., 2011
Authenticit y	Cattle and buffalo milk	Duplex qPCR	Mitochondrial D loop region of cattle, buffalo (126 bp, 226 bp)	0.15 ng buffalo, 0.04 ng cattle DNA; 0.1% adulteration of cow and buffalo milk	De et al., 2011
	Beef meat	qPCR	Bovine-specific cytochrome b gene (*cytb*) (116 bp)	35 pg bovine DNA	Zhang et al., 2007
Bacterial toxin	Staphylococcal enterotoxins in food samples	qPCR	*Sea, seb, sec, sed* genes of *S. aureus*	ND	Trničikova et al., 2010
	Toxinogenic *B. cereus*	Multiplex qPCR	Genes of toxins (*nheA, hblD* and *cytK1*) and emesis (*ces*)	10 CFU/g after overnight enrichment	Wehrle et al., 2010

Table 3. Application of PCR for the detection and quantification of different trace components in foods. LOD, limit of detection in artificially spiked food samples; ND, not defined.

geographic regions. MLVA assays were successfully used for typing *L. monocytogenes* strains using six specific genetic loci (Chen et al., 2011). The MLVA assay discriminated between outbreak isolates and unrelated food, animal and environmental isolates, with identical MLVA patterns seen for known outbreak-related isolates. The typing of *L. monocytogenes* strains with MLVA was also optimised for direct application to food samples. Differentiation of *S. aureus* strains isolated from raw milk and dairy products with MLVA using six tandem repeat loci grouped the strains into seven clusters that revealed clear genomic variability among the strains tested. MLVA assays with eight genetic loci were also developed for *Brucella melitensis* and *Brucella abortus*, which has enabled strain identification and the establishment of source of infection in several cases (Rees et al., 2009).

In the implementation of different typing methods, different parameters need to be considered; i.e. stability, discriminatory power, typing ability, reproducibility and agreement (Belkum et al., 2007). For practical reasons, the cost and availability of equipment also need to be considered.

2.5 PCR as a tool for analysis of trace components in foods

In recent years, PCR technology has been brought into use in other areas of the analysis of foods, such as for the authenticity of food, for food allergens, and for the indirect determination of bacterial toxins and mycotoxins. PCR offers possibilities for these food analyses as the DNA target for this reaction is a very stable and long-live molecule that is present in all organisms. The main problem for these assays is the preparation of the food sample for the analysis, as the concentrations of these unwanted compounds are usually very low. Thus the DNA extraction method has to be very effective to provide a relatively high yield of the target DNA for PCR. The other problem is the standards that are needed as control samples in all cases where qPCR is applied. However, some examples of recently applied assays are listed in Table 3.

3. Conclusion

PCR as a new technique that since its development in 1983 has reached many areas in a short period of time, including that of food analysis. Multiple use of PCR has been most pronounced in the field of food microbiology, although in recent years, PCR has been increasingly used in other areas, such as food hygiene, food toxicology and food analysis. Therefore, this chapter can only provide a brief summary of the various studies and applications of PCR and qPCR in the food industry.

4. References

Abufera, U.; Bhugaloo-Vial, P.; Issack, M. I. & Jaufeerally-Fakim, Y. (2009). Molecular characterization of *Salmonella* isolates by REP-PCR and RAPD analysis. *Infection, Genetics and Evolution*, Vol. 9, No. 3, (May, 2009), pp. 322-327

Agersborg, A.; Dahl, R. & Martinez, I. (1997). Sample preparation and DNA extraction procedures for polymerase chain reaction identification of *Listeria monocytogenes* in seafoods. *International Journal of Food Microbiology*, Vol. 35, No. 3, (December, 1997), pp. 275-280

Akiba, M.; Kusumoto, M. & Iwata, T. (2011). Rapid identification of *Salmonella enterica* serovars, Typhimurium, Choleraesuis, Infantis, Hadar, Enteritidis, Dublin and Gallinarum, by multiplex PCR. *Journal of Microbiological Methods*, Vol. 85, No. 1, (April, 2011), pp. 9–15

Al-Soud, W. A. & Rådström, P. (2000). Effects of amplification facilitators on diagnostic PCR in the presence of blood, feces, and meat. *Journal of Clinical Microbiology*, Vol. 38, No. 12, (December, 2000), pp. 4463–4470

Aprodu, I.; Walcher, G.; Schelin, J.; Hein, I.; Norling, B.; Rådström, P.; Nicolau, A. & Wagner, M. (2011). Advanced sample preparation for the molecular quantification of *Staphylococcus aureus* in artificially and naturally contaminated milk. *International Journal of Food Microbiology*, Vol. 145, No. 1, (March, 2011), pp. 61–65

Arlorio, M.; Cereti, E.; Coïsson, J.D.; Travaglia, F. & Martelli, A. (2007). Detection of hazelnut (*Corylus* spp.) in processed foods using real-time PCR. *Food Control*, Vol. 18, No. 2, (February, 2007), pp. 140–148

Bai, J.; Shi, X. & Nagaraja, T.G. (2010). A multiplex PCR procedure for the detection of six major virulence genes in *Escherichia coli* O157:H7. *Journal of Microbiological Methods*, Vol., 82, No. 1, (July, 2010), pp. 85–89

Bailey, J. S. (1998). Detection of *Salmonella* cells within 24 to 26 hours in poultry samples with the polymerase chain reaction BAX system. *Journal of Food Protection*, Vol. 61, No. 7, July 1998), pp. 792–795

Bartlett, J. M. S. & Stirling, D. (2003). A short history of the polymerase chain reaction. In Methods in Molecular Biology, PCR Protocols, *Humana Press*, Vol. 226, 2 edition., (2003), pp. 3–6

Beutler, E.; Gelbart, T. & Kuhl, W. (1990). Interference of heparin with the polymerase chain reaction. *BioTechniques*, Vol. 9, No. 2, (August, 1990), pp. 166

Bickley, J.; Short, J.K.; McDowell, D.G. & Parkes, H.C. (1996). Polymerase chain reaction (PCR) detection of *Listeria monocytogenes* in diluted milk and reversal of PCR inhibition caused by calcium ions. *Letters in Applied Microbiology*, Vol. 22, No. 2, (February, 1996), pp. 153–158

Blatter, S.; Giezendanner, N.; Stephan, R. & Zweifel, C. (2010). Phenotypic and molecular typing of *Listeria monocytogenes* isolated from the processing environment and products of a sandwich-producing plant. Food Control, Vol. 21, No. 11, (April, 2010), pp. 1519–1523

Brock biology of microorganisms. 2000. 9th ed. Madigan M. T.; Martinko J. M. & Parker J. (eds.). Upper Saddle River, Prentice Hall International, pp. 33-40; 172-177

Brock, T. D. & Boylen, K. L. (1973). Presence of thermophilic bacteria in laundry and domestic hot-water heaters. *Applied Microbiology*, Vol. 25, No. 1, (June, 1973), pp. 72-76

Bustin, S. A. & Nolan, T. (2004). Pitfalls of quantitative real-time reverse-transcription polymerase chain reaction. *Journal of Biomolecular Techniques*, Vol. 15, No. 3, (September, 2004), pp. 155-166

Bustin, S. A. (2004). A-Z of quantitative PCR. International University line, La Jolla, California, (2004), pp. 5-26, 87-112, 244-245

Cabanas, R. ; Bragulat, M. R. ; Abarca, M. L. ; Castella, G. & Cabanes, F. J. (2008). Occurrence of *Penicillium verrucosum* in retail wheat flours from the Spanish market. *Food*

Chen, S.; Li, J.; Saleh-Lakha, S.; Allen, V. & Odumeru, J. (2011). Multiple-locus variable number of tandem repeat analysis (MLVA) of *Listeria monocytogenes* directly in food samples. *International Journal of Food Microbiology,* Vol. 148, No. 1, (July, 2011), pp. 8–14

Chien, A.; Edgar, D. B. &, Trela, J. M. (1976) Deoxyribonucleic acid polymerase from the extreme thermophile *Thermus aquaticus. Journal of Bacteriology,* Vol. 127, No. 3, (September, 1976), pp. 1550–1557

Dahlenborg, M.; Borch, E. & Rådström, P. (2001). Development of a combined selection and enrichment PCR procedure for *Clostridium botulinum* types B, E and F and its use to determine prevalence in fecal samples from slaughter pigs. *Applied and Environmental Microbiology,* Vol. 67, No. 10, (October, 2001), pp. 4781–4789

De Medici, D., L.; Croci, E.; Delibato, S.; Di Pasquale, E. & Toti. L. (2003). Evaluation of DNA extraction methods for use in combination with SYBR Green I real-time PCR to detect *Salmonella enterica* serotype Enteritidis in poultry. *Applied and Environmental Microbiology,* Vol. 69, No. 6, (June, 2001), pp. 3456–3461

de Wet, S. C.; Denman, S. E.; Sly, L. & McSweeney, C. S. (2008). An improved method for RNA extraction from carcass samples for detection of viable *Escherichia coli* O157:H7 by reverse-transcriptase polymerase chain reaction. *Letters in Applied Microbiology,* Vol. 47, No. 5, (November, 2008), pp. 399-404

De, S.; Brahma, B.; Polley, S.; Mukherjee, A.; Banerjee, D.; Gohaina, M.; Singh, K. P.; Singh, R.; Datta, R. K. & Goswami, S. L. (2011). Simplex and duplex PCR assays for species specific identification of cattle and buffalo milk and cheese. *Food Control,* Vol. 22, No. 5, (May, 2011), pp. 690-696

Deschaght, P.; Van Simaey, L.; Decat, E.; Van Mechelen, E.; Brisse, S. & Vaneechoutte, M. (2010). Rapid genotyping of *Achromobacter xylosoxidans, Acinetobacter baumannii, Klebsiella pneumoniae, Pseudomonas aeruginosa* and *Stenotrophomonas maltophilia* isolates using melting curve analysis of RAPD-generated DNA fragments (McRAPD). *Research in Microbiology,* Vol. 162, No. 4, (May, 2011), pp. 386-392

Elhariry, H.; Bahobial,A. A. & Gherbawy, Y. (2011). Genotypic identification of *Penicillium expansum* and the role of processing on patulin presence in juice. *Food and Chemical Toxicology,* Vol. 49, No. 4, (April, 2011), pp. 941–946

Estrada, C. S. M. L.; Velázquez, L. C.; Escudero, M. E.; Favier, G. I.; Lazarte, V. & Stefanini de Guzmán, A. M. (2011). Pulsed field, PCR ribotyping and multiplex PCR analysis of *Yersinia enterocolitica* strains isolated from meat food in San Luis Argentina. *Food Microbiology,* Vol. 28, No. 1, (February, 2011), pp. 21-28

Fahle, G. A. & Fischer, S. H. (2000). Comparison of six commercial DNA extraction kits for recovery of cytomegalovirus DNA from spiked human specimens. *Journal of Clinical Microbiology,* Vol. 38, No. 10, (October, 2000), pp. 3860–3863

Freise, J.; Gerard, H. C.; Bunke, T.; Whittum-Hudson, J. A.; Zeidler, H.; Köhler, L.; Hudson, A. P. & Kuipers, J. G. (2001). Optimised sample DNA preparation for detection of *Chlamydia trachomatis* in synovial tissue by polymerase chain reaction and ligase chain reaction. *Annals of the Rheumatic Diseases,* Vol. 60, No. 2, (February, 2001), pp. 140–145

Fuchs, M.; Cichna-Markl, M. & Hochegger, R. (2012).Development and validation of a novel real-time PCR method for the detection of celery (*Apium graveolens*) in food. *Food Chemistry,* Vol. 130, No. 1, (January, 2012), pp. 189–195

Galan, A. M. G.; Brohée, M.; de Andrade Silva, E.; van Hengel, A. J. & Chassaigne, H. (2011). Development of a real-time PCR method for the simultaneous detection of soya and lupin mitochondrial DNA as markers for the presence of allergens in processed food. *Food Chemistry*, Vol. 127, No. 2, (July, 2011), pp. 834–841

Germini, A.; Masola, A.; Carnevali, P. & Marchelli, R. (2009). Simultaneous detection of *Escherichia coli* O175:H7, *Salmonella* spp., and *Listeria monocytogenes* by multiplex PCR. *Food Control*, Vol. 20, No. 8, (August, 2009), pp. 733–738

Hahm, B. K. ; Maldonadob, Y. ; Schreiberc, E.; Bhuniab, A. K. & Nakatsu, C. H. (2003). Subtyping of foodborne and environmental isolates of *Escherichia coli* by multiplex-PCR, rep-PCR, PFGE, ribotyping and AFLP. *Journal of Microbiological Methods*, Vol. 53, No. 3, (June, 2003), pp. 387- 399

Hallier-Soulier, S. & Guillot, E. (1999). An immunomagnetic separation polymerase chain reaction assay for rapid and ultra-sensitive detection of *Cryptosporidium parvum* in drinking water. *FEMS Microbiological Letters*, Vol. 176, No. 2, (June, 1999), pp. 285–289

Healy, M.; Huong, J.; Schrock, R ; Manry, J.; Renwick, A ; Nieto, R ; Woods, C.; Versalovic, J. &. Lupski, J. (2005). Microbial DNA typing by automated repetitive-sequence-based PCR. *Journal of Clinical Microbiology*, Vol.43, No.1, (October, 2005), pp. 199–207

Higuchi, R.; Dollinger, G.; Walsh, P. S. & Griffith, R. (1992). Simultaneous amplification and detection of specific DNA sequences. *Biotechnology*, Vol. 10, No. 4, (April, 1992), pp. 413–417

Higuchi, R.; Fockler, C.; Dollinger, G. & Watson, R. (1993). Kinetic PCR: Real time monitoring of DNA amplification reactions. Biotechnology, Vol. 11, No. 119, (September, 1993), pp. 1026–1030

Hill, W. E. (1996). The polymerase chain reaction: applications for the detection of foodborne pathogens. *Critical Reviews in Food Science and Nutrition*, Vol. 36, No. 1-2, (January, 1996), pp. 123-173

Holland, P. M.; Abramson, R. D.; Watson, R. & Gelfand, D. H. (1991). Detection of specific polymerase chain reaction product by utilizing the 5' to 3' exonuclease activity of *Thermus aquaticus* DNA polymerase. Proceedings of the National Academy of Sciences USA, Vol. 88, No. 16, (August, 1991), pp. 7276–7280

Hoppe, B. L.; Conti-Tronconi, B. M. & Horton, R. M. (1992). Gel-loading dyes compatible with PCR. *BioTechniques*, Vol. 12, No. 5, (May, 1992), pp. 679–680

Hudson, J. A.; Lake, R. J.; Savill, M. G.; Scholes P. & McCormick, R. E. (2001). Rapid detection of *Listeria monocytogenes* in ham samples using immunomagnetic separation followed by polymerase chain reaction. *Journal of Applied Microbiology*, Vol. 90, No. 4, (April, 2001), pp. 614–621

Jaimes, C. P.; Aristizábal, F. A.; Bernal, M. M.; Suárez, Z. R. & Montoya, D. (2006). AFLP fingerprinting of Colombian *Clostridium* spp. strains, multivariate data analysis and its taxonomical implications. *Journal of Microbiological Methods*, Vol. 67, No. 1, (Ocrober, 2006), pp. 64–69

Jeršek, B.; Gilot, P.; Gubina, M.; Klun, N.; Mehle, J.; Tcherneva, E.; Rijpens, N. & Herman, L. Typing of *Listeria monocytogenes* strains by repetitive element sequence-based PCR. *Journal of Clinical Microbiology*, Vol. 37, No. 1, (January, 1999), pp. 103-109

Jeršek, B.; Tcherneva, E.; Rijpens, N. & Herman L. (1996). Repetitive element sequence-based PCR for species and strain discrimination in the genus *Listeria*. *Letters in Applied Microbiology*, Vol. 23, No. 1, (July, 1996), pp. 55-60

Jothikumar, N.; Kang, G. & Hill, V. R. (2009). Broadly reactive TaqMan® assay for real-time RT-PCR detection of rotavirus in clinical and environmental samples. *Journal of Virological Methods*, Vol. 155, No. 2, (February, 2009), pp. 126–131

Kainz, P. 2000. The PCR plateau phase–Towards an understanding of its limitations. *Biochimica et Biophysica Acta*, Vol. 1494, No. 1-2, (November, 2000), pp. 23–27

Kanki, M.; Sakata, J.; Taguchi, M.; Kumeda, Y.; Ishibashi, M.; Kawai, T.; Kawatsu, K.; Yamasaki, W.; Inoue, K. & Miyahara, M. (2009). Effect of sample preparation and bacterial concentration on *Salmonella enterica* detection in poultry meat using culture methods and PCR assaying of preenrichment broths. *Food Microbiology*, Vol. 26, No. 1, (February, 2009), pp. 1-3

Keim, P.; Price, L. B.; Klevytska, A. M.; Smith, K. L.; Schupp, J. M.; Okinaka, R.; Jackson, P.J. & Hugh-Jones, M. E. (2000). Multiple-locus variable-number tandem repeat analysis reveals genetic relationships within *Bacillus anthracis*. *Journal of Bacteriology*, Vol. 182, No. 10, (May, 2000), pp. 2928–2936

Kérouanton, A.; Marault, M.; Petit, L.; Grout, J.; Dao, T. T. & Brisabois, A. (2010). Evaluation of a multiplex PCR assay as an alternative method for *Listeria monocytogenes* serotyping. *Journal of Microbiological Methods*, Vol. 80, No.2, (February, 2010), pp. 134–137

Kim, C. H.; Khan, M.; Morin, D. E.; Hurley, W. L.; Tripathy, D. N.; Kehrli M.Jr.; Oluoch A. O. & Kakoma, I. (2001). Optimization of the PCR for detection of *Staphylococcus aureus nuc* gene in bovine milk. *Journal of Dairy Scienece*, Vol. 84, No. 1, (January, 2001), pp. 74–83

Kim, D. M.; Chung, S. H. & Chun, H. S. 2011. Multiplex PCR assay for the detection of aflatoxigenic and non-aflatoxigenic fungi in *meju*, a Korean fermented soybean food starter. *Food Microbiology*, Vol. 28, No. 7, (October, 2011), pp. 1402-1408

King, C. E.; Debruynel, R.; Kuch, M; Schwarz, C. & Poinar, H. N. (2009). A quantitative approach to detect and overcome PCR inhibition in ancient DNA extracts. *Research Reports*, Vol. 47, No. 5, (November, 2009), pp. 941-949

Klančnik, A.; Smole Možina, S. & Jeršek, B. (2003). The effect of DNA preparation procedure on sensitivity of PCR detection of *Listeria monocytogenes*. *Zbornik Biotehehniške fakultete*, Vol. 81, No. 1, (May, 2003), pp. 65-73

Krämer, N.; Löfström, C.; Vigre, H.; Hoorfar, J.; Bunge, C. & Malorny, B. (2011). A novel strategy to obtain quantitative data for modelling: Combined enrichment and real-time PCR for enumeration of salmonellae from pig carcasses. *International Journal of Food Microbiology*, Vol. 145, No. 1, (March, 2011), pp. 86–95

Kreader, C. A. (1996). Relief of amplification inhibition in PCR with bovine serum albumin or T4 gene 32 protein. *Applied and Environmental Microbiology*, Vol. 62, No. 3, (March, 1996), pp. 1102–1106

Kure, C. F.; Skaara, I.; Holst-Jensena, A. & Abeln, E. C. A. (2003). The use of AFLP to relate cheese-contaminating *Penicillium* strains to specific points in the production plants. *International Journal of Food Microbiology*, Vol. 83, No. 2, (June, 2003), pp. 195–204

Lantz, P. G.; Abu Al-Soud, W.; Knutsson, R.; Hahn-Hägerdal, B. & Rådström, P. (2000). Biotechnical use of the polymerase chain reaction for microbiological analysis of biological samples. *Biotechnology Annual Review,* Vol. 5, pp. 87–130

Lantz, P. G.; Hahn-Hägerdal, B. & Rådström, P. (1994). Sample preparation methods in PCR-based detection of food pathogens. *Trends in Food Science and Technology,* Vol. 5, No. 12, (December, 1994), pp. 384-389

Lantz, P. G.; Matsson, M.; Wadström, T. & Rådström, P. (1997). Removal of PCR inhibitors from human faecal samples through the use of an aqueous two phase system for sample preparation prior to PCR. *Journal of Microbiological Methods,* Vol. 28, No. 3, (March, 1997), pp. 159–167

Lantz, P. G.; Tjerneld, F.; Hahn-Hägerdal, B. & Rådström, P. (1996). Use of aqueous two-phase systems in sample preparation for polymerase chain reactionbased detection of microorganisms. *Journal of Chromatography. B, Biomedical Applications,* Vol. 680, No. 1-2, (May, 1996), pp. 165–170

Lawyer, F. C.; Stoffels, S.; Saiki, R. K.; Chang, S. Y.; Landre, P. A.; Ambrason, R. D. & Gelfand, D. H. (1993). High-level expression, purification, and enzymatic characterization of full-length *Thermus aquaticus* DNA polymerase. *PCR Methods Appl.,* Vol. 2, No. 4, (May, 1993), pp. 275–287

Lee, J. L. & Levin, R. E. (2011). Detection of 5 CFU/g of *Escherichia coli* O157:H7 on lettuce using activated charcoal and real-time PCR without enrichment. *Food Microbiology,* Vol. 28, No. 3, (May, 2011), pp. 562-567

Lee, M. D. & Fairchild, A. (2006). Sample Preparation for PCR, In: PCR Methods in Foods, J. Maurer (Ed.), 41-50, Springer, ISBN-10: 0-387-28264-5, Athens, GA, United States of America, 24.10.2011, Available from http://www.springerlink.com/content/k4172w2414867830

Leung, K. T.; Mackereth, R.; Tien, Y.-C. & Topp, E. (2004). A comparison of AFLP and ERIC-PCR analyses for discriminating *Escherichia coli* from cattle, pig and human sources. *FEMS Microbiology Ecology,* Vol. 47, No. 1, (January, 2004), pp. 111-119

Lindqvist, R.; Norling, B. & Lambertz, S. T. (1997). A rapid sample preparation method for PCR detection of food pathogens based on buoyant density centrifugation. *Letters in Applied Microbiology,* Vol. 24, No. 4, (April, 1997), pp. 306–310

Löfström, C.; Schelin, J.; Norling, B.; Vigre, H.; Hoorfar, J. & Rådström, P. (2010). Culture independent quantification of *Salmonella enterica* in carcass gauze swabs by flotation prior to real-time PCR. *International Journal of Food Microbiology,* Vol.145, No. 1, (March, 2010), pp. 103-109

Logan, J., Edwards, K. &, Saunders, N. (2009). Real-Time PCR: Current Technology and Applications Caister Academic Press. Editor: Applied and Functional Genomics, Health Protection Agency, London. (2009), pp. 284

Lomonaco, S.; Nucera D.; Parisi, A.; Normanno, G. & Bottero, M. T. (2011). Comparison of two AFLP methods and PFGE using strains of *Listeria monocytogenes* isolated from environmental and food samples obtained from Piedmont, Italy. *International Journal of Food Microbiology,* Vol. 149, No. 2, (September, 2011), pp. 177–182

Lukue, M. I.; Rodríguez, A.; Andrade, M. J.; Gordillo, R.; Rodríguez, M. & Córdoba J. J. (2011). Development of a PCR protocol to detect patulin producing moulds in food products. *Food Control,* Vol. 22, No. 12, (December, 2011), pp. 1831-1838

Lyamichev, V.; Brow, M. A. D. & Dahlberg, J. E. (1993). Structure-specific endonucleolytic cleavage of nucleic acids by eubacterial DNA polymerases. *Science*, Vol. 260 No. 5109, (May, 1993), pp. 778–783

Martínez-Blanch, J. F.; Sánchez, G.; Garay, E. & Aznar, R. (2009). Development of a real-time PCR assay for detection and quantification of enterotoxigenic members of *Bacillus cereus* group in food samples. *International Journal of Food Microbiology*, Vol. 135, No. 30, (July, 2009) pp.15-21

Mayrl, E.; Roeder, B.; Mester, P.; Wagner, M. & Rossmanith, P. (2009). Broad range evaluation of the matrix solubilization (matrix lysis) strategy for direct enumeration of foodborne pathogens by nucleic acids technologies. *Journal of Food Protection*, Vol. 72, No. 6, (June, 2009), pp. 1225–1233

McKillip, J. L.; Jaykus, L. A. & Drake, M. A. (2000). A comparison of methods for the detection of *Escherichia coli* O157:H7 from artificially contaminated dairy products sing PCR. *Journal of Applied Microbiology*, Vol. 89, No. 1, (July, 2000), pp. 49–55

McKnight, I. C.; Eiroa, M. N. U.; Sant'Ana, A. S. & Massaguer, P.R. (2010). *Alicyclobacillus acidoterrestris* in pasteurized exotic Brazilian fruit juices: Isolation, genotypic characterization and heat resistance. *Food Microbiology*, Vol. 27, No. 8, (December, 2010), pp. 1016-1022

Opel, K. L.; Chung, D. & McCord, B. R. (2010). A study of PCR inhibition mechanisms using real time PCR. *Journal of Forensic Sciences*, Vol. 55, No. 1, (January, 2010), pp. 25–33

Perry, L.; Heard, P.; Kane, M.; Kim, M.; Savikhin, S.; Domínguez, W. & Applegate, B. (2007). Application of multiplex polymerase chain reaction to the detection of pathogens in food. *Journal of Rapid Methods and Automation in Microbiology*, Vol. 15, No. 2, (June, 2007), pp. 176–198

Piskernik, S.; Klančnik, A.; Toplak, N.; Kovač, M. & Jeršek, B. (2010). Rapid detection of *Escherichia coli* O157:H7 in food using enrichment and real-time polymerase chain reaction. *Journal of Food and Nutrition Research*, Vol. 49, No. 2, pp. 78–84

Pomp, D. & Medrano, J. F. (1991). Organic solvents as facilitators of polymerase chain reaction. *Biotechniques*, Vol. 10, No. 1, (January, 1991), pp. 58–59

Postollec, F.; Falentin, H.; Pavan, S.; Combrisson. J. & Sohier, D. (2011). Recent advances in quantitative PCR (qPCR) applications in food microbiology. *Food Microbiology*, Vol. 28, No. 5, (August, 2011), pp. 50-54

Rådström, P.; Knutsson, R.; Wolffs, P.; Lovenklev, M. & Lofstrom, C. (2004). Pre-PCR processing strategies to generate PCR-compatible samples. *Molecular Biotechnology*, Vol. 26, No. 2, (February, 2004), pp. 133–146

Rådström, P.; Löfström, C.; Lövenklev, M.; Knutsson, R. & Wolffs, P. *(*2008*)* Strategies for overcoming PCR inhibition. *Cold Spring Harbor Protocols (doi:10.1101/pdb.top20)*

Ramakers, C.; Ruijter, J. M.; Deprez, R. H. & Moorman. A. F. (2003). Assumption-free analysis of quantitative real-time polymerase chain reaction (PCR) data. *Neuroscience Letters*, Vol. 339, No. 1, (March, 2003), pp. 62-66

Raymaekers, M.; Smets, R.; Maes, B. & Cartuyvels, R. (2009). Checklist for optimization and validation of real-time PCR assays. *Journal of Clinical Laboratory Analysis*, Vol. 23, No. 3, (May, 2009), pp. 145–151

Rees, R. K.; Graves, M.; Caton, N.; Ely, J. M. & Probert, W. S. (2009). Single tube identification and strain typing of *Brucella melitensis* by multiplex PCR. *Journal of Microbiological Methods*, Vol. 78, No. 1, (July, 2009), pp. 66–70

Robyt, J. F. & White, B. J. (1990). Biochemical Techniques Theory and Practice. Waveland Press, Inc., (1990), pp. 129-156

Rossen, L.; Nørskov, P.; Holmstrøm, K. & Rasmussen, O.F. (1992). Inhibition of PCR by components of food samples, microbial diagnostic assays and DNA extraction solutions. *International Journal of Food Microbiology,* Vol. 17, No. 1, (September, 1990), pp. 37–45

Rutledge, R. G. & Côté, C. (2003). Mathematics of quantitative kinetic PCR and the application of standard curves. *Nucleic Acids Research,* Vol. 31, No. 16, (May, 2009), pp. 1-6,

Saiki, R. K.; Gelfand, D. H.; Stoffel, S.; Scharf, S. J.; Higuchi, R.; Horn, G. T., Mullis, K. B. & Erlich, H. A. (1988). Primer-directed enzymatic amplification of DNA with a thermostable DNA polymerase. *Science,* Vol. 239, No. 4839, (January, 1988), pp. 487–491

Saiki, R. K.; Scharf, S.; Faloona, F.; Mullis, K. B.; Horn, G. T.; Erlich, H. A. & Arnheim, N. (1985). Enzymatic amplification of beta-globin genomic sequences and restriction site analysis for diagnosis of sickle cell anemia. *Science,* Vol. 230, No. 4732, (December, 1985), pp. 1350–1354

Sakai, F.; Takemoto; A.; Watanabe; S.; Aoyama; K.; Ohkubo; T.; Yanahira; S.; Igarashi; H.; Kozaki; S.; Hiramatsu; K. & Ito; T. (2010). Multiplex PCRs for assignment of Staphylocoagulase types and subtypes of type VI Staphylocoagulase. *Journal of Microbiological Methods,* Vol. 75, No. 2, (October, 2010), pp. 312–317

Sambrook J. & Russel D. W. (2001). Molecular Cloning: A Laboratory Manual 3rd Edition, *Cold Spring Harbor Laboratory Press.* Cold Spring Harbor, NY, (2001), pp. 5.4-5.14

Savelkoul; P. H. M.; Aarts; H. J. M.; DeHaas; J.; Dijkshoorn; L.; Duim; B.; Otsen; M. Rademarker; L. W.; Schouls; L. & Lenstra; A. (1999). Amplified-Fragment Length Polymorphism Analysis: the State of an Art. *Journal of Clinical Microbiology,* Vol. 37; No. 10; (October, 1999), pp. 3083-3091

Silva, D. S. P.; Canato, T.; Magnani, M.; Alves, J.; Hirooka, E. Y.; Rocha, T. C. & Moreira de Oliveira. T. C. R. M. (2011). Multiplex PCR for the simultaneous detection of *Salmonella* spp. & *Salmonella* Enteritidis in food. *International Journal of Food Science and Technology,* Vol. 46, No. 7, (July, 2011), pp. 1502–1507

Steffan, R. J. & Atlas R. M. (1991). Polymerase chain reaction - Applications in enviromental microbiology. *Annual Review of Microbiology,* Vol. 45, (October, 1991), pp. 137-161

Stewart, D. & Gendel, S. M. (1998). Specificity of the bax polymerase chain reaction system for detection of the foodborne pathogen *Listeria monocytogenes. Journal of AOAC International,* Vol. 81, No. 4, (July-August, 1998), pp. 817–822

Stock, P.; Vanderberg, J.; Glazer, I. & Boemare, N. (2009). 1.6.2. Primers development and virus identification strategies. *Insect Pathogens: Molecular Approaches and Techniques.* CAB International, (2009), pp. 22

Thisted Lambertz, S.; Lindqvist, R.; Ballagi-Pordány, A. & Danielsson-Tham, M.-L. (2000). A combined culture and PCR method for detection of pathogenic *Yersinia enterocolitica* in food. *International Journal of Food Microbiology,* Vol. 57, No. 1-2, (June, 2000), pp. 63–73

Trnčíková, T.; Hrušková, V.; Oravcová, K; Pangallo, D. & Kaclíková, K. (2009). Rapid and sensitive detection of *Staphylococcus aureus* in food using selective enrichment and real-time PCR targeting a new gene marker. *Food Analytical Methods*, Vol. 2, No. 4, (December, 2009), pp. 241-250

Trnčíková, T.; Piskernik, S.; Kaclikova, E.; Smole-Možina, S.; Kuchta, T. & Jeršek, B. (2010). Characterization of *Staphylococcus aureus* strains isolated from food produced in Slovakia and Slovenia with regard to the presence of genes encoding for enterotoxins. *Journal of Food and Nutrition Research*, Vol. 49, No. 4, pp. 215-220

Tyler, K. D.; Wang, G.; Tyler, D. & Johnson,W. M. (1997). Factors affecting reliability and reproducibility of amplification-based DNA fingerprinting of representative bacterial pathogens. *Journal of Clinical Microbiology*, Vol. 35; No.2, (February, 1997), pp. 339-346

Ulve, V. M.; Monnet, C.; Valence, F.; Fauquant, J.; Falentin, H. & Lortal, S. (2008). RNA extraction from cheese for analysis of *in situ* gene expression of *Lactococcus lactis*. *Journal of Applied Microbiology*, Vol. 105, No. 5, (November, 2008), pp. 1327-1333

Uyttendaele, M.; van Boxstael, S. & Debevere, J. (1999). PCR assay for detection of the *E. coli* O157:H7 *eae*-gene and effect of the sample preparation method on PCR detection of heat-killed *E. coli* O157:H7 in ground beef. *International Journal of Food Microbiology*, Vol. 52, No. 1-2, (November, 1999), pp. 85–95

van Belkum, A.; Sluijuter, M.; de Groot, R.; Verbrugh, H. & Hermans, P. W. (1996). Novel BOX repeat PCR assay for high-resolution typing of *Streptococcus pneumoniae* strains. *Journal of Clinical Microbiology*, Vol. 34, No. 5, (May, 1996), pp. 1176–1179

van Belkum, A.; Tassios, P. T.; Dijkshoorn, L.; Haeggman, S.; Cookson, B.; Fry, N. K.; Fussing, V.; Green, J.; Feil, E.; Gerner-Smidt, P.; Brisse, S. & Struelens, M. (2007). Guidelines for the validation and application of typing methods for use in bacterial epidemiology. *Clinical Microbiology and Infectious Diseases*, No. 13, Suppl. 3, (October, 2007), pp. 1–46

Versalovic, J.; Koeuth, T. & Lupski, J. R. (1991). Distribution of repetitive DNA sequences in eubacteria and application to fingerprinting of bacterial genomes. *Nucleic Acids Research*, Vol. 19; No. 24, (November, 1991), pp. 6823 -6831

Warren, B. R.; Yuk, H. G. & Schneider, K. R. (2007). Detection of *Salmonella* by flow-through immunocapture real-time PCR in selected foods within 8 hours. *Journal of FoodProtection*, Vol. 70, No. 4, (April, 2007), pp. 1002–1006

Wehrle, E.; Didier, A.; Moravek, M.; Dietrich, R. & Märtlbauer, E. (2010). Detection of *Bacillus cereus* with enteropathogenic potential by multiplex real-time PCR based on SYBR green I. *Molecular and Cellular Probes*, Vol. 24, No. 3, (June, 2010), pp. 124-130, doi:10.1016/j.mcp.2009.11.004

Wehrle; E.; Moravek; M.; Dietrich; R.; Bürk; C.; Didie;r A. & Märtlbauer; E. (2009). Comparison of multiplex PCR; enzyme immunoassay and cell culture methods for the detection of enterotoxinogenic *Bacillus cereus*. *Journal of Microbiological Methods*, Vol. 78, No. 3, (September, 2009), pp. 265–270

Weyant, R. S.; Edmonds, P. & Swaminathan, B. (1990). Effect of ionic and nonionic detergents on the *Taq* polymerase. *BioTechniques*, Vol. 9, No. 3, (September. 1990), pp. 308–309

Whitehouse, C. A. & Hottel, H. E. (2007). Comparison of five commercial DNA extraction kits for the recovery of *Francisella tularensis* DNA from spiked soil samples. *Molecular and Cellular Probes*, Vol. 21, No. 2, (April, 2007), pp. 92–96

Williams, J. G.; Kubelik, A. R.; Livak, K. J.; Rafalski, J. A. & Tingey, S. V. (1990). DNA polymorphisms amplified by arbitrary primers are useful as genetic markers. *Nucleic Acids Research*, Vol. 8, No. 22, (October, 1990), pp. 6531–6535

Wilson, I. G. (1997). Minireview: Inhibition and facilitation of nucleic acid amplification. *Applied and Environmental Microbiology*, Vol. 63, No. 10, (November, 1997), pp. 3741–3751

Wolffs, P. F. G.; Glencross, K.; Norling, B. & Griffiths, M. W. (2007). Simultaneous quantification of pathogenic *Campylobacter* and *Salmonella* in chicken rinse fluid by a flotation and real-time multiplex PCR procedure. *International Journal of Food Microbiology*, Vol. 117, No. 1, (June, 2007), pp. 50-54

Wolffs, P. F. G.; Norling, B. & Rådström, P. (2004). Rapid quantification of *Yersinia enterocolitica* in pork samples by a novel sample preparation method, flotation, prior to real-time PCR. *Journal of Clinical Microbiology*, Vol. 42, No. 3, (March, 2004), pp. 1042–1047

Zhang, C.-L.; Fowler, M. R.; Scott, N. W.; Lawson, G. & Slater, A. (2007). A TaqMan real-time PCR system for the identification and quantification of bovine DNA in meats, milks and cheeses. *Food Control*, Vol. 18, No. 9, (September, 2007), pp. 1149–1158

11

PCR in Disease Diagnosis of WND

Asifa Majeed, Abdul Khaliq Naveed, Natasha Rehman and Suhail Razak
Dept. of Biochemistry and Molecular Biology, College of Medical Sciences, National University of Sciences and Technology, Rawalpindi Pakistan

1. Introduction

Polymerase chain reaction (PCR) invented by Karry Mullis in 1983 is a cornerstone of molecular biology techniques. Polymerase chain reaction (PCR) is used to generate millions of copies of a DNA sequence in a short interval. PCR is commonly used for a variety of applications including diagnosis of genetic disease, cloning, QTLR analysis, forensic analysis, diagnosis of infectious diseases. Polymerase chain reaction becomes indispensable technique in medical sciences these days for the diagnosis of infectious diseases. Molecular genetic testing has made possible to identify the mutations in first degree relatives of index case even in the absence of clinical and biochemical presentations of symptoms. PCR is a basic step in molecular genetic testing.

Wilson disease is a genetic disorder of copper metabolism with hepatic or neuropsychiatric presentations [1]. The copper-transporting P-type ATPase, *ATP7B* gene was identified in 1993 (ATP7B; OMIM 606882) [2] and found responsible for WND. Copper is a nutritional trace element and play indispensable role in variety of biological reactions [3, 4]. The homeostasis of copper is maintained by liver. It regulates excretion of copper into bile and into secretary pathway. Ceruloplasmin protein is abundant in blood and plays functional role in copper transport. Copper excretes from hepatocytes as ceruloplasmin or through bile [5, 6]. The human Cu-ATPases regulate the intracellular copper homeostasis. The copper is transported from cytosol to secretory pathway using energy released by the hydrolysis of ATP and supply the copper for various copper-dependent biological process and enzymes. The biosynthesis of copper-dependent ferroxidase ceruloplasmin is dependent on *ATP7B*. In addition, Cu-ATPases play part in export of excess copper out of the cell [7, 8, 9].

Liver is involved in the copper homeostasis and remove excess copper via bile [10] through the activity of *ATP7B* transporter. Therefore defect in *ATP7B* gene results in disturbance of copper homeostasis and caused WND [2, 11] due the copper accumulation in liver, brain and cornea.

ATP7B is present in the Golgi & trans-Golgi Network (TGN), travel to the vesicular compartment. There, *ATP7B* delivers copper to ceruloplasmin in the cell which is copper binding protein [12, 13]. *ATP7B* gene contains six copper binding domains, transduction domain, cation and phosphorylation domain, nucleotide-binding domain and eight hydrophobic transmembrane sequences [7]. The Wilson disease gene *ATP7B* was localized at chromosome 13 on q14.3 band and cloned in 1993. *ATP7B* gene comprised of 21 exons which encodes 1465 amino acid residues [2, 14, 15].

The WND diagnosis is complex due to variable disease onset and clinical symptoms. Clinical symptoms of the disease include hepatic, neurologic and psychiatric disturbances [16]. WND patients with hepatic manifestation may present asymptomatic to mild hepatomegaly, cirrhosis, acute hepatitis and jaundice. Haemolysis can be present in acute liver failure [17, 18, 19].

Neurological symptoms mostly appeared in second decade of life due to the toxic copper accumulation in brain that damages the nerve cells. This leads to hypokinetic speech, tremor, dystonia and later dysphagia, mutism and Parkinsonism. Hepatic and neurological presentations in combination have been observed in 50% WND patients [20, 21].

Psychiatric symptoms have been seen in the early stages of WND including incompatible behavior, irritability, depression and cognitive impairment [22, 23].

Therefore, molecular testing of WND [21, 24] has served as a very useful approach for presymptomatic disease diagnosis in the absence of clinical symptoms. In present study, we investigated presymptomatic WND in siblings of two index case of Wilson disease.

2. Material and methods

The informed consent was obtained from all subjects and study was approved from ethnic committee of the institution.

2.1 Subjects

Two families with WND patients were enrolled for present study. The age of patients was between 7-11 years and of siblings was between 2-5 years. The children of family 1 were born to non-consanguineous parents. History of patient belongs to family I was described earlier [25]. The children of family II were born to consanguineous parents. Past family history was negative for presence of WND. However, liver disease was reported in family II. Patients were enrolled for mutational analysis based on clinical diagnosis. Siblings were screened for presymptomatic WND. The patients of family II were 11 & 8 years old. The laboratory investigations of respective patients were described in table 1 & 2. The siblings of index patients were included without clinical data. Patients of this family were presented with hepatic manifestation. The 11 year old boy was also patient of hepatitis C.

2.2 Mutational analysis

DNA was extracted from peripheral blood of all subjects by standard phenol/chloroform extraction method [26]. The quality plus quantity of DNA was checked through gel electrophoresis and spectrophotometer. PCR was done in 30µl volume containing 200ng genomic DNA, 1X *Taq* buffer, 200µM dNTPs mixture, 2.5 pmol of both primers, 1unit of *Taq* polymerase. The PCR conditions were optimized for 11 exons of *ATP7B* gene. The PCR products were purified through PCR purification kit (Genomed GmbH Inc). For sequencing, 0.1-0.5ng of PCR product was used as template and sequencing PCR was performed with quickstart DTCS kit (Beckman Coulter). The sequencing PCR program was comprised of 30 cycles: 96°C for 20 seconds, 20 seconds at respective annealing temperature and 60°C for 4 min. Salt precipitation method was used to remove unincorporated dye terminators as described by manufacturer. The sequencing was performed with forward and reverse primers on CEQ8000 Genetic Analyzer (Beckman Coulter). The sequences were compared for the detection of mutation through BioEdit Sequence Alignment Editor ver 7.0.9.0.

3. Results and discussion

Wilson disease is a recessive autosomal disorder caused by increased accumulation of copper in liver, brain, cornea [5]. The prevalence of WND is around 1:30,000 with carrier frequency 1:90 [27] while 4% carrier frequency is also reported [28]. Clinical presentation of disease is variable and mostly appears between ages 5 to 35 with rare case of onset in 2 to 72 years [29, 30]. The diagnosis of WND is based on presence of KF ring, low plasma ceruloplasmin, elevated urinary copper and liver copper concentration [1, 24]. WND is caused due to defect in *ATP7B* gene, which is copper transporter. The worldwide data revealed the population specific pattern of *ATP7B* gene mutations. The most common mutation found in Europe, American and Greece population was H1069Q in exon 14. About 50–80% of WND patients from these countries carry at least one allele with this mutation with an allele frequency ranging between 30 and 70% [28, 31, 32]. Mutational analysis of *ATP7B* gene has been extensively carried out in Chinese population and showed a high prevalence of WND. Mutations have been detected in all exons except 21. Most mutations were found in exons 8, 12, 13 and 16 accounts for 74.0% of the reported WND alleles. The most frequent WND mutations were p.Arg778Leu and p.Pro992Leu, which account for 50.43% of all the reported WND alleles in Chinese population [33, 34]. The R778L mutation was also frequent in Korean and Taiwan population with an allele frequency of 20-35% & 55.4% [35, 36, 37]. In addition to R778L mutation at exon 8, hotspot for *ATP7B* mutation in exon 12 were also detected in Taiwan WND patients [38] where 9.62% of all mutations occurred.

The spectrum of *ATP7B* gene mutations in our population is yet to be studied. In present study, the siblings of index patients of WND were screened for mutations in *ATP7B* gene. We enrolled two families for genetic testing and novel mutation was described previously in one patient [25]. The past history of patients revealed the sudden onset of disease while no WND patient was reported earlier in these families. Based on current family history, siblings were screened for presympotomatic WND. The exon 15 and 19 were amplified at 55°C. The exons 16, 17, & 21 were optimized at 64°C. Same annealing temperatures were used for sequencing PCR. Phenotypic data of subjects of family 1 was found normal. No laboratory investigations were performed at the time of genetic testing. The family members of patient were screened without relying on clinical data. The patients and siblings were born to non-consanguineous parents. The parents were also found normal in genetic testing.

The family-II was also presented with same history. The family did not have any past WND history. The two child of respective family were declared WND patient. The phenotypic presentations of sibling and parents were found normal. However, occurrence of liver diseases (details not available) was reported in three generations. The parents were not reported any type of liver disease. The Hb level of sibling was reported below normal range when he was 3years old. We encountered problem in collecting blood sample from family-II. Therefore we reamplified each exon through PCR with same product to get product in sufficient quantity. This step reduced the chance for loss of sample. Because subjects were belonged to remote area and access to them was not feasible.

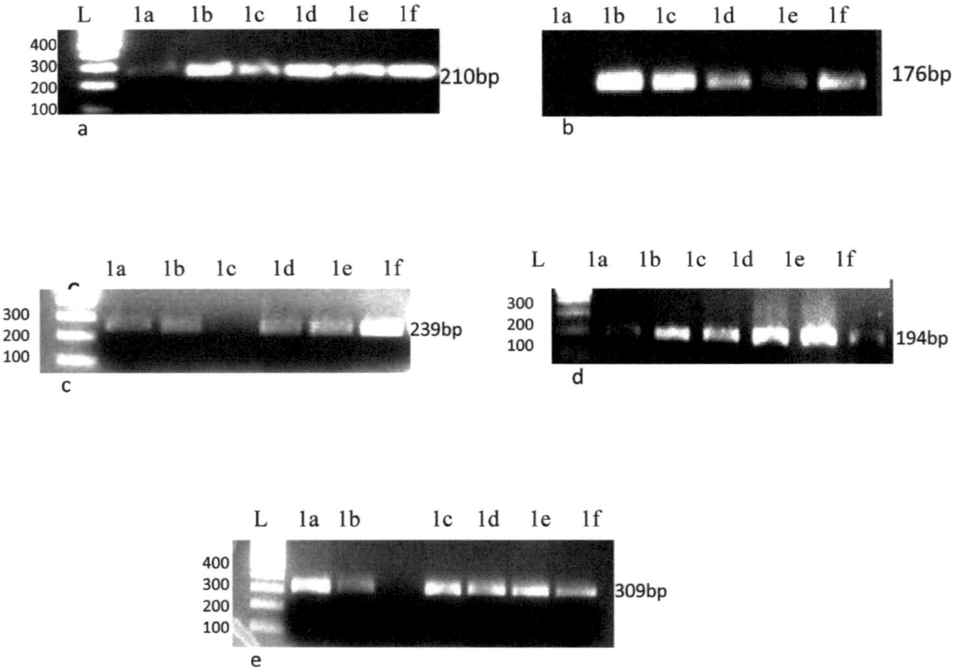

Fig. 1. PCR amplification of samples of family 1
a) Exon-15, b)Exon-16, c)Exon-17, d)Exon-19, e)Exon-21

Patients	CP	U-Cu	ALT
Normal	**<20mg/dl**	**>100mg/24h**	**7-45 U/L**
8a	6.8 mg/dl	1796mg/24h	40 U/L
8f	20 mg/dl	1000mg/24h	200 U/L

Table 1. Biochemical analysis of WND patients of family II

Patients	Hb	Total Bilirubin	AST	ALP	Serum Albumin
Normal	**13-17 g/dl**	**0.3-1.2 mg/dl**	**7-45U/L**	**98-279 U/L**	**3.5-5.5 g/dl**
8a	8.9 g/dl	1.8 mg/dl	1400 U/L	569 U/L	2.3 g/dl
8f	5 g/dl	30 mg/dl	450 U/L	350 U/L	2.2 g/dl

Table 2. Clinical features of WND patients of family II

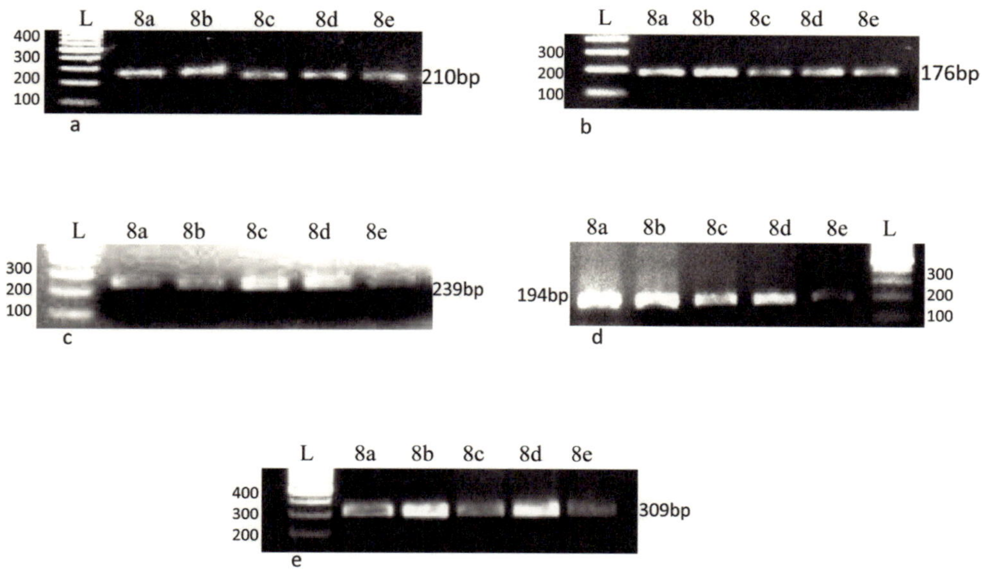

Fig. 2. PCR amplification of samples of family II
a) Exon-15, b)Exon-16, c)Exon-17, d)Exon-19, e)Exon-21

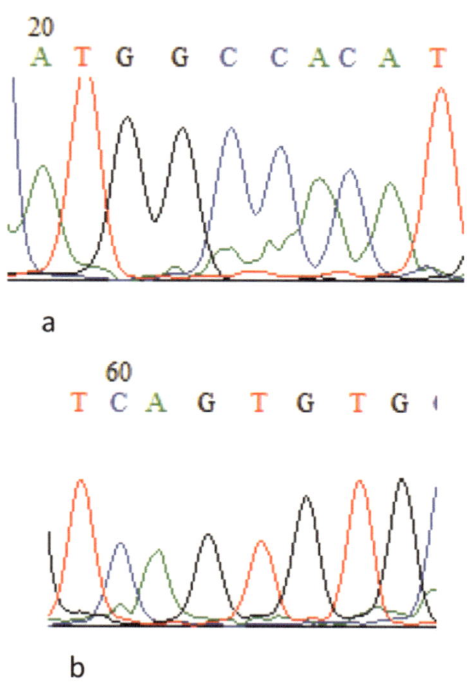

Fig. 3. Sequencing of *ATP7B* gene
a) Family 1, b) Family II

The phenotypic presentations of sibling were found normal. Fig 1 & Fig 2 shows the PCR amplification of respective exons. In mutational analysis, subjects of both families were detected negative for any mutation on exons 15, 16, 17, 19 & 21 (Fig 3).

The prospective of our study was the screening of carrier in respective families and inclusion of genetic testing for earlier diagnosis of WND. PCR and gene sequencing are the basic steps of genetic testing. We have not found prevalent mutation in our Wilson disease patients. Screening the siblings is a grueling task when common mutation was not identified. Genotyping of *ATP7B* gene in families of WND patients was first time carried out in our country. Genetic testing is useful tool for the screening of presymptomatic WND case or carriers in the family of index patient. It could help in genetic counseling of respective family based on molecular diagnosis. We have also found that optimization of parameters for PCR and sequencing is indispensible for effective screening. The result's integrity dependent on quantity and quality of template and is a paramount factors in PCR. Presymptomatic diagnosis through genetic analysis could help to stop progression & prevention of the disease and timely treatment. The Taqman allelic discrimination reported a valid technique for efficient screening of common mutation in index patient and sibs [39]. Single nucleotide polymorphism (SNP) have also been studied in combination with the prevalent mutations for WND diagnosis and evaluated as a comprehensive strategy for the detection presymptomatic or carrier sibs of WND patients [40].

4. Conclusion

This case has provided a base to establish a system for genetic testing for the earlier diagnosis of disease and detection of heterozygote carrier. PCR based genetic testing using different approaches like multiplex PCR, SNP markers and Taqman assay will turn into a cost effective screening. These results further more have provided a platform for haplotype analysis. Presymptomatic diagnosed patient can undergo regular treatment to prevent disease progression and onset.

5. References

[1] Roberts EA, Schilsky ML. Diagnosis and treatment of Wilson disease: an update. Hepat 2008;47(6): 2089-111.

[2] Bull PC, Thomas GR, Rommens JM, Frobes JR, Cox DW. The Wilson disease gene is a putative copper transporting p-type ATPase similar to the Menkes gene. Nat Gen 1993;5: 327-337

[3] Prohaska JR. Role of copper transporters in copper homeostasis. Amer J Clin Nutr 2008; 88(3):826S-829S

[4] Lallioti V, Muruais G, Tsuchiya Y, Pulido D, Sandoval IV. Molecular mechanisms of copper homeostasis. Front Biosci 2009;14:4878–4903

[5] Nicholas AV, Ann PG, Richard B P, Kipros G, James C. The multi-layered regulation of copper translocating P-type ATPases. Biometals 2009;22:177–190

[6] Linder MC, Wooten L, Cerveza P, Cotton S, Shulze R, Lomeli N. Copper transport. Am J Clin Nutr 1998;67(5):965S-971S

[7] Lutsenko S, Erik S, Shane L, Shinde U. Biochemical basis of regulation of human copper-transporting ATPases. Arch Bioch Biophy 2007;463(2):134-148

[8] Bie P, Muller P, Wijmenga C, Klomp LWJ. Molecular pathogenesis of Wilson and Menkes disease: correlation of mutations with molecular defects and disease phenotypes. J Med Gen 2007; 44: 673-688
[9] Hellman NE, Kono S, Mancini GM, Hoogeboom AJ, DeJong GJ, Gitlin JD. Mechanisms of copper incorporation into human ceruloplasmin. J Biol Chem 2002;277:46632–8
[10] Wijmenga C, Klomp LW. Molecular regulation of copper excretion in the liver. Proc Nutr Soc 2004;63:31–9
[11] Langner C, Denk H. Wilson disease. Virchows Arch 2004; 445:111–8.
[12] Bartee MY, Lutsenko S. Hepatic copper-transporting ATPase *ATP7B*: function and inactivation at the molecular and cellular level. Biometals 2007;20:627–637
[13] Guo Y, Nyasa, L, Braiterman LT, Hubbard AL. NH2-terminal signals in *ATP7B* cu-ATPase mediate its Cu-dependent anterograde traffic in polarized hepatic cells. Am J Physiol Gastr Liver Phy 2005; 289:G904–16.
[14] Tanzi RE, Petrrukhin K, Chernov I, Pellequer JL, Wasco W, Ross, B. The Wilson disease gene is a copper transporting ATPase with homology to Menkes disease gene. Nat Genet 1993;5:344-350
[15] Yamaguchi Y, Heiny ME, Gitlin JD. Isolation and characterization of a human liver cDNA as a candidate gene for wilson disease. Biochem. Biophys. Res. Commun 1993;197:271–277
[16] Ferenci P, Caca K, Loudianos G, Mieli-Vergani G, Tanner S, Sternlieb I, Schilsky M, Cox D, Berr F. Diagnosis and phenotypic classification of Wilson disease. Liver Int 2003 Jun; 23(3):139-42.
[17] Merle U, Schaefer M, Ferenci P, Stremmel W. Clinical presentation, diagnosis and long-term outcome of Wilson's disease: a cohort study. Gut 2007; 56(1):115-20.
[18] El-Youssef M. Wilson disease. Mayo Clin Proc 2003 Sep; 78(9):1126-36.
[19] Sharma S, Toppo A, Rath B, Harbhajanka A, Lalita Jyotsna P. Hemolytic Anemia as a Presenting Feature of Wilson's Disease: A Case Report. Indian J Hematol Blood Transfus 2010 Sep; 26(3):101-2.
[20] Singh P, Ahluwalia A, Saggar K, Grewal CS. Wilson's disease: MRI features. J Pediatr Neurosci 2011 Jan; 6(1):27-8.
[21] Medici V, Mirante VG, Fassati LR, Pompili M, Forti D, Del Gaudio M, Trevisan CP, Cillo U, Sturniolo GC, Fragiuoli S,Monotematica AISF. Liver transplantation for Wilson's disease: The burden of neurological and psychiatric disorders. Liver Transpl 2005 Sep; 11(9):1056-63.
[22] Sahoo MK, Avasthi A, Sahoo M, Modi M, Biswas P. Psychiatric manifestations of Wilson's disease and treatment with electroconvulsive therapy. Indian J Psychiatry 2010 Jan; 52(1):66-8.
[23] Benhamla T, Tirouche YD, Abaoub-Germain A, Theodore F. The onset of psychiatric disorders and Wilson's disease. Encephale 2007 Dec; 33(6):924-32.
[24] Ala A, Borjigin J, Rochwarger A, Schilsky M. Wilson disease in septuagenarian siblings: raising the bar for diagnosis. Hepatology 2005;41:668–670
[25] Naveed AK, Majeed A, Mansoor S. Spectrum of *ATP7B* gene mutations in Pakistani wilson disease patients: A novel mutation is associated with severe hepatic and neurological complication. Inter J Biol 2010;2(1):117-122
[26] Sambrook J. Russell D. Molecular Cloning: A laboratory Manual 3rdedition. Cold Spring Harbor laboratory press ;2001

[27] Scheinberg I, Sternlieb I. Wilson disease.In: Lloyd H, Smith J, editors. Major problems in Internal medicine. Philadelphia: Saunders; 1984. pp 23

[28] Loudianos G, Kostic V, Solinas P, Lovicu M, Dessi V, Svetel M, Major T, Cao A. Delineation of the spectrum of Wilson disease mutations in the Greek population and the identification of six novel mutations. Genet Test 2000;4:399–402.

[29] Perri RE, Hahn SH, Ferber MJ, Kamath PS. Wilson disease keeping the bar for diagnosis raised. Hep 2005;42:974

[30] Wilson DC, Phillips MJ, Cox DW, Roberts EA. Severe hepatic Wilson's disease in preschool-aged children. J Pediatr 2000;137:719–722

[31] Caca K, Ferenci P, Kuhn HJ. High prevalence of the H1069Q mutation in East German patients with Wilson disease: rapid detection of mutations by limited sequencing and phenotype-genotype analysis. J Hepatol 2001; 35:575–581

[32] Olivarez L, Caggana M, Pass KA, Ferguson P, Brewer GJ. Estimate of the frequency of Wilson's disease in the US Caucasian population: a mutation analysis approach. Ann Hum Genet 2001; 65:459–463

[33] Li XH, Lu Y, Yun L, Fu QC, Xu J, Zang GQ, Zhou F, Yu D, Han Y, Zhang D, Gong QM, Lu ZM, Kong XF, Wang JS, Zhang XX. Clinical and molecular characterization of Wilson's disease in China: identification of 14 novel mutations. BMC Med Gen 2011; 12:6

[34] Wang LH, Huang YQ, Shang X, Su QX, Xiong F, Yu QY, Lin HP, Wei ZS, Hong MF, Xu XM. Mutation analysis of 73 southern Chinese Wilson's disease patients: identification of 10 novel mutations and its clinical correlation. J Hum Genet 2011;56(9):660-665

[35] Park HD, Ki CS, Lee SY, Kim JW. Carrier frequency of the R778L, A874V, and N1270S mutations in the *ATP7B* gene in a Korean population. Clin Genet 2009; 75:405-7

[36] Yoo HW. Identification of novel mutations and the three most common mutations in the human *ATP7B* gene of Korean patients with Wilson disease. Genet Med 2002; 4: 43S-48S.

[37] Wan L, Tsai CH, Tsai Y, Hsu CM, Lee CC, Tsai FJ. Mutation analysis of Taiwanese Wilson disease patients. Biochem Biophys Res Commu 2006; 345:734-8.

[38] Wan L, Tsai CH, Hsu CM, Huang CC, Yang CC, Liao CC, Wu CC, Hsu YA, Lee CC, Liu SC, Lin WD, Tsai FJ. Mutation analysis and characterization of alternative splice variants of the Wilson disease gene *ATP7B*. Hepatol 2010; 52(5):1662-1670.

[39] Zappu A, Lepori BM, Incollu S, Noli CSa, De Virgiliis S, Cao A, Loudianos G. Development of TaqMan allelic specific discrimination assay for detection of the most common Sardinian Wilson's disease mutations. Implications for genetic screening. Mol Cell Prob 2010; 24:233-235.

[40] Gupta A, Maulik M, Nasipuri P, Chattopadhyay I, Das KS, Gangopadhyay KP. The Indian Genome Variation Consortium,3 and Kunal Ray1Molecular Diagnosis of Wilson Disease Using Prevalent Mutations and Informative Single-Nucleotide Polymorphism Markers Clin Chem 2007;53:9:1601–1608.

12

Real-Time PCR for Gene Expression Analysis

Akin Yilmaz, Hacer Ilke Onen, Ebru Alp and Sevda Menevse
Department of Medical Biology and Genetics, Faculty of Medicine,
Gazi University, Ankara
Turkey

1. Introduction

After discovery of polymerase chain reaction (PCR) by Dr. Kary Mullis in 1983, several different types of PCR have been invented and continually improved upon over the years. One of them called "Real-time PCR" or "fluorescence based PCR" allows us to quantitate nucleic acids obtained from cells or tissues, to compare the variable states of infection, to detect chromosomal translocations, to genotype single nucleotide polymorphisms, to determine gene expression level of samples and so on. For the detection and quantification of nucleic acids, Real-time PCR has become the most accurate and sensitive method. Quantitative measurement of specific gene expression using quantitative PCR (qPCR) is necessary for understanding basic cellular mechanisms and detecting of alteration in gene expression levels in response to specific biological stimuli (e.g., growth factor or pharmacological agent) (Bustin, 2000; Bustin, 2002). Quantification of nucleic acids has been significantly simplified by the development of the Real-time PCR technique (Bustin, 2002; Huggett et al., 2005). It is mostly used for two reasons: either as a primary investigative tool to determine gene expression or as a secondary tool to validate the results of DNA microarrays (Valasek & Repa, 2005).

2. Basic principles of PCR and Real-time PCR

PCR is an easy and quick *in vitro* method to amplify any target DNA fragment (Powledge, 2004). In PCR, DNA polymerase enzymes are used for the amplification of specific DNA fragments. For this purpose, the most commonly used DNA polymerase is called *Taq* DNA polymerase, isolated from thermophillic bacteria, *Thermus aquaticus*. Another enzyme named *Pfu* DNA polymerase, isolated from *Pyrococcus furiosus*, is also used for PCR because of its higher fidelity during amplification of DNA. These two enzymes are heat stable and DNA dependent DNA polymerases. They can synthesize new DNA strand using a DNA template in the presence of primers, deoxyribonucleotide triphosphates (dNTPs), Mg^{2+} and proper buffer system (Old & Primrose, 1994; Valasek & Repa, 2005).

The PCR involves two oligonucleotide primers which flank the DNA sequence that is to be amplified. The primers hybridize to opposite strands of the DNA after it has been denatured, and are oriented so that DNA synthesis by the polymerase proceeds through the region between the two primers. PCR involves three steps: denaturation, primer hybridization or annealing and extension. During the extension, polymerase creates two double stranded target regions, each of which can again be denatured ready for a second cycle of hybridization and extension (Fig. 1A). The third cycle produces two double-stranded molecules that comprise precisely the target region in double-stranded form. As shown in Fig. 1B, by repeated cycles of heat denaturation, primer hybridization and extension, there follows a rapid exponential accumulation of specific target fragment of DNA (Old & Primrose, 1994).

If the reaction runs with perfect efficiency (100%), there will be two fold increases in target DNA fragment after each cycle of PCR. For example; after n cycles of PCR, the copy number of the target fragment will be 2^n. In practice, reactions, however, do not work with perfect efficiency as reactants within PCR mixture are depleted after many cycles, and then the reaction will reach a plateau phase in which no change of the amount of the product (Gibson et al., 1996; Heid et al., 1996; Valasek & Repa, 2005).

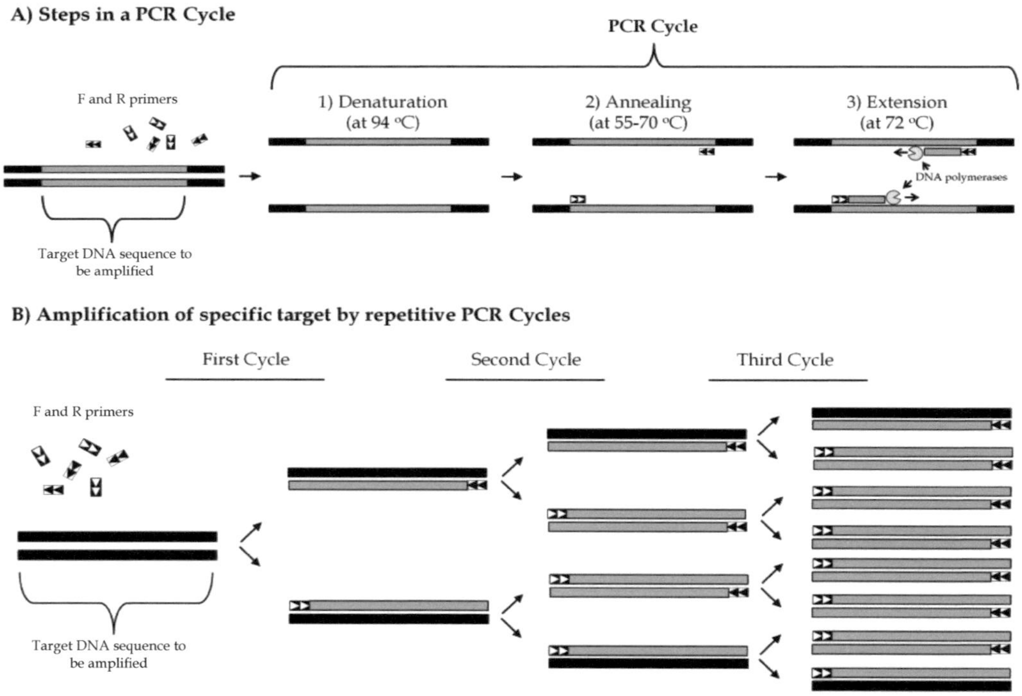

Fig. 1. Schematic representation of PCR. A: Steps in a PCR cycle.
B: Exponential amplification of specific target by repetitive PCR cycles.

PCR can be divided into four phases: the linear ground phase, exponential phase, log-linear phase and plateau phase (Fig. 2) (Tichopad et al. 2003). The advantage of using fluorogenic dyes in the Real-time PCR experiments is to visualize these phases during the reaction. Real-time PCR exploits the fact that the quantity of PCR products in exponential phase is in proportion to the quantity of initial template under ideal conditions (Gibson et al., 1996; Heid et al., 1996).

At the linear ground phase, (usually the first 10-15 cycles), fluorescence emission produced at each cycle has not been higher than the background. Thus, it is obscured by the background fluorescence. This fluorescence is calculated at linear ground phase (Tichopad et al., 2003).

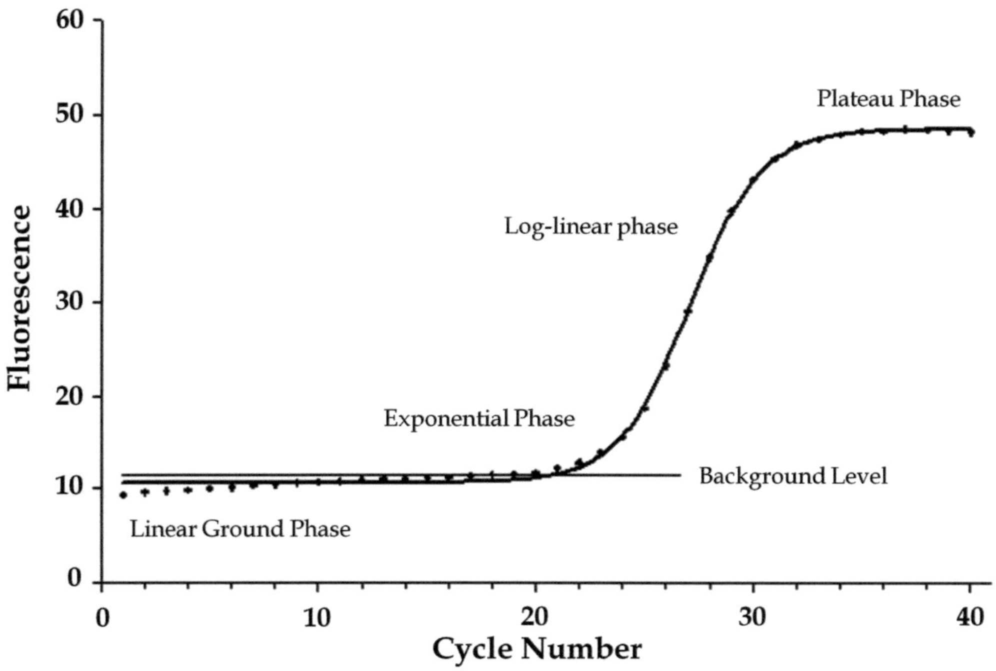

Fig. 2. The four phases of PCR shown in a plot of fluorescence signal versus cycle number.

At exponential phase, the amount of fluorescence reaches a threshold where it can be detected as significantly stronger than the background fluorescence signal. The cycle in which this detection happened is known as threshold cycle (Ct) in ABI PRISM® literature (Applied Biosystems, Foster City, USA) or crossing point (Cp) in LightCycler® literature (Roche Applied Science, Indianapolis, USA) (Tichopad et al. 2003). This value will be called as Ct throughout the text. Ct is a very important point of Real-time PCR because this value represents the amount of target gene found in the sample and it is used to calculate experimental results. If the amount of target sequence is high in the sample, the reaction reaches exponential phase more quickly and thus, the cycle (or Ct value) in which the amount of fluorescence reaches a threshold will be lower for this sample (Heid et al., 1996).

In log-linear phase, the reaction takes place with linear efficiency and increase in florescence signal in every cycle is occured. As stated before, in the last phase of reaction, plateau phase, reactants become limited and exponential product accumulation do not occur anymore (Gibson et al., 1996; Heid et al., 1996; Tichopad et al. 2003).

It is important to underlie that the quantity of PCR products in exponential phase correlates to the quantity of initial template under an ideal conditions (Gibson et al., 1996; Heid et al., 1996). Because the reaction efficiently accomplishes DNA amplification only up to a certain quantity before the plateau effect, it is not possible to reliably calculate the amount of starting DNA by quantifying the amount of product at the end of reaction. After reaction, similar amounts of amplified DNA in the samples that contain different amount of a specific target DNA sequence is found because of plateau effect. Thus, any distinct correlation between samples is lost. Real-time PCR solves this problem by measuring product formation during exponential phase since efficient amplifications occur early in the reaction process (Valasek & Repa, 2005).

3. Template preparation

The first step in gene expression studies is isolation of the high quality RNA from samples. RNA is chemically less stable than DNA so that maintaining RNA integrity in an aqueous solution and protection of RNA against degradation is very important (Fraga et al., 2008). There are numerous protocols and commercially available kits for isolating total RNA and/or mRNA from different tissue samples and some of them are tissue specific (Bustin, 2002; Fraga et al., 2008).

Inhibitory components present frequently in biological samples may cause a significant reduction in the sensitivity and kinetics of Real-time PCR (Radstrom et al., 2004). These inhibitors may originate from reagents used during nucleic acid extraction or co-purified components from the biological sample, for example bile salts, urea, heme, heparin or IgG (Nolan et al., 2006). The presence of any inhibitors of polymerase activity in both reverse transcription and Real-time PCR steps should be considered crucially as many biological samples contain inhibitors for the polymerases (Smith et al., 2003). Inhibitors affect the experiment in two ways by generating incorrect quantitative results or creating false-negative results. The presence of inhibitors within biological samples can be checked by various methods (Nolan et al., 2006).

Differences in mRNA expression patterns at the cellular level may also be masked because of by variability resulting from RNA samples extracted from complex tissue specimens. Such tissues contain variable subpopulations of cells of different lineage at different stages of differentiation. Moreover, malign tissue specimens may also consist of normal cells such as epithelial, stromal, immune or vascular cells. Thus, Real-time PCR data obtained from such a mixed sample is an average of different cell populations. To solve this problem, cell sorting technique can be used for enriching specific cell populations using flow cytometry or antibody-coated beads for blood samples (Deggerdal & Larsen 1997; Raaijmakers et al., 2002;). However, there is no practical way of sorting cells without affecting the expression profile of the sample for solid tissue biopsies. This variability may be partly excluded after tumor and normal tissue samples checked by pathologist view before starting Real-time PCR experiment. Moreover, the introduction of laser capture microdissection (LCM) technique promises to address this particular problem. By using this technique, target mRNA levels can be reported conveniently as copies per area or cells dissected (Fink et al., 1998).

Furthermore, total RNA extracted tissue specimens are usually contaminated with DNA. If the tissue has high DNA content, DNase I treatment is necessary to eliminate residual DNA. If the samples are to be DNase-treated, it is compulsory to remove DNase before cDNA synthesis (Bustin, 2002). After isolation, RNA should be stored at -80°C.

Traditionally, the ratio of absorbance at 260 nm and 280 nm or analysis of the rRNA bands on agarose gels are used to determine the purity of RNA. OD 260/280 ratio higher than 1.8 is accepted as proper for downstream applications (Manchester, 1996). RNA is considered of high quality when the ratio of 28S:18S bands is about 2.0. Nowadays, Agilent BioAnalyser, Ribogreen, NanoDrop and BioRad Experion are used for this purpose (Nolan et al., 2006). NanoDrop ND-1000 spectrophotometer only needs 1 µl of sample and can be used with concentrations as low as 2 µg ml^{-1}.

Agilent 2100 Bioanalyzer and BioRad Experion are used for the quality control of RNA. These instruments use a lab on a chip approach to perform capillary electrophoresis (Nolan et al., 2006) These instrument softwares calculate a numerical value: RNA integrity number (RIN) on the 2100 Bioanalyzer and quality index (RQI) on the Experion. A RQI/RIN of 1 exhibits nearly fragmented and degraded RNA and a RQI/RIN of 10 exhibits intact and non-degraded RNA (Schroeder et al., 2006). RNA quality score (RIN or RQI) higher than five is determined as good total RNA quality, moreover, higher than eight is perfect total RNA for gene expression studies (Fleige & Pfaffl, 2006).

After isolation of total RNA and evaluating its integrity and purity, cDNA synthesis can be simply made using commercially available kits starting with equal amounts of RNA samples. Moreover, cDNA can be synthesized using random primers, oligo(dT), target gene specific primers or a combination of oligo(dT) and random primers.

4. Real- time PCR primer design

Optimal primers are essential to insure that only a single PCR product is amplified. In order to avoid non-specific PCR products, primers should not have high sequence similarity with other sequences. This can be checked using the Basic Local Alignment Search Tool (BLAST) from the National Center for Biotechnology Information (http://blast.ncbi.nlm.nih.gov/Blast.cgi). Primers containing 16-28 nucleotides are enough for successful PCR amplification. GC content of the primer should be between 35%-65% (Wang & Seed, 2006).

In addition to general rules used for designing common PCR primers, some important parameters should be considered for Real-time PCR amplification. Primers should be designed to give product size of 100-200 bp. Primer melting temperatures (Tm) should be 60–65 °C. Intron spanning primer pair should be preferred in order to prevent potential signals from genomic DNA contamination in the sample. Finally, if oligo(dT) is used for priming in reverse transcription, primers should be located within 1000 bp of the 3' end of mRNA (Wang et al., 2006).

There are some free online tools or commercially available softwares which can be used for primer design if the parameters described above are provided. The selected list of useful web resources and some commercial programs is given in table 1.

Website or Software Name	Specification	URL
Primer3	Picking primer and hybridization probes	http://frodo.wi.mit.edu/primer3/input.htm
Primer-BLAST	For making primers. It uses Primer3 to design primers and then submits them to BLAST search	http://www.ncbi.nlm.nih.gov/tools/primer-blast/
PrimerBank	Public database of Real-time PCR primers. Contains over 306.800 primer mostly for human and mouse	http://pga.mgh.harvard.edu/primerbank/
RTPrimer DB	Public database for primer and probe sequences used in real-time PCR assays. Contains 8309 real-time PCR assays for 5740 genes.	http://www.rtprimerdb.org/
Real-time PCR Primer Sets	Primer and Probe database, mostly for SYBR green assays	http://www.realtimeprimers.org/
OligoCalc	Calculates oligonucleotide properties	http://www.basic.northwestern.edu/biotools/oligocalc.html
Universal Probe Library from Roche Applied Science	Designing primers and UPL hydrolysis probe	www.universalprobelibrary.com or http://www.roche-applied-science.com
Primer Express	Designing primers and TaqMan probes	www.appliedbiosystems.com
Beacon Designer	Real-time PCR primers and probes	http://www.premierbiosoft.com
Primer Premier	Primer Design	http://www.premierbiosoft.com

Table 1. Some free online tools or commercially available softwares

5. Real-time PCR detection chemistries

In the following section we will focus on the detection chemistries which deviates Real-time PCR from conventional PCR assays. Real-time PCR detection chemistries can be classified into sequence non-specific or specific detection chemistries.

5.1 Sequence non-specific detection

The principle of the sequence non-specific detection is the use of DNA binding fluorogenic dyes. This method is not affected when the presence of variations (i.e. single nucleotide polymorphisms or SNPs) on the target sequence. Moreover, less specialist knowledge is required as compared to the design of fluorogenic oligoprobes (Komurian-Pradel et al., 2001). DNA binding dyes are also inexpensive and simple to use (VanGuilder et al., 2008).

The first dye used as DNA binding fluorophore was ethidium bromide (Higuchi et al., 1993; Wittwer et al., 1997), and other dyes such as SYBR Green I, YO-PRO and BEBO have been also used (Ishiguro et al., 1995; Tseng et al., 1997; Morrison et al., 1998; Bengtsson et al., 2003). All these dyes fluoresce when binding with double-stranded DNA (dsDNA) and this dsDNA-dye complex is revealed by a suitable wavelength of light. Thus, observing the amplification of any dsDNA template is possible during reaction (Fig. 3).

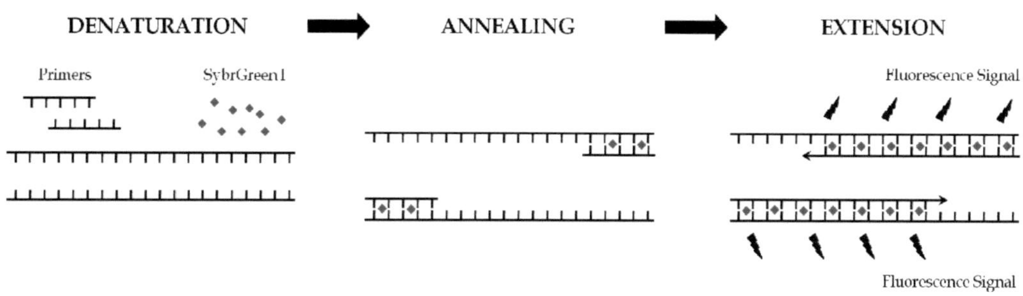

Fig. 3. Detection principle of SYBRGreen I dye during PCR amplification.

SYBR Green I is the most frequently used dsDNA specific dye in Real-time PCR. It is a cyanine dye with an asymmetric structure. The binding affinity of SYBR Green I to dsDNA is 100 times higher than that of ethidium bromide (Wittwer et al., 1997; Morrison et al., 1998). After SYBR Green I binds to dsDNA, it emits 1000-fold greater fluorescence as compared to unbound dyes (Wittwer et al., 1997).

As the amount of double-stranded amplification product increases during reaction, the amount of dye that can bind and fluoresce, also increases. Thus, the fluorescence signal elevates proportionally to dsDNA concentration (Wittwer et al., 1997). However, these dyes also bind primer-dimers, commonly occur during reaction, and non-specific PCR products. This non-specific dsDNA-dye interaction can cause misinterpretation of the results. That is why these dyes provide sensitive but not specific detection (Espy et al., 2006). However, this problem can be solved using melting curve analysis. Instruments performing a melting curve analysis to determine the Tm allow detection of accumulation of different products based upon the G+C% content and length of the amplification product (Espy et al., 2006).

After melting curve analysis, if two or more peaks are present, it means that there are more than one amplified products in the reaction and thus no specific amplification for a single DNA sequence is occurred (Valasek & Repa, 2005).

5.2 Sequence specific detection

Development of fluorescent probe technology allows us to perform sensitive and specific detection with Real-time PCR. Mainly, there are three types of probes used in the reaction although they have distinct molecular structure and dyes attached. These probes can be grouped as follows: hybridization probes, hydrolysis probes and hairpin probes. All detection methods using fluorescent probe technology rely on a process referred to as fluorescence resonance energy transfer (FRET) in which the transfer of light energy between two adjacent dye molecules occurs (Espy et al., 2006). However, both hydrolysis and hybridization probes depend on FRET to change fluorescence emission intensity, the energy transfer works in opposite manners in these two chemistries. While FRET reduces fluorescence intensity in hydrolysis probes, it increases intensity in hybridization probes (Wong & Medrano, 2005).

5.2.1 Hybridization probes

One or two hybridization probes can be used in a reaction (Bernard & Wittwer, 2000). In an assay utilizing two hybridization probes, they bind to target sequence in close proximity to each other in a head-to-tail arrangement (Wittwer et al., 1997a; Wong & Medrano, 2005;). The upstream probe carries an acceptor (or quencher) dye on its 3' end the second probe or downstream probe is labeled with a donor (or reporter) dye on 5' end (Wittwer et al., 1997; Bustin, 2000; Wong & Medrano, 2005). On the other hand, in one probe method, the upstream primer is labeled with an acceptor dye on the 3' end instead of labeling probe. Thus, labeled primer replaces the function of one of the probes used two hybridization probe method (Wong & Medrano, 2005). In both cases, the energy transfer depends on the distance between two dye molecules. Because of the distance between two dyes in solution, donor dye emits only background fluorescence (Bustin, 2000). When the probes hybridize to their complementary sequence, this binding brings the two dyes in close proximity to one another and FRET occurs at high efficiency. Since, a fluorescent signal is detected only as a result of two independent probes hybridizing to their correct target sequence, increasing amounts of measured fluorescence is proportional to the amount of DNA synthesized during the PCR reaction. Moreover, as the probes are not hydrolyzed, fluorescence signal is reversible and allows the generation of melting curves (Bustin, 2000) (Fig. 4).

5.2.2 Hydrolysis probes

Hydrolysis probes (also known as TaqMan probes or 5' nuclease assay) contain a fluorescent reporter dye at its 5' end and quencher dye at its 3' end (Wong & Medrano, 2005; VanGuilder et al., 2008). If the probe is unbound, reporter and quencher dyes are maintained in close proximity, which allows the quencher to reduce the reporter fluorescence intensity by FRET, and thus no reporter fluorescence is detected (Bustin, 2000) (Fig. 4).

On the other hand, the probe binds to the target sequence, when the specific PCR product is generated. It remains hybridized while the polymerase extends the primer until the enzyme reaches the hybridized probe. Then the 5'-3' exonuclease activity of DNA polymerase degrades the probe during extension step of the PCR (Heid et al., 1996). 5'- exonuclease activity of the polymerase releases the 5' reporter dye from the quenching effect of the 3' dye and this release is detected as an increase in fluorescence intensity (Heid et al. 1996; Bustin, 2000; VanGuilder et al., 2008) (Fig. 4).

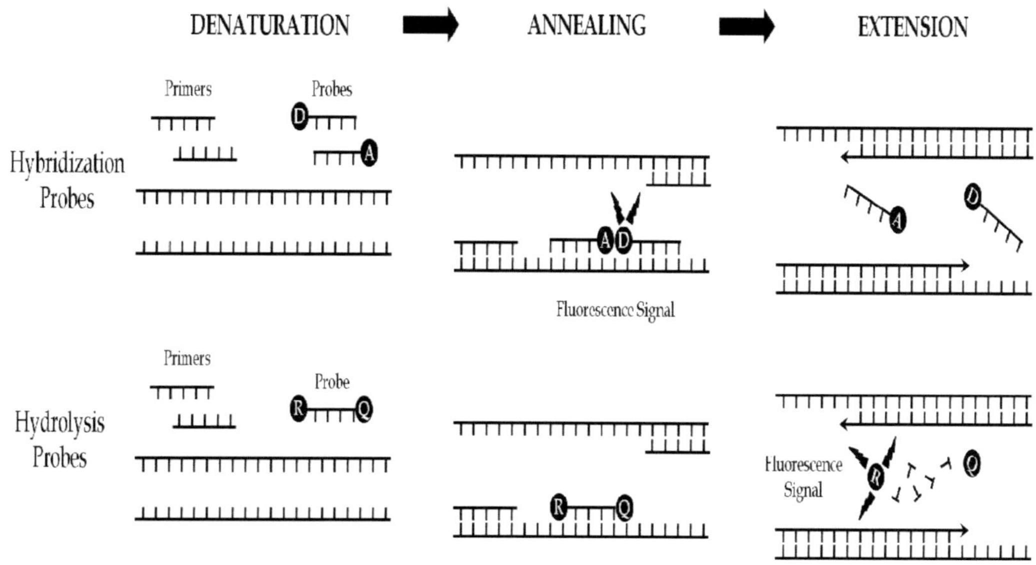

Fig. 4. Detection of nucleic acids using hybridization and hydrolysis probes in Real-time PCR.

Hydrolysis probes commonly are in structure of nucleic acids, however, recently developed, Locked Nucleic Acids (LNA) containing hydrolysis probes are commercially available from Roche Applied Science under the name of Universal Probe Library (UPL) probes and can be accessed online from the site given in Table 1. LNAs are DNA nucleotide analogues with increased binding strengths compared to standard DNA nucleotides. In order to maintain the specificity and Tm, LNA bases are incorporated in each UPL probes (www.universalprobelibrary.com).

5.2.3 Hairpin probes

When compared with linear DNA probes, hairpin or stem-loop DNA probes have an increased specifity of target recognition. Hairpin DNA probes are single-stranded oligonucleotides and contain a sequence complementary to the target that is flanked by self-complementary target unrelated termini. Invention of hairpin probes is let to view hybridization process in real-time. They are widely used in different applications and two major factors are responsible for such broad applications of these DNA probes: Enhanced specificity of the probe-target interaction and the possibility of closed-tube real-time monitoring formats (Broude, 2005). There are several types of hairpin probes commercially available including molecular beacons, scorpions, LUX™ fluorogenic primers and Sunrise™ Primers.

5.2.3.1 Molecular beacons

Molecular beacons are a class of hairpin probes and first developed in 1996 (Tyagi & Kramer 1996). Sequence-specific loop region is located between two inverted repeats which form stem of the hairpin by complementary base pairing (Tyagi & Kramer 1996; Bonnet et al., 1999). Reporter and quencher dyes are linked to each end of the molecular beacon. In solution, reporter's fluorescence is effectively reduced via contact quenching. When probe binds to the target sequence, reporter and quencher dyes separates, resulting in increased fluorescence (Tan et al., 2000) (Fig. 5A). The fluorescence of the probe increases 100-fold when it binds to its target (Bonnet et al., 1999).

5.2.3.2 Scorpions

Scorpion primers are bi-functional molecules because probe sequence is covalently linked to primer. Probe sequence forms stem-loop structure and contains fluorophore at 5'-end which is quenched by a moiety attached to the 3'-end of the loop. The probe is linked to the 5'-end of a primer via a nonamplifiable monomer or DNA polymerase blocker (Whitcombe et al., 1999). The probe part of the scorpion is complementary to the extension product of the attached primer. After extension step in a PCR, the probe will bind to the extended part of the primer when the complementary strands are separated in the denaturation step of the next cycle. Therefore, scorpion primers generate self-probing PCR products (Whitcombe et al., 1999; Thelwell et al., 2000) (Fig. 5B).

5.2.3.3 LUX™ fluorogenic primers

Light Upon eXtension (LUX) primers (Invitrogen, Carlsbad, CA, USA) are self-quenched single-fluorophore labeled primers. It is designed to be self-quenched with secondary structure of the its 3' end (Nazarenko et al., 2002a). This secondary structure reduces initial fluorescence to a minimal amount until the primer is incorporated into a double-stranded PCR product (Nazarenko et al., 2002b). This incorporation causes an unfolding of the primer which abolishes the self-quenching and thus an increase in fluorescence occurs (Kusser, 2006) (Fig. 5C). LUX primers are designed to have a G or C 3'-terminal nucleotide and fluophore attached to the second or third base (Thymine nucleotide) from the 3' end. It also has five to seven nucleotide 5'-tail that is complementary to the 3' end of the primer. Such a design of the primer allows the molecule to form a blunt-end hairpin structure with low fluorescence at temperatures below its Tm. Various fluorescent dyes can be used, allowing the potential for multiplex assays to simultaneously quantitate multiple genes (Nazarenko et al., 2002b).

5.2.3.4 Sunrise™ primers

Sunrise primers consist of a dual-labeled (reporter and quencher) hairpin loop on the 5' end. Similar to the scorpions, their 3' end is used as a PCR primer by the polymerase (Nazarenko et al., 1997). However, the probe part of the Scorpion is complementary to the extension product of the attached primer. On the other hand, the probe sequence of the Sunrise primers is complementary to the hybridized strand of the primer. After integration of the Sunrise primer into the newly formed PCR product, the reporter and quencher locate far enough which allow reporter emission (Wong & Medrano, 2005) (Fig. 5D). It is important to consider that Sunrise primers may produce fluorescence signals due to nonspecific products and primer-dimers.

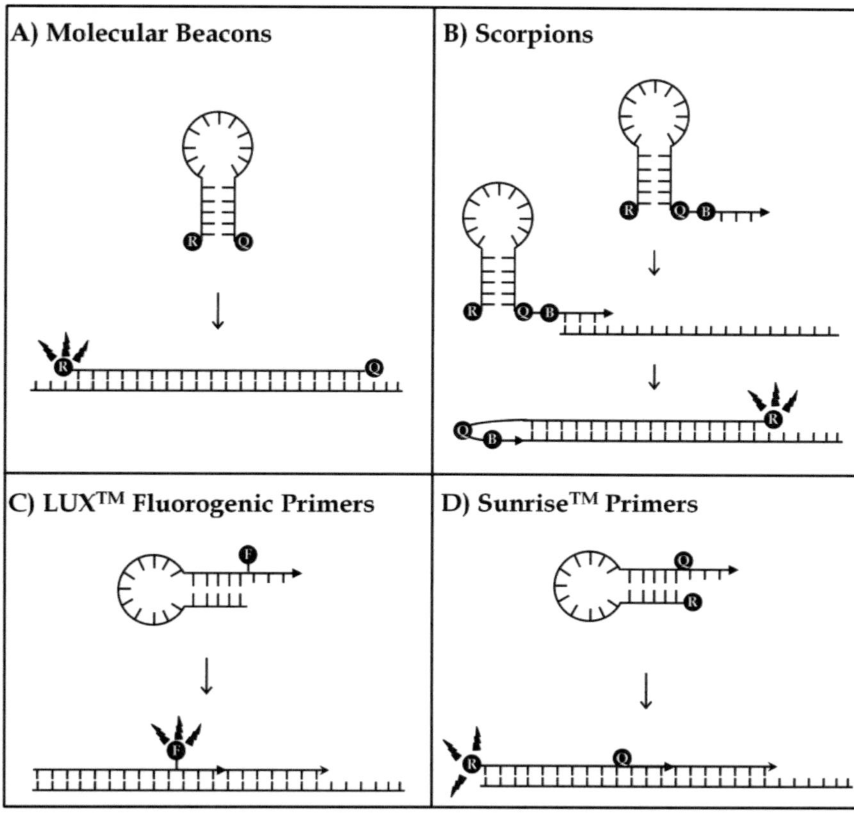

Fig. 5. Representative diagram showing hairpin probes and their principles of detection. A: Molecular beacons. B: Scorpions. C: LUX™ fluorogenic primers and D: Sunrise™ primers.

5.2.3.5 Other detection chemistries

Recently, newly developed detection chemistries, which will not be discussed in further detail here, have been introduced for Real-time PCR. Like hybridization and hydrolysis probes, these new systems all rely on the FRET principle. Although the list of these new chemistries is rapidly growing, some of them are minor groove-binding probes (MGB probes), ResonSense probes, Hy-beacon probes, Light-up probes, Simple probes, Lion probes, AllGlo probes, Displacement probes and etc (Overbergh et al., 2010). Some of these probes contain synthetic nucleic acid analogs as in the case of Light-up probes.

Light-up probes are peptide nucleic acids (PNAs) oligonucleotide. They are linked with thiazole orange as the fluorophore and no quencher is required (Svanvik et al., 2000). PNA molecules have a backbone with peptide like covalent bonds and exocyclic bases. When light-up probes hybridize with specific target DNA, the resulting duplex or triplex structures elicit increased fluorescence of the fluorophore (Isacsson et al., 2000).

6. Real time quantification

The quantification strategy is an important factor for detecting of mRNA gene expression level. Quantification of mRNA transcription can be measured by absolute or relative quantitative Real-time PCR (Souazé et al., 1996; Pfaffl, 2001a; Bustin, 2002). Absolute quantification is an analysis method to accurately measure the copy number of a target sequence (in picograms or nanograms of DNA or RNA) in the sample, while relative quantification provides relative changes in mRNA expression levels as a ratio of the amount of initial target sequence between control and analysed samples (Souazé et al., 1996; Pfaffl, 2001a; Fraga et al., 2008). Thus, relative quantification simply allows us to determine the fold changes between sample and control. If the purpose is accurately measuring the copy number of a target sequence, absolute quantification strategy which requires standards of known copy number, should be performed. Moreover, these standards should be amplified in the same run (Peirson et al., 2003).

Both approaches are generally used but relative quantification requires less set up time and easier to perform than absolute quantification because a standard curve is not essential (Livak, 2001; Pfaffl 2004b; Fraga et al., 2008). Furthermore, it is commonly not necessary to know the absolute amount of mRNA in biological applications examining gene expression (Bustin, 2002; Huggett et al., 2005).

6.1 Absolute quantification

This approach is more precise but often more labor-intensive (Bustin & Nolan, 2004a). Absolute quantification requires a standard calibration curve using serially diluted standards of known concentrations for highly specific, sensitive and reproducible data (Reischl & Kochanowski, 1995; Heid et al., 1996; Bustin, 2000; Pfaffl & Hageleit, 2001; Pfaffl, 2001b; Fraga et al., 2008). Linear relationship between Ct and initial amounts of total RNA or cDNA using standart curve allows the detection of unknowns' concentration based on their Ct values (Heid et al., 1996).

In this method, all standards and samples have equal amplification efficiency. It is necessary to control the efficiency of the Real-time PCR reaction to quantify mRNA levels (Fraga et al., 2008). Real-time PCR amplification efficiencies for calibration curve and target cDNA must have identical reverse transcription efficiency to provide a valid standard for mRNA quantification (Pfaffl & Hageleit, 2001). The amplification efficiencies of the standard and unknown target sequence should be approximately equal and the concentration of the serial dilutions should be within the range of the unknown(s) in order to ensure correct results.

The standard and target sequence should have the same primer binding sites and produce a product of approximately the same size and sequence (Fraga et al., 2008). The standard can be based on known concentrations of double-stranded DNA (dsDNA), single-stranded DNA (ssDNA), commercially synthesized big oligonucleotide and complementary RNA (cRNA) bearing the target sequence (Reischl & Kochanowski, 1995; Morrison et al., 1998; Bustin 2000; Kainz, 2000; Pfaffl & Hageleit, 2001; Pfaffl et al., 2001c; Wong & Medrano, 2005). DNA standards can be synthesized by cloning the target sequence into a plasmid (Gerard et al., 1998), purifying a conventional PCR product, or directly synthesizing the target nucleic acid (Liss, 2002). These standards have a property of larger quantification range, greater sensitivity, more reproducibility and higher stability than RNA standards (Pfaffl, 2004b).

However, DNA standards are generally not possible to use as a standard for absolute quantitation of RNA because there is no control for the efficiency of the reverse transcription step. (Livak, 2001; Wong & Medrano, 2005). Therefore, RNA molecules are strongly recommended as standards for quantification of RNA (Real Time PCR, 2010).

In standard preparation for RNA quantitation, an in vitro-transcribed sense RNA transcript is generated, the sample is digested with RNase-free DNase so that the RNA is quantitated accurately. Because any significant DNA contamination will result in inaccurate quantification (Bustin, 2000, 2002). A recombinant RNA (recRNA) can be synthesized *in vitro* by cloning the DNA of the gene of interest (GOI) into a suitable vector, containing typically SP6, T3, or T7 phage RNA polymerase promoters (Gibson et al., 1996; Pfaffl & Hageleit, 2001; Pfaffl, 2001b; Fronhoffs et al., 2002; Fraga et al., 2008). Several commercial kits are available that facilitate the production of RNA from these vectors. After in vitro transcribed RNA (standard RNA) is synthesized, the standart concentration is measured on a spectrophotometer and converted the absorbance to a 'target copy number per µg RNA' (Bustin, 2000). Once the standard has been accurately quantified, it is serially diluted in increments of 5- to 10-fold and each dilution should be run in triplicate (Fraga et al., 2008). The dilutions should be made over the range of copy numbers that include the likely amount of target mRNA expected to be present in the experimental samples to maximize accuracy (Bustin, 2000; Fraga et al., 2008). After dilutions generated, known amounts of RNAs are converted to cDNA for subsequent Real-time PCR. A standart curve is created by plotting the average Ct values (inversely proportional to the log of the initial copy number) from each dilution versus the absolute amount of standard present in the sample (Higuchi et al., 1993; Bustin, 2000; Fraga et al., 2008).

As shown in Fig. 6, the copy numbers of sample RNAs can be calculated via comparison of samples' Ct values to this standard curve after real time amplification (Bustin, 2000). Under certain circumstances, absolute quantification models can also be normalized using suitable and unregulated references or housekeeping genes (Pfaffl, 2004b). Standard design, production, determination of the exact standard concentration, and stability over long storage time are tortuous and can be difficult (Bustin & Nolan, 2004a). In addition, the primary limitation to this approach is the necessity of obtaining an independent reliable standard for each gene. Moreover, running concurrent standard curves during each assay is needed.

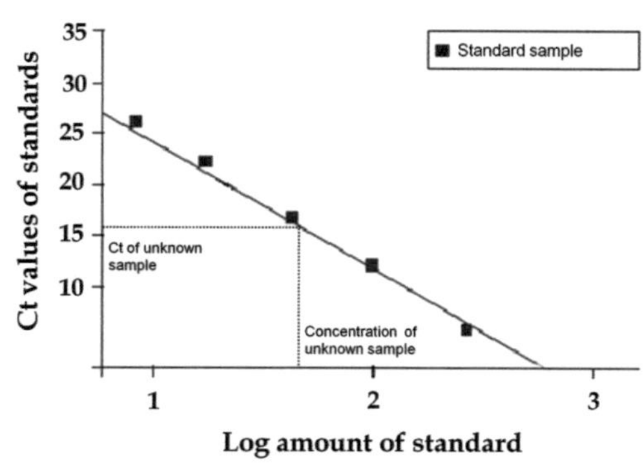

Fig. 6. Absolute quantification with standard curve. It shows determination of concentration of unknown sample.

6.2 Relative quantification

Relative quantification determines the changes in steady-state mRNA levels of GOI in response to different treatments (e.g., control versus experimental) or state of the tissue (e.g., infected versus uninfected samples, different developmental states or benign versus malign tissue) (Pfaffl et al., 2002a; Bustin &Nolan, 2004a; Valasek & Repa, 2005;). The advantage of using a relative quantification approach is that standards with known concentrations are not required so that there is no need for generating a standard calibration curve (Pfaffl 2004b; Fraga et al., 2008). In this approach, only relative changes can be determined because of the unknown internal standard quantity (Valasek & Repa, 2005;). This may not pose a problem for more research projects because absolute value of mRNA is commonly not necessary and fold change is adequate for investigating physiological changes in gene expression levels in many biological applications (Bustin, 2002; Pfaffl, 2004b; Huggett etal., 2005; Valasek & Repa, 2005). During relative quantification, amounts of target and reference gene's (sometimes called a housekeeping gene or internal control) are determined within the same sample. Housekeeping gene selection is an important issue and has been discussed in separate title (see housekeeping gene selection). After reaction, the Ct ratio between each target and the reference gene is calculated (Real-time PCR, 2010). The housekeeping gene which helps to normalize the data for experimental error, can be co-amplified in the same tube in a multiplex assay or can be amplified in a separate tube (Wittwer et al., 2001; Pfaffl et al., 2002a; Fraga et al., 2008).

6.3 Amplification efficiency

Amplification efficiency is an important factor for accurate relative quantification. In an optimal PCR reaction (100% efficient), every amplicon will be replicated and the amount of product will double after each cycle and a plot of copy number versus cycle number produces a line. However, if the reaction is only 90% efficient, the amount of product will not double after each cycle and the slope of the plot will be less than the same plot assuming 100% efficiency (Fraga et al., 2008). As mentioned earlier, the amplification efficiency is assumed an ideal or 1 (Gibson et al., 1996). However, small efficiency differences between target and reference gene result in inaccurate expression ratio (over or under initial mRNA amount instead of real). Difference in PCR efficiency (ΔE) of 3% ($\Delta E = 0.03$) between target gene and reference gene generates a falsely calculated expression ratio of 47% in case of $E_{target} < E_{ref}$ and 209% in case of $E_{target} > E_{ref}$ after 25 PCR cycles (Pfaffl et al., 2002b; Rasmussen, 2001). The amplification efficiencies of the target gene and housekeeping gene are preferred to be the same so that relative expression values for the target gene in samples are accurately compared (Schmittgen et al, 2000; Fraga et al., 2008). However, it is difficult to achieve identical amplification efficiencies in all PCRs. Therefore, lack of an appropirate correction factor might result in overestimation of the target gene's starting concentration (Liu & Saint, 2002a).

Traditionally, the amplification efficiencies of a genes (for example; target or reference genes) can be determined by preparing a 10-fold dilution series from a reference RNA or cDNA sample and by plotting the Ct as a function of log[10] concentration of template. The slope of the resulting trend line (S) will be a clue of the PCR efficiency. Simply, amplification efficiency of a reaction is calculated using data collected from a standard curve plot with the following formula (Rasmussen, 2001):

$$\text{Exponential amplification} = 10^{(-1/S)} \quad (1)$$

$$E = [10^{(-1/S)}]-1 \quad (2)$$

In above formulas, "E" refers to the efficiency of the reaction and "S" refers to the slope of the standart curve plot generated by Ct value versus the log of the input template amount.

A slope of -3.32 indicates the PCR reaction is 100% efficient. When a slope value is between −3.6 and −3.1, amplification efficiency ranges from 90% to 110% (e. g., E = 0.9 − 1.1). The appropriate efficiency of the PCR should be 90-110%. (Rasmussen, 2001; Tichopad et al., 2003). Theoretically, 3.3 cycles are required in order to increase the amplicon concentration 10-fold when a PCR reaction proceeding at 100% efficiency. Additionally, a Ct alteration of 1 between different samples corresponds to a 2-fold changes in starting material (Fraga et al., 2008).

Amplification efficiency depends on many factors, such as efficiency of primer annealing, the length of the amplicon, GC content of the amplicon and sample impurities (McDowell et al., 1998). These factors affect primer binding, the melting point of the target sequence, and the processivity of the DNA polymerase. (Wiesner, 1992). Therefore, the target gene and the reference gene's amplification efficiencies are usually found different. Thus, determination of the amplification efficiencies of analysed genes should be done carefully in Real-time PCR assays.

6.4 Data analysis methods and software applications

The Ct values obtained from Real-time PCR analysis need to be converted using different procedure in order to make valid comparisons (Fraga et al., 2008). Besides, the classical Real-time PCR parameters (i.e. primer design, RNA quality, reverse transcription and polymerase performances), Real-time PCR data processing can influence or even change the final results. Analysis of Real-time PCR data can be either of absolute levels (i.e., numbers of copies of a specific RNA per sample) or relative levels.

In absolute quantification, the Ct value for each sample should be compared with standard curve to extrapolate the starting concentration. (Wilkening & Bader, 2004; Fraga et al., 2008; Vanguilder et al., 2008). Besides, the absolute gene quantification strategy, the relative expression strategy compares GOI in relation to a reference gene, is commonly used by the academic research community.

To date, several mathematical models using for calculating relative expression ratio (R) or fold induction have been developed and they are based on the comparison of the distinct cycle differences (Meijerink et al., 2001; Pfaffl, 2001a; Liu and Saint, 2002b). Two types of relative quantification models are available and used generally;

A) Relative quantification without efficiency correction or the Comparative Ct method; The comparative Ct method is a mathematical model based on the delta-Ct (ΔCt) (Wittwer et al., 2001) or delta-delta-Ct (ΔΔCt) values in most applications, described by Livak and Schmittgen (Livak & Schmittgen, 2001) without efficiency correction. In this model, an optimal doubling of the target sequence during each performed Real-time PCR cycle is assumed (Winer et al., 1999; Livak, 2001; Livak & Schmittgen, 2001). This analysis can be performed in two ways; Non-normalized expression (also known as ΔCt method) and normalized expression (also known as ΔΔCt method).

A1) Non-normalized Expression (ΔCt) method; In relative quantification, a comparision is made with the gene expressed in the sample to that of the same gene expressed in the control. Ct values are non-normalized using housekeeping gene, but normalization is accomplished via equal loading of samples. Quantitation is performed relative to the control by subtracting the Ct value of the control gene from Ct of the sample gene (ΔCt). The fold difference of target gene in sample and control is calculated by using the resulting differences in cycle number (ΔCt) as the exponent of the base 2 (due to the doubling function of PCR) as given below in eq. 3 and 4.

$$R = 2^{\Delta Ct} \tag{3}$$

$$R = 2^{[Ct\ sample\ -\ Ct\ control]} \tag{4}$$

A2) Normalized Expression (ΔΔCT) method; In this approach, loading differences are eliminated. Moreover, the Ct values of both the control and the samples for target gene are normalized to an appropriate housekeeping or reference gene. This method also known as $2^{-\Delta\Delta Ct}$ method, where ΔΔCt = ΔCt sample - ΔCt control. Formulas are given below in eq.5 and 6.

$$R = 2^{-\Delta\Delta Ct} \tag{5}$$

$$R = 2^{-[\Delta Ct\ sample\ -\ \Delta Ct\ control]} \tag{6}$$

$$\Delta Ct\ (sample) = Ct\ target\ gene - Ct\ reference\ gene$$

$$\Delta Ct\ (control) = Ct\ target\ gene - Ct\ reference\ gene$$

$$\Delta\Delta Ct = \Delta Ct\ (sample) - \Delta Ct\ (control)$$

The reaction is rigorously optimized and the PCR product size should be kept small (less than 150 bp) (Marino et al., 2003; Wong & Medrano, 2005). Comparative Ct method can be chosen when assaying a large number of samples because the standart curve is unnecessary (Wong & Medrano, 2005).

This model is acceptable for a first approximation of the crude expression ratio. However, efficiency (E) corrected models are useful to obtained reliable relative expression data (Pfaffl et al., 2009).

B) Relative quantification with efficiency correction or Pfaffl model: Pfaffl developed a mathematical formula widely used for the relative quantification of gene expression in Real-time PCR (Pfaffl, 2001a). This model combines gene quantification and normalization with an amplification efficiency of the target and reference genes. This calculation can be based on one sample (Souazé et al., 1996;; LightCycler® Relative Quantification Software, 2001) or multiple samples (Pfaffl, 2001a,, 2004b) and their formulas are given in Eqs. 7-8 and 9, respectively. Reactions for the determination of efficiencies of the genes should be run in a 5 or 10-fold serially diluted sample.

$$R = \frac{\left(E_{target}\right)^{\Delta Ct_{target}(control - sample)}}{\left(E_{ref}\right)^{\Delta Ct_{ref}(control - sample)}} \tag{7}$$

$$R = \frac{\left(E_{ref}\right)^{Ctsample}}{\left(E_{target}\right)^{Ctsample}} \div \frac{\left(E_{ref}\right)^{Ctcontrol}}{\left(E_{target}\right)^{Ctcontrol}} \tag{8}$$

$$R = \frac{\left(E_{target}\right)^{\Delta Ct_{target}(MEANcontrol - MEANsample)}}{\left(E_{ref}\right)^{\Delta Ct_{ref}(MEANcontrol - MEANsample)}} \tag{9}$$

In new approaches, multiple reference genes is used to obtain more stable and reliable results (Vandesompele et al., 2002). An efficiency corrected calculation models, based on multiple samples and reference genes (so-called REF index), should consist of at least three reference genes (eq. 10) (Pfaffl, 2004b).

$$R = \frac{\left(E_{target}\right)^{\Delta Ct_{target}(MEANcontrol - MEANsample)}}{\left(E_{ref} index\right)^{\Delta Ct_{ref\ index}(MEAN\ control - MEAN\ sample)}} \tag{10}$$

Analysis of the raw data in precise mathematical and statistical manner should be performed rationally in gene expression analysis. Various software tools and excel spreadsheets are available to calculate the raw data. The LightCycler relative expression software (Roche Applied Science), Q-Gene (Muller et al., 2002), qBASE (Hellemans et al., 2007), SoFar (Wilhelm et al., 2003), DART (Peirson et al., 2003), qPCR-DAMS (Jin et al., 2006) and REST software applications (Pfaffl et al., 2002b) can be used for calculation. Only Q-Gene (Muller et al., 2002) and REST (Pfaffl et al., 2002b) software packages are freely available. Q-Gene uses a paired or an unpaired Student's t test, a Mann-Whitney U-test, or Wilcoxon signed-rank test (Muller et al., 2002).

The REST software established in 2002 performs Pair-Wise Fixed Reallocation Randomization Test which repeatedly and randomly reallocates at least 2000 times the observed Ct values to the two groups (Pfaffl et al., 2002b; Pfaffl et al., 2004b). Two new version of REST software package (REST 2008 and REST 2009) were developed by Pfaffl and co-workers and the single run efficiency is implemented in REST 2008 as well as multiple reference gene normalization. In REST 2009, randomization algorithms have been improved to obtain better confidence intervals and more accurate p values. Moreover, the best fit for the standard curve is used for the determination of the efficiency and it is used in the randomization process.

7. Housekeeping gene selection

The proper housekeeping gene (HKG) is continuously expressed in all cell types and tissues (Thellin et al., 1999). Additionally, the expression level of a suitable reference gene should be stable and is not affected by the biologic and experimental condition or by the disease state (Vandesompele et al., 2002). Nevertheless, there is no universal housekeeping gene having invariable expression under all these circumstances (Thellin et al., 1999). Therefore, choosing a stable housekeeping gene is crucial for the accurate interpretation of gene expression data. (Zhang et al., 2005). Furthermore, using more than one HKG is recommended for the convenient results. The most frequently used housekeeping genes involved β-actin (ACTB), glyceraldehyde-3-phosphate dehydrogenase (GAPDH), hypoxanthine phosphoribosyl transferase 1 (HPRT1), β2-microglobulin (B2M), phosphoglycerokinase1 (PGK1), cyclophilin A (CPA) and 18S rRNA.

GeNorm (Vandesompele et al., 2002) (available at http://medgen.ugent.be/»jvdesomp/genorm/), Normfinder (Andersen et al., 2004) (available at http://www.mdl.dk/publicationsnormfinder.htm) and Bestkeeper (Pfaffl et al.2004a) (available at http://www.gene-quantification.de/bestkeeper. html) programs are used to determination of housekeeping gene mRNA expression stability.

There are numerous studies on the selection of the proper reference gene in many different tissues and cell types. Calcagno et al. suggested that plasma membrane calcium-ATPase 4 (PMCA4) is a suitable reference gene for normalization of gene expression for polytopic membrane proteins including transporters, ATPases and receptors (Calcagno et al., 2006). Cicinnati et al. showed that hydroxymethyl-bilane synthase (HMBS) and GAPDH are good reference genes for normalizing gene expression data between paired tumoral and adjacent non-tumoral tissues derived from patients with human hepatocellular carcinoma (HCC) (Cicinnati et al., 2008). It is shown that TATA box binding protein (TBP) and HPRT1 are the most reliable reference genes for q-PCR normalization in HBV related HCCs' matched tumor and non-tumor tissue samples (Gao et al., 2008; Fu et al., 2009). The cyclophilin A gene [peptidylprolyl isomerase A gene (PPIA)] is recommended as a housekeeping gene for gene expression studies in atopic human airway epithelial cells (AEC) of asthmatics (He et al., 2008). Penna et al. suggested that the use of two reference genes [Eukaryotic translation initiation factor 4A2 (EIF4A2) and Cytochrome c-1 (CYC1)] is proper for the normalization of the RT-qPCR data in human brain tissues (Penna et al., 2011). Pfister et al. demonstrated that the ribosomal protein L37A (RPL37A) is the most ideal housekeeping gene in meningiomas and their normal control tissue arachnoidea, dura mater and normal brain. The use of the combination of RPL37A and eukaryotic translation initiation factor 2B, subunit 1 alpha (EIF2B1) housekeeping genes is also recommended (Pfister et al., 2011). In another study, it is shown that the best choice of a reference gene for expression studies on astrocytomas is GAPDH. If two genes are used for gene normalization, authors recommend the combination of ribosomal protein, large, P0 (RPLP0) and histone cluster 1 (H3F). (Gresner et al., 2011).

Silver et al. showed that GAPDH is the most suitable HKG in reticulocyte studies (Silver et al., 2006). It is shown that succinate dehydrogenase complex subunit A (SDHA) is the best individual reference gene in neonatal human epidermal keratinocytes after UVB exposure. Also, SDHA and PGK1 were designated as the best combination (Balogh et al., 2008).

8. Normalization

Data normalization is a further major step for quantification of target gene expression in Real-time PCR (Pfaffl 2001a; Bustin, 2002). Appropriate normalization strategies are required to correct errors in Real-time PCR(Huggett et al., 2005; Wong & Medrano, 2005). These errors can be originated from a number of factors (variation in RNA integrity, sample-to-sample variation, PCR efficiency differences, cDNA sample loading variation etc.) (Karge et al., 1998; Mannhalter et al., 2000). Performing a normalization strategy is also crucial to control for the amount of starting material, variation of amplification efficiencies and differences between samples. However, this remains the most intractable problem for real-time quantification (Thellin et al., 1999). Starting material usually varies in tissue mass, cell number or experimental treatment. mRNA levels can be standardized to cell number under ideal conditions in *in vitro* model. Ensuring similar tissue volume or weight appear to be straightforward, but this type of normalization is not possible because it can be difficult to ensure that different samples contain the same cellular material (Vandesompele et al., 2002).

Several strategies can be chosen for normalising Real-time PCR data including reference gene selection, similarity of sample size and quality of RNA (Huggett et al., 2005). Precise quantification and good quality of RNA is essential prior to reverse transcription (Bustin & Nolan, 2004b). Data normalization can be carried out against an endogenous unregulated reference gene transcript or against total cellular RNA content (molecules/g total RNA and concentrations/g total RNA) but normalization to total RNA is unreliable. Because knowlegde about the total RNA content or even about the mRNA concentrations of the cells can not be accurately determined (Bustin, 2000, 2002). Normalising strategy using a housekeeping gene is a simple and convenient method for correction of sample-to-sample variation in Real-time PCR. Target and housekeeping gene expression levels should be within a similar range. For example, HPRT gene expression is very low in most human tissues so that this gene is only suitable for the normalization of lowly expressed target genes (Huggett et al., 2005).

Although it is best to start with the same amount of RNA concentration in cDNA synthesis step, sometimes this can not be achieved due to pipetting errors. Such an error can be partly controlled by using reference genes (Huggett et al., 2005).

9. Conclusion

In summary, qPCR is rapid, cost-effective, accurate, sensitive, reliable and reproducible method so that this technology has become a routine and robust approach for nucleic acid-based diagnostics and research area. It is frequently used for the analysis of gene expression profiles, the discovery of novel and surrogate molecular biomarkers of disease and validation of microarray data. Real-time PCR technique is preferred by numerous research labs around the world. While convenient normalisation and choosing an appropriate housekeeping gene are critical for obtaining biologically relevant results, an ideal normalisation remains to be answered in a satisfactory manner.

10. References

Andersen, C. L., Jensen, J. L., & Ørntoft, T. F. (2004). Normalization of real-time quantitative reverse transcription-PCR data: a model-based variance estimation approach to identify genes suited for normalization, applied to bladder and colon cancer data sets. *Cancer Res.*, 64, 15, 5245-5250.

Balogh, A., Paragh, G. Jr., Juhász, A., Köbling, T., Törocsik, D., Mikó, E., Varga, V., Emri, G., Horkay, I., Scholtz, B., & Remenyik, E. (2008). Reference genes for quantitative real time PCR in UVB irradiated keratinocytes. *J. Photochem. Photobiol. B.,* 93, 3, 133-139.

Bengtsson, M., Karlsson, H. J., Westman, G., & Kubista, M. (2003) A new minor groove binding asymmetric cyanine reporter dye for real-time PCR. *Nucleic Acids Res.*, 31, 8, e45.

Bernard, P.S. & Wittwer, C. T. (2000). Homogeneous amplification and variant detection by fluorescent hybridization probes. *Clin. Chem.*, 46, 147-148.

Bonnet, G., Tyagi, S., Libchaber, A., & Kramer, F. R. (1999). Thermodynamic basis of the enhanced specificity of structured DNA probes. *Proc. Natl. Acad. Sci. U. S. A.*, 96, 11, 6171-6176.

Broude, N. E. (2005). Molecular Beacons and Other Hairpin Probes. *Encyclopedia of Diagnostic Genomics and Proteomics*, 846-850 Marcel Dekker, Inc., New York.

Bustin, S. A. (2000). Absolute quantification of mRNA using real-time reverse transcription polymerase chain reaction assays. *J. Mol. Endocrinol.,* 25, 2, 169-193.

Bustin, S. A. (2002). Quantification of mRNA using real-time reverse transcription PCR (RT-PCR): trends and problems. *J. Mol. Endocrinol.,* 29, 1, 23-39.

Bustin, S. A., & Nolan, T. (2004b). Pitfalls of quantitative real-time reversetranscription polymerase chain reaction. *J. Biomol. Tech.*, 15, 155–166.

Bustin, S.A., & Nolan, T. (2004a). Analysis of mRNA expression by real-time PCR. In: *Real-Time PCR: An Essential Guide,* Edwards, K., Logan J., & Saunders, N., eds., pp. 125-184. Horizon Bioscience, Norfolk, U.K.

Calcagno, A. M., Chewning, K. J., Wu, C. P., & Ambudkar, S. V. (2006). Plasma membrane calcium ATPase (PMCA4): a housekeeper for RT-PCR relative quantification of polytopic membrane proteins. *BMC Mol. Biol.*, 7, 29.

Cicinnati, V. R., Shen, Q., Sotiropoulos, G. C., Radtke, A., Gerken, G., & Beckebaum, S. (2008). Validation of putative reference genes for gene expression studies in human hepatocellular carcinoma using real-time quantitative RT-PCR. *BMC Cancer.*, 8, 350.

Deggerdal, A., & Larsen, F. (1997). Rapid isolation of PCR-ready DNA from blood, bone marrow and cultured cells, based on paramagnetic beads. *Biotechniques*, 22, 3, 554-557.

Espy, M. J., Uhl, J. R., Sloan, L. M., Buckwalter, S. P., Jones, M. F., Vetter, E. A., Yao, J. D., Wengenack, N. L., Rosenblatt, J. E., Cockerill, F. R. 3rd., & Smith, T. F. (2006). Real-time PCR in clinical microbiology: applications for routine laboratory testing. *Clin. Microbiol. Rev.*, 19, 1, 165-256.

Fink, L., Seeger, W., Ermert, L., Hänze, J., Stahl, U., Grimminger, F., Kummer, W., Bohle, R. M. (1998). Real-time quantitative RT-PCR after laser-assisted cell picking. *Nat. Med.*, 4, 11, 1329-1333.

Fleige, S., & Pfaffl, M. W. (2006). RNA integrity and the effect on the real-time qRT-PCR performance. *Mol. Aspects Med.*, 27, 2-3, 126-139.

Fraga, D., Meulia, T., & Fenster, S. (2008). Real-Time PCR. In: *Current Protocols Essential Laboratory Techniques,* Gallagher, S. R., & Wiley, E. A. (Eds), 10.3.1–10.3.34, John Wiley and Sons, Inc. Retrieved from http://onlinelibrary.wiley.com/doi/10.1002/9780470089941.et1003s00/full

Fronhoffs, S., Totzke, G., Stier, S., Wernert, N., Rothe, M., Brüning, T., Koch, B., Sachinidis, A., Vetter, H., & Ko, Y. (2002). A method for the rapid construction of cRNA standard curves in quantitative real-time reverse transcription polymerase chain reaction. *Mol. Cell. Probes.*, 16, 2, 99-110.

Fu, L. Y., Jia, H. L., Dong, Q. Z., Wu, J. C., Zhao, Y., Zhou, H. J., Ren, N., Ye, Q. H., & Qin, L. X. (2009). Suitable reference genes for real-time PCR in human HBV-related hepatocellular carcinoma with different clinical prognoses. *BMC Cancer*, 9, 49.

Gao, Q., Wang, X. Y., Fan, J., Qiu, S. J., Zhou, J., Shi, Y. H., Xiao, Y. S., Xu, Y., Huang, X. W., & Sun, J. (2008). Selection of reference genes for real-time PCR in human hepatocellular carcinoma tissues. *J. Cancer Res. Clin. Oncol.*, 134, 9, 979-986.

Gerard, C.J., Olsson, K., Ramanathan, R., Reading, C., & Hanania, E. G. (1998). Improved quantitation of minimal residual disease in multiple myeloma using real-time polymerase chain reaction and plasmid-DNA complementarity determining region III standards. *Cancer Res.*, 58, 17, 3957-3964.

Gibson, U. E., Heid, C. A., & Williams, P. M. (1996). A novel method for real time quantitative RT-PCR. *Genome Res.*, 6, 10, 995-1001.

Goidin, D., Mamessier, A., Staquet, M. J., Schmitt, D., & Berthier-Vergnes, O., (2001). Ribosomal 18S RNA prevails over glyceraldehyde-3-phosphate dehydrogenase and beta-actin genes as internal standard for quantitative comparison of mRNA levels in invasive and noninvasive human melanoma cell subpopulations. *Anal. Biochem.*, 295, 1, 17-21.

Gresner, S. M., Golanska, E., Kulczycka-Wojdala, D., Jaskolski, D. J., Papierz, W., & Liberski, P. P. (2011). Selection of reference genes for gene expression studies in astrocytomas. *Anal. Biochem.*, 408, 1, 163-5.

Haberhausen, G., Pinsl, J., Kuhn, C. C., & Markert-Hahn, C. (1988) Comparative study of different standardization concepts in quantitative competitive reverse transcription-PCR assays. *J. Clin. Microbiol.*, 3, 628-633.

He, J. Q., Sandford, A. J., Wang, I. M., Stepaniants, S., Knight, D. A., Kicic, A., Stick, S. M., & Paré, P. D. (2008). Selection of housekeeping genes for real-time PCR in atopic human bronchial epithelial cells. *Eur. Respir. J.*, 32, 3, 755-762.

Heid, C. A., Stevens, J., Livak, K. J., & Williams, P. M. (1996). Real time quantitative PCR. *Genome Res.*, 6, 10, 986-994.

Hellemans, J., Mortier, G., De Paepe, A., Speleman, F., & Vandesompele, J. (2007). qBase relative quantification framework and software for management and automated analysis of real-time quantitative PCR data. *Genome Biol.*, 8, 2, R19.

Higuchi, R., Fockler, C., Dollinger, G., & Watson, R. (1993). Kinetic PCR analysis: real-time monitoring of DNA amplification reactions. *Biotechnology*, 11, 9, 1026-1030.

Huggett, J., Dheda, K., Bustin, S., & Zumla, A. (2005). Real-time RT-PCR normalization; strategies and considerations. *Genes Immun.*, 6, 4, 279-284.

Isacsson. J., Cao, H., Ohlsson, L., Nordgren, S., Svanvik, N., Westman, G., Kubista, M., Sjöback, R., & Sehlstedt, U. (2000). Rapid and specific detection of PCR products using light-up probes. *Mol. Cell. Probes*, 14, 5, 321-328.

Ishiguro, T., Saitoh, J., Yawata, H., Yamagishi, H., Iwasaki, S., & Mitoma, Y. (1995). Homogeneous quantitative assay of hepatitis C virus RNA by polymerase chain reaction in the presence of a fluorescent intercalater. *Anal. Biochem.*, 229, 207–213.

Jin, N., He, K., & Liu, L. (2006) qPCR-DAMS: a database tool to analyze, manage, and store both relative and absolute quantitative real-time PCR data. *Physiol. Genomics.*, 25, 3, 525-527.

Kainz, P. (2000). The PCR plateau phase - towards an understanding of its limitations. *Biochim. Biophys. Acta.,* 1494, 1-2, 23-27.

Karge, W. H., Schaeferi E. J., & Ordovas, J. M. (1998). Quantification of mRNA by polymerase chain reaction (PCR) using an internal standard and a nonradioactive detection method. *Methods Mol. Biol.,* 110, 43-61.

Karrer, E. E., Lincoln, J. E., Hogenhout, S., Bennett, A. B., Bostock, R. M., Martineau, B., Lucas W. J., Gilchrist, D. G., & Alexander, D. (1995). In situ isolation of mRNA from individual plant cells: creation of cell-specific cDNA libraries. *Proc Natl Acad Sci U S A.,* 92, 9, 3814-3818.

Komurian-Pradel, F., Paranhos-Baccalà, G., Sodoyer, M., Chevallier, P., Mandrand, B., Lotteau,V., & André, P. (2001). Quantitation of HCV RNA using real-time PCR and fluorimetry. *J. Virol. Methods,* 95, 111–119.

Kusser, W. (2006). Use of self-quenched, fluorogenic LUX primers for gene expression profiling. *Methods Mol Biol.,* 335, 115-133.

Liss, B. (2002). Improved quantitative real-time RT-PCR for expression profiling of individual cells. *Nucleic Acids Res.,* 30, 17, e89.

Liu, W., & Saint, D. A. (2002a). A new quantitative method of real time reverse transcription polymerase chain reaction assay based on simulation of polymerase chain reaction kinetics. *Anal. Biochem.,* 302, 1, 52-59.

Liu, W., & Saint, D. A. (2002b). Validation of a quantitative method for real time PCR kinetics. *Biochem. Biophys. Res. Commun.,* 294, 2, 347-353.

Livak, K. J., & Schmittgen, T. D. (2001) Analysis of relative gene expression data using real-time quantitative PCR and the 2(-Delta Delta C(T)) Method. *Methods,* 25, 4, 402-408.

Livak, K.J., (2001). Relative quantification of gene expression, ABI Prism 7700 Sequence detection System User Bulletin #2;. http://docs.appliedbiosystems.com/pebiodocs/04303859.pdf

Manchester, K. L. (1996). Use of UV methods for measurement of protein and nucleic acid concentrations. *Biotechniques,* 20, 6, 968-970.

Mannhalter, C., Koizar, D., & Mitterbauer, G., (2000). Evaluation of RNA isolation methods and reference genes for RT-PCR analyses of rare target RNA. *Clin. Chem. Lab. Med.,* 38, 171-177.

Marino, J.H., Cook, P., Miller P. S. (2003). Accurate and statistically verified quantification of relative mRNA abundances using SYBR Green I and real-time RT-PCR. *J. Immunol. Methods,* 283, 291-306.

McDowell, D. G., Burns, N. A., & Parkes, H. C. (1998). Localised sequence regions possessing high melting temperatures prevent the amplification of a DNA mimic in competitive PCR. *Nucleic Acids Res.,* 26, 14, 3340-3347.

Meijerink, J., Mandigers, C., van de Locht, L., Tonnissen, E., Goodsaid, F., & Raemaekers, J. (2001). A novel method to compensate for differential amplification efficiencies between patient DNA samples in quantitative real-time PCR. *J. Mol. Diagn.,* 3, 55-61.

Morrison, T. B., Weis, J. J., & Wittwer, C. T. (1998). Quantification of low-copy transcripts by continuous SYBR Green I monitoring during amplification. *Biotechniques,* 24, 6, 954-962.

Muller, P. Y., Janovjak, H., Miserez, A. R., & Dobbie, Z. (2002). Processing of gene expression data generated by quantitative real-time RT-PCR. *Biotechniques, 32, 6,* 1372-1378.

Nazarenko, I., Lowe, B., Darfler, M., Ikonomi, P., Schuster, D., & Rashtchian, A. (2002a). Multiplex quantitative PCR using self-quenched primers labeled with a single fluorophore. *Nucleic Acids Res.*, 30, e37.

Nazarenko, I., Pires, R., Lowe, B., Obaidy, M., & Rashtchian, A. (2002b). Effect of primary and secondary structure of oligodeoxyribonucleotides on the fluorescent properties of conjugated dyes. *Nucleic Acids Res.*, 30, 2089–2195.

Nazarenko, I.A., Bhatnagar, S. K., & Hohman, R. J. (1997). A closed tube format for amplification and detection of DNA based on energy transfer. *Nucleic Acids Res.*, 25, 2516-2521.

Nolan, T., Hands, R. E., & Bustin, S. A. (2006). Quantification of mRNA using real-time RT-PCR. *Nat. Protoc.*, 1, 3, 1559-1582.

Old, R. W. & Primrose, S. B. (1994). *Principles of gene manipulation* (5th edition), Blackwell Science, 0-632-03712-1, Oxford, UK.

Overbergh, L., Giulietti A., Valckx, D., & Mathieu C. (2010). Real time polymerase chain reaction, In: *Molecular diagnostics*, Patrinos G.P., & Ansorge, W., pp. 87-105, 978-0-12-374537-8, Academic Press, London, UK.

Peirson, S. N., Butler, J. N., & Foster, R. G. (2003). Experimental validation of novel and conventional approaches to quantitative real-time PCR data analysis. *Nucleic Acids Res.*, 31, 14, e73.

Penna, I., Vella, S., Gigoni, A., Russo, C., Cancedda, R., & Pagano, A. (2011). Selection of candidate housekeeping genes for normalization in human postmortem brain samples. *Int. J. Mol. Sci.*, 12, 9, 5461-5470.

Pfaffl M. W. (2006). Relative quantification, In: *Real-time PCR*, Dorak T. M., pp. 63-87, International University Line, 0-4153-7734-X, New York, USA.

Pfaffl M. W., (2004a) Quantification strategies in real-time PCR, In: *A-Z of quantitative PCR*, Bustin, S.A., pp. 87–112, International University Line, 0963681788, CA, USA.

Pfaffl, M. W. (2001a). A new mathematical model for relative quantification in real-time RT-PCR. *Nucleic Acids Res.*, 29, 9, e45.

Pfaffl, M. W., & Hageleit, M. (2001). Validities of mRNA quantification using recombinant RNA and recombinant DNA external calibration curves in real-time RT-PCR. *Biotechnology Letters*, 23, 4, 275-282.

Pfaffl, M. W., Georgieva, T. M., Georgiev, I. P., Ontsouka, E., Hageleit, M., & Blum, J. W. (2002a). Real-time RT-PCR quantification of insulin-like growth factor (IGF)-1, IGF-1 receptor, IGF-2, IGF-2 receptor, insulin receptor, growth hormone receptor, IGF-binding proteins 1, 2 and 3 in the bovine species. *Domest. Anim. Endocrinol.*, 22, 2, 91-102.

Pfaffl, M. W., Tichopad, A., Prgomet, C., & Neuvians, T. P. (2004a). Determination of stable housekeeping genes, differentially regulated target genes and sample integrity: BestKeeper--Excel-based tool using pair-wise correlations. *Biotechnol Lett.*, 26, 6, 509-515.

Pfaffl, M. W., Vandesompele, J., & Kubista M. (2009). Data analysis software, In: Real-time PCR: Current Technology and Applications, Logan, J., Edwards K., & Saunders N., pp. 65-83, Caister Academic Press, 978-1-90-44-55-39-4, Norfolk, UK.

Pfaffl, M.W., (2001b). Development and validation of an externally standardised quantitative Insulin like growth factor-1 (IGF-1) RT-PCR using LightCycler SYBR Green I technology. In: *Rapid Cycle Real-time PCR, Methods and Applications,* Meuer,

S., Wittwer, C., and Nakagawara, K., pp. 281-191, Springer Press, 3-540-66736-9, Heidelberg.

Pfaffl, M.W., Horgan, G.W., & Dempfle, L. (2002b). Relative expression software tool (REST) for group-wise comparison and statistical analysis of relative expression results in real-time PCR. *Nucleic Acids Res.*, 30, 9, e36.

Pfaffl, M.W., Lange, I.G., Daxenberger, A., & Meyer, H. H. (2001c). Tissue-specific expression pattern of estrogen receptors (ER): quantification of ER alpha and ER beta mRNA with real-time RT-PCR. *APMIS,* 109, 5, 345-355.

Pfister, C., Tatabiga, M. S., & Roser, F. (2011). Selection of suitable reference genes for quantitative real-time polymerase chain reaction in human meningiomas and arachnoidea. *BMC Res. Notes.*, 4, 275.

Powledge, T. M. (2004). The polymerase chain reaction. *Adv. Physiol. Educ.*, 28, 1-4, 44-50.

Raaijmakers, M. H, van Emst, L., de Witte, T., Mensink, E., & Raymakers, R. A. (2002). Quantitative assessment of gene expression in highly purified hematopoietic cells using real-time reverse transcriptase polymerase chain reaction. *Exp. Hematol.*, 30, 5, 481-487.

Radstrom, P., Knutsson, R., Wolffs, P., Lovenklev, M. & Lofstrom, C. (2004). Pre-PCR processing: strategies to generate PCR-compatible samples. *Mol. Biotechnol.,* 26, 133–146.

Rasmussen, R. (2001). Quantification on the LightCycler. In: *Rapid Cycle Real-time PCR, Methods and Applications,* Meuer, S., Wittwer, C., and Nakagawara, K., pp. 281-191, Springer Press, 3-540-66736-9, Heidelberg.

Real-time PCR Brochure, Qiagen. (2010). Critical Factors for Successful Real-Time PCR, 04 Nov 2011, http://www.qiagen.com/literature/render.aspx?id=23490

Reischl, U., & Kochanowski, B. (1995). Quantitative PCR. A survey of the present technology. *Mol. Biotechnol.*, 3, 1, 55-71.

Schmittgen, T. D., Zakrajsek, B. A., Mills, A. G., Gorn, V., Singer, M. J., & Reed, M. W. (2000) Quantitative reverse transcription-polymerase chain reaction to study mRNA decay: comparison of endpoint and real-time methods. *Anal. Biochem.*, 285, 2, 194-204.

Schroeder, A., Mueller, O., Stocker, S., Salowsky, R., Leiber, M., Gassmann, M., Lightfoot, S., Menzel, W., Granzow, M., & Ragg, T. (2006). The RIN: an RNA integrity number for assigning integrity values to RNA measurements. *BMC Mol. Biol.*, 7, 3.

Silver, N., Best, S., Jiang, J., & Thein, S. L. (2006). Selection of housekeeping genes for gene expression studies in human reticulocytes using real-time PCR. *BMC Mol. Biol.,* 7, 33.

Smith, R. D, Brown, B, Ikonomi, P, & Schechter, A. N. (2003). Exogenous reference RNA for normalization of real-time quantitative PCR. *Biotechniques,* 34, 1, 88-91.

Souazé, F., Ntodou-Thomé, A., Tran, C. Y., Rostène, W., & Forgez, P. (1996). Quantitative RT-PCR: limits and accuracy. *Biotechniques*, 21, 2, 280-285.

Stahlberg, A., Kubista, M., & Pfaffl, M. (2004). Comparison of reverse transcriptases in gene expression analysis. *Clin. Chem.*, 50, 1678–1680.

Svanvik, N., Westman, G., Wang, D., & Kubista, M. (2000). Light-up probes: Thiazole orange conjugated peptide nucleic acid for detection of target nucleic acid in homogeneoussolution. *Anal. Biochem.*, 81, 26–35.

Tan, W., Fang, X., Li, J., & Liu, X. (2000). Molecular beacons: a novel DNA probe for nucleic acid and protein studies. *Chemistry*, 6, 7, 1107-1111.

Thellin, O., Zorzi, W., Lakaye, B., De Borman, B., Coumans, B., Hennen, G., Grisar, T., Igout, A., & Heinen, E. (1999). Housekeeping genes as internal standards: use and limits. *J. Biotechnol.*, 75, 2-3, 291-5.

Thelwell, N., Millington, S., Solinas, A., Booth, J., & Brown, T. (2000) Mode of action and application of Scorpion primers to mutation detection. *Nucleic Acids Res.*, 28, 19, 3752-3761.

Tichopad, A., Dilger, M., Schwarz, G., & Pfaffl, M. W. (2003). Standardized determination of real-time PCR efficiency from a single reaction set-up. *Nucleic Acids Res.*, 31, 20, e122.

Tseng, S. Y., Macool, D., Elliott, V., Tice, G., Jackson, R., Barbour, M., & Amorese, D. (1997). An homogeneous fluorescence polymerase chain reaction assay to identify Salmonella. *Anal. Biochem.*, 245, 207–212.

Tyagi, S., & Kramer, F. R. (1996). Molecular beacons: probes that fluoresce upon hybridization. *Nat. Biotechnol.*, 14, 303–308.

Valasek, M. A., Repa, J. J. (2005) The power of real-time PCR. *Adv Physiol Educ.*, 29, 3, 151-159.

Vandesompele, J., De Preter, K., Pattyn, F., Poppe, B., Van Roy, N., De Paepe, A., & Speleman, F. (2002). Accurate normalization of real-time quantitative RT-PCR data by geometric averaging of multiple internal control genes. *Genome Biol.*, 3, 7, 0034.1-003411.

VanGuilder, H. D., Vrana, K. E., & Freeman, W. M. (2008). Twenty-five years of quantitative PCR for gene expression analysis. *Biotechniques,* 44, 5, 619-626.

Wang, X., & Seed, B. (2006). High-throughput primer and probe design, In: *Real-time PCR*, Dorak T. M., pp. 93-106, International University Line, 0-4153-7734-X, New York, USA.

Wang, Y., Zhu, W., & Levy, D. E. (2006). Nuclear and cytoplasmic mRNA quantification by SYBR green based real-time RT-PCR. *Methods.* 39, 4, 356-362.

Whitcombe, D., Theaker, J., Guy, S. P., Brown, T., & Little, S. (1999). Detection of PCR products using self-probing amplicons and fluorescence. *Nat Biotechnol.*, 17, 8, 804-807.

Wiesner, R. J. (1992). Direct quantification of picomolar concentrations of mRNAs by mathematical analysis of a reverse transcription/exponential polymerase chain reaction assay. *Nucleic Acids Res.*, 20, 21, 5863-5864.

Wilhelm, J., Pingoud, A., & Hahn, M. (2003). Validation of an algorithm for automatic quantification of nucleic acid copy numbers by real-time polymerase chain reaction. *Anal Biochem.*, 317, 2, 218-225.

Wilkening, S., & Bader, A. (2004). Quantitative real-time polymerase chain reaction: methodical analysis and mathematical model. *J. Biomol. Tech., 15,* 107-111.

Winer, J., Jung, C. K., Shackel, I., & Williams, P. M. (1999). Development and validation of real-time quantitative reverse transcriptase-polymerase chain reaction for monitoring gene expression in cardiac myocytes in vitro. *Anal. Biochem.* 270, 1, 41-49.

Wittwer, C. T., Herrmann, M. G., Gundry, C. N., & Elenitoba-Johnson, K. S. (2001). Real-time multiplex PCR assays. *Methods,* 25, 4, 430-442.

Wittwer, C. T., Herrmann, M. G., Moss, A. A., Rasmussen, R. P. (1997). Continuous fluorescence monitoring of rapid cycle DNA amplification. *Biotechniques,* 22, 1, 130-138.

Wong, M. L., & Medrano, J. F. (2005). Real-time PCR for mRNA quantitation. *Biotechniques,* 39, 1, 75-85.

Zhang, X., Ding, L., & Sandford, A. J. (2005). Selection of reference genes for gene expression studies in human neutrophils by real-time PCR. *BMC Mol. Biol.*, 6, 1, 4.

13

Recent Advances and Applications of Transgenic Animal Technology

Xiangyang Miao
Institute of Animal Sciences, Chinese Academy of Agricultural Sciences
China

1. Introduction

Transgenic animal technology is one of the fastest growing biotechnology areas. It is used to integrate exogenous genes into the animal genome by genetic engineering technology so that these genes can be expressed and inherited by offspring. The transgenic efficiency and precise control of gene expression are the key limiting factors in the production of transgenic animals. A variety of transgenic techniques are available, each of which has its own advantages and disadvantages and needs further study because of unresolved technical and safety issues. Further studies will allow transgenic technology to explore gene function, animal genetic improvement, bioreactors, animal disease models, and organ transplantation. This article reviews the recent developments in animal gene transfer techniques, including microinjection method, sperm vector method, Embryonic stem cell, somatic cell nuclear transplantation method, retroviral vector method, germ line stem cell mediated method to improve efficiency, gene targeting to improve accuracy, RNA interference-mediated gene silencing technology, zinc-finger nucleases-gene targeting technique and induced pluripotent stem cell technology. These new transgenic techniques can provide a better platform to develop transgenic animals for breeding new animal varieties, and promote the development of medical sciences, livestock production, and other fields.

2. Microinjection

In the past 20 years, DNA microinjection has become the most widely applied method for gene transfer in animals. The mouse was the first animal to undergo successful gene transfer using DNA microinjection. This method involves: 1) transfer of a desired gene construct (of a single gene or a combination of genes that are recombined and then cloned) from another member of the same species or from a different species into the pronucleus of a reproductive cell; 2) *in vitro* culture of the manipulated cells to develop to a specific embryonic phase; and 3) then transfer of the embryonic cells to the recipient female.

Microinjection equipments include microscopes and micromanipulators. Various microscope configurations from upright to inverted styles made by different companies (e.g. Leica, Zeiss, Nikon, Olympus) afford excellent differential interference contrast. Micromanipulator systems are grouped into either air-driven systems (e.g. Nikon, Zeiss, Eppendorf) or oil-driven hydraulic systems. Microinjection needles and slides are also needed during experiment.

Steps for getting transgene production and evaluation from DNA microinjection are: collection of fertilized eggs from superovulated donors, injection of interested genes into male pronuclei, surgical transfer of 20-25 eggs into oviduct of pseudopregnant recipients that carry eggs to term, and PCR or southern blot analyses to detect offspring carrying the transgenes.

Transgenic frequencies obtained from pronuclear microinjection are about 5%-30%. Some factors influence transgenesis production. The studies in mouse done by Brinster et al. (1985) showed that: 1) linear DNA fragments integrated with greater efficiency than circular or supercoiled DNA; 2) transgene DNA should be injected in low amounts othervise it had toxic effects on the embryos; 3) nuclear injection was dramatically more efficient than cytoplasmic injection. In general, good results may be obtained with equipment systems, injector's experience and skills, and the technique. A major advantage of this method is its applicability to a wide variety of species.

3. Sperm-mediated gene transfer

The finding that mature spermatozoa act as vectors of genetic materials, not only for their own genome, but also for exogenous DNA molecules, has suggested a strategy for animal transgenesis alternative to DNA microinjection. Exploiting this possibility, protocols for sperm-mediated gene transfer (SMGT) have been developed in a variety of animal species with extremely variable results.

In 1989 Lavitrano et al. described a simple and efficient technique, sperm-mediated gene transfer to produce transgenic mice. In this technique, DNA was mixed with sperm cells before *in vitro*. 30% of offspring mouse were integrated foreign DNA. In subsequent years, however, many successful reports of SMGT have been published.

The basic principle of sperm-mediated gene transfer is quite straightforward: seminal plasma-free sperm cells are suspended in the appropriate medium, and then incubated with DNA. The resultant DNA-carrying sperms are then used to fertilize eggs, via *in vitro* fertilization or artificial insemination or, in the case of aquatic animals, via waterborne (natural) fertilization. To improve transgenic efficiency, 'augmentation' techniques such as electroporation or liposomes to 'force' sperm to capture transgenes were used in some studies.

Here we briefly introduced methods of SMGT. 1) Sperm cells directly incubated with exogenous DNA. Lavitrano et al. (1989) first incubated mouse sperms with DNA to integrate foreign genes into the germ cells and produce transgenic mice. But as low efficiency of the method, now researchers use this method in combination with other methods of sperm carrier techniques to obtain transgenes. 2) Transfection of DNA into the sperm cells mediated by liposomes. These cationic lipids interact with the negatively charged nucleic acid molecules forming complexes that the nucleic acid is coated by the lipids. The positive outer surface of the complex can then associate with the negatively charged cell membrane, allowing the internalization of the nucleic acid. 3) Electroporation-mediated import of foreign DNA into sperms. High voltage electric field can induce a temporary reversible membrane permeability changes, which allowes the foreign DNA into the cells more easily. Studies showed that this method can improve the transgenic efficiency up to 22% of the embryos in pigs, cattle, chickens and other animals. However, sperm cells underwent electroporation have two aspects. On one hand, some channels temporarily opened on cell membranes are conducive to the entry of foreign gene; on the other hand, the shock is also an injury to cells that causes irreversible damage to sperm motility. 4) Adenoviral vector mediated gene transfer. Farre et al. (1999) tested the ability of adenoviral vectors to transfer DNA into boar spermatozoa and to offspring. Exposure of spermatozoa to adenovirus bearing the E. coli *lacZ* gene resulted in the transfer of the gene to the head of the spermatozoa. Of the 2-to 8-cell embryos obtained after *in vitro* fertilization with adenovirus-exposed sperm, 21.7% expressed the LacZ product. Four out of 56 piglets (about 7%) obtained foreign gene in PCR analyses after artificial insemination with adenovirus-exposed spermatozoa. SMGT is based on the intrinsic ability of sperm cells to bind and internalise exogenous DNA and to transfer it into the egg at fertilisation as illustrated in Fig. 1.

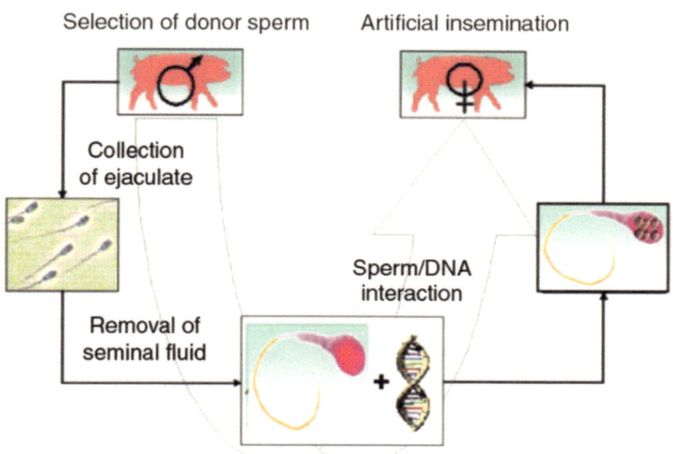

Fig. 1. Sperm-mediated gene transfer in the pig.

Mechanisms of nucleic acid uptaken by sperm are not known very well. It is suggested that the interaction of exogenous molecules may trigger an endogenous reverse transcriptase (RT) activity in spermatozoa. Such RT activity is able to reverse transcribe exogenous RNA molecules (specifically, the human poliovirus RNA genome was used in the study that provided the first set of evidence) into cDNA, which are transferred to embryos following *in vitro* fertiliszation (Giordano et al., 2000). That finding suggested for the first time that a sperm endogenous RT is implicated in the generation of newly reverse-transcribed sequences and, more generally, established the notion that the retrotransposon/ retroviral machinery is involved in SMGT.

Now it is well established that spermatozoa can play a role in transgenesis in all species. Their ability to take up exogenous DNA molecules can be exploited to transmit novel genetic information to the offspring after fertilization. This potential is highlighted by the recent development of SMGT variant protocols.

4. Embryonic stem cell-mediated gene transfer

This method involves isolation of totipotent stem cells, which are undifferentiated cells that have the potential to differentiate into any type of cells (somatic and germ cells) and therefore to give rise to a complete organism. Then the desired DNA sequences are inserts into the genome of embryonic stem (ES) cells cultured *in vitro* by homologous recombination. The cells containing the desired DNA are incorporated into the host's embryo, resulting in a chimeric animal. This technique is of particular importance for the study of the genetic control of developmental processes and works well in mice.

It has the advantage of allowing precise targeting of defined mutations in the gene via homologous recombination. Based on the resultant function of the targeted gene, gene targeting methods have opened two lines of investigation: the gene knock-out (KO) to disrupt the existing gene, and the gene knock-in (KI) to insert a functional new gene.

Stem cells of the embryo can be classified into three types: embryonal carcinoma (EC) cells, embryonic stem (ES) cells, and primordial germ cells or embryonic germ (PGCs) cells. EC cells are derived from the stem cells of teratocarcinomas; ES cells are derived from the inner cell mass of blastocysts; and PGCs are derived from the posterior third of the embryo. Here, we focus on ES cells, which are undifferentiated, pluripotent and usually are expected to have the capacity to produce both gametes and all somatic-cell lineages. Up to now, two methods of producing transgenic mice are widely used (Figure 2).

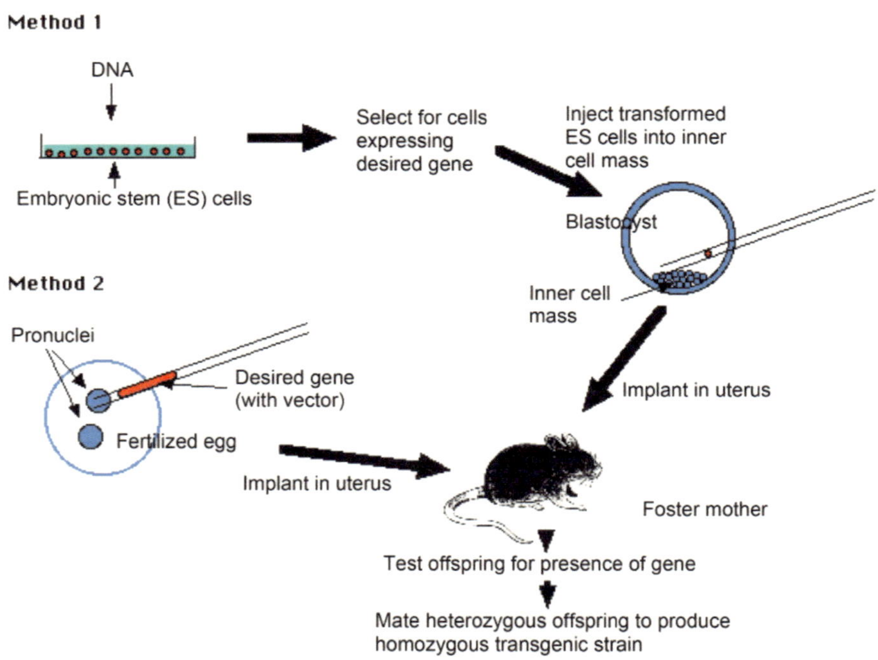

Fig. 2. Two methods of producing transgenic mice are widely used. Method 1, transforming embryonic stem cells (ES cells) growing in tissue culture with the desired DNA; Method 2, injecting the desired gene into the pronucleus of a fertilized mouse egg.

4.1 Establishment of ES cells

ES cells were first established by two laboratories independently in 1981 (Evans and Kaufman, 1981; Martin, 1981). They changed hormone levels, blastocyst was cultured to obtain delay in its development. 4 to 6-day blastocyst inner cell mass (ICM) was separated, then co-cultured on the mitomycin C-treated Unlimited lines of fibroblasts (STO) feeder layer (Feeder). After cell proliferation and inhibition of differentiation of passage, the first mouse undifferentiated ES cell lines were established. It was shown that mouse ES cells by blastocyst injection could widely induce variety of organizations involved in the formation of chimeric animals at rate of 61%.

Currently, mouse fibroblasts unlimited line (STO) or mouse embryonic fibroblasts (Mouse Embryonic Fibroblast, namely MEF) was widely used to prepare the feeder layer. These cells can secrete fibroblast growth factor (FGF), differentiation inhibitory factor (DIA), white leukemia inhibitory factor (LIF) and other substances. They also promote the growth and colonization of ES cells and suppress their differentiation. The STO or MEF is treated with mitomycin vitamin C or other mitotic inhibitor, early embryos or PGCs is cultured on the feeder layer and ES cells can be obtained. Rat liver cell conditioned medium (buffalo liver conditioned medium) is another wildly used differentiation inhibiting medium. In addition, sheep oviduct epithelial (oTE), goat oviduct epithelial (cTE), sheep uterine epithelium (oUE), goat uterine epithelium (cUE), bovine granulosa cells (bG), bovine uterine fibroblasts (bUF) and fetal bovine testes, kidney and liver fibroblasts (fbTF, fbKF, fbLF), etc., are also used as feeder layers in laboratories.

Several conditions must be met for isolating ES cells. 1) Present cells in culture must be undifferentiated and pluripotent. 2) The pluripotent cells must be deprived of differentiation signals in culture. 3) The cells must be stimulated, or at least be allowed, to proliferate.

4.2 Characteristics of ES cells

ES cells are small, aggregated and unpolarized cells forming islands on the feeder layers and have large nucleoli and a high nucleo-cytoplasmic rate. Cell markers have been used to characterize undifferentiated or differentiated ES cells. Alkaline phosphatase is equivalent to the cell surface nonspecific alkaline phosphatase of the inner cell mass of the mouse blastocyst. Other markers are surface glycolipids (ECMA-7), embryoglycans (SSEA-1) and so on.

As gene expression characteristics of ES cells, they express all the "house-keeping" genes involved in the machinery of cell cycling and some receptors to factors which allow them escaping cycling and differentiating.

4.3 ES cells for transgenesis

ES cells are highly efficient materials for animal cloning. With the development of chimeric nuclear transfer technology and production technology, ES cells are widely used in animal cloning. Proliferation of ES cells derived from donor as the nucleus, which in chimeric germline, produce following generations by sperm or egg cell proliferation.

Introduction of foreign DNA by electroporation into ES cells is very efficient. The DNA integrates into genome at a frequency of -10^{-3}, then numerous markers are used for efficient selection. The gene-modified cell clones are introduced back into preimplantation stage embryos, either by blastocyst injection or by morula aggregation, to produce chimeras.

There are two main methods of embryonic stem cell-mediated gene transfer: one is gene trap, another is gene targeting. We will discuss in the following section.

5. Somatic cell nuclear transfer

ESCs are totipotent in development, capable of limitless passage and proliferation, and have become the ideal cells for gene targeting. However, in many species, especially the large agricultural animals, ESCs have not been successfully isolated or cultured. For these animals, somatic cells are easily obtained in large numbers and can be cultured *in vitro*.

Somatic cell nuclear transfer (SCNT) is a technique for cloning. The nucleus is removed from a healthy egg. The enucleated egg becomes the host for a nucleus that is transplanted from another cell, such as a skin cell. The resulting embryo can be used to generate embryonic stem cells with a genetic match to the nucleus donor (therapeutic cloning), or can be implanted into a surrogate mother to create a cloned individual, such as Dolly the sheep (reproductive cloning) (Figure 3).

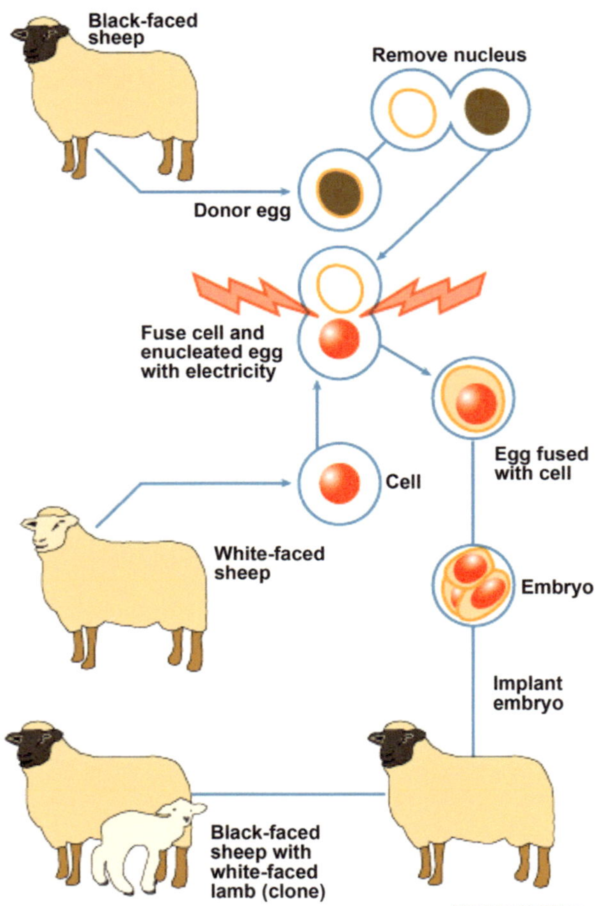

Fig. 3. Diagram of the nuclear transfer procedure that produced the dolly sheep.

Cloning by nuclear transfer from adult somatic cells is a remarkable demonstration of developmental plasticity. When a nucleus is placed in oocyte cytoplasm, the changes in chromatin structure that govern differentiation can be reversed, and the changed nucleus can control the development of oocyte to term.

Dolly was cloned by SCNT with a nucleus from a cultured mammary gland cell. In the Edinburgh experiment, three types of cells were used as karyoplasts: (i) mammary epithelium, (ii) fetal fibroblasts, and (iii) embryo-derived cells. The efficiencies of fusion were relatively high for all three cell types, which were 63.8%, 84.7% and 82.8% respectively.

5.1 The main methods of nuclear transfer

According to the different donor cells, somatic cell nuclear transfer divided into early embryo nuclear transplantation, nuclear transfer and differentiated embryonic stem cell nuclear transfer. From the present point of view, there are two somatic cell nuclear transfer technology: one is Roslin technology, another is Honolulu technique. Compared with previous techniques, the major breakthrough of Roslin method is the use of blood starvation method, which enables the proliferation of cultured cells temporarily in the G0 phase week. To ensure that the donor nucleus and development of the cytoplasm of recipient cells synchronized, electric pulse method was used. The fusion and activation of donor nucleus and enucleated oocyte were triggered by oocyte.

In Honolulu technique, a slight modification was made in the direct use of G0 phase or G1 phase of somatic cell nuclear as donor to avoid serum starvation. Donor nucleus was transferred into the oocyte cytoplasm and stayed for some time (6h). Then the oocyte was activated and stimulated to proliferation with strontium ions (chemical activation). More details of nuclear transplantation technology include the following steps:

5.1.1 Choose of recipient cells and the nuclear removing

MII oocytes are the most use of nuclear transplantation donor, but different laboratories used different activation time: before the activation of nuclear transfer, nuclear transfer when activated, delayed activation of nuclear transfer, and so on. No matter which methods using the blastomeres of early embryos, ES cells or somatic cell, nuclear transfer ported to this type of recipient, all get the offspring.

Nuclear removing from oocytes are mainly done by mechanical and chemical methods. Earlier mechanical method is the use of glass needle under the microscope, including the blind absorption and fluorescence dye staining under UV to remove nuclear. In 1998, Wakayama et al. first applied piezoelectric microinjection (PEM) to remove nuclear. Because of its high frequency of vibration control, more easily through the zona pellucida, and the small damage to the fertilized egg, the success rate to get transgenes was greatly improved. Compared with the mechanical method, chemical method is much simpler and easier to remove nuclear. In the early time, researchers mainly used early etopside (ethylene grapes pyran sugar) and cycloheximide treatment. However, the success rate was low. Recently Gasparrini et al. (2003) chose the decarboxylation youthful acid base to successfully remove the nucleus from the mouse oocyte. But whether this method is applicable to other species needs to further experiments.

5.1.2 The choice of donor cells and transplantation

Currently cells mainly used in the nuclear donor are: cumulus cells, testicular pillar cells, sperm cells, brain cells, fetal or adult fibroblasts, breast cells, embryonic stem cells (ES cells) and so on. The synchronization of donor and recipient cell cycle is an important factor affecting the development of nuclear transfer. Early studies suggest that donor cells in the G0 is essential to the success of somatic cell nuclear transplantation, and that cells in the G0 phase can be selected or artificially induced. However, the recent discovery indicated that the nuclear transfer from donor cells in G1 or G2 or M phase was also a success way. Meanwhile, the study also found that in mice, somatic cell nuclear transfer had a greater efficiency of transplantation.

There are two ways to transplant donor nuclei: fusion and injection. Fusion includes the chemical fusion, sendai virus mediated fusion and electro-fusion. And injection includes glass needle injection and piezoelectric microinjection.

5.1.3 Oocyte activation

Mature oocytes as recipients are lack of nuclear migration and the fertilization, so they must be manually activated to promote their further development. The activation methods are currently mostly used with electrical activation, ethanol, ionomycin, calcium ionophore A23187, chlorine strontium, 3-phosphatidylinositol (IP3), sperm extract and so on. These methods are often used in combination or with protein synthesis inhibitors (actinomycetes ketone, puromycin), serine threonine protein kinase inhibitor DMAP. However, studies have found that all of these methods can only lead to increased concentration of intracellular calcium in oocytes, and calcium can not form vibration, which might be the reason of low efficiency of nuclear transfer.

5.2 Application of SCNT

This technology will be helpful in understanding the most important issues such as nuclear matter interaction, the nucleus division and reprogramming, changes in mitochondria of reconstructed embryos, cell aging. Nuclear transfer technology also provides a powerful tool to wildlife, endangered species' protection.

In agricultural area, this technology will further enrich the quality of breedings. Nuclear transfer technology can be used to accelerate the breeding process and expand population within the effective number in a short time. Both nuclear transfer and gene targeting can be used to modify target genes and produce new varieties with superior traits (such as increased fecundity, increased milk yield, strengthen resistance to disease, etc.).

In the medical field, SCNT was used to clone tissues and organs for patient transplant, such as the treatment of Parkinson's disease, etc. Meanwhile, the combination of gene targeting and nuclear transfer technology established various animal models of human disease for medical and pharmaceutical research.

But lots of questions of nuclear transplantation in mammals are still not well solved, including cytoplast aging, cytoplast cell-cycle stage, activation procedure, source of karyoplasts and its differentiation, karyoplast cell cycle stage, serial transfer, karyoplast: cytoplast (nucleocytoplasmic interactions), species specific differences. Instead, we have aimed at highlighting some of the unanswered questions relating to somatic cell cloning that will require resolution before the procedure becomes useful for practical purposes.

6. Retrovirus-mediated gene transfer

A retrovirus is a virus that carries its genetic material in the form of RNA rather than DNA. In this method, retroviruses are used as vectors to transfer genetic material into the host cell, resulting in a chimera, an organism consisting of tissues or parts of diverse genetic constitution. Chimeras are inbred for as many as 20 generations until homozygous (carrying the desired transgene in every cell) transgenic offspring are born.

Retrovirus-mediated expression cloning was developed in mid 1990s. The most important features of retrovirus as vectors are the technical ease and effectiveness of gene transfer and target cells specificity. When cells are infected by retroviruses, the resultant viral DNA, after reverse transcription and integration, becomes a part of the host cell genome to be maintained for the life of the host cell (Ponder, 2001). It is also reported that DNase hypersensitive regions are the preferred targets for retrovirus integration. Unlike DNA microinjection, integration of a viral gene does not seem to induce rearrangements of the host genome, except for a short duplication at the site of integration (Jaenisch, 1988).

6.1 Retrovirus biology and Retroviral vector design

Retroviruses are animal viruses contain two positive-strand RNA genomes. The word "retro" means, when the virus vectors infect a host cell, the viral RNA is reverse transcribed in the cytoplasm making linear double-stranded DNA. This dsDNA then is transported into host cell mucleus and integrates into a chromosome directly with no change of its original linear form.

The retrovirus genome can be divided into trans- and cis- acting sequences. Trans-acting protein-coding genes are *gas, pol* and *env*. The *gas* gene encodes the structural components of the virus. The *pol* gene encodes the RNA-dependent DNA polymerase (reverse-transcriptase), integrase for the integration of reverse-transcribed viral DNA into the host cell chromosome, and the protease for posttranslational cleavage of viral proteins. The *env* gene encodes the surface envelope glycoproteins.

Cis region are located at the 5′ and 3′ ends of the genome. 1) The long terminal repeat (LTR) contains transcription and integration signals. 2) Primer binding site and polypurine sequence are for reverse transcription. 3) Posttranscriptional splicing sites include splicing donor and acceptor sites along with two short fragments within the viral intron for *env* mRNA production. 4) E signal is for encapsidation of murine leukemia virus (MLV) and reticuloendotheliosis virus (REV), respectively.

The viral vector based on Moloney murine leukemia virus (Mo-MLV) is the most widely used retroviral vector systems. Mo-MLV retroviral vector system is constructed based on the Mo-MLV genes for packaging, reverse transcription and integration of the required cis-acting elements and trans-acting protein coding sequence of separation, respectively, as well as recombinant retroviral vector elements and packaging cell line. In the molecule level of recombinant retroviral vector, the foreign gene replace the original virus structural protein coding region, but essential components, the virus replication, transcription and packaging sequences are preserved. Retrovirus structural protein coding genes are provided in trans from the packaging cells. Target cells are infected with this virus particles, so that the target gene stably integrated in the genome or chromosomes of target cells, in order to achieve the transfer of foreign genes.

6.2 Packing cell lines

Packing cells are designed to synthetize all retroviral proteins necessary for the production infectious particles. The purpose of a packing cell is to provide Gag, Pol, and Env protein to the retroviral vectors having no trans-acting sequences. NIH3T3 cells transformed with appropriate MLV genes are the most popular system for gene transfer in mammalian cells.

In earlier packaging cell construction, NIH3T3 cells were transfected with the *gag*, *pol*, and *env* genes of MLV in a single transcriptional unit, causing production of replication-competent helper virus. In the later work, decreased the possibility of homologous recombination were made. For instance, in ampli-GPE packaging cells, the 5′ LTR promoter replaced with mouse metallothionein promoter in controlling of the *gag*, *pol*, and *env* genes. Then the PG13 packing cell line, the BOSC23 packing cell line, and the 293GP/VSV-G packing cell line, have the advantage of low replication competent virus production, high DNA transfection efficiency and wide host cell ranges.

Retroviral vectors have been mainly used in somatic transgenesis for gene expression studies by using reporter genes, cell lineage, and for antisense sequences inhibition of gene expression in specific cell types.

6.3 Problems of retroviral vectors in gene transfer

The maximum size for reverse transcription of each vector is about 10kb, which may affect the expression level in transgenic animals. Another problem is the recombination, which is production of replication competent retrovirus from virus-producing cells. So nowdays, reducing the homologous sequences between DNAs for packing cells and vector and by using different plasmids to separate different genes gets over this problem.

7. Germ line stem cell technique

7.1 Spermatogonial stem cell technique

Spermatogonial stem cells (SSCs) are a population of cells in mammalian testes that have high potential to self-renew and differentiate similarly to embryonic stem cells. SSCs transplantation is a recently developed animal reproduction technique that involves the injection of *in vitro* cultured spermatogonial stem cells from an age-matched male donor into the seminiferous tubule of age-matched host animal to produce germ cells. During the *in vitro* culture of spermatogonial stem cells, positive spermatogonial stem cells that will be transferred can be screened and, thus, the transgenic efficacy can be significantly enhanced.

7.1.1 Origin of spermatogonial stem cells

Spermatogonial stem cells derive from the birth of the original sex cells, the original sex cells are from primordial germ cells (primordial germ cells, PGCs). PGCs are generated in early embryo development in mammals, processing in earlier period independent of other cell lines on a small group of cells. They have been integrated into the base from the yolk sac formed after the intestine. After along intestinal active migration, and migration on the way to proliferate in pregnancy 10.5 d, PGCS eventually arrives at the genital ridge to form a gonad. PGCs then differentiate and form germ cell precursor cells known as the the original sex cells. In the male animals, the original cells of strong mitotic activity are not stationary for a long time, until the animal was born. In mice, spermatogonial stem cells appear in the course of 6 d after birth, the earliest spermatogonial stem cells appeared probably after birth 3-4 d. The specific time for the original cell into spermatogonial stem cells in other species is unclear. Livestock may take a few months, and spiritual may take about a few years. Some studies have shown that there are two types of primary cells in neonatal mouse testis, one directly differentiates into spermatogonia and completes the first form of spermatogenesis, and another form of spermatogonial stem cells, in the later time provides the basis for spermatogenesis.

7.1.2 Spermatogonial stem cell proliferation and differentiation

Spermatogonial stem cells are located in the testis seminiferous tubule basement membrane, and are round, less cytoplasm. Their nucleus were round or slightly oval, often dominant with chromatin, little heterochromatin. They have abundant of cytoplasmic ribosomal cores, mitochondria. Based on cell arrangement characteristics, mice type A spermatogonia can be divided into three types: A single spermatogonia (As); A paired spermatogonia (Apr); and 8, 16 or 32 cells of A aligned spermatogonia (Aal). As having stem cell activity, As, Apr, Aal are referred as undifferentiated spermatogonia. After division, daughter cells of As separate from each other as two new As stem cells, or due to incomplete cytokinesis, two daughter cells form Apr by cytoplasmic bridges between connected to each other. Under normal circumstances, about half of the As cell divides and forms Apr cells, while the other half through the proliferation renews themselves and keeps the number of stem cells. Apr cells, by four further division, form 8, 16 or 32 cells Aal-based original cells, which then divide into the Al type A spermatogonia. After over six consecutive division, Al type A spermatogonia differentiate into type A2 spermatogonia, which in turn gradually differentiate from the A2 → A3 → A4 → In → B, and finally into B Type A spermatogonia. Proliferation and differentiation of stem cells generally are subject to balance to their microenvironment (niche) of the regulation. Current study suggests that, spermatogonial stem cell micro-environment is seminiferous tubules and interstitial blood vessels around the area. As /Apr / Aal tends to distribution there, and will move out of this area when they differentiate into A1 spermatogonial cells.

GDNF from supporting (Sertoli) cells is one of the most important cytokines regulating spermatogonial stem cell proliferation. Studies have shown that over-expression GDNF expression in mice leads to accumulate undifferentiated spermatogonia. When the GDNF$^{-/+}$ mouse spermatogonial stem cells is exhausted, the process of spermatogenesis is damaged. More importantly, GDNF has become a necessary factor for culturing spermatogonial stem cells. GDNF binds to GFRA1, induces activation of RET, and then recruits other molecules to the RET intracellular domain. The recent study showed that BCL6 and Etv5 genes were regulated by GDNF. BCL6 and Etv5 knockout mice are showing spermatogenesis and sertoli cell degeneration syndrome phenotype. In addition, genetic models in mice also observed that two non GDNF regulated genes: Plzf and Taf4b, play an important role in maintaining the proliferation of spermatogonial stem cells in body. Plzf knockout mice with aging, accompanied by the original cell degeneration, and gradually lose the structure of the seminiferous tubule. And Plzf - / - mouse spermatogonial cell transplantation can not be re-formed seminiferous epithelium. Taf4b knockout mice become sterile at 3 months, also appear cell syndrome phenotype.

There are three important regulatory points in the differentiation of spermatogonial cells: 1) the change between As and Apr; 2) Aal changes to the A1 and A1 to B spermatogonial cell transformation; 3) Apr spermatogonia without completing cytokinesis. The cytoplasmic bridge connecting two daughter cells is considered to be first visible sign of spermatogonia differentiation. However, little is known about how to control differentiation of spermatogonia process. Current studies showed that Vitamin A (RA), c-Kit and other genes in the germ cell differentiation play an important role. If RA defects, only undifferentiated testicular spermatogonia exist, when RA renew, Aal spermatogonia after block of re-entered the cell cycle differentiate into A1 spermatogonia. RA can induce cultured undifferentiated spermatogonia to express large amount of Stra8 and c-Kit, and these two genes are considered as the molecular markers for spermatogonial stem cells starting to differentiate. C-kit oncogene is the original W locus, encoding the tyrosine kinase receptor and its ligand is stem cell factor (SCF). C-kit point mutation in male mice results in the initial stage of spermatogenesis DNA synthesis disorder, no DNA synthesis in the process of Aal to A1 differentiation, and complete infertility.

7.1.3 Pluripotency of spermatogonial stem cell

It has long been considered a single spermatogonial stem cells can only differentiate into sperm-specific manner. But in recent years, studies have shown that the original stem cells *in vitro* can be induced to become pluripotent stem cells. In 2003, Kanatsu-Shinohara first began research in this area and found that spermatogonial stem cells isolated from newborn mouse could produce embryonic-like stem cells (embryonic stem cell-like, ES-like) cells when cultured *in vitro*. These ES-like cells further cultured *in vitro* produced ES-like clones, which formed teratomas when injected into mice. Subsequently a number of scientific researchers discovered that adult mouse spermatogonial stem cells ccould be induced into ES-like pluripotent stem cells named multipotent adult germline stem cells (maGSCs). Ko et al. (2009) recently made an outstanding contribution by first proving that maGSCs cells was indeed changing from spermatogonial stem cells, and differentiated and formed the three germ layers both *in vitro* and *in vivo,* and the reproductive system could transfer to the next generation. Other study also showed that spermatogonial stem cells were very malleable and could direct transdifferentiation into other cell types. In 2007, Boulang et al. tried spermatogonial stem cell transplantation to the breast, and achieved the breast epithelium *in vivo*. In 2009, U.S. scientists fusioned spermatogonial stem cells and appropriate cells, and then grafted the hybrids into the body. Results showed that spermatogonial stem cells could differentiate directly into the prostate epithelium, uterine epithelium and skin epithelium in newborn mouse. Although the molecular mechanisms of the transition of spermatogonial stem cells to pluripotent cells are not clear, scientists successfully induced spermatogonial stem cells into pluripotent stem cells in the adult testis, which suggested that spermatogonial stem cells will become an important source of pluripotent stem cells in medical field.

7.2 Primordial germ cell technique

Primordial germ cells (PGCs) refer to the ancestral cells that can develop into sperm or ovum cells. PGCs reside in the recipient's gonads, migrate and proliferate in the recipient embryonic gonads. Because PGCs at different stages of development can serve as transgenic recipient cells (Honaramooz etal., 2011), transgenic studies using PGCs as vectors are simple, highly effective and will likely gain favor for the production of transgenic animals. Development of Mouse Embryonic Primordial Germ Cells was shown in Figure 4.

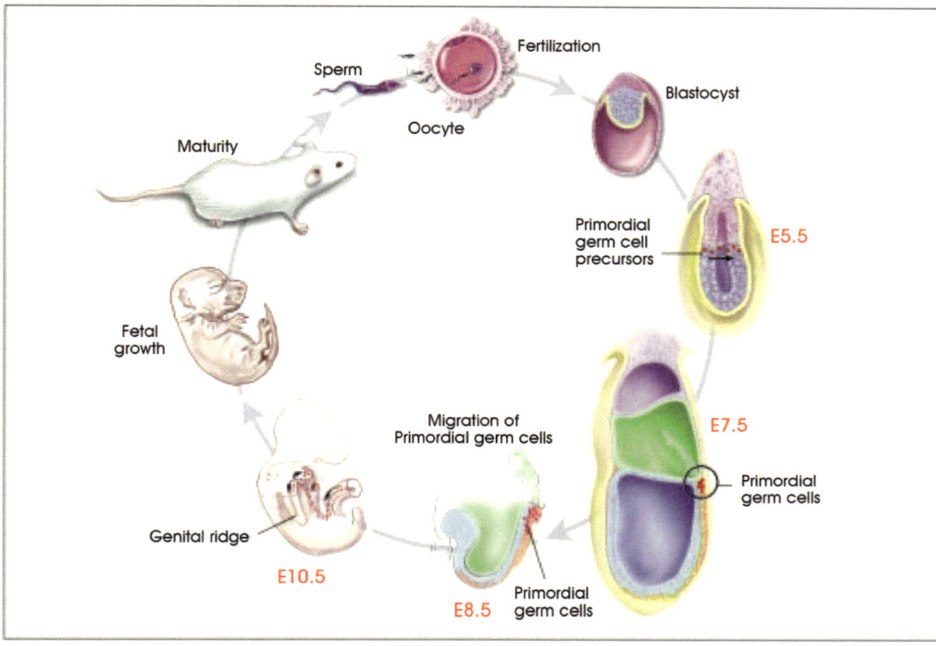

Fig. 4. Development of Mouse Embryonic Primordial Germ Cells.

7.2.1 Origin and characteristics of PGCs

The PGCs originate in the epiblast of the stage X blastoderm. The fusion of an egg and a sperm, at fertilization gives rise to a zygote that is totipotent and capable of giving rise to all embryonic lineages. In the early embryo, PGCs can be distinguished from the somatic cells at 7 days post coitus (d.p.c), because they express an alkaline phosphatase izozyme, tissue non-specific alkaline phosphatase (TNAP).

PGCs are ordinarily identified through morphological characteristics, e.g. large size (12-20um in diameter), large eccentrically placed nuclei with prominent, often fragmented nucleoli. Also some histochemical markers such as periodic acid-schiff (PAS), which stains for glycogen, or immunohistochemical markers such as EMA-1 and SSEA-1 (stage-specific embryonic antigen 1), which recognize cell-surface carbohydrate epitopes, are used for identifing PGCs.

Recent studies demonstrated that the normal progression of the germ cell lineage during gonadogenesis involved a delicate balance of primordial germ cell survival and death factors generated by surrounding somatic cells. This balance operates in a different fashion in females and males. The tuning primordial germ cell specification in the wall of the yolk sac, migration through the hindgut and dorsal mesentery, and colonization in the urogenital ridges involves the temporal and spatial activation of the following signaling pathways. Primordial germ cell specification involves bone morphogenetic proteins 2, 4 and 8b, and their migration is facilitated by the c-kit receptor-ligand duet. When colonization occurs: (1) neuregulin-b ligand is expressed and binds to an ErbB2-ErbB3 receptor tyrosine kinase heterodimer on primordial germ cells; (2) Vasa, an ortholog of the Drosophila gene vasa, a member of an ATP-dependent RNA helicase of the DEAD (Asp-Glu-Ala-Asp)-box family protein is also expressed by primordial germ cells; (3) Bcl-x (cell survival factor) and Bax (cell death factor) join forces to modulate the first burst of primordial germ cell apoptosis; (4) Cadherins, integrins, and disintegrins bring together primordial germ cells and somatic cells to organize testis and ovary. Information on other inducers of primordial cell survival, such as teratoma (TER) factor, is beginning to emerge.

7.2.2 Transgenesis using PGCs

Several methods of inserting DNA into PGCs are available. Mueller et al. (1999) isolated PGCs from pig fetuses, generated hemizygous transgenic cells for a human growth hormone (hGH) gene, and obtained chimeric pigs following blastocyst injection of transgenic porcine PGCs. Van de Lavoir et al. (2006) targeted chicken PGCs with a GFP gene construct, and transplanted the targeted cells into primordial gonads of stage XIII–XV chicken embryos that had been incubated for 3 days. A total of eight male chicks were obtained. Once fully matured, seven sired transgenic chicks carried and expressed the foreign GFP gene. These results indicated that it is feasible to use transgenic or gene targeted PGCs to produce transgenic animals, even in large animals. In early study, only retroviral infection of germinal crescent PGCs had been successfully produced transgenic chicken. DNA complexed with liposomes provided a convenient method both in situ and *in vitro*. Recently, electroporation of germinal crescent or blood PGCs or gonagal tissues was used as transfection method.

8. Gene targeting

Gene targeting is a technology to specifically modify a particular gene in the chromosome through homologous recombination and the integration of extrinsic gene into the specific target site. Gene targeting technology overcomes random integration events and, therefore, is ideal for the modification and reconstruction of biological genetic materials.

The advent of pronuclear injection, ES cells, and gene knockout technology led to the generation of mice harboring gain-of-function or loss-of-function mutations.

The integrase family consisted of 28 proteins from bacteria, phage, and yeast that have a common invariant His-Arg-Tyr triad. These proteins have the function of DNA recognition, synapsis, cleavage strand exchange, and relegation. There are four wildly used site-specific recombination system in eukaryotic applications: 1) Cre-loxP from bacteriophage P1, 2) FLP-FRT from plasmid of Saccharomyces cerevisiae, 3) R-RS from Zygosaccharomyces rouxii, 4) Gin-Gix from bacteriophage Mu. The Cre-loxP and FLP-FRT systems have been developed as wildly applied tools in Drosophila and mouse genetics.

8.1 Cre-loxP system

Cre recombinase, a P1 phage enzyme belongs to λ Int super-gene family. Cre recombinase gene is 1029 bp and codes for a 38kDa protein that is a 343 amino acid monomeric protein. The enzyme is λ Int super-gene family. It not only has the catalytic activity, but also is similar to restriction enzyme by recognizing specific DNA sequences, which are loxP sites, and deleting the gene sequence between the loxP sites. Without the aid of any auxiliary factors, the efficiency of the reorganization of DNA by Cre recombinase is 70%. Cre recombinase can affect the structure of DNA in a variety of substrates, such as linear, circular or supercoiled DNA. It is a site-specific recombination enzyme, can mediate two LoxP sites (sequence)-specific recombination, the gene sequence between the LoxP sites are deleted or reorganized.

LoxP (locus of X-over P1) sequence from the P1 phage, has two 13bp inverted repeat sequences and an interval of 8bp common form. 8bp interval LoxP sequence also defines direction. Cre catalyzes DNA strand exchange in the process of covalent binding to DNA, 13bp repeat sequence is the reverse by Cre enzyme-binding domain. A model experiment in genetics using the Cre-lox system was shown in Figure 5.

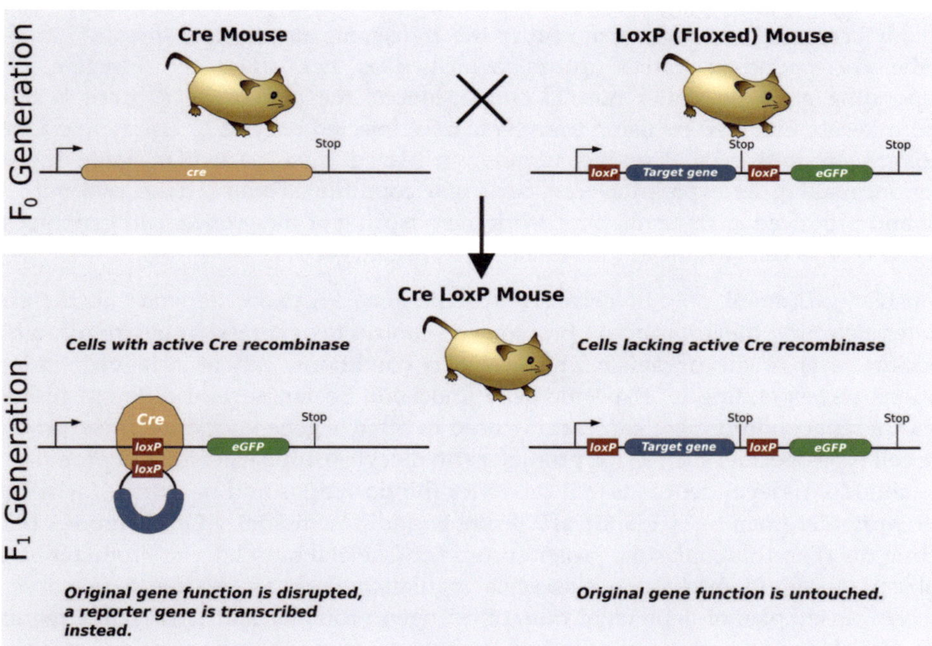

Fig. 5. A model experiment in genetics using the Cre-lox system.

8.2 Cre-LoxP system characteristics

Cre recombinase mediated recombination between two LoxP sites in the restructuring, which is a dynamic, reversible process divided into three cases: 1) if the two LoxP sites located in a DNA chain, and in the same direction, Cre recombinase can effectively excise the sequence between two LoxP sites; 2) if the two LoxP sites located in a DNA chain, but in the opposite direction, Cre recombinase can lead to two LoxP sites between the sequence inversion; 3) if the two LoxP sites are located in two different strands of DNA or chromosomes, Cre enzyme can mediate the exchange of two strands of DNA or chromosomal translocations. In addition, Cre/LoxP can not only recognize the two 13bp inverted repeat sequences and 8bp interval region, but also a 13bp inverted repeat sequence of the interval or 8bp able when changes occurred and recombined. Using this feature, people can build a LoxP site carrier as needed transformation sequences for specific gene mutations or repair, to increase the scope of application of the system.

8.3 Cre-LoxP and transgenic models

The Cre / LoxP system is wildly used in transgenic technologies in a loss-of-function model.

The use of Cre / LoxP system to achieve knockout a particular gene *in vivo* under certain conditions needs two transgenic mice. The first mice commonly are obtained with embryonic stem cell technology. In both ends of each gene locus contains a loxP sequence, then this sequence inserts into embryonic stem cells by homologous recombination, replaced the original genome sequence. After this treatment, the embryonic stem cells are re-implanted into pseudo-pregnant mouse uterus, to re-develop into a complete embryo, eventually becoming a transgenic mouse. In this transgenic mouse, loxP sites are introduced into the corresponding gene's intron, which does not affect the function of the corresponding gene, so under normal circumstances, the phenotype of mice is normal. Second mice are obtained by using transgenic mice injected oocytes or embryonic stem cell technology. In this mice, Cre recombinase is placed in a particular gene under the regulation, making its expression in a particular condition. Finally, these two mice were mated and produced at the same time. With the offspring of these mice, two genotypes in a particular type of cells results in the absence of a specific gene.

Obviously, in different cells or organs a specific gene knockout depends on the chosen promoter. Selecting the appropriate promoter to control the expression of Cre recombinase in specific parts of the organism under certain conditions, can be achieved under the conditions corresponding to a specific gene knockout. So far, several different promoters under different conditions are successfully used to achieve gene knockout. These promoters can be cell type-specific, such as lck promoter (thymocytes), alphaA crystallin promoter (eye lens), calmodulin-dependent kinase II promoter (hippocampus and neocortex), whey acidic protein promoter (mammary gland), aP2 promoter (adipose tissue), AQP2 promoter (kidney collecting duct) and sarcoplasmic protein promoter (skeletal muscle), etc. Promoter can also be subject to certain exogenous chemicals regulation such as interferon response Mxl promoter, Mosey phenol-dependent mutant estrogen promoter and tetracycline regulation system. Regulation of exogenous gene knockout can be avoided in the early embryo because the abnormal gene function may have side-effects.

9. Gene silencing mediated by RNA interference

RNA interference (RNAi) is the silencing of specific gene expression mediated by the formation of double-stranded RNA that results in the inhibition of gene expression by degrading mRNA. Therefore, RNAi can achieve the spatiotemporality and reversibility of gene expression modulation.

Gene silencing has been found to be an important methods of regulating gene expression. In gene transfer studies, it is easier to insert a foreign gene into an animal genome than to remove an existing gene, unless by a knock-out technique. Gene knock-out is complex, difficult and irreversible, once the gene knocked out, it cannot be recovered. The development of RNA interference (RNAi) technology, through blocking of gene expression or cleavage of the expressed mRNA, allows specific, partial and reversible knock-down of the desired gene, and can also achieve spatio-temporal regulation of gene expression.

The introduction of 21 nt small RNA, which is fully or partially complementary to an endogenous gene, into animal cells or tissues will interfere with the expression of the endogenous gene, or trigger the cleavage of the expressed mRNA. Consequently, RNAi impairs gene function or alters an animal trait. For example, Acosta et al. (2005) microinjected myostatin small interference RNA (siRNA) into fertilized eggs of zebra fish, which resulted in virtually eliminating myostatin mRNA, reducing the inhibitory effect of myostatin on muscle growth and creating muscle hypergrowth fish. Such microinjected siRNA molecules, although they might be carried over to progenies and remain functional, were not integrated into the genome and not inheritable. Dann et al. (2006) microinjected a rat fertilized egg with a lentiviral vector that directed expression of a short hairpin RNA (shRNA) of the Dazl gene, which is normally expressed by germ cells and is critical for fertility. Result showed that Dazl protein was substantially depressed in the testes of pups, and the pups were sterile. This RNA interference effect by the transferred gene was inherited for at least three generations.

By using the Cre–loxP system and combining RNA polymerase promoter sequence and complementary sequence of the targeted gene, a new method of spatio-temporal RNA interference was invented (Ventura et al., 2004; Yu and McMahon, 2006). However, this gene targeting based technique is not reversible. Turning the RNAi on and off at will should not involve the genomic modification, but may be achieved through the regulation of transcription of the RNA interfering gene. Kistner et al. (1996) developed a transcription factor (tTA) which was only activated in the presence of tetracycline. Such a transcription factor was transferred into the mouse genome and, in the presence of tetracycline, bound with the specific promoter, tetracycline response element (TRE), and activated transcription of the downstream gene of interest. Dickins et al. (2007) combined the TRE with the RNAi sequence, and obtained transgenic mice having the hybrid gene. Using a proper mating system, transgenic mice with both the TRE-RNAi and tTA transgenes were prepared. Administration of tetracycline to such mice activated the transcription factor tTA, and subsequently activated transcription of the RNAi gene, which silenced the expression of the target gene (Dickins et al., 2007). Withdrawal of tetracycline terminated the expression of the RNAi gene and the interference with the target gene, thus achieving reversible RNA interference. Using the reversible RNAi theory to create transgenic animals, so to reversibly silence or knock down genes, may help to change reversibly physiological activities of animals and even humans.

10. Zinc-finger nucleases – Gene targeting technique

Recently the emergence of the zinc finger nuclease (ZFN) technique signified a qualitative leap in gene targeting techniques. ZFN is comprised of one DNA binding domain and one non-specific endonuclease domain. ZFN can bind and cut DNA at specific sites, introduce double-stranded DNA break at a specific location, transfer extrinsic DNA by induction of the endogenous DNA repair procedure, homology-directed repair or non-homology terminal junction, and modify the cellular endogenous gene. This technique breaks through the limitation of gene targeting efficacy which is enhanced by five orders of magnitude.

Hence the development of ZFN-mediated gene targeting technology provides molecular biologists with the ability to site-specifically modify mammalian genomes, including the human genome, via the homology-directed repair of a targeted genomic double-strand break (DSB). ZFNs are showing promise as reagents that can create gene-specific DSBs. ZFNs are artificial proteins by fusing a specific zinc finger DNA-binding domain with a nonspecific endonuclease domain from the *Fok*I restriction enzyme. ZFNs can create specific DSBs *in vitro*. Some studies showed that DSBs stimulate gene targeting or homologous recombination in *Xenopus* oocytes, *Drosophila melanogaster*, even in plants. In mammalian cells, model ZFNs stimulate gene targeting by a factor of several thousand, as observed using a green fluorescent protein (GFP) reporter system. In addition, ZFNs have been designed to recognize an endogenous gene (IL2RG) and stimulate gene targeting at one allele of the endogenous IL2RG locus in 11% of the cells and at both alleles in 6.5% of the cells.

Mechanism of DSB by ZFNs requires: 1) two different ZFN monomers to bind to their adjacent cognate sites on DNA; 2) the Fokl nuclease domains to dimerize to form the active catalytic site for the induction of the DSB.

ZFN-mediated gene transformation has been successfully achieved in different kinds of cells from diverse species, for example, frog oocytes, nematodes, zebra fish, mice, rats and humans. Through this approach, the endogenous gene modification efficiency reached significant high (>10%).

10.1 Application of ZFN introduction of foreign genes or fragments of target gene mutation

Up to now, there are very few published papers describing the use of ZFNs to stimulate the targeting of natural sites in mammalian cells. For instance, two fundamental biological processes: DNA recognition by C2H2 zinc-finger proteins and homology-directed repair of DNA double-strand breaks were used in Urnov et al. (2005) research. Zinc-finger proteins recognizing a unique chromosomal site can be fused to a nuclease domain, and a double-strand break induced by the resulting zinc-finger nuclease can create specific sequence alterations by stimulating homologous recombination between the chromosome and an extrachromosomal DNA donor. Result showed that zinc-finger nucleases designed against mutation in the IL2Rgamma gene in an X-linked severe combined immune deficiency (SCID) yielded more than 18% gene-modified human cells without selection.

Foley et al. (2009) adapted this technology to create targeted mutations in the zebrafish germ line. ZFNs were engineered that recognize sequences in the zebrafish ortholog of the vascular endothelial growth factor-2 receptor, *kdr* (also known as *kdra*). Co-injection of mRNAs encoding these ZFNs into one-cell-stage zebrafish embryos led to mutagenic lesions at the target site that were transmitted through the germ line with high frequency. The use of engineered ZFNs to introduce heritable mutations into a genome obviates the need for embryonic stem cell lines and should be applicable to most animal species for which early-stage embryos are easily accessible.

So far the development of ZFN technology has gradually matured. The technology is based on the ZFN specifically identifiable target DNA. Therefore, the future of ZFP study will focus on finding more highly specific ZFP, ZFP and the optimal combination, which can greatly reduce the workload of experimental design and validation.

11. Induced pluripotent stem cell technique

Induced pluripotent stem cells (iPS) are a cell type that the differentiated body cell is transfected with several transcriptional factors and is re-programmed as an embryonic-like stem cell. Similarly, iPS has the totipotency of self-renew and differentiation like embryonic stem cells.

iPS cells can be used as target cells for transgenes with extrinsic genes through transgenic techniques or be genetically modified through gene targeting or gene knock-out. The iPS cells can be used as donor cells for somatic cell nuclei and fused with suitable recipient somatic cells to produce transgenic animals.

iPSCs are adult cells that have been genetically reprogrammed to an embryonic stem cell–like state by being forced to express genes and factors important for maintaining the defining properties of embryonic stem cells. Although these cells meet the defining criteria for pluripotent stem cells, it is not known if iPSCs and embryonic stem cells differ in clinically significant ways. Mouse iPSCs were first reported in 2006, and human iPSCs were first reported in late 2007. Mouse iPSCs demonstrate important characteristics of pluripotent stem cells, including expressing stem cell markers, forming tumors containing cells from all three germ layers, and being able to contribute to many different tissues when injected into mouse embryos at a very early stage in development. Human iPSCs also express stem cell markers and are capable of generating cells characteristic of all three germ layers. The Figure 6 addresses each of iPSC steps in detail.

11.1 Choice of reprogramming factors

Direct reprogramming was initially performed in mouse fibroblasts through retroviral transduction of 24 candidate genes that were all implicated in the establishment and maintenance of the pluripotent state. Four transcription factors, Oct4 (Pou5f1), Sox2, c-Myc, and Klf4, play important roles in process.

However, with further research, it found that in certain circumstances, four transcription factors Oct4, Sox2, Nanog, Lin28 could be used to reprogramm human fibroblasts into iPSCs. Similarly, in rat fibroblasts transformed by renumbering process, we can use Sox1 or Sox3 alternative Sox2, Klf4 can also be replaced with Klf2. According to the target cells to different levels of gene expression, transcription factor for re-programming can also be adjusted, such as neural stem cells (NSCs) themselves high expression of Sox2 and c-Myc, only Oct4 can re-compiled NSCs into iPSCs, only Oct4 and Sox2 are enough to induce human umbilical cord blood stem cells into iPSCs. In reprogramming human fibroblast to iPSCs, no c-Myc factor involved in can also get successful result, indicating that c-Myc is not the necessary iPSCs transcription factor of human fibroblasts. This finding has important meaning because c-Myc is a transcription factor highly expressed in tumor cells, import of exogenous human c-Myc may activate the new iPSCs into tumor cells at inappropriate time.

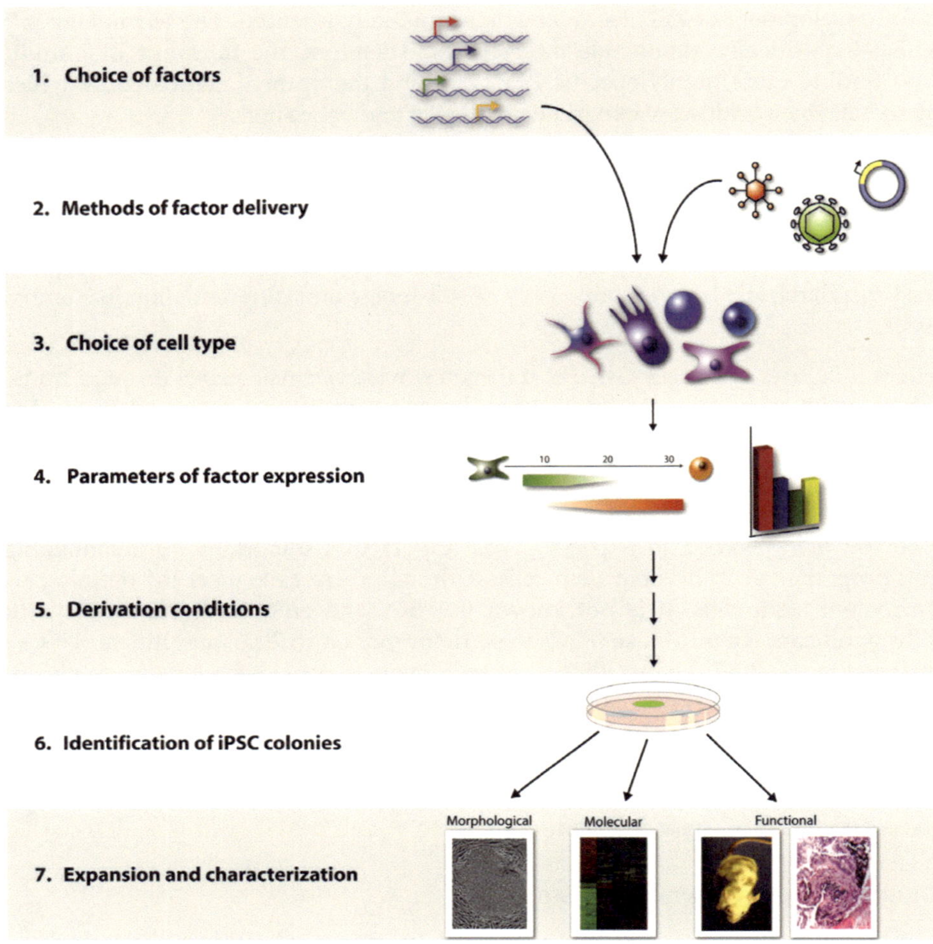

Fig. 6. Overview of the iPSC Derivation Process.

So far, it is difficult to effectively control a variety of transcription factor gene expression parameters. Transcription factor genes in a variety of programming over the entire role of process, status, and mechanisms are poorly understood. Generally speaking, it seems that Oct4 is essential, while the other three main factors are to promote or strengthen the effectiveness of Oct4, which can be replaced.

While the original suite of four factors remains the standard for direct reprogramming, a lot of small molecules and additional factors have been reported to enhance the reprogramming process and/or functionally replace the role of some of the transcription factors. For example: valproic acid, a histone deacetylase inhibitor, enhances reprogramming efficiency with four factors (O/S/M/K) in mouse fibroblasts, restores reprogramming efficiency in mouse fibroblasts without c-Myc (O/S/K only), permits reprogramming of human fibroblasts treated with OCT4 and SOX2, though at extremely low efficiency. Others are: 5-azacytidine; shRNA against Dnmt1, BIX01294, BayK8644, Wnt3a, siRNA against p53 and Utf1 cDNA.

11.2 Methods of factor delivery

Once the transcription factors are selected to transfer to target cells, delivery methods is another key to success. From the initial use of retroviral vectors to Lentiviral vector, these carriers can ensure the expression pattern of transcription factors in transformation of cells. However, this viral vector may change or even increase the potential of cell differentiation. For iPSCs safely used in clinical, non-integrated carriers must be taken. The new carrier called Cre-recombinant virus can be renumbered 5 cases for Parkinson's disease skin fibroblast dimensional cells. This vector has the advantage that you can remove one of the transcription factor when iPSCs turn to success. PiggyBac transposon renumbering systems are also used. When the piggyBac transposon re-compiled successfully, then transposase expression transposon is removed to obtain iPSCs without exogenous viral vectors or gene transcription.

The use of integrating viruses for iPSC induction has represented a major roadblock in the pursuit of clinically relevant applications. For HIV-based lentivirus method, it is constitutive, transduction of both dividing and nondividing cells, temporal control over factor expression. But lower expression levels than integrated form is main disadvantage.

In short, in order to be more conducive to clinical iPSCs, the carrier of transcription factors becomes hot spots in the field of iPSCs research.

11.3 Choice of cell type

As the capacity of skin fibroblasts derived easily, and culture conditions similar to ESCs, skin fibroblasts can be used as trophoblast cells. But fibroblasts are not the only option, a variety of somatic cells, such as stomach, liver, pancreas, blood cells and bone marrow stromal cells, neural progenitor cells, and even in the adult division at the end of the angle stromal cells have been successfully re-compiled into the iPSC.

Several factors must be considered in determining the optimal cell type for a given application: 1) the ease at which reprogramming factors can be introduced, which varies both by cell type and delivery approach; 2) the availability and ease of derivation of the given cell type; and 3) the age and source of the cell.

Although additional research is needed, iPSCs are already useful tools for drug development and modeling of diseases, and scientists hope to use them in transplantation medicine. Viruses are currently used to introduce the reprogramming factors into adult cells, and this process must be carefully controlled and tested before the technique can lead to useful treatments for humans. In animal studies, the virus used to introduce the stem cell factors sometimes causes cancers. Researchers are currently investigating non-viral delivery strategies. In any case, this breakthrough discovery has created a powerful new way to "de-differentiate" cells whose developmental fates had been previously assumed to be determined. In addition, tissues derived from iPSCs will be a nearly identical match to the cell donor and thus probably avoid rejection by the immune system. The iPSC strategy creates pluripotent stem cells that, together with studies of other types of pluripotent stem cells, will help researchers learn how to reprogram cells to repair damaged tissues in the human body.

12. Applications

Transgenic animals have potentially broad application for the improvement of animal production quality, the enhancement of production capacity, the studies of human disease models and the production of biomedical materials.

The benefits of these animals to human welfare can be grouped into the following areas:

12.1 Agricultural applications

The application of biotechnology to farm animals has the potential to benefit both humans and animals in significant ways.

a. Breeding: Farmers have always used selective breeding to produce animals that exhibit desired traits (e.g., increased milk production, high growth rate). Traditional breeding is a time-consuming, difficult task. When technology using molecular biology was developed, it became possible to develop traits in animals in a shorter time and with more precision. In addition, it offers the farmer an easy way to increase yields.
   Take ES cell technology as an example, chimeric nuclear transfer technology and production technology is improving, as ES cells are widely used in animal cloning. Proliferation of ES cells derived from donor as the nucleus, produced cloned animals. ES cells in germline chimeric, then develop into sperm or eggs to produce offspring. Animal cloning technology can produce excellent breeding, combination of genes and their high proportion in the population in short time.
b. Quality: Transgenic cows exist that produce more milk or milk with less lactose or cholesterol, pigs and cattle that have more meat on them, and sheep that grow more wool. In the past, farmers used growth hormones to spur the development of animals but this technique was problematic, especially since residue of the hormones remained in the animal product.

At present the production of transgenic animals in low efficiency is one of the main problems. The results of the testing work are carried out at the individual level. Using ES cells as a carrier, directed transformation of ES cells, the integration of inserted genes, expression level and stability of interested genes can be screened. The work is carried out at the cellular level, which is easy to obtain stable cell line with expression of satisfaction, accessing to the target gene carrying the transgene for animals. One success story is artificial insemination: the use of this technology from 1950s to 1990s in US, increased the average milk production per cow over 300%.

12.2 Medical applications

a. Xenotransplantation: Transplant organs may soon come from transgenic animals. Transgenic pigs may provide the transplant organs needed to alleviate the shortfall. Currently, xenotransplantation is hampered by a pig protein that can cause donor rejection but research is underway to remove the pig protein and replace it with a human protein.
   For organ and tissue transplantation, which is known as a "species of daughter cells ", for the clinical organization, organ transplantation offers great amount of material knockout cells. U.S. ACT companies put the nucleus of human skin into bovine oocytes without the genetic information, nurturing issued totipotency cell. If they could be

successfully used in clinical, in future, many difficult diseases such as Parkinson's disease will be cured.

b. Nutritional supplements and pharmaceuticals: Milk-producing transgenic animals are especially useful for medicines. Products such as insulin, growth hormone, and blood anti-clotting factors may soon be or have already been obtained from the milk of transgenic cows, sheep, or goats. Research is also underway to manufacture milk through transgenesis for treatment of debilitating diseases such as phenylketonuria (PKU), hereditary emphysema, and cystic fibrosis.
   ES cell culture techniques are used in some special body, then the cost can be a huge improvement. For example, some special drugs (interferon, antithrombin, erythropoietin and other biological systems agents or genetically modified), in body fluids from animals (milk, blood, etc.) or tissue extract achieve the body of the animal drug production factory.
c. Human gene therapy: A transgenic cow exists that produces a substance to help human red cells grow. Human gene therapy involves in adding a normal copy of a gene (transgene) to the genome of a person carrying defective copies of the gene. The potential for treatments for the 5,000 named genetic diseases is huge and transgenic animals could play a role.

The most current human serious medical diseases are cancer, genetic diseases, including birth defects, These diseases are caused by abnormal cell transformation and differentiation, such as Lesch, Nyhan. Fully understanding the process of cell differentiation and development will be able to cure the diseases. Many scientists have established many mouse disease models, and expressed human disease gene in mice for further treatment of human disease. For example, U.S. National Institute of Molecular Neurology Laboratory used mice ESC to induce neuroepithelial cells, implanted them into the brain, and got a large number of small conflicts like cells and glial cells. It can be envisaged to treat multiple sclerosis diseases.

13. Problems and prospects

Transgenic animals have potentially broad application for the improvement of animal production quality, the enhancement of productivity, the studies of human disease models and the production of pharmaceuticals. However, there are many pressing problems that need to be resolved for transgenic animal studies.

a. Dietary and food safety concerns: Food safety of bioengineered products is always a significant public topic. For the transgenic animals, some of the foreign gene and its promoter sequences from the virus may occur in the recipient animals. Homologous recombination or integration may cause the formation of new virus. Foreign gene inserted in the chromosome locus may also result in different genetic changes in different degrees, causing unintended effects. Transgenic animals may also increase the risk of zoonotic disease, and cause human allergic reactions.
b. Environmental impacts: If transgenic animals are in the external environment and mating with wildlife, foreign gene may spread, which results in changing the species composition of the original genes, causing confusion in species resources. It may also lead to the loss of the wild allele, resulting in a decline in genetic diversity. Once released into the environment, transgenic animals can disrupt the ecological balance of species, genetic diversity of threatened species. For example, once the transgenic fishes are into ponds or rivers and out of control, they may affect the balance of ecology.

c. Respect for life and "unnaturalness" of genetic engineering: Ethical concern has also been discussed about the "unnaturalness" of genetic engineering and the ways it might devalue nature and commercialize life. Here we quoted Strachan Donnelley's view:

"Animal biotechnology, inspired by of often genuine and legitimate desires to meet human and animal need and interests, must beware that it does not pre-empt 'nature natural' in the minds and hearts of us human beings and replace it with its own 'nature contrived'...This would be the end of us as seekers after 'living' natural norms and ways of being human, and given the press of our present technological powers, no doubt the end of nature's richness and goodness itself. This would decidedly be a double moral disaster and irresponsibility."

With the fast development of animal gene transfer technology, scientists had well improved the efficiency of making transgenic animals as well as the control of the transgene. Combination of gene targeting with somatic cell cloning or RNAi techniques had created a powerful platform for preparation of transgenic animals. However, cloning was still a highly unpredictable laboratory protocol, which existed uncertainty results in the experiments. These questions deserved each scientist careful attention.

14. Conclusion

Transgenic animal techniques have developed rapidly and provided more and improved platforms for the preparation of transgenic animals since their emergence. These techniques provide an entirely new pathway for the accurate modulation of genes. In addition, transgenic animal research may provide the tools for a series of research hotspots like microRNA function and iPS cells. All of these developments will provide new ideas and bring forth important changes in fields like medicine, health and livestock improvement. In particular, the economic and social benefits from the production of bioreactors, drug production, and organ culture for human transplantation will be great.

In summary, this review has attempted to present a comprehensive comparison of the currently available transgenic animal technology. Transgenic animal research involves consistent exploration and creation, and searching simple, reliable and efficient transgenic techniques is the key for transgenic animals. It is conceivable that the development of more simple and novel animal transgenic techniques will lead to more transgenic animals and related products that will likely improve our livelihood and wellbeing.

15. References

Aasen, T.; Raya, A.; Barrero, M.J.; Garreta, E.; Consiglio, A.; Gonzalez, F.; Vassena, R.; Billic, J.; Pekarik, V. & Tiscornia, G. (2008). Efficient and rapid generation of induced pluripotent stem cells from human keratinocytes. *Nat. Biotechnol.,* Vol.26, No.11, (October 2008), pp. 1276–1284, ISSN 1087-0156.

Abremski, K.E. & Hoess, R.H. (1992). Evidence for a second conserved arginine residue in the integrase family of recombination proteins. *Protein Eng.,* Vol.5, No.1, (October 1991), pp. 87-91, ISSN 1741-0134.

Acosta, J; Carpio, Y; Borroto, I; Gonzalez, O & Estrada, M.P. (2005). Myostatin gene silenced by RNAi show a zebrafish giant phenotype. *J Biotechnol.,* Vol.119, No.4, (October 2005), pp. 324–331, ISSN 0168-1656.

Aponte, P.M; van Bragt, M.P.; de Rooij, D.G. & van Pelt, A.M. (2005). Spermatogonial stem cells: characteristics and experimental possibilities. *APMIS*, Vol.113, No.11-12, (November 2005), pp. 727-742, ISSN 1600-0463.

Baubonis, W. & Sauer, B. (1993). Genomic targeting with purified Cre recombinase. *Nucleic Acids Res.*, Vol.21, No.9, (May 1993), pp. 2025-2029, ISSN 1362-4962.

Blelloch, R.; Venere, M.; Yen, J. & Ramalho-Santos, M. (2007). Generation of induced pluripotent stem cells in the absence of drug selection. *Cell Stem Cell*, Vol.1, No.3, (September 2007), pp. 245–247, ISSN 1934-5909.

Bosnali, M. & Edenhofer, F. (2008). Generation of transducible versions of transcription factors Oct4 and Sox2. *Biol. Chem.*, Vol.389, No.7, (July 2008), pp. 851–861, ISSN 1437-4315.

Boulanger, C.A.; Mack, D.L.; Booth, B.W. & Smith, G.H. (2007). Interaction with the mammary microenvironment redirects spermatogenic cell fate *in vivo*. *Proc. Natl. Acad. Sci.*, Vol.104, No.10, (March 2007), pp. 3871-3876, ISSN 1091-6490.

Brinster, R.L.; Chen, H.Y.; Trumbauer, M.E.; Yagle, M.K. & Palmiter, R.D. (1985). Factors affecting the efficiency of introducing foreign DNA into mice by microinjection eggs. *Proc. Natl. Acad. Sci.*, Vol.82, No.13, (July 1985), pp. 4438-4442, ISSN 1091-6490.

Chen, C.; Ouyang, W.; Grigura, V.; Zhou, Q.; Carnes, K.; Lim, H.; Zhao, G.Q.; Arber, S.; Kurplos, N.; Murphy, T.L.; Cheng, A.M.; Hassell, J.A; Chandrashekar, M.C.; Hess, R.A. & Murphy, K.M. (2005). ERM is required for transcriptional control of the spermatogonial stem cell niche. *Nature*, Vol.436, No.7053, (August 2005), pp. 1030-1034, ISSN 0028-0836.

Cibelli, J.B.; Stice, S.L.; Golueke, P.J.; Kane, J.J.; Jerry, J.; Blackwell, C.; Deleon, F.A. &, Robl, J.M. (1998). Transgenic bovine chimeric offspring produced from somatic cell-derived stem-like cells. *Nature Biotechnol.*, Vol.16, No.7, (July 1998), pp. 642–646, ISSN 1087-0156.

Costoya, J.A.; Hobbs, R.M.; Barna, M. Cattoretti, G.; Manova, K.; Sukhwani, M.; Orwiq, K.E.; Wolgemuth, D.J. & Pandolfi, P.P. (2004). Essential role of Plzf in maintenance of spermatogonial stem cells. *Nat Genet.*, Vol.36, No.6, (June 2004), pp. 653-659, ISSN 1061-4036.

Dai, Y.F.; Vaught, T.D. & Boone, J. (2002). Targeted disruption of the alpha-1,3-galactosyltransferase gene in cloned pigs. *Nature Biotechnol.*, Vol.20, No.3, (March 2002), pp. 251–255, ISSN 1087-0156.

Dann, C.T.; Alvarado, A.L.; Hammer, R.E. & Garbers, D.L. (2006). Heritable and stable gene knockdown in rats. *Proc. Natl. Acad. Sci.*, Vol.103, No.30, (July 2006), pp. 11246–11251, ISSN 1091-6490.

de Rooij, D.G. (2001). Proliferation and differentiation of spermatogonial stem cells. *Reproduction*, Vol.121, No.3, (March 2001), pp. 347-354, ISSN 1470-1626.

Dexter, M. & Allen, T. (1992). Haematopoiesis. Multi-talented stem cells? *Nature*, Vol.360, No.6406, (December 1992), pp. 709-710, ISSN 0028-0836.

Dickins, R.A.; McJunkin, K.; Hernando, E.; Premsrirut, P.K.; Krizhanovsky, V.; Burgess, D.J.; Kim, S.Y.; Cordon-Cardo, C.; Zender, L.; Hannon, G.J.; Lowe, S.W. (2007). Tissuespecific and reversible RNA interference in transgenic mice. *Nat Genet.*, Vol.39, No.7, (July 2007), pp. 914–921, ISSN 1061-4036.

Evans, M.J. & Kaufman, M.H. (1981). Establishment in culture of plurpotential cells from mouse embryos. *Nature*, Vol.292, No.2819, (July 1981), pp. 154-156, ISSN 0028-0836.

Farre, L.; Riqau, T.; Garcia-Rocha, M.; Canal, M.; Gomez-Foix, A.M. & Rodriquez-Gil, J.E. (1999). Adenovirus-mediated introduction of DNA into pig sperm and offspring. *Mol. Reprod.Dev.*, Vol.53, No.2, (June 1999), pp. 149-158, ISSN 1098-2795.

Foley, J.E.; Yeh, J.R.J; Maeder, M.L.; Reyon, D.; Sander, J.D.; Pe-terson, R.T. & Joung, J.K. (2009). Rapid mutation of endogenous ze-brafish genes using zinc finger nucleases made by Oli-gomerized Pool ENgineering (OPEN). PLoS One, Vol.4, No.2, (February 2009), pp. e4348, ISSN 1932-6203.

Fulka, J. Jr.; First, N.L.; Loi, P. & Moor, R.M. (1998). Cloning by somatic cell nuclear transfer. *BioEssays*, Vol.20, No.10, (October 1998), pp. 847-851, ISSN 1521-1878.

Fulka, J. Jr.; First, N.L.; Moor, R.M. (1996). Nuclear transplantation in mammals: Remodelling of transplanted nuclei under the influence of maturation promoting factor. *BioEssays*, Vol.18, No.10, (October 1996), pp. 835–840, ISSN 1521-1878.

Gasparrini, B.; Gao, S.; Ainslie, A.; Fltcher, J.; McGarry, M.; Ritchie, W.A.; Springbett, A.J.; Overstrom, E.W.; Wilmut, I.; DeSousa, P.A. (2003). Cloned mice derived from embryonic stem cell karyoplasts and activated cytoplasts prepared by induced enucleation. *Biol Reprod*, Vol.68, No.4, (April 2003), pp. 1259-1266, ISSN 0006-3363.

Giordano, R.; Magnano, A.R.; Zaccagnini, G.; Pittoggi, C. & Moscufo, N. (2000). Reverse transcriptase activity in mature spermatozoa of mouse. *J Cell Biol.*, Vol.148, No.6, (March 2000), pp. 1107–1113, ISSN 1540-8140.

Guan, K.; Nayernia, K.; Maier, L.S.; Wagner, S.; Dressel, R.; Lee, J.H.; Nolte, J.; Wolf, F.; Li, M.Y.; Engel, W. & Hasenfuss, G. (2006). Pluripotency of spermatogonial stem cells from adult mouse testis. *Nature*, Vol.440, No.7088, (April 2006), pp. 1199-203, ISSN 0028-0836.

Huangfu, D.; Maehr, R.; Guo, W.; Eijkelenboom, A.; Snitow, M.; Chen, A.E. & Melton, D.A. (2008a). Induction of pluripotent stem cells by defined factors is greatly improved by small-molecule compounds. *Nat. Biotechnol.*, Vol.26, No.7, (July 2008), pp. 795–797, ISSN 1087-0156.

Huangfu, D.; Osafune, K.; Maehr, R.; Guo, W.; Eijkelenboom, A.; Chen, S.; Muhlestein, W. & Melton, D.A. (2008b). Induction of pluripotent stem cells from primary human fibroblasts with only Oct4 and Sox2. *Nat. Biotechnol.* Vol.26, No.11, (November 2008), pp. 1269–1275, ISSN 1087-0156.

Jaenisch, R. (1988). Transgenic animals. *Science*, Vol.240, No.4858, (November 1998), pp. 1468-1474, ISSN 0036-8075.

Kim, J.B.; Zaehres, H.; Wu, G.; Gentile, L.; Ko, K.; Sebastiano, V.; Arauzo- Bravo, M.J.; Ruau, D.; Han, D.W. & Zenke, M. (2008). Pluripotent stem cells induced from adult neural stem cells by reprogramming with two factors. *Nature*, Vol.454, No.7204, (July 2008), pp. 646–650, ISSN 0028-0836.

Kitamura, T.; Koshino, Y.; Shibata, F.; Oki, T.; Nakajima, H.; Nosaka, T. & Kumagai, H. (2003). Retrovirus-mediated gene transfer and expression cloning: powerful tools in functional genomics. *Exp Hematol.*, Vol.31, No.11, (November 2003), pp. 1007-1014, ISSN 0301-472X.

Ko, K.; Tapia, N.; Wu, G.; Kim, J.B.; Bravo, M.J.; Sasse, P.; Glaser, T.; Han, D.W.; Hausdorfer, K.; Sebastiano, V.; Stehling, M.; Fleischmann, B.K.; Brustle, O.; Zenke, M. U. & Scholer, H.R. (2009). Induction of pluripotency in adult unipotent germline stem cells. *Cell Stem Cell*, Vol.5, No.1, (July 2009), pp. 87-96, ISSN 1934-5909.

Kubota, H.; Avarbock, M.R. & Brinster, R.L. (2004). Growth factors essential for self-renewal and expansion of mouse spermatogonialmstem cells. *Proc Natl Acad Sci*, Vol.101, No.47, (November 2004), pp. 16489-16494, ISSN 1091-6490.

Kubota, H.; Avarbock, M.R. & Brinster RL. (2003). Spermatogonial stem cells share some, but not all, phenotypic and functional characteristics with other stem cells. *Proc Natl Acad Sci*, Vol.100, No.11, (May 2003), pp. 6487-6492, ISSN 1091-6490.

Kuroiwa, Y.; Kasinathan, P. & Matsushita H. (2004). Sequential targeting of the genes encoding immunoglobulin-m and prion protein in cattle. *Nat Genet.*, Vol.36, No.7, (July 2004), pp. 775–780, ISSN 1061-4036.

Lavitrano, M.; Busnelli, M.; Cerrito, M.G.; Giovannoni, R.; Manzini, S.; Vargiolu, A. (2006) Sperm-mediated gene transfer. *Reproduction Fertility and Development*, Vol.18, No.2, (January 2006), pp. 19–23, ISSN 1031-3613.

Lavitrano, M.; Camaioni, A.; Fazio, V.M.; Dolci, S., Farace M.G. & Spadafora, C. (1989). Sperm cells as vectors for introducing foreign DNA into eggs: Genetic transformation of mice. *Cell*, Vol.57, No.5, (June 1989), pp. 717-723, ISSN 0092-8674.

Maherali, N. & Hochedlinger, K. (2008). Guidelines and techniques for the generation of induced pluripotent stem cells. *Cell Stem Cell*, Vol. 3, No. 6, (December 2008), pp. 595-605, ISSN 1934-5909.

Martin, G.R. (1981). Isolation of a pluripotent cell lines from early mouse embryos cultured in medium conditioned by teratocarcinoma stem cells. *Proc. Natl. Acad. Sci.*, vol.78, No. 12, (December 1981), pp. 7634-7638, ISSN 1091-6490.

McCreath, K.J.; Howcroft, J.; Campbell, K.H.; Colman, A.; Schnieke, A.E. & Kind, A.J. (2000). Production of gene-targeted sheep by nuclear transfer from cultured somatic cells. *Nature*, Vol. 405, No. 6790, (June 2000), pp. 1066–1069, ISSN 0028-0836.

McLaren, A. (2003). Primordial germ cells in the mouse. *Dev Biol.*, Vol. 262, No. 1, (October 2003), pp. 1-15, ISSN 0012-1606.

McLean, D.J.; Friel, P.J.; Johnston, D.S. & Griswold, M.D. (2003). Characterization of spermatogonial stem cell maturation and differentiation in neonatal mice. *Biol Reprod*, Vol. 69, No. 6, (September 2003), pp. 2085-2091, ISSN 0006-3363.

Mikkelsen, T.S.; Hanna, J.; Zhang, X.; Ku, M.; Wernig, M.; Schorderet, P.; Bernstein, B.E.; Jaenisch, R.; Lander, E.S. & Meissner, A. (2008). Dissecting direct reprogramming through integrative genomic analysis. *Nature* Vol. 454, No. 7200, (May 2008), pp. 49–55, ISSN 0028-0836.

Mueller, S.; Prelle, K.; Rieger, N.; Petznek, H.; Lassnig, C.; Luksch, U.; Aigner, B.; Baetscher, M.; Wolf, E.; Mueller, M. & Brem, G. (1999). Chimeric pigs following blastocyst injection of transgenic porcine primordial germ cells. *Mol Reprod Dev.*, Vol. 54, No. 3, (November 1999), pp. 244–254, ISSN 1040-452X.

Oatley, J.M. & Brinster, R.L. (2008). Regulation of spermatogonial stem cell self-renewal in mammals. *Annu Rev Cell Dev Biol*, Vol. 24, (June 2008), pp. 263-286, ISSN 1081-0706.

Petolino, J.F.; Worden ,A.; Curlee, K.; Connell, J.; Strange Moynahan, T.L. & Larsen C, Russell. (2010). Zine finger nuclease-mediated transgene deletion. *Plant Mol Biol*, Vol. 73, No. 6, (May 2010), pp. 617–628, ISSN 0167-4412.

Phillips, B.T.; Gassei, K.; Orwig, K.E. (2010). Spermatogonial stem cell regulation and spermatogenesis. *Philos Trans R Soc Lond B Biol Sci*, Vol. 365, No. 1546, (May 2010), pp. 1663-1678, ISSN 0962-8436.

Porteus, M.H. & Baltimore, D. (2003). Chimeric nucleases stimulate gene targeting in human cells. *Science*, Vol. 300, No. 5620, (May 2003), pp. 763, ISSN 0036-8075.

Seandel, M.; James, D.; Shmelkov, S.V,; Falciatori, I.; Kim, J.; Chavala, S.; Scherr, D.S.; Zhang, F.; Torres, R.; Gale, N.W.; Yancopoulos, G.D.; Murphy, A.; Valenzuela, D.M. Hobbs, R.M.; Pandolfi, P.P. & Rafii, S. (2007). Generation of functional multipotent adult stem cells from GPR125+ germline progenitors. *Nature*, Vol. 449, No. 7160, (September 2007), pp. 346-50, ISSN 0028-0836.

Smith, K. & Spadafora, C. (2005). Sperm-mediated gene transfer: application and implications. *Bioessays*, Vol. 27, No. 5, (May 2005), pp. 551-562, ISSN 0265-9247.

Stadtfeld, M.; Brennand, K. & Hochedlinger, K. (2008a). Reprogramming of pancreatic beta cells into induced pluripotent stem cells. *Curr. Biol.*,Vol. 18, No. 12, (May 2008), pp. 890–894, ISSN 0960-9822.

Stadtfeld, M.; Maherali, N.; Breault, D.T. & Hochedlinger, K. (2008b). Defining molecular cornerstones during fibroblast to iPS cell reprogramming in mouse. *Cell Stem Cell*, Vol.2, No. 3, (February 2008), pp. 230–240, ISSN 1934-5909.

Stadtfeld, M.; Nagaya, M.; Utikal, J.; Weir, G. & Hochedlinger, K. (2008c). Induced Pluripotent Stem Cells Generated Without Viral Integration. *Science*, Vol. 322, No. 5903, (September 2008), pp. 945–949, ISSN 0036-8075.

Streckfuss-Bomeke, K.; Vlasov, A.; Hulsmann, S.; Yin, D.; Nayernia, K.; Engel, W.; Hasenfuss, G. & Guan, K. (2009). Generation of functional neurons and glia from multipotent adult mouse germ-line stem cells. *Stem Cell Res*, Vol. 2, No. 2, (March 2009), pp. 139-154, ISSN 1873-5061.

Takahashi, K. & Yamanaka, S. (2006). Induction of pluripotent stem cells from mouse embryonic and adult fibroblast cultures by defined factors. *Cell*, Vol. 126, No. 4, (August 2006), pp. 663–676, ISSN 0092-8674.

Tegelenbosch, R.A.; de Rooij, D.G. (1993). A quantitative study of spermatogonial multiplication and stem cell renewal in the C3H/101 F1 hybrid mouse. *Mutat Res*, Vol. 290, No. 2, (December 1993), pp. 193-200, ISSN 0027-5107.

Tong-Starksen, S.E. & Peterlin, B.M. (1990). Mechanism of retroviral transcriptional activation. *Semin. Virol.* 1, pp. 215-227.

Urnov, F.D.; Miller, J.C.; Lee, Y.L.; Beausejour, C.M.; Rock, J.M.; Augustus, S; Jamieson A.C.; Porteus, M.H.; Gregory, P.D. & Holmes, M.C. (2005). Highly efficient endogenous human gene correction using designed zinc-finger nucleases. *Nature*, Vol. 435, No. 7042, (April 2005), pp. 646–651, ISSN 0028-0836.

Van de Lavoir, M.C.; Diamond, J.H. & Leighton, P.A. (2006). Germline transmission of genetically modified primordial germ cells. *Nature*, Vol. 441, No. 6907, (April 2006), pp. 766–769, ISSN 0028-0836.

Wakayama, T.; Perry, A.C.F.; Zuccotti, M.; Johnson, K.R. & Yanagimachi, R. (1998), Full-term development of mice from enucleated oocytes injected with cumulus cell nuclei. *Nature*, Vol. 394, No. 6691, (June 1998), pp. 369-374, ISSN 0028-0836.

Willadsen, S.M. (1986). Nuclear transplantation in sheep embryos. *Nature*, Vol. 320, No. 6057, (March 1986), pp. 63–65, ISSN 0028-0836.

Wilmut, I.; Beaujean, N.; de Sousa, P.A.; Dinnyes, A.; King, T.J.; Paterson, L.A.; Wells, D.N. & Young, L.E. (2002). Somatic cell nuclear transfer. *Nature*, Vol. 419, No. 6907, (October 2002), pp. 583-586, ISSN 0028-0836.

Yoshida, S.; Sukeno, M. & Nabeshima, Y. (2007). A Vasculature-Associated Niche for Undifferentiated Spermatogonia in the Mouse Testis. *Science*, Vol. 317, No. 5845, (August 2007), pp. 1722-1726, ISSN 0036-8075.

Yoshida, S. (2010). Stem cells in mammalian spermatogenesis. *Dev Growth Differ*, Vol.52, No.3, (January 2010), pp. 311-317, ISSN 0012-1592.

14

Measuring of DNA Damage by Quantitative PCR

Ayse Gul Mutlu
Mehmet Akif Ersoy University, Arts and Sciences Faculty, Department of Biology, Burdur Turkey

1. Introduction

1.1 QPCR; principles and development

PCR is an in vitro method of nucleic acid synthesis by which a particular segment of DNA can be specifically replicated. It involves two oligonucleotide primer that flank the DNA fragment to be amplified and repeated cycles of heat denaturation of the DNA, annealing of the primers to their complementary sequences, and extension of the annealed primers with DNA polymerase. Successive cycles of amplification essentially double the amount of the target DNA synthesized in the previous cycle (1).

Recent advances of the fluorometric dyes allow the very sensitive and quick quantitation of DNA. Before the invention of fluorometric quantitative PCR (QPCR) method the researchers who measured a gene's amount, have used the different methods like competetive PCR, solid phase assays, HPLC, dot blot or immunoassay (2). Many of the applications of real-time Q-PCR include measuring mRNA expression levels, DNA copy number, transgene copy number and expression analysis, allelic discrimination, and measuring viral titers (3).

The detection of gene-spesific damage and repair has been studied in nuclear and mitochondrial DNA by the use of southern analysis. But this method requires knowledge of the restriction sites flanking the damaged site, the use of large quantities of DNA, and incision of DNA lesions with a spesific endonuclease (4). Govan and collegues has reported a new approach to measuring of DNA damage in 1990 (5). This PCR based quantitative technique has been improved by Kalinowski and collegues (6). Principle of this analysis is that lesions present in the DNA, block the progression of any thermostable polymerase on the template. So the DNA amplification decreases in the damaged template when compared to the undamaged DNA (4). QPCR is a suitable method for the measuring damage and repair in the subgene level functional units like promotor regions, exons and introns (7). Method also useful to determining DNA damage and repair that originated by the genotoxic agents and oxidative stress (8,9,10). The method capable of detect 1 lesion/10^5 nucleotides from as little as 5 ng of total genomic DNA (4).

DNA extraction, pre quantitation of DNA template, PCR amplification and quantitation of PCR products are crucial for success of the application (Figure 1). Quantity and quality of the DNA sample is important. We use mini column based kits for DNA extraction in our laboratory. These extraction kits and carefully pipetting, minimize the artificial DNA damages. Pico Green dsDNA quantitation kit is used for both template DNA quantitation and the analysis of PCR products as fluorometrically 485 nm excitation, 530 nm emission. Pico Green and SYBR green are substantially more sensitive for quantifying DNA concentrations than ethidium bromide and some other fluorimetric dyes (11). Initial DNA template quantity in the all PCR tubes must be the same. mtDNA damage is quantified by comparing the relative efficiency of amplification of long fragments of DNA and normalizing this to gene copy numbers by the amplification of smaller fragments, which have a statistically negligible likelihood of containing damaged bases. To calculate relative amplification, the long QPCR values are divided by the corresponding short QPCR results to account for potential copy number differences between samples. Decreased relative amplification is an indicator of the damaged DNA (4, 12).

DNA isolation

↓

Quantitation of the isolated DNA (template quantitation)

↓

PCR

(Starting amounts of the template DNA must be same in all of the samples)

↓

PCR products quantitation

↓

Calculation of the relative amplification

Fig. 1. Flowchart of the QPCR assay for measuring of DNA damage

2. Optimization of the assay; the crucial steps

Crucial step of the QPCR is PCR optimization. Thermal conditions, especially annealing temperature must be optimized. Extention temperature may be lower for long PCR amplifications. mtDNA amplification may needs some adjuvants. We use in our laboratory DMSO (%4) for improve the efficiency of the PCR reaction. Various authors recommend DMSO and glycerol to improve amplification efficiency (higher amount of product) and specificity (no unspecific products) of PCR, when used in concentrations varying between 5%-10% (vol/vol). However, in the multiplex reactions, these adjuvants give conflicting results. For example, 5% DMSO improve the amplification of some products and decrease the amount of others. There are similar results with 5% glycerol. Therefore, the usefulness of these adjuvants needs to be tested in each case. Also BSA may increase the efficiency of the PCR (13).

Hot start PCR improve specifity of PCR reaction. Hot start PCR is reported to minimize nontarget amplification and the formation of primer-dimer (14) .

Optimization might involve changes in nucleic acids preparation, in primer usage, in buffer usage and in cycling parameters. One of the recent developments in PCR optimization is to recognize the importance of eliminating some undesired hybridization events that often happen in the first cycle and can carry potentially devastating effects. Theoretically, if the amplification precedes with an efficiency of 100%, the amount of amplicons is doubling at each cycle. However, in most PCR procedures, the overall efficiency is less than 100% and a typical amplification runs with a constant efficiency of about 70-80% from the 15th cycle to the 30th cycle, depending on the amount of starting material. The increase in the amount of amplicons stays exponential only for a limited number of cycles, after which the amplification rate reaches a plateau. The factors that contribute to this plateau phenomenon include substrate saturation of enzyme, product strand reannealing, and incomplete product strand separation. In this latter phase, the quantitated amount of amplified product is no longer proportional to the starting amount of target molecules. Therefore, to make PCR suitable in quantitative settings, it is imperative that a balance be found between a constant efficiency and an exponential phase in the amplification process. This will ultimately depend on the number of cycles, on the amount of targets in the starting material, and on the system of detection and quantitation of the amplified product (15).

We run a 50% template control and a nontemplate control in PCR in our laboratory. 50% template control should given a 50% reduction of the amplification signal (values between 40%-60% reduction are acceptable). The nontemplate control would detect contamination with spurious DNA or PCR products (4).

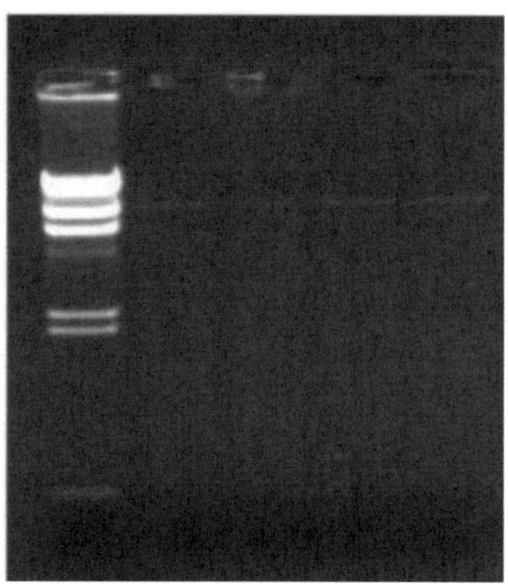

Fig. 2. PCR band of 10 kb mtDNA fragment of Mus musculus (Balb C). (Band 1: λ *Hind III* digest marker DNA)

3. Measuring of mtDNA damage on mice

We study mtDNA damage by QPCR method in different organisms like fruit flies, mice and snails in our laboratory. In our research that we used mice, oxidative mtDNA damage that created by cigarette smoke and protective effects of VitE and selenium was investigated (Figure 3).

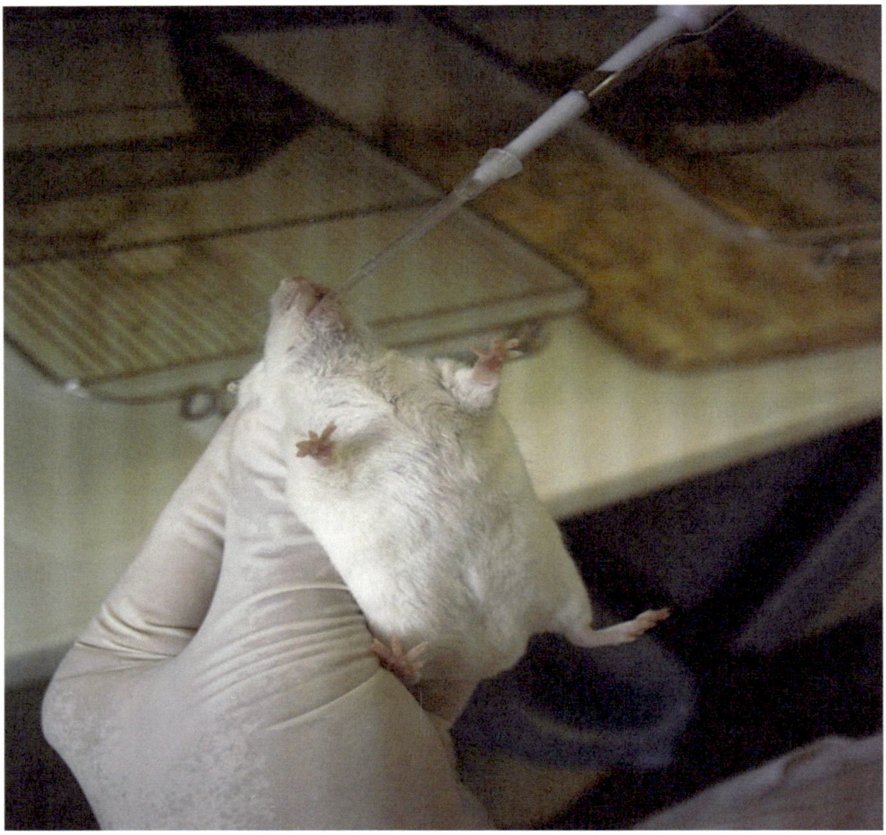

Fig. 3. Vitamin supplementation

DNA damage that is originated by cigarette smoke in various organs is declared by some research (16,17). Tobacco smoking contains many thousands of chemicals including a plethora of mutagens. Many carcinogens undergo metabolic activation in mammalian tissues to reactive intermediates that interact with and modify informational macromolecules, such as DNA with potentially mutagenic consequences (18). PAHs (Polycyclic aromatic hydrocarbons) cause irreversible DNA damage via covalent binding or oxidation (19). However genetic damage reflecting individual exposure and susceptibility to PAH may play a role in disease development (20). Tobacco smoke contains major classes of carcinogens that include PAHs, aromatic amines and tobaccospecific nitrosamines. In addition, toxic compounds such as formaldehyde, acetaldehyde, acrolein, short-lived radicals and reactive oxygen intermediates generated by redox cycling from catechol and hydroquinone and nitric oxide (NO) may also contribute to the toxic and carcinogenic effects of tobacco smoke. Direct DNAdamaging compounds that are present in cigarette smoke (CS) have previously been reported to include reactive oxygen intermediates, peroxynitrite, ethylating agents and unidentified compounds (21).

Many carcinogens in the cigarette smoke like PAHs, nitrosamine and cisplatin bind mitochondrial DNA (mtDNA) preferentially (22). The antioxidants are used frequently as food supplements may be effective to preventing cigarette smoke damage on mtDNA. Damages that are created by CS may be prevented by vitamin E (Vit E) and selenium (Se) which are powerful antioxidants.

Genomic DNA mini column kit (SIGMA) was used for total DNA isolation according to the technical bulletin. We used Pico Green dsDNA quantitation kit for both template DNA quantitation and the analysis of PCR products as fluorometrically 485 nm excitation, 530 nm emission (23). A crucial step of quantitative PCR is the concentration of the DNA sample. In fact, the accuracy of the assay relies on initial template quantity because all of the samples must have exactly the same amount of DNA. The Pico Green dye has not only proved efficient in regarding to template quantitation but also to PCR product analysis (10). Hot Start ready mix Taq (SIGMA) were used for PCR. In this mix, taq polymerase combines the performance enhancements of Taq antibody for hot start. When the temperature is raised above 70°C in the first denaturation step of the cycling process, the complex dissociates and the polymerase becomes fully active. DMSO as 4% of total volume and 20 ng of template total DNA were added into the each PCR tube. Mouse 117 bp Mouse 117 bp mtDNA fragment (small fragment) primers were:

13597 5′- CCC AGC TAC TAC CAT CAT TCA AGT- 3′

13688 5′- GAT GGT TTG GGA GAT TGG TTG ATG T- 3′ (Table 1)

Mus musculus primers for long fragment (10085 bp): 3278 5′- GCC AGC CTG ACC CAT AGC CAT AAT AT- 3′ 13337 5′- GAG AGA TTT TAT GGG TGT AAT GCG G- 3′ (4)
Mus musculus primers for small fragment (117 bp): 13597 5′- CCC AGC TAC TAC CAT CAT TCA AGT- 3′ 13688 5′- GAT GGT TTG GGA GAT TGG TTG ATG T- 3′ (4)
Drosophila primers for long fragment (10629 bp): 1880 5′- ATGGTGGAGCTTCAGTTGATTT - 3′ 12487 5′- CAACCTTTTTGTGATGCGATTA - 3′
Drosophila primers for small fragment (100 bp): 11426 5′- TAAGAAAATTCCGAGGGATTCA - 3′ 11504 5′- GGTCGAGCTCCAATTCAAGTTA - 3′

Table 1. QPCR primers for measuring of mtDNA damage in *Mus musculus* and *Drosophila melanogaster*

Thermal conditions for *Mus musculus* long fragment (10085 bp): 75ºC for 2 min 95ºC for 1 min **94°C for 15 sec** **59°C for 30 sec → 21 cycles** **65 °C for 11 min** 72ºC for 10 min
Thermal conditions for *Mus musculus* small fragment (117 bp): 75 ºC for 2 min 95 ºC for 15 sec **94°C for 30 sec** **50°C for 45 sec → 19 cycles** **72 °C for 45 sec.** 72ºC for 10 min
Thermal conditions for *Drosophila* long fragment (10629 bp): 75ºC for 1 min 95ºC for 1 min **94°C for 15 sec** **52°C for 45 sec → 21 cycles** **65 °C for 11 min** 68ºC for 10 min.
Thermal conditions for *Drosophila* small fragment (100 bp): 75 ºC for 2 min 95 ºC for 15 sec **94°C for 30 sec** **55°C for 45 sec → 21 cycles** **72 °C for 45 sec** 72ºC for 10 min

Table 2. Thermal conditions for QPCR in *Mus musculus* and *Drosophila melanogaster*

Mouse 10 kb mtDNA fragment (Figure 2) primers were:

3278 5'- GCC AGC CTG ACC CAT AGC CAT AAT AT- 3'

13337 5'- GAG AGA TTT TAT GGG TGT AAT GCG G- 3'

(4).

For long fragment PCR amplification, DNA was denatured initially at 75°C for 2 min and 95°C for 1 min, and then the reaction underwent 21 PCR cycles of 94°C for 15 sec, 59°C for 30 sec, and 65 °C for 11 min. Final extension was allowed to proceed at 72°C for 10 min (Table 2). For small fragment PCR amplification, DNA was denatured initially at 75 °C for 2 min and 95 °C for 15 sec, and then the reaction underwent 19 PCR cycles of 94°C for 30 sec, 50°C for 45 sec, and 72 °C for 45 sec. Final extension was allowed to proceed at 72°C for 10 min (23).

We were always run a 50% template control and a nontemplate control in PCR. To calculate relative amplification, the long QPCR values were divided by the corresponding short QPCR results to account for potential copy number differences between samples (mtDNA/total DNA value may be different in 20 ng template total DNA of each PCR tube) (3,4,10,23). The copy number results not indicate the damage.

We detected mtDNA damage in the mouse heart succesfully. According to these relative amplification results "cigarette smoke" application group was significantly different from all other groups. mtDNA damage was significantly higher in the cigarette smoke group than the other groups. However "Cigarette Smoke+Vitamin E+Selenium" group had lowest mean damage (23,24).

4. Measuring of oxidative mtDNA damage and copy number on Drosophila

The free radical theory of aging postulates that aging changes are caused by free radical reactions. Aging is the progressive accumulation of changes with time that are responsible for the ever-increasing likelihood of disease and death. These irreversible changes are attributed to the aging process. This process is now the major cause of death in the developed countries. The aging process may be due to free radical reactions (25). The free radical theory of aging posits that the accumulation of macromolecular damage induced by toxic reactive oxygen species (ROS) plays a central role in the aging process. The mitochondria are the principal generator of ROS during the conversion of molecular oxygen to energy production where approximately 0.4% to 4% of the molecular oxygen metabolized by the mitochondrial electron transport chain is converted to ROS (26). Cellular damage caused by radicals may induce cancer, neurodegeneration and autoimmun disease (27). Toxic materials may produce ROS and generate oxidative damage on mitochondrial DNA (mtDNA) (23). mtDNA damages may trigger mitochondrial dysfunction (28). Damage to mtDNA could be potentially more important than deletions in nDNA, because the entire mitochondrial genome codes for genes that are expressed while nDNA contains a large amount of non-transcribed sequences. Also, mtDNA, unlike nDNA, is continuously replicated, even in terminally differentiated cells, such as neurons and cardiomyocytes; hence, somatic mtDNA damage potentially causes more adverse effects on cellular functions than does somatic nDNA damage (29).

Cereals naturally contain a wide variety of polyphenols such as the hydroxycinnamic acids, ferulic, vanillic, and *p*-coumaric acids which show a strong antioxidant power and may help to protect from oxidative stress and, therefore, can decrease the risk of contracting many diseases. Flavonoids are present in small quantities, even though their numerous biological effects and their implications for inflammation and chronic diseases have been widely described. The mechanisms of action of polyphenols go beyond the modulation of oxidative stress-related pathways (30).

Wheat is an important component of the human diet. But the distribution of phytochemicals (total phenolics, flavonoids, ferulic acid, and carotenoids) and hydrophilic and lipophilic antioxidant activity in milled fractions (endosperm and bran/germ) are different each other. Different milled fractions of wheat have different profiles of both hydrophilic and lipophilic phytochemicals. Total phenolic content of bran/germ fractions is 15−18-fold higher than that of endosperm fractions. Hydrophilic antioxidant activity of bran/germ samples is 13−27-fold higher than that of the respective endosperm samples. Similarly, lipophilic antioxidant activity is 28−89-fold higher in the bran/germ fractions. In whole-wheat flour, the bran/germ fraction contribute 83% of the total phenolic content, 79% of the total flavonoid content, 51% of the total lutein, 78% of the total zeaxanthin, 42% of the total β-cryptoxanthin, 85% of the total hydrophilic antioxidant activity, and 94% of the total lipophilic antioxidant activity (31).

Aim of our study was investigate the effects of a wheat germ rich diet on oxidative mtDNA damage, mtDNA copy number and antioxidant enzyme activities in the aging process of *Drosophila* (32).

Genomic DNA kits (invitrogen) were used for total DNA isolation according to the technical bulletin. İnvitrogen (Molecular Probes) Pico Green dsDNA quantitation dye and QUBIT 2.0 fluorometer were used for both template DNA quantitation and the analysis of PCR products as fluorometrically (Figure 4). DMSO as 4% of total volume and 5 ng of template total DNA were added into the each PCR tube.

Primers for Drosophila mtDNA 100bp fragment were designed as;

11426 5′- TAAGAAAATTCCGAGGGATTCA - 3′

11525 5′- GGTCGAGCTCCAATTCAAGTTA - 3′

Primers for Drosophila mtDNA 10629 bp fragment were designed as;

1880 5′- ATGGTGGAGCTTCAGTTGATTT - 3′

12508 5′- CAACCTTTTTGTGATGCGATTA - 3′ (Table 1)

For long fragment PCR amplification, DNA was denatured initially at 75°C for 1 min and 95°C for 1 min, and then the reaction underwent 21 PCR cycles of 94°C for 15 sec, 52°C for 45 sec, and 65 °C for 11 min. Final extension was allowed to proceed at 68°C for 10 min (Table 2).

For small fragment PCR amplification, DNA was denatured initially at 75 °C for 2 min and 95 °C for 15 sec, and then the reaction underwent 21 PCR cycles of 94°C for 30 sec, 55°C for 45 sec, and 72 °C for 45 sec. Final extension was allowed to proceed at 72°C for 10 min.

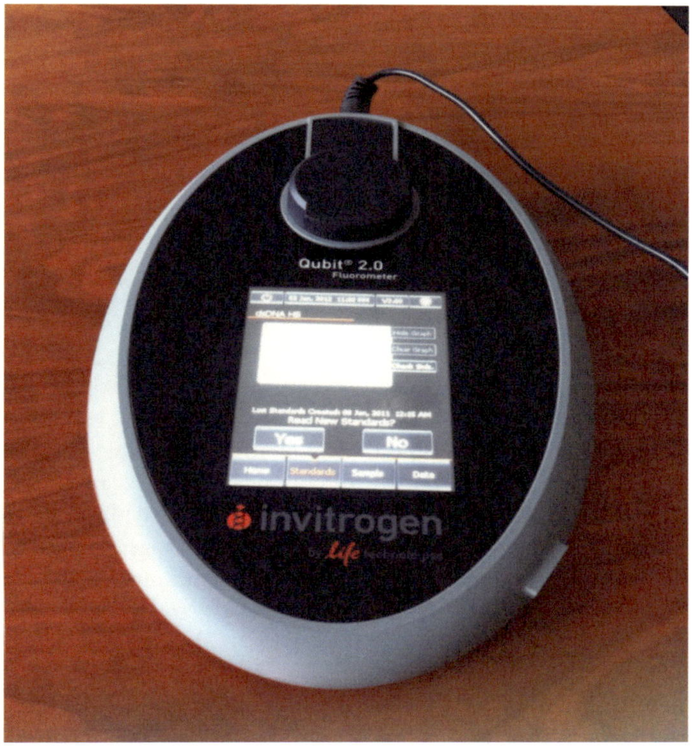

Fig. 4. QUBIT 2.0 fluorometer were used for both template DNA quantitation and the analysis of PCR products as fluorometrically

5. Conclusions

QPCR is a suitable method for the measuring damage and repair in the subgene level functional units like promotor regions, exons and introns (7). Recent advances of the fluorometric dyes allow the very sensitive and quick quantitation of DNA. Before the invention of fluorometric quantitative PCR (QPCR) method, the researchers who measured a gene's amount, have used the different methods like competetive PCR, solid phase assays, HPLC, dot blot or immunoassay (2). Many of the applications of real-time Q-PCR include measuring mRNA expression levels, DNA copy number, transgene copy number and expression analysis, allelic discrimination, and measuring viral titers (3). Method also useful to determining DNA damage and repair that originated by the genotoxic agents and oxidative stress (8,9,10). Crucial step of the QPCR is PCR optimization. Thermal conditions, especially annealing temperature must be optimized. Important points of the optimization:

1. Determination of annealing temperature
2. Optimization of he extention temperature (Extention temperature may be lower for long PCR amplifications)
3. Adjuvants (if necessary)
4. Hot start PCR (minimize nontarget amplification and the formation of primer-dimer)
5. Determination of cycling number

6. Running of %50 template and nontemplate controls in PCR (50% template control should given a 50% reduction of the amplification signal -values between 40%-60% reduction are acceptable-. The nontemplate control would detect contamination with spurious DNA or PCR products)

We detected mtDNA damage that originated by the genotoxic agents, oxidative stress and age, above mentioned conditions in our various studies (23,24,32). Also, QPCR method is suitable for the nutritional studies and some cancer researches.

6. References

[1] Saiki RK, Amplification of genomic DNA. PCR Protocols: a guide to methods and applications (eds. Innis, Gelfand, Sninsky, White). Academic Press, California. 1990.

[2] Wong A, Cortopassi G, Reproducible QPCR of mitochondrial and nuclear DNA copy number using the LightCycler. Mitochondrial DNA methods and Protocols (eds.Copeland). pp.129-149, Humana Press Inc, Totawa, NJ. 2002.

[3] Ginzinger DG, 2002. Gene quantification using real-time quantitative PCR: An emerging technology hits the mainstream. Experimental Hematology, 30(6): 503-512.

[4] Santos JH, Mandavilli BS, Van Houten B, Measuring oxidative mtDNA damage and repair using QPCR. Mitochondrial DNA methods and Protocols (eds.Copeland). pp.159-176, Humana Press Inc, Totawa, NJ. 2002.

[5] Govan HL, Valles-Ayoub Y, Braun J, 1990. Fine-mapping of DNA damage and repair in specific genomic segments. Nucleic Acids Research, 18 (13):3823-3830.

[6] Kalinowski DP, Illenye S, Van Houten B, 1992. Analysis of DNA damage and repair in murine leukemia L1210 cells using a QPCR assay. Nucleic Acids Research, 20(13):3485-3494.

[7] Grimaldi KA, Bingham JP, Souhami RL, Hartley JA, 1994. DNA damage by anticancer reagents and its repair: mapping in cells in the subgene level with QPCR reaction. Anal Biochem, 222(1):236-242.

[8] Van Houten B, Cheng S, Chen Y, 2000. Measuring gene spesific nucleotide excision repair in human cells using quantitative amplification of long targets from nanogram quantities of DNA. Mutat Res, 25; 460(2):81-94.

[9] Ayala-Torres S, Chen Y, Suoblada T, Rosenblatt J, Van Houten B, 2000. Analysis of gene spesific DNA damage and repair using QPCR. Methods, 22: 135-147.

[10] Yakes FM, Van Houten B, 1997. Mitochondrial DNA damage is more extensive and persists longer than nuclear DNA damage in human cells following oxidative stres. Proc Natl Acad Sci USA, 94: 514-519.

[11] Rengarajan K, Cristol SM, Mehta M, Nickerson JM, 2002. Quantifying DNA concentrations using fluorimetry: A comparison of fluorophores. Molecular Vision, 8: 416-421.

[12] Venkatraman A, Landar A, Davis AJ, Chamlee L, Sandersoni T, Kim H, Page G, Pompilius M, Ballinger S, Darley-UsmarV, Bailey SM, 2004. Modification of the mitochondrial proteome in response to the stres of ethanol-dependent hepatoxicity. J Biol Chem, 279: 22092-22101.

[13] O. Henegariu, N.A. Heerema, S.R. Dlouhy, G.H. Vance and P.H. Vogt, 1997. Multiplex PCR: Critical Parameters and Step-by-Step Protocol. BioTechniques, 23: 504-511.

[14] Erlich HA, Gelfand D, Sninsky JJ, 1991. Recent Advances in the Polymerase Chain Reaction. Science, 252: 1643-1651.

[15] Ferre F, 1992. Quantitative or semi-quantitative PCR: reality versus myth. Genome Research, 2:1-9.

[16] Izzotti A, Balanksy RM, Blagoeva PM, Mircheva Z, Tulimiero L, Cartiglia C, De Flora S, 1998. DNA alterations in rat organs after chronic exposure to cigarette smoke and/or ethanol digestion. FASEB J, 12: 753-758.

[17] Izzotti A, Bagnasco M, D'Agostini F, Cartiglia C, Lubet RA, Kelloff G, De Flora S, 1999. Formation and persistence of nucleotide alterations in rats exposed whole-body to environmental cigarette smoke. Carcinogenesis, 20: 1499-1506.

[18] Phillips DH, 2002. Smoking related DNA and protein adducts in human tissues. Carcinogenesis, 23: 1979- 2004.

[19] Gelboin HV, 1980. Benzo α pyrene metabolism, activation and carcinogenesis: role and regulation of mixed-function oxidases and related enzymes. Physiol Rev, 60:1107–66.

[20] Rundle A, Tang D, Hibshoosh H, Estabrook A, Schnabel F, Cao W, Grumet S, Perera FP, 2000. The relationship between genetic damage from polycyclic aromatic hydrocarbons in breast tissue and breast cancer. Carcinogenesis, 21: 1281-1289.

[21] Yang Q, Hergenhahn M, Weninger A, Bartsch H, 1999. Cigarette smoke induces direct DNA damage in the human B-lymphoid cell line Raji. Carcinogenesis, 20: 1769-1775.

[22] Sawyer DE, Van Houten B. (1999) Repair of DNA damage in mitochondria. Mutat Res, 434: 161- 176.

[23] Mutlu AG, Fiskin K, 2009. Can Vitamin E and Selenium Prevent Cigarette Smoke-Derived Oxidative mtDNA Damage? Turkish Journal of Biochemistry, 34 (3); 167-172.

[24] Mutlu AG., Fiskin K, 2009. Oxidative mtDNA damage in the heart tissue of cigarette smoke exposed mice and protective effects of Vitamın E and Selenium. IUBMB Life. 61: 328-329. (III. International Congress of Molecular Medicine, İstanbul, 5-8 May 2009).

[25] Harman D, 2006. Free Radicals in Aging. Moll Cell Biochem, 84: 155-61.

[26] Lim H, Bodmer R and Perrin L, 2006. Drosophila aging 2005-2006. Exp Gerontol, 41: 1213-1216.

[27] Rodriguez C, Mayo JC, Sainz RM, Antolin I, Herrera F, Martin V and Reiter RJ, 2004. Regulation of antioxidant enzymes: a significant role for melatonin. J Pineal Res, 36: 1-9.

[28] Lesnefsky EJ, Moghaddas S, Tandler B, Kerner J and Hoppel CL, 2001. Mitochondrial dysfunction in cardiac disease: ischemia –reperfusion, aging and heart failure. J Mol Cell Cardiol, 33: 1065-1089.

[29] Liang F-Q and Godley BF, 2003. Oxidative stres induced mtDNA damage in human retinal pigment epithelial cells: a possible mechanism for RPE aging and ge-related macular degeneration. Experimental Eye Research, 76: 397-403.

[30] Alvarez P, Alvarado C, Puerto M, Schlumberger A, Jimenez L and De la Fuente M, 2006. Improvement of leucocyte functions in prematurely aging mice after five weeks of diet supplementation with polyphenol-rich cereals. Nutrition, 22: 913-921.

[31] Adom KK, Sorrells ME and Liu RH, 2005. Phytochemicals and antioxidant activity of milled fractions of different wheat varieties. J Agric Food Chem, 53: 2297-2306.

[32] Mutlu AG, 2011. Bugday embriyosunca zengin bir diyetin, Drosophila'nın yaslanma surecinde, oksidatif mtDNA hasarı, mtDNA kopya sayısı, ve antioksidan enzim aktiviteleri üzerine etkileri. Türkiye Klinikleri Tıp Bilimleri Dergisi, 31(6): 132.

15

Detection of *Apple Chlorotic Leaf Spot Virus* in Tissues of Pear Using *In Situ* RT-PCR and Primed *In Situ* Labeling

Na Liu, Jianxin Niu* and Ying Zhao

Department of Horticultural, Agricultural College of Shihezi University, Shihezi People's Republic of China

1. Introduction

Apple chlorotic leaf spot virus (ACLSV) is the type member of the Trichovirus genus, the family *Flexiviridae* (Martelli et al., 1994; Adams et al., 2004) and is known to infect most pome and stone fruit tree species, including apple, peach, pear, plum, almond, cherry and apricot (Lister, 1970; Németh, 1986). ACLSV has a worldwide distribution and induces a large variety of symptoms in sensitive fruit trees (Németh, 1986; Dunez & Delbos, 1988; Desvignes & Boyé, 1989). However, In Japan, this virus is one of the causative agents of topworking disease and induces lethal decline in apple trees grown on Maruba kaido (Malus prunifolia var. ringo) rootstocks (Yanase, 1974). Other severe symptoms of stone fruit trees in Europe caused by ACLSV including bark split and pseudopox in plum, bark split in cherry, pseudopox and graft incompatibility in apricot and ring pattern mosaic in pear (Dunez et al., 1972; Desvignes & Boyé, 1989; Cieślińska et al., 1995; Jelkmann & Kunze, 1995). ACLSV has very flexuous filamentous particles, approximately 640 to 760 nm in length and consisting of a single-stranded positive-sense RNA with Mr of $2{\cdot}48 \times 10^6$ and multiple copies of a 22 kDa coat protein (CP) (Yoshikawa & Takahashi, 1988).

In situ detection techniques allow specific nucleic acid sequences to be exposed in morphologically preserved tissue sections. In combination with immunocytochemistry, *in situ* detection can relate microscopic topological information to gene activity at the transcript or protein levels in specific tissues. In certain cases, they also can provide increased specificity and more rapid analyses. *In situ* reverse transcription polymerase chain reaction (RT-PCR) is a molecular biological-cytological method. *In situ* RT-PCR combined the sensitiveness of PCR amplification with spatial localization of products to monitor the appearance of specific transcripts in the tissue sections. Therefore, *in situ* RT-PCR defined a powerful tool for the low abundance transcript detection (Pesquet et al., 2004). Hasse et al. (1990) first reported the *in situ* PCR technology, which combined the strong points of PCR and *in situ* hybridization. It was widely used for all kinds of disease and genetic studies in human and animal (Gressens & Martin, 1994; Staskus et al., 1991; Nuovo et al., 1991; Bagasra et al., 1992; Cohen, 1996; Chen & Fuggle, 1993; Höfler et al., 1995). The first application of *in situ* RT-PCR for the plant tissue was reported by Woo et al. (1995). Most recently, this

* Corresponding Author

method had not been used to a large extent in plants (Greer et al., 1991; Johansen, 1997; Matsuda et al., 1997).

The primed *in situ* labeling (PRINS) procedure is a fast and efficient alternative to conventional fluorescence *in situ* hybridization for nucleic acid detection. According to the PRINS method, laboratory-synthesized oligonucleotide probes are used instead of cloned DNA for the *in situ* localization of individual genes. The PRINS primers are annealed to complementary target sequences on tissues and are extended in the presence of labeled nucleotides (Koch et al., 1995) utilizing *Taq* DNA polymerase. Since its introduction, the PRINS protocol has been continuously optimized, and numerous applications have been developed (Thomas et al., 2001; Yan et al., 2001; Xu et al., 2002; Tharapel & Wachtel, 2006a, 2006b; Wachtel & Tharapel, 2006; Kaczmarek et al., 2007). The technique has thus proved to be a useful tool for *in situ* screening, and has become a simple and efficient complement to conventional and molecular cytogenetic methods.

In this paper, we optimized the *in situ* RT-PCR and PRINS method for increased sensitivity to localize the virus in plant tissues with ACLSV. Based on this research, through observing distribution of amplified cDNA in tissues, we can analysis the virus infection. In this way, it can provide a new approach to detection virus in fruit trees, as well as investigate the formation, distribution and transformation of virus and produce innocuity fruit trees.

2. Materials

2.1 Virus sources

Leaves were collected from Korla pear in Shayidong commercial orchard of Korla, Xinjiang, China. Virus-free healthy leaves were used as negative controls.

2.2 Reagents and enzymes

Taq DNA Polymerase, dNTPs, dATP, dGTP, dCTP, dTTP, PMD19-T were all purchased from TakaRa (China); M-MLV Reverse Transcriptase, T4 DNA ligase were from Fermentas (USA); TIANprep Mini Plasmid Kit and TIANgel Midi purification Kit were from TIANGEN (China); SuperScript II RNase H-Reverse Transcriptase were from Invitrogen (EU); Proteinase K were from Merk (Germany); Digoxigenin-11- dUTP, alkaline phosphatase labeled anti-digoxin, anti-digoxin- fluorescence, Ribonuclease inhibitor, DNaseI were purchased from ROCH (USA); Nitro blue tetrazolium chloride (NBT)/5-bromo-4-chloro- 3-indolylphosphate (BCIP) were purchased from Shanghai Sangon (China); others were all analysis purity made in China. *E. coli* DH5α as preserved strains were stored at Biotechnology Laboratory of Horticultural Department, Agriculture College, Shihezi University, China.

2.3 Primer design

The sequences were amplified by *in situ* RT-PCR reaction with specific primers, which were designed according to the cDNA sequence of ACLSV (Sato et al., 1993). Primer sequences are as follows: forward primer (P3) 5′-GGCAACCCTGGAACAGA-3′ and the reverse primer (P4) 5′-CAGACCCTTATTGAAG TCGAA-3′.

The sequences were amplified by PRINS reaction with specific primers, which were designed according to the cDNA sequence of ACLSV from GenBank D14996 (Table 1). A

Blast search of the primer sequences showed that they were specific for their intended targets.

Primer	Primer Sequence (5'-3')	Annealing Temp (°C)
acls Pa 1	CTTTACGAGCCCATTTCTTGCC	61.5
acls Ps 1	GAACATAGCGATACAGGGGACC	60.3
acls Pa 2	TGCCTCACACACTTGGCGGAG	60.6
acls Ps 2	CGATACAGGGGACCTCGGAAC	61.5
acls Pa 3	GCCTTTACGAGCCCATTTCTTG	59.5
acls Ps 3	AGGGGACCTCGGAACAAACAG	60.5
acls Pa 4	GTACAAAAGAGGTTTGTGAAG	54.2
acls Ps 4	GTGCTGGTGGAGGTGAAATC	57.4
acls Pa 5	CAATCTGAAGGAGGTAGTCGGT	56.4
acls Ps 5	TTCAGGCGTAGTAGAAAAGAGG	57.7

Table 1. Oligonucleotide primers used to PRINS

3. Methods

3.1 Total RNA extraction and RT-PCR

Total RNAs were extracted from phloem infected by ACLSV. The 200 mg fresh Pear phloem tissue were grinded in liquide nitrogen for a fine powder and transferred to a 1.5 mL eppendorf tube which has added 800 μL extraction buffer (50 mmol·L^{-1} Tris-Cl pH 8.0, 140 mmol·L^{-1} NaCl, 10 mmol·L^{-1} EDTA, 4% SDS, 3% PVP, 15% ethanol, 5% β-mercaptoethanol), well mixed by invertion of the tube. Added 500 μL Tris-saturated phenol (pH 8.0): chloroform: isoamyl alcohol (25: 24: 1) to the tube, sepaeated by centrifugation at 12 000 rpm for 15 min at 4°C. Transferred the supernatant by hand-suction to a fresh tube and mixed with an equal volume of Tris-saturated phenol (pH 8.0): chloroform: isoamyl alcohol (25: 24: 1), followed by centrifugation at 12 000 rpm at 4 °C for 15 min. The supernatant was transferred to a fresh tube and mixed with an equal volume of chloroform: isoamyl alcohol (24: 1) and then centrifugation at 12 000 rpm at 4°C for 10 min. Transferred the supernatant to a fresh tube and added 2.0 volumes of LiCl. Precipitated at –20°C for 2-3 h. RNA was separated by centrifugation at 12 000 rpm for 15 min at 4°C. Removed the supernatant by hand-suction, washed the pellet two times by 70% ethanol, air-dry at room temperature. Suspended the pellet in 20-30 μL of TE solution or DEPC-treated sterile water and analysed it immediately by electrophoresis or stored at - 20°C.

The reverse transcription mixture contained 1.0 μL specific reverse primer and 5.0 μL of total RNA and 9.5 μL of ddH_2O. The mixture was kept at 70°C for 5 min, and then immediately transferred to ice for 5 min. Then 2.5 μL of dNTPs (10 mM each), 5.0 μL of 5×M-MLV buffer, 1.0 μL of RNasin ribonuclease inhibitor (40 U μL^{-1}), 1.0 μL of M-MLV reverse transcriptase (200 U μL^{-1}) and made the total volume of 25.0 μL. The mixture was incubated at 42°C for 1 h.

PCR reaction volumes were 20.0 µL, and contained 2.0 µL of 10×PCR buffer, 0.5 µL of dNTPs (each 10 mM), 2.0 µL of primers, 2.0 µL of cDNA, 0.2 µL (5U μL^{-1}) *Tap* DNA polymerase and 13.3 µL of ddH_2O. PCR was carried out with an initial denaturation of 4 min at 94°C, followed by 35 cycles of 30s, 94°C; 30s, 55°C; 1 min, 72°C; and then by a final elongation step of 7 min at 72°C.

3.2 Cloning and sequencing

The amplified PCR products were gel purified and extracted using TIANgel Midi Purification Kit (TIANGEN, China). The purified DNA fragments were ligated into the PMD19-T vector (TaKaRa Biotechnology, China) following the manufacturer's instruction, and used to transform *E. coli* DH5α. The positive clones were confirmed by PCR and restriction enzyme digestion before sequencing. Two clones from independent PCR reactions were sequenced from both directions.

3.3 Tissue embedding and preparation of slide

1. Slide disposal: After rinsed, ultrasonic cleaned and high temperature baked, the slide must be pre-prepared with poly-L-lysine for 5 min, and then incubated it at 26°C overnight, sealed and stored at room temperature for use within 10 d.
2. Tissues fixation: Leaves were cut into small pieces (3×2 mm) and rinsed the tissues in 4% paraformaldehyde immediately for 1h at room temperature with gentle shaking.
3. Dehydration: Washed the tissues in PBS buffer two times (5 min each), immersed the tissues in series of concentration of ethanol (50%, 70%, 85%, 95% and 100%) for 1h, respectively, at room temperature.
4. Transparences: Put the tissues into pure alcohol: xylene (1: 1) and pure xylene for 1 h, respectively, at room temperature.
5. Low-temperature wax infiltration: Put the tissues into the container which contained transparence and paraffin, covered the container with lid, and incubated at 38°C overnight.
6. High-temperature wax infiltration: Removed the lid, and put the container into incubator at 58°C, and then changed the pure paraffin three times for 2 h each.
7. Paraffin-embedding: Pour melted paraffin wax to pre-folded carton for embedding.
8. Sectioning: Tissue sections (2-16 µm) were obtained by a conventional rotary microtome. If very thin sections were required, a retracting rotary microtome should be used to avoid the compression of the tissue block by the up-stroke of the knife and sections should be mounted onto poly-L-lysine-coated pre-prepared slides.
9. Stretched section: Wax sections needed to be stretched before adhesion to the glass slide. Sections were lifted onto a layer of de-gassed water on a slide held on a warmed flat plate (45°C). Once the sections was stretched, drained away the excess water and left the slide into incubator at 40°C, overnight, the section has dried onto the slide, stored at -20°C.

3.4 Pretreatment of slides

1. De-waxed: Removed the slides from the refrigerator, put the slide into the oven incubated for 1-3 h, at 60°C in order to melt paraffin. Rinsed the slide in xylene for 5 min

and transferred to ethanol for 5 min, repeated more times until the paraffin was completely removed, then left the slide at room temperature for air-dry.

2. Proteinase K treatment: Added 1 μg·mL^{-1} Proteinase K digested 10-45 min at 37°C, stopped reaction by washings for 5 min in PBS buffer and transferred to DEPC-treated sterile water for 5 min at room temperature, then air-dry.
3. DNaseI treatment: For each slide, 4.0 μL 10×DNase I buffer, 4.0 μL DNase I (10 U·μL^{-1}), 1.0 μL Ribonuclease inhibitor (40 U·μL^{-1}) and DEPC water added to 20.0 μL in a 0.5 mL microtube. Applied the reaction solution onto the slide and put it into humidified chamber and incubated at 37°C overnight.
4. Wash the slide two times in DEPC-treated sterile water for 5 min each and in alcohol for 5 min at room temperature.

3.5 *In situ* reverse transcription reaction

For each slide, 4.0 μL 5×Frist-Strand Buffer ($MgCl^{2+}$ 15 mM), 2.0 μL dNTPs (10 mM each), 1.0 μL RNasin (40 U·μL^{-1}), 1.0 μL Antisense primer (20 μM), 2.0 μL DDT (0.1 M), 1.0 μL SuperScript II RT (200 U·μL^{-1}), and DEPC water added to 20.0 μL in a 0.5 mL microtube. Applied the reaction solution onto the slide and put it into a humidified chamber and incubated at 42°C for 1 h, then inactived at 92°C for 1 min. Washed the slide two times for 5 min each in distilled water at room temperature.

3.6 *In situ* RT-PCR detection

3.6.1 *In situ* RT-PCR reaction

The reaction was consisted of 2.5 μL 10 × PCR buffer (Mg^{2+} free), 0.5 μL dNTP (10 mmol·μL^{-1}), 1.0 μL each primer (20 pmol·μL-1), 2.5 μL Dig-11-dUTP (1 nmol·μL^{-1}), 1.0 μL *Taq* DNA polymerase (2.5 U·μL^{-1}) and distilled water to 25.0 μL. Mounted the slide with genic frame, added the reaction solution, and covered the slide with a cover slip ,then put the slide on the flate bloke of the thermocycler. Cycling parameters consisted of 94°C for 3 min, 94°C for 2 min and 35 cycles of a two-step PCR with an annealing temperature of 56°C for 1 min. Removed the cover slip and inactivated at 94°C for 2 min. Washed the slid two times for 10 min each in washing buffer with gentle shaking. Several slides were used as negative controls for each *in situ* RT-PCR experiment. One slide was healthy plant, the other slides were amplified without primers, *Taq* DNA polymerase, or RT step.

3.6.2 Immunoenzymatic detection

1. Mounted the slide with 100 μL blocking buffer (100 mmol·L^{-1} Tris-HCl, pH 7.5, 150 mmol·L^{-1} NaCl, and 3% BSA). Incubated the slide in a humidified chamber at 37°C for 30min. Drained the blocking buffer from the slide.
2. Added anti-Dig-alkaline phosphatase (1: 100 in blocking buffer), and incubated the slide in a humidified chamber for 30 min at room temperature.
3. Stopped the reaction by rinsing the slide with washing buffer (100 mmol·L^{-1} Tris-HCl, pH 7.5, 150 mmol·L^{-1} NaCl) two times for 10 min each at room temperature with gentle shaking.
4. Developed the color reaction by adding 100 μL of NBT/BCIP solution to the slide and incubated the slide in a humidified chamber for 60 min in the dark at room temperature. Then rinsed the slide with water to stop the reaction.

5. Rinsed the slide in series of concentration of ethanol, 50%, 70%, 85%, 95%, and 100% for 2 min, respectively, at room temperature for dehydration.
6. Put the slide into pure xylene for 3 min for transparent.
7. Covered the section with the cover slip using mounting solution, air-dry. Then the sections were ready for data recording, which could view under bright field microscopy through stained with Alcian Blue.

3.7 PRINS detection

3.7.1 PRINS reaction

1. Immersed slides in 0.02 N HCl for 20 min.
2. Denature the samples by immersing them in 70% formamide/2×SSC, at 72°C for 2 min.
3. Dehydrate the slides in a series (70%, 90%, and 100%) of ice-cold ethanol washes (4°C) before allowing them to air-dry.
4. Prepare reaction mixture in a final volume of 25.0 μL consisted of specific primers (20 μM) 10.0 μL, 0.1% BSA 2.5 μL, 0.2 mM dNTPs 2.5 μL (each), 0.02 mM dTTP 1.0 μL, 0.02 mM Dig-11-dUTP 3.0 μL, *Taq* buffer 2.5 μL, *Taq* DNA polymerase (2.5 U μL^{-1}) 1.0 μL and distilled water to 25.0 μL. Kept the mix on ice during preparation and until application to the slide.
5. Reaction mixture incubated at annealing temperature and incubated the denatured the slide for 7 min at annealing temperature. Applied the reaction mixture and covered the working area of the slide completely with a 22 × 22 cover slip on the denatured the slide, and then transferred to the heating block of the thermal.
6. Set up the PRINS program and start the reaction. The program was carried out on a programmable thermal cycler equipped with a flat plate for slides. The program consisted of one cycle of 9 min at annealing temperature with an additional 30 min at 72°C for extension.
7. After extension, the slide was removed from cycler, the cover slip was removed, and the slide washed in NE solution (500 mM NaCl, 50 mM EDTA, pH 8.0) at 72°C for 5 min, and transferred the slide to 4×SSC/0.2% Tween-20 at 50°C for 5 min to stop the reaction.

3.7.2 Visualization of PRINS products

1. For each slide, added 10 μg mL^{-1} avidin-Rhodamine and 20 μg mL^{-1} anti-digoxigenin-FITC.
2. Placed slides in a humidified chamber for 30 min at room temperature, worked in the dark as much as possible to avoid fluorescence bleaching.
3. The slide was rinsed in preheated solutions (1×PBS/0.2%Tween-20, 37°C; 0.5×PBS/0.2% Tween-20, 37°C; 0.2×PBS/0.2%Tween-20, 37°C) for 5min, respectively, air-dried.
4. Mounted the slide with 3μg mL^{-1} of DAPI/antifade solution under a 22×22 coverslip counterstained for 10min, in dark.
5. Let the excess mounting medium dry. Approximately 1 h, permanently seal the slide with nail polish. Slide can be maintained at 4°C until scored.

3.7.3 Signal detection and image analysis

Olympus BX51 fluorescence microscope system was adopted for this process. This system contained Olympus UPlanFI 100×/1.30 Oil ∞/0.17 C1field lens, pass band filter with

DAPI/FITC/Rhodamine, AxioCam Camera module and Video Test-FISH 4.0 image analysis system.

4. Results

4.1 Detection ACLSV by RT-PCR

Total RNA were extract from the phloem of pear which were infected with ACLSV, first strand cDNA synthesis was obtained by reverse transcription using specific primer and 358 bp fragment was amplified by P3/P4 primers as shown in Figure. 1. The purified DNA fragments were ligated into the PMD19-T vector and transformed into *E. coli* DH5α. The positive clones were confirmed by PCR and restriction enzyme digestion before sequencing.

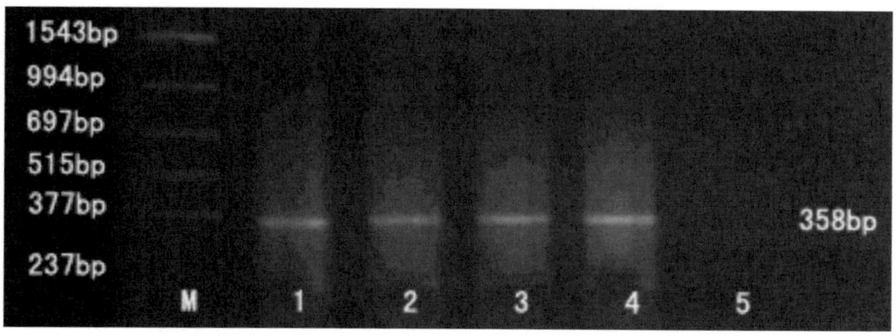

Fig. 1. The productions of RT-PCR of ACLSV
M: Marker; 1-4: productions; 5: negative control

4.2 Detection the reliability of alkaline phosphatase chromogenic system

The slide were digested by 1μg·mL^{-1} Proteinase K for 20 min at 37°C, and incubated at 37°C overnight with DNase I. Washed the slide two times for 10 min each in PBS buffer. Mounted the slide with blocking buffer and incubated at 37°C for 30min. Added anti-Dig-alkaline phosphatase (1:100 in blocking buffer) and incubated the slide in a moist chamber for 60 min at room temperature, then washed the slide two times for 10 min each in PBS buffer at room temperature with gentle shaking. Added NBT/BCIP solution to the slide and incubated the slide in a humidified chamber for 60 min in the dark at room temperature. The result showed that sections were not stained.

4.3 The effect of treatment with proteinase K

After treated with Proteinase K treatment for 10 min or 15 min, the organization performed a piece of blue, which indicated that Proteinase K digested inadequately. Morphology was fuzzy when digested for 30 min or 40 min, illustrating excessive digestion. Proteinase K treatment 20 min was more moderate.

4.4 The effect of RT-component concentration

The results showed there was no signal when RNasin was less than 0.2 U·μL^{-1}, and it was enhanced with the increased RNasin. The concentration of dNTPs was above 0.4 mmol·L^{-1}, the signal was appeared; the concentration of SuperScript II ranged from 0.1U·μL^{-1} to 1.3

U·μL^{-1} and the signal was enhanced with the increase concentration of SuperScript II; the concentration of primers above 0.9 μmol·L^{-1} were effective, less than 0.8 μmol·L^{-1} could not synthesized sufficient quantities of cDNA and above 1.2 μmol·L^{-1} could produce non-specific product.

4.5 The effect of other factors

The result showed that positive signals were appeared on the slide only when the annealing temperature at 56°C, which indicated that the suitable temperature was 56°C. Amplification with 10-20 cycles, the signals were not appeared, 25 cycles appeared weaker blue signal, 30-35 cycles showed stronger signals, which demonstrated that fewer cycles led to lower

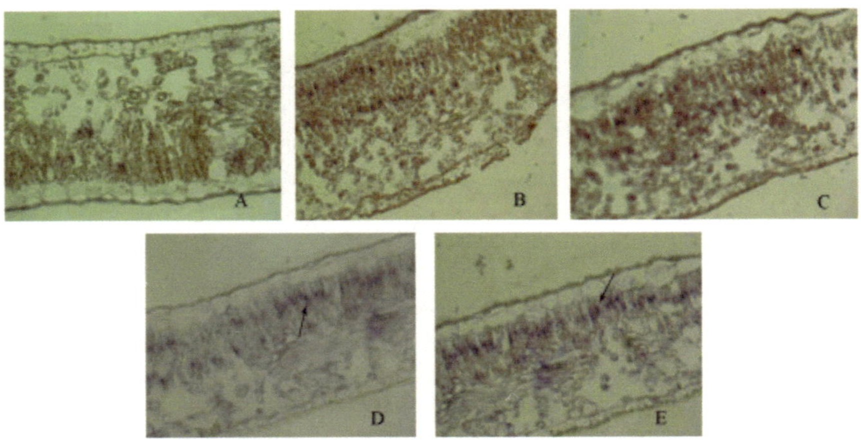

Fig. 2. The effect of cycle number on *In situ* RT-PCR
A: 10 cycles; B: 15 cycles; C: 20 cycles; D: 25 cycles; E: 30 cycles

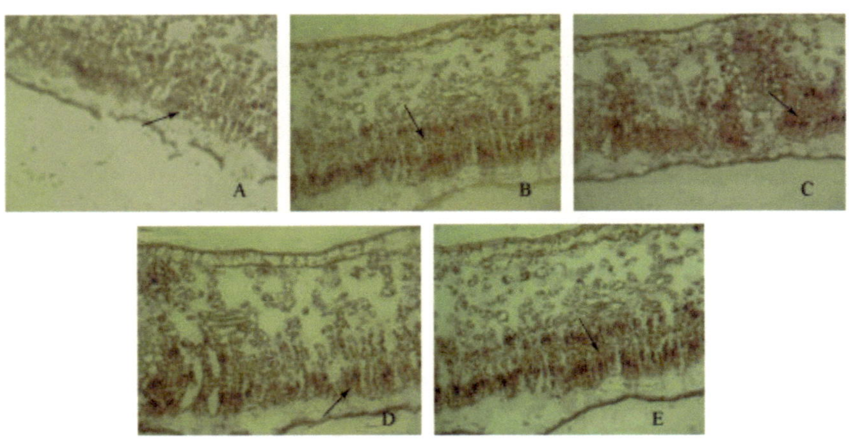

Fig. 3. The effect of the different *Taq* DNA polymerase concentration on the detection of *In situ* RT-PCR
A: 2 U·100 μL^{-1}; B: 4 U·100 μL^{-1}; C: 6 U·100 μL^{-1}; D: 8 U·100 μL^{-1}; E: 10 U ·100 μL^{-1}

synthesis (Figure. 2). The concentration of *Taq* DNA polymerase with 2 U·100μL^{-1}-10 U·100μL^{-1} could satisfy amplification and showed stronger signals, which indicated that the suitable concentration of *Taq* DNA polymerase was 2 U·100μL^{-1} (Figure. 3).

4.6 PRINS-Rhodamine staining

Applied PRINS-Rhodamine staining detected ACLSV showed that the infected leaves of pear tissues were presented red fluorescence positive signals (Fig. 4, A~D, arrows showing the locations), which were consistent with the results of *In situ* RT-PCR detection (Niu et al., 2007). Healthy leaves and infected leaves without SuperScript II RT, fluorescent antibody and *Taq* DNA polymerase, did not present red fluorescence signals (Fig. 5, E~H).

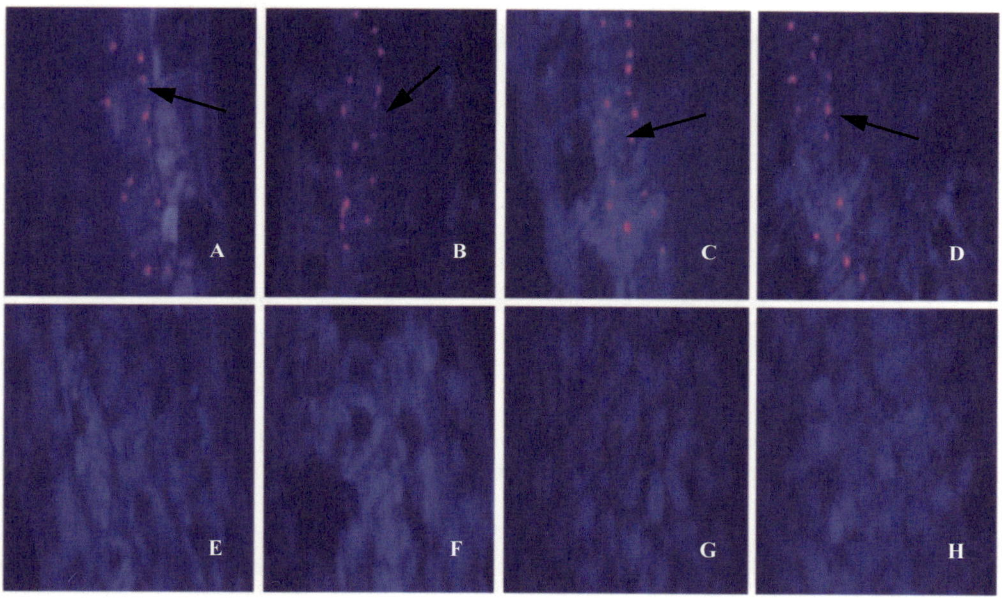

Fig. 4. PRINS-Rhodamine staining results of ACLSV in pear tissues
A-D: Labeled results of virus infected pear leaves from the same positions of different trees; E: Labeled results of healthy pear leave (control); F-H: PRINS-Rhodamine staining results of ACLSV in pear tissues (control: Left out of SuperScript II RT, fluorescence antibody, *Taq* enzyme).

4.7 PRINS-FITC staining

FITC fluorochrome was more sensitive to the temperature and pH, and the efficiency was lower than Rhodamine staining, and the results showed inconspicuous signals. Applied PRINS-FITC staining detected ACLSV showed that the infected leaves of pear tissues were presented green fluorescence positive signals (Fig. 5, A~D, arrows showing the locations), which were consistent with the results of *In situ* RT-PCR detection (Niu et al., 2007). Healthy leaves and infected leaves without SuperScript II RT, fluorescent antibody and *Taq* DNA polymerase, did not present red fluorescence signals (Fig. 2, E~H).

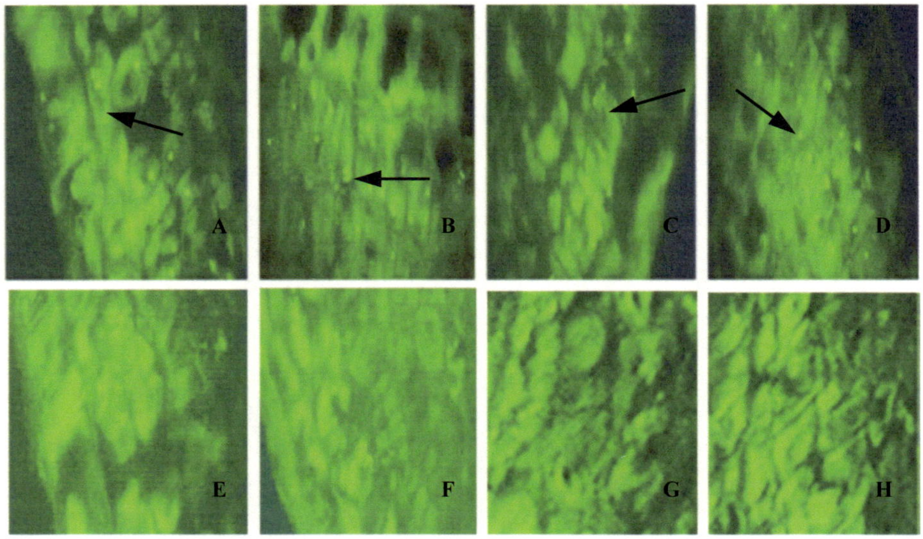

Fig. 5. PRINS-FITC staining results of ACLSV in pear tissues
A-D: Labeled results of virus infected pear leaves from the same positions of different trees; E: Labeled results of healthy pear leave (control); F-H: PRINS- Rhodamine staining results of ACLSV in pear tissues (control: Left out of SuperScript II RT, fluorescence antibody, *Taq* enzyme).

5. Discussion

The study is based on virus RNA as a template to reverse transcription cDNA and *in situ* amplification. Before amplification, the slides treatment with DNA exonuclease without RNA enzyme overnight digest the original genomic DNA in tissues which can eliminating DNA fragment decorated by polymerase which could form false-positive amplification (Long et al., 1993). In our studies, the known virus-free material of pear tree used as the negative control did not appear specificity of fluorescence signals. Negative control without SuperScript II RT, fluorescence antibody, *Taq* enzyme showed the same result of virus-free material. Signals did not display without RT steps indicated that the products were amplified by cDNA, which excluded the possible of experimental reagents cross produced fluorescent complex and attached to the tissue surface induced fluorescence signals. In our studies, ACLSV of leave sections of Korla Pear were detected by *in situ* RT-PCR and PRINS, the results showed that the positive materials were found obviously alcian blue and fluorescence signals in mesophyll cells, while the negative control tissue did not appear. It was indicated that ACLSV mainly distributed in the palisade tissue of the mesophyll cells, and the same results as *in situ* RT-PCR detection (Niu et al., 2007). In addition, the results showed that the thickness of section had a great influence on detection. Thin slices can easy to cause the tissues were not complete, and the cell of thick slices were multiple and overlapping, which unfavorable for observing, and seriously affect the detection results. So, in order to obtain desire results of detection, the 4-6 μm of sections were used.

Because of the *in situ* amplified cDNA in tissues, we must consider the number of primers to use. A single primer would not allow a strong enough signal for fluorescent detection.

However, too many primers would likely lead to primer-dimers or non-specific hybridization. In PRINS reaction system, primer extensions strictly followed the principle of complementary base pairing, and ensure the specificity labeling. Synthesis of labeled DNA will remain in the amplified position and not diffusion. In this study, we used five specific primers for PRINS, and achieved clearly fluorescence signals.

Terkelsen et al., (1993) using repeated primed *in situ* labeling (repeated-PRINS). This change of strategy results in a localized accumulation of sequence-specific labeled DNA, resulting in up to a 15-fold amplification of the signal as compared to the standard PRINS method. Ni et al., (1998) results showed that the repeated-PRINS technology could to enhance the signal; however, repeated heat denaturation and extension process for long time which induced the cell loss normal forms. In our study, we pretreatment species with appropriate concentration of protease K, and the optimal time of proteinase K digestion was necessary. The tissues slices were treated with proteinase K for 10, 20, 30, and 45 min. The best results were achieved after 20 min of the proteinase K digestion. The morphology of the tissue was well retained, and interpretation of results was unambiguous. The signal was recognized as fluorescence-signals the site of the label. The 10 min durations turned out to be too short and led to lack of signal. The extension of the reaction time up to 45 min produced morphological distortions to the point that interpretation of results became impossible. In addition, our research showed that increasing the ratio of dTTP and labeled-dUTP could improve the signal intensity. In general, the ratio of dTTP and labeled-dUTP was 1: 1 could generate enough strong signals. We increased the dTTP and labeled-dUTP concentration ratio to 1: 3 generated strong signals.

In this study, two fluorescence labeling were used, FITC and Rhodamine, respectively. Fluorescent-FITC was used *in situ* labeling showed sensitive on PH and easy to decay. In the conditions of susceptible pH or strong UV irradiation, the fluorescence excitation rapid decay. In addition, increase the times of washing, the tissues were more easily damaged and higher backgrounds were obtained. Therefore, on the basis of complete elution, appropriate to reduce washing processing steps were necessary.

Primed *in situ* labeling (PRINS) of nucleic acids was developed as an alternative to traditionally used fluorescence *in situ* hybridization (FISH). PRINS is based on sequence-specific annealing of unlabelled oligonucleotide primer under stringent conditions to the DNA of denaturated. Compared to FISH, PRINS is faster and does not require preparation of labeled probes, the process costs much less in terms of reagents (Velagelati et al., 1998; Tharapel & Kadandale, 2002; Pellestor et al., 2002), and hybridization signal is stronger, more specific and easy to control. In addition, we believe that this modified PRINS technique can have very meaningful applications in molecular cytogenetics. It can be used for the visualization and mapping of genetic loci on chromosomes, and for detection of the presence or absence of small DNA segments involved in genetic diseases. PRINS will have a more extensive application prospects in plant virus detection.

6. Conclusions

ACLSV of leave sections of Korla Pear were detected by *in situ* RT-PCR and PRINS, and the positive materials were found obviously alcian blue and fluorescence signals in mesophyll cells. The results showed that *in situ* RT-PCR and PRINS, which had two staining methods of PRINS-FITC and PRINS-Rhodamine, could get good detection results in which the parts have viruses showed alcian blue, green and red fluorescence light, respectively. Therefore, primed *in situ* labeling technique can be perfectly used for virus *in situ* detection of fruit trees, and it is also a rapid, simple and reliable *in situ* detection method.

7. Acknowledgements

This stsudy was supported by National Natural Science Foundation of China (30360066), the National Key Technologies R&D Program of China (2003BA546C), the Foundation Science and Technology Commission Xinjiang Production and Construction Crops, China (NKB02SDXNK01 SW) and Natural Science and Technology Innovation of Shihezi University, China (ZRKX200707).

8. References

Adams, M J.; Antoniw, J F.; Bar-Joseph, M.; Brunt, A A.; Candresse, T.; Foster, G D.; Martelli, G P.; Milne, R G. & Fauquet, C M. (2004). The new plant virus family Flexiviridae and assessment of molecular criteria for species demarcation. *Archives of Virology*, Vol.149, No.5, (May 2004), pp. 1045-1060, ISSN 0304-8608

Bagasra, O.; Hauptman, S. P.; Lischner, H. W.; Sachs, M, & Pomerantz, R. J. (1992). Detection of human immunodeficiency virus type 1 provirus in mononuclear cells by in situ polymerase chain reaction. *The New England Journal of Medicine*, Vol.326, No.21, (May 1992), pp. 1385-1391, ISSN 0028-4793

Chen, R. H. & Fuggle, S. V. (1993). In situ cDNA polymerase chain reaction. A novel technique for detecting mRNA expression. *American Journal of Pathology*, Vol.143, No.6, (December 1993), pp. 1527-1533, ISSN 0002-9440

Cieślińska, M.; Malinowski, T. & Zawadzka, B J. (1995). Studies on several strains of Apple chlorotic leaf spot virus (ACLSV) isolated from different fruit tree species. *Acta Horticulturae*, Vol.386, No.89, (July 1995), pp. 63-71, ISSN 0567-7572

Cohen, N. S. (1996). Intracellular localization of the mRNAs of argininosuccinate synthetase and argininosuccinate lyase around liver mitochondria, visualized by high-resolution in situ reverse transcription-polymerase chain reaction. *Journal of Cellular Biocemistry*, Vol.61, No.1, (April 1996), pp. 81-96, ISSN 1097-4644

Desvignes, J C. & Boyé, R. (1989). Different diseases caused by the Chlorotic leaf spot virus on the fruit trees. *Acta Horticulturae*, Vol.235, No.49, (April 1989), pp. 31-38, ISSN 0567-7572

Dunez, J.; Marenaud, G.; Delbos, R. P. & Lansac, M. (1972). Variability of symptoms induced by the *Apple chlorotic leaf spot* (CLSV). A type of CLSV probably responsible for bark split disease of prune trees, *Plant Disease Reporter*, Vol.56, No.4, (April 1995), pp. 293-295, ISSN 0032-0811

Greer, C. E.; Peterson, S. L.; Kiviat, N. B. & Manos M. M. (1991). PCR amplification from paraffin- embedded tissues. Effects of fixative and fixation time. *American Journal of Clinical Pathology*, Vol.95, No.2, (February 1991), pp. 117-124, ISSN 0002-9173

Gressens, P. & Martin, J. R. (1994). In situ polymerase chain reaction: Localization of HSV-2 DNA sequences in infections of the nervous system. *Journal of Virological Methods*, Vol.46, No.1, (January 1994), pp. 61-83, ISSN 0166-0934

Dunez, J. & Delbos, R. 1988, Apple chlorotic leaf spot virus, In: *European handbook of plant diseases*, I.M. Smith, (Ed.), 5-7, ISBN 0-632-01222-0, Britain

Haase, A. T.; Retzl, E. F. & Staskus, K.A. (1990). Amplification and detection of lentiviral DNA inside Cells. *Proceeding of the National Academy of Sciences of the United States of America*, Vol.87, No.13, (February 1990), pp. 4971-4975, ISSN 0027-8424

Harrer, T.; Schwinger, E. & Mennicke, K. (2001). A new technique for cyclic in situ amplification and a case report about amplification of a single copy gene sequence in human metaphase chromosomes through PCR-PR INS. *Human Mutation*, Vol. 17, No.2, (February 2001), pp. 131-140, ISSN 1059-7794

Höfler, H.; Pütz, B.; Mueller, J. D.; Neubert, W.; Sutter, G. & Gais, P. (1995). In situ amplification of measles virus RNA by the self-sustained sequence replication reaction. *Laboratory Investigation*, Vol.73, No.4, (October 1995), pp. 577-585, ISSN 0023-6837

Jelkmann, W. & Kunze, L. (1995). Plum pseudopox in German prune after infection with an isolate of Apple chlorotic leafspot virus causing plum line pattern. *Acta Horticulturae*, Vol.386, No.89, (July 1995), pp. 122-125, ISSN 0567-7572

Johansen, B. (1997). In situ PCR on plant material with sub-cellular resolution. *Annals of Botany*, Vol.80, No.5, (November 1997), pp. 697-700, ISSN 3035-7364

Kaczmarek, A.; Naganowska, B. & Wolko, B. (2007). PRINS and C-PRINS: Promising tools for the physical mapping of the lupin genome. *Cellular & Molecular Biology Letters*, Vol.12, No.1, (March 2007), pp. 16-24, ISSN 1425-8153

Koch, J.; Hindkjær, J.; Kølvraa, S. & Bolund, L. (1995). Construction of a panel of chromosome-specific oligonucleotide probes (PRINS-primers) useful for the identification of individual human chromosomes in situ. *Cytogenetics and Cell Genetics*, Vol.71, No.2, (June 1995), pp. 142–147, ISSN 0301-0171

Long, A. A.; Komminoth, P.; Lee, E. & Wolfe, H. J. (1993). Comparison of indirect and direct in-situ polymerase chain reaction in cell preparations and tissue sections. *Histochemistry*, Vol.99, No.2, (February 1993), pp. 151-162, ISSN 0948-6143

Lister, R M. (1970). Apple chlorotic leaf spot virus. In. *CMI/AAB Descriptions of Plant Viruses*, A.J. Gobbs (Ed.), 30, Commonwealth Mycological Institute, Kew, Surrey, England

Martelli, G. P.; Candresse, T. & Namba, S. (1994). *Trichovirus*, a new genus of plant viruses. *Archives of Virology*, Vol.134, No.3-4, (September 1994), pp. 451-455, ISSN 0304-8608

Németh, M. V. (1986). *Virus, Mycoplasma and Rickettsia diseases of fruit trees,* M. V. Németh, (Ed.), Akadémiai KiadóBudapest

Ni, B.; Li, H.; Zou, Y. H. & Wu, S. L. (1998). Use of the primed in situ labeling technique for a rapid detection of chromosomes X, 18. *Chinese Journal of Medical Genetics*, Vol.15, No.6, (December 1998), pp. 373-375, ISSN 1003-9406

Niu, J X.; Zhou, M. S.; Ma, B. G.; Zhao, Y. & Liu, H. (2007). Detection of Apple chlorotic leaf sopt virus in pear by in situ RT-PCR. *Acta Horticulturae Sinica*, Vol.34, No.1, (February 2007), pp. 53-58, ISSN 0513-353X

Nuovo, G. J.; MacConnell, P.; Forde, A. & Delvenne P. (1991). Detection of human papillomavirus DNA in formalin-fixed tissues by in situ hybridization after amplification by polymerase chain reaction. *American Journal of Pathology*, Vol.139, No.4, (October 1991), pp. 847-854, ISSN 0002-9440

Pellestor, F.; Imbert, I. & Andréo, B. (2002a). Rapid chromosome detection by PRINS in human sperm. *American journal of Medical Genetics*, Vol.107, No.2, (January 2002), pp. 109-114, ISSN 1552-4833

Pesquet, E., Barbier, O., Ranocha, P., Jauneau, A. & Goffner, D. (2004). Multiple gene detection by in situ RT-PCR in isolated plant cells and tissues. *The Plant Journal*, Vol.39, No.6, (September 2004), pp. 947–959, ISSN 1365-313X

Sato, K.; Yoshikawa, N. & Takahashi, T. (1993). Complete nucleotide sequence of the genome of an apple isolate of apple chlorotic leaf spot virus. *Journal of General Virology*, Vol.74, No.9, (September 1993), pp. 1927-1931, ISSN 0022-1317

Staskus, K. A.; Couch, L.; Bitterman, P.; Retzel, E. F.; Zupancic, M. & Haase, A. T. (1991). In situ amplification of visna virus DNA in tissue sections reveals a reservoir of latently infected cells. *Microbial Pathogenesis*, Vol.11, No.1, (July 1991), pp. 67-76, ISSN 0882-4010

Tharapel, S. A. & Kadandale, J. S. (2002). Primed in situ labeling (PRINS) for evaluation of gene deletions in cancer. *American journal of Medical Genetics*, Vol.107, No.2, (January 2002), pp. 123-126, ISSN 1552-4833

Tharapel, A. T. & Wachtel, S. S. (2006a). PRINS for mapping single-copy genes, In: *Gene Mapping, Discovery, and Expression*, M. Bina, (Ed.), 59-67, ISBN 1-58829-575-3, United States of America

Tharapel, A. T. & Wachtel, S. S. (2006b). PRINS for the detection of gene deletions in cancer, In: *PRINS and In Situ PCR Protocols*, F. Pellestor, (Ed.), 105-113, ISBN 1-58829-549-4, United States of America

Terkelsen, C.; Koch, J.; Kølvraa, S.; Hindkjr, J.; Pedersen, S. & Bolund, L. (1993). Repeated primer in situ labelling: Formation and labeling of specific DNA sequences in chromosomes and nuclei. *Cytogenetics and Cell Genetics*, Vol.63, No.4, (December 1993), pp. 235–237, ISSN 0301-0171

Velagelati, G. V.; Shulman, L. P.; Phillips, O. P.; Tharapel, S. A. & Tharapel, A. T. (1998). Primed in situ labeling for rapid prenatal diagnosis. *American journal of Obstetrics & Gynecology*, Vol. 178, No.6, (June 2001), pp. 1311-1320, ISSN ISSN: 0002-9378

Wachtel, S. S. & Tharapel, A. T. (2006). PRINS for the detection of unique sequences, In: *PRINS and In Situ PCR Protocols*, F. Pellestor, (Ed.), 33-40, ISBN 1-58829-549-4, United States of America

Woo, H. H., Brigham, L. A. & Hawes, M. C. (1995). In-cell RT-PCR in a single, detached plant cell. *Plant Molecular Biology Reporter*, Vol.13, No.4, (December 1995), pp. 355–362, ISSN 0734-9640

Xu, J.; Wang, A. & Chen, J M. (2002). New developments in primed in situ labeling (PRINS) methods and applications. *Journal of Clinical Medicine in Practice*, Vol.6, No.3, (February 2002), pp. 169-175, ISSN 1672-2353

Yan, J.; Bronsard, M. & Drouin, R. (2001). Creating a new color by omission of 3′ end blocking step for simultaneous detection of different chromosomes in multi-PRINS technique. *Chromosoma*, Vol.109, No.8, (February 2001), pp. 565-570, ISSN 0009-5915

Yanase, H. (1974). Studies on apple latent viruses in Japan, *Bulletin of the Fruit Tree Research Station, Japan* Series C1, pp. 47-109 ISSN 0916-5851 0385-2326

Yoshikawa, N. & Takahashi, T. (1988). Properties of RNAs and proteins of apple stem grooving and apple chlorotic leaf spot viruses. *Journal of General Virology*, Vol.69, No.1, (January 1988), pp. 241-245, ISSN 0022-1317

16

Application of PCR Technologies to Humans, Animals, Plants and Pathogens from Central Africa

Ouwe Missi Oukem-Boyer Odile[1], Migot-Nabias Florence[2], Born Céline[3], Aubouy Agnès[4] and Nkenfou Céline[1]
[1]*Chantal Biya International Reference Centre for Research on Prevention and Management of HIV/AIDS (CIRCB), Yaoundé,*
[2]*Institut de Recherche pour le Developpement (IRD), UMR 216 (IRD/UPD) Faculté de Pharmacie, Paris,*
[3]*Institut de Recherche pour le Developpement (IRD), UMR 152, Université Paul Sabatier, Toulouse,*
[4]*University of Stellenbosch, Department of Botany and Zoology, Stellenbosch,*
[1]*Cameroon*
[2,3]*France*
[4]*South Africa*

1. Introduction

The Central African region, also called Atlantic Equatorial Africa, harbors one of the biggest worldwide biodiversity. It is true for human, with a great diversity of ethnic groups, but also for animals, plants, and microorganisms including pathogen species. Although this region is lagging behind in various domains, few research centers and laboratories have been able to develop sophisticated research work for diagnostics, fundamental research, and operational research, using polymerase chain reaction (PCR) techniques. This present paper intends to give an overview of the use of PCR technology in Central Africa and its various applications in the field of genetics, phylogeography, ecology, botany, and infectious diseases, which may have a broad impact on interspecies relationships, diagnostics of diseases, environment and biodiversity.

We will successively describe the main research findings in humans, animals, plants and pathogens from Central Africa, and show how the PCR has allowed scientists from this region to contribute significantly to generalized knowledge in these fields. Then, we'll discuss opportunities and challenges in conducting such kind of research in these particular limited-resources settings before concluding this chapter.

2. Humans

Since the nineties, the extensive use of molecular techniques has contributed to deepen the knowledge on human genetics. In most studies related to Central Africa, such methodologies have often been used in the context of immunogenetics or genetic epidemiology of infectious diseases. The host genetic background is as important as immunity in the individual fight against infections. These studies were a fabulous opportunity to investigate the richness and extreme diversity of the genetic polymorphisms that characterize populations from Central Africa.

2.1 HLA characterization

The major histocompatibility complex (MHC) is one of the most polymorphic genetic systems of many species, including human leukocyte antigen (HLA) in humans. The class I and class II MHC genes encode cell-surface heterodimers that play an important role in antigen presentation, tolerance, and self/non-self recognition. The HLA molecules bind intracellularly processed antigenic peptides, forming complexes that are the ligands of the antigen receptors of T lymphocytes. In addition, the class I and class II histocompatibility antigens play an important role in allogeneic transplantation. Matching for the alleles at the class I and class II MHC loci impacts the outcome of both solid-organ and hematopoietic stem cell allogeneic transplants.

The HLA class II typing of 167 unrelated Gabonese individuals living in the village of Dienga, located in the South-East of Gabon (province of the Haut-Ogooué) was assessed by polymerase chain reaction-restriction fragment length polymorphism (PCR-RFLP) [2]. All individuals belonged to the Banzabi ethnic group, which represents the second most important population grouping in Gabon after the Fang, with 55,000 to 60,000 individuals living in an area of 32,000 km^2. At the date of realization, in 1996, restriction endonuclease mapping of the PCR products provided profiles that allowed identification of 135 major alleles or groups of alleles among the 184 known DRB1 alleles [3]. Similarly, 9, 24 and 53 major alleles or groups of alleles were recognizable out of a total of 19, 35 and 83 DQA1, DQB1 and DPB1 alleles respectively, so far reported in the literature. For each locus, the PCR-RFLP identified alleles include all major alleles, while unidentifiable alleles were corresponding to rare and newly described alleles. The most frequent alleles at each locus were DRB1*1501–3 (0.31), DQA1*0102 (0.50), DQB1*0602 (0.42) and DPB1*0402 (0.29). The estimation of the haplotype frequencies as well as the observation of the segregation of several haplotypes using additional HLA typing of relatives, revealed that the three-locus haplotype DRB1*1501–3-DQA1*0102-DQB1*0602 was found at the highest frequency (0.31) among these individuals. This haplotype is not typically African and has already been described in Caucasians, but its presence at high frequency is exclusive to populations originating from Central Africa, and can thus be designated as a particular genetic marker of these populations. On the other hand, the absence in the Gabonese Banzabi group of DRB1*04 and the concomitant predominance at equal prevalence rates of DRB1*02 and DRB1*05, conforms to the other sub-Saharan population groups which have already been typed for their DR1-DR10 allospecificities [4]. Similarly, the predominant alleles observed at the DQA1, DQB1 and DPB1 loci studied have already been described in other sub-Saharan populations [5]. As an example, the determination of DRB1-DQA1-DQB1 haplotype frequencies for 230 Gabonese individuals belonging to tribes as different as Fang, Kele, Myene, Punu, Sira and Tsogo, revealed, as for the Banzabi group, the highest frequency (0.24) for the DRB1*15/16-DQA1*0102-DQB1*0602 haplotype [6]. The same predominant haplotype was observed with a high frequency of 0.27 among 126 healthy individuals in Cameroon, by means of a determination by high-resolution PCR using sequence-specific oligonucleotide probes (PCR-SSOP) and/or DNA sequencing [7].

Few studies investigated the extensive allelic diversity in the class I loci (to date, more than 250 HLA-A, 500 HLA-B, and 120 HLA-C alleles) by means of molecular methods among populations of Central Africa [5]. In populations as geographically close as Cameroonians (Yaoundé) [8] and Gabonese (Dienga, South-East of Gabon) [9], the two most frequently detected HLA-A and HLA-B allele families diverged, illustrating the patchwork representation of the different genetic backgrounds (Cameroon: HLA-A*23, A*29, HLA-B*53 and B*58; Gabon: HLA-A*19, A*10, HLA-B*17 and B*70). In Cameroon, where populations are very heterogeneous in their origin, culture and language, the most frequently encountered HLA-A, HLA-B and HLA-C alleles differed in four ethnic groups distributed from the north to the south of the country, reflecting the complex migrations and admixtures that occurred in this area located in the borders of Central and west Africa, before that populations settled [10].

2.2 Red blood cell polymorphisms

Red blood cell polymorphisms are frequently found in areas where malaria is currently or was historically endemic. This observation led to the idea that some of these polymorphisms might provide a relative advantage for survival [11]. The best-characterized polymorphism in this context is the sickle cell trait (HbAS), comprising heterozygous carriage of hemoglobin (Hb) S, which results from a valine substitution for glutamic acid at position 6 of the hemoglobin β chain. HbAS provides carriers with a high degree of protection against severe *Plasmodium falciparum* malaria during early life, which explains the relatively high penetrance of this mutation— in some areas reaching 30%—in sub-Saharan African communities exposed to high rates of infection with *P. falciparum* [12]. The mutation in the homozygous state (HbSS) leads to the disease referred to as "sickle cell anemia," a life-threatening condition that usually results in early death [13, 14]. HbAS in such populations thus exemplifies a balanced polymorphism that confers a selective advantage to the heterozygote [15]. Molecular determination of the HbS carriage is assessed by PCR-RFLP, where a 369-bp segment of the codon 6 in the beta-globine gene, encompassing the A>T substitution, is amplified, before being digested with the restriction endonuclease *DdeI*.

In sub-Saharan populations, the ABO blood group distribution is in large part dominated by the O blood group, with prevalence rates of at least 50%. Strong hypotheses favor a selection pressure exerted by the plasmodial parasite on its host cell, and include i) the worldwide distribution of the ABO blood groups with a type O predominance in malarious regions of the world [16], ii) the fact that *Plasmodium falciparum* has substantially affected the human genome and was present when the ABO polymorphisms arose [17], iii) the associations of ABO blood groups and clinical outcome of malaria with the observation of a degree of protection conferred by blood group O against severe courses of the disease [18] and iv) the potential role that erythrocyte surface antigens may play in cytoadhesion of infected erythrocytes to micro vessel endothelia and in parasite invasion [19]. No molecular method is used for the determination of ABO blood groups, as hematological methods (Beth-Vincent and Simonin techniques) are both simple and robust.

G6PD is a cytoplasmic enzyme allowing cells to withstand oxidant stress. It is encoded by one of the most polymorphic genes in humans, located on the X chromosome. In Africa, G6PD is represented by three major variants, G6PD B (normal), G6PD A (90% enzyme activity) and G6PD A- (12% enzyme activity) [20]. The location of the G6PD gene on the X chromosome and the subsequent variable X-chromosome inactivation implies that the expression of G6PD deficiency differs markedly among heterozygous females and therefore that these females do not constitute a homogeneous group [21]. PCR-RFLP is used for the molecular determination of the predominant G6PD A- variant in sub-Saharan Africa: mutation 376 A>G responsible for the G6PD A electrophoretic mobility and mutation 202 G>A responsible for the A- deficiency, are determined by PCR amplification of exons 5 and 4 respectively, followed by restriction enzyme analysis, using *FokI* (376 A>G mutation) and *NlaIII* (202 G>A mutation). However, the 376 A>G mutation may also be associated with other deleterious mutations such as 542 A>T (G6PD Santamaria), 680 G>T or 968 T>C, revealed after electrophoretic migration of digested amplified products with *BspEI*, *BstNI* and *NciI* respectively.

Table 1 presents data obtained among healthy individuals in order to avoid distribution bias due to selection of genetic traits by secularly settled diseases such as malaria. No HbSS individual was recorded in the studies gathered in this Table, because of an age range beyond the life expectancy of most HbSS patients in developing countries. Since the G6PD A and B variants have almost the same enzyme activity, the patients were stratified into groups with normal (female BB, AB, AA and male B and A genotypes), heterozygous (female A-B and A-A genotypes) and homo-/hemi-zygous (female A-A- and male A- genotypes) state, corresponding to decreasing levels of G6PD enzymatic activity. Some research teams have extensively studied erythrocyte polymorphisms in relation to malaria morbidity, among children hospitalized at the Albert Schweitzer Hospital from Lambaréné, in the Moyen Ogooué province of Gabon. As these genetic traits strongly influence the distribution of the clinical pattern of malaria, their frequency distribution is not representative of the whole population, and therefore they could not be reported in Table 1.

Erythrocyte polymorphisms	**Prevalence rate (%)**		
	Gabon (Dienga)	**Cameroun (Ebolowa)**	**Republic of Congo (Brazzaville)**
ABO blood groups:	N = 279 [22] [23]	N = 1,007 [24]	N = 13,045 [27]
Group O	54	51	53
Group A	27	24	22
Group B	17	19	21
Group AB	2	6	4
HbS genotypes:	N = 279 [22] [23]	N = 240 [25]	N = 868 [28]
Hb AA	77	81	80
Hb AS	23	19	20
Hb SS	0	0	0
G6PD state:	N = 271 M & F [22] [23]	N = 561 M [26]	N = 398 M & F [29]
- Normal (genotypes BB, AB, AA, B & A)	78	93	68
- Heterozygous (genotypes A-B & A-A)	13	0	21
- Homo-/hemi-zygous (genotypes A-A- & A-)	9	7	11

M: males; F: females.

Table 1. Erythrocyte polymorphisms among healthy individuals from Central Africa

Other erythrocyte polymorphisms characterize the sub-Saharan populations, including Central Africans. It is the case of the alpha-thalassemia, which consists in the deletion of 1, 2, 3 or the 4 genes encoding the alpha chain of the globin. Several forms of alpha-thalassemia are distributed worldwide, and the form encountered in sub-Saharan Africa resides in a gene deletion of 3.7 kb ($-\alpha^{3.7}$ type), which generates the formation of a functional hybrid gene. A PCR amplification strategy using three primers allows to determine the normal ($\alpha\alpha/\alpha\alpha$), heterozygous ($-\alpha^{3.7}/\alpha\alpha$) and homozygous ($-\alpha^{3.7}/-\alpha^{3.7}$) state as well as the $--/-\alpha^{3.7}$ form (H haemoglobin) [30]. The prevalence of α^{+}-thalassemia in Africa ranges from 5 to 50%, according to a gradient from North Africa to equatorial Africa and from South Africa to equatorial Africa: so, the highest prevalence rates are reached in the Central African Republic [31] and in a Bantu population from the republic of Congo [32]. Different erythrocyte polymorphisms may coexist in the same individual, as the results of advantageous interactions. Namely, a beneficial effect of α^{+}-thalassemia on the hematological characteristics of sickle-cell anemia patients has been found, in accordance with the observation in HbAS individuals of decreasing values of HbS quantification accompanying decreasing numbers of α-globin genes (from 4 to 2) [32].

2.3 Innate immunity

For the needs of malaria-linked studies, polymorphisms of some products of the inflammatory response have been investigated among populations from Central African countries.

Mannose binding lectin (MBL) is a member of the collectin family of proteins, which are components of the innate immune system, acting therefore against multiple pathogenic organisms. MBL is thought to be more effective at an early age, before effective acquired immune responses have developed, and low plasma concentrations of non-functional MBL have been attributed to mutations in the first exon of the MBL gene: $MBL_{IVS\text{-}I\text{-}5}$ G>A, MBL_{54} G>A and MBL_{57} G>A. PCR-RFLP determination may be performed, using *NlaIII (*codon 52), *BanI* (codon 54) and *MboII* (codon 57) endonucleases. At least one MBP gene mutation was present in 34% of a Gabonese population sample (Banzabi), with an overall gene frequency of 0.03, 0.02 and 0.18 mutations at codons 52, 54 and 57, respectively [22, 25]. There are other published MBL2 genotyping techniques, based on sequence-specific PCR, denaturing gradient gel electrophoresis of PCR-amplified fragments, real-time PCR with the hybridization of sequence-specific probes and sequence-based typing. A new strategy that combines sequence-specific PCR and sequence-based typing (Haplotype Specific Sequencing or HSS) was recently improved and allowed identification of 14 MBL allele-specific fragments (located in the promoter and exon 1) among Gabonese individuals [33].

Inducible nitric oxide synthase 2 (NOS2) is the critical enzyme involved in the synthesis of nitric oxide, a short-lived molecule with diverse functions including antimalarial activity, that can also cause damage to the host cell. The most investigated polymorphism is located in the promoter region of NOS2, and concerns the point mutation $NOS2_{-954}$ G>C, which is associated with an increased production of NOS2. By the means of a PCR amplification followed by enzymatic digestion with *BsaI*, this point mutation was found in 18% of Gabonese individuals from the Banzabi ethnic group, mainly in the heterozygous state [22, 25]. A similar high prevalence was found in another Gabonese population group, recruited in Lambaréné [34].

Tumor necrosis factor α (TNF-α) is a proinflammatory cytokine that provides rapid host defense against infection but is detrimental or fatal in excess. The main studied polymorphisms are located in the promoter region of the gene and are TNFα_{-308} G>A and TNFα_{-238} G>A base substitutions. These two polymorphisms have not been related to any variation in cytokine production, but may serve as markers for a functional polymorphism elsewhere in the TNF-α gene. Indeed, the TNFα_{376} A allele (G>A substitution), which is frequently found in linkage disequilibrium with TNFα_{-238} A allele, is related to enhanced secretion of TNF and might be responsible for increased antigen- or T-cell mediated B-cell stimulation and proliferation [35]. Molecular determination is assessed by PCR-RFLP using *NcoI* (-308), *AlwI* (-238) and *FokI* (376) restriction endonucleases. Prevalence rates of 22% (TNFα_{-308} A allele) and 17% (TNFα_{-238} A allele) were found in a Gabonese population (Banzabi), mainly in the heterozygous state [22, 25].

Haptoglobin (Hp) is an acute-phase protein that binds irreversibly to hemoglobin (Hb), enabling its safe and rapid clearance. Therefore, Hp has an important protective role in hemolytic disease because it greatly reduces the oxidative and peroxidative potential of free Hb. Haptoglobin exists in three phenotypic forms: Hp1-1, 2-1, and 2-2, which are encoded by two co-dominant alleles, Hp^1 and Hp^2. A fourth phenotype HpO, referred to as hypo- or an-haptoglobinaemia has been reported to be the predominant phenotype in West Africa. Functional differences between the different Hp phenotypes have been reported, the ability to bind Hb being in the order of 1-1 > 2-1 > 2-2. The gene frequencies of different Hp phenotypes show marked geographical differences as well as large variations among different ethnic groups. Hp genotypes determined by PCR in 511 Gabonese children from the village of Bakoumba (South-East of Gabon), distributed into 36.5%, 47.6% and 15.9% for Hp1-1, Hp2-1 and Hp2-2 respectively [36]. In South-West Cameroon, the genotype distribution among 98 pregnant women was 53% for Hp1-1, 22% for Hp2-1 and 25% for Hp2-2 [37].

2.4 Polymorphism of the cytochrome P450 superfamily

The DNA samples of the Gabonese individuals from the Banzabi ethnic group already described [2] entered a dataset of DNA samples from European (French Caucasians), African (Senegalese), South American (Peruvians) and North African (Tunisians) populations, in order to evaluate the inter-ethnic variations in the genetic polymorphism of several components of the cytochrome P450 superfamily (CYP) which gathers a large and diverse group of enzymes (Table 2). The function of most CYP enzymes is to catalyze the oxidation of organic substances. Their substrates include metabolic intermediates such as lipids and steroidal hormones, as well as xenobiotic substances such as drugs and other toxic chemicals. The investigation of the variable number of tandem repeat (VNTR) polymorphism of the human prostacyclin synthase gene (CYP8A1) revealed a particular distribution of the nine characterized alleles in the Gabonese population group, which may be associated with a more frequent and severe form of hypertension found in some Black populations [38]. The frequencies of three single nucleotide polymorphisms occurring in the CYP2A13 were determined by PCR-single strand conformational polymorphism (PCR-SSCP) (578C>T (Arg^{101}Stop)) and PCR-RFLP (3375C>T (Arg^{257}Cys) and 720C>G (3'-untranslated region)) and were respectively 0, 15.3 and 20.8 among the Gabonese group, differing from those of other groups under comparison: these marked inter-ethnic variations in an enzyme involved in the metabolism of compounds provided by the use of tobacco, have consequences on the susceptibility to lung cancer [39]. More precisely, it appears that black populations could present a higher deficit in CYP2A13 activity compared with other population groups, compatible with a reduced risk for smoking-related lung adenocarcinoma. In the same way, a frameshift mutation, responsible for the synthesis of a truncated protein of the CYP2F1, which activity in lung tissue is linked to carcinogenic effects, was mostly represented in the Gabonese population sample [40]. The genetic polymorphism of the CYP3A5 enzyme, implicated in the metabolism of chemotherapeutic agents but also toxins, was analyzed using a PCR-SSCP strategy, leading to the observation of great inter-ethnic differences in the distribution of a maximum of 17 alleles, some of them being linked to the synthesis of a non functional enzyme. According to the determination of the CYP3A5 predicted phenotype, Gabonese individuals were the most numerous (90.0%) to express a complete and functional CYP3A5 protein compared to French Caucasians (10.4%) and Tunisians (30.0%) [41]. The CYP4A11 enzyme is involved in the regulation of the blood pressure in the kidney, and an 8590T>C mutation has been associated to an increased prevalence of hypertension. Using PCR-SSCP and nucleotide sequence analysis, the frequency of this mutation was found lower in Gabonese compared to other investigated African population groups (Tunisians, Senegalese) [42]. Lastly, 3 single nucleotide polymorphisms (SNPs) affecting the human type II inosine monophosphate dehydrogenase (IMPDH2) gene have been determined by PCR-SSCP. This enzyme participates in the metabolism of purines and constitutes a target for antiviral drugs. It resulted that African

P450 Tissue location	**Clinical implication**	**Gene polymorphism**	**DNA samples origin (n)**	**Reference**
CYP8A1 Ovary, heart, skeletal muscle, lung and prostate	Pathogenesis of vascular diseases	9 VNTRs in the 5′-proximal regulatory region of the *CYP8A1* gene	European (78 French Caucasians); African (50 Gabonese and 50 Tunisians)	[38]
CYP2A13 Lung tissue	Susceptibility of tobacco-related tumorigenesis	3 SNPs: 578C>T (exon 2), 3375C>T (exon 5) and 720C>G (3′UTR)	European (52 French Caucasians); African (36 Gabonese and 48 Tunisians)	[39]
CYP3A5 Liver	Metabolism of chemotherapeutic agents and toxins	17 SNPs on the 13 exons of the *CYP3A5* gene	European (51 French Caucasians); African (36 Gabonese and 36 Tunisians)	[41]
CYP2F1 Lung tissue	Metabolism of pneumotoxicants with carcinogenic effects	Frameshift mutation in *CYP2F1* exon 2 (c.14_15insC)	European (90 French Caucasians); African (32 Gabonese, 37 Tunisians and 75 Senegalese)	[40]
CYP4A11 Liver and kidney	Regulation of blood pressure in the kidney	1 SNP on *CYP4A22*-exon 11: 8590T>C	European (99 French Caucasians); African (36 Gabonese, 53 Tunisians and 50 Senegalese); South American (60 Peruvians)	[42]

VNTR: variable number of tandem repeats; SNP: single nucleotide polymorphism; 3′UTR: 3′ untranslated region

Table 2. Genetic polymorphisms in enzymes of the cytochrome P450 superfamily (CYP), in diverse populations including Gabonese

population groups (Tunisians, Gabonese, and Senegalese) presented a higher IMPDH2 activity than Caucasians, with implications for the dose requirement of IMPDH2 inhibitors administered to patients [43].

This compilation of genetic data on populations from Central Africa is far from being exhaustive. As an example, the genetic polymorphism of Toll-Like Receptors (TLR) is to date extensively explored in order to deepen the understanding of the first steps of the immune recognition. Also, cytokines that regulate adaptive immune responses (humoral immunity and cell-mediated immunity) may present inter-individual genetic variations such as it is the case for IL-2, IL-4, IL-5, IFN-gamma, TGF-beta, LT-alpha or IL-13. Finally, increasing information is generated every day thanks to equipments (such as real-time PCR systems or DNA sequencers) that allow handling simultaneously a great number of biological samples. Altogether, this review of genetic data gathered during the last twenty years among Central African populations, illustrates in which point Africa, which is thought to be the homeland of all modern humans, is the most genetically diverse region of the world.

3. Animals

Methods used to infer the respective role of historical, environmental and evolutionary processes on animal distribution are related to the molecular ecology field and, as such, very similar to those employed to study plant dynamic (see section 4.). For animal, sequence of genes of mitochondrial DNA (mtDNA) such as cytochrome b or control region genes are largely used in phylogenetic and phylogeographic studies. The evolutionary pace of mitochondrial genomes being relatively fast, mtDNA sequences can also be used in population genetics study even if nuclear markers (microsatellites, SNP, etc.) provide a higher level of information.

3.1 Species identification from fecal pellets

The inability to correctly identify species and determine their proportional abundance in the wild is of real conservation concern, not only for species management but also in the regulation of illegal trade. However, estimating species abundance using classical ecological methods based on direct observation is very challenging in Central Africa. Indirect methods based on animal tracks, especially fecal pellets have been proposed; however pellets of parapatric related species are sometimes very similar and difficult to use to reliably differentiate species in the field. To address this problem, a PCR-based method has been proposed to differentiate Central African artiodactyls species and especially duikers (*Cephalophus*) from their fecal pellets [44]. In this purpose, a mtDNA sequence database was compiled from all forest *Cephalophus* species and other similarly sized, sympatric *Tragelaphus*, *Neotragus* and *Hyemoschus* species. The tree-based approach proposed by the authors is reliable to recover most species identity from Central African duikers.

3.2 Rivers are playing a major role in genetic differentiation for large primates in central Africa

For both Gorillas (*Gorilla gorilla*; [45, 46]) and Mandrills (*Mandrillus sphinx*; [47]) phylogeographic studies based on mtDNA (for both species) and microsatellite (only for Gorilla) markers have shown that rivers hamper gene flow among populations and have a major role in partitioning the species diversity. For Mandrills, the Ogooué river (Gabon) separates populations in Cameroon and northern Gabon from those in southern Gabon [47]. For Gorilla, rivers are more permeable and allow limited admixture among populations separated by waterways [45]. Anthony et al. also showed that like for plant species (see section 4) past vicariance events and Pleistocene refugia played an important role in shaping genetic diversity of current Gorilla populations [45].

3.3 Central African elephants: Forest or savannah elephants?

Despite their morphology typical from forest elephants, a genetic study based on mtDNA [48] shows that Central African elephants are sharing their history with both forest and savannah elephants from Western Africa. It also gives evidence that Central African forest populations show lower genetic diversity than those in savannahs, and infers a recent population expansion. These results do not support the separation of African elephants into two evolutionary lineages (forest and savannah). The demographic history of African elephants seems more complex, with a combination of multiple refugial mitochondrial lineages and recurrent hybridization among them rendering a simple forest/savannah elephant split inapplicable to modern African elephant populations.

4. Plants

4.1 Methods and approaches

This paragraph is giving on overview of approaches and methods related to the molecular ecology field and used to study natural or human-induced dynamic of plant species in Central Africa. Acknowledging the past history of the Central African forest domain is crucial for our understanding of spatial and temporal evolution of species living throughout the region.

Historical processes responsible for the contemporary distributions of individuals can be studied within the field of historical biogeography or phylogeography. For phylogeographic studies the distribution of genetic lineages within or among closely related species is considered throughout the geographical space and current patterns are interpreted in light of past vicariance events, population bottleneck, survival in glacial refugia and/or colonization routes [49, 50, 51]. This approach can be combined with landscape genetic methods to respectively infer impact of historical and environmental processes on the distribution of the genetic diversity. Landscape genetic methods allow to correlate the distribution of the genetic diversity with environmental parameters and to reveal, for example, the impact of topographic features on gene flow or the role of soil heterogeneity in structuring the genetic diversity [52]. At finer scales, classical population genetic approaches address the role of additional evolutionary forces (drift, dispersal, mutation, mating system, etc.) in shaping current patterns. All these genetic-based approaches belong to the molecular ecology field and are combined to address questions linked to the natural species dynamic or more importantly, questions linked to the survival of threatened species facing forest fragmentation, logging activities, etc.

All these approaches primary necessitate analyses of the genetic diversity at individual level. In this purpose, various techniques based on PCR are used. Different genetic markers can be chosen based on their respective evolutionary properties. For analyses of large-scale patterns, sequences of cytoplasmic DNA (ctDNA) like chloroplastic DNA (cpDNA) for plants are chosen. Cytoplamic DNA are haploid, non-recombining (or recombination events are rare) and generally characterized by uniparental inheritance (chloroplasts are generally maternally inherited for angiosperm, paternally for gymnosperm plant species). These markers allow inference in genealogical histories of individuals, populations and/or species. It is however highly recommended to combine cytoplasmic with nuclear markers for intraspecific phylogeographic studies because of the uniparental inheritance of ctDNA. It is especially true for species with sex-biased dispersal capacities. For instance, cpDNA would show a very strong spatial structure for tree with heavy barochore (dispersed by gravity) seeds whereas nuclear genes dispersed by both seed and anemophilous (transported by wind) pollen, would not reveal any spatial structure. Therefore, sequences from nuclear genes could provide valuable information in phylogeographic assessments. They are nonetheless more complicated to analyze because of i) the difficulty to isolate haplotype from diploid organisms, ii) intragenic recombination and iii) the relatively slow pace of sequence evolution at most nuclear loci. Other nuclear PCR-based genetic markers such as microsatellites, AFLP (Amplified Fragment Length Polymorphism), RAPD (Random Amplification of Polymorphic DNA) or SNPs (described in section 2.4) are also used for phylogeographic studies, most of them being particularly valuable for population genetic studies.

4.2 Importance of the past climatic changes in shaping pattern of genetic diversity in Central Africa

The Lower Guinea forest domain (the Atlantic coastal forest distributed from Nigeria to Congo) has undergone major distribution range shifts during the Quaternary, but few studies have investigated their impact on the genetic diversity of plant species. Several phylogeographic studies using either cpDNA polymorphism [52, 53, 54, 55, 56, 57] and/or nuclear markers such as RAPD [58] and microsatellite markers [53, 59, 60] have recently been published, considering Central African trees as model species, to give insight into the historical biogeography of the region. For most of the studied species, the genetic diversity is very spatially structured throughout the species distribution giving strong phylogeographic signals. These results show that the Central African rainforest domain was very fragmented during the cool and dry periods from the Last Glacial Maximum period at the end of the Pleistocene (20000-13000 years before present) and more recently during the Little Ice Age (about 4000-2500 years before present). During these periods, most tree species and probably forest species in general, only survived in a reduced number of isolated populations in areas where environmental conditions remained suitable. The question is now to test for the presence of forest refugia in Central Africa, in other words: did forest-species all survived in the same areas? In this case, effort for the conservation of these areas must be treated with the highest priority as refugia may play a major role in the survival of forest-species, while climate is changing, probably in buffering effect of the fluctuations. First results show that some refugia were shared among several tree species with one main refugium in the North and one in the South of the thermal equator (e.g. *Milicia excelsa* in [53], *Erythrophleum suaveolens* in [55], *Irvingia gabonensis* in [56], *Distemonanthus benthamianus* in [60]. Other species managed to survive in additional areas with at least four remaining populations for *Aucoumea klaineana* in Gabon [59]. More species covering all functional groups (pioneer, understorey, long-lived, etc.) must be studied to be able to infer general trends to allow predictions about impact of the Global Climate Change on species distribution.

4.3 Importance of species life history traits in the maintenance of genetic diversity

At finer scale, microsatellite loci were used to infer species dispersal ability of threatened tree species. *Baillonella toxisperma* Pierre Sapotaceae is a very low-density tree. The species is insect-pollinated and its seeds are dispersed by animals, including elephants. Using spatial genetic structure analyses, Ndiade-Bourobou et al. were able to demonstrate that dispersal distances were uncommonly high and able to connect trees present in very low density throughout forest [61]. This process allows the maintenance of high genetic diversity in reducing inbreeding effect and assures as such the survival of the species. This equilibrium is very vulnerable as both tree and animal-vectors densities have dramatically dropped due to additional effects of logging, hunting and poaching activities. For *Aucoumea klaineanea* Pierre Burseraceae, a highly logged tree species in Gabon, Born et al. show that dispersal distance is very limited and that founder effects associated with colonization processes are avoided by the homogeneity in reproductive success in adults [62]. Their results also suggest [63] that reduced density of trees and/or forest opening is balanced by higher gene dispersal distances. This result is linked with dispersal syndromes of the species that locally contribute to the maintenance of the genetic diversity.

5. Pathogens

A lot of diseases of animal origin and their rapid spread and possible transmission to humans (HIV/AIDS, Ebola, Avian Influenza, etc.) can pose a threat to human health. Tools have evolved from simple serological screenings to specific amplification using conventional or Real Time PCR methods, hence allowing more suitable diagnostic methods for early stage detection, identification and characterization of emerging or re-emerging pathogens. We'll successively take examples of pathogens infecting i) humans (parasites, viruses, bacteria, in section 5.1), non-human primates and other animals (section 5.2), and finally pathogens of plants (section 5.3).

5.1 Pathogens in humans

5.1.1 Parasites

Health in Central Africa is triggered by malaria, the most studied human parasite. Malaria transmission remains holoendemic in Central Africa in spite of decades of efforts in implementation/operational research. Other parasitic diseases are of utmost importance in term of public health, as human African trypanosomiasis (or sleeping sickness), filariasis, intestinal parasites, schistosomiasis, toxoplasmosis and amibiasis; however, they are all considered as neglected diseases. The PCR techniques contribute to the diagnostic of these infections. These techniques also improve our understanding of the physiopathology of these diseases through basic research. PCR indubitably helps to diagnose more efficiently and to find new therapeutic strategies.

5.1.1.1 PCR and diagnostic for human parasites in Central Africa

The Table 3 shows a few examples of PCR-based diagnostics for human parasites, although these techniques are not the gold standard for diagnosis of human parasites. The high cost of the PCR-based techniques is mainly mentioned as inconvenient. New diagnostic techniques should be implemented once it's demonstrated that the balance cost/benefit is lower than 1. First, the technique must be feasible in routine laboratories in terms of equipment and training of local agents. Secondly, the new technique has to offer a benefit in terms of clinical treatment of the patients. This clinical benefit may result in a better specificity and sensitivity, and in a reduced time to diagnosis. The improvement of sensitivity allows the detection of sub-microscopic infections, as detailed in the chapter of this book titled "Submicroscopic infections of human parasitic diseases" by Touré-Ndouo.

The main advantages of diagnosis by PCR for human parasites from Central Africa are both the higher specificity and the small amounts of blood or tissue required. The specificity of DNA sequences offers a simple tool to distinguish species. As an example, the species spectrum of intestinal parasites involved in hospitalized AIDS patients was determined in the Democratic Republic of the Congo [64]. Opportunistic infections were detected by PCR, as *Cryptosporidium* sp., *Enterocytozoon bieneusi*, *Isospora belli* and *Encephalitozoon intestinalis*. The other intestinal parasites detected by PCR were *Entamoeba histolytica*, *Entamoeba dispar*, *Ascoris lumbricoides*, *Giardia intestinalis*, hookworm, *Trichiuris trichiura*, *Enterobius vermicularis*, and *Schistosoma mansoni*. Furthermore, the PCR-based diagnostic is quite more sensitive than microscopic examination, which is sometimes not sufficient to differentiate various parasite species. This is clearly the case for filariasis [65] and schistosomiasis [66]. In human sleeping sickness, PCR on blood allows avoiding painful lumbar punctures and was proposed as a less-invasive alternative to replace the cerebrospinal fluid examination. However, in this case, PCR is a good tool for primodiagnostic but cannot be used for post-treatment follow-up. Indeed, the high sensitivity of PCR leads to detection of persisting DNA in blood of patients even after successful treatment [67].

<table>
<tr><th></th><th>Se.*</th><th>Spe.*</th><th>Advantage</th><th>Inconvenient</th><th>Ref. technique</th><th>Reference</th></tr>
<tr><td>Plasmodium spp (qPCR)§</td><td>99.40%</td><td>90.90%</td><td>Limit of detection greatly reduced</td><td>High cost</td><td>Microscopy examination of thick and thin blood smears</td><td>[70]</td></tr>
<tr><td>T. brucei gambiense in blood by PCR</td><td>88.40%</td><td>99.20%</td><td>Non invasive</td><td>Not suitable for follow-up</td><td>Microscopic analysis of the CSF</td><td>[67]</td></tr>
<tr><td>L. loa, M. perstans and W. bancrofti by nested PCR</td><td>100%$</td><td>100%$</td><td>High se. and spe. for 3 filariosis co-endemic</td><td>Cost</td><td>Knott's concentration and microscopic examination</td><td>[65]</td></tr>
<tr><td>S. mansoni in fecal samples by qPCR</td><td>86.50%</td><td>100%</td><td rowspan="2">High spe. to distinguish species</td><td rowspan="2">High cost; Not intended for routine diagnostic</td><td>Microscopic examination of Kato</td><td rowspan="2">[66]</td></tr>
<tr><td>S. haematobium in fecal samples by qPCR</td><td>82.80%</td><td>100%</td><td>Microscopic examination of filtrated urine samples</td></tr>
</table>

*Se. sensitivity, Spe. Specificity, CSF cerebrospinal fluid
§ qPCR, quantitative polymerase chain reaction
$ 30% of samples not done by PCR

Table 3. Efficiency and characteristics of PCR-based diagnostic in several endemic human parasitosis that are prevalent in Central Africa

Malaria constitutes one of the major public health problems in Central Africa. As *Plasmodium falciparum* infection is deadly when untreated in children and pregnant women, its diagnostic has to be accurate and fast. At hospital level, where many malaria diagnostics are performed a day, cost/benefit may be convincing and PCR-based diagnostic may be implemented. However, the benefits linked to PCR-based diagnosis for malaria are the identification of the different *Plasmodium* species and a lower detection limit. This is not necessarily clinically relevant. In addition, the existence of alternative diagnostic techniques as rapid diagnostic tests (RDTs) based on immunochromatographic assays to detect specific *Plasmodium* antigens that are recommended by the WHO, increases the cost/benefit ratio for PCR [68, 69].

Finally, PCR-based diagnosis is a very good tool for epidemiological survey. It still needs improvement in terms of cost, feasibility and quickness to deserve an implementation in Central African routine laboratories.

5.1.1.2 PCR and research on human parasites in Central Africa

As malaria is the most prevalent infection in Central Africa with the higher mortality incidence, this part will focus on malaria. The aim of this part is to point out the central role of PCR techniques in malaria research performed in Central Africa, without providing an exhausting list of its applications. The Figure 1 summarizes the research applications in the malaria field related to PCR-based techniques.

Fundamental research

The link between fundamental and operational research is tight, particularly for pathologies like malaria that need field studies to confirm hypotheses. Molecular epidemiology for malaria parasite is an example of this tight link. The study of SNPs related to drug resistance in *P. falciparum* on a genome-wide scale in a diversity of strains from Africa provides information on the frequency of the studied SNPs. If drug resistance requires several SNPs and those naturally occurring SNPs are rare in most genes, it may last years for the parasite to acquire a drug resistant phenotype. So, it is important to know whether *P. falciparum* genome presents low or high level of SNPs in endemic areas. However, the generation of new *P. falciparum* variants encoding for different levels of SNPs can result of tandem repeats of similar sequences (called RATs) that could undergo slip-strand mispairing. Replication slippage or deletion mechanisms lead to the apparition or lost of different RATs. Interestingly, the high frequency of RATs close to drug resistance or immune response target sequences can result in a fast increase of important SNPs (reviewed in [71]).

The development of new diagnostics for malaria is also dependant of PCR-based techniques. The first RDTs for malaria were supplied more than 15 years ago. Some of them are based on immunochromatographic detection of *P. falciparum* histidine-rich protein 2 (PfHRP2), using monoclonal antibodies. PfHRP2 is an abundant circulating protein easily detectable in the blood of patients. However, some studies reported variable test performances. In that way, complementary studies were necessary to compare the PfHRP2 sequences from several parasite strains and the potential consequences on the performance of PfHR2-based RDTs. The genetic diversity of the *pfhrp2* gene was studied in isolates originating from 19 countries including Central African countries and the relationship between the *pfhrp2* diversities and the sensitivities of PfHRP2-based RDTs was assessed [72]. The results indicated that 2 types of repeats in the DNA sequence of PfHRP2 were predictive of RDT detection sensitivity with 87.5% accuracy. These results pointed out the importance of the genetic background of the parasites and their diversity in the different geographic endemic areas.

Parasite antigen diversity studies at the molecular level are also performed for vaccine research. *P. falciparum* erythrocyte membrane protein 1 (PfEMP1) is a major vaccine target as evidence supports the central role of PfEMP1 in the development of a protective acquired immunity in children and pregnant women living in high level endemic areas. However, PfEMP1 undergoes a serious problem. PfEMP1 is highly polymorphic and encoded by a gene family of 50-60 *var* genes. To identify specific *var* genes or domain structuring these genes and related to protective immunity, many molecular studies were done and are still currently performed, all based on the basic molecular technique, PCR. In pregnancy-associated malaria, some studies showed that the *var* gene expressed called *var2csa* is relatively conserved. A comparative study showed that Duffy binding-like domains from placental parasites from Gabon and Cameroon shared 85%-99% amino-acid identities, confirming the conserved nature of placental variants [73]. This demonstration of sequence conservation in PfEMP1 DNA and its implication in the binding to chondroitin sulfate A (CSA) and to the pathology was clearly relevant to vaccine development for pregnancy-associated malaria. Today, it is largely recognized that the parasite ligand mediating CSA binding and causing malaria in pregnancy is VAR2CSA, a member of PfEMP1 family, and that it is a promising target for vaccine design. Recent researches focus on the molecular variability of *var2csa* in field isolates and on the immune response induced by different domains of the protein. Vaccine research largely depends on immunological studies, as this is clearly the case with the example of PfEMP1. However, PCR is not the favorite technique for such studies unlike flow cytometry or Enzyme Linked Immunosorbent Assay (ELISA). For immunological topics related to malaria, PCR is mainly used in studies on human genetic markers linked to malaria protection (see section 2 of this chapter).

Operational research

The evaluation of the therapeutic and control strategies implemented to fight against malaria constitutes operational research. First, PCR has become an essential technique for the evaluation of antimalarial treatment efficiency. Historically, *in vivo* resistance of *P. falciparum* to antimalarial drugs was classified into three grades, RI (low), RII (intermediate), and RIII (high) [74]. Since 2002, therapeutic failures are divided in early and late treatment failures (ETF, LTF), and LTF includes late clinical failures and late parasitological failures [75]. Both classifications are based on follow-up studies of parasitemia in patients treated with antimalarial treatments. Usually, follow-ups last 28 days, but are now extended to 42 days with the use of artemisinin-based treatment combinations (ACT) [75]. The classification relies on the reappearance or not of parasites during the follow-up. In highly endemic areas for malaria, the reappearance of parasites may be linked to the persistence of the initial infection, or to a new infection that occurred during the follow-up (the incubation time for *P. falciparum* is 7 to 10 days). A first study was performed in Central Africa in Gabon to demonstrate the great advantage of PCR to distinguish recrudescent *P. falciparum* clones from new ones, in studies of antimalarial treatment efficacy [76]. The technique involves amplification by PCR of regions of 3 highly polymorphic parasite genes, merozoïte surface protein-1 (*msp-1*), *msp-2* and glutamate-rich protein (*glurp*). Through this study, the authors showed that 39% of RI resistant cases were in fact due to new infections. Today, PCR genotyping is systematically included in treatment efficacy studies [75].

The implementation of therapeutic strategies for malaria in a specific area has an impact on the deployment of parasite resistance to the drug used. It is of high importance to study the development of parasite resistance in malaria endemic areas, in order to suggest new policies once treatments become inefficient. PCR is definitely the basic tool to perform such studies once molecular mechanisms of resistance have been demonstrated through more fundamental research. Sulfadoxine-pyrimethamine (SP) treatment has been used for a long time as second-line treatment for uncomplicated malaria in case of chloroquine treatment failure. The parasite mechanisms of resistance to SP have been well described and result in SNPs located on *Pfdhfr* and *Pfdhps* genes that appear in a few years following the implementation of such molecules. PCR followed by sequencing is the usual technique to study the rate of these mutations. In Gabon, Congo and Cameroon, the rate of *Pfdhfr* and *Pfdhps* mutations has been followed for years and constituted serious arguments to search other alternative treatments to chloroquine [77, 78, 79]. Since the era of ACT has begun, research teams based in Central Africa also use PCR-based techniques to follow the emergence of molecular markers related to the resistance to artemisinin-based molecules [80, 81].

Malaria prevention is also carried out through the use of insecticide treated materials or indoor residual spraying in Central Africa. This strategy has some implications on the spread of pyrethroid resistance in *Anopheles gambiae* and this has become a major concern in Africa. A PCR-RFLP assay was developed in Cameroon to follow two SNPs in the gene encoding subunit 2 of the sodium channel, also called the knockdown (*kdr*) mutations [82]. Since that time, studies to follow the situation of insecticide resistance are performed. In Gabon, both *kdr-e* and *kdr-w* alleles were shown to be present at high frequency in the *Anopheles gambiae* population. Of course, these results have implications for the effectiveness of the current vector control programmes that are based on pyrethroid-impregnated bed nets [83].

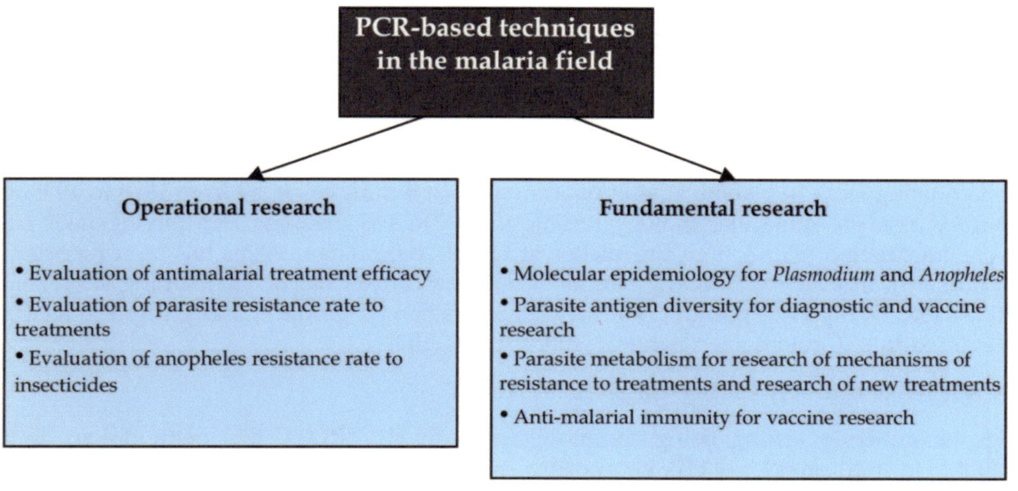

Fig. 1. The use of PCR-based techniques in the malaria field for operational and fundamental research

5.1.2 Viruses

This part will describe how the PCR-based techniques have been applied to many viruses infecting humans living in Central Africa, such as Human Immunodeficiency Virus (HIV), Human T cell Leukemia Virus (HTLV), Influenza virus, Hepatitis virus, and Ebola virus, for their origin, circulation, diversity, diagnosis, surveillance, and/or monitoring. Table 4 gives several examples of pathogens infecting humans in Central Africa, which have benefited from PCR technologies, with a particular emphasis on viruses.

5.1.2.1 Human Immunodeficiency Virus (HIV)

Central Africa has been described as the "epicenter of the HIV pandemic"[84]. Scores of articles have used PCR methods to report findings related to the viral diversity of HIV in this region, emergence of new strains [85] and recombinant forms [86], emergence of resistance to antiretroviral drugs [87], and challenges encountered for the genotyping tests because of the broad diversity of HIV strains [88]. In this section we'll explain the usefulness of PCR in i) the identification of various HIV strains found in Central Africa, ii) the early diagnosis of HIV, especially in exposed infants, iii) the management of infected patients, iv) implementation research and finally, we'll underline the need of an African AIDS vaccine.

PCR has help in the discovery and description of the virus

Since the discovery of HIV in the early 80s by Montagnier and Gallo, many strains, types, subtypes, circulating recombinant forms (CRFs) and unique recombinant forms (URFs) have been described and characterized in patients from the Central African region. The discovery of new HIV variants occurred by atypical serological reaction, and confirmation was obtained by simple PCR, nested PCR, heteroduplex mobility assay (HMA) (see Box 1) or sequencing. Particularly, full-length genomes sequencing has been instrumental in the characterization of new HIV CRFs, such as HIV-1 CRF 25_cpx [89] and CRF 22_01A1 [86, 90] in Cameroon. Obviously, the characterization of all these variants has an impact on HIV diagnosis, treatment and vaccine development, especially for the HIV-infected individuals leaving in Central Africa. The genetic diversity of HIV-1 group M in the republic of Congo was described and documented [91]. This was achieved using specific PCR coupled to HMA techniques of the *env* and *gag* genes (see Box 1). In Equatorial Guinea, Hunt et al. described the variability of HIV-1 group O, while Peeters et al. performed a wider study of HIV-1 group O distribution in Africa [92, 93, 94]. Although ELISA was mainly used in this latter study, indeterminate cases were solved using PCR. In Gabon, a great quantity of HIV strains collected from 1986 to 1994 was characterized by molecular biology techniques (PCR, sequencing); then phylogenetic trees were constructed [95]. A high prevalence of HIV-1 recombinant forms has been reported in Gabon [96]. In Cameroon, many studies have been carried out on genotyping subtypes of HIV-1 [86, 97, 98, 99]. Recently, new HIV-1 groups named group N and group P have been identified from Cameroonian patients [100, 101, 102, 103].

PCR is used routinely for the diagnosis of HIV

Despite antibody testing being commonly used in HIV RDTs, this methodology is not suitable in children born of HIV seropositive mothers during the first 15 to 18 months of life. The reason is that maternal antibodies transferred to the infant during pregnancy or breastfeeding persist up to 18 months and could give false positive results. Therefore, detection of proviral DNA by PCR is recommended for the early diagnosis in HIV-exposed infants. Detection of HIV proviral DNA is performed using the Roche Amplicor HIV-1 DNA commercial test, which is so far considered as the gold standard. This test reveals an HIV-1 infection within neonates and infants from 6 weeks of life and beyond. This test targets the *gag* gene during amplification where a fragment of 120bp is amplified and then, detection is based on ELISA. The kit is stored at 4°C and was especially designed for HIV-1 group M. Blood samples are collected as Dried Blood Spots (DBS), which have already been used for nationwide HIV prevalence survey in Africa [104]. More than 305,000 children in 34 countries worldwide have been offered early infant diagnosis (EID) and antiretroviral treatment thanks to the Clinton HIV/AIDS initiative (CHAI) and UNICEF, both managing the funds from UNITAID. The Amplicor HIV-1 DNA commercial test is currently used in several laboratories throughout Africa, and Cameroon is probably the leading country in Central Africa with a well-developed national EID programme, implemented by the Ministry of Public Health in the 10 regions of the country since 2007 [105].

PCR allows the management of HIV infection

Two main tests employing PCR techniques are useful for the biological follow-up of HIV-1-infected individuals i) the viral load (VL), which uses RNA PCR and ii) the resistance testing, which consists in amplification of specific viral fragments and sequencing. Viral load is mostly useful to follow the progression of the disease and for therapeutic monitoring as well. According to the commercial kits that are currently available, products of amplification can either be detected at the end of the reaction or while they accumulate in a real time manner. The lack of a commercially available viral load assay for HIV-2 is a concern for the proper management of patients infected with HIV-2 strains [106]. The resistance testing is actually an HIV-1 genotyping assay where the protease and the reverse transcriptase conserved regions of the *pol* gene are amplified and sequenced, as described by Fokam et al. [107]. Only two commercial tests approved by the Food and Drug Administration are currently available, and have been used widely to follow-up patients under antiretroviral treatment [108, 109, 110] and to report drug resistance mutations in HIV-1 reverse transcriptase or protease [109, 111, 112]. However, such commercial kits are very expensive for resource-limited countries like those of Central Africa and also their performance is questionable because of the great diversity of strains found in that region. For these reasons, an in-house genotyping assay has been developed in Cameroon recently and it is considered as more performant and cost effective than commercial kits [107].

The heteroduplex mobility analysis (HMA) is a molecular biology technique based on PCR amplification then followed by polyacrylamide gel electrophoresis analysis. This method has been first used for the subtype determination of HIV-1 group M envelope sequences, but has been recently developed for *gag* gene sequences.

Principle of the HMA test:
Heteroduplexes are formed with uncharacterized DNA fragments and known DNA sequences (as reference) included in the kit. Importantly, *env* gene fragments of uncharacterized DNA fragments are amplified by nested PCR whereas the reference sequences are obtained by direct amplification of plasmids from the kit.
Mobility of such heteroduplexes is analyzed on polyacrylamide gels. The closest is the unknown DNA sequence with the reference sequence; the fastest is the mobility of the heteroduplex on the polyacrylamide gel.

The HMA technique has been used to characterize HIV strains from Cameroon [1].

Box 1. Heteroduplex Mobility Analysis

The use of PCR in implementation research

Implementation research is essential for the control of infectious diseases of poverty [113]. Although PCR technologies are sophisticated and require a certain level of technical expertise and facilities that are usually not available and not affordable in poor-resources settings, implementation research studies can help to find alternative solutions. For example, the fact that DBS can replace blood samples advantageously has been instrumental in increasing access to HIV diagnosis in exposed infants living in remote settings, through the implementation and scale-up of the EID program [105]. Equally, DBS can improve the biological follow-up of HIV-1-infected individuals, both for the VL quantification and the resistance testing. Indeed, DBS, which can be collected on sites, transported and tested after a long-term storage, are suitable for the differed quantification of HIV-1 RNA, thus allowing people living with HIV/AIDS in rural areas to have access to this sophisticated test [114]. On another hand, implementation of resistance testing on DBS is in progress in Africa [115, 116] and will soon benefit HIV-1-infected patients living far from urban areas in Central Africa [108]. While waiting for the development of point of care assays, DBS appear to be a good alternative for the monitoring of HIV-1-infected people in remote settings (reviewed in [117]). However, the transport of samples and the return of results remains challenging, and need additional implementation research.

Back to the sites

Central Africa could be the ideal place where an AIDS vaccine could be designed, because of the great diversity of strains that are found in this region. However, when the scientific community is reflecting on how simian immunodeficiency virus infections hosted by African nonhuman primates could help in designing an AIDS vaccine for example, Central African scientists are absent [118]. This situation should change and African institutions, supported by their government, should advocate strongly for and invest in an African AIDS vaccine. To this end, the African AIDS Vaccine Partnership (AAVP) intends to promote cutting-edge research for the development of an African HIV vaccine [119]. In addition, the European Developing Clinical Trial Partnership (EDCTP) is supporting several African institutions from Gabon, Congo and Cameroon to build capacity for the conduct of future HIV/AIDS clinical trials [120] and is advocating for support from governments.

5.1.2.2 Human T cell Lymphotropic Virus (HTLV)

Central Africa is one of the few regions of the world where HTLV type 1 (HTLV-1) is highly endemic, as reviewed by Gessain & Mahieux [121]. Sequencing of HTLV-1 focuses on the gene *env* and the long terminal repeat fragments [122]. Molecular studies have demonstrated that the several molecular subtypes (genotypes) are related to the geographical origin and not to the disease. For example, while the subtype A is considered as cosmopolitan, the subtype B is mainly found in Central Africa (Democratic Republic of Congo, Gabon, and Cameroon). The subtype D has also been described in individuals from Cameroon, Gabon, Central African Republic, but less frequently than the subtype B, and more specifically in Pygmies. New subtypes (E and F) would be equally present in this region [121]. Interestingly, the first complete nucleotide sequence of HTLV type 2 (HTLV-2) has been obtained in a 44-year-old male living in a rural area of Gabon, by using nested PCR [123]. However, HTLV-2 does not seem to be as prevalent as HTLV-1 in this region since in a recent epidemiological survey performed on 907 pregnant women, only one case of HTLV-2 was reported [122]. In Cameroon however, HTLV-2 seroprevalence was 2.5% in Bakola Pygmies, but no HTLV-2 infection was found in Bantus [124]. HTLV type 3 (HTLV-3) and HTLV type 4 (HTLV-4) have been recently identified in primate hunters in Central Africa. Real-time PCR quantitative assays have been developed in the USA and allow detecting as few as 10 copies of HTLV-3 or HTLV-4 sequences of the gene *pol* in a small amount of DNA from human peripheral blood lymphocytes [125]. However, a new method using a single tube, multiplex, real time PCR has been developed at the Centre International de Recherches Médicales de Franceville (CIRMF), Gabon, which allows detecting HTLV-1, HTLV-2 and HTLV-3 simultaneously [126]. This new PCR-based technique could be of valuable use for epidemiological studies in countries where those viruses are prevalent.

5.1.2.3 Influenza virus

Despite influenza surveillance was increasing worldwide, developing countries in general and Central Africa in particular paid very little attention to the 2009 pandemic. Very recently however, samples from patients living with influenza-like illness in Yaounde, Cameroon were analyzed with various techniques including real time reverse transcription-polymerase chain reaction (RT-PCR) thus allowing the detection and subtyping of influenza A (H1N1 and H3N2) and B viruses from these patients [127]. Because of the H1N1 influenza A pandemic, Cameroon entered in a global surveillance network and received a laboratory equipped with a robust PCR platform for diagnosing influenza viruses in remote settings [128].

5.1.2.4 Hepatitis viruses

Hepatitis B virus (HBV) and hepatitis C virus (HCV) are endemic in the Central African region. Since the last two decades, the use of PCR techniques and phylogenetic analysis has led to characterize the genotype distribution of HCV in this area. The RNA is amplified by RT-PCR and nested PCR and the primers commonly used are specific to the 5'UTR and NS5B regions. In Cameroon, genotypes 1 and 4 are the most prevalent, but highly heterogeneous, with 5 subtypes 1, 4 subtypes 4 and unclassified subtypes, while the genotype 2 prevalence is low, with homogeneous sequences [129, 130]. Further work has help to understand the history of the HCV epidemic in Cameroon, where mass therapeutic or vaccine campaigns would have contributed to the spread of this infection during the colonial era [131]. In Gabon and Central African Republic, the predominance of the heterogeneous genotype 4 has been reported [132, 133, 134]. Equally, few HBV genotype studies have been conducted Central Africa. Makuwa et al. reported the identification of HBV-A3 in rural Gabon [135], while this genotype is co-circulating with HBV-E among Pygmies in Cameroon [136]. More recently, a pilot study was conducted in the village of Dienga, Gabon (previously described in section 2.1) with the aim of looking at potential interactions between HBV, HCV and *P. falciparum* infections, which are all very prevalent in this region [137]. In this study, HCV chronic carrier were identified by ELISA and by qualitative RT-PCR amplification of the 5' non coding region, and *P. falciparum* infection were assessed by microscopic examination and in case of negative result, by PCR targeting the gene encoding *P. falciparum* SSUrRNA, previously described by Snounou et al. [138]. Interestingly, these results showed that HCV infection may lead to slower emergence of *P. falciparum* in blood [137]. Other studies have demonstrated the usefulness of the PCR as a tool for the description of the molecular diversity of other less known/marginal viruses in this region, such as hepatitis delta virus in Cameroon [139] and in Gabon [140], or hepatitis GB-C/HG virus and TT virus in Gabon [141].

5.1.2.5 Ebola virus

Since the first declaration of deaths due to Ebola virus in Zaïre in 1976, the Central African region has been particularly affected by repeated Ebola outbreaks, which affected populations from Gabon and Republic of Congo in addition to the Democratic Republic of Congo. However, publications on the detection of Ebola virus in humans by molecular studies such as RT-PCR are scarce. The first reason is that infected patients have been reluctant to any type of invasive sampling method. The second is that for cultural reasons, families strongly refuse that researchers collect postmortem skin biopsies [142]. By analyzing few serum samples and less invasive specimens such as oral fluid samples, Formenty et al. could detect Ebola virus by RT-PCR and compare the two types of specimens [142]. This RT-PCR method has been developed, implemented and evaluated for diagnostics purposes at the CIRMF in Gabon, where a tremendous work is being done in the field of Ebola and other hemorrhagic fevers [143]. It is clear that the RT-PCR is the most appropriate tool not only to diagnose the infection in patients at a very early stage, but also to follow-up recovering patients [144]. Of note, studies were more easily carried out in animals, where important findings using PCR technologies were reported (see section 5.2).

In conclusion to this section on viruses, it is important to mention that new random priming methods adapted from the sequence independent single primer amplification (SISPA) technology are now available, and could be used to sequence whole genomes of all sorts of (known or unknown) RNA and DNA viruses [145]. This methodology, together with molecular clock analyses are needed to better understand the origin, circulation and diversity of all the viruses present in Central African populations.

5.1.3 Bacteria

In a review on the molecular epidemiology of bacterial infections in sub-Saharan infections, almost no information is reported from Central Africa [156]. Recently, molecular epidemiology methods have been applied to the genetic typing of *Mycobacterium tuberculosis* complex strains, the etiologic agents of tuberculosis, whose incidence is increasing dramatically in sub-Saharan Africa [157]. In 1993, a novel typing method called spoligotyping has been described [158]. This PCR-based method uses the DNA polymorphism of *M. tuberculosis* complex strains to detect and differentiate clinical isolates simultaneously, and allows their genotypic classification [159]. Briefly, this method aims at analyzing the so called DVR regions, which is composed of direct repeat (DR) regions, in which variable repeat sequences are inserted [160]. Spoligotyping, which is frequently compared to the conventional and more powerful RFLP method, remains a useful tool for genotyping clinical isolates in various epidemiological settings. In Cameroon, Niobe-Eyangoh et al. have used spoligotyping for analysis of hundreds of *M. tuberculosis* complex isolates from patients living in the West region [155]. This technique, which is considered as rapid, simple, and cost-effective, has been found accurate and easy to implement in that country, where the distribution of *M. tuberculosis* complex strains remains however still poorly documented, as well as the rest of Central Africa (see Table 4).

5.2 Pathogens in animals

Non-human primates from Central Africa have been extensively studied because it has been found that they are naturally infected with viruses or parasites similar to those affecting humans. The fact that humans are living in permanent contact with wild animals through hunting and butchering can explain transmission of pathogens from animals to humans.

Pathogen-genotype	Group/ Subtype	Regions (specific group)	Technique	Zone of amplification	References	Reviews
HIV-1	M/A,C, D, G, H, F, J, K, CRF01-AE	DRC	PCR & HMA	*env* V3-V5 region	[91]	[93] [146] [117]
	M/CRFs	Cameroon	Nested PCR	*gag*, *pol*, *env* genes	[86]	
	M/CRFs	South Est Gabon	PCR	*pol* gene	[147]	
	N	Cameroon	PCR	LTR-*gag*, *pol-vif*, *env* genes, entire genome	[101]	
	O	Cameroon, Equatorial Guinea	PCR Nested PCR	LTR-*gag*, *pol-vif*, *env* genes, entire genome	[94, 148]	
	P	Cameroon	RT PCR	*pol* integrase and *env* fragments	[100, 102]	
HIV-2		Equatorial Guinea	nested PCR	*pol* gene	[149]	
HTLV-1	A	Congo, DRC, Chad	nested PCR, PCR multiplex, real time PCR	gene *env* and LTR, gene *tax*	[150] [122] [126]	[121]
	B	DRC, Gabon, Cameroon, CAR				
	D	Cameroon, Gabon (Pygmies)				
	E	DRC				
	F	Gabon				
HTLV-2	Gab, B	Gabon Cameroon (Bakola Pygmies)	nested PCR, PCR, multiplex, real time PCR	entire proviral genome, gene *env* and LTR, gene *tax*, Long Terminal Repeats	[123] [122] [126] [124]	
HTLV-3		Gabon, Cameroon	multiplex, real time PCR, nested PCR	gene *tax* genes *tax* and *pol*	[126] [151]	[152]

Pathogen-genotype	Group/ Subtype	Regions (specific group)	Technique	Zone of amplification	References	Reviews
HTLV-4		South East Cameroon	nested PCR	gene *tax* genes *tax* and *pol*	[151]	
Influenza A Influenza B	H1N1 H3N2 B/Victoria/2/ 87 lineage and B/Yagamata/1 6/88 lineage	Cameroon	RT PCR	HA NA and M sequences	[127]	
HCV-1	1a, 1b, 1c, 1e, 1h, 1l	Cameroon South-West CAR	RT PCR & nested PCR	NS5b gene NS5b and E2 regions 5'UTR region	[129, 131, 132, 133]	
HCV-2	2f	Cameroon South-West CAR				
HCV-4	4e, 4f, 4k, 4c 4r, 4t, 4p, unclassified	Cameroon, South-West CAR, Gabon				
HBV	A3 E	Gabon, DRC, Cameroon (Pygmies) Cameroon (Pygmies)	Semi nested PCR	HBs (surface) gene	[135, 136]	
Ebola		DRC, Gabon, Congo	RT PCR	RNA polymerase L and NP genes	[142, 153]	[154] [143]
Mycobacterium	*tuberculosis*	Cameroon	spoligotyping	DVR region	[155]	
Plasmodium	*falciparum*	Gabon	PCR	SSUrRNA gene	[137, 138]	Touré-Ndouo, 2011 (chapter in this book)

HIV: Human Immunodeficiency Virus, HTLV: Human T cell Leukemia Virus, HCV: Hepatitis Virus C, HBV: Hepatitis Virus B, LTR: Long Terminal Repeats, CAR: the Central African Republic, DRC: Democratic Republic of Congo

Table 4. Examples of pathogens infecting humans in Central Africa, which have benefited from PCR technologies

5.2.1 Pathogens in non-human primates

A substantial proportion of wild-living primates in Central Africa are naturally infected with Simian Immunodeficiency Viruses (SIVs) [161, 162, 163], Simian T-cell Lymphotropic Viruses (STLVs) [164, 165, 166, 167], Simian Foamy Viruses (SFV) [168] and also Hepatitis B Viruses (HBV) [169].

SIVs are lentiviruses that are found naturally in a great variety of nonhuman primates from Equatorial Africa, including but no restricted to chimpanzees (SIVcpz), mandrills, (SIVmnd-1 and SIVmnd-2), drills (SIVdrl), talapoin monkeys (SIVtal), sun tailed monkeys (SIVsun), African green monkeys (SIVagm), red-capped mangabeys (SIVrcm) (see [162, 163, 170] and [171] for review). The evolutionary origins of these related viruses have been studied by amplification of the *gag*, *pol*, and *env* genes, and by construction and analysis of evolutionary trees. Sequence analysis of the entire genome and phylogenetic analyses have led to the identification of distinct primate lentivirus lineages in which most of the SIV strains described so far can be classified (see [171] and Table 5). The example of SIVs illustrates how the PCR techniques have been instrumental in the characterization of new strains of pathogens in non-human primates of Central Africa. As previously mentioned for animals (see section 3) phylogeographic studies have been equally carried out for pathogens. In mandrills for example, the two types of viruses appear to be geographically distributed, since SIVmnd-1 was found in mandrills from central and southern Gabon whereas SIVmnd-2 was identified in northern and western Gabon, as well as in Cameroon [163].

Other examples of pathogens in non-human primates from Central Africa could have been used, like the STLVs (the simian counterpart of HTLVs), the SFVs and/or HBV, which similarly to SIVs have been described and characterized with molecular techniques including PCR. With no pretention of being exhaustive, the Table 5 summarizes several examples of pathogens found in animals from this region, with the technique used, the gene amplified, and appropriate references for more details. Of note, molecular techniques adapted to non-invasive fecal samples have been pivotal to identify simian viruses in quite a number of species, especially in case of wild living primates.

These findings from Central Africa on pathogens in non-human primates together with those reported in humans, give a more comprehensive picture of the relationship between simian viruses and their counterpart in humans.

Indeed, the use of PCR related technologies and the clustering of sequences has helped to understand that i) cross species transmission of viruses (from non-human primates to humans) occurred in Central Africa through highly exposed population such as hunters and people handling primates as bush meat [164] and ii) species barriers could be easier to cross over than geographic barriers [165]. Taken together, these observations reveal that the risk of emergence of new viral diseases in Central Africa is still latent.

Similarly, various species of *Plasmodium*, including *P. falciparum* have been found in great apes (chimpanzees and gorillas) in Central Africa [172, 173]. If blood samples are not suitable for systematical analyses in primates, especially in case of wild primates; the identification of *Plasmodium* by PCR has been facilitated by the use of fecal primate samples, which are also broadly collected for the identification of simian viruses (see above). The identification of new species of *Plasmodium*, such as *P. gaboni*, which infects chimpanzees and *P. GorA* and *P. GorB*, which infect gorillas, has help to obtain a more comprehensive view of the phylogenetic relationships among *Plasmodium* species [173].

Pathogen-genotype	Subtype/ lineage	Regions (animals)	Technique	Zone of amplification	References	Reviews
Plasmodium	*gaboni*	Gabon (chimpanzees)	PCR	complete mitochondrial genome (including *Cyt b*, *Cox I* and *Cox III* genes)	[172]	[177]
	GorA *GorB*	Gabon (wild chimpanzees, wild gorilla, captive wild-born gorilla)	Plasmodium-specific PCR assay	mitochondrial *cytochrome b* gene	[173]	
	falciparum	Gabon (wild chimpanzees, gorilla)		nuclear and mitochondrial genomes	[177]	
SIV	SIVmnd-1 SIVmnd-2	Gabon (mandrills), Cameroon (mandrills)	PCR	entire genome	[178] [163]	[171]
	SIVtal	Cameroon (talapoin monkeys)	PCR	entire genome	[162]	
	SIVsun	Gabon (wild-caught sun tailed monkey)	PCR	entire genome	[161]	
	SIVrcm	Gabon (red capped mangabeys); Nigeria/Cameroon border (red-capped mangabeys)	PCR	entire genome	[170] [179]	
	SIVcpz	Cameroon, Gabon, DRC (chimpanzees)	PCR	entire genome	[180] [181] [182]	
STLV-1	D, F	Cameroon (agile mangabeys, mustached monkeys, talapoins, gorilla, mandrills, African green monkeys, agile mangabeys, and crested mona and greater spot-nosed monkeys); Gabon (mandrills)	Discriminatory PCR	LTR & *env* sequences	[164] [165]	
STLV-2		DRC (wild-living bonobos)	Generic PCR	*tax* gene	[183]	
STLV-3		Cameroon (agile mangabeys)	Discriminatory PCR	LTR & *env* sequences	[164]	

Pathogen-genotype	Subtype/lineage	Regions (animals)	Technique	Zone of amplification	References	Reviews
SFV	SFVcpz	Gabon, Cameroon (chimpanzees); Cameroon, CAR, Gabon, Republic of Congo, DRC (wild chimpanzees); Gabon (wild and semi-free ranging captive mandrills)	nested PCR RT PCR	integrase and LTR region *gag*, *pol*-RT and *pol*-IN LTR	[184] [185] [168]	
Ebola		Gabon (Fruit bats)	PCR	RNA polymerase	[153]	
Influenza	H5N1	Northern Cameroon (ducks)	PCR	NA sequences	[176]	

SIV: Simian Immunodeficiency Virus, STLV: Simian T cell Lymphotropic Virus, SFV: Simian Foamy Virus, LTR: Long Terminal Repeats, CAR: the Central African Republic, DRC: Democratic Republic of Congo

Table 5. Examples of pathogens infecting animals of Central Africa that have benefited from PCR technologies

By sequencing the complete mitochondrial gene or at least a part of the cytochrome b, and Bayesian or maximum-likelihood methods, phylogenetic analyses can be performed, hence allowing a better understanding of the origins and evolution of malaria parasites and possibly transmission between apes and humans [172].

5.2.2 Pathogens in other animal species

Apart from non-human primates, other animals from the Central African region have been studied for their possible implication in the life cycle of viruses causing hemorrhagic fever like Ebola or Marburg, which are both affecting great apes and humans. For example, sequences of Ebola were detected by PCR in small rodents and shrews, suggesting that common terrestrial small mammals living in peripheral forest areas may play a role in the life cycle of the Ebola virus [174]. More recently, Ebola and Marburg viruses were found in symptomless infected fruit bats in Central Africa, indicating that these animals could therefore act as the natural reservoir of these both viruses [153, 175].

In the context of outbreaks of highly pathogenic avian influenza, ducks from the far north region of Cameroon were found to host a highly pathogenic avian influenza subtype H5N1, whose sequence was closely related to H5N1 isolates reported in other African countries [176].

5.3 Pathogens in plants

For plant pathogen, PCR-based techniques are essentially used in two purposes: i) to identify pathogen species, comparing pathogen sequences to known pathogen sequence libraries or ii) to characterize pathogen colonization dynamic. One example of each application is summarized below.

5.3.1 Which fungi are attacking Central African *Terminalia* species?

Begoude et al. collected fungal inoculum on *Terminalia* in Cameroon to identify which pathogens are threatening these highly logged tree species. They compared DNA sequence for the ITS and tef 1-alpha gene regions to known pathogen libraries and showed that the majority of isolates are from the *Lasiodiplodia* genus [186].

5.3.2 The colonization dynamic of *Mycosphaerella fijiensis* in Cameroon

Dispersal processes of fungal plant pathogens can be inferred from analyses of spatial genetic structures resulting from recent range expansions. The fungus *Mycosphaerella fijiensis*, pathogenic on banana, is an example of a recent worldwide epidemic and is currently threatening Cameroonian banana plantations. Halkett et al. collected fungal isolates in Cameroon and analyzed them using 19 microsatellite markers. They demonstrated that large gene flows are linking populations even separated by long distances, through dense banana plantations, and so ensuring stable demographic regime and promoting efficient colonization dynamic of the fungus [187].

6. Opportunities and challenges

Some of the few research institutes and molecular biology laboratories that have been mainly involved in the findings reported above are the CIRMF (Franceville, Gabon), which is equipped with BSL3 and BSL4 facility, and the CIRCB (Yaounde, Cameroon), among others. Despite the amount of work and publications that have been generated from the Central African region, institutions and scientists involved in molecular biology research in Central Africa are facing several problems including procurement, maintenance, human resources, capacity building and ethics–related issues.

Obtaining the valuable results depends on multiple factors including methodology of sampling, processing, storage and shipment of samples to laboratory with respect of maintain of the cold chain. As described above, problems related to sampling were well circumvented with animals. Indeed, by using shed hair or feces, which are non invasive methods of sampling, phylogenetic analyses have allowed a better understanding of the evolutionary history of gorillas [46] mandrills [47] or elephants [48]. Equally, a number of simian viruses have been characterized in fecal samples, which is more convenient, especially in case of wild-living primates. In these contexts, new reagents such as the RNA later® have been very helpful to stabilize and protect RNA in fresh collected specimens, hence allowing an extended period of storage before processing the samples. In humans, the collection of samples via DBS is simple, convenient, and cost effective. Transportation does not require any cold chain, and storage is easier than samples obtained from whole blood. In the field of HIV, DBS are advantageous for the biological follow-up of infected patients living in remote areas [117].

Another issue, which has to be taken into consideration, is related to the issue of the quality control and quality assurance, which need permanent improvement and capacity building efforts. Due to limited resources and equipment, and possibly because the culture of research is still dramatically lacking in most of sub-Saharan African countries [188], only few laboratories have obtained certification and the roadmap to accreditation is still far ahead. Therefore there is an urgent need that institutions from Central Africa participate more in laboratory accreditation programs, with the goal of seeking lab accreditation and excellence in general. For example, the World Health Organization (WHO)-AFRO and the Center for Disease Control Global AIDS program have launched recently an accreditation program for quality improvement of African laboratories for HIV monitoring. However, such programs will also improve the monitoring of HIV-TB coinfected patients, and by extension, the follow-up of patients suffering from other diseases, such as malaria or any neglected disease. Equally, support from the EDCTP is currently helping African institutions -grouped in regional Networks of Excellence- to conduct future clinical trials in the four regions of sub-Saharan Africa. To achieve this goal, a lot of efforts have been put into building capacity of young African scientists and laboratories, which have to meet international standards and respect good clinical and laboratory practices [120].

Studies reported here have been carried out mainly in the framework of collaborative research with institutions from the North. However, DNA samples are often kept abroad, with the partners, without any signed material transfer agreement. In some other cases, African scientists and institutions from the region are not associated to the work and/or publications. The researcher's community has to be aware of avoiding the "banking" of DNA from African populations outside from Africa, mutualising benefits with the concerned populations and scientific partners as well as respecting ethical issues, such as establishing a fair partnership between African scientists and scientist from the North. The lack of these aspects have been demonstrated in a recent bibliometric review on human genetic studies performed during the two last decades in Cameroon [189]. Recently, the

African Society of Human Genetics launched the Human Heredity and Health in Africa (H3Africa) initiative, with the support of the National Institutes of Health and the Wellcome Trust (see http://h3africa.org/). The aim of this initiative, which was first discussed at the Yaoundé meeting in March 2009, is to conduct genomics-based research projects in Africa in order to better understand health and diseases in various African populations and to identify possible populations that are at risk of developing a specific disease. To this end, various calls for proposals have been launched, in which African institutions will take the leading role. One of these calls is the H3 Africa biorepository grant, which will address the need of biobanking samples in Africa for Africa. This H3Africa programme gives a lot of hope that capacity building and ethics-related will be soon addressed in favor of African institutions and African scientists and other scientists living in Africa, and that partnerships will eventually result in true win-win collaborations.

7. Conclusion

The contribution of PCR technologies to humans, animals, plants and pathogens from Central Africa is considerable, hence allowing the discovery of new species of plants and pathogens in this region, particularly in Gabon (see http://www.cirmf.org/en/publications). The richness of animals, plants, and pathogens is unquestionable and the Central African region is notorious for its great biodiversity.

In this chapter, a great number of PCR-based techniques have been described, including but not limited to PCR-restriction fragment length polymorphism, PCR using sequence-specific oligonucleotide probes, combination of sequence-specific PCR and sequence-based typing also called Haplotype Specific Sequencing, PCR-single strand conformational polymorphism, reverse transcriptase PCR, sequence independent single primer amplification technology, nested and semi-nested PCR, quantitative PCR, real time PCR, PCR multiplex, Heteroduplex Mobility Analysis, and spoligotyping. Applied to humans, these techniques have contributed significantly to increase the knowledge on human genetics, through immunogenetics and genetics epidemiology of infectious diseases. Particularly, a great number of molecular studies describe the genetic polymorphism of the various populations and ethnic groups living in this region (section 2). Applied to wild animals and non-invasive samples such as shed hair or feces, PCR technologies have for example facilitated the identification of related species, which are not easy to differentiate just by direct observation as done by ecologists, by using mitochondrial DNA (section 3). Applied to plants, PCR-based methods have contributed to a better understanding of spatial and temporal evolution of species found in that region, including colonization routes, and tree densities than can be modified because of activities of humans in that region (section 4). Finally, application of PCR technologies has been reported for pathogens infecting humans, animals and plants (section 5). Parasites, viruses, and bacteria that are prevalent in humans, non-human primates and other animal species, and fungal plant pathogens have been discovered and characterized through PCR-based techniques.

The PCR-generated knowledge is benefiting to a broad range of disciplines, such as genetics, molecular ecology, phylogeography, botany, evolution, molecular epidemiology, and infectious diseases, amongst others.

Altogether, these finding have contributed to a better understanding of the relationship between humans from Central Africa and their environment (animals, plants and pathogens), and particularly the inter relationship between species. Indubitably, this will be of help for a better management of resources at the global level. In addition, progresses have been made in fundamental research, operational research, and research applied to diagnostics and monitoring of infected individuals.

Challenges in conducting PCR-based research are procurement and storage of reagents and blood samples due to the cold chain, maintenance of equipment, as well as human resources, capacity-building and ethics-related issues. However, new initiatives such as those launched by the African Society of Human Genetics (H3 Africa), the AAVP (promoting an African AIDS Vaccine), and the EDCTP (supporting regional Networks of Excellence for the future conduct of clinical trials) are real opportunities for the scientific community that is working in Africa, to perform cutting-edge research where sophisticated molecular biology laboratories and bioinformatics platforms will be created/renovated and will complement each other.

In conclusion, despite a challenging research environment and though the paucity of facilities, scientists from Central Africa have brought a significant contribution to the scientific community, through PCR-related technologies. Collaborative research with northern partners has been fruitful and must be always conducted while keeping in mind a fair partnership and authorship. PCR-based research is increasing significantly in Central Africa and must be recognized at the level of the scientific community.

8. Acknowledgments

This paper has voluntarily been written by female scientists only, who have personally contributed to some of the findings presented in this chapter. All authors and individuals acknowledged below have been working or are currently working in Central Africa, particularly in Cameroon (at the CIRCB, Yaoundé and/or University of Buea) and Gabon (at the CIRMF, Franceville). We acknowledge Dr Mireille Bawe Johnson, Cardiff University, Biodiversity and Ecological Processes Group, Cardiff, UK, Dr Maria Makuwa, Laboratory Coordinator and Administrator at the Global Viral Forecasting Initiative (GVFI)/Institut National de Recherche Biomedicale (INRB), Kinshasa, Democratic Republic of Congo and Dr Lucy M. Ndip, head of the laboratory for Emerging Infectious Diseases, University of Buea, Cameroon for their contribution during the pre submission of this chapter. We also want to thank Dr Michaela Müller-Trutwin for her advice and Dr Sandrine Souquiere, for the critical reading of the manuscript. Finally, we are grateful to Mrs Clemence Rochelle Akoumba for her kind assistance in collecting some of the full papers referenced below, and Mrs Nchangwi Syntia Munung for her great help in managing references in Endnote®. Odile Ouwe Missi Oukem-Boyer is member of the Central Africa Network for Tuberculosis, HIV/AIDS, and Malaria (CANTAM) and of the Initiative to Strengthen Health Research Capacity in Africa (ISHReCA).

9. References

[1] Pasquier, C., et al. (2001). HIV-1 subtyping using phylogenetic analysis of pol gene sequences. *J Virol Methods*, 94, (1-2), 45-54, ISSN 0166-0934 (Print).

[2] Migot-Nabias, F., et al. (1999). HLA class II polymorphism in a Gabonese Banzabi population. *Tissue Antigens*, 53, (6), 580-585, ISSN 0001-2815 (Print).

[3] Bodmer, J. G., et al. (1997). Nomenclature for factors of the HLA system, 1996. *Tissue Antigens*, 49, (3 Pt 2), 297-321, ISSN 0001-2815 (Print).

[4] Tiercy, J. M., et al. (1992). HLA-DR polymorphism in a Senegalese Mandenka population: DNA oligotyping and population genetics of DRB1 specificities. *Am J Hum Genet*, 51, (3), 592-608, ISSN 0002-9297 (Print).

[5] Hammond MG, d. T. E., Sanchez-Mazas A, Andrien M, Coluzzi M, de Pablo MR, de Stefano G, Kaplan C, Kennedy LJ, Louie L, Migot F. (1997). HLA in sub-Saharan Africa (12th IHCW SSAF Report). . *Genetic diversity of HLA. Functional and medical implications.*, I, 345-353, ISSN 2-84254-003-4.

[6] Schnittger, L., et al. (1997). HLA DRB1-DQA1-DQB1 haplotype diversity in two African populations. *Tissue Antigens*, 50, (5), 546-551, ISSN 0001-2815 (Print).

[7] Pimtanothai, N., et al. (2001). HLA-DR and -DQ polymorphism in Cameroon. *Tissue Antigens*, 58, (1), 1-8, ISSN 0001-2815 (Print).

[8] Ellis, J. M., et al. (2000). Diversity is demonstrated in class I HLA-A and HLA-B alleles in Cameroon, Africa: description of HLA-A*03012, *2612, *3006 and HLA-B*1403, *4016, *4703. *Tissue Antigens*, 56, (4), 291-302, ISSN 0001-2815 (Print).

[9] Migot-Nabias, F., et al. (2001). HLA alleles in relation to specific immunity to liver stage antigen-1 from plasmodium falciparum in Gabon. *Genes Immun*, 2, (1), 4-10, ISSN 1466-4879 (Print).

[10] Spinola, H., et al. (2011). HLA Class-I Diversity in Cameroon: Evidence for a North-South Structure of Genetic Variation and Relationships with African Populations. *Ann Hum Genet*, 75, (6), 665-677, ISSN 0003-4800 (Print).

[11] Weatherall, D. J. (1987). Common genetic disorders of the red cell and the 'malaria hypothesis'. *Ann Trop Med Parasitol*, 81, (5), 539-548, ISSN 0003-4983 (Print).

[12] Lell, B., et al. (1999). The role of red blood cell polymorphisms in resistance and susceptibility to malaria. *Clin Infect Dis*, 28, (4), 794-799, ISSN 1058-4838 (Print).

[13] Bunn, H. F. (1997). Pathogenesis and treatment of sickle cell disease. *N Engl J Med*, 337, (11), 762-769, ISSN 0028-4793 (Print).

[14] Serjeant, G. R. (1997). Sickle-cell disease. *Lancet*, 350, (9079), 725-730, ISSN 0140-6736 (Print).

[15] Allison, A. C. (1954). Protection afforded by sickle-cell trait against subtertian malareal infection. *Br Med J*, 1, (4857), 290-294, ISSN 0007-1447 (Print).

[16] Cserti, C. M. and W. H. Dzik. (2007). The ABO blood group system and Plasmodium falciparum malaria. *Blood*, 110, (7), 2250-2258, ISSN 0006-4971 (Print).

[17] Kwiatkowski, D. P. (2005). How malaria has affected the human genome and what human genetics can teach us about malaria. *Am J Hum Genet*, 77, (2), 171-192, ISSN 0002-9297 (Print).

[18] Fischer, P. R. and P. Boone. (1998). Short report: severe malaria associated with blood group. *Am J Trop Med Hyg*, 58, (1), 122-123, ISSN 0002-9637 (Print).

[19] Udomsangpetch, R., et al. (1993). The effects of hemoglobin genotype and ABO blood group on the formation of rosettes by Plasmodium falciparum-infected red blood cells. *Am J Trop Med Hyg*, 48, (2), 149-153, ISSN 0002-9637 (Print).

[20] Ruwende, C. and A. Hill. (1998). Glucose-6-phosphate dehydrogenase deficiency and malaria. *J Mol Med (Berl)*, 76, (8), 581-588, ISSN 0946-2716 (Print).

[21] Beutler, E., et al. (1962). The normal human female as a mosaic of X-chromosome activity: studies using the gene for C-6-PD-deficiency as a marker. *Proceeding of the National Academy of Sciences of the United States of America*, 48, 9-16, ISSN 0027-8424 (Print).

[22] Migot-Nabias, F., et al. (2000). Human genetic factors related to susceptibility to mild malaria in Gabon. *Genes Immun*, 1, (7), 435-441, ISSN 1466-4879 (Print).
[23] Tagny, C. T., et al. (2009). [The erythrocyte phenotype in ABO and Rh blood groups in blood donors and blood recipients in a hospital setting of Cameroon: adapting supply to demand]. *Rev Med Brux*, 30, (3), 159-162, ISSN 0035-3639 (Print).
[24] Carme, B., et al. (1989). Clinical and biological study of Loa loa filariasis in Congolese. *Am J Trop Med Hyg*, 41, (3), 331-337, ISSN 0002-9637 (Print).
[25] Mombo, L. E., et al. (2003). Human genetic polymorphisms and asymptomatic Plasmodium falciparum malaria in Gabonese schoolchildren. *Am J Trop Med Hyg*, 68, (2), 186-190, ISSN 0002-9637 (Print).
[26] Le Hesran, J. Y., et al. (1999). Longitudinal study of Plasmodium falciparum infection and immune responses in infants with or without the sickle cell trait. *Int J Epidemiol*, 28, (4), 793-798, ISSN 0300-5771 (Print).
[27] Michel, R., et al. (1981). [Plasmodium falciparum and drepanocytic gene in Popular Republic of Congo. I. Prevalence of malaria and drepanocytic trait among school children in Brazzaville area (author's transl)]. *Med Trop (Mars)*, 41, (4), 403-412, ISSN 0025-682X (Print).
[28] Bernstein, S. C., et al. (1980). Population studies in Cameroon: hemoglobin S, glucose-6-phosphate dehydrogenase deficiency and falciparum malaria. *Hum Hered*, 30, (4), 251-258, ISSN 0001-5652 (Print).
[29] Bouanga, J. C., et al. (1998). Glucose-6-phosphate dehydrogenase deficiency and homozygous sickle cell disease in Congo. *Hum Hered*, 48, (4), 192-197, ISSN 0001-5652 (Print).
[30] Dode, C., et al. (1993). Rapid analysis of -alpha 3.7 thalassaemia and alpha alpha alpha anti 3.7 triplication by enzymatic amplification analysis. *Br J Haematol*, 83, (1), 105-111, ISSN 0007-1048 (Print).
[31] Pagnier, J., et al. (1984). alpha-Thalassemia among sickle cell anemia patients in various African populations. *Hum Genet*, 68, (4), 318-319, ISSN 0340-6717 (Print).
[32] Mouele, R., et al. (2000). alpha-thalassemia in Bantu population from Congo-Brazzaville: its interaction with sickle cell anemia. *Hum Hered*, 50, (2), 118-125, ISSN 0001-5652 (Print).
[33] Boldt, A. B., et al. (2009). Haplotype specific-sequencing reveals MBL2 association with asymptomatic Plasmodium falciparum infection. *Malar J*, 8, 97, ISSN 1475-2875 (Electronic).
[34] Kun, J. F., et al. (2001). Nitric oxide synthase 2 (Lambarene) (G-954C), increased nitric oxide production, and protection against malaria. *J Infect Dis*, 184, (3), 330-336, ISSN 0022-1899 (Print).
[35] Boussiotis, V. A., et al. (1994). Tumor necrosis factor alpha is an autocrine growth factor for normal human B cells. *Proc Natl Acad Sci U S A*, 91, (15), 7007-7011, ISSN 0027-8424 (Print).
[36] Fowkes, F. J., et al. (2006). Association of haptoglobin levels with age, parasite density, and haptoglobin genotype in a malaria-endemic area of Gabon. *Am J Trop Med Hyg*, 74, (1), 26-30, ISSN 0002-9637 (Print).
[37] Minang, J. T., et al. (2004). Haptoglobin phenotypes and malaria infection in pregnant women at delivery in western Cameroon. *Acta Trop*, 90, (1), 107-114, ISSN 0001-706X (Print).
[38] Chevalier, D., et al. (2002). Sequence analysis, frequency and ethnic distribution of VNTR polymorphism in the 5'-untranslated region of the human prostacyclin

synthase gene (CYP8A1). *Prostaglandins Other Lipid Mediat*, 70, (1-2), 31-37, ISSN 1098-8823 (Print).

[39] Cauffiez, C., et al. (2005). CYP2A13 genetic polymorphism in French Caucasian, Gabonese and Tunisian populations. *Xenobiotica*, 35, (7), 661-669, ISSN 0049-8254 (Print).

[40] Tournel, G., et al. (2007). CYP2F1 genetic polymorphism: identification of interethnic variations. *Xenobiotica*, 37, (12), 1433-1438, ISSN 0049-8254 (Print).

[41] Quaranta, S., et al. (2006). Ethnic differences in the distribution of CYP3A5 gene polymorphisms. *Xenobiotica*, 36, (12), 1191-1200, ISSN 0049-8254 (Print).

[42] Lino Cardenas, C. L., et al. (2011). Arachidonic acid omega-hydroxylase CYP4A11: inter-ethnic variations in the 8590T>C loss-of-function variant. *Mol Biol Rep*, ISSN 0301-4851 (Print).

[43] Garat, A., et al. (2011). Inter-ethnic variability of three functional polymorphisms affecting the IMPDH2 gene. *Mol Biol Rep*, 38, (8), 5185-5188, ISSN 0301-4851 (Print).

[44] Ntie, S., et al. (2010). A molecular diagnostic for identifying central African forest artiodactyls from faecal pellets. *Animal Conservation*, 13, (1), 80-93, ISSN 1367-9430 (Print).

[45] Anthony, N. M., et al. (2007). The role of Pleistocene refugia and rivers in shaping gorilla genetic diversity in central Africa. *Proceedings of the National Academy of Sciences of the United States of America*, 104, (51), 20432-20436, ISSN 0027-8424 (Print).

[46] Clifford, S. L., et al. (2004). Mitochondrial DNA phylogeography of western lowland gorillas (Gorilla gorilla gorilla). *Molecular Ecology*, 13, (6), 1567-1567, ISSN 0962-1083 (Print).

[47] Telfer, P. T., et al. (2003). Molecular evidence for deep phylogenetic divergence in Mandrillus sphinx. *Molecular Ecology*, 12, (7), 2019-2024, ISSN 0962-1083 (Print).

[48] Johnson, M. B., et al. (2007). Complex phylogeographic history of central African forest elephants and its implications for taxonomy. *BMC Evolutionary Biology*, 7, ISSN 1471-2148 (Electronic).

[49] Avise, J. C. (2009). Phylogeography: retrospect and prospect. *Journal of Biogeography*, 36, (1), 3-15, ISSN 0305-0270 (Print).

[50] Avise, J. C., et al. (1998). Speciation durations and Pleistocene effects on vertebrate phylogeography. *Proceedings of the Royal Society of London Series B-Biological Sciences*, 265, (1407), 1707-1712, ISSN 0962-8452 (Print).

[51] Hewitt, G. (2000). The genetic legacy of the Quaternary ice ages. *Nature*, 405, (6789), 907-913, ISSN 0028-0836 (Print).

[52] Manel, S., et al. (2003). Landscape genetics: combining landscape ecology and population genetics. *Trends in Ecology & Evolution*, 18, (4), 189-197, ISSN 0169-5347 (Print).

[53] Dainou, K., et al. (2010). Forest refugia revisited: nSSRs and cpDNA sequences support historical isolation in a wide-spread African tree with high colonization capacity, Milicia excelsa (Moraceae). *Molecular Ecology*, 19, (20), 4462-4477, ISSN 0962-1083 (Print).

[54] Dauby, G., et al. (2010). Chloroplast DNA Polymorphism and Phylogeography of a Central African Tree Species Widespread in Mature Rainforests: *Greenwayodendron suaveolens* (Annonaceae). *Tropical Plant Biology*, 3, 4-13, ISSN 1935-9756 (Print).

[55] Duminil, J., et al. (2010). CpDNA-based species identification and phylogeography: application to African tropical tree species. *Molecular Ecology*, 19, (24), 5469-5483, ISSN 0962-1083 (Print).

[56] Lowe, A. J., et al. (2010). Testing putative African tropical forest refugia using chloroplast and nuclear DNA phylogeography. *Tropical Plant Biology*, 3, (1), 50-58, ISSN 1935-9756 (Print).

[57] Muloko-Ntoutoume, N., et al. (2000). Chloroplast DNA variation in a rainforest tree (*Aucoumea klaineana*, Burseraceae) in Gabon. *Molecular Ecology*, 9, (3), 359-363, ISSN 0962-1083 (Print).

[58] Lowe, A. J., et al. (2000). Conservation genetics of bush mango from central/west Africa: implications from random amplified polymorphic DNA analysis. *Molecular Ecology*, 9, (7), 831-841, ISSN 0962-1083 (Print).

[59] Born, C., et al. (2011). Insights into the biogeographical history of the Lower Guinea Forest Domain: evidence for the role of refugia in the intraspecific differentiation of Aucoumea klaineana. *Molecular Ecology*, 20, (1), 131-142, ISSN 0962-1083 (Print).

[60] Debout, G. D., et al. (2011). Population history and gene dispersal inferred from spatial genetic structure of a Central African timber tree, Distemonanthus benthamianus (Caesalpinioideae). *Heredity*, 106, (1), 88-99, ISSN 0018-067X (Print).

[61] Ndiade-Bourobou, D., et al. (2010). Long-distance seed and pollen dispersal inferred from spatial genetic structure in the very low-density rainforest tree, Baillonella toxisperma Pierre, in Central Africa. *Molecular Ecology*, 19, (22), 4949-4962, ISSN 0962-1083 (Print).

[62] Born, C., et al. (2008). Colonization processes and the maintenance of genetic diversity: insights from a pioneer rainforest tree, Aucoumea klaineana. *Proceedings of the Royal Society B-Biological Sciences*, 275, (1647), 2171-2179, ISSN 0962-8452 (Print).

[63] Born, C., et al. (2008). Small-scale spatial genetic structure in the Central African rainforest tree species *Aucoumea klaineana*: a stepwise approach to infer the impact of limited gene dispersal, population history and habitat fragmentation. *Molecular Ecology*, 17, (8), 2041-2050, ISSN 0962-1083 (Print).

[64] Wumba, R., et al. (2010). Intestinal parasites infections in hospitalized AIDS patients in Kinshasa, Democratic Republic of Congo. *Parasite*, 17, (4), 321-328, ISSN 1252-607X (Print).

[65] Jimenez, M., et al. (2011). Detection and discrimination of Loa loa, Mansonella perstans and Wuchereria bancrofti by PCR-RFLP and nested-PCR of ribosomal DNA ITS1 region. *Exp Parasitol*, 127, (1), 282-286, ISSN 0014-4894 (Print).

[66] ten Hove, R. J., et al. (2008). Multiplex real-time PCR for the detection and quantification of Schistosoma mansoni and S. haematobium infection in stool samples collected in northern Senegal. *Trans R Soc Trop Med Hyg*, 102, (2), 179-185, ISSN 0035-9203 (Print).

[67] Deborggraeve, S., et al. (2011). Diagnostic accuracy of PCR in gambiense sleeping sickness diagnosis, staging and post-treatment follow-up: a 2-year longitudinal study. *PLoS Negl Trop Dis*, 5, (2), e972, ISSN 1935-2727 (Print).

[68] WHO (2003). *Malaria Rapid Diagnosis Making it Work: Informal Consultation on Field Trials and Quality Assurance on Malaria Rapid Diagnostic Tests*, Retrieved from http://www.who.int/malaria/publications/atoz/rdt2/en/index.html.

[69] WHO (2006). *The Use of Malaria Rapid Diagnostic Tests*. World Health Organization, Regional Office for the Western Pacific, ISBN 92 9061 204 5, Manila.

[70] Khairnar, K., et al. (2009). Multiplex real-time quantitative PCR, microscopy and rapid diagnostic immuno-chromatographic tests for the detection of Plasmodium spp: performance, limit of detection analysis and quality assurance. *Malar J*, 8, 284, ISSN 1475-2875 (Electronic).

[71] Hartl, D. L., et al. (2002). The paradoxical population genetics of Plasmodium falciparum. *Trends Parasitol*, 18, (6), 266-272, ISSN 1471-4922 (Print).

[72] Baker, J., et al. (2005). Genetic diversity of Plasmodium falciparum histidine-rich protein 2 (PfHRP2) and its effect on the performance of PfHRP2-based rapid diagnostic tests. *J Infect Dis*, 192, (5), 870-877, ISSN 0022-1899 (Print).

[73] Khattab, A., et al. (2003). Common surface-antigen var genes of limited diversity expressed by Plasmodium falciparum placental isolates separated by time and space. *J Infect Dis*, 187, (3), 477-483, ISSN 0022-1899 (Print).

[74] WHO (1986). *Chemotherapy of Malaria*. World Health Organization, ISBN 92 4 140127 3, Geneva.

[75] WHO (2002). *Monitoring antimalarial drug resistance, Report of a WHO consultation*, Retrieved from http://www.rollbackmalaria.org/cmc_upload/0/000/015/800/200239.pdf.

[76] Ranford-Cartwright, L. C., et al. (1997). Molecular analysis of recrudescent parasites in a Plasmodium falciparum drug efficacy trial in Gabon. *Trans R Soc Trop Med Hyg*, 91, (6), 719-724, ISSN 0035-9203 (Print).

[77] Mombo-Ngoma, G., et al. (2011). High prevalence of dhfr triple mutant and correlation with high rates of sulphadoxine-pyrimethamine treatment failures in vivo in Gabonese children. *Malar J*, 10, 123, ISSN 1475-2875 (Electronic).

[78] Ndounga, M., et al. (2007). Therapeutic efficacy of sulfadoxine-pyrimethamine and the prevalence of molecular markers of resistance in under 5-year olds in Brazzaville, Congo. *Trop Med Int Health*, 12, (10), 1164-1171, ISSN 1360-2276 (Print).

[79] Tahar, R. and L. K. Basco. (2006). Molecular epidemiology of malaria in Cameroon. XXII. Geographic mapping and distribution of Plasmodium falciparum dihydrofolate reductase (dhfr) mutant alleles. *Am J Trop Med Hyg*, 75, (3), 396-401, ISSN 0002-9637 (Print).

[80] Lekana-Douki, J. B., et al. (2011). Increased prevalence of the Plasmodium falciparum Pfmdr1 86N genotype among field isolates from Franceville, Gabon after replacement of chloroquine by artemether-lumefantrine and artesunate-mefloquine. *Infect Genet Evol*, 11, (2), 512-517, ISSN 1567-1348 (Print).

[81] Menard, D., et al. (2006). Frequency distribution of antimalarial drug-resistant alleles among isolates of Plasmodium falciparum in Bangui, Central African Republic. *Am J Trop Med Hyg*, 74, (2), 205-210, ISSN 0002-9637 (Print).

[82] Etang, J., et al. (2006). First report of knockdown mutations in the malaria vector Anopheles gambiae from Cameroon. *Am J Trop Med Hyg*, 74, (5), 795-797, ISSN 0002-9637 (Print).

[83] Mourou, J. R., et al. (2010). Malaria transmission and insecticide resistance of Anopheles gambiae in Libreville and Port-Gentil, Gabon. *Malar J*, 9, 321, ISSN 1475-2875 (Electronic).

[84] Powell, R., et al. (2010). The Evolution of HIV-1 Diversity in Rural Cameroon and its Implications in Vaccine Design and Trials. *Viruses*, 2, (2), 639-654, ISSN 1999-4915 (Print).

[85] Lee, S., et al. (2007). Detection of emerging HIV variants in blood donors from urban areas of Cameroon. *AIDS Res Hum Retroviruses*, 23, (10), 1262-1267, ISSN 0889-2229 (Print).

[86] Ragupathy, V., et al. (2011). Identification of new, emerging HIV-1 unique recombinant forms and drug resistant viruses circulating in Cameroon. *Virol J*, 8, 185, ISSN 1743-422X (Electronic).

[87] Djoko, C. F., et al. (2010). HIV type 1 pol gene diversity and genotypic antiretroviral drug resistance mutations in Malabo, Equatorial Guinea. *AIDS Res Hum Retroviruses*, 26, (9), 1027-1031, ISSN 0889-2229 (Print).

[88] Aghokeng, A. F., et al. (2011). High failure rate of the ViroSeq HIV-1 genotyping system for drug resistance testing in Cameroon, a country with broad HIV-1 genetic diversity. *J Clin Microbiol*, 49, (4), 1635-1641, ISSN 0095-1137 (Print).

[89] Luk, K. C., et al. (2008). Near full-length genome characterization of an HIV type 1 CRF25_cpx strain from Cameroon. *AIDS Res Hum Retroviruses*, 24, (10), 1309-1314, ISSN 0889-2229 (Print).

[90] Zhao, J., et al. (2010). Identification and genetic characterization of a novel CRF22_01A1 recombinant form of HIV type 1 in Cameroon. *AIDS Res Hum Retroviruses*, 26, (9), 1033-1045, ISSN 0889-2229 (Print).

[91] Vidal, N., et al. (2000). Unprecedented degree of human immunodeficiency virus type 1 (HIV-1) group M genetic diversity in the Democratic Republic of Congo suggests that the HIV-1 pandemic originated in Central Africa. *J Virol*, 74, (22), 10498-10507, ISSN 0022-538X (Print).

[92] Hunt, J. C., et al. (1997). Envelope sequence variability and serologic characterization of HIV type 1 group O isolates from equatorial guinea. *AIDS Res Hum Retroviruses*, 13, (12), 995-1005, ISSN 0889-2229 (Print).

[93] Peeters, M., et al. (1997). Geographical distribution of HIV-1 group O viruses in Africa. *AIDS*, 11, (4), 493-498, ISSN 0269-9370 (Print).

[94] Roques, P., et al. (2002). Phylogenetic analysis of 49 newly derived HIV-1 group O strains: high viral diversity but no group M-like subtype structure. *Virology*, 302, (2), 259-273, ISSN 0042-6822 (Print).

[95] Delaporte, E., et al. (1996). Epidemiological and molecular characteristics of HIV infection in Gabon, 1986-1994. *AIDS*, 10, (8), 903-910, ISSN 0269-9370 (Print).

[96] Pandrea, I., et al. (2002). Analysis of partial pol and env sequences indicates a high prevalence of HIV type 1 recombinant strains circulating in Gabon. *AIDS Res Hum Retroviruses*, 18, (15), 1103-1116, ISSN 0889-2229 (Print).

[97] Carr, J. K., et al. (2001). The AG recombinant IbNG and novel strains of group M HIV-1 are common in Cameroon. *Virology*, 286, (1), 168-181, ISSN 0042-6822 (Print).

[98] Fonjungo, P. N., et al. (2000). Molecular screening for HIV-1 group N and simian immunodeficiency virus cpz-like virus infections in Cameroon. *AIDS*, 14, (6), 750-752, ISSN 0269-9370 (Print).

[99] Nkengasong, J. N., et al. (1994). Genotypic subtypes of HIV-1 in Cameroon. *AIDS*, 8, (10), 1405-1412, ISSN 0269-9370 (Print).

[100] Plantier, J. C., et al. (2009). A new human immunodeficiency virus derived from gorillas. *Nat Med*, 15, (8), 871-872, ISSN 1078-8956 (Print).

[101] Simon, F., et al. (1998). Identification of a new human immunodeficiency virus type 1 distinct from group M and group O. *Nat Med*, 4, (9), 1032-1037, ISSN 1078-8956 (Print).

[102] Vallari, A., et al. (2011). Confirmation of putative HIV-1 group P in Cameroon. *J Virol*, 85, (3), 1403-1407, ISSN 0022-538X (Print).

[103] Yamaguchi, J., et al. (2006). Identification of HIV type 1 group N infections in a husband and wife in Cameroon: viral genome sequences provide evidence for horizontal transmission. *AIDS Res Hum Retroviruses*, 22, (1), 83-92, ISSN 0889-2229 (Print).

[104] Ouwe-Missi-Oukem-Boyer, O. N., et al. (2005). [The use of dried blood spots for HIV-antibody testing in Sahel]. *Bull Soc Pathol Exot*, 98, (5), 343-346, ISSN 0037-9085 (Print).

[105] Nkenfou, C. N., et al. (2011). Implementation of HIV Early Infant Diagnosis and HIV-1 RNA viral load determination on Dried Blood Spots in Cameroon: challenges and propositions. *AIDS Res Hum Retroviruses*, 2011 Jul 27. [Epub ahead of print] ISSN 0889-2229 (Print).

[106] Damond, F., et al. (2008). Quality control assessment of human immunodeficiency virus type 2 (HIV-2) viral load quantification assays: results from an international collaboration on HIV-2 infection in 2006. *J Clin Microbiol*, 46, (6), 2088-2091, ISSN 0095-1137 (Print).

[107] Fokam, J., et al. (2011). Performance evaluation of an in-house human immunodeficiency virus type-1 protease-reverse transcriptase genotyping assay in Cameroon. *Arch Virol*, 156, (7), 1235-1243, ISSN 0304-8608 (Print).

[108] Charpentier, C., et al. (2011). Virological failure and HIV type 1 drug resistance profiles among patients followed-up in private sector, Douala, Cameroon. *AIDS Res Hum Retroviruses*, 27, (2), 221-230, ISSN 0889-2229 (Print).

[109] Fokam, J., et al. (2011). Drug Resistance Among Drug-naive and First-line Antiretroviral Treatment-failing Children in Cameroon. *Pediatr Infect Dis J*, 30(12): 1062-1068, ISSN 0891-3668 (Print).

[110] Laurent, C., et al. (2011). Monitoring of HIV viral loads, CD4 cell counts, and clinical assessments versus clinical monitoring alone for antiretroviral therapy in rural district hospitals in Cameroon (Stratall ANRS 12110/ESTHER): a randomised non-inferiority trial. *Lancet Infect Dis*, 11, (11), 825-833, ISSN 1473-3099 (Print).

[111] Burda, S. T., et al. (2010). HIV-1 reverse transcriptase drug-resistance mutations in chronically infected individuals receiving or naive to HAART in Cameroon. *J Med Virol*, 82, (2), 187-196, ISSN 0146-6615 (Print).

[112] Soria, A., et al. (2009). Resistance profiles after different periods of exposure to a first-line antiretroviral regimen in a Cameroonian cohort of HIV type-1-infected patients. *Antivir Ther*, 14, (3), 339-347, ISSN 1359-6535 (Print).

[113] WHO/TDR. (2011) *Implementation research for the control of infectious diseases of poverty*, Retrieved from http://apps.who.int/tdr/svc/publications/tdr-research-publications/access_report.

[114] Mbida, A. D., et al. (2009). Measure of viral load by using the Abbott Real-Time HIV-1 assay on dried blood and plasma spot specimens collected in 2 rural dispensaries in Cameroon. *J Acquir Immune Defic Syndr*, 52, (1), 9-16, ISSN 1525-4135 (Print).

[115] Johannessen, A. (2010). Quantification of HIV-1 RNA on dried blood spots. *AIDS*, 24, (3), 475-476, ISSN 0269-9370 (Print).

[116] Ziemniak, C., et al. (2011). Use of Dried Blood Spot Samples and In-house Assays to Identify Antiretroviral Drug Resistance in HIV-Infected Children in Resource-Constrained Settings. *J Clin Microbiol*, 49(12): 4077-4082, ISSN 0095-1137 (Print).

[117] Johannessen, A. (2010). Dried blood spots in HIV monitoring: applications in resource-limited settings. *Bioanalysis*, 2, (11), 1893-1908, ISSN 1757-6180 (Print).

[118] Sodora, D. L., et al. (2009). Toward an AIDS vaccine: lessons from natural simian immunodeficiency virus infections of African nonhuman primate hosts. *Nat Med*, 15, (8), 861-865, ISSN 1078-8956 (Print).

[119] Kaleebu, P., et al. (2008). African AIDS vaccine programme for a coordinated and collaborative vaccine development effort on the continent. *PLoS Med*, 5, (12), e236, ISSN 1549-1277 (Print).

[120] Dolgin, E. (2010). African networks launch to boost clinical trial capacity. *Nat Med*, 16, (1), 8, ISSN 1078-8956 (Print).

[121] Gessain, A. and R. Mahieux. (2000). [Epidemiology, origin and genetic diversity of HTLV-1 retrovirus and STLV-1 simian affiliated retrovirus]. *Bull Soc Pathol Exot*, 93, (3), 163-171, ISSN 0037-9085 (Print).

[122] Etenna, S. L., et al. (2008). New insights into prevalence, genetic diversity, and proviral load of human T-cell leukemia virus types 1 and 2 in pregnant women in Gabon in equatorial central Africa. *J Clin Microbiol*, 46, (11), 3607-3614, ISSN 0095-1137 (Print).

[123] Letourneur, F., et al. (1998). Complete nucleotide sequence of an African human T-lymphotropic virus type II subtype b isolate (HTLV-II-Gab): molecular and phylogenetic analysis. *J Gen Virol*, 79 (Pt 2), 269-277, ISSN 0022-1317 (Print).

[124] Mauclere, P., et al. (2011). HTLV-2B strains, similar to those found in several Amerindian tribes, are endemic in central African Bakola Pygmies. *J Infect Dis*, 203, (9), 1316-1323, ISSN 0022-1899 (Print).

[125] Duong, Y. T., et al. (2008). Short communication: Absence of evidence of HTLV-3 and HTLV-4 in patients with large granular lymphocyte (LGL) leukemia. *AIDS Res Hum Retroviruses*, 24, (12), 1503-1505, ISSN 0889-2229 (Print).

[126] Besson, G. and M. Kazanji. (2009). One-step, multiplex, real-time PCR assay with molecular beacon probes for simultaneous detection, differentiation, and quantification of human T-cell leukemia virus types 1, 2, and 3. *J Clin Microbiol*, 47, (4), 1129-1135, ISSN 0095-1137 (Print).

[127] Njouom, R., et al. (2010). Circulation of human influenza viruses and emergence of Oseltamivir-resistant A(H1N1) viruses in Cameroon, Central Africa. *BMC Infect Dis*, 10, 56, ISSN 1471-2334 (Electronic).

[128] Burke, R. L., et al. (2011). Department of Defense influenza and other respiratory disease surveillance during the 2009 pandemic. *BMC Public Health*, 11 Suppl 2, S6, ISSN 1471-2458 (Electronic).

[129] Njouom, R., et al. (2003). High rate of hepatitis C virus infection and predominance of genotype 4 among elderly inhabitants of a remote village of the rain forest of South Cameroon. *J Med Virol*, 71, (2), 219-225, ISSN 0146-6615 (Print).

[130] Pasquier, C., et al. (2005). Distribution and heterogeneity of hepatitis C genotypes in hepatitis patients in Cameroon. *J Med Virol*, 77, (3), 390-398, ISSN 0146-6615 (Print).

[131] Njouom, R., et al. (2007). The hepatitis C virus epidemic in Cameroon: genetic evidence for rapid transmission between 1920 and 1960. *Infect Genet Evol*, 7, (3), 361-367, ISSN 1567-1348 (Print).

[132] Ndong-Atome, G. R., et al. (2008). High prevalence of hepatitis C virus infection and predominance of genotype 4 in rural Gabon. *J Med Virol*, 80, (9), 1581-1587, ISSN 0146-6615 (Print).

[133] Njouom, R., et al. (2009). Predominance of hepatitis C virus genotype 4 infection and rapid transmission between 1935 and 1965 in the Central African Republic. *J Gen Virol*, 90, (Pt 10), 2452-2456, ISSN 0022-1317 (Print).

[134] Xu, L. Z., et al. (1994). Hepatitis C virus genotype 4 is highly prevalent in central Africa (Gabon). *J Gen Virol*, 75 (Pt 9), 2393-2398, ISSN 0022-1317 (Print).

[135] Makuwa, M., et al. (2006). Identification of hepatitis B virus subgenotype A3 in rural Gabon. *J Med Virol*, 78, (9), 1175-1184, ISSN 0146-6615 (Print).

[136] Foupouapouognigni, Y., et al. (2011). Hepatitis B and C virus infections in the three Pygmy groups in Cameroon. *J Clin Microbiol*, 49, (2), 737-740, ISSN 0095-1137 (Print).

[137] Ouwe-Missi-Oukem-Boyer, O., et al. (2011). Hepatitis C virus infection may lead to slower emergence of P. falciparum in blood. *PLoS One*, 6, (1), e16034, ISSN 1932-6203 (Electronic).

[138] Snounou, G., et al. (1993). High sensitivity of detection of human malaria parasites by the use of nested polymerase chain reaction. *Mol Biochem Parasitol*, 61, (2), 315-320, ISSN 0166-6851 (Print).

[139] Foupouapouognigni, Y., et al. (2011). High prevalence and predominance of hepatitis delta virus genotype 1 infection in Cameroon. *J Clin Microbiol*, 49, (3), 1162-1164, ISSN 0095-1137 (Print).

[140] Makuwa, M., et al. (2009). Prevalence and molecular diversity of hepatitis B virus and hepatitis delta virus in urban and rural populations in northern Gabon in central Africa. *J Clin Microbiol*, 47, (7), 2265-2268, ISSN 0095-1137 (Print).

[141] Tuveri, R., et al. (2000). Prevalence and genetic variants of hepatitis GB-C/HG and TT viruses in Gabon, equatorial Africa. *Am J Trop Med Hyg*, 63, (3-4), 192-198, ISSN 0002-9637 (Print).

[142] Formenty, P., et al. (2006). Detection of Ebola virus in oral fluid specimens during outbreaks of Ebola virus hemorrhagic fever in the Republic of Congo. *Clin Infect Dis*, 42, (11), 1521-1526, ISSN 1058-4838 (Print).

[143] Leroy, E., et al. (2011). [Ebola and Marburg hemorrhagic fever viruses: update on filoviruses]. *Med Trop (Mars)*, 71, (2), 111-121, ISSN 0025-682X (Print).

[144] Leroy, E. M., et al. (2000). Diagnosis of Ebola haemorrhagic fever by RT-PCR in an epidemic setting. *J Med Virol*, 60, (4), 463-467, ISSN 0146-6615 (Print).

[145] Djikeng, A., et al. (2008). Viral genome sequencing by random priming methods. *BMC Genomics*, 9, 5, ISSN 1471-2164 (Electronic).

[146] Creek, T. L., et al. (2007). Infant human immunodeficiency virus diagnosis in resource-limited settings: issues, technologies, and country experiences. *Am J Obstet Gynecol*, 197, (3 Suppl), S64-71, ISSN 0002-9378 (Print).

[147] Caron, M., et al. (2008). Human immunodeficiency virus type 1 seroprevalence and antiretroviral drug resistance-associated mutations in miners in Gabon, central Africa. *AIDS Res Hum Retroviruses*, 24, (9), 1225-1228, ISSN 0889-2229 (Print).

[148] Mas, A., et al. (1999). Phylogeny of HIV type 1 group O isolates based on env gene sequences. *AIDS Res Hum Retroviruses*, 15, (8), 769-773, ISSN 0889-2229 (Print).

[149] Heredia, A., et al. (1997). Evidence of HIV-2 infection in Equatorial Guinea (central Africa): partial genetic analysis of a B subtype virus. *AIDS Res Hum Retroviruses*, 13, (5), 439-440, ISSN 0889-2229 (Print).

[150] Mahieux, R., et al. (1997). Molecular epidemiology of 58 new African human T-cell leukemia virus type 1 (HTLV-1) strains: identification of a new and distinct HTLV-1 molecular subtype in Central Africa and in Pygmies. *J Virol*, 71, 1317-1333, ISSN 0022-538X (Print).

[151] Wolfe, N. D., et al. (2005). Emergence of unique primate T-lymphotropic viruses among central African bushmeat hunters. *Proceeding of the National Academy of Sciences of the United States of America*, 102, (22), 7994-7999, ISSN 0027-8424 (Print).

[152] Mahieux, R. and A. Gessain. (2011). HTLV-3/STLV-3 and HTLV-4 Viruses: Discovery, Epidemiology, Serology and Molecular Aspects. *Viruses*, 3, (7), 1074-1090, ISSN 1999-4915 (Print).

[153] Leroy, E. M., et al. (2005). Fruit bats as reservoirs of Ebola virus. *Nature*, 438, (7068), 575-576, ISSN 0028-0836 (Print).

[154] Morvan, J. M., et al. (2000). [Forest ecosystems and Ebola virus]. *Bull Soc Pathol Exot*, 93, (3), 172-175, ISSN 0037-9085 (Print).

[155] Niobe-Eyangoh, S. N., et al. (2003). Genetic biodiversity of Mycobacterium tuberculosis complex strains from patients with pulmonary tuberculosis in Cameroon. *J Clin Microbiol*, 41, (6), 2547-2553, ISSN 0095-1137 (Print).

[156] Picard, B. (2000). [Molecular epidemiology of large bacterial endemics in Sub-Saharan Africa]. *Bull Soc Pathol Exot*, 93, (3), 219-223, ISSN 0037-9085 (Print).
[157] Raviglione, M. C., et al. (1995). Global epidemiology of tuberculosis. Morbidity and mortality of a worldwide epidemic. *JAMA*, 273, (3), 220-226, ISSN 0098-7484 (Print).
[158] Groenen, P. M., et al. (1993). Nature of DNA polymorphism in the direct repeat cluster of Mycobacterium tuberculosis; application for strain differentiation by a novel typing method. *Mol Microbiol*, 10, (5), 1057-1065, ISSN 0950-382X (Print).
[159] Kamerbeek, J., et al. (1997). Simultaneous detection and strain differentiation of Mycobacterium tuberculosis for diagnosis and epidemiology. *J Clin Microbiol*, 35, (4), 907-914, ISSN 0095-1137 (Print).
[160] Warren, R. M., et al. (2002). Use of spoligotyping for accurate classification of recurrent tuberculosis. *J Clin Microbiol*, 40, (10), 3851-3853, ISSN 0095-1137 (Print).
[161] Liegeois, F., et al. (2011). Full-Length Genome Sequence of a Simian Immunodeficiency Virus from a Wild-Captured Sun-Tailed Monkey in Gabon Provides Evidence for a Species-Specific Monophyletic SIVsun Lineage. *AIDS Res Hum Retroviruses*, 27(11): 1237-1241, ISSN 0889-2229 (Print).
[162] Liegeois, F., et al. (2006). Molecular characterization of a novel simian immunodeficiency virus lineage (SIVtal) from northern talapoins (Miopithecus ogouensis). *Virology*, 349, (1), 55-65, ISSN 0042-6822 (Print).
[163] Souquiere, S., et al. (2001). Wild Mandrillus sphinx are carriers of two types of lentivirus. *J Virol*, 75, (15), 7086-7096, ISSN 0022-538X (Print).
[164] Courgnaud, V., et al. (2004). Simian T-cell leukemia virus (STLV) infection in wild primate populations in Cameroon: evidence for dual STLV type 1 and type 3 infection in agile mangabeys (Cercocebus agilis). *J Virol*, 78, (9), 4700-4709, ISSN 0022-538X (Print).
[165] Makuwa, M., et al. (2004). Two distinct STLV-1 subtypes infecting Mandrillus sphinx follow the geographic distribution of their hosts. *AIDS Res Hum Retroviruses*, 20, (10), 1137-1143, ISSN 0889-2229 (Print).
[166] Sintasath, D. M., et al. (2009). Simian T-lymphotropic virus diversity among nonhuman primates, Cameroon. *Emerg Infect Dis*, 15, (2), 175-184, ISSN 1080-6040 (Print).
[167] Sintasath, D. M., et al. (2009). Genetic characterization of the complete genome of a highly divergent simian T-lymphotropic virus (STLV) type 3 from a wild Cercopithecus mona monkey. *Retrovirology*, 6, 97, ISSN 1742-4690 (Electronic).
[168] Mouinga-Ondeme, A., et al. (2010). Two distinct variants of simian foamy virus in naturally infected mandrills (Mandrillus sphinx) and cross-species transmission to humans. *Retrovirology*, 7, 105, ISSN 1742-4690 (Electronic).
[169] Makuwa, M., et al. (2005). Identification of hepatitis B virus genome in faecal sample from wild living chimpanzee (Pan troglodytes troglodytes) in Gabon. *J Clin Virol*, 34 Suppl 1, S83-88, ISSN 1386-6532 (Print).
[170] Georges-Courbot, M. C., et al. (1998). Natural infection of a household pet red-capped mangabey (Cercocebus torquatus torquatus) with a new simian immunodeficiency virus. *J Virol*, 72, (1), 600-608, ISSN 0022-538X (Print).
[171] Peeters, M., et al. (2008). [Genetic diversity and phylogeographic distribution of SIV: how to understand the origin of HIV]. *Med Sci (Paris)*, 24, (6-7), 621-628, ISSN 0767-0974 (Print).
[172] Ollomo, B., et al. (2009). A new malaria agent in African hominids. *PLoS Pathog*, 5, (5), e1000446, ISSN 1553-7366 (Print).

[173] Prugnolle, F., et al. (2010). African great apes are natural hosts of multiple related malaria species, including Plasmodium falciparum. *Proceeding of the National Academy of Sciences of the United States of America*, 107, (4), 1458-1463, ISSN 0027-8424 (Print).

[174] Morvan, J. M., et al. (1999). Identification of Ebola virus sequences present as RNA or DNA in organs of terrestrial small mammals of the Central African Republic. *Microbes Infect*, 1, (14), 1193-1201, ISSN 1286-4579 (Print).

[175] Towner, J. S., et al. (2007). Marburg virus infection detected in a common African bat. *PLoS One*, 2, (1), e764, ISSN 1932-6203 (Electronic).

[176] Njouom, R., et al. (2008). Highly pathogenic avian influenza virus subtype H5N1 in ducks in the Northern part of Cameroon. *Vet Microbiol*, 130, (3-4), 380-384, ISSN 0378-1135 (Print).

[177] Prugnolle, F., et al. (2011). Plasmodium falciparum is not as lonely as previously considered. *Virulence*, 2, (1), 71-76, ISSN 2150-5594 (Print).

[178] Georges-Courbot, M. C., et al. (1996). Occurrence and frequency of transmission of naturally occurring simian retroviral infections (SIV, STLV, and SRV) at the CIRMF Primate Center, Gabon. *J Med Primatol*, 25, (5), 313-326, ISSN 0047-2565 (Print).

[179] Beer, B. E., et al. (2001). Characterization of novel simian immunodeficiency viruses from red-capped mangabeys from Nigeria (SIVrcmNG409 and -NG411). *J Virol*, 75, (24), 12014-12027, ISSN 0022-538X (Print).

[180] Gao, F., et al. (1999). Origin of HIV-1 in the chimpanzee Pan troglodytes troglodytes. *Nature*, 397, (6718), 436-441, ISSN 0028-0836 (Print).

[181] Huet, T., et al. (1990). Genetic organization of a chimpanzee lentivirus related to HIV-1. *Nature*, 345, (6273), 356-359, ISSN 0028-0836 (Print).

[182] Vanden Haesevelde, M. M., et al. (1996). Sequence analysis of a highly divergent HIV-1-related lentivirus isolated from a wild captured chimpanzee. *Virology*, 221, (2), 346-350, ISSN 0042-6822 (Print).

[183] Ahuka-Mundeke, S., et al. (2011). Identification and Molecular Characterization of New Simian T Cell Lymphotropic Viruses in Nonhuman Primates Bushmeat from the Democratic Republic of Congo. *AIDS Res Hum Retroviruses*, 2011 Sep 14. [Epub ahead of print], ISSN 0889-2229 (Print).

[184] Calattini, S., et al. (2006). Detection and molecular characterization of foamy viruses in Central African chimpanzees of the Pan troglodytes troglodytes and Pan troglodytes vellerosus subspecies. *J Med Primatol*, 35, (2), 59-66, ISSN 0047-2565 (Print).

[185] Liu, W., et al. (2008). Molecular ecology and natural history of simian foamy virus infection in wild-living chimpanzees. *PLoS Pathog*, 4, (7), e1000097, ISSN 1553-7366 (Print).

[186] Begoude, B. A. D., et al. (2011). The pathogenic potential of endophytic Botryosphaeriaceous fungi on Terminalia species in Cameroon. *Forest Pathology*, 41, (4), 281-292, ISSN 1437-4781 (Print).

[187] Halkett, F., et al. (2010). Genetic discontinuities and disequilibria in recently established populations of the plant pathogenic fungus Mycosphaerella fijiensis. *Molecular Ecology*, 19, (18), 3909-3923, ISSN 0962-1083 (Print).

[188] Ntoumi, F. (2011). The ant who learned to be an elephant. *Science*, 333, (6051), 1824-1825, ISSN 0036-8075 (Print).

[189] Wonkam, A., et al. (2011). Ethics of Human Genetic Studies in Sub-Saharan Africa: The Case of Cameroon through a Bibliometric Analysis. *Dev World Bioeth*, 11(3): 120-127, ISSN 1471-8731 (Print).

17

Study of Mycobacterium Tuberculosis by Molecular Methods in Northeast Mexico

H. W. Araujo-Torres[1,3], J. A. Narváez-Zapata[2], M. G. Castillo-Álvarez[1], MS. Puga-Hernández[4], J. Flores-Gracia[5] and M. A. Reyes-López[1,*]

[1]*Conservation Medicine Lab., Centro de Biotecnología Genómica del Instituto Politécnico Nacional, Cd. Reynosa, Tamps,*
[2]*Industrial Biotechnology Lab., Centro de Biotecnología Genómica del Instituto Politécnico Nacional, Cd. Reynosa, Tamps,*
[3]*Centro de Investigación en Ciencia Aplicada y Tecnología Avanzada del Instituto Politécnico Nacional, Altamira, Tamps,*
[4]*Laboratorio Estatal de Salud Pública de Tamaulipas, Cd. Victoria, Tamps,*
[5]*Instituto Tecnológico de Ciudad Victoria, Cd. Victoria, Tamps*
México

1. Introduction

One third of the world population is afected by TB and one million people did die this year 2011 in undeveloped countries (Venkatesh et al., 2011). In Tamaulipas, a Northern State of Mexico and a border state between USA and Mexico, frequency is 26.9 new TB cases per 100,000 people, twice of national rate of 12.85 cases per 100,000 people (Ferrer et al., 2010). Only on the border of Tamaulipas about 320 cases are diagnosed each year. Many of these cases correspond to people from other states of Mexico, probably by geographic position and by migration problematic of this study zone (Fitchett et al., 2011). Only 92% of the treated population are cured mainly because much of these people are poor and whose nutritional status directly affects the possibility of quick recovery (SSA, 2009).

The long presence of this disease has increased the need to know specifically which *Mycobacterium tuberculosis* strains are circulating in the region. Additionally, it is necessary to know the antibiotic/susceptibility profile of these strains since many of them acquire resistance against the traditional antibiotics along time.

In general, the diagnostic of this disease is traditionally conducted by using gold standard techniques focused to identify the presence of *M. tuberculosis* in clinical specimen of humans or cattle. These techniques included the strain of microorganism in Ziehl-Neelsen and culture in Lowenstein-Jensen medium (Cadmus et al., 2011), both regarded as reference techniques in the diagnosis of TB. Differentiation among mycobacteria of the *M. tuberculosis* complex (MTC) and other than MTC (NMTC) is accomplished by applying biochemical tests: niacin production, catalase activity, thermostable at 68 ° C and reduction of nitrate.

* Corresponding Author

Actually, detection for *M. tuberculosis* has been shorted due mainly to application of molecular methods directly to clinical samples. Usually, the detection of this bacterium takes 2 to 4 weeks (Marhöfer et al., 2011). Some molecular techniques are already on the market, being the most commonly used *AMPLICOR M. tuberculosis PCR test* (Roche), *M. tuberculosis* Direct test (MTDT) (GenProbe) and LCX *M. tuberculosis* assay (Abbott). In addition, PCR amplification of ribosomal sequences (Ribotyping) or amplification of repetitive intragenic consensus sequences (i.e. ERIC-PCR, spoligotyping, MIRU-VNTR), among others, are the usual methods used to specifically discriminate among different *M. tuberculosis* strains (Rodwell et al., 2010; Pang et al., 2011).

The use of molecular approach is also applied in the analysis of the antibiotic resistance of these isolates. In this sense, molecular detection of specific mutations in genes involved with drug resistance has successfully been applied in the identification of these (i.e. detection of mutations on *rpo*B, *kat*G, *mab*A, etc.) (Sala & Hartkoorn, 2011).

As an example of the conjunction of the background described above, this chapter briefly presents a work of potential tuberculosis patient samples, to which mycobacteria were isolated to determine whether any of them were resistant to some antibiotics and if it could be grouped by health districts of Tamaulipas and also grouped the isolates strains in MTC and NMTC, and identify them, potentially.

Therefore, the main aim of this study was determinated by using a molecular approach the specific *M. tuberculosis* strains presents in the region and identify specific mutations in these strains related with drug resistance. The information here generated helps to take epidemiological decisions aimed to control and to prevent this disease in the Northeast of Tamaulipas.

1.1 TB statistical

This main issue is due to produce a resistance against antibiotics used traditionally to control the disease. The first antibiotic against TB was created in the 40's decade. Consequently, the incidence of this disease declined in the following decades, especially in developed countries. However, in the last 20 years it has been observed an increase in the TB cases around of the world, particularly due to generation of new *M. tuberculosis* strains with resistance of the traditional antibiotics, or multidrug resistance (Yew et al., 2010). The death associated to TB may increase in undeveloped countries since some others diseases as the HIV may duplicate the death frequency in patients with both diseases (Havlir & Barnes 1999; Sonnenberg et al., 2005). Given this situation the TB was declared as global emergency by WHO (WHO 1993).

1.2 Situation of TB in northern Mexico

The Northeast state of Tamaulipas exhibits a high peak of occurrence of TB with regarding at Mexican rate and the frequency of this disease has remained stable during the last 10 years (Ferrer et al., 2010). Besides, this region is a natural corridor for exporting of cattle between Mexico and EE.UU. Therefore, both countries have commitment to keep safe their borders (Fitchett et al., 2011). Among the factors that may partially explain it, the high migrations rate reported in this region with people from other Mexican states (mainly

Veracruz, Coahuila, Mexico city, among others) and other Central America countries (i.e. Guatemala, Honduras, among others).

One of the reasons of this migration may be high number of manufacturing factories that offer a high number of jobs and that appeal and wait to travel. In addition, many of these people only remain of 4 to 5 years here and wait to travel to U.S.A. As it was previously mentioned, many of these people are poor and with low nutritional status and their lifestyle (i.e. drug or alcohol consumes) could prompted the TB disease (Wagner et al., 2011). Therefore, TB will become a big issue between Mexico and USA. From here, both countries have agreements on health and security cooperation. These agreements include the fast detection of this disease and the discrimination among *M. tuberculosis* strains (Fitchett et al, 2011).

1.3 Multidrug resistant in Mycobacterium tuberculosis

Recently, it has been reported an increase in the TB cases around of the world, particularly due to generation of new *M. tuberculosis* strains with resistance to traditional antibiotics, or multidrug resistance (Sougakoff, 2011). In 2007, the 14th edition of the Merck list shows 30 different anti-TB drugs, many analogues or prodrugs of antibiotics, as the first line of defense against this disease. In Mexico, the antibiotics most commonly used are the rifampicin (RIF), the isoniazid (INH), the pyrazinamide (PZA), the streptomycin (STR) and the ethambutol (EMB) (Borrell, Gagneux., 2011).

Worldwide, rifampicin is the drug mostly in the control of this bacterium (Connell et al., 2011). These antibiotics are not enough to halt the emergence and spread of multidrug resistant (MDR) strains causing a serious problem for the TB control and increasing public health problems (Zumi, et al. 2001). This have prompts the development of fast and reliable diagnostic process to detect, to discriminate, and to evaluate resistance of *M. tuberculosis* strains against main drugs.

1.4 Molecular approaches

Molecular biology has allowed detection of DNA or RNA sequence of different mycobacteria. An example of these approaches is using probes. Probes were prepared from nucleic acid sequences complementary to the DNA or RNA sequences from different species (including *M. tuberculosis, M. avium, M. kansasii, M. gordonae.*), which may be labeled with radioactive isotopes (hot probes) or chromogenic substances (cold probes). The gene probe is capable of binding or hybridizing with a homologous fragment of the study sample, which has been previously denatured by physical means. Hybridization of the probe to its complementary fragment is easily detected with addition of a marker. The main advantages of these techniques are fast and specific. Its disadvantages high cost and that many probes cannot identify species within the MTC.

Typing techniques based on amplification of nucleic acids by PCR provide a fast and reliable approach to obtain genetic information about bacteria or microorganism groups. Molecular typing methods for tuberculosis are based on that those infected by strains of *M. tuberculosis* have the same genotype (genetic fingerprinting) and are epidemiologically related, while those infected with different genotypes (unique patterns) are not.

Among the techniques of molecular biology that are currently used, as: ribotyping, the PCR amplification of repetitive extragenic palindromic sequences (REP-PCR) and the repetitive intragenic consensus sequences of *Enterobacteriaceae* (ERIC-PCR). These techniques can also be used in clinical studies to establish patterns of colonization and to identify sources of transmission of infectious microorganisms, which may contribute to a better understanding of the epidemiology and pathogenesis thereby helping to develop disease prevention strategies (Struelens, MESGEM, 1996).

1.5 Ribotyping

Ribotyping technique applied in the diagnostic of diseases has been used for differentiation of bacterial serotypes involved with the occurrence of outbreaks. Additionally, this technique has an extended use in the study of nosocomial fungus (Pavlic and Griffiths, 2009). Ribotyping is also used to study the ecology, the genotypic variation and the transmission of Streptococcus mutants from person to person (Alam et al., 1999). The patterns are simplified to ribotyping, making visible the DNA fragments containing parts or all of ribosomal genes, sometimes detected bacterial serotypes (Pavlic and Griffiths, 2009).

Because of the epidemiological and clinical importance of some bacterial strains such as *M. tuberculosis*, it is interesting the application of related techniques like typing by PCR, in breaking through in a better understanding of the ecology and epidemiology of these bacteria. Some studies show this approach to evaluate the discriminatory power of different methods for genotyping of MTC isolates, they compared the performance of i) IS6110 DNA fingerprint, ii) spoligotyping and iii) 24-loci MIRU-VNTR (mycobacterial interspersed repetitive units - variable number of tandem repeats) typing in a long term study on the epidemiology of tuberculosis (TB) in Schleswig-Holstein, the most-northern federal state of Germany (Roetzer et al, 2011), other group studied the clustered cases identified using a population-based universal molecular epidemiology strategy over a 5-year period. Clonal variants of the reference strain defining the cluster were found in 9 (12%) of the 74 clusters identified after the genotyping of 612 M. tuberculosis isolates by IS6110 restriction fragment length polymorphism analysis and mycobacterial interspersed repetitive units-variable-number tandem repeat typing. Clusters with microevolution events were epidemiologically supported and involved 4 to 9 cases diagnosed over a 1- to 5-year period (Pérez-Lago et al, 2011), another study was to compare polymerase chain reaction (PCR)-based methods--spoligotyping and mycobacterial interspersed repetitive units (MIRU) typing--with the gold-standard IS6110 restriction fragment length polymorphism (RFLP) analysis in 101 isolates of Mycobacterium tuberculosis to determine the genetic diversity of M. tuberculosis clinical isolates from Delhi, North India (Varma-Basil et al 2011) and finally, a study where Forty three isoniazid (INH)-resistant *M. tuberculosis* isolates were characterized on the basis of the most common INH associated mutations, katG315 and mabA -15C→T, and phenotypic properties (i.e. MIC of INH, resistance associated pattern, and catalase activity). Typing for resistance mutations was performed by Multiplex Allele-Specific PCR and sequencing reaction (Soudani et al, 2011).

1.6 ERIC-PCR

Amplification of enterobacterial repetitive intergenic consensus by PCR (ERIC-PCR) has only been used sporadically to detect mycobacteria. ERIC sequences are repetitive elements of 126 bp that appear to be restricted only to transcribed regions of chromosome. Its position in the genome appears to be different in different species. As any technique, ERIC-PCR is used as typing, to study the clonal relationship in various Gram-negative bacteria such as *Acinetobacter baumannii*. The DNA patterns obtained with the ERIC-PCR are usually less complex than those generated by other techniques such as REP-PCR. The technique is quick and easy to perform, and provides highly reproducible results (Gillings & Holley 1997.).

Additionally, the presence of ERIC sequences has been detected in genome of *M. tuberculosis* (Sechi et al, 1998). Studies showed that the level of differentiation obtained by ERIC-PCR is superior to that obtained by the RFLP-IS6110 genetic profile comparable to that obtained by (*GTG*) *5-PCR* fingerprinting (PCR-GTG). The use of the PCR-GTG, a repetitive marker in the *M. tuberculosis* chromosome with an IS6110 sequence has been successfully applied to a PCR-based fingerprinting method. This method is fast and sensitive and can be applied to the study of the epidemiology of infections caused by *M. tuberculosis* and therapeutic implications for health, particularly when the IS6110 RFLP-DNA profile does not provide any help.

2. Objective

The aim of this study was to conduct a molecular characterization of mycobacteria strains by typing and drug resistance gene mutations from samples of potential TB patients in Northeast Mexico.

3. Methods

Two strategies were conducted to study the samples isolated from patients with probable TB clinical diagnosis from the State Public Health Laboratory of Tamaulipas (LESPT) from The State of Tamaulipas, MX. The first one was the identification of strains as belonged or not to MTC. Second one was to detect mutations on the genes related to drug resistance to major antibiotics against *M. tuberculosis.*

3.1 Samples and cultures

Specimens included in this study were collected over a period of 16 months (October 2008 to January 2010) from acid-fast bacilli AFB-positive sputum obtained from the State Public Health Laboratory (LESPT). Basically, LESPT concentrates most of the TB cases from Tamaulipas. All the samples were taken under the informed consent of the patients. . In addition, a structured test was used to obtain standard demographic and epidemiologic data of the patients. Two sputum consecutive specimens were collected from each individual. These samples were mixed with 1% cetylpyridinium chloride and immediately transported to the LESPT where they were stored at 4o C (Kent and Kubica, 1985). All strains cultured were identified to species level by standard microbiological procedures in the LESPT.

3.2 DNA extraction

Samples were first lysed (tissue samples were mechanically disrupted) and proteins simultaneously denatured in the appropriate lysis buffer. QIAGEN Proteinase K was then added and after a suitable incubation period, lysates were loaded onto the QIAGEN Genomic-tip. DNA binds to the column while other cell constituents passed through. Following a wash step to remove any remaining contaminants, pure, high-molecular-weight DNA was eluted and precipitated with isopropanol. Hands-on time for the complete procedure was just 45 minutes for samples.

Bacterial strains obtained from patients with TB were preliminary analyzed by an antibiogram test to verify if these strains exhibit some class of antibiotic resistant. Approximately, One hundred consecutive strains were selected to further molecular characterization. Bacteria selected were growth in solid Lowenstein-Jensen and 7H9 Middlebrook broths supplemented with 10% (vol/vol) of oleic acid-albumin-dextrose-catalase. After that, the samples were incubated for at least 8 weeks. DNA from bacterial samples was obtained from those grew strains by used the QIAGEN kit (QIAGEN) of according to manufacturing instructions

3.3 Molecular detection of *M. tuberculosis*

The following primers were used (Yeboah-Manu et al. 2001): spacer region-specific primers, spacer region 33 specific (5'ACACCGACATGACGGCGG3') and spacer region 34 specific (5'CGACGGTGTGGGCGAGG3'); IS6110 (5'GGACAACGCCGAATTGCG'3 and 5'TAGGCGTCGGTGACAAAGGCCAC'3), and Mycobacterium genus-specific TB11 (sequence 5'ACCAACGATGGTGTGTCCAT3') and TB12 (sequence 5'CTTGTCGAACCGCATACCCT3'). Expected PCR products are 550, 439, and 172 and 99 bp, respectively.

PCR mixtures contained 20 µl of 2× PCR mix, 10 µL of primer mix with each primer at 0.66 pmol/µL, 0.2 µL of Taq polymerase enzyme (Roche Diagnostics), and 10 µL of extracted DNA. The PCR conditions were 95°C for 3 min; 30 cycles of 95°C for 20 s, 65°C for 30 s, and 72°C for 30 s; and 72°C for 7 min. After PCR, the products were analyzed by electrophoresis in agarose gel.

3.4 Typing methods

For ribotyping, the standardization of PCR was done using three sets of primers to amplify the 16S region (Table 1). Note: The primers 16S and 16S R F were used for amplification of 16S Mycobacterium.

Primer	Sequence	Size (nt)	Tm (°C)	Reference
R1 R2	5'-TTGTACACACCGCCCGTCA-3' 5'-GAAACATCTAATACCT-3'	19 16	62.3 46.5	Sechi A, et al, 1998.
16S F 16S R	5'AGAGTTTGATCCTGGCTC-3' 5'-CGGGAACGTATTCACCG-3'	18 17	57.62 59.61	Strom et al, 2002.
P13P F P11P R	5'-GAGGAAGGTGGGGATGACGT-3' 5'-AGGCCCGGAACGTATTCAC-3'	20 19	64 60	Sorrell et al, 1996.

Table 1. Primers used for amplification of 16S Ribosomal DNA of Mycobacterium.

A set of chosen primers, which amplified for desired sequences are shown below (Table 1). (Strom et al, 2002).

Ribotyping by PCR was performed with two primers complementary to conserved regions. The sequences of the primers were described on Table 1. Amplifications were carried out in a final volume of 25 µl. Twenty five cycles of amplification were performed, with each cycle consisting of 2 min of denaturation at 94°C, 45 seconds of annealing at 62°C, and 1 min at 72°C. The last cycle consisted of a 7 min extension at 72°C. The amplification products were visualized after electrophoresis at 90 V for 90 min in a 2% agarose gel, and the gel was stained with SYBER Gold (Invitrogen).

3.5 ERIC-PCR

For ERIC-PCR, a pair of primers (Sechi et al, 1998) used and their characteristics are described below (Table 2).

Amplification reactions were performed in a volume of 50 µl with final amounts of 1 U of Taq polymerase, 20 mM Tris-HCl (pH 8.3), 50 mM KCl, 1.5 mM MgCl2, and 200 µM of deoxynucleoside triphosphate (Gibco, BRL, Life Technology, Paisley, United Kingdom). The reaction mixtures were then incubated for 5 min at 95°C, followed by 35 cycles of 94°C for 30 s, Touch-down (47-57°C), and 65°C for 4 min and a final extension at 70°C for 7 min. The amplification products were visualized after electrophoresis at 90 V for 90 min in a 2% agarose gel, and the gel was stained with SYBER Gold (Invitrogen).

Primer	Nucleotide sequence	Size (nt)	Tm(°C)
ERIC 1R	5'-ATGTAAGCT CCT GGGGATTCAC-3'	22	62.7
ERIC 2	5'-AAGTAAGTGACT GGGGTGAGCG-3'	22	64.5

Table 2. Primers for ERIC-PCR

3.6 Gene drug resistant analysis

Eight pairs of PCR primers (PR1 to PR16) were used to simultaneously amplify regions of eight genes associated with resistance to six antituberculosis drugs. In addition, eight pairs (PR17 to PR32) of internal PCR primers were then used to determine the DNA sequences of these genes (Table 3 and 4)

3.7 Sequencing

PCR products obtained from only 36 out of 100 bacterial strains for ERIC-PCR and 15 bacteria drug resistant were purified with an EXO-SAP. Components were supplemented with gold buffer (Applied Biosystem) and sequenced on an Applied Biosystem 310 Genetic analyzer (ABI Prism 310 Genetic analyzer), using big dye terminator cycle sequencing Ready Kit (Applied Biosystem).

For Drug resistant, the purified samples were analyzed with the ABI PRISM 3100 Genetic Analyzer (Applied Biosystems). The DNA sequences are collected and edited with Data Collection software version 1.01 and Sequencing Analysis version 3.7 (Applied Biosystems) and compared with those of M. tuberculosis H37Rv (GenBank access no. NC_000962) with the program Geneious version 4.5.4 (Software Development Biomatters Ltd).

Additionally, for ribotyping, sequence of 16S Ribosomal DNA of mycobacterial determined in an ABI Prism® 3130. Obtained sequences were analyzed in NCBI database using BlastN analysis (http://www.ncbi.nlm.nih.gov/blast/Blast.cgi). Alignment editing of 16S sequences was performed by Chromas Lite 2.0, BioEdit Sequence Alignment Editor Version 7.0.4.1 and CLC Sequence Viewer Version 6.1 software.

Finally, the comparison between isolated Mycobacteria and reported Mycobacterium tuberculosis strains were made by a microbial identification and phylogenetic analysis of obtained data using MEGA4 Software Version 4.0.2. Tree topologies were determined by methods of Minimum evolution criterion and Maximum Parsimony, with a value of reliability, "Bootstrap" of 100 replications for phylogenetic analysis.

4. Results and conclusion

Male population in Tamaulipas is the most affected by TB with 61% of the isolates evaluated in this work. Geographical distribution of infected people represented a greater proportion in Central and South of the State with 52% and 45% of isolates evaluated, respectively.

Region	Sequences	Position	Size (bp)
rpoB	PR1 (forward) 5-CCGCGATCAAGGAGTTCTTC-3 PR2 (reverse) 5-ACACGATCTCGTCGCTAACC-3	1256-1275 1570-1551	315
katG	PR3 (forward) 5-GTGCCCGAGCAACACCCACCCATTACAGAAAC -3 PR4 (reverse) 5-TCAGCGCACGTCGAACCTGTCGAG-3	1-32 223-2200	2,223
mabA	PR5 (forward) 5-ACATACCTGCTGCGCAATTC-3 PR6 (reverse) 5-GCATACGAATACGCCGAGAT-3	-217 a -198 1145-1126	1,362
embB	PR7 (forward) 5-CCGACCACGCTGAAACTGCTGGCGAT-3 PR8 (reverse) 5-GCCTGGTGCATACCGAGCAGCATAG-3	640-665 3387-3303	2,748
pncA	PR9 (forward) 5-GGCGTCATGGACCCTATATC-3 PR10 (reverse) 5-CAACAGTTCATCCCGGTTC-3	-80 a -61 590-572	670
rpsL	PR11 (forward) 5-CCAACCATCCAGCAGCTGGT-3 PR12 (reverse) 5-GTCGAGAGCCCGCTTGAGGG-3	4-23 575-556	572
rrs	PR13 (forward) 5-AAACCTCTTTCACCATCGAC-3 PR14 (reverse) 5-GTATCCATTGATGCTCGCAA-3	428-447 1756-1737	1,329
gyrA	PR15 (forward) 5-GATGACAGACACGACGTTGC-3 PR16 (reverse) 5-GGGCTTCGGTGTACCTCAT-3	1-19 397-379	398

Table 3. Primers for multiplex-PCR (Sekiguchi et al. 2007)

Thirty-seven out of 40 samples were analyzed by 16S gene sequences, 34 of them were grouped in the MTC, and the 3 remaining sequences were integrated into the NMTC (Figure 1). Moreover, from 37 sequences analyzed, only 12 of these showed polymorphisms on a segment of 250 nucleotides with an average size of 750 nucleotides for each sequence. Of these 12 sequences, only 3 isolates were grouped in NMTC, the 9 remaining isolated strains show polymorphism in their nucleotide sequences belong to the MTC. Two sequences of isolates tested showed 100% and 98% identity respectively with the species of *M. fortuitum* according to our analysis in the NCBI database (strain *M. fortuitum* 16S gene with accession number DQ973806.1 and strain *M. fortuitum* 16S gene with accession number AY457066.1, respectively).

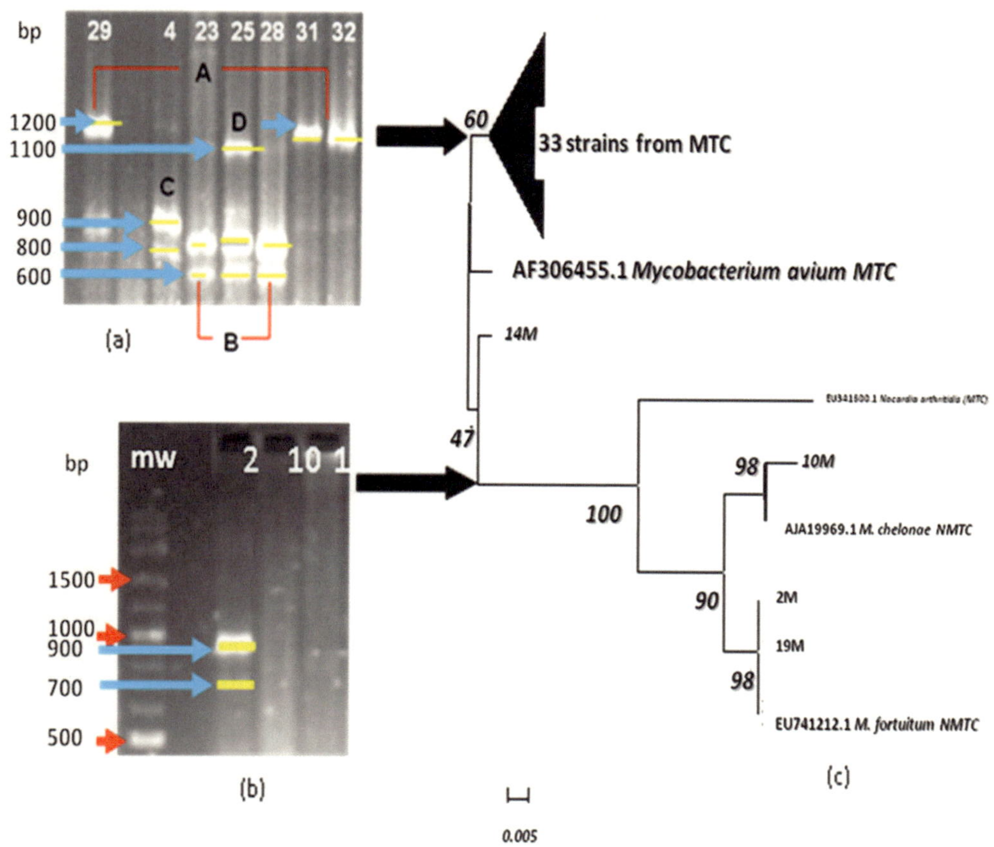

Fig. 1. Comparison of genetic profiles from isolated mycobacteria by ERIC-PCR VS Ribotyping (a) In picture, letters A, B, C, and D show four different genetic profiles of ERIC-PCR grouped in 7 isolates of mycobacteria. Arrow on direction of figure (c) indicates 7 isolated strains are part of the MTC. (b) Image shows ERIC-PCR amplifications of 3 isolates (samples 2M, 10M, 19M) clustered in NMTC. (c) Phylogenetic tree based on comparison of 16S gene mycobacterium species from MTC and NMTC. *Mycobacterium spp* sample.

A third sequence of one isolated strain showed 100% identity with *M. chelonae* (*M. chelonae* strain T9 with AM884324.1 access number). In this sense, it is useful to mention that LESPT identified these strains as *M. tuberculosis* based only on their microbiological and biochemical results. It is important to mention that LESPT does not conducted molecular analysis to identify their samples.

One out of the two identified strains was *M. fortuitum*, made by sequencing, but no for microbiology, since this was identified as *M. tuberculosis* by LESPT. The other isolated itself coincided with both techniques.This gives us a different result, although microbiology diagnostic and taken at this time, we do believe strongly that this is due to *M. fortuitum.*

Gene	Sequences	Position
rpoB	PR17 5-TACGGCGTTTCGATGAAC-3 (complementary strand)	1529–1512
katG	PR18 5-ACGTAGATCAGCCCCATCTG-3 (complementary strand) PR19 5-GAGCCCGATGAGGTCTATTG-3 PR20 5-CCGATCTATGAGCGGATCAC-3 PR21 5-GAACAAACCGACGTGGAATC-3	689–670 574–593 1162–1181 1729–1748
mabA	PR22 5-ACATACCTGCTGCGCAATTC-3	-217 a -198
embB	PR23 5-ACGCTGAAACTGCTGGCGAT-3 PR24 5-GTCATCCTGACCGTGGTGTT-3 PR25 5-GGTGGGCAGGATGAGGTAGT-3 (complementary strand) PR26 5-CACAATCTTTTTCGCCCTGT-3 PR27 5-GCGTGGTATCTCCTGCCTAAG-3	646–665 1462–1481 1596–1577 2007–2026 2581–2601
pncA	PR28 5-GGCGTCATGGACCCTATATC-3	-80 -61
rpsL	PR29 5-CCAACCATCCAGCAGCTGGT-3	4–23
Rrs	PR30 5-CAGGTAAGGTTCTTCGCGTTG-3 (complementary strand) PR31 5-GTTCGGATCGGGGTCTGCAA-3	979–959 1291–1310
gyrA	PR32 5-GATGACAGACACGACGTTGC-3	1–19

Table 4. Primers for sequencing (Sekiguchi et al. 2007)

The use of the reference strain H37Rv of *M. tuberculosis* sequence served as a reference or guide for clustering of the isolates studied, since in the phylogenetic analysis, the type strain H37Rv was integrated with the 34 isolated MTC, thus reaffirming the phylogenetic relationship of isolates tested with the species *M. tuberculosis* (Figure 1). On the other hand, the reference sequence of *Nocardia arthritidis* (No. Access EU841600) showed no relation with the 37 evaluated strains in phylogenetic analysis, being totally separated from the two complexes formed, MTC and NMTC. This comparison is done, because of *Nocardia* is also acid-resistant and can be confused with *Mycobacteria* on microscopic analysis, hence the importance of making the comparison. Then, It should be mentioned that the identification of mycobacteria isolated from the 16S sequences was proved to be an appropriate strategy to establish the level of genetic relatedness among isolates studied and know how related isolates were isolated or if these could be separated into MTC and NMTC.

ERIC-PCR technique gave 4 different genetic profiles for mycobacteria (Figure 1). It should be emphasized that three of these genetic profiles are consistent with those reported in molecular epidemiology studies by amplifying sequences ERIC (Sechi et al, 1998). From 34 isolates clustered in the complex of M. tuberculosis, seven of these 4 make up the genetic profiles obtained by ERIC-PCR. Then they expect the rest of the isolates (27 isolates) that make up this complex terms grouped in the 4 genetic profiles (A, B, C and D) obtained by ERIC-PCR.

However, one aspect to consider in the results obtained in this work is that of obtaining the 4 different genetic profiles amplified by ERIC-PCR, they did not allow to discriminate among species of MTC and NMTC as it was expected, those profiles or genetic patterns of 3 isolated strains have been totally different from other profiles of 7 isolated strains (Figure 1), all of 10 isolated strains are in MTC.

Regarding to genetic relatedness of 40 isolates of mycobacteria studied, phylogenetic analysis of 16S gene sequences showed 37 sequences, which formed two groups. The first group of MTC consisting of 34 isolates and the second group resulted in NMTC consisting of 3 isolates. The percentages of identity were from 98% to 100% for isolates clustered in both complexes.

The analysis on relationship between the isolates studied and their geographical origin revealed that the Mycobacterium tuberculosis complex is distributed both in the central and south of the state of Tamaulipas, MX. Meanwhile, species such as *M. fortuitum* and *M. chelonae* are only found circulating in the central region of Tamaulipas.

It should be noted that the isolates studied are only samples originating from the central and southern Tamaulipas. No isolated strains were obtained from northern part of Tamaulipas, which would have complemented the results of this research. For example, as mentioned before in those border states (US-Mexico border) there is a great number of people and cattle to move from different parts of country and abroad, which could suggest existence of different strains of *M. tuberculosis* in Tamaulipas, and even the presence of other species found in the central and southern Tamaulipas, MX, allowing us to know not only that isolated strains in each region are preferably, but also known how those isolated strains are circulating all around Tamaulipas.

Note this work was limited to samples of LESPT; however the proposal will be to analyze samples from all health districts in Tamaulipas, and analyze samples of other states, such as Veracruz and/or Coahuila. Additionally, we will seek to refine the ERIC-PCR and implementing the MIRU-VNTR and spoligotyping for more complete diagnosis. In addition, arrangements are made between the County LESPT and McAllen, Texas, United States, to have samples (about 14,000 isolates) identified and stored in the United States from Tamaulipas.

Finally, the preliminary results were shown (Table 5), where mutations, insertions, transversions, and transitions were found. In general, the mutations obtained did not alter the chemical or structural composition of proteins that confer resistance to an antibiotic to the mycobacteria and their regions sequenced. In these particular cases, we selected to work with isolated strains were resistant to antibiotics commonly administered in Mexico, the results obtained for the case of pyrazinamide, a silent mutation was found, so that the resistance exhibited by the bacteria should be caused by mutations on the sequenced region. In the case of isoniazid *mab*A gene, we found an insertion within the gene that could be the cause of resistance exhibited.

These results indicate that DNA sequencing-based method was effective for detection of MDR strains. However, when novel mutations in drug resistance-related genes are detected by the method, it is essential to also perform drug susceptibility testing, because novel mutation may not be associated with drug resistance.

Gene	Number of sample	Changes	
		Nucleotide	Protein
*rpo*B	2	CGG-TGG	R476W
rrs	2	TGG-AGG	W193R
*rps*L	2	AAA-AAG	K121K
*pnc*A	2	GGT-GGC	G75G
*mab*A	2	702 T insertion -15 C-T upstream	
*gyr*A	2	GAG-CAG	E21Q

Table 5. Relationship of changes found in the sequences of the genes of interest.

5. Conclusion

In conclusion, two strategies were carried out to study the samples isolated from patients with TB diagnostic of LESPT from Tamaulipas, MX. The first was the identification of isolates and determine if these isolates belonged or not to MTC. Second, to determinate if mutations in primary sequences of genes related to resistance to major antibiotics used to kill mycobacteria in Tamaulipas, could be detected.

For the first part of the study, there were used 3 strategies, a multiplex-PCR, ERIC-PCR, and ribotyping. For the second direct amplification of 16S DNA region was performed.

Multiplex-PCR for 99% of the samples coincided with the microbiological results, identifying *M. tuberculosis*, primary. In the case of ERIC-PCR, the samples could be grouped into 4 different groups; however it could differentiate between MTC and NMTC. Finally, ribotyping produced promising results by discriminating the isolated strains and identifying 99% as *M. tuberculosis*.

Finally, the results indicate that DNA sequencing-based method was effective for detection of MDR strains. However, when novel mutations in drug resistance-related genes are detected by the method, it is essential to also perform drug susceptibility testing, because novel mutations are not always associated with drug resistance.

6. Perspectives

This kind of work will answer other questions: is it necessary ribotyping before or after ERIC-PCR or multiplex-PCR and it is important to recognize each species of Mycobacteria to understand if TB strains would circulate all around Tamaulipas and if those ones would be or get in USA too? In a few years we will understand this phenomenon; meanwhile this chapter makes the first approach to understand how TB strains are moving and if those strains are or not drug resistant on a border State between USA and Mexico. The present investigation continues, pending to sequence regions of resistance to pyrazinamide and ethambutol, which are largest genes.

7. Acknowledgments

Financing of this research was by Instituto Politécnico Nacional (projects SIP 20090679 and SIP 20100504). Narváez-Zapata J. A. and Reyes-López M. A. are fellows of CONACYT-SNI,

COFAA, and EDI from Instituto Politécnico Nacional. Authors are also grateful for the support received from Network of Drug Development and Diagnostic Methods (RED FARMED) from CONACyT. Finally, authors really appreciated the samples submitted for QFB. Norma Alicia Villareal Reyes from LESPT.

8. References

Alam S., Brailsford S., Whiley R. & Beighton D. (1999). PCR-Based Methods for Genotyping Viridans Group Streptococci. Journal of Clinical Microbiology. Vol. 37, No. 9. p. 2772-2776.

Borrell S, Gagneux S. (2011). Strain diversity, epistasis and the evolution of drug resistance in Mycobacterium tuberculosis. Clin Microbiol Infect. Jun;17(6):815-20

Cadmus SI, Falodun OI, Fagade OE. (2011). Methods of sputum decontamination with emphasis on local tuberculosis laboratories. Afr J Med Med Sci. 40(1):5-14

Connell DW, Berry M, Cooke G, Kon OM. (2011).Update on tuberculosis: TB in the early 21st century. Eur Respir Rev. 20(120):71-84

Ferrer G, Acuna-Villaorduna C, Escobedo M, Vlasich E, Rivera M. (2010). Outcomes of multidrug-resistant tuberculosis among binational cases in El Paso, Texas. Am J Trop Med Hyg. 83(5):1056-8

Fitchett JR, Vallecillo AJ, Espitia C. (2011). Tuberculosis transmission across the United States-Mexico border. Rev Panam Salud Publica. 29(1):57-60

Gillings, M. and Holley, M. (1997). Repetitive element PCR fingerprinting (rep-PCR) using enterobacterial repetitive intergenic consensus (ERIC) primers is not necessarily directed at ERIC elements. The Society for Applied Bacteriology, Letters in Applied Microbiology. No. 25. p. 17-21.

Havlir, D.V. and Barnes, P.F. (1999) Tuberculosis in patients with human immunodeficiency virus infection. N. Engl. J. Med. 340, 367–373

Kent, P. T., and G. P. Kubica. (1985). Public health mycobacteriology: a guide for the level III laboratory. U.S. Department of Health and Human Services, Centers for Disease Control, Atlanta, Ga.

Marhöfer, RJ, Oellien F, Selzer PM. (2011). Drug discovery and the use of computational approaches for infectious diseases. Future Med Chem. 3(8):1011-25

Pang Y, Zhou Y, Wang S, Lu J, Lu B, He G, Wang L, Zhao Y. (2011). A novel method based on high resolution melting (HRM) analysis for MIRU-VNTR genotyping of Mycobacterium tuberculosis. J Microbiol Methods. 86(3):291-7

Pavlic M, Griffiths MW. (2009). Principles, applications, and limitations of automated ribotyping as a rapid method in food safety. Foodborne Pathog Dis. 6(9):1047-55

Pérez-Lago L, Herranz M, Martínez Lirola M; on behalf of the INDAL-TB Group, Bouza E, García de Viedma D. 2011. Characterization of Microevolution Events in Mycobacterium tuberculosis Strains Involved in Recent Transmission Clusters. J Clin Microbiol. 49(11):3771-3776.),

Rodwell TC, Kapasi AJ, Moore M, Milian-Suazo F, Harris B, Guerrero LP, Moser K, Strathdee SA, Garfein RS. (2010). Tracing the origins of Mycobacterium bovis tuberculosis in humans in the USA to cattle in Mexico using spoligotyping. Int J Infect Dis.;14 Suppl 3:e129-35

Roetzer A, Schuback S, Diel R, Gasau F, Ubben T, di Nauta A, Richter E, Rüsch-Gerdes S, Niemann S. 2011.Evaluation of Mycobacterium tuberculosis typing methods in a four-year study in Schleswig-Holstein, Northern Germany. J Clin Microbiol. Oct 12 [Epub ahead of print])

Sala C, Hartkoorn RC. (2011). Tuberculosis drugs: new candidates and how to find more. Future Microbiol. 6(6):617-33

Sechi A., Zanetti S., Dupré I., Delogu G. and Fadda G. (1998). Enterobacterial Repetitive Consensus Sequences as Molecular Targets for Typing of Mycobacterium tuberculosis Strains. Journal of Clinical Microbiology. Vol. 3, No. 1. p. 128-132.

Sekiguchi J, Miyoshi-Akiyama T, Augustynowicz-Kopeć E, Zwolska Z, Kirikae F, Toyota E, Kobayashi I, Morita K, Kudo K, Kato S, Kuratsuji T, Mori T, Kirikae T. (2007). Detection of multidrug resistance in Mycobacterium tuberculosis. J Clin Microbiol. 45(1):179-92.

Sonnenberg, P. et al. (2005) How soon after infection with HIV does the risk of tuberculosis start to increase? A retrospective cohort study in South African gold miners. J. Infect. Dis. 191, 150–158

Sorrell TC, Chen SC, Ruma P, Meyer W, Pfeiffer TJ, Ellis DH, Brownlee AG. (1996). Concordance of clinical and environmental isolates of Cryptococcus neoformans var. gattii by random amplification of polymorphic DNA analysis and PCR fingerprinting. J Clin Microbiol. 34(5):1253-60.

Soudani A, Hadjfredj S, Zribi M, Messaadi F, Messaoud T, Masmoudi A, Zribi M, Fendri C. Genotypic and phenotypic characteristics of tunisian isoniazid-resistant Mycobacterium tuberculosis strains. 2011. J Microbiol. 49(3):413-7).

Sougakoff W. (2011). Molecular epidemiology of multidrug-resistant strains of Mycobacterium tuberculosis. Clin Microbiol Infect. 17(6):800-5

SSA. (2009). Secretaría de Salud. Informe México 2009.

Ström K, Sjögren J, Broberg A, Schnürer J. (2002). Lactobacillus plantarum MiLAB 393 Produces the Antifungal Cyclic Dipeptides Cyclo(L-Phe-L-Pro) and Cyclo(L-Phe-trans-4-OH-L-Pro) and 3-Phenyllactic Acid. Appl Environ Microbiol. 68(9):4322-7.

Struelens, J., and Members of the European Study Group on Epidemiological Markers (MESGEM). (1996). Consensus guidelines for Appropriate use and evaluation of microbial epidemiologic typing systems. Clin Microbiology and Infect, 2 (1): 2-11.

Varma-Basil M, Kumar S, Arora J, Angrup A, Zozio T, Banavaliker JN, Singh UB, Rastogi N, Bose M. 2011. Comparison of spoligotyping, mycobacterial interspersed repetitive units typing and IS6110-RFLP in a study of genotypic diversity of Mycobacterium tuberculosis in Delhi, North India. Mem Inst Oswaldo Cruz. 106(5):524-35)

Venkatesh KK, Swaminathan S, Andrews JR, Mayer KH. (2011). Tuberculosis and HIV co-infection: screening and treatment strategies. Drugs. 71(9):1133-52

Wagner KD, Pollini RA, Patterson TL, Lozada R, Ojeda VD, Brouwer KC, Vera A, Volkmann TA, Strathdee SA. (2011). Cross-border drug injection relationships among injection drug users in Tijuana, Mexico. Drug Alcohol Depend. 113(2-3):236-41

Yeboah-Manu D, Yates MD, Wilson SM. (2001). Application of a simple multiplex PCR to aid in routine work of the mycobacterium reference laboratory. J Clin Microbiol. 39(11):4166-8.

Yew WW, Sotgiu G, Migliori GB. (2010). Update in tuberculosis and nontuberculous mycobacterial disease Am J Respir Crit Care Med. 2011 Jul 15;184(2):180-5

18

Development of a Molecular Platform for GMO Detection in Food and Feed on the Basis of "Combinatory qPCR" Technology

Sylvia Broeders[1], Nina Papazova[1],
Marc Van den Bulcke[2] and Nancy Roosens[1]
[1]*Wetenschappelijk Instituut Volksgezondheid, Institut Scientifique de Santé Publique, Platform Biotechnology and Molecular Biology,*
[2]*European Commission, Joint Research Centre, Institute for Health and Consumer Protection, Molecular Biology and Genomics Unit*
[1]*Belgium*
[2]*Italy*

1. Introduction

Fifteen years after the first commercialisation of biotech crops, the global area of their cultivation comprises more than one billion hectares. The increase in the area between 1996 and 2010 is 87-fold which makes biotech crops the fastest adopted technology in modern agriculture (James, 2010).

In 2010, 184 Genetically Modified (GM - see glossary) events, representing 24 crops have already received worldwide regulatory approval. To date, 29 countries have cultivated GM crops, whereas 59 countries have granted regulatory approvals for their import for food and feed use and release into the environment. The six main countries cultivating GM crops are USA, Brazil, Argentina, India, Canada and China. In the EU the cultivation area of biotech crops amounts only 0,1%of the cultivation area reaching 125 million hectares in 25 countries (Stein & Rodriguez-Cerezo, 2009). The most important biotech crop is soybean (50% of the biotech crops cultivation area), followed by maize (31%), cotton (14%) and oilseed rape (4%) (James, 2010).

Herbicide tolerance and insect resistance are the main traits used in the first generation of GM crops. After 2009, many GM events conferring novel traits have entered the regulatory system. New traits were introduced in soybean, maize, cotton and oilseed rape. The second generation of traits comprises altered crop composition, new herbicide tolerances, virus and nematode resistance and abiotic stress tolerance. Furthermore, new crops such as potato and rice were approved in different countries (Stein & Rodriguez-Cerezo, 2009). Moreover, gene stacking is a trend that is likely to increase in the near future. There are new events containing up to four stacked traits in the regulatory pipeline. A maize stacked event containing up to eight traits is in an advanced research and development stage (Dow AgroSciences SmartStax® platform; James, 2010).

In the EU until 2010, 39 events were authorised for import and processing in food and feed and two for cultivation. This includes 23 maize events from which 12 containing double and triple stacked traits, seven cotton events from which two containing stacked traits, four oilseed rape events, three soybean events, one potato and one sugar beet event. A detailed list of the EU-authorised GM events per crop with their main traits is presented in table 1.

Another tendency is that new GM events are not solely developed and commercialised by international biotech companies anymore, but also by scientific governmental institutions. Many of these GM events are commercialised by Asian national research centres (e.g. China, India) and are intended for the local markets. However, as many food and feed materials are imported in the EU from third party countries, events that are not submitted for authorisation in the EU (unauthorised GMO or UGM) might accidentally end up into in the food and feed chain (Stein & Rodriguez-Cerezo, 2009).

In reaction to the public concern about the presence of Genetically Modified Organisms (GMO - see glossary) in the food chain, many countries have adopted a specific legislation with respect to the introduction of GMO on their market. The legislation requirements vary from country to country, but there are some common elements such as case by case safety assessment, distinction between contained use and release into the environment and a distinction between cultivation and use as raw products in processing. Commonly recognised is the concept of substantial equivalence (Shauzu, 2001). In many regulatory systems tolerances or labelling thresholds, varying between 0.9 and 5%, were introduced.

The EU legislation on GMO is complex and consists of several core elements: a pre-authorisation safety assessment, use of a labelling threshold, strict requirements for traceability of the GM products along the food chain and post-market monitoring. Labelling and traceability of new GM products are regulated mainly under Commission Regulations 1829/2003 and 1830/2003. For all events submitted under EC/1829/2003 a safety assessment is performed by the European Food Safety Authority (EFSA- see glossary). Food, feed and environmental risks are evaluated based on the data provided by the company requesting authorisation of a GM product. The food and feed safety assessment includes several issues such as allergenicity, toxicology, nutritional characteristics and post-market monitoring of the GM food and feed. The environmental risk assessment includes evaluation of the potential of gene transfer, interaction of the GM plant with target and non-target organisms and monitoring (EFSA, 2011).

A very important issue is the molecular characterisation of the GM event. The objective of this characterisation is to obtain information on the introduced trait or genetic modification and to assess if unintended effects due to the genetic modification have taken place (Organisation for Economic Co-operation and Development [OECD], 2010). The molecular characterisation is an evaluation of relevant scientific data on the transformation process and vector constructs used, inserted transgenic sequences, copy number of the inserts, presence of partial copies, expression of the transgenic protein, stability and the inheritance of the transgenic insert (EFSA, 2011). The information on the elements introduced in the GMO as well as the sequence information on the junction regions between the plant genome and the transgenic insert are an essential part as they are related to the development of detection methods.

Transformation event (Unique identifier)	**Trait**	**Transformation event (Unique identifier)**	**Trait**
		Maize single events	
Bt11 (SYN-BT Ø11-1)	Insect resistance Herbicide tolerance (glufosinate)	**DAS59122 (DAS-59122-7)**	Insect resistance (Coleopteran insects) Herbicide tolerance (glufosinate)
DAS1507 (DAS-Ø15Ø7-1)	Insect resistance (Lepidopteran insects) Herbicide tolerance (glufosinate)	**GA21 (MON-ØØØ21-9)**	Herbicide tolerance (glyfosate)
MON810 (MON-ØØ81Ø-6)	Insect resistance (Lepidopteran insects)	**MON863 (MON-ØØ863-5)**	Insect resistance (Coleopteran insects)
T25 (ACS-ZMØØ3-2)	Herbicide tolerance (glufosinate)	**NK603 (MON-ØØ6Ø3-6)**	Herbicide tolerance (glyfosate)
MON88017 (MON-88Ø17-3)	Insect resistance (Coleopteran insects) Herbicide tolerance (glyfosate)	**MIR604 (SYN-IR6Ø4-5)**	Insect resistance (Coleopteran insects)
MON89034 (MON-89Ø34-3)	Insect resistance (Lepidopteran insects)	***Bt176 (SYN-EV176-9)***	*Insect resistance (European corn borer)* *Herbicide tolerance (glufosinate)*
3272 maize (SYN-E3272-5)	*Altered composition (increased a-amilase content)*	***MIR162 (SYN-IR162-4)***	*Insect resistance (Lepidopteran insects)*
98140 (DP-098140-6)	*Herbicide tolerance (ALS-inhibiting herbicides)*		
		Maize stacked events	
DAS1507xNK603 (DAS-Ø15Ø7-1xMON-ØØ6Ø3-6)	Insect resistance (Coleopteran insects) Double herbicide tolerance (glufosinate and glyfosate)	**NK603xMON810 (MON-ØØ6Ø3-6 x MON-ØØ81Ø-6)**	Insect resistance (Lepidopteran insects) Herbicide tolerance (glyfosate)
DAS59122xNK603 (DAS-59122-7xMON-ØØ6Ø3-6)	Insect resistance (Coleopteran insects) Double herbicide tolerance (glufosinate and glyfosate)	**MON863xMON810 (MON-ØØ863-5 x MON-ØØ81Ø-6)**	Double insect resistance (Lepidopteran and Coleopteran insects)
Bt11xGA21 (SYN-BTØ11-1xMON-ØØØ21-9)	Insect resistance (Lepidopteran insects) Double herbicide tolerance (glufosinate and glyfosate)	**MON863xNK603 (MON-ØØ863-5 x MON-ØØ6Ø3-6)**	Insect resistance (Coleopteran insects) Herbicide tolerance (glyfosate)

Transformation event (Unique identifier)	Trait	Transformation event (Unique identifier)	Trait
MON88017xMON810 (MON-88Ø17-3xMON-ØØ81Ø-6)	Double insect resistance (Lepidopteran and Coleopteran insects) Herbicide tolerance (glyfosate)	**MON89034xNK603 (MON-89Ø34-3x MON-ØØ6Ø3-6)**	Insect resistance (Lepidopteran) Herbicide tolerance (glyfosate)
DAS1507xDAS59122 (DAS-Ø15Ø7x DAS-59122-7)	Double insect resistance (Lepidopteran and Coleopteran insects) Herbicide tolerance (glufosinate)	**MON89034xMON88017 (MON-89Ø34-3x MON-88Ø17-3)**	Double insect resistance (Lepidopteran and Coleopteran insects) Herbicide tolerance (glyfosate)
MON863xMON810XNK 603 (MON-ØØ863-5xMON-ØØ81Ø-6xMON-ØØ6Ø3-6)	Double insect resistance (Lepidopteran and Coleopteran insects) Herbicide tolerance (glyfosate)	**DAS59122xDAS1507xN K603 (DAS-59122-7xDAS-Ø15Ø7xMON-ØØ6Ø3-6)**	Double insect resistance (Lepidopteran and Coleopteran insects) Double herbicide tolerance (glyfosate and glufosinate)
GA21xMON810 (MON-ØØØ21-9 x MON-ØØ81Ø-6)	*Insect resistance (Lepidopteran insects) Herbicide tolerance (glyfosate)*		
	Cotton single events		
MON1445 (MON-Ø1445-2)	Herbicide tolerance (glyfosate)	**MON15985 (MON-15985-7)**	Insect resistance (Lepidopteran insects)
MON531 (MON-ØØ531-6)	Insect resistance	**LLcotton25 (ACS-GHØØ1-3)**	Herbicide tolerance (glufosinate)
GHB614 (BCS-GHØØ2-5)	Herbicide tolerance (glyfosate)		
	Cotton stacked events		
MON15985xMON1445 (MON-15985-7 x MON-Ø1445-2)	Insect resistance (Lepidopteran insects) Herbicide tolerance (glyfosate)	**MON531xMON1445 (MON-ØØ531-6 x MON-Ø1445-2)**	Insect resistance Herbicide tolerance (glyfosate)
281-24-236/3006-210-23 (DAS-24236-5 x DAS-21Ø23-5)	*Insect resistance (Lepidopteran insects) Herbicide tolerance (glufosinate)*		
	Oilseed rape single events		
GT73 (MON-ØØØ73-7)	Herbicide tolerance (glyfosate)	**T45 (ACS-BNØØ8-2)**	Herbicide tolerance (glufosinate)
Ms8, Rf3, MS8xRf3 (ACS-BNØØ5-8ACS-BNØØ3-6ACS-BNØØ5-8 x ACS-BN003-6)	Herbicide tolerance (glufosinate) Fertility restoration	*Ms1, Rf1, Ms1xRf1 (ACS-BNØØ4-7 ACS-BNØØ1-4 ACS-BNØØ4-7xACS-BNØØ1-4)*	Herbicide tolerance (glufosinate) Fertility restoration

Transformation event (Unique identifier)	Trait	Transformation event (Unique identifier)	Trait
Ms1, Rf2, Ms1xRf2 (ACS-BNØØ4-7 ACS-BNØØ2-5 ACS-BNØØ4-7xACS-BNØØ2-5)	Herbicide tolerance (glufosinate) Fertility restoration	*Topas 19/2 (ACS-BNØØ7-1)*	Herbicide tolerance (glufosinate)
Soybean single events			
GTS40-3-2 (MON-Ø4Ø32-6)	Herbicide tolerance (glyfosate)	**A2704-12 (ACS-GMØØ5-3)**	Herbicide tolerance (glufosinate)
MON89788 (MON-89788-1)	Herbicide tolerance (glyfosate)	***356043 (DP-356043-5)***	*Double herbicide tolerance (glyfosate and ALS-inhibiting herbicides)*
305423 (DP-305423-1)	*High oleic acid content*	***A5547-127 (ACS-GM006-4)***	*Herbicide tolerance (glufosinate)*
MON87701 (MON-877Ø1-2)	*Insect resistance (Lepidopteran insects)*		
Potato single events			
EH92-527-1 (BPS-25271-9)	Low amylase content		
Sugar beet single events			
H7-1 (KM-ØØØ71-4)	Herbicide tolerance (glyfosate)		
Rice single events			
LLrice62 (ACS-OSØØ2-5)	*Herbicide tolerance (glufosinate)*		

Table 1. GM events authorised in the EU and events under under EC/619/2011 (*in italic*).

A labelling threshold of 0,9% is set up for all authorised GM events in the EU. Food and feed products containing GM events above this threshold have to be labelled as 'containing GMO'. The existence of a labelling threshold requires development of a system for GMO detection and quantification. Several types of methods exist, primarily bioassays, both protein-based (immunological) and DNA-based (mainly based on the Polymerase Chain Reaction (PCR) technology). The protein assays are based on the immunological reaction between the target protein and the specific antibody coupled with colorimetric detection (Holst-Jensen, 2009). Practical applications are the ELISA test or flow strip tests, which are widely used in testing of seed or grain materials. For instance, the United States Department of Agriculture- Grain Inspection, Packers and Stockyards (USDA-GIPSA, 2011) has certified several protein-based rapid kits for detection of biotech-derived grain/oilseeds. However, sensitivity and reliable quantification are often a problem for the immunological assays, due to for example low protein expression. Additionally, proteins are instable and nearly impossible to be reliably detected in processed products. Therefore, the DNA-based methods provide a reliable alternative for detection. In the European Union (EU), the detection of GMO is based on DNA and the recommended technique is real-time PCR. Moreover, this technique also provides the possibility for quantification of the GM target. In this context it is recommended to express the GM percentage as a ratio between the GM copy numbers and taxon-specific copy numbers (Commission Recommendation EC/787/2004).

The GMO detection policy in the EU is based on two important elements: availability of validated methods for detection and availability of Certified Reference Materials (CRM - see glossary). According to the EU legislation before a new GMO is approved to be released on the market a validated event-specific detection method should be available. The event-specific methods are developed by the company submitting the GMO for authorisation. The company has to develop a method complying with the acceptance criteria described in the document "Definition of Minimum Performance Requirements for analytical methods of GM testing" (ENGL, 2008) developed by the European Network of GMO Laboratories (ENGL - see glossary). The ENGL is a consortium of National Reference Laboratories (NRL - see glossary) assisting the European Union Reference Laboratory for GM Food and Feed (EU-RL GMFF - see glossary) by providing scientific expertise. The EU-RL is responsible for testing and validation of the method submitted by the applicant. Upon validation the method is published on the EU-RL web site (http://gmo-crl.jrc.ec.europa.eu/) and made available for further use in the control laboratories involved in GMO testing.

In addition to detection methods, the EU legislation requires availability of Certified Reference Materials for the authorised events (EC/641/2004; EC/1829/2003). The CRM for GM testing are produced by the EC-JRC Institute for Reference Materials and Measurements (IRMM, BE) and the American Oil Chemists' Society (AOCS, USA) and usually are powder or leaf DNA extract with a certified content of the GM event.

The GM testing laboratories have to verify that they are capable to achieve the method performance characteristics before using it for routine analyses by performing in house validation by testing the relevant validation parameters as described in the guidance document (ENGL, 2011). Additionally, the control laboratories must be accredited under ISO 17025 (2005) or another equivalent international standard (Commission Regulation EC/1981/2006).

Although the EU legislation regulates the availability of event-specific methods for GMO detection, other methods such as construct-specific (recognising the GM constructs with which several events are transformed) or element-specific (detecting the elements present in many GMO) methods are used in the control laboratories in order to perform the analysis. These methods are subject to development and introduction of the laboratories themselves: there are no official guidelines describing how to validate such methods and which parameters have to be assessed.

The increasing GM cultivation worldwide and the number of authorisations in the EU and elsewhere pose a significant challenge to the control laboratories. They have to be able to apply all official methods for GM detection of authorised events. A second problem, are the asynchronous approvals of GM events in the EU and third party countries which can lead to low level presence of non-authorised GMO in food and feed. The recently adopted Commission Regulation EC/619/2011 regulates the presence of events which are pending for authorisation or withdrawn from the market in feed and for which methods for detection and reference materials (RM - see glossary) are available (table 1).

Given the fact that an increasing number of events have to be analysed in order to comply with the legislation requirements, the control laboratories need to develop analytical approaches (platforms) which allow them to perform the analyses in a fast, cost and time-efficient manner.

2. Plant DNA extraction and its impact on GMO detection

2.1 Introduction

In view of the EU legislation on GMO commercialisation and the fact that GM events are being authorised, it is mandatory to have control on the products being used and brought onto the market in the EU. Hereto, detection of GM events in food and feed samples is necessary to decide on the conformity of a sample. To enable this detection, real-time PCR (qPCR) is to date the method of choice. For this purpose, DNA needs to be extracted from the sample under analysis. In this process it is important to obtain not only enough DNA to perform the necessary qPCR reaction(s) (part 3) but also DNA of high quality (i.e. purity and integrity). As PCR is an enzymatic reaction, it is kinetically sensitive and the presence of other substances in the reaction may affect the PCR efficiency by for example impairing the binding of the primers to the target sequence in the genomic DNA. Such interference can have an impact on the GMO analysis cascade, especially on the last step namely the GMO quantification.

It has indeed been shown (Corbisier et al., 2007) that the quality of the DNA used in the qPCR has an important influence on the GM% obtained. Depending on the DNA extraction method used and the degree of purity of the extracted genomic DNA (gDNA), a deviating GM% was recorded. An interlaboratory study designed for the maize event MON 810, further demonstrated a significant influence of the DNA extraction method on the measurement results when using the construct-specific qPCR method while this impact was not seen when the event-specific detection method was utilised (Charels et al., 2007). It must thus be noted that even using 'pure' materials such as reference materials, DNA extraction is not so straightforward and that attention should be paid to the choice of the applied extraction method. This becomes even more important for enforcement laboratories as they mainly have to deal with processed and mixed samples. In this respect, Peano et al. (2004) reported the effect of treatment (mechanical, technological, chemical) of a sample in combination with the applied extraction method on the quality of the gDNA. When the feed and food product showed extensive fragmentation, due to a certain treatment during the preparation, the detection of these DNA fragments was dependant on the kit used for DNA extraction. Furthermore, Bellocchi et al. (2010) demonstrated that the result of a quantification experiment may be affected by the DNA extraction method employed unless DNA extracts that do not comply with previously set criteria were removed from the GM% calculations.

This highlights the importance of taking into account different parameters when using a modular approach (Holst-Jensen & Berdal, 2004). It is necessary to set up criteria for DNA quantity, purity, integrity and inhibition prior to using the extracted DNA in the qPCR reactions and to choose an appropriate DNA extraction method. Furthermore, attention should be paid to the fact that different targets might not be affected in the same way by impurities or co-extracted substances. Both Corbisier et al. (2007) and Cankar et al. (2006) demonstrated that this would impair in a strong way the final result. If, in a GMO quantification the two targets (i.e. the transgene and the taxon-specific element) do not behave in the same way and the PCR efficiencies are deviating too much, the obtained GM% would be biased.

It should also be noted that the extraction method used has a double impact on GMO quantification as not only the sample needs to be extracted but also the CRM. As the DNA extracted from the CRM powder will be used to construct the calibration curve in the quantification experiment it should also be free of inhibitors as this otherwise will affect the PCR efficiency. DNA extracted from the CRM powder needs to be pure and free of inhibitors to obtain a curve falling within the ENGL criteria (ENGL, 2011). Additionally, the PCR efficiencies for the calibrant and the sample should be the same to obtain reliable quantification. As this is not always the case, controls such as dilutions of the sample to evaluate inhibition, should be included in the reaction (point 2.2).

Although many DNA extraction protocols are quite user friendly and many extraction kits exist, their downstream application in qPCR is not clear-cut and additional evaluation of the quality of the extracted gDNA is necessary as well as assessment of the presence of possible PCR inhibitors.

2.2 Assessment of DNA yield, purity, integrity and inhibition

The determination of the DNA concentration in an extract is not straightforward and different techniques exist. The obtained **DNA yield** after extraction can, for example, be determined using spectrophotometry (UV). This determination is based on the absorbance of nucleic acids at a wavelength of 260 nm. It is a method that has been used commonly for the estimation of the concentration of nucleic acids in a range of applications (Sambrook & Russell, 2001). Although it is a fast and simple method, it allows only determination of the concentration in a range of 5 to 50 µg/ml. Another drawback of this method is the fact that it is not specific for double stranded DNA (dsDNA) but also detects RNA and single stranded DNA (ssDNA) molecules (Gallagher, 2011). Additionally, substances like proteins and phenolics also absorb between 220 and 340 nm and can thus interfere with the measurement.

Alternatively, fluorimetry can be used to determine the concentration of the extracted gDNA in the solution (Singer et al., 1997). This method uses a dye that fluoresces upon intercalating in the dsDNA such as the PicoGreen (Molecular Probes). This enables a more specific measurement of the dsDNA amount present in an extract as there is no binding with interfering proteins and only a limited interaction with RNA and ssDNA. This method is more sensitive than UV measurements permitting to work with samples with lower concentrations in a linear range of 0,05 to 1 µg/ml (Singer et al., 1997). The method is reliable and well introduced in GMO testing laboratories. It should however be noted that a standard curve using lambda DNA needs to be prepared which requests a little more time. Furthermore it has been observed that the presence of various compounds have an effect on the accuracy of PicoGreen-based measurements (Singer et al., 1997; Holden et al., 2009 – see below).

A deviation between the concentration obtained by UV measurement and fluorimetry is often seen (Holden et al., 2009), especially for highly processed products (Bellocchi et al., 2010). This may be due to the fact that short or single stranded nucleic acid fragments interfere more with UV than with the PicoGreen dye. It has been proven that the fluorescence signal decreases with increasing length of sonication time (and thus fragmentation) showing the inability of the PicoGreen dye to bind with single stranded fragments (Georgiou & Papapostolou 2006; Holden et al., 2009; Shokere et al., 2009). One of the possible sources of single-stranded DNA may be denaturation of DNA during the drying phase after ethanol precipitation, the final step in many extraction protocols (Svaren et al., 1996). Utilizing spectrophotometry to quantify the DNA in an extract may thus lead to overestimation of the concentration.

Although one should determine the concentration of an extract to ensure that the DNA amount in a quantification reaction is above the limit of quantification (LOQ -part 5), the exact DNA concentration is of less importance. As the determination of the GM content of a sample relies on a relative calculation (ratio transgene copies versus endogene copies - part 3), it is imperative that a same amount of DNA is engaged in both qPCR reactions necessary in quantification, i.e. the event-specific and taxon-specific qPCR methods, whereas the exact amount engaged is of lesser importance. Carrying out both reactions in a single well, i.e. performing a duplex reaction would thus be a good solution.

When using spectrophotometry, additional to measurements at 260 nm, also measurements at wavelengths of 230 and 280 nm may be done. The **purity** of the DNA can then be assessed using the absorbance ratios A260/280 and A260/230. The A260/280 ratio gives an idea of the occurrence of residual proteins. On the other hand, the A260/230 ratio gives an indication on the presence of carbohydrates. In an ideal situation, both ratios should tend to 2,0 (Glasel 1995; Manchester 1995). Any deviation could indicate the presence of co-extracted materials that can impair the availability of the DNA for hybridisation with the primers and thus affect the PCR efficiency.

Another important aspect is the **integrity** or intactness of the gDNA (degradation). When the DNA becomes fragmented, the GM target which is less abundant (compared to the endogene) might fall below the quantification limit of the qPCR method. It is evident that this has a practical consequence on the correct quantification of the target. One must thus ensure that the average length of the extracted DNA molecules is longer than the size of the amplicon. To avoid that degradation of the DNA impairs the GMO quantification, the methods are generally designed to amplify sequences ranging in size between 70 and 100 bp. However, one should take into account the minimum length of an amplicon necessary to allow binding of the oligonucleotides (two primers in SYBR®Green chemistry, two primers and one probe used in TaqMan® chemistry). To this purpose for example MGB probes (Kutyavin et al., 2000) can be used to allow even shorter sequences that are stable and have an elevated melting temperature. Further, the amplicon sizes for the endogene and transgene target should not differ too much as shorter fragments are more efficiently amplified than longer ones. This difference in amplification efficiencies will have an impact on the correctness of the quantification reaction. The intactness of the extracted DNA can be assessed using agarose gel electrophoresis with ethidium bromide staining or an alternative. This technique also allows observing if any RNA has been co-extracted.

Knowledge of the presence of co-extracted substances and RNA and the existence of fragmented DNA in the extract is however not sufficient. It is known that **PCR inhibitors** are one of the most important influencing factors of the reliability of quantification (Bickley & Hopkins, 1999). It is thus important to know the impact of these molecules, present in the solution, on the GM quantification. Hereto, a preliminary inhibition test should be performed to evaluate their possible effect on the PCR efficiency. In this view, it is important to check if both targets of the quantification reaction (i.e. endogene and transgene) are equally affected by the presence of the inhibitors. If this is not the case, it would influence the detection of the real number of targets and thus lead to a deviating result (Corbisier et al., 2007).

There are several ways to study the presence of inhibition in a qPCR reaction. It is for example possible to include Internal Amplification Controls (IAC; Nolan et al., 2006; Burggraf & Olgemoller, 2004) or to add a positive control nucleic acid to the sample (Cloud et al., 2003). Further, mathematical algorithms can provide a measure of PCR efficiency from analysis of the amplification curves (Tichopad et al., 2003; Ramakers et al., 2003; Liu and Saint, 2003; Lievens et al., 2011). A simple alternative is the use of dilution series to assess the impact of inhibitory substances on the PCR reaction.

Recently, the ENGL released a document wherein they describe an approach to evaluate inhibition of a PCR reaction (ENGL, 2011). To this purpose the gDNA is serially diluted and each dilution is measured in duplicate using the validated qPCR method that will be applied for quantification. According to the previously published ENGL document (2008), the difference between the measured and theoretical C_t value should not exceed 0,5 C_t to exclude inhibition. In practice, four four-fold dilutions (from 1/4 till 1/256) need to be prepared from a stock solution. Both the dilutions and the stock are subsequently analysed in qPCR. This yields five qPCR results: the undiluted sample and the four (four-fold) dilutions. Using the latter, a curve is constructed by regressing the C_t values against the log of the dilution factor. This relation then allows the calculation (extrapolation) of a theoretical C_t value for the undiluted sample. Subsequently, this 'extrapolated' C_t value is compared with the measured value: there should be no more than 0,5 difference. Additionally, the regression line should comply with the following criteria: the slope must be between -3,6 and -3,1 and the linearity (R^2) must be equal or above 0,98.

A practical adaptation of this method is being used in the WIV-ISP-GMOlab. A series of dilutions is made from the gDNA under investigation and each dilution is analyzed using qPCR. Subsequently it is assumed that the last dilution contains the least inhibitors as the co-extracted substance will be diluted together with the DNA and will be below inhibitory concentration. The theoretical/expected C_t can be calculated for the other dilutions using knowledge of the dilution factors (e.g. a dilution of 2 corresponds to a C_t difference of 1). If the difference between the measured and theoretical C_t is equal or below 0,5, inhibition can be excluded. It must be noted that a difference of 0,5 for the highest concentration can be considered as an indication of inhibition. If this is observed for lower concentrations (more diluted samples) it is more probable that it comes from a dilution or pipeting mistake as it is unlikely that a low concentration would show inhibition that is not seen in the more concentrated solution.

These experiments and criteria should be set up by the laboratories prior to the quantification qPCR reaction to ensure correct quantification of a GM event in a sample. It should hereby be noted that also the DNA extracted from the CRM, used to construct the calibration curve, should be subjected to an inhibition test. Furthermore, these criteria should be evaluated for each DNA extraction method in combination with at least the most common matrices.

2.3 Evaluation of DNA extraction methods

Samples under investigation in GMO detection can vary to a great extend in the context of composition (single ingredient versus mixture), texture (solid versus liquid) and matrix (different plant species, processed versus raw material). The use of one universal DNA extraction method can thus difficultly be envisaged. The choice of an appropriate extraction procedure suitable for a particular sample matrix is thus a prerequisite for successful qPCR analysis. It must however be noted that this is not always straightforward as enforcement laboratories are not necessarily informed on the ingredients present in the sample under investigation.

The C-hexadecyl-Trimethyl-Ammonium-Bromide ('CTAB') extraction method is widely used in the enforcement laboratories for GMO detection (Pietsch et al., 1997). The method starts with lysis of the cells to release all contents. Addition of RNase and Proteinase K allows removal of respectively RNA and proteins. The ionic detergent CTAB forms an insoluble complex with the nucleic acids. The polyphenolic compounds, polysaccharides and other components remain in the supernatant and can be washed away. The DNA is released from the pellet by raising the salt content and is then concentrated by alcohol precipitation. It can be used for a variety of matrices such as maize, oilseed rape, potato and rice. The DNA yield is in most cases sufficient to conduct the necessary qPCR steps. However, the purity of the DNA solution is not always satisfactory. Yet, it is one of the more suitable methods for processed food and feed. In any case, an inhibition test is always advisable. In the GMOlab, inhibition is sometimes seen with very complex matrices such as processed feed products and liquid samples. The protocol is also less efficient for some rice containing materials. One of the drawbacks of the CTAB method is that the procedure is quite time-consuming as it contains different steps of incubation and centrifugation and also an overnight step necessary to ensure that the DNA pellet is completely dissolved. The method further requires some pre-extraction manipulations such as the preparation of specific buffers. It should also be noted that residues of the CTAB buffer can interfere with the PicoGreen dye and impair a correct measurement of the DNA concentration. It was observed that the magnitude of the effect of the CTAB detergent was in inverse proportion to the amount of DNA in the assay (Holden et al., 2009).

The CTAB extraction method can alternatively be combined with an extra purification step. Hereto a Genomic-Tip 20 column can be used (QIAGEN). This is an anion-exchange chromatography column to which the DNA fragments will be bound by electrostatic interactions between the negatively charged phosphate groups of the DNA and the positively charged resin. Upon subsequent washing steps, the impurities are removed while the DNA remains bound to the column. Finally the DNA is eluted and precipitated with alcohol. The method is very efficient for DNA extraction from soybean and cotton matrices which are more difficult to extract using the classic CTAB extraction method. For cotton powders for example, this is also the method recommended by the EU-RL (http://gmo-crl.jrc.ec.europa.eu/summaries/281-3006%20Cotton_DNAExtr.pdf). Utilizing this alternative procedure, solutions of higher purity can be obtained although the DNA yields are lower. However, they are in most cases still sufficient to perform all necessary qPCR analyses. Due to the purification of the gDNA on the column, these extracts are most often free of inhibitors. As for the classic CTAB method, specific buffers need to be made and an overnight step has to be incorporated to allow the pellet to dissolve. Additionally, the Genomic-Tip 20 columns and buffers that need to be purchased tend to be rather expensive.

A big advantage of the CTAB and CTAB-Tip20 methods is that there is no restriction on the sample intake. This allows the laboratories to easily scale up the extraction protocol. This is for example very convenient for the extraction of gDNA from CRM to ensure sufficient

DNA for validation of methods. The production of large batches of CRM DNA allows the laboratory to have a tested material readily available for several subsequent experiments. Also for several samples such a scaling up is sometimes necessary as the DNA content of some samples may be very low (due to for instance processing).

To reduce the time of DNA extraction, several kits are commercially available. Different companies offer their own DNA extraction kit which is mostly based on isolation of the gDNA using a silica-based method. Usually these kits deliver very fast gDNA and are easy to handle. A drawback of these kits is that often the sample intake is limited which has an impact on the final DNA yield. In, for example, the Wizard Genomic DNA Purification Kit (Promega), a maximum intake of 20 mg is allowed. It is thus necessary to pool several extracts to obtain a sufficient DNA amount for the subsequent qPCR analysis. In addition, when using DNA extracted with this kit, fluctuations in PCR efficiencies upon repetitions were observed which could lead to over- or underestimation of the GMO content (Cankar et al., 2006). Moreover, when comparing the PCR efficiencies of different amplicons, the gDNA extracted with the Wizard kit showed a high dispersion of the data.

The GENESpin kit (Eurofins GeneScan) is one of the few kits where an indication for possible scaling up of the system is given. According to the manufacturers, the kit would be suitable for several food samples such as cakes, bread, sausages,... They also indicate adapted protocols for liquid and powdered hygroscopic samples.

Furthermore, it should be noted that the kits are not always suitable for the extraction of DNA from all matrices. The DNeasy plant kits (QIAGEN) for example, are very efficient kits for the purification of DNA from fresh material (leaves, roots,...) but are less suited for powder materials. Corbisier et al. (2007) showed in their pilot study that this kit yielded a DNA concentration that was twice as low in comparison to the CTAB method. However, using this protocol relatively pure extracts were obtained. In the same study, it was observed that the Nippon Gene GM Quicker protocol (Diagenode), although a low yield and purity was achieved, delivered DNA which was less contaminated by RNA in comparison to the other procedures used.

The situation is even more complicated when it comes to DNA extraction of real-life samples. These not only can contain different species but also additional substances that affect DNA extraction. One such example is the presence of lecithin. This substance is often used in bakery products and as emulgator, stabilisator or anti-oxidant. Additionally, some products such as soybeans contain natural lecithin. As soybean is widely used in food and feed materials and Roundup Ready Soybean is one of the most cultivated GM crops (James, 2010), GMO detection laboratories often have to deal with this product. Wurz et al. (1998) presented an efficient extraction protocol for the isolation of soybean DNA from soy lecithin and showed its application in downstream qPCR. This method can thus be used for extraction of DNA from products such as soymilk and soy sauce.

Last but not least, it should be taken into account that the same product (e.g. bread) can have a different composition when produced by different procedures and can thus contain different substances that could affect the efficiency of the PCR. Even when taking for example only soybean products into account, the PCR efficiency is very much dependant on the nature of the product (Cankar et al., 2006). It was reported that for example DNA extracted with the DNeasy kit (QIAGEN) from a soybean feed sample revealed a higher inhibition effect on the transgene compared to the endogene although that for other samples such as the CRM, soybean milk and tortilla chips this was not observed.

It is thus advisable to validate an extraction method for different matrices. And although the extraction method is validated for a certain matrix, one should keep in mind that gDNA extracted from different samples is not necessarily equally suitable for quantitative analysis. Considering this, it is worthwhile for a GM detection laboratory to put some effort in the evaluation of the different existing extraction protocols in combination with the variety of samples that need to be analysed in GMO detection. And subsequently to chose the extraction method that is the most suitable to remove potential compounds such as lipids, polysaccharides and phenolics that could otherwise impair the PCR efficiency.

2.4 Conclusion

GM quantification is performed in different steps in which DNA extraction is the first one. This pre-PCR phase is of great importance for the trueness of the quantification result. The DNA extracted from different materials should be evaluated for yield, purity and integrity before performing the qPCR experiment. Furthermore, the DNA solution should be assessed for the presence of inhibitors and their impact on the two targets of the quantification i.e. the endogene and transgene. It is clear that these parameters not only have to be evaluated for the sample under investigation but also for the gDNA extracted from the Certified Reference Material used as a calibrant. Both the sample and CRM DNA need to meet the set criteria to ensure reliable quantification. Seen the diversity of products and matrices that need to be analysed by GM testing laboratories, several DNA extraction protocols exist including home-made buffers and kits. It is obvious, that the extraction protocol to be used needs to be evaluated and that the gDNA extracted has to pass the requirements set by the laboratories before it is used in subsequent PCR analysis. In addition to the choice of the DNA extraction method, thought should also be given to the method used to determine the concentration of the extracted DNA.

In general, the validated DNA extraction protocols used in routine such as the CTAB method are valid for different matrices. However, when dealing with a complex matrix it is important to verify the quality of the DNA. As the extraction method may in some cases have an influence on the GM content, optimalisation of the extraction procedure may be needed. Furthermore, the presence of inhibitors should be checked as they may impair the efficiency of the PCR reaction and thus influence the quantification of GM events in a sample. Hereto, the impact of co-extracted substances and products used in the extraction protocol should be evaluated on the sample, the CRM and the two targets under investigation. If a considerable inhibitory effect is observed, further DNA purification should be performed.

3. Description of the structure of a transgenic insert and the type of DNA sequence used for qPCR analysis

3.1 Introduction

All the GM events currently on the EU market are plants in which a piece of foreign DNA has been introduced into the genome. This piece of DNA generally consists of a regulatory promoter region, a coding sequence and a terminator (Fig. 1) and is called the transgenic construct or insert. To introduce this construct into the plant genome, genetic engineering techniques (Darbani et al., 2008), such as *Agrobacterium*-mediated transformation and particle bombardment, are being used. Hereto the transgene is cloned in a plasmid for example between two specific and unique sequences (T-DNA borders).

For *Agrobacterium*-mediated transformation, the plasmid carrying the transgene is introduced into this bacterium. Further, the intrinsic properties of this soil bacterium are used to incorporate the transgenic construct into the plant genome: the bacterium namely infects the plant and transfers the T-DNA part of the plasmid to the plant genome. In this way the transgene is stably inherited in the subsequent generations (Chilton et al., 1977). Different explants such as leaves (Horsch et al., 1985), roots (Valvekens et al., 1988), embryos (Hensel et al., 2009), ovules (Holme et al., 2006) and microspores (Kumlehn et al., 2006) can be used for transformation. In particle bombardment, gold or tungsten particles are coated with the plasmid containing the transgene (Kikkert et al., 2004). Subsequently, these particles are fired onto the explants with high voltage allowing the incorporation of the transgene into the plant genome. Compared to *Agrobacterium*-mediated transformation, particle bombardment more often leads to multiple inserts of the transgenic construct into the genome.

The detection of this transgenic insert forms the basis of the EU legislation concerning the introduction of GMO onto the market and thus requests the development of GMO detection methods. This detection is carried out by enforcement laboratories and the method of choice is real-time PCR (qPCR). At WIV-ISP, a GMO detection platform, allowing the verification of the presence of GM material in food and feed samples was developed. The platform consists of

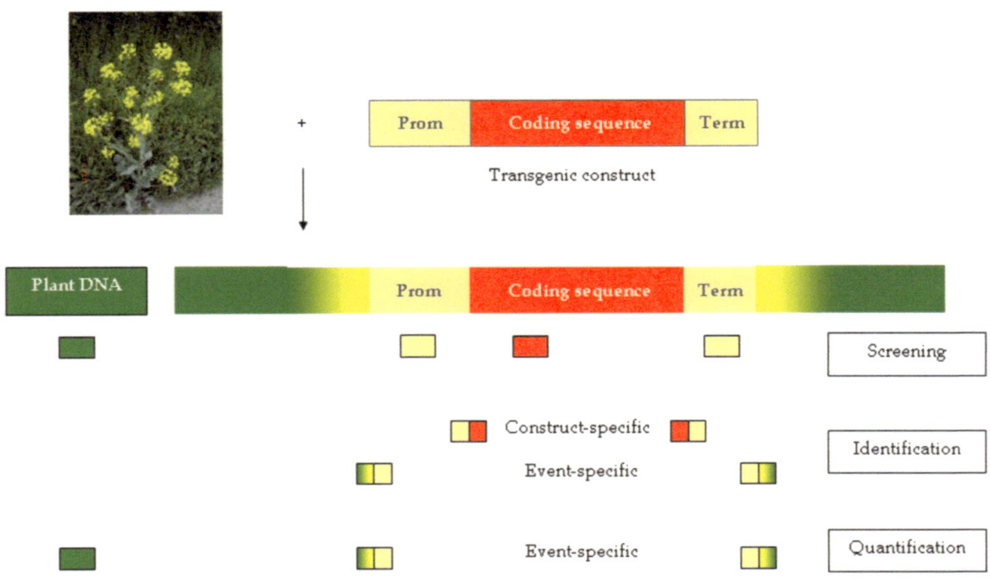

Fig. 1. Plant transformation and type of sequence targeted by the different steps in qPCR analysis.
In screening, a sequence inside one of the elements of the transgenic construct is targeted. A construct-specific method used for the identification of the GMO targets the junction between two elements within the transgenic construct. An event-specific method, used in identification and quantification of a GM event, targets the junction between the transgenic insert and the plant genome DNA.

a preparative step namely DNA extraction (part 2) and three consequent qPCR steps namely screening, identification and quantification (Fig. 2). Hereto, in-house developed and validated SYBR®Green screening methods (part 4) are combined with EU-RL validated TaqMan® event-specific methods (part 5). In each step of the qPCR analysis, a different part of the transgenic construct is being targeted. The region in the construct targeted by the method is linked with the specificity of the method. By using a more specific method in each subsequent step, it is possible to gradually narrow down the possibilities to a specific GM event.

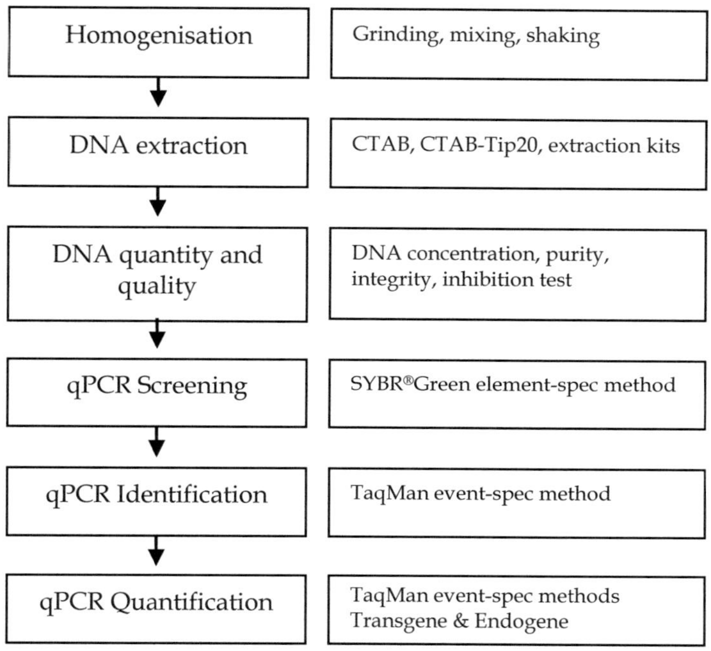

Fig. 2. Flowchart of the analysis steps in GMO detection

In support of these analyses, a matrix-based approach called CoSYPS (Combinatory SYBR®Green qPCR Screening) has been developed (Van den Bulcke et al., 2010). This approach relies on the integration of the analytical results obtained for a sample in a mathematical Decision Support System and the application of a "prime-number"-based algorithm (part 6). Based on the outcome of the screening results of a set of markers in a sample, the system will identify which GM events are possibly present in a sample.

3.2 GMO screening methods

After DNA extraction, screening is the next crucial step in GMO detection. In view of the growing number of GM events introduced on the market and new upcoming traits, screening methods will become more and more important and necessary to enable the discrimination between the different GMO. Testing for each possible GM event separately would namely become too expensive and labour-intensive.

A screening method usually targets a sequence inside one of the elements of the transgenic construct (Fig. 1). Seen the fact that the elements that are used in transgenic constructs are recurrent, detection of a single element often does not confer high specificity and, as a consequence, does not allow deciding on which GM event might be present. A combination of different screening markers is therefore necessary to get a better idea of the possible GM events occurring in a sample. This allows the reduction of the number of identifications to be performed.

To date several screening methods for the detection of GM materials in food and feed samples have already been published. These methods often target the Cauliflower Mosaic Virus 35S promoter (p35S) and/or the *Agrobacterium tumefaciens* nopaline synthase terminator (tNOS) seen the fact that these elements are the most represented in the EU authorised GM events. From the twenty four authorised events, nineteen events contain the p35S target, fifteen the tNOS element and eleven combine both markers (GMO Compass website; Agbios website). Additionally, methods for the detection of herbicide tolerance (HT) genes used in transgenic constructs have been reported. These mainly target two classes of HT sequences: the bacterial phosphinotricin-*N*-acetyltransferases from *Streptomyces viridochromogenes* (*pat*) and from *Streptomyces hygroscopicus* (*bar*) (Wehrmann et al., 1996), and the 5-enolpyruvylshikimate-3-phosphate synthase (*epsps*) from *Agrobacterium tumefaciens* strain CP4 or from plant origin (*in casu* petunia) (Kishore et al., 1988; Padgette et al., 1996). Apart from herbicide tolerance, the GM events currently on the market are transformed with insect resistance traits. Hereto the *Bacillus thuringiensis* endotoxin encoding genes (e.g. the *cryIAb/Ac*) are being used and detection methods have been developed (Bravo et al., 2007). It should however be noted that the above-mentioned methods are mostly either end-point detection on agarose gel or real-time qPCR using TaqMan® chemistry (Hamels et al., 2009; Raymond et al., 2010; Nadal et al., 2009; Prins et al., 2008). Development of screening methods using the SYBR®Green qPCR technology only started recently (Barbau-Piednoir et al., 2010; Barbau-Piednoir et al., 2011; Mbongolo Mbella et al., 2011) although this approach offers a number of advantages over the TaqMan chemistry. The use of melting temperature analysis for instance allows detection of the expected target but also allows distinction between closely-related elements, which is important in the evaluation of the specificity of the method. But more important for enforcement laboratories is the fact that SYBR®Green methods do not require the use of fluorescent labelled oligonucleotides which is much more cost effective.

In view of the growing amount of GM events and the lack of cost-effective screening methods, the WIV-ISP platform puts a major effort in the development of an extensive number of qPCR SYBR®Green screening methods. They form a unique combination targeting different elements within the transgenic construct in addition to plant sequences and are gathered in the patented CoSYPS matrix (Combinatory SYBR®Green qPCR Screening; Van den Bulcke et al., 2010). The methods used to build the CoSYPS were in-house developed and validated (part 4). They are used together with the CoSYPS matrix in the routine analysis of food and feed samples in the GMOlab under ISO 17025 accreditation. To cover the increasing number of GM events and to add discriminative power to the CoSYPS system, new screening methods are being developed on a regular basis and are subsequently being introduced in the CoSYPS (part 6) after in-house validation.

The in-house developed methods target different types of DNA elements (table 2). Firstly, a screening method aiming to target the chloroplastic *rbcl* gene (plant kingdom marker) was developed. This element will permit to decide on the presence of vegetative DNA in an unknown sample. Secondly, methods that detect plant taxon-specific sequences (Mbongolo

Method name	**Target**	**Fragment size (bp)**	**Reference**
Plant kingdom marker			
Rbcl	Ribulose-1,5-biphosphate carboxylase oxygenase	95	Mbongolo Mbella et al., 2011
Plant taxon-specific methods			
Lectin	Lectin gene of soybean (*Glycine max* L.)	81	Mbongolo Mbella et al., 2011
Adh	Alcohol dehydrogenase I gene from maize (*Zea mays* L.)	83	Mbongolo Mbella et al., 2011
Cru	Cruciferin gene from oilseed rape (*Brassica napus*)	85	Mbongolo Mbella et al., 2011
PLD	Phospholipase D gene from rice (*Oryza sativa*)	80	Mbongolo Mbella et al., 2011
Sad 1	Stearoyl-acyl carrier protein desaturase gene of cotton (Gossypium genus)	107	Mbongolo Mbella et al., 2011
Glu3	Glutamine synthetase gene from sugar beet (*Beta vulgaris*)	118	Mbongolo Mbella et al., 2011
Methods specific for generic element			
p35S	Promoter of the 35S Cauliflower Mosaic Virus	75	Barbau-Piednoir et al., 2010
tNOS	Terminator of the nopaline synthase gene	69	Barbau-Piednoir et al., 2010
pFMV	Promoter of the 34S Figworth Mosaic Virus	79	Broeders et al., (in preparation)
pNOS	Promoter of the nopaline synthase gene	75	Broeders et al., (in preparation)
t35S	Terminator of the Cauliflower Mosaic Virus	107	Broeders et al., (in preparation)
Methods specific for GM elements			
CryIAb	Gene encoding the *Bacillus thuringiensis* δ-endotoxin (insect resistance)	73	Barbau-Piednoir et al., 2011
Cry3Bb	Gene encoding the *Bacillus thuringiensis* δ-endotoxin (insect resistance)	105	Broeders et al., (personal communication)
Pat	Phosphinotricin-*N*-acetyltransferases gene from *Streptomyces viridochromogenes*	109	Barbau-Piednoir et al., 2011
Bar	Phosphinotricin-*N*-acetyltransferases gene from *Streptomyces hygroscopicus*	69	Barbau-Piednoir et al., 2011
EPSPS-CP4	5-enolpyruvylshikimate-3-phosphate synthase gene from *Agrobacterium tumefasciens* strain CP4	108	Barbau-Piednoir et al., 2011
P35S discriminating method			
CRT	Reverse transcriptase gene from the Cauliflower Mosaic Virus	94	Papazova et al., (in preparation)

Table 2. List of SYBR®Green screening methods developed and validated by the GMOlab.

Mbella et al., 2011) have been developed. These methods target the main GM commodity crops such as soybean, maize, oilseed rape, cotton, sugar beet and rice. They make it possible determining the species composition of the sample and allow a first discrimination of GM events (e.g. the presence of a soybean GM event can be excluded if the soybean taxon-specific marker is negative). Thirdly, methods specific for GM generic elements were developed (Barbau-Piednoir et al., 2010). These are elements that are included in many transgenic constructs used in commercial GM plants. Such elements are represented by promoter and terminator sequences such as the Cauliflower Mosaic Virus promoter (p35S) and the *Agrobacterium tumefaciens* nopaline synthase terminator (tNOS). Adding the information from the qPCR experiments targeting these generic elements gives a first idea of the putative presence of a GM event in the sample. However, seen these elements are widespread in the transgenic constructs currently used, they do not contain enough discriminative power to sufficiently reduce the number of possible GM events present. These elements need thus, in a fourth step, to be combined with methods targeting other GM specific elements such as herbicide tolerance and insect resistance genes (e.g. Cry genes, bar, pat). Such methods have also been developed and were recently published (Barbau-Piednoir et al., 2011). Last but not least, a marker was developed to be able to discriminate between the p35S present in a GM event and the one due to possible natural presence of the Cauliflower Mosaic Virus from which the transgenic sequence was originally taken (the so-called donor organism). The combination of the results of the eighteen markers, currently used in routine, will allow defining the putative GM events present in a sample. Utilizing the CoSYPS to this purpose, a list of possible events to be identified will be obtained. Additionally, the use of the various markers in combination with the CoSYPS is a powerful tool in the detection of unauthorised GMO (UGM) events. In principle, the elements that are positive in the screening qPCR should be covered by the EU authorised events (EC/1829/2003) or the GM events included in the 'Low Level Presence' legislation (EC/619/2011). If this is not the case, one might suspect the presence of an unauthorised event in the sample.

For each of the screening methods developed and validated at the WIV-ISP-GMOlab, the corresponding amplicon is cloned in a pUC18 background. These plasmids, called Sybricons, are submitted under "Safe Deposit" at the BCCM (Ghent, BE). They can be used to determine the nominal T_m value of the target and further utilized as positive controls in routine analysis.

In addition to the 18 SYBR®Green screening markers, the GMOlab applies two markers in TaqMan® chemistry for the detection of potato (UGPase) and linseed (SAD).

3.3 GMO identification methods

Based on the outcome of the screening step, a second phase will be necessary namely identification of the GM event.

Identification methods are directed to the detection of a specific GM event. These qPCR methods, contrary to the screening methods, use TaqMan® chemistry. They can be either construct-specific or event-specific qPCR methods. A construct-specific method targets the junction between two elements within the transgenic construct. They are thus directed to the sequence covering a part of the promoter and coding sequence or of the coding sequence and the terminator (Fig. 1). Event-specific methods, in contrast, target the junction between the transgenic insert and the plant genome DNA. They are thus designed to cover part of the sequence of the plant and the promoter or of the terminator and the plant DNA (Fig. 1).

As the location of the transgenic insert into the plant genome is unique, the event-specific methods are specific to a sole GM event. Indeed, one and the same construct can be inserted into the genome of different plant species and will not be discriminated by using a construct-specific method alone whereas the plant-insert junction, targeted by the event-specific method, will be unique. This makes the event-specific methods the technique of choice in GMO identification. These methods are in fact part of the GM quantification methods available. They are laid down by the GM Company together with the request for GM authorisation. Subsequently the EU-RL validates them in a ring trial in which the NRL for GMO detection participate. Once the validated method is published and a CRM is available, the enforcement laboratories need to be able to implement the method in their laboratory (part 5). The construct-specific methods, on the other hand, can be in-house developed methods, methods developed by research groups or the qPCR methods that are published by the EU-RL for quantification of GM events. As they are less specific than the event-specific methods, they have a less discriminative power and are thus not recommended. However, for some GM events (e.g. rice GM events) no other methods exist to date.

At the GMOlab, the coming out of the different identifications are gathered in a Decision Support System (part 6) which will further indicate at which level a specific GM event is present. Only if the GMO is found at quantifiable levels (i.e. above the limit of quantification), a third step will be involved namely quantification of the GM event.

3.4 GMO quantification methods

In this last step in the process of GMO detection, the amount of the present GM event will be determined. This quantification is necessary to assess the compliance of a sample with the 0,9% labelling threshold (EC/1829/2003) and the recently voted 'Low Level Presence' (LLP) legislation (EC/619/2011).

Quantification of a GM event in a sample relies on the relative determination of the number of copies of the transgene in relation to the number of copies of the endogene (i.e. the taxon-specific sequence). Hereto a combination of a GM event-specific method and a taxon-specific method will be used. Both methods need to be provided by the GM plant developing companies when requesting EU authorisation and are subsequently validated by the EU-RL. Each laboratory involved in GMO detection needs then to verify in-house if the method complies with the set acceptance criteria before to use it in routine analysis of samples (part 5).

The result of GMO quantification is expressed as a GM mass percentage in relation to the ingredient for authorised events and in relation to the GM material for the LLP events. This result is reported to the competent authorities who will decide if the sample is conform to the legislations or not.

3.5 Conclusion

As the number of GM events being introduced on the market is rapidly increasing, screening will become a necessary first step in GMO detection. Additionally, an intensive screening provides an indication on the presence of GM material originating from unauthorised and unapproved GMO. Indeed, countries that produce GM plants only for local consumption will not request for EU authorisation but these crops might still "escape" and end up in the EU food chain. As a consequence also the detection of these UGM will become a major task of enforcement laboratories.

The GMO platform developed by the WIV-ISP-GMOlab allows detection of authorised GM events as well as UGM in a cost- and time effective manner. It consists of a preparative DNA extraction step and three consecutive qPCR steps. The CoSYPS system, including in-house developed SYBR®Green screening methods, forms an innovative tool in GMO detection allowing reducing the number of identifications to be carried out. The TaqMan® identification further allows a narrowing down of the GM events present to a specific GMO and quantification permits the determination of the GM content.

4. Development and validation of a qualitative qPCR method in view of its application for screening purposes in the WIV-ISP GMO detection platform

4.1 Introduction

As described previously, in order to face the rapidly increasing number of GMO in food and feed products, new methods facilitating an initial screening of analytical samples is needed. Therefore, one of the major objectives of the molecular platform at WIV-ISP is to develop qualitative screening methods targeting either new genetic elements commonly found in transgenic constructs or species frequently used in food and feed in view of rationalizing GMO detection.

The methods developed are singleplex qPCR, based on SYBR®Green chemistry. Additionally, the methods are designed to work under uniform conditions (primer concentrations, PCR program) in order to facilitate their simultaneous application in a 96-well plate format. These SYBR®Green methods were in-house validated in order to be applied under ISO 17025 accreditation. As there is no 'golden standard' for the validation of qualitative methods related to GMO detection, enforcement laboratories need to decide which parameters need to be evaluated in the validation. In addition, the laboratories have to set their own criteria based on the guidance document for quantitative qPCR methods.

Part 4.3 of this chapter focuses on the method validation criteria and proposes a pragmatic approach for the in-house validation of singleplex real-time PCR qualitative methods. This proposal is mainly based on the recently adopted Codex Alimentarius guidelines on performance criteria and validation of methods for GMO analysis (Codex, 2010), and on the minimum performance requirements for methods for GMO testing set forward by the ENGL (ENGL, 2008). During the in-house validation critical values are determined for the screening methods to be introduced in the Decision Support System currently used in the routine analyses, namely the CoSYPS (part 6).

4.2 Development of SYBR®Green methods for screening purposes

The first step of method development is to determine the screening qPCR target. Targets for screening can be any element present in the transgenic construct inserted in authorised or unauthorised GMO and taxon-specific sequences. Application of the screening approach requires development of many targets in order to cover the growing range of GM events. Selection of the methods to be developed is based on a number of priorities. Firstly, methods targeting the main commodity crops used in transformation events are of high importance. Secondly, priority is given to transgenic elements frequently occurring in EU authorised GM events in addition to targets that provide an extra discriminative power. Thirdly, other important transgenic elements occurring in unauthorised GM events which might be necessary to test for by the enforcement laboratories should be targeted.

The development of a new screening method depends on several prerequisites: information on the elements inserted in a GM event, their copy number and the nucleotide sequence of the inserted element. Information on the elements of the transgenic construct inserted in a GMO can be obtained from publicly available dossiers submitted by the applicant for authorisation or patent databases. This information is usually available after the authorisation is granted or after the competent authorities have given a positive advice. Important information sources are the GMO crop database of the Centre for Environmental Risk Assessment (CERA) (http://www.cera-gmc.org/?action=gm_crop_database) and the GMO database on authorisations and approval of GMO in the EU (http://www.gmo-compass.org/eng/gmo/db/). The nucleotide sequences are available in public databases such as the National Center for Biotechnology Information (NCBI) (http://www.ncbi.nlm.nih.gov/), patent databases and scientific publications. One must however take care when using the information present in these databases as for example Single Nucleotide Polymorphisms (SNP) may exist in the sequence of the elements inserted in different GM events (Morisset et al., 2009). Therefore, the information in the public databases is not always completely reliable and more than one source should be consulted.

Additionally, variations in the sequences used to design taxon-specific assays exists as for instance SNP can occur between the varieties of one plant species (Broothaerts et al., 2008; Papazova et al., 2010). The difficulty here is that information on the nucleotide sequence in different plant varieties is not available. This problem can be partially solved by designing the SYBR®Green primers on basis of existing TaqMan® taxon-specific assays for which experimental tests have been performed. Presence of SNP in the primer annealing sites can lead to a false negative result and to the conclusion that an event containing this target is not present when the assay is applied to an unknown sample (Broothaerts et al., 2008; Papazova et al., 2010).

Upon selection of the suitable sequence different primer pairs are designed by using appropriate bioinformatic tools. One of the most widely used programs is Primer3 (Rozen & Scaletzky, 2001). These primer pairs are further assessed *in silico* for their specificity. This can be done by means of bioinformatic tools such as the primer search module in the EMBOSS bioinformatic platform, BLAST searches etc. For transgenic elements, this theoretical specificity test is performed using sequences from authorised GM events. If the primers target a reference taxon-specific sequence, it should be tested if they are specific for the target taxon and do not amplify closely related species. Here, the criteria for specificity for reference assays of the event-specific quantification methods also apply (part 5).

As the goal is to use all the methods simultaneously under uniform conditions, particular attention is paid on the amplicon size and the primer annealing temperature (T_m) when developing the primers. Amplicons with a size lower than 100 bp are preferred although the size for real-time PCR amplicons can be as large as 250 bp. For qPCR detection smaller amplicons are favoured in order to avoid lack of amplification due to the possible fragmented status of the DNA in the sample (part 2). In addition, the melting temperature of the primers should be around 60°C according to the general requirements for qPCR primers (www.appliedbiosystems.com). The formation of primer dimers and hairpins should be checked and primer pairs showing this feature should be excluded for further analysis.

4.3 Validation of a SYBR®Green screening method

The in-house validation of a SYBR®Green screening method is based on the determination of several method characteristics that are required for the validation of event-specific quantitative methods (ENGL, 2008 - part 5), namely applicability, practicability, specificity, Limit of Detection (LOD), Limit of Quantification (LOQ) and precision (RSDr%). The definitions of these parameters can be found in the glossary. The GMOlab has developed its own experimental set up in order to assess these parameters. Upon validation the results are evaluated and if they meet the acceptance criteria the method can be used under accreditation. Additionally, the critical values which are introduced in the CoSYPS (part 6) are determined during the in-house validation.

The method is **applicable** when it detects the target in the respective GMO for which it was designed. To test this aspect of a method a list of GM events containing the target (positive samples) and events not containing the target (negative samples) is made. Usually, this list is limited to GM events which are authorised and for which (certified) reference materials are available. If possible, different matrices (e.g. gDNA, pDNA, raw material, processed material,…) are included and different GM concentrations are used. Further the applicability of the methods is assessed by screening certified reference materials which are used in the GMOlab for validation and calibration purposes.

The **practicability** of the SYBR®Green screening methods follows directly from the fact that all methods have been developed in-house. During the development, the use of the same conditions (qPCR program, reaction volume, …) and qPCR instruments have been taken into account. This will thus allow using all methods in a same run during routine analysis of a sample.

The **specificity** of the method is first assessed *in silico* (part 4.2) and further experimentally. The screening method should be specific for the target for which it is developed and should not be homologous and give an amplification product with other sequences. The specificity is experimentally tested on all materials to which the analysis can be applied. The GM events or taxa containing the target should give a positive amplification signal, while the ones which do not contain it should give no amplification signal. An amplification signal is considered positive when a C_t value and a melting curve analysis are recorded. Absence of amplification is considered when either no C_t is recorded or when a C_t value at least 10 C_t higher than the one of the positive samples is measured. To assess the nominal T_m value, a plasmid containing the construct under analysis may be used.

As the screening methods developed and validated at the GMOlab are based on the SYBR®Green detection chemistry, the melting temperature of the amplicon is an important parameter related to the specificity of the method. The melting temperature (T_m) of a DNA sequence is dependent on a large number of factors, among which the ionic conditions in the sample solution, the DNA nature (sequence, secondary structure, etc.) and the starting concentration of the DNA molecule (Hillen et al., 1981; Rouzina & Bloomfield, 2001). Moreover different qPCR instruments tend to measure slightly different values for a given amplicon (due to differences in heating block control, mathematical integration, extrapolation, etc.). The variation of the T_m follows a normal distribution and the T_m of the method is calculated as the average T_m from all the data obtained during validation. Additionally, a T_m confidence interval is calculated ($T_m \pm 3$ standard deviations) which is used further to decide whether the correct target has been amplified (part 6). The T_m and its confidence interval can be updated regularly by adding data from analysis of routine samples to the existing dataset.

Using the data from the *in silico* and experimental specificity tests, mostly only one primer pair is selected for determination of the method sensitivity (LOD and LOQ) and repeatability.

To assess the **sensitivity** of the developed method, a GM event containing the target is used (usually a CRM with a known GM%). It should however be noted that the GM-specific CRM are certified for the content of a specific GM event and not for the content of the screening target (promoter, coding sequence, terminator). This demonstrates that the preliminary information on the elements inserted in a GM event and their copy number is crucial in order to estimate the correct copy number of the target. For taxon-specific markers, this assessment can be done using a wild type (non-GM) material. The LOD and the LOQ are determined on basis of serial dilutions starting from at least 2000 target copies until the theoretical zero copy numbers. Each of the dilutions is run in six replicates.

The **LOD** is set up at the level where less than 5% false negatives are observed (Codex Alimentarius, 2009). As it is not feasible to perform the analysis on a large number of PCR replicates, six repeats are run per dilution point. If all six repeats are positive, this means that 95% of the time a positive sample will indeed be detected. Therefore the LOD of the screening method is set at the haploid genome copy level at which all six replicates provide a specific positive signal (n = 6; 6/6 specific signals) (AFNOR XP V 03-020-2).

The **LOQ** is defined as the target copy number with a similar positive PCR result (expressed as C_t value) upon six-fold measurement of the target sequence in the same DNA sample with a minor standard deviation (SD_{Ct}<0,5) (AFNOR XP V 03-020-2). A screening target is in principle not quantified, but the LOQ can give an idea about the content of the target in an unknown sample.

Additionally, the **precision** (inter-run repeatability) of the method is determined. In practice this is done by calculating the relative repeatability standard deviation (RSDr%) on each of the dilutions used to determine the LOD and LOQ. Hereto, the experiment is performed under repeatability conditions (in a short period of time, on the same qPCR instrument by the same operator) in four independent runs. The RSDr% is calculated according to the ISO 5725-2. The method is accepted as valid when the RSDr% is below 25%.

4.4 Conclusion

As, to date, no instructions on the development and validation of screening methods are available, the laboratories need to set up their own experimental plan and criteria. At the WIV-ISP-GMOlab, development and validation of SYBR®Green methods for screening purposes is done in a harmonized way to allow applying the methods in a single qPCR run. The parameters evaluated, the way to perform this assessment and the acceptance criteria are based on previously published documents (ENGL, 2008; Codex Alimentarius, 2009; AFNOR XP V 03-020-2).

Upon evaluation of all the necessary parameters and their accordance with the set criteria, a validation dossier is established. The LOD, LOQ (expressed as a C_t value) and the T_m interval are introduced into the CoSYPS Decision Support System and serve as decision values to conclude if a sample is positive for the target or not (part 6). Subsequently the method is implemented in routine GMO detection under ISO 17025.

5. Validation of a qPCR method for GMO quantification and its implementation in a routine laboratory under ISO 17025 accreditation

5.1 Introduction into the legal context

Regulation (EC) 1829/2003 on genetically modified food and feed defines that food and feed products containing or derived from GMO must be labelled. The labelling requirements do not apply to food and feed containing GMO in a proportion not higher than 0,9% of the ingredients, provided that this presence is adventitious or technically unavoidable. Moreover, the recently adopted "Low Level Presence" Commission Regulation (EC/619/2011) requires a reliable quantification at a level of 0.1%. Member States are responsible for monitoring the GMO content of products and compliance with GMO labelling requirements. In this context, the enforcement of the EU legislation on GMO labelling requires GMO detection methods that are sound, precise and robust. It is, therefore, an essential requirement to use validated methods for GMO detection and quantification. Only in this manner it can be assured that independent control laboratories achieve comparable analysis results and are able to fulfil regulatory tasks (JRC, 2010).

The submission and validation of a GMO detection method is an integral part of the regulatory and approval process for GM food and feed to be placed on the market (EC/1829/2003). This Commission Regulation states that the application for authorisation should include, amongst others, "methods for detection, sampling and identification of the transformation event". As a consequence, the biotech companies have to provide detection protocols and control samples to validate the event-specific method to the EU-RL GMFF. These methods should be based on the real-time PCR technology (EC/787/2004). In view of the European harmonisation and standardisation of methods for sampling, detection, identification and quantification of GMO, the EU-RL has published a list of parameters to be tested and their acceptance criteria in the a document "Definition of minimum performance requirements for analytical methods of GMO testing" (ENGL, 2008).

A GM event cannot be authorised in the EU before a relevant detection method has been validated. The method validation process is conducted by the European Commission's Joint Research Centre (JRC) in its capacity as European Union Reference Laboratory for GM Food and Feed, and is assisted in its task by the European Network of GMO Laboratories. Commission Regulation EC/882/2004 establishes that analytical methods used for food and feed control must be verified by control laboratories before their use (JRC, 2010). In practice, after testing of the material and protocol, the JRC distributes the sample material and corresponding reagents to the participating laboratories in a ring trial. The validation ring trials are organised according to the requirements set up in ISO 5725 and following the IUPAC protocol (IUPAC, 1995). In such a collaborative validation trial, the EU-RL is assisted by the National Reference Laboratories (NRL) which are assigned as official control laboratories at national level (EC/882/2004). The NRL have to be accredited under ISO 17025 standard. Usually there are 12-13 participating laboratories, randomly selected from all available NRL. The validation ring trial aims at determining the method performance characteristics.

In this way the submitted method is evaluated with regard to the validation criteria. Failure to meet these criteria leads to rejection of the method and consequently to a delay in the authorisation of the GMO. Upon acceptance, the EU-RL GMFF prepares a validation report with the results of the study and the validated protocol. These are submitted to the European Food Safety Authority (EFSA) and are subsequently published on the EU-RL GMFF official website. Upon publication the validated methods become official methods. The method validation thus provides the enforcement laboratories with standardised and harmonised methods applicable in official GMO detection.

5.2 Evaluated parameters for newly developed event-specific methods for GMO quantification

5.2.1 Evaluation of method performance characteristics by the EURL-GMFF

The requirements for method **specificity** are laid down in the legislation. The method submitted has to be event-specific (based on the specific sequence of the plant-transgenic construct junction, part 3) and should detect only the specific GMO submitted for authorisation to be useful for unequivocal detection/identification/quantification of the GM event (EC/641/2004). To demonstrate that the method is event-specific, it has to be tested against all GM events from the applicant which are currently authorised in different parts of the world and against those still in development.

As the submitted methods are quantitative, they also include a reference taxon-specific assay. The specificity of this assay should also be tested. For taxon-specific assays the target should be preferably a unique sequence present in a single copy in the target plant genome. The copy number and the specificity have to be assessed *in silico* by using BLAST (http://blast.ncbi.nlm.nih.gov/Blast.cgi) searches against known databases. In addition, the taxon-specific target should not show amplification signals with close relatives or taxa of the most important food crops. Usually, the different biotech companies develop their own taxon-specific method and test it on a range of taxa selected by them. This can pose several problems for the laboratories applying the methods. Firstly, there is no standard list of taxa and varieties to be included in the test. Ideally, the reference assay should be tested on a large range of varieties covering the existing natural variation within the taxon in order to assure that it will amplify any material from the plant species targeted by the method. Secondly, the existence of more than one reference system for events of the same plant taxon requires the use of several reference assays in quantification, which increases the costs of the analysis by the laboratory. In this context the requirements for the specificity of the taxon-specific reference assays should be made more precise and harmonisation in the methods used for different GM events is needed.

Information on the **applicability** of the method should be provided. This includes information on the scope of the method. In addition, information on known interferences with other analytes and the applicability to certain matrices should be supplied.

The **practicability** of the method should be demonstrated. For instance, methods where the reference and the event-specific assays are run on different PCR plates or under different PCR cycling conditions are less practicable and would be time and cost consuming when applied in a routine laboratory.

Besides these criteria, other parameters related to the method performance are assessed namely the **dynamic range, linearity, amplification efficiency, LOD and LOQ, trueness, precision and robustness**. The definitions of all parameters can be found in the glossary.

5.2.2 Evaluation of method performance characteristics, performed by the analysis of the results of the inter-laboratory collaborative trial

Once the EU-RL GMFF has made a scientific evaluation of the method based on the performance of the above-mentioned parameters (as provided by the method developer), it organizes a validation ring trial (concerning dynamic range, precision, relative reproducibility standard deviation and trueness). The participating laboratories receive the necessary samples and reagents and a detailed experimental protocol. It should be noted that the purpose of the ring trial is to assess the performance of the method and not of the laboratory. Therefore each participant has to follow the experimental procedure strictly. The results obtained by the laboratories are expressed as GM% for each tested level. These results are further scrutinised for outliers by the EU-RL GMFF using statistical methods recommended by ISO 5725. In addition, the mean value is calculated for each GM level analysed. Based on the parameters assessed during the ring trial, a conclusion is made on the compliance of the method with the ENGL method acceptance criteria and if it can be considered applicable in regard to the requirements of EC/641/2004.

5.3 Implementation of a validated event-specific method in a testing laboratory

When the interlaboratory validation study is completed and the method is considered as applicable, the method is ready to be implemented in routine testing laboratories like the GMOlab.

On the one hand, Commission Regulation EC/882/2004 states that official laboratories shall be accredited according to the ISO 17025 standard. An ISO 17025 accreditation, under a fixed or flexible scope, implies that "the laboratory shall confirm that it can properly operate standard methods before introducing the tests for calibrations". On the other hand, according to the same regulation, it is the task of the EU-RL GMFF to provide the NRL with details of analytical methods, including reference methods. In this context, guidelines for implementation of the validated methods in the routine laboratory are set up by the ENGL in the document "Verification of analytical methods for GMO testing when implementing interlaboratory validated methods" (ENGL, 2011). These guidelines reflect the requirements set up in the document "Definition of the Minimum Performance Requirements for analytical methods of GMO testing" (ENGL, 2008), but also give additional guidance on how to design the experimental set up and to calculate the required values. In practice the laboratories have to design the quantification experiment in which two or three GM levels are quantified and the parameters described hereunder have to be assessed.

Dynamic range, R^2 coefficient and amplification efficiency: these parameters can be calculated simultaneously from calibration curves when testing other parameters (trueness and precision). For each target, the average values of at least two calibration curves should be taken. The dynamic range should be tested between $1/10^{th}$ of the threshold value and 5 times this value i.e. between 0,09% and 4,5% for the 0,9% labelling threshold. The PCR efficiency should be between 90 and 100% and the R^2 coefficient needs to be equal or above 0,98 to have a linear curve.

Trueness should be determined at a level close to the level set in the legislation (0,9%) or according to the intended use of the method and additionally at a level close to the LOQ. The trueness can be measured using a CRM or if not available on a sample from a proficiency test (PT). To comply with the acceptance criterion, the measured value should not deviate more than 25% from the true value. In the case of a PT sample a z-score in the range of (-2;2) should have been obtained.

The **Relative Repeatability Standard Deviation (RSDr)** should be calculated on at least 16 single test results obtained under repeatability conditions. Repeatability should be available for all tested GM levels. The RSDr needs to be equal or below 25% to be acceptable.

Furthermore, the enforcement laboratory should estimate the **sensitivity** of the method. Hereto, four parameters can be calculated. The *Relative LOQ* (LOQ_{rel}) is estimated at low concentration(s) of positive material e.g. 0,1%. The LOQ_{rel} is set at this level if the RSDr is below 25%. The *Absolute LOQ* (LOQ_{abs}) is estimated by measuring dilution series of low copy numbers of the target. The LOQ_{abs} is set as the last dilution where the RSDr is lower than 25%.

The *Relative LOD* (LOD_{rel}) is estimated using ten replicates of a positive control material with a low GM level. The LOD_{rel} is set at this level if the ten replicates show a positive amplification. The *Absolute LOD* (LOD_{abs}) is estimated as the copy number at which not more than 5% false negatives are obtained. In practice this is performed by evaluating ten PCR replicates of low copy number of the target. The LOD_{abs} is set at this level if the ten replicates score positive.

5.4 Conclusion

A GMO quantification method filed by the biotech companies together with the application for authorisation follows different steps. Firstly, the developer needs to provide information on the performance of the method. Hereto, he needs to evaluate different parameters as laid down in the ENGL document (ENGL, 2008). Secondly, the EU-RL GMFF evaluates the submitted information and decides whether the dossier is in compliance with the set criteria. Thirdly, the EU-RL organises a ring trial to validate the method. Hereto it gets the support of the different NRL that participate in the validation. Fourthly, the enforcement laboratories need to assess a number of parameters before to implement the method in their laboratory for routine analysis under ISO 17025 accreditation.

At WIV-ISP-GMOlab, the assessed parameters and the data obtained during the in-house verification are gathered in a validation dossier. The event-specific method is in a first time used as a qualitative identification method in the second step of GMO analysis. The critical parameters determined during the in-house validation for these methods are the LOD_{abs} and LOQ_{abs}. These parameters, expressed as C_t values, are introduced into the DSS and serve as a threshold to decide if the GM event is present in the sample and in case of presence if it is quantifiable.

For quantification methods, no real DSS exists but different parameters are evaluated at each use in routine analysis and have to be in compliance with the set criteria. In a first step, the parameters of the calibration curves of the event-specific and the taxon-specific method (linearity, slope, PCR efficiency) are evaluated. Additionally, control samples (0,1% and 1%) are quantified and the result has to fulfil the acceptance criterion for trueness. In this way the obtained quantitative results for unknown samples are validated.

6. Introduction of the qPCR methods in the Decision Support System (DSS)

6.1 General strategy

As described before (part 3), to cover the broadest GMO spectra, SYBR®Green qPCR methods have been developed and validated in the GMO detection platform. In this context, it rapidly becomes tedious in routine analyses to manually combine all the screening results in order to decide which GMO are potentially present in a sample. Therefore, in support to the qPCR data, a simple mathematical model has been developed to automatically calculate the possible presences in a product based on the outcome of the qPCR screening analysis (Van den Bulcke et al., 2010). The CoSYPS, standing for Combinatory SYBR®Green qPCR screening, represents a novel tool for GMO analysis based on the SYBR®Green qPCR technology. Using this decision support system alone is not sufficient. The suspected GM events need to be specifically identified in a second step, using e.g. the EU-RL Taqman® event-specific qPCR method(s). In a third step, the positively identified GM events are quantified to asses if their content complies or not with the 0,9% labelling threshold (EC1830/2003).

This newly developed tool is a versatile, cost-effective and time-efficient approach in assessing the GMO presence in analytical samples and can be applied in routine analysis for enforcement purposes. The full system has been patent protected (Van den Bulcke et al., 2008).

Here the construction, functioning and the theoretical basis of the CoSYPS will be described. Further explanation on the mathematical functioning of the CoSYPS may be found in the recently published paper "A theoretical introduction to "Combinatory SYBR®Green qPCR screening", a matrix-based approach for the detection of materials derived from genetically modified plants" (Van den Bulcke et al., 2010).

6.2 Screening for GMO candidates by CoSYPS analysis

The CoSYPS is based on the determination of the presence of certain element(s) originating from GMO and plant taxa frequently occurring in food and feed products. Hereto, SYBR®Green qPCR analysis of gDNA extracted from the product is performed, using primer pairs targeting different (multiple) discriminatory marker amplicons (part 3 and table 2).

During the SYBR®Green qPCR analysis of the sample, two critical qPCR parameters are recorded for each method used: the C_t and T_m values. Within the Decision Support System the obtained values are then compared to the LOD (expressed as a C_t value - see glossary) determined in the validation of the qPCR screening method and the nominal T_m value of the amplicon (see glossary). Both parameters are used as decision criteria for the analysis and are incorporated as such in the CoSYPS Decision Support System.

In a first step, the CoSYPS algorithm compares the measured C_t and the T_m values for each screening element with the corresponding "decision values" in the DSS. The latter values are determined during the in-house validation of the method (part 4). A signal generated in SYBR®Green qPCR analysis for a sample is considered as positive by the CoSYPS when an exponential amplification below the C_t value of the LOD (+ 1 C_t) is obtained and the amplicon has a T_m value that falls within the determined T_m confidence interval (part 4). In agreement with the decision principles of the ISO norm 24276 (twice positive, twice negative), all decisions within the CoSYPS are based on the extraction and analysis of two distinct representative sub-extracts and eventually confirmed by a third analysis in case of ambiguous results (one positive, one negative). Therefore, a sample is positive for a specific screening element when the C_t and T_m results are unambiguously for both sub-extracts. Any positive signal obtained with a SYBR®Green qPCR method targeting a particular GM element indicates that a GMO comprising this target could be present in the sample. When several GMO contain the same target, a positive result generated by this screening method indicates that potentially all these GMO may be present in the sample. However, when multiple targets are present in a GMO and the CoSYPS contains methods for each of these targets, all targets present in that GMO must be positive to conclude that this GMO might be present.

The second step in the CoSYPS algorithm is based on a mathematical model. A unique prime number (a prime number is a natural number that has exactly two distinct natural number divisors: 1 and itself) is associated with each particular screening method. When the sample is considered positive for a certain screening element, this specific prime number is assigned to the sample. When it is considered negative, the number 1 (neutral element in multiplication) is assigned. By multiplying all assigned values, the algorithm calculates the "Gödel prime product" (GPP_{sample}) of the sample (the product of the prime numbers corresponding to the positive scoring screening methods). In a similar way each GMO can be represented by a product of the different prime numbers corresponding to the elements belonging to the GMO. This product is designed as the "Gödel prime product" (GPP_{GMO}) of the GMO and represents a "mathematical tag" for this GMO. Note that several GMO can be associated with a same GPP product as they comprise the same genetic elements.

The third step of the CoSYPS is based on the fact that, as a consequence of the nature of prime numbers, the division of the GPP by any of the prime numbers used in the generation of the GPP is an integer. Therefore the presence of a target in a GMO can be mathematically traced by generating this fraction: the program makes the ratio between the GPP_{sample} and the GPP_{GMO} to identify which GMO could be present in the sample (the division generates an integer).

Consequently, on the basis of the positive signal(s) obtained during the screening for each specific SYBR®Green qPCR method, the specific prime number assigned to each method is scored by the CoSYPS. The multiplication of these prime numbers allows the CoSYPS to calculate the GPP for the analysed sample. From this number, the CoSYPS can select all the potential GMO present in the sample by a series of simple divisions.

6.3 Integration of an event-specific method in the Decision Support System and interpretation

On the basis of outcome of the CoSYPS analysis a set of candidate GMO which could possibly reside within the product can be identified. In order to confirm the presence of a certain GM event in this product, event-specific Taqman® qPCR analysis is performed in a next step by applying methods validated and published by the EU-RL (http://gmo-crl.jrc.ec.europa.eu).

During the sample analysis, the C_t value obtained as outcome of the event-specific qPCR is recorded. This C_t value is compared to the LOD and LOQ (as determined during the verification of the identification method in the laboratory - part 5). These values were previously introduced in the Decision Support System.

A GM event is considered detectable by the DSS when an exponential amplification below the C_t value of the LOD (+ 1 C_t) is obtained. The LOD was obtained under repeatability conditions (part 4).

To conclude which GM events are effectively present and identified in the sample, the DSS retains all prime numbers of the GM event with a C_t value below the C_t value of the LOD (+

1 C_t) threshold level. The C_t value is also compared with the LOQ + 1 C_t to decide if the GM event is present at a quantifiable level.

If no authorised GMO can explain the presence of a set of screening targets, it can be concluded that the sample contains one or more unassigned targets. The unassigned signals are mostly due to unauthorised GMO or donor organisms (bacterial, viral and plant sources of transgenic elements). In such cases more complex analysis like DNA walking, DNA sequencing has to be performed outside of the routine to elucidate their origin.

6.4 Practical case

As an example, the accredited SYBR®Green qPCR methods available in a qPCR platform for GMO detection and their associated prime numbers are p35S, tNOS, pNOS, t35S, CryIAb, PAT/pat, CP4, PAT/bar for the transgenic elements and ADH1, LEC and CRU for the taxon-specific markers (table 3a). The elements targeted by these methods can be found in part 3 table 2.

During the screening analysis a positive signal (correct T_m and Ct < C_t of LOD + 1 C_t) is found for the p35S, tNOS, CryIAb, t35S , PAT/pat and ADH1 elements while no positive signal was obtained for pNOS, CP4, PAT/bar and the other species-specific targets (table 3b–step1). For each positive screening marker (p35S, tNOS, CryIAb, t35S and PAT/pat and ADH1) the specific prime number is assigned to each of the corresponding methods. As the pNOS, CP4, PAT/bar targets and the other taxon-specific markers are considered as negative the assigned number for all of these methods is 1. The "Gödel prime product" of the sample (= 1057485) is calculated by multiplying all the assigned prime numbers (table 3b-step2). The CoSYPS will compare this GPP_{sample} with the GPP of all GM events that have previously been introduced in the system. The example is given here for four GM events.

The transgenic MON 810 and T25 events are described as a function of three transgenic elements (p35S, tNOS, CryIAb) and (p35S, t35S, PAT/pat) respectively and one maize-specific (ADH1). The GA21 maize is covered by the tNOS and maize–specific element. The GTS40-3-2 event is defined by three transgenic elements (p35S, tNOS, CP4) and the soybean endogen (LEC). Consequently, the "Gödel prime product" of the MON 810, T25, GA21 and GTS40-3-2 are 5655 (= 3 X 5 X 13 X 29), 16269 (= 3 X 11 X17 X 29), 145 (= 5 X 29) and 8835 (= 3 X 5 X 19 X 31) respectively (table 3b-step 3).

To assess which GMO are potentially present in the sample, the "Gödel prime product" of the sample is divided by the GPP of each GMO (table 3b-step 3). The result is an integer only for MON 810, T25 and GA21. From the screening analysis, the CoSYPS thus predicts that MON 810, T25 and GA21 are potentially present while GTS40-3-2 is not. As a consequence MON 810, GA21 and T25 have to be further analysed with the event-specific method to confirm their presence.

In order to confirm the presence of MON 810, T25 and GA21 in the sample product, the event-specific qPCR analyses are performed. The results (expressed as C_t values) confirm the presence of MON 810 and T25 while GA21 is not detectable (table 3c). The C_t values obtained are compared with the LOQ + 1 C_t of each method and show that only MON 810 can be quantified. Finally the GM% of this event will be compared to the labelling threshold (0,9% mass per ingredient) in order to conclude on the conformity of the sample.

a. Accredited SYBR®Green qPCR available

Screening methods	p35S	pNOS	t35S	tNOS	CryIAb	PAT/pat	PAT/bar	CP4	ADH1	LEC	CRU
Prime numbers	3	7	11	5	13	17	23	19	29	31	37

b. CoSYPS algorithm and screening methods

CoSYPs step1 - Sample analysis "+" is assigned when value < LOD+1 C_t; " –" is assigned when value > LOD + 1 C_t											
Subsample 1	+	-	+	+	+	+	-	-	+	-	-
Subsample 2	+	-	+	+	+	+	-	-	+	-	-
CoSYPS step 2 - Calculation of the Gödel prime product of the sample											
Product of	3	1	11	5	13	17	1	1	29	1	1
GPP			1057485								
CoSYPS step 3 - Assessment of potential GMO present in the sample											
GMO	GPP of GMO		GPP_{sample}/ GPP_{GMO}			Decision					
MON810	5655		187			Confirmation by event-specific Taqman method					
T25	16269		65			Confirmation by event-specific Taqman method					
GA21	145		7293			Confirmation by event-specific Taqman method					
GTS 40-3-2	8835		119,69			No confirmation by event-specific Taqman method					

c. DSS and confirmation by event-specific Taqman method

Taqman				
	MON 810	T25	GTS 40-3-2	GA21
Subsample 1	LOD > C_t < LOQ	LOD > C_t > LOQ		LOD < C_t > LOQ
Subsample 2	LOD > C_t < LOQ	LOD > C_t > LOQ		LOD < C_t > LOQ
Results	*Present Quantifiable*	*Present*		*Not detectable*

Table 3. Mathematical functioning of the CoSYPS, allowing demonstrating the possible presence of a set of GMO in a product based on the outcome of a qPCR screening analysis.

6.5 Conclusion

By combining the results of the screening analysis, the CoSYPS allows to decide in a fast way which GM events are possibly present in the sample under analysis. The use of the mathematical algorithm, which compares the GPP_{sample} and GPP_{GMO}, excludes the need for manual calculations and comparisons. The only thing that needs to be done by the operator is the preliminary introduction of the critical values (C_t corresponding to the LOD and LOQ, T_m) obtained during method validation in the system. Further, in identification, the obtained results for a sample are compared with the LOD and LOQ values determined during in-house validation of the event-specific methods. From this comparison, the Decision Support System will indicate which GM events are present and at which level and thus allow deciding which GMO needs to be quantified in the sample. This Decision Support System, developed and patented by the WIV-ISP-GMOlab is thus a very efficient, user friendly and cost-saving tool in GMO detection.

7. Conclusion

In the near future, the number and the diversity of GM crops will continue to increase, as well as the requests for authorisation for their import for food and feed in the EU. Beside the notifications of GM events produced by multinational biotech companies, many GM events will be developed by universities, national research centres and small private companies. Thus, the chance for accidental occurrence of unapproved GMO in the EU food and feed chain trough importation will be higher. As the EU's general policy supports strong commitment to consumer protection and freedom of choice, and therefore mandatory product labelling, the development of sensitive, reliable but also cost-effective and flexible strategies for the detection of GMO in products through establishment of molecular platforms will become more and more crucial.

The GMO detection platform developed at WIV-ISP consists of a pre-PCR step namely DNA extraction and three consecutive qPCR phases. In this view, the choice of efficient methods to extract good quality DNA, in particular for processed food and feed, is a critical factor. A pre-PCR evaluation of the extracted gDNA is necessary as well as setting criteria for the purity and integrity of the DNA. Furthermore, the presence of PCR inhibitors is a major obstacle for efficient amplification in qPCR. This step may even become more important as the number of GM plant taxa becomes larger. Developing simple standard methods for genomic DNA extraction minimizing inhibition will therefore be the key for providing concordant results when using qPCR techniques.

Due to the broad range of GMO that my occur in the EU food and feed chains, the use of screening strategies only based on the 35S promoter of the Cauliflower Mosaic Virus (p35S) and the nopaline synthase terminator of *Agrobacterium tumefaciens* (tNOS) followed by the analysis of the sample with event-specific EU validated methods by the enforcement laboratories will become insufficient. As a consequence, new methods focusing on an intensive screening analysis need to be developed.

At the present time several high-tech strategies like multiplex PCR and consecutive detection and identification of the amplification products using micro-arrays (Chaouachi et al., 2008, Morisset et al., 2008, Hamels et al., 2009) or PCR combined with capillary electrophoresis (Nadal et al., 2009) have been proposed to deal with this discriminative problem and the broad diversity of GMO. However, at the present time, these technologies require additional costly equipment and investments in technical support. Furthermore, they need technological optimalisation as they show a high background at low target level. These difficulties make them less suitable for routine or enforcement purposes.

Contrary to the above-mentioned technically complex strategies, our approach based on numerous singleplex qPCR-based methods developed to function under the same reaction conditions combined with the informatics decision support tool CoSYPS may in the future represent a very effective alternative. This newly developed tool is considered as a versatile, cost-effective and time-efficient platform assessing the GMO presence in analytical samples. In addition, it functions in routine analysis for enforcement purposes in a commonly applied 96-well plate qPCR format.

In the future, the research of the molecular platform of the WIV-ISP will focus on the development of more discriminative SYBR®Green qPCR screening methods to cover the broad range of GMO and UGM and thus to improve the resolution of the system. Particular importance will be given to their use in a modular approach associated with a decision tree cascade. Moreover, our strategy aiming at developing harmonised SYBR®Green qPCR screening methods incorporated in the Combinatory SYBR®Green qPCR Screening (CoSYPS) system has a potential to be applied in other scientific fields than GMO detection. The application of this strategy for food borne pathogenic bacteria is now under development in our team.

8. Glossary

Amplification Efficiency

The amplification efficiency is the rate of amplification that leads to a theoretical slope of -3,32 with an efficiency of 100% in each cycle. The efficiency of the reaction can be calculated by the following equation:

$$Efficiency = 10^{\left(\frac{-1}{slope}\right)} - 1 \quad (1)$$

Applicability

Applicability is the description of analytes, matrices and concentrations to which the method can be applied.

Certified Reference Material (CRM)

A Certified Reference Material is a reference material characterized by a metrologically valid procedure for one or more specified properties, accompanied by a certificate that provides the value of the specified property, its associated uncertainty, and a statement of metrological traceability.

Correlation Coefficient (R^2)

The R^2 coefficient is the correlation coefficient of a (calibration) curve obtained by linear regression analysis.

Dynamic Range

The dynamic range is the range of concentrations over which the method performs in a linear manner with an acceptable level of trueness and precision.

European Food Safety Authority (EFSA)

EFSA is an agency of the EU that provides independent scientific advice and communication on all matters concerning food and feed safety.

European Network of GMO Laboratories (ENGL)

The European Network of GMO Laboratories is a platform of EU experts that plays an eminent role in the development, harmonisation and standardisation of means and methods for sampling, detection, identification and quantification of Genetically Modified Organisms (GMO) or derived products in a wide variety of matrices, covering seeds, grains, food, feed and environmental samples. The network was inaugurated in Brussels on December

4th 2002 and it currently consists of more than 100 national enforcement laboratories, representing all 27 EU Member States plus Norway and Switzerland. Its plenary meetings are open to particular observers, such as to representatives from Acceding and Candidate Countries.

European Union Reference Laboratory for GM Food and Feed (EU-RL GMFF)

The core task of the EU-RL GMFF is the scientific assessment and validation of detection methods for GM Food and Feed as part of the EU authorisation procedure. The Joint Research Centre (JRC) of the European Commission and, more precisely, the Molecular Biology and Genomics Unit of the Institute for Health and Consumer Protection (IHCP), has been given the mandate for the operation of the EU-RL GMFF. Activities are carried out in close collaboration with European Network of GMO Laboratories (ENGL).

Genetically Modified (GM) event

A GM event refers to the unique DNA recombination event that took place in one plant cell, which was then used to generate entire transgenic plants

Genetically Modified Organism (GMO)

A Genetically Modified Organism is officially defined in the EU legislation as "organisms, not from human origin, in which the genetic material (DNA) has been altered in a way that does not occur naturally by mating and/or natural recombination"

Limit of Detection (LOD)

The limit of detection is the lowest amount or concentration of analyte in a sample, which can be reliably detected but not necessarily quantified, as demonstrated by single-laboratory validation.

Limit of Quantification (LOQ)

The limit of quantification is the lowest amount or concentration of analyte in a sample that can be reliably quantified with an acceptable level of precision and accuracy.

Melting temperature (T_m)

The melting temperature is the temperature at which 50% of the DNA is single stranded.

National Reference Laboratory (NRL)

A National Reference Laboratory on GMO operates in the frame of Commission Regulation EC/1829/2003 on GM Food and Feed and Commission regulation EC/1830/2003 on labelling and traceability of GMO. It assists the EU-RL and the NRL from the different member states are gathered in the ENGL.

Practicability

Practicability is the ease of operations, the feasibility and efficiency of implementation, the associated unitary costs (e.g. cost/sample) of the method.

Precision - Relative Repeatability Standard Deviation (RSD_r%)

The relative repeatability standard deviation is the relative standard deviation of test results obtained under repeatability conditions. Repeatability conditions are conditions where test results are obtained with the same method, on identical test items, in the same laboratory, by the same operator, using the same equipment within short intervals of time.

Precision – Relative Reproducibility Standard Deviation ($RSD_R\%$)

The relative reproducibility standard deviation is the relative standard deviation of test results obtained under reproducibility conditions. Reproducibility conditions are conditions where the test results are obtained with the same method, on identical test items, in different laboratories, with different operators, using different equipment. Reproducibility standard deviation describes the inter-laboratory variation.

Reference material (RM)

A Reference Material is a material that is sufficiently homogeneous and stable with respect to one or more specified properties, which has been established to be fit for its intended use in a measurement process.

Robustness

The robustness of a method is a measure of its capacity to remain unaffected by small, but deliberate deviations from the experimental conditions described in the procedure.

Specificity

Specificity is a property of a method to respond exclusively to the characteristic or analyte of interest.

Threshold cycle (C_t)

The threshold cycle reflects the cycle number at which the fluorescence generated within a reaction crosses the threshold. It is inversely correlated to the logarithm of the initial copy number. The C_t value assigned to a particular well thus reflects the point during the reaction at which a sufficient number of amplicons has been accumulated.

Trueness

The trueness is defined as the closeness of agreement between the average value obtained from a large series of test results and an accepted reference value. The measure of trueness is usually expressed in terms of bias.

9. Acknowledgement

The authors like to acknowledge Lievens Antoon for his critical review of the manuscript and De Keersmaecker Sigrid for her help with figures and layout. This work of the GMOlab was financially supported by four projects: the British Food Standard Agency (FSA, contract G03032) and the German Federal Office of Consumer Protection and Food Safety (BVL) through the Project GMOseek, under the European ERA-NET consortium SAFEFOODERA; the GMODETEC project (RT-06/6) of the Belgian Federal Ministry "Health, Food Chain Safety and Environment'; the European Commission through the Integrated Project Co-Extra (Contract No. 007158), under the 6th Framework Program and by the Belgian Science Policy projects SPSD I & II of the Belgian Federal Ministry of Science Policy.

10. References

Agbios web site: http://www.agbios.com/dbase.php

Barbau-Piednoir, E., Lievens, A., Mbongolo-Mbella, G., Roosens, N., Sneyers, M., Leunda-Casi, A. & Van den Bulcke, M. (2010). SYBR®Green qPCR screening methods for the presence of "35S promoter" and "NOS terminator" elements in food and feed products. *European Food Research Technology*, Vol. 230, No. 3, pp. 383-393, 1438-2377

Barbau-Piednoir, E., Lievens, A., Vandermassen, E., Mbongolo-Mbella, G., Leunda-Casi, A., Roosens, N., Sneyers, M. & Van den Bulcke, M. (2011). Four new SYBR®Green qPCR screening methods for the detection of Roundup Ready®, LibertyLink®, and CryIAb traits in genetically modified products. *European Food Research Technology*. Vol. 234, No. 1, pp. 13-23, doi: 10.1007/s00217-011-1605-7

Bellocchi, G., De Giacomo, M., Foti, N., Mazzara, M., Palmaccio, E., Savini, C., Di Domenicantonio, C., Onori R. & Van den Eede, G. (2010). Testing the interaction between analytical modules: an example with Roundup Ready® soybean line GTS 40-3-2. *BMC Biotechnology*, Vol. 10, No. 55, doi:10.1186/1472-6750-10-55

Bickley, J. & Hopkins, D. (1999). Inhibitors and enhancers of PCR, In: *Analytical Molecular Biology: Quality and Validation*, Saunders GC, Parkes HC, pp. 81-102, Royal Society of Chemistry, Thomas Graham House, ISBN 0854044728, Cambridge UK 1234

Bravo, A., Gill, S.S. & Soberon, M. (2007). Mode of action of *Bacillus thuringiensis* Cry and Cyt toxins and their potential for insect control. *Toxicon*, Vol. 49, No. 4, pp. 423-435, 0041-0101

Broothaerts, W., Corbisier, P., Schimmel, H., Trapmann, S., Vincent, S. & Emons, H. (2008). A single nucleotide polymorphism (SNP839) in the *adh1* reference gene affects the quantitation of genetically modified maize (*Zea mays* L.). *Journal of Agricultural and Food Chemistry*, Vol. 56, No. 19, pp. 8825-8831, doi: 10.1021/jf801636d

Burggraf, S. & Olgemoller, B. (2004). Simple technique for internal control of real-time amplification assays. *Clinical Chemistry*, Vol. 50, No. 5, pp. 819-825, 0009-9147

Cankar, K., Štebih, D., Dreo, T., Žel, J. & Gruden, K. (2006). Critical points of DNA quantification by real-time PCR – effects of DNA extraction method and sample matrix on quantification of genetically modified organisms. *BMC Biotechnology*, Vol. 6, No. 36, doi:10.1186/1472-6750-6-37

Center for Environmental Risk Assessment (CERA) GMO crop database: http://www.cera-gmc.org/?action=gm_crop_database

Chaouachi, M., Chupeau. G., Bernard, A., Mckhann, H., Romaniuk, M., Giancola, S., Laval, V., Bertheau, Y. & Brunel, D. (2008). A high-throughput multiplex method adapted for GMO detection. *Journal of Agricultural and Food Chemistry*, Vol. 56, pp. 11596–11606

Charels, D., Broeders, S., Corbisier, P., Trapmann, S., Schimmel, H., Linsinger, T. & Emons, H. (2007). Toward Metrological Traceability for DNA Fragment Ratios in GM Quantification. 2.Systematic Study of Parameters Influencing the Quantitative Determination of MON 810 Corn by Real-Time PCR. *Journal of Agricultural and Food Chemistry*, Vol. 55, No. 9, pp. 3258-3267, doi: 10.1021/jf062932d-8893

Chilton, M.D., Drummond, M.H., Merio, D.J., Sciaky, D., Montoya, A.L., Gordon, M.P. & Nester, E.W. (1977). Stable incorporation of plasmid DNA into higher plant cells: The molecular basis of crown gall tumorigenesis. *Cell*, Vol. 11, No. 2, pp. 263-271, 0092-8674

Cloud, J.L., Hymas, W.C., Turlak, A., Croft, A., Reischl, U., Daly, J.A. & Carroll, K.C. (2003). Description of a multiplex *Bordetella pertussis* and *Bordetella parapertussis* LightCycler PCR assay with inhibition control. *Diagnostic Microbiological infectious Diseases*, Vol. 46, No. 3, pp. 189-195, 0732

Codex Alimentarius Commission - Joint FAO/WHO Food Standards Programme - Codex Committee on Methods of Analysis and Sampling, Thirtieth Session (9-13 March 2009). Proposed Draft Guidelines on Criteria for Methods for the Detection and Identification of Foods Derived from Biotechnology. CX/MAS 09/30/8; Available at ftp://ftp.fao.org/codex/ccmas30/ma30_08e.pdf (assessed on 26/10/2011).

Codex Committee On Methods Of Analysis And Sampling. (2010). Guidelines On Performance Criteria And Validation Of Methods For Detection, Identification And Quantification Of Specific DNA Sequences And Specific Proteins In Foods. *Codex alimentarius commission – WHO*. Rome.

Commission Recommendation (EC) No 787/2004 of 4 October 2004 on technical guidance for sampling and detection of genetically modified organisms and material produced from genetically modified organisms as or in products in the context of Regulation (EC) No 1830/2003. *Official Journal of the European Union*, L 348, pp. 12-16

Commission Regulation (EC) No 641/2004 of 6 April 2004 on detailed rules for the implementation of Regulation (EC) No 1829/2003 of the European Parliament and of the Council as regards the application for the authorisation of new genetically modified food and feed, the notification of existing products and adventitious or technically unavoidable presence of genetically modified material which has benefited from a favourable risk evaluation. *Official Journal of the European Union*, L 102, p.14-25

Commission Regulation (EC) No 619/2011 of 24 June 2011 laying down the methods of sampling and analysis for the official control of feed as regards presence of genetically modified material for which an authorisation procedure is pending or the authorisation of which has expired. *Official Journal of the European Union*, L166, p.9-15

Commission Regulation (EC) No 882/2004 of the European Parliament and of the council of 29 April 2004 on official controls performed to ensure the verification of compliance with feed and food law, animal health and animal welfare rules. *Official Journal of the European Union*, L 191, p.1-141

Commission Regulation (EC) No 1829/2003 of the European Parliament and of the Council of 22 September 2003 on genetically modified food and feed. *Official Journal of the European Union*, L 268, pp.1-23

Commission Regulation (EC) No 1830/2003 of the European Parliament and of the Council of 22 September 2003 concerning the traceability and labelling of genetically modified organisms and the traceability of food and feed products produced from genetically modified organisms and amending Directive 2001/18/EC. *Official Journal of the European Union*, L 268, pp. 24-28

Commission Regulation (EC) No 1981/2006 of 22 December 2006 on detailed rules for the implementation of Article 32 of Regulation (EC) No 1829/2003 of the European Parliament and of the Council as regards the Community reference laboratory for genetically modified organisms. *Official Journal of the European Union*, L 368, pp.99-109

Corbisier, P., Broothaerts, W., Gioria, S., Schimmel, H., Burns, M., Baoutina, A, Emslie, K.R., Furui, S., Kurosawa, Y., Holden, M., Kim, H-H;, Lee, Y., Kawaharasaki, M., Sin D. & Wang J. (2007). Toward Metrological Traceability for DNA Fragment Ratios in GM Quantification. 1. Effect of DNA Extraction Methods on the Quantitative Determination of Bt176 Corn by Real-Time PCR. *Journal of Agricultural and Food Chemistry*, Vol. 55, No. 9, pp. 3249-3257, 0021-8561

Darbani, B.; Farajnia, S.; Toorchi, M.; Zakerbostanabad, S.; Noeparvar S. & Stewart Jr., C.N. (2008). DNA-delivery Methods to Produce Transgenic Plants. *Biotechnology*, Vol. 7, No. 3, pp. 385-402, 1682-296X

European Food Safety Authority (EFSA) Pannel on Genetically Modified Organisms (GMOs). (2011). SCIENTIFIC OPINION Guidance for risk assessment of food and feed from genetically modified plants. *EFSA Journal*, Vol. 9, No. 5, pp. 2150

European Network of GMO Laboratories. (2008). Definition of minimum performance requirements for analytical methods of GMO testing. V.13/10/2008. Available at http://gmo-crl.jrc.ec.europa.eu/guidancedocs.htm

European Network of GMO Laboratories (ENGL). (2011). Verification of analytical methods for GMO testing when implementing interlaboratory validated methods. *EUR24790-EN*

European Union Reference Laboratory report: Sampling and DNA extraction of cotton seeds Report from the Validation of the "CTAB/Genomic-tip 20" method for DNA extraction from ground cotton seeds: http://gmo-crl.jrc.ec.europa.eu/summaries/281-3006%20Cotton_DNAExtr.pdf

European Union Reference Laboratory web site: http://gmo-crl.jrc.ec.europa.eu/

Gallagher, S.R. (2011). Quantitation of DNA and RNA with Absorption and Fluorescence Spectroscopy. In: *Current Protocols in Molecular Biology*. Ausubel, F.M., John Wiley & Sons Inc.: New York 2011, Appendix 3D

Georgiou, C.D. & Papapostolou, I. (2006) Assay for the quantification of intact/fragmented genomic DNA. *Analytical Biochemistry*, Vol. 358, No. 2, pp. 247-256, doi:10.1016/j.ab.2006.07.035

Glasel, J.A. (1995). Validity of nucleic acid purities monitored by 260 nm/280 nm absorbance ratios. *Biotechniques*, Vol. 18, No. 1, pp. 62-63

GMO compass web site: http://www.gmo-compass.org

Hamels, S., Glouden, T., Gillard, K., Mazzara, M., Debode, F., Foti, N., Sneyers, M., Nuez, T.E., Pla, M., Berben, G., Moens, W., Bertheau, Y., Audeon, C., Van den Eede, G. & Remacle, J. (2009). A PCR-microarray method for the screening of genetically modified organisms. *European Food Research Technology*, Vol. 228, pp. 531-541

Hensel, G., Kastner, C., Oleszczuk, S., Riechen, J. & Kumlehn, J. (2009). *Agrobacterium*-Mediated Gene Transfer to Cereal Crop Plants: Current Protocols for Barley, Wheat, Triticale, and Maize. *International Journal of Plant Genomics*, doi: 10.1155/2009/835608

Hillen, W., Goodman, T.C. & Wells, R.D. (1981). Salt dependence and thermodynamic interpretation of the thermal denaturation of small DNA restriction fragments. *Nucleic Acid Research*, Vol. 9, No. 2, pp.415-436

Holden, M.J., Haynes, R.J., Rabb, S.A., Satija, N., Yang, K. & Blasic Jr., J.R. (2009). Factors Affecting Quantification of Total DNA by UV Spectroscopy and PicoGreen Fluorescence. *Journal of Agricultural and Food Chemistry*, Vol. 57, No. 16, pp. 7221-7226, 0021-8561

Holme, I.B., Brinch-Pederson, H., Lange, M., & Holm, P.B. (2006). Transformation of barley (*Hordeum vulgare* L.) by *Agrobacterium tumefaciens* infection of in vitro cultured ovules. *Plant Cell Reports*, Vol. 25, No. 12, pp. 1325-1335

Holst-Jensen, A. & Berdal, K.G. (2004). The modular analytical procedure and validation approach, and the units of measurement for genetically modified materials in foods and feeds. *Journal of AOAC International*, Vol. 87, No. 4, pp. 927-936.

Holst-Jensen, A. (2009). Testing for genetically modified organisms (GMOs): Past, present and future perspectives. *Biotechnology Advances*, Vol. 27, No. 6, pp.1071-1082, doi:10.1016/j.biotechadv.2009.05.025

Horsch, R.B., Fry, J.E., Hoffman, N.L., Eichholtz, D., Rogers, S.G. & Fraley, R.T. (1985). A simple and general method for transferring genes into plants. *Science*, Vol. 227, No. 4691, pp. 1229-1231, doi: 10.1126/science.227.4691.1229

International Standard ISO 5725-2. (1994).Accuracy (trueness and precision) of measurement methods and results - Part 2 (11994). International Organisation for Standardisation, Genève, Switzerland

International Standard ISO/IEC 17025. (2005). General requirements for the competence of testing and calibration laboratories. International Organisation for Standardisation, Genève, Switzerland

IUPAC protocol for the Design, Conduct and Interpretation of Method Performance Studies. (1995). *Pure and Applied Chemistry*, Vol. 67, pp.331-343

James, C. (2010). Global Status of Commercialized Biotech/GM Crops: 2010. Executive summary. *ISAA brief*, No. 42.

JRC web site: Development and harmonisation of GMO detection methods - Overview on EU activities (2010) http://ihcp.jrc.ec.europa.eu/our_activities/gmo/development-and-harmonisation-of-gmo-detection-methods-an-overview-of-eu-activities

Kikkert, J.R., Vidal, J.R. & Reisch, B.I. (2004). Stable Transformation of Plant Cells by Particle Bombardment/Biolistics. Methods in Molecular Biology, In: *Transgenic Plants: Methods and Protocols* Vol 286, Edited by L Peña, Human Press Inc., Totowa, NJ

Kishore, G., Shah, D., Padgette, S., Dells-Cioppa, G., Gasser, C., Re, D., Hironak, C., Taylor, M., Wibbenmeyer; J., Eichholtz, D., Hayford, M., Hoffmann, N., Delannay, X., Horsch, R., Klee, H., Roger, S., Rochester, D., Brundage, L., Sanders, P. & Fraley, R.T. (1988). 5-Enolpyruvylshikimate 3-Phosphate Synthase. In: *Biochemistry to Genetic Engineering of Glyphosate Tolerance*, pp. 37-48, American Chemical Society, patent 4971908

Kumlehn, J., Serazetdinova, L., Hensel, G., Becker, D. & Loerz, H. (2006). Genetic transformation of barley (Hordeum vulgare L.) via infection of androgenic pollen cultures with *Agrobacterium tumefaciens*. *Plant Biotechnology Journal*, Vol. 4, No. 2, pp. 251-261, 1467-7652

Kutyavin, I.V., Afonina, I.A., Mills, A., Gorn, V.V., Lukhtanov, E.A., Belousov, E.S., Singer, M.J., Walburger, D.K., Lokhov, S.G., Gall, A.A., Dempcy, R., Reed, M.W., Meyer, R.B. & Hedgpeth, J. (2000). 3′-minor groove binder-DNA probes increase sequence specificity at extension temperatures. *Nucleic Acids Research*, Vol. 28, No. 2, pp. 655-661, doi: 10.1093/nar/28.2.655

Lievens, A.J., Van Aelst, S., Van den Bulcke, M. & Goetghebeur, E. (2011). Enhanced analysis of real-time PCR data by using a variable efficiency model: FPK-PCR. *Nucleic Acids Research*, Vol. 5, pp. 1-15, doi:10.1093/nar/gkr775

Liu, W. & Saint, D.A. (2003). Validation of a quantitative method for real-time PCR kinetics. *Biochemical and Biophysical Research Communication*, Vol. 294, pp. 347-353

Manchester, K.L. (1995). Value of A260/A280 ratios for measurement of purity of nucleic acids. *Biotechniques*, Vol. 19, No. 2, pp. 208-209

Mbongolo Mbella, E. G., Lievens, A., Barbau-Piednoir, E., Sneyers, M., Leunda-Casi, A., Roosens, N. & Van den Bulcke, M. (2011). SYBR®Geen qPCR methods for detection of endogenous reference genes in commodity crops: a step ahead in combinatory screening of genetically modified crops in food and feed products. *European Food Research Technology*, Vol. 232, pp. 485-196, doi: 10.1007/s00217-010-1408-2

Morisset, D., Dobnik, D., Hamels, S., Zel, J. & Gruden, K. (2008). NAIMA: target amplification strategy allowing quantitative on chip detection of GMOs. *Nucleic Acids Research*, Vol. 36, No. 18, doi:10.1093/nar/gkn524

Morisset, D., Demšar, T., Gruden, K., Vojvoda, J., Štebih, D. & Žel, J. (2009). Detection of genetically modified organisms—closing the gaps. *Nature Biotechnology*, Vol. 27, pp. 700 - 701, doi:10.1038/nbt0809-700

Nadal, A., Esteve, T. & Pla, M. (2009). Multiplex polymerase chain reaction-capillary gel electrophoresis: a promising tool for GMO screening--assay for simultaneous detection of five genetically modified cotton events and species. *Journal AOAC International*, Vol. 92, No. 3, pp. 765–777

National Center for Biotechnology Information (NCBI): http://www.ncbi.nlm.nih.gov/

Nolan, T., Hands, R.E., Ogunkolade, W. & Bustin, S.A. (2006). SPUD: A quantitative PCR assay for the detection of inhibitors in nucleic acid preparations. *Analytical Biochemistry*, Vol. 351, pp. 308-310, doi:10.1016/j.ab.2006.01.051

Norm AFNOR XP V 03-020-2, 09/200, ISSN 0335-3931

Organisation for Economic Co-operation and Development (OECD). (2010). Molecular Characterisation of Plants Derived from Modern Biotechnology. Safety assessment of Transgenic Organisms: OECD Consensus documents, Vol. 3, pp. 305-319. OECD Publishing, ISBN 978-92-64-09542

Padgette, S.R., Re, D.B., Barry, G.F., Eichholtz, D.E., Delannay, X., Fuchs, R.L., Kishore, G.M. & Fraley, R.T. (1996). New weed control opportunities: development of soybeans with a Roundup Ready gene. In: *Herbicide-resistant crops: agricultural, environmental, economic, regulatory, and technical aspects*, pp. 53-84

Papazova, N., Zhang, D., Gruden, K., Vojvoda, J., Yang, L., Buh, M., Blejec, A., Fouilloux, S., De Loose, M. & Taverniers, I. (2010). Evaluation of the reliability of maize reference assays for GMO quantification. *Analytical and Bioanalytical Chemistry*, Vol. 396, pp. 2189–2201

Peano, C., Samson, M. C., Palmieri L., Gulli, M. & Marmiroli, N. (2004). Qualitative and Quantitative Evaluation of the Genomic DNA Extracted from GMO and Non-GMO Foodstuffs with Four Different Extraction Methods. *Journal of Agricultural and Food Chemistry*, Vol. 52, No. 23, pp. 6962-6968, doi: 10.1021/jf040008i

Pietsch, K., Waiblinger, H.U., Brodmann, P. & Wurz, A., (1997). Screeningverfahren zur Identifizierung ‚gentechnisch veränderter' pflanzlicher Lebensmittel. *Deutsche Lebensmittel-Rundschau*, Vol. 93, No. 2, pp. 35-58, 0012-0413

Prins, T.W., van Dijk, J.P., Beenen, H.G., Van Hoef, A.M.A., Voorhuijzen, M.M., Schoen, C.D., Aarts, H.J.M. & Kok, E.J. (2008). Optimised padlock probe ligation and microarray detection of multiple (non-authorised) GMOs in a single reaction. *BMC Genomics*, Vol. 9, No. 584, doi:10.1186/1471-2164-9-584

Ramakers, C., Ruijter, J.M., Deprez, R.H. & Moorman, A.F. (2003). Assumption-free analysis of quantitative real-time polymerase chain reaction (PCR) data. *Neuroscience Letter*, Vol. 339, No. 1, pp. 62-66, doi:10.1016/S0304-3940(02)01423-4

Raymond, P., Gendron, L., Khalf, P.S., Dibley, K.L., Bhat, S., Xie, V.R.D., Partis, L., Moreau, M.E., Dollard, C., Coté, M.J., Laberge, S. & Emslie, K.R. (2010). Detection and identification of multiple genetically modified events using DNA insert fingerprinting. *Analytical and Bioanalytical Chemistry*, Vol. 396, No. 6, pp. 2091-2102, doi: 10.1007/s00216-009-3295-6

Rouzina, L. & Bloomfield, V.A. (2001). Force-Induced melting of the DNA double Helix. 2. Effects of Solution Conditions. *Biophysical Journal*, Vol. 80, No. 2, pp. 894-900, 0006-3495/01/02/894/07

Rozen, S. & Skaletsky, H. (2000). Primer3 on the WWW for general users and for biologist programmers. In: Krawetz S, Misener S (eds) *Bioinformatics Methods and Protocols: Methods in Molecular Biology*. Humana Press, Totowa, NJ, pp 365-386.

Sambrook, J. & Russell, D.W. (2001). Molecular Cloning, a laboratory manual, third edition, Vol 1, 2 & 3, Cold Spring Harbor Laboratory Press.

Shauzu, M. (2001). The concept of substantial equivalence in safety assessment of foods derived from genetically modified organisms. *AgBiotechNet 2000*, Vol. 2, April 2001, ABN 044.

Shokere, L.A., Holden, M.J. & Jenkins, J.R. (2009). Comparison of fluorometric and spectrophotometric DNA quantification for real-time quantitative PCR of degraded DNA. *Food Control*, Vol. 20, No. 4, pp. 391-401, 0956-7135

Singer, V.L., Jones, L.J., yue, L.J. & Haugland, R.P. (1997). Characterization of PicoGreen reagent and development of fluorescence-based solution assay for double stranded DNA quantitation. *Analytical Biochemistry*, Vol. 249, pp. 223-238

Stein, A. & Rodriguez-Cerezo, E. (2009). The global pipeline of new GM crops. Implications of asynchroneous approval for international trade. *EU23846-EN*.

Svaren, J., Inagami, S., Lovegren, E. & Chakley, R. (1987). DNA denatures upon drying after ethanol precipitation. *Nucleic Acids Research*, Vol. 15, No. 21, pp. 8739-8753

Tichopad, A., Dilger, M., Schwarz, G. & Pfaffl, M.W. (2003). Standardized determination of real-time PCR efficiency from single reaction set-up. *Nucleic Acids Research*, Vol. 31, No. 20, el22, doi: 10.1093/nar/gng122

United States Department of Agriculture - Grain Inspection, Packers and Stockyards (USDA-GIPSA). (2011). Directive 9181.2. Performance verification of qualitative mycotoxin and biotech rapid test kits. Available from http://www.gipsa.usda.gov/fgis/standproc/biotech.html (assessed on 25/10/2011).

Valvekens, D., Van Montagu, M. & Van Lijsbettens, M. (1988). *Agrobacterium-tumefaciens*-mediated transformation of *Arabidopsis thaliana* root explants by using kanamyscin selection. *Proceedings of the National Academy of Sciences USA*, Vol. 85, No. 15, pp. 5536-5540

Van den Bulcke, M., Lievens, A., Leunda, A., Mbongolo Mbella, E., Barbau-Piednoir, E. & Sneyers, M. (2008). Transgenic plant event detection. PATENT PCT/EP2008/051059

Van den Bulcke, M., Lievens, A., Barbau-Piednoir, E., Mbongolo Mbella, G., Roosens, N., Sneyers, M. & Leunda Casi, A. (2010). A theoretical introduction to "Combinatory SYBR®Green qPCR Screening", a matrix-based approach for the detection of materials derived from genetically modified plants. *Analytical Bioanalytical Chemistry*, Vol. 396, No. 6, pp. 2113-2123, DOI 10.1007/s00216-009-3286-7

Wehrmann, A., Van, V.A., Opsomer, C., Botterman, J. & Schulz, A. (1996). The similarities of bar and pat gene products make them equally applicable for plant engineers. *Nature Biotechnology*, Vol. 14, No. 10, pp. 1274-1278, doi:10.1038/nbt1096-1274

Wurz, A., Rüggeberg, H., Brodmann, P., Waiblinger, H.U. & Pietsch, K., (1998). DNA-Extraktionmethode für den nachweis gentechnisch veränderter Soja in Sojalecithin. *Deutsche Lebensmittel-Rundschau*, Vol. 94, No. 5, pp. 159-161

19

Overview of Real-Time PCR Principles

Morteza Seifi[1,*], Asghar Ghasemi[1], Siamak Heidarzadeh[2], Mahmood Khosravi[3], Atefeh Namipashaki[4], Vahid Mehri Soofiany[5], Ali Alizadeh Khosroshahi[6] and Nasim Danaei[7]

[1]Laboratory of Genetics, Legal Medicine Organization of Tabriz, Tabriz,
[2]Division of Microbiology, School of Public Health, Tehran University of Medical Sciences, Tehran,
[3]Hematology Department of Medicine Faculty, Guilan University of Medical Sciences, Rasht,
[4]Department of Biotechnology, School of Allied Medical Sciences, Tehran University of Medical Sciences, Tehran,
[5]Faculty of Medicine, Shahid Behesti University of Medical Sciences, Tehran,
[6]Jarrah Pasha Medicine Faculty of Istanbul, Istanbul,
[7]Department of Health and Nutrition, Tabriz University of Medical Sciences, Tabriz
[1,2,3,4,5,7]Iran
[6]Turkey

1. Introduction

Real-time PCR is based on the revolutionary method of PCR, developed by Kary Mullis in the 1980s, which allows researchers to amplify specific pieces of DNA more than a billion-fold (Saiki, Scharf et al. 1985; Mullis and Faloona 1987; Mullis 1990). PCR-based strategies have propelled molecular biology forward by enabling researchers to manipulate DNA more easily, thereby facilitating both common procedures, such as cloning, and huge endeavors such as the Human Genome Project (Olson, Hood et al. 1989; Ausubel, Brent et al. 2005). Real-time PCR represents yet another technological leap forward that has opened up new and powerful applications for researchers throughout the world. This is in part because the enormous sensitivity of PCR has been coupled to the precision afforded by "real-time" monitoring of PCR products as they are generated (Valasek and Repa 2005).

Higuchi and co-workers (Higuchi, Dollinger et al. 1992; Higuchi, Fockler et al. 1993) at Roche Molecular Systems and Chiron accomplished the first demonstration of real-time PCR. By including a common fluorescent dye called ethidium bromide (EtBr) in the PCR and running the reaction under ultraviolet light, which causes EtBr to fluoresce, they could visualize and record the accumulation of DNA with a video camera. It has been known since 1966 that EtBr increases its fluorescence upon binding of nucleic acids (Le Pecq and Paoletti 1966), but only by combining this fluorescent chemistry with PCR and real-time videography could real-time PCR be born as it was in the early 1990s. Subsequently, this

* Corresponding Author

technology quickly matured into a competitive market, becoming commercially widespread and scientifically influential (Valasek and Repa 2005).

Real-time PCR instrumentation was first made commercially available by Applied Biosystems in 1996, after which several other companies added new machines to the market. Presently, Applied Biosystems, BioGene, Bioneer, Bio-Rad, Cepheid, Corbett Research, Idaho Technology, MJ Research, Roche Applied Science, and Stratagene all offer instrumentation lines for real-time PCR (BioInformatics 2003).

Widespread use has also resulted in a multiplicity of names for the technology, each with a different shade of meaning. Real-time PCR simply refers to amplification of DNA (by PCR) that ismonitored while the amplification is occurring. The benefit of this real-time capability is that it allows the researcher to better determine the amount of starting DNA in the sample before the amplification by PCR. Present day real-time methods generally involve fluorogenic probes that "light up" to show the amount of DNA present at each cycle of PCR. "Kinetic PCR" refers to this process as well. "Quantitative PCR" refers to the ability to quantify the starting amount of a specific sequence of DNA. This term predates real-time PCR because it can refer to any PCR procedure, including earlier gel-based end-point assays, that attempts to quantify the starting amount of nucleic acid. Rarely, one might see the term "quantitative fluorescent PCR" to designate that the quantification was accomplished via measuring output from a fluorogenic probe, although this is redundant because all of the present chemistries for real-time PCR are fluorescent. In addition, if reverse transcriptase enzymes are used before PCR amplification in any of the above situations, then "RT-PCR" replaces "PCR" in the term. Today, the two most common terms, real-time and quantitative, are often used interchangeably or in combination, because real-time PCR is quickly becoming the method of choice to quantify nucleic acids (Valasek and Repa 2005).

The basic goal of real-time PCR is to precisely distinguish and measure specific nucleic acid sequences in a sample even if there is only a very small quantity. Real-time PCR amplifies a specific target sequence ina sample then monitors theamplification progress using fluorescent technology. During amplification, how quickly the fluorescent signal reaches a threshold level correlates with the amount of original target sequence, thereby enabling quantification. In addition, the final product can be further characterized by subjecting it to increasing temperatures to determine when the double-stranded product "melts." This melting point is a unique property dependent on product length and nucleotide composition. To accomplish these tasks, conventional PCR has been coupled to state-of-the-art fluorescent chemistries and instrumentation to becomereal-time PCR (Valasek and Repa 2005).

2. The chemistries of real-time PCR

Today fluorescence is exclusively used as the detection method in real-time PCR. Both sequence specific probes and non-specific labels are available as reporters. In his initial work Higuchi used the common nucleic acid stain ethidium bromide, which becomes fluorescent upon intercalating into DNA (Higuchi, Dollinger et al. 1992). Classical intercalators, however, interfere with the polymerase reaction, and asymmetric cyanine dyes such as SYBR Green I and BEBO have become more popular (Bengtsson, Karlsson et al. 2003; Zipper, Brunner et al. 2004). Asymmetric cyanines have two aromatic systems containing nitrogen, one of which is positively charged, connected by amethine bridge. These dyes have virtually no fluorescence when they are free in solution due to vibrations engaging both aromatic systems, which convert electronic excitation energy into heat that dissipates to the surrounding solvent. On the other hand the dyes become brightly fluorescent when they bind to DNA, presumably to the minor groove, and rotation around the methine bond is restricted (Nygren, Svanvik et al. 1998). In PCR the fluorescence of these dyes increases with the amount of double stranded product formed, though not strictly in proportion because the dye fluorescence depends on the dye: base binding ratio, which decreases during the course of the reaction. The dye fluorescence depends also to some degree on the DNA sequence. But a certain amount of a particular double-stranded DNA target, in the absence of significant amounts of other double-stranded DNAs, gives rise to the same fluorescence every time. Hence, the dyes are excellent for quantitative real-time PCR when samples are compared at the same level of fluorescence in absence of interfering DNA. Although minor groove binding dyes show preference for runs of AT base-pairs (Jansen, Norde´n et al. 1993), asymmetric cyanines are considered sequence non-specific reporters in real-time PCR. They give rise to fluorescence signal in the presence of any double stranded DNA including undesired primer–dimer products. Primer–dimer formation interferes with the formation of specific products because of competition of the two reactions for reagents and may lead to erroneous readouts. It is therefore good practice to control for primer–dimer formation. This can be done by melting curve analysis after completing the PCR. The temperature is then gradually increased and the fluorescence is measured as function of temperature. The fluorescence decreases gradually with increasing temperature because of increased thermal motion which allows for more internal rotation in the bound dye (Nygren, Svanvik et al. 1998). However, when the temperature is reached at which the double stranded DNA strand separates the dye comes off and the fluorescence drops abruptly (Ririe, Rasmussen et al. 1997). This temperature, referred to as the melting temperature, Tm, is easiest determined as the maximum of the negative first derivative of the melting curve. Since primer–dimer products typically are shorter than the targeted product, they melt at a lower temperature and their presence is easily recognized by melting curve analysis (Kubista, Andrade et al. 2006).

Labeled primers and probes are based on nucleic acids or some of their synthetic analogues such as the peptide nucleic acids (PNA) (Egholm, Buchardt et al. 1992) and the locked nucleic acids (LNA) (Costa, Ernault et al. 2004). The dye labels are of two kinds: (i) fluorophores with intrinsically strong fluorescence, such as fluorescein and rhodamine derivatives (Sjöback, Nygren et al. 1995), which through structural design are brought into contact with a quencher molecule, and (ii) fluorophores that change their fluorescence properties upon binding nucleic acids. Examples of probes with two dyes are the hydrolysis probes, popularly called Taqman probes (Holland, Abramson et al. 1991), which can be based either on regular oligonucleotides or on LNA (Braasch and Corey 2001), Molecular Beacons (Tyagi and Kramer 1996; Tyagi, Bratu et al. 1998), Hybridization probes (Caplin, Rasmussen et al. 1999), and the Lion probes (http://www.biotools.net). The dyes form a donor–acceptor pair, where the donor dye is excited and transfers its energy to the acceptor molecule if it is in proximity. Originally the acceptor molecule was also a dye, but today quencher molecules are more popular (Wilson and Johansson 2003). Energy transfer and quenching are distance dependent and structural rearrangement of the probe, or, in the case of hydrolysis probes, degradation, change the distance between the donor and acceptor and, hence, the fluorescence of the system (Kubista, Andrade et al. 2006).

Probes based on a single dye, whose fluorescence changes upon binding target DNA include the LightUp probes (Svanvik, Westman et al. 2000), AllGlo probes (http://www.allelogic.com), Displacement probes (Li, Qingge et al. 2002), and the Simple probes (http://www.idahotech.com/itbiochem/simpleprobes.html).

Chemical modifications and alterations of the oligonucleotide backbone are employed in some probes to improve the binding properties to the target template. This makes it possible to use shorter probes, which is advantageous for the detection of targets with short conserved regions such as retroviruses. LightUp probes have a neutral peptide nucleic acid (PNA) backbone that binds to DNA with greater affinity than normal oligonucleotides (Kubista, Andrade et al. 2006).

The LightUp probes are 10–12 bases, which is short compared to normal oligonucleotide probes that are usually at least 25 bases (http://www.lightup.se). LNA-probes make use of modified nucleotides to enhance binding affinity. MGBprobes are hydrolysis probes with a minor groove binding molecule attached to the end of the probe to increase affinity for DNA, which makes it possible to use shorter probes (Kutyavin, Afonina et al. 2000). Examples of modified primers include: Scorpion primers (Whitcombe, Theaker et al. 1999), LUX primers (Nazarenko, Lowe et al. 2002), Ampliflour primers (Uehara, Nardone et al. 1999), and the QZyme system (BD QZymeTM Assays for Quantitative PCR, 2003). As long as a single target is detected per sample there is not much of a difference in using a dye or a probe. Assay specificity is in both cases determined by the primers. Probes do not detect primer–dimer products, but using non-optimized probe assays is hiding the problem under the rug. If primer–dimers form they cause problems whether they are seen or not. In probe based assays, particularly when high CT values are obtained, one should verify the absence of competing primer–dimer products (Kubista, Andrade et al. 2006). The traditional way is by gel electrophoresis. Recently, an alternative approach was proposed based on the BOXTO dye. BOXTO is a sequence non-specific doublestranded DNA binding dye that has distinct spectral characteristics to fluorescein and can be used in combination with FAM based probes. The BOXTO and the probe signals are detected in different channels of the real-time PCR instrument. While the probe reflects formation of the targeted product as usual, the BOXTO dye also reports the presence of any competing primer–dimer products, which can be identified by melting curve analysis (Lind, Stahlberg et al. 2006). The great advantage of probes is for multiplexing, where several products are amplified in the same tube and detected in parallel (Wittwer, Herrmann et al. 2001). Today multiplexing is mainly used to relate expression of reporter genes to that of an exogenous control gene in diagnostic applications (Mackya 2004), and for single nucleotide polymorphism (SNP) and mutation detection studies (Mhlanga and Malmberg 2001; Mattarucchi, Marsoni et al. 2005). Multiplex assays are more difficult to design because when products accumulate the parallel PCR reactions compete for reagents. To minimize competition limiting amounts of primers must be used. Also, primer design is harder, because complementarity must be avoided between all the primers. Multiplex assays can be based either on probes or on labeled primers, where labeled primers usually give rise to signal from primer–dimer products, while probes do not. The different probing technologies have their advantages and limitations. Dyes are cheaper than probes but they do not distinguish between products. Hairpin forming probes have the highest specificity, because the formation of the hairpin competes with the binding to mismatched targets. This makes them most suitable for SNP and multi-site variation (MSV) analysis (Bonnet, Tyagi et al. 1999). Hydrolysis probes require two-step PCR to function properly, which is not optimal for the polymerase reaction, and short amplicons are necessary to obtain reasonable amplification efficiencies. But they are easier to design than hairpin forming probes and an 80% success rate was recently reported (Kubista 2004).

In summary, a 'good'probe, independent of chemistry, should have low background fluorescence, high fluorescence upon target formation (high signal to noise ratio), and high target specificity. The dyes'excitation and emission spectra are important parameters to consider when designing multiplex reactions. Spectral overlap in excitation and emission should be minimized to keep cross-talk to a minimum (Kubista, Andrade et al. 2006).

2.1 SYBR green I

SYBR green I binds to the minor groove of dsDNA, emitting 1,000-fold greater fluorescence than when it is free in solution (Wittwer, Herrmann et al. 1997). Therefore, the greater the amount of dsDNA present in the reaction tube, the greater the amount of DNA binding and fluorescent signal from SYBR green I. Thus any amplification of DNA in the reaction tube is measured (Valasek and Repa 2005).

2.2 BEBO

The minor groove binding asymmetric cyanine dye BEBO is tested as sequence nonspecific label in real-time PCR. The Fluorescence intensity of BEBO increases upon binding to double-stranded DNA allowing emission to be measured at the end of the elongation phase in the PCR cycle. BEBO concentrations between 0.1 and 0.4 mM generated sufficient Fluorescence signal without inhibiting the PCR. A comparison with the commonly used reporter dye SYBR Green I shows that the two dyes behave similarly in all important aspects. The dye has absorbance and emission wavelengths that can be detected on the FAM channel on most common real-time PCR platforms, and shows a strong fluorescence increase when bound to dsDNA. BEBO can be used as an unspecific dye for real-time PCR applications or other applications where staining of dsDNA is wanted (Bengtsson, Karlsson et al. 2003).

2.3 BOXTO

The unsymmetrical cyanine dyes BOXTO and its positive divalent derivative BOXTO-PRO were studied as real-time PCR reporting fluorescent dyes and compared to SYBR GREEN I (SG). Unmodified BOXTO showed no inhibitory effects on real-time PCR, while BOXTO-PRO showed complete inhibition, sufficient fluorescent signal was acquired when 0.5–1.0 µM BOXTO was used with RotorGene and iCycler platforms. Statistical analysis showed that there is no significant difference between the efficiency and dynamic range of BOXTO and SG (Ahmad 2007).

2.4 5' nuclease (TaqMan) probes

Hydrolysis probes (also called 5/-nuclease probes because the 5/-exonuclease activity of DNA polymerase cleaves the probe) offer an alternative approach to the problem of specificity. These are likely the most widely used fluorogenic probe format (Mackay 2004) and are exemplified by TaqMan probes. In terms of structure, hydrolysis probes are sequence- specific dually fluorophore-labeled DNA oligonucleotides (Valasek and Repa 2005). One fluorophore is termed the quencher and the other is the reporter. When the quencher and reporter are in close proximity, that is, they are both attached to the same short oligonucleotide; the quencher absorbs the signal from the reporter (Valasek and Repa 2005). This is an example of fluorescence resonance energy transfer (also called Forster transfer) in which energy is transferred from a "donor" (the reporter) to an "acceptor" (the quencher) fluorophore. During amplification, the oligonucleotide is broken apart by the action of DNA polymerase (5/-nuclease activity) and the reporter and quencher separate, allowing the reporter's energy and fluorescent signal to be liberated. Thus destruction or hydrolysis of the oligonucleotide results in an increase of reporter signal and corresponds with the specific amplification of DNA (Valasek and Repa 2005). Examples of common quencher fluorophores include TAMRA, DABCYL, and BHQ, whereas reporters are more numerous (e.g., FAM, VIC, NED, etc). Hydrolysis probes afford similar precision as SYBR green I (Wilhelm, Pingoud et al. 2003), but they give greater insurance regarding the specificity because only sequence-specific amplification is measured. In addition, hydrolysis probes allow for simple identification of point mutations within the amplicon using melting curve analysis (Valasek and Repa 2005).

2.5 Molecular beacons

Molecular beacons are similar to TaqMan probes but are not designed to be cleaved by the 5' nuclease activity of Taq polymerase. These probes have a fluorescent dye on the 5' end and a quencher dye on the 3' end of the oligonucleotide probe. A region at each end of the molecular beacon probe is designed to be complementary to itself, so at low temperatures, the ends anneal, creating a hairpin structure. This integral annealing property positions the two dyes in close proximity, quenching the fluorescence from the reporter dye (Espy, Uhl et al. 2006). The central region of the probe is designed to be complementary to a region of the PCR amplification product. At high temperatures, both the PCR amplification product and probe are single stranded. As the temperature of the PCR is lowered, the central region of the molecular beacon probe binds to the PCR product and forces the separation of the fluorescent reporter dye from the quenching dye. The effects of the quencher dye are obviated and a light signal from the reporter dye can be detected. If no PCR amplification product is available for binding, the probe reanneals to itself, forcing the reporter dye and quencher dye together, preventing fluorescent signal (Espy, Uhl et al. 2006). Typically, a single molecular beacon is used for detection of a PCR amplification product and multiple beacon probes with different reporter dyes are used for single nucleotide polymorphism detection. By selection of appropriate PCR temperatures and/or extension of the probe length, molecular beacons will bind to the target PCR product when an unknown nucleotide polymorphism is present but at a slight cost of reduced specificity. There is not a specific temperature thermocycling requirement for molecular beacons, so temperature optimization of the PCR is simplified (Espy, Uhl et al. 2006).

2.6 FRET hybridization probes

FRET hybridization probes, also referred to as LightCyclerprobes; represent a third type of probe detection format commonly used with real-time PCR testing platforms. FRET hybridization probes are two DNA probes designed to anneal next to each other in a head-to-tail configuration on the PCR product. The upstream probe has a fluorescent dye on the 3' end and the downstream probe has an acceptor dye on the 5' end. If both probes anneal to the target PCR product, fluorescence from the 3' dye is absorbed by the adjacent acceptor dye on the 5' end of the second probe. The second dye is excited and emits light at a third wavelength and this third wavelength is detected. If the two dyes do not align together because there is no specific DNA for them to bind, then FRET does not occur between the two dyes because the distances between the dyes are too great. A design detail of FRET hybridization probes is the 3' end of the second (downstream) probe is phosphorylated to prevent it from being used as a primer by Taq during PCR amplification. The two probes encompass a region of 40 to 50 DNA base pairs, providing exquisite specificity (Espy, Uhl et al. 2006). FRET hybridization probe technology permits melting curve analysis of the amplification product. If the temperature is slowly raised, eventually the probes will no longer be able to anneal to the target PCR product and the FRET signal will be lost. The temperature at which half the FRET signal is lost is referred to as the melting temperature of the probe system (Espy, Uhl et al. 2006). The Tm depends on the guanine plus cytosine content and oligonucleotide length. In contrast to TaqMan probes, a single nucleotide polymorphism in the target DNA under a hybridization FRET probe will still generate a signal, but the melting curve will display a lower Tm. The lowered Tm can be characteristic for a specific polymorphism underneath the probes; however, a lowered Tm can also be the result of any sequence difference under the probes. The target PCR product is detected and the altered Tm informs the user there is a difference in the sequence being detected. Generally, more than three base pair differences under a FRET hybridization probe prevent hybridization at typical annealing temperatures and are not detected (Espy, Uhl et al. 2006). This trait of FRET hybridization probes is advantageous in cases where the genome of the organism is known to mutate at a high frequency, such as with viruses. When a single or limited number (<3) of known polymorphisms occur between two similar targets, FRET hybridization probes can also be used for discriminating strains of organisms (Espy, Uhl et al. 2006). Like molecular beacons, there is not a specific thermocycling temperature requirement for FRET hybridization probes. Molecular beacons and FRET hybridization probes, unlike TaqMan probes, are both recycled (conserved) in each round of PCR temperature cycle. Also, for Molecular beacons and FRET hybridization probes, unlike TaqMan probes, fluorescent signal does not accumulate as PCR product accumulates after each PCR cycle (Espy, Uhl et al. 2006).

2.7 Scorpions

Scorpions combine the detection probe with the upstream PCR primer (Whitcombe, Theaker et al. 1999) and consist of a fluorophore on the 5′ end, followed by a complementary stem-loop structure (also containing the specific probe sequence), quencher dye, DNA polymerase blocker (a nonamplifiable monomer that prevents DNA polymerase extension), and finally a PCR primer on the 3′ end. The probe sequence contained within the hairpin allows the scorpion to anneal to the template strand, which separates the quencher for the fluorophore and results in increased fluorescence. Because sequence-specific priming and probing is a unimolecular event, scorpions perform better than bimolecular methods under conditions of rapid cycling such as the LightCycler (Thelwell, Millington et al. 2000). Cycling is performed at a temperature optimal for DNA polymerase activity instead of the reduced temperature necessary for the 5′ nuclease assay. Scorpions are specific enough for allele discrimination and may be multiplexed easily (Thelwell, Millington et al. 2000). The scorpion chemistry has been improved with the creation of duplex scorpions in which the reporter dye/probe and quencher fragment are on separate, complementary molecules (Solinas, Brown et al. 2001). The duplex scorpions still bind in a unimolecular event, but because the reporter and quenchers are on separate molecules, they yield greater signal intensity because the reporter and quencher can separate completely (Wong and Medrano 2005).

2.8 Sunrise™ primers

Created by Oncor (Gaithersburg, MD, USA), Sunrise primers are similar to scorpions in that they combine both the PCR primer and detection mechanism in the same molecule (Nazarenko, Bhatnagar et al. 1997). These probes consist of a dual-labeled (reporter and quencher fluorophores) hairpin loop on the 5' end, with the 3' end acting as the PCR primer. When unbound, the hairpin is intact, causing reporter quenching via FRET. Upon integration into the newly formed PCR product, the reporter and quencher are held far enough apart to allow reporter emission (Wong and Medrano 2005).

2.9 LUX™ fluorogenic primers

Light upon extension (LUX) primers (Invitrogen, Carlsbad, CA, USA) are self-quenched single-fluorophore labeled primers almost identical to Sunrise primers. However, rather than using a quencher fluorophore, the secondary structure of the 3' end reduces initial fluorescence to a minimal amount (Nazarenko, Lowe et al. 2002). Because this chemistry does not require a quencher dye, it is much less expensive than dual-labeled probes. While this system relies on only two oligonucleotides for specificity, unlike the SYBR Green I platform in which a dissociation curve is used to detect erroneous amplification, no such convenient detection exists for the LUX platform. Agarose gels must be run to ensure the presence of a single PCR product, a step that is extremely important not only for the LUX primers but also for the Sunrise primers and scorpions because PCR priming and probe binding are not independent in these chemistries (Wong and Medrano 2005).

2.10 Light-up probes

Light-up probes are peptide nucleic acids (PNAs) that use thiazole orange as the fluorophor. Upon hybridisation with DNA, duplex or triplex structures are formed with increased fluorescence intensity of the fluorophor. A quencher is not required. This technique is limited by unspecific fluorescence, which increases during PCR and therefore restricts the achievable sensitivity (Isacsson, Cao et al. 2000; Svanvik, Stahlberg et al. 2000;Svanvik, Westman et al. 2000). Some other formats use the increasing quench as indicator for product accumulation (Crockett and Wittwer 2001; Kurata, Kanagawa et al. 2001). In this case, the fluorescence is quenched by a guanine residue of the PCR product. These probes are comparatively inexpensive and easy to construct; however, measurement of the decrease of a signal is problematic, especially during the early exponential phase in which only very few probes are quenched (Wilhelm and Pingoud 2003).

2.11 Eclipse probe

qPCR assays using an Eclipse probe employ two primers and a sequence-specific oligonucleotide probe. The probe is complementary to a sequence within the amplicon and contains a fluorescent reporter at the 3' end, a quencher at the 5' end, and a minor groove binder (Bio-Rad Laboratories 2006). The unhybridized probe adopts a conformation that brings the reporter and quencher together, quenching the reporter. During the annealing step of PCR, the probe hybridizes to the target with the help of the minor groove binder. The probe thus becomes linearized, separating the reporter and quencher and allowing the reporter to fluoresce. The resulting fluorescent signal is proportional to the amount of amplified product in the sample (Bio-Rad Laboratories 2006).

2.12 Amplifluor primer

qPCR assays using Amplifluor chemistry employ two target-specific primers and one universal primer called the UniPrimer. The first target-specific primer contains a 5' extension sequence called the Z-sequence that is also found at the 3' end of the UniPrimer. The UniPrimer forms a hairpin structure (Bio-Rad Laboratories 2006). A fluorescent reporter and a quencher are attached at the 5' and the 3' ends of the stem structure, respectively. In the hairpin conformation, the reporter fluorescence is quenched due to its proximity to the quencher. During the first amplification cycle, the first target-specific primer (with the Z-sequence) hybridizes to the template and is extended. During the second amplification cycle, the second target-specific primer is used to synthesize a new target template that contains a sequence complementary to the Z-sequence. The product from the second amplification cycle can then serve as the template for the UniPrimer. In the third amplification cycle, the extended UniPrimer serves as a template for the next amplification cycle (Bio-Rad Laboratories 2006). In the fourth cycle, extension of the template through the hairpin region of the UniPrimer causes the UniPrimer to open up and adopt a linear configuration, which allows the reporter to fluoresce. Exponential amplification using the second target-specific primer and the UniPrimer occurs in subsequent amplification cycles. The resulting fluorescent signal is proportional to the amount of amplified product in the sample (Bio-Rad Laboratories 2006).

2.13 BD QZyme primer

qPCR assays using BD QZyme primers employ a target-specific zymogene primer, a target-specific reverse primer, and a universal oligonucleotide substrate. The oligonucleotide contains a fluorescent reporter on the 5' end and a quencher on the 3' end. When oligonucleotide substrate is intact, the fluorescence of the reporter is quenched by the quencher due to their proximity (Bio-Rad Laboratories 2006). The zymogene primer contains a sequence that encodes a catalytic DNA. During the first amplification cycle, the zymogene primer is extended. In the second cycle, the product of the first cycle is used as the template by the target-specific reverse primer, which is extended to create a new target sequence containing a catalytic DNA region. In the subsequent annealing step, the fluorescently labeled oligonucleotide substrate hybridizes to the catalytic DNA sequence and is cleaved. This cleavage separates the reporter from the quencher, resulting in a fluorescent signal that is proportional to the amount of amplified product in the sample (Bio-Rad Laboratories 2006).

3. Design and optimization of SYBR Green I reactions

A SYBR Green I assay uses a pair of PCR primers that amplifies a specific region within the target sequence of interest and includes SYBR Green 1 for detecting the amplified product. The steps for developing a SYBR Green I assay are:

1. Primer design and amplicon design

2. Assay validation and optimization

A) Primer and Amplicon Design:

A successful real-time PCR reaction requires efficient and specific amplification of the product. Both primers and target sequence can affect this efficiency. Therefore, care must be taken when choosing a target sequence and designing primers. A number of free and commercially available software programs are available for this purpose. One popular web-based program for primer design is Primer3 (http://frodo.wi.mit.edu/cgi-bin/primer3/primer3_www.cgi). A commercially available program such as Beacon Designer software performs both primer design and amplicon selection (Bio-Rad Laboratories 2006).

B) Guidelines of amplicon design:

1. Design amplicon to be 75–200 bp.Shorteramplicons are typically amplified with higher efficiency. An amplicon should be at least 75 bp to easily distinguish it from any primer-dimers that might form
2. Avoid secondary structure if possible. Use programs such as mfold http://www.bioinfo.rpi.edu/applications/mfold/) to predict whether an amplicon will form any secondary structure at annealing temperature. See Real-Time PCR: General Considerations (Bio-Rad bulletin 2593) for more details
3. Avoid templates with long (>4) repeats of single bases
4. Maintain a GC content of 50–60%

C) Parameters of primer design:

1. Design primers with a GC content of 50–60%
2. Maintain a melting temperature (T_m) between 50°C and 65°C. We calculate T_m values using the nearest-neighbor method with values of 50 mM for salt concentration and 300 nM for oligonucleotide concentration
3. Avoid secondary structure; adjust primer locations outside of the target sequence secondary structure if required
4. Avoid repeats of Gs or Cs longer than three bases
5. Place Gs and Cs on ends of primers
6. Check sequence of forward and reverse primers to ensure no 3' complementarity (avoid primer-dimer formation)
7. Verify specificity using tools such as the Basic Local Alignment Search Tool (http://www.ncbi.nlm.nih.gov/blast/)

D) Assay Validation and Optimization:

Components a SYBR Green I qPCR reaction:

1. PCR master mix with SYBR Green I
2. Template
3. Primers

Preformulated real-time PCR master mixes containing buffer, DNA polymerase, dNTPs, and SYBR Green I dye are available from several vendors.

Optimized SYBR Green I qPCR reactions should be sensitive and specific and should exhibit good amplification efficiency over a broad dynamic range (Bio-Rad Laboratories 2006).

Steps of to determine the performance of your SYBR Green I qPCR assay:

a. Identify the optimal annealing temperature for your assay
b. Construct a standard curve to evaluate assay performance

E) Annealing Temperature Optimization:

The optimal annealing temperature can easily be assessed on qPCR instruments that have a temperature gradient feature, such as the MiniOpticon™, MyiQ™, DNA Engine Opticon®, Opticon™ 2, iCycleriQ®, Chromo4™, and iQ™5 systems.

A gradient feature allows you to test a range of annealing temperatures simultaneously, so optimization reactions can be performed in a single experiment.

To find the optimal annealing temperature for reaction, recommend testing a range of annealing temperatures above and below the calculated T_m of the primers.

Because SYBR Green I binds to all dsDNA, it is necessary to check the specificity of your qPCR assay by analyzing the reaction product(s). To do this, use the melt-curve function on your real-time instrument and also run products on an agarose gel. An optimized SYBR Green I qPCR reaction should have a single peak in the melt curve, corresponding to the single band on the agarose gel.

Nonspecific products that may have been co-amplified with the specific product can be identified by melt-curve analysis. In this example, the specific product is the peak with a T_m of 89°C and corresponds to the upper band on the gel. The nonspecific product is the peak with a T_m of 78°C and corresponds to the lower band in the gel. By comparing the gel image with the melt curve, you can identify peaks in the melt curve that correspond to specific product, additional nonspecific bands, and primer-dimers. If nonspecific products such as primer-dimers are detected by melt-curve analysis, recommend that redesign primers (Bio-Rad Laboratories 2006).

F) Assay Performance Evaluation Using Standard Curves:

The efficiency, reproducibility, and dynamic range of a SYBR Green I assay can be determined by constructing a standard curve using serial dilutions of a known template. The efficiency of the assay should be 90–105%, the R^2 of the standard curve should be >0.980 or $r > |-0.990|$, and the C_T values of the replicates should be similar.

It is important to note that the range of template concentrations used for the standard curve must encompass the entire range of template concentration of the test samples to show that results from the test samples are within the linear dynamic range of the assay. If the test samples give results outside of the range of the standard curves, one of the following must be performed:

1. Construct a wider standard curve covering the test sample concentrations and perform analysis to ensure that the assay is linear in that new range
2. If the test samples give a lower C_T than the highest concentration of standards used in the standard curve, repeat the assay using diluted test samples
3. If the test samples give a higher C_T than the lowest concentration of standards used in the standard curve, repeat the assay using larger amounts of the test samples

4. Design and optimization of TaqManProbe reactions

A TaqMan assay uses a pair of PCR primers and a dual-labeled target-specific fluorescent probe. The steps for developing a TaqMan assay are:

a. Primer and probe design
b. Assay validation and optimization

A. Primer and Probe Design:

As with any qPCR reaction, TaqMan-based assays require efficient and specific amplification of the product. Typically, the primers are designed to have an annealing temperature between 55 and 60°C. We recommend using software such as Beacon Designer for designing your TaqMan primers and TaqMan probe.. Because the dual-labeled probe is the most costly component of a TaqMan assay, suggested that order the two primers and validate their performance using SYBR Green I before ordering the dual-labeled probe.

The TaqMan probe should have a T_m 5–10°C higher than that of the primers. In most cases, the probe should be <30 nucleotides and must not contain a G at its 5' end because this could quench the fluorescent signal even after hydrolysis. Choose a sequence within the target that has a GC content of 30–80%, and design the probe to anneal to the strand that has more Gs than Cs (so the probe contains more Cs than Gs).

An important aspect of designing a TaqMan probe is reporter and quencher selection. We recommend using FAM-labeled probes when designing singleplex reactions, because they are inexpensive and readily available, perform well, and can be detected by all instruments currently on the market.

Another important consideration for obtaining accurate real-time qPCR data is probe quality. Even a perfectly designed probe can fail if the probe is improperly synthesized or purified. Improper removal of uncoupled fluorescent label, inefficient coupling, and/or poor quenching can produce high fluorescent background or noise. A low signal-to-noise ratio results in decreased sensitivity and a smaller linear dynamic range. Two probes with identical sequences and identical fluorophore labels can be measurably different when synthesized by different suppliers or even at different times by the same supplier.

B. Assay Validation and Optimization:

A TaqMan probe-based qPCR reaction contains the following components:

1. PCR master mix
2. Template
3. Primers
4. Probe(s)

Preformulated PCR master mixes containing buffer, DNA polymerase, and dNTPs are commercially available from several vendors. For TaqMan assays, we recommend using iQ™ supermix with 300 nM of each of the two primers and 200 nM of probe(s). TaqMan assays require careful attention to temperature conditions. A typical TaqMan protocol contains a denaturation step followed by a combined annealing and extension step at 55–60°C, instead of the traditional three-step PCR cycle of denaturation, annealing, and extension. This is to ensure that the probe remains bound to its target during primer extension. Typical TaqMan probes for nucleic acid quantification are designed to have a T_m of 60–70°C. An optimized TaqMan assay should be sensitive and specific, and should exhibit good amplification efficiency over a broad dynamic range.

In short, construct a standard curve using dilutions of a known template and use this curve to determine the efficiency of the assay along with R^2 or r of the regression line. The efficiency of the reaction should be between 90 and 105%, the R^2 should be >0.980 or r > |-0.990|, and the replicates should give similar C_T values. If the assay performs within these specifications, you are ready to start your experiment. If the assay performs outside these specifications, we suggest that you redesign your primers and TaqMan probe. It is important to note that the range of template concentrations used for the standard curve must encompass the entire range of template concentrations of the test samples to demonstrate that results from the test samples are within the dynamic range of the assay (Bio-Rad Laboratories 2006). If test samples give results outside the range of the standard curve, one of the three following steps must be performed:

1. Construct a wider standard curve covering the test sample concentrations and perform analysis to ensure that the assay is linear in that new range
2. If the test samples give a lower C_T than the highest concentration of standards used in the standard curve, repeat the assay using diluted test samples
3. If the test samples give a higher C_T than the lowest concentration of standards used in the standard curve, repeat the assay using larger amounts of the test samples

5. The instrumentation of real-time PCR

A critical requirement for real-time PCR technology is the ability to detect the fluorescent signal and record the progress of the PCR. Because fluorescent chemistries require both a specific input of energy for excitation and a detection of a particular emission wavelength, the instrumentation must be able to do both simultaneously and at the desired wavelengths. Thus the chemistries and instrumentation are intimately linked (Valasek and Repa 2005).

At present, there are three basic ways in which real-time instrumentation can supply the excitation energy for fluorophores: by lamp, light-emitting diode (LED), or laser. Lamps are classified as broad-spectrum emission devices, whereas LEDs and lasers are narrow spectrum. Instruments that utilize lamps (tungsten halogen or quartz tungsten halogen) may also include filters to restrict the emitted light to specific excitation wavelengths. Instruments using lamps include Applied Biosystem's ABI Prism 7000, Stratagene's Mx4000 and Mx3000P, and Bio-Rad's iCycleriQ. LED systems include Roche's LightCycler, Cepheid's SmartCycler, Corbett's Rotor-Gene, and MJ Research's DNA Engine Opticon 2. The ABI Prism 7900HT is the sole machine to use a laser for excitation (Valasek and Repa 2005). To collect data, the emission energies must also be detected at the appropriate wavelengths. Detectors include charge-coupled device cameras, photomultiplier tubes, or other types of photodetectors. Narrow wavelength filters or channels are generally employed to allow only the desired wavelength(s) to pass to the photodetector to be measured. Usually, multiple discrete wavelengths can be measured at once, which allows for multiplexing, i.e., running multiple assays in a single reaction tube (Valasek and Repa 2005). Another portion of the instrumentation consists of a thermocycler to carry out PCR. Of particular importance for real-time PCR is the ability of the thermocycler to maintain a consistent temperature among all sample wells, as any differences in temperature could lead to different PCR amplification efficiencies. This is accomplished by using a heating block (Peltier based or resistive), heated air, or a combination of the two. As one might expect, heating blocks generally change temperature more slowly than heated air, resulting in longer thermocycling times. For example, Roche's LightCycler models utilizing heated air can perform 40 cycles in 30 min, whereas Applied Biosystem's ABI Prism 7900HT utilizing a Peltier-based heating block take s 1 h 45 min (Valasek and Repa 2005). Real-time instrumentation certainly would not be complete without appropriate computer hardware and data-acquisition and analysis software. Software platforms try to simplify analysis of real-time PCR data by offering graphical output of assay results including amplification and dissociation (melting point) curves. The amplification curve gives data regarding the kinetics of amplification of the target sequence, whereas the dissociation curve reveals the characteristics of the final amplified product (Valasek and Repa 2005).

5.1 Comparison of the different systems

Essentially, each real time PCR instrument consists of a computer-controlled thermocycler integrated with fluorescent detection system and dedicated software to analyze the result. Some systems can detect four different wave lengths (I-cycler, Mx4000 [stratagene] and Smart Cycler®, Version 2.0 Light Cycler®) whereas others can detect two different wavelengths (Light Cycler®). The Light Cycler® and Smart Cycler® are capable of performing rapid-cycle real time PCR because the reaction is set-up in capillaries or especially designated tubes. Both have optimized heating- cooling characteristic. A complete amplification protocol can be performed in 30-45 minutes (Myi ; Giulietti, Overbergh et al. 2001; Soheili and Samiei 2005). The Smart Cycler® is a combination of 16 individual, one tube real time PCR units. It is capable of performing a different PCR program on each of 16 reaction tubes. This is very useful for a rapid optimization of the assay as many variables can be tested at the same time. The Bio-Rad I-cycler IQ® instrument can perform real time amplification with a temperature gradient for specific PCR steps, allowing the optimization of real time PCR assay. The spectrofluorometers in the thermal cycler have a number of differences. Laser-based systems are tuned to excite each fluorophore at a specific wavelength and provide maximum efficiency. Lamp-based systems provide a broad excitation range that can be filtered to work with a number of fluorophores. The laser source not only gives brighter illumination to the fluorophore signal, but also produces less background noise(Myi ; Giulietti, Overbergh et al. 2001; Soheili and Samiei 2005).

In conclusion, real time PCR is a powerful advancement of the basic PCR technique. The important steps in deciding which particular assay format to use are related to the type of data required. The requirement for a research laboratory is quite distinct from those of a diagnostic laboratory. For the latter, probe confirmation of the PCR product is an essential part of the assay, whereas SYBR green detection may be sufficient for many other applications such as quantifying expression of a gene. All of the real-time PCR machines analyzed are capable of detecting PCR product in real time and a specific assay can be made optionally on every system. However, there are some decisions to be made when selecting among different formats. The choice of system is dependent on individual laboratory needs (Myi ; Giulietti, Overbergh et al. 2001; Soheili and Samiei 2005). Considering diagnostic applications, the Light Cycler® or Smart Cycler® may obtain faster results for urgent assays.

This could reduce the time of analysis to result from 3-4 hours to 1.5 hours. On the other hand, if sensitivity is the most important issue, these machines, with their smaller reaction volume and consequently lower sensitivity, wouldn't be the first choice. The ABI 7700 and Bio-Rad -I-Cycler IQ® have a 96 well format, enabling higher throughput than other systems. The 384-well plates, as designed by ABI for use in the 7900 HT system, can further enhance through put. For diagnostic application, internal control of nucleic acid isolation and PCR inhibition, it is essential to obtain valid results. This can be achieved using the system that enables multi-color detection, such as the I-Cycler IQ® and the Smart Cycler®. Recently, a multi-color format of the Light Cycler® is also present in market. Multiplex real-time PCRs can be developed for three different targets and an internal control by using the four detection wavelengths possible in multicolor detection. As a matter of fact, the choice of which real time system to use depends on the range of application required. To achieve meaningful results, each assay must be validated and optimized for the particular system chosen (Myi ; Giulietti, Overbergh et al. 2001; Soheili and Samiei 2005).

6. Advantages of real-time PCR quantitation

There are many methods in molecular biology for measuring quantities of target nucleic acid sequences. However, most of these methods exhibit one or more of the following shortcomings: they are time consuming, labor intensive, insufficiently sensitive, non-quantitative, require the use of radioactivity, or have a substantial probability of cross contamination (Reischl, Wittwer et al. 2002). These methods include but are not limited to Northern and Southern hybridizations, HPLC, scintillation proximity assay, PCR-ELISA, RNase protection assay, in situ hybridization, and various gel electrophoresis PCR end-point systems (Valasek and Repa 2005). Real-time PCR has distinct advantages over these earlier methods for several reasons. Perhaps the most important is its ability to quantify nucleic acids over an extraordinarily wide dynamic range (at least 5 log units). This is coupled to extreme sensitivity, allowing the detection of less than five copies (perhaps only one copy in some cases) of a target sequence, making it possible to analyze small samples like clinical biopsies or miniscule lysates from laser capture microdissection. With appropriate internal standards and calculations, mean variation coefficients are 1–2%, allowing reproducible analysis of subtle gene expression changes even at low levels of expression (Klein 2002; Luu-The, Paquet et al. 2005). In addition, all real-time platforms are relatively quick, with some affording high-throughput automation. Finally, real-time PCR is performed in a closed reaction vessel that requires no post-PCR manipulations, thereby minimizing the chances for cross contamination in the laboratory (Valasek and Repa 2005).

7. Limitations of real-time PCR quantitation

There are several limitations to real-time PCR methods. The majority of these are present in all PCR or RT-PCR-based techniques. Real-time PCR is susceptible to PCR inhibition bycompounds present in certain biological samples. For example, clinical and forensic uses for real-timePCR may be affected by inhibitors found in certain body fluidssuch as hemoglobin or urea (Wilson 1997). Food microbiological applications may encounter organic and phenolic inhibitors (Wilson 1997). To circumvent this problem, alternative DNA polymerases (e.g., Tfl, Pwo, Tth, etc.) that are resistant to particular inhibitors can be used. Other limitations primarily concern real-time PCR- based analysis of gene expression

(Bustin 2000; Bustin 2002;Bustin and Nolan 2004). Because of the necessary use of RNA in an extra enzymatic step, more problems have the opportunity to occur. RNA itself is extremely labile compared with DNA, and therefore isolation must be carefully performed to ensure both the integrity of the RNA itself and the removal of contaminating nucleases, genomic DNA, and RT or PCR inhibitors. This can be a problem with any sample source, but clinical samples are of special concern because inconsistencies in sample size, collection, storage, and transport can lead to a variable quality of RNA templates. Conversion of RNA to cDNA during the RT reaction is also subject to variability because multiple reverse transcriptase enzymes with different characteristics exist, and different classes of oligonucleotides (e.g., random, poly-dT, or gene specific primers) can be used to prime RT (Valasek and Repa 2005). Probably the largest present limitation of real-time PCR, however, is not inherent in the technology but rather resides in human error: improper assay development, incorrect data analysis, or unwarranted conclusions. In our experience using real-time PCR for gene expression analysis, real-time PCR primer sets must be designed and validated by stringent criteria to ensure specificity and accuracy of the results. For microbiology, false positives or negatives must be considered when designing an assay to detect pathogens. Amplification and melting curves must be visually inspected while independent calculations based on these curves should be double-checked for accuracy. Real-time PCR gene expression analysis measures mRNA levels and, therefore, only suggests possible changes in protein levels or function rather than demonstrating them. And although there is a tight connection between gene expression and gene product function (Brown and Botstein 1999)(8), this is certainly not always the case, and formal demonstration may be needed for a given research project. Of course, conclusions based on data derived from real-time PCR are best utilized when the biological context is well understood (Bustin 2002).

8. Types of real-time quantification

1. Absolute Quantitation

Absolute quantitation uses serially diluted standards of known concentrations to generate a standard curve. The standard curve produces a linear relationship between Ct and initial amounts of total RNA or cDNA, allowing the determination of the concentration of unknowns based on their Ct values (Heid, Stevens et al. 1996). This method assumes all standards and samples have approximately equal amplification efficiencies (Souaze, Ntodou-Thome et al. 1996). In addition, the concentration of serial dilutions should encompass the levels in the experimental samples and stay within the range of accurately quantifiable and detectable levels specific for both the real-time PCR machine and assay.The PCR standard is a fragment of double-stranded DNA (dsDNA), single-stranded DNA (ssDNA), or cRNA bearing the target sequence (Wong and Medrano 2005). A simple protocol for constructing a cRNA standard for one-step PCR can be found in Fronhoffs et al. (Fronhoffs, Totzke et al. 2002), while a DNA standard for two-step real-time PCR can be synthesized by cloning the target sequence into a plasmid (Gerard, Olsson et al. 1998), purifying a conventional PCR product (Liss 2002), or directly synthesizing the target nucleic acid. The standard used must be a pure species. DNA standards have been shown to have a larger quantification range and greater sensitivity, reproducibility, and stability than RNA standards (Pfaffl, Tichopad et al. 2004). However, a DNA standard cannot be used for a one-step real-time RT-PCR due to the absence of a control for the reverse transcription efficiency (Giulietti, Overbergh et al. 2001).

2. Relative quantitation

During relative quantitation, changes in sample gene expression are measured based on either an external standard or a reference sample, also known as a calibrator (Livak and Schmittgen 2001). When using a calibrator, the results are expressed as a target/reference ratio. There are numerous mathematical models available to calculate the mean normalized gene expression from relative quantitationassays. Depending onthe method employed, these can yield different results and thus discrepant measures of standard error (Liu and Saint 2002; Muller, Janovjak et al. 2002). Table 1 shows a comparison of the different methods, with an explanation of each method to follow (Wong and Medrano 2005).

9. Quantitative analyses

For quantitative analysis, the amplification curves are evaluated. The amplification process is monitored either through the fluorescence of dsDNA-specific dyes (like SYBR Green I) or ofsequence-specific probes. Each curve consists of at least three distinct phases: 1) an initial lag phase in which no product accumulation can be measured, 2) an exponential phase, and 3) a plateau phase (Wilhelm and Pingoud 2003). The exponential phase in principle could be extrapolated to the start of the reaction (Cycle 0) to calculate the template copy number, but the error would be too high. The template copy number can be estimated with greater precision from the number of cycles needed for the signal to reach an arbitrary threshold. The threshold must intersect the signal curve in its exponential phase, in which the signal increase correlates with product accumulation. The intersection point is the so-called threshold value (C_T) or crossing point (C_P). This point may be between two successive cycles (i.e. it may be a fractional number). For exact quantifications, the efficiency of the amplification reaction must be known. It is crucial that the amplification efficiencies of standards and unknowns are identical (Wilhelm and Pingoud 2003). The efficiency can be estimated from the C_T values of samples with known template concentrations ('standards') as described below (Wilhelm and Pingoud 2003).

During the exponential phase, the signal *S*can be described by Equation 1:

$$S = pN_0\varepsilon^c \qquad (1)$$

where*p*is a proportionality factor to relate PCR product concentration and signal intensity, $N_0$is the amount of template, $\varepsilon$is the amplification efficiency ($1 \leq \varepsilon \leq 2$; $\varepsilon = 2$ means 100% efficiency) and *c*is the cycle number.

Solving for c results in Equation 2:

$$c = -(\log\varepsilon)^{-1}(\log N_0 + \log p - \log S) \qquad (2)$$

With $m = -(\log\varepsilon)^{-1}$ and $b = -(\log\varepsilon)^{-1}(\log p - \log S)$, Equation 2 simplifies to Equation 3:

$$c = m\log N_0 + b \qquad (3)$$

This equation describes the linear relationship between the $C_T$values determined and the log of the template concentration (N_0). The parameters *m*and *b*can be determined by a regression analysis of the C_T values of the standards. When solved for N_0, this equation serves as a calibration curve for the calculation of the unknowns according to Equation 4:

$$N_0 = 10^{(C_T - b)/m} \qquad (4)$$

Methods (Reference)	Amplification Efficiency Correction	Amplification Efficiency Calculation	Amplification Efficiency Assumptions	Automated Excel-Based Program
Standard Curve (31)	no	standard curve	no experimental sample variation	no
Comparative C_t ($2^{-\Delta\Delta Ct}$) (21)	yes	standard curve	reference = target	no
Pfaffl et al. (26)	yes	standard curve	sample = control	REST[a]
Q-Gene (23)	yes	standard curve	sample = control	Q-Gene[b]
Gentle et al. (7)	yes	raw data	researcher defines log-linear phase	no
Liu and Saint (22)	yes	raw data	reference and target genes can have different efficiencies	no
DART-PCR (30)	yes	raw data	statistically defined log-linear phase	DART-PCR[c]

C_t, cycle threshold, DART-PCR, data analysis for real-time PCR; REST, relative expression software tool.

[a]www.gene-quantification.info
[b]www.BioTechniques.com
[c]nar.oupjournals.org/cgi/content/full/31/14/e73/DC1

Table 1. Characteristics of Relative Quantitation Methods

Detection Chemistries	Specificity	Multiplex Capability	Specific Oligonucleotide Required	Allelic Discrimination	Cost
DNA Binding Dyes	two PCR primers	No	No	No	$
Hybridization Probe Four Oligonucleotide Method	two PCR primers; two specific probes	Yes	Yes	Yes	$$$
Hybridization Probe Three Oligonucleotide Method	two PCR primers; one specific probe	Yes	Yes	Yes	$$$
Hydrolysis Probes	two PCR primers; one specific probe	Yes	Yes	Yes	$$$
Molecular Beacons	two PCR primers; one specific probe	Yes	Yes	Yes	$$$
Scorpions	one PCR primer; one primer/probe	Yes	Yes	Yes	$$$
Sunrise Primers	two PCR primers	Yes	Yes	Yes	$$$
LUX Primers	two PCR primers	Yes	Yes	No	$$

$$$, very expensive; $$, moderately expensive; $, inexpensive. LUX, light upon extension.

Table 2. Characteristics of Detection Chemistries

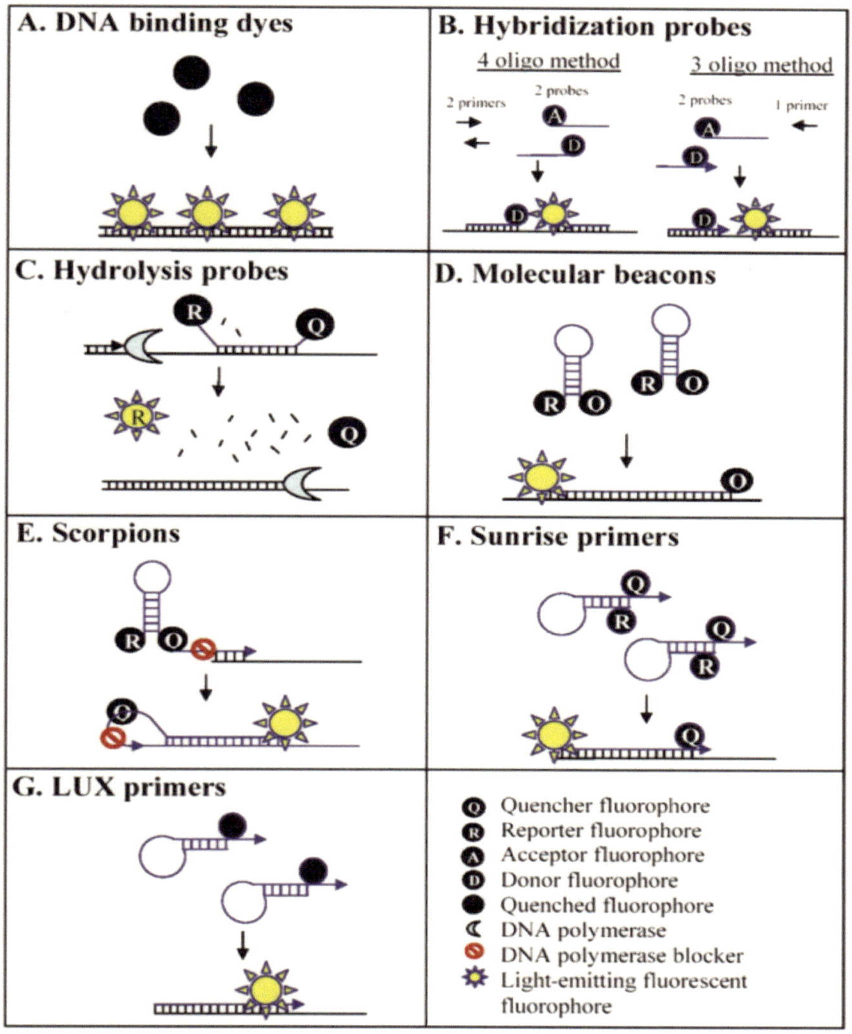

Fig. 1. Real-time PCR detection chemistries. Probe sequences are shown in blue while target DNA sequences are shown in black. Primers are indicated by horizontal arrowheads. Not all unlabeled PCR primers are shown. Oligo, oligonucleotide.

The efficiency can be calculated from the parameter m by using Equation 5:

$$\varepsilon = 10^{-1/m} \tag{5}$$

By inserting ε back into Equation 4, one obtains Equation 6:

$$N_0 = \varepsilon^{(b-C_T)} \tag{6}$$

The maximum value for ε is 2.0 (i.e. the amount of product is doubled in each cycle). The experimental value for ε usually varies between 1.5 and 1.9. Lower efficiencies limit the sensitivity of the assay but allow quantifications with higher precisions. Therefore, reactions should be optimized for high efficiency. The effect of the efficiency on the precision, however, is not pronounced.

With more than six orders of magnitude, the dynamic range of this procedure is extraordinarily high (Marcucci, Livak et al. 1998; Verhagen, Willemse et al. 2000;Sails, Fox et al. 2003). The accuracy of this technique is limited by the precision of the determination of the C_T values. The error of the C_T values results from the signal noise and the C_T calculation method. In highly optimized assays, standard errors of less than ± 0.2 cycles can be achieved. By assuming an amplification efficiency of 2 (i.e. 100 %), this implies that the minimum relative error for the quantification is about 10- 20%. The effects of different analysis and calculation methods and the effects of amplification-independent signal trends on the accuracy and precision of quantifications by realtime PCR are described in detail in papers by Lui et al. and Wilhelm et al (Liu and Saint 2002; Wilhelm, Pingoud et al. 2003).

Quantification is relative to the standard used. Only when the absolute concentration of the template molecules in the standard sample is known can the results be absolute. However, in most cases, determination of absolute concentrations is not required. That real-time PCR allows absolute quantification is demonstrated in principle by the reported determination of genome sizes (Wilhelm, Pingoud et al. 2003).

All quantifications by PCR are relative, either to a standard or to a reference gene. Interestingly, Equation 6 nicely illustrates the relative character of the quantifications using a dilution series of a standard; the meaning of the parameter b is the expected C_T value of a sample with 'one' copy (or any other unit as defined by the operator). The difference of this value minus the C_T value determined for the unknown sample ($\Delta C_T = b - C_T$) is a direct measure for the relative difference in template concentrations of the unknown and standard (Wilhelm and Pingoud 2003).

To analyse relative changes in transcript levels, the chosen standard is usually a reference transcript, for example from a housekeeping gene, itself with unknown template concentration. The calculation of ΔC_T values between reference and sample transcript in a reference and a test sample then provides a simple tool to estimate relative changes. The derivation, assumptions and applications of the so-called $2^{\Delta\Delta CT}$ method are described elsewhere by Livak et al (Livak and Schmittgen 2001). The results of this method are only semiquantitative because the efficiency ε is assumed to be 2.0 in all experiments and for all templates, which is at best an optimistic estimate. More precise results are obtained with a procedure introduced by Pfaffl et al., which includes a measured value for ε (Pfaffl 2001; Wilhelm and Pingoud 2003).

In general, care must also be taken for accurate quantifications with external standardization, especially with respect to polymerase inhibitors, which may be present in differentconcentrations in the unknowns and standards. This problem is circumvented by internal standardization. Here, an analytically distinguishable standard template ('competitor') is added to the sample and co-amplified in the same reaction (Gilliland, Perrin et al. 1990; Goerke, Bayer et al. 2001). The direct and simultaneous quantitative analysis of both products in realtime PCR also poses problems. These difficulties are mostly due to the fact that different fluorophors have to be used to distinguish the sequences of competitor and sample. As a result of different FRET and quantum efficiencies, the C_T values obtained for competitor and sample are not directly comparable.

The problem of where to set the threshold makes relativequantifications difficult if not impossible. However, a simple trickcan be used to combine the advantages of both methods: thereaction mixtures are prepared in duplicate (Gibson, Heid et al. 1996). To one of thesemixtures, the probe specific for the competitor sequence isadded, whereas the probe specific for the sample sequence isadded to the other mixture. This process is carried out for a seriesof reactions with different amounts of competitor added. Withthis procedure, two calibration lines are obtained and theintersection of the two lines is the equivalence point (Wilhelm and Pingoud 2003).

10. Melting curve analyses

Melting curves represent the temperature dependence of the fluorescence. They are recorded subsequent to the amplification of the target sequence by PCR. The detection can be performed either with dsDNA-specific dyes like SYBR Green I or with sequence-specific probes such as the molecular beacons and the hybridisation probes (scorpion and sunrise primers cannot be used for melting curve analysis because they are integrated into the PCR products; TaqMan probes cannot be used for melting curve analyses either, since their signal generation depends on the hydrolysis of the probe). Melting curves of sequence-specific probes are used for genotyping, resolving single base mismatches between target sequence and probe (Lay and Wittwer 1997; Whitcombe, Brownie et al. 1998), whereas SYBR Green I is used most frequently for product characterization (Ririe, Rasmussen et al. 1997). It has been reported that melting curves measured with SYBR Green I can also be utilized for genotyping of insertion/deletion polymorphisms and of single nucleotide polymorphisms (SNPs) (Akey, Sosnoski et al. 2001; Lin, Tseng et al. 2001).

In melting curves, the signal decreases gradually as a result of a temperature-dependent quench and more abruptly at a certain temperature because of the melting of the products

(dsDNA or ssDNA/probe hybrid). The melting temperature (T_m) of a product is defined as the temperature at which the steepest decrease of signal occurs. This can be identified conveniently as the peak value(s) (global or local maxima) in the negative derivative of the melting curve. Additionally, the area under the curve (AUC) of the peaks is proportional to the amount of product. Therefore, melting curve analysis may be used for quantifications with internal standardization when the T_m values of sample and competitor products are significantly different (Al-Robaiy, Rupf et al. 2001). However, well-performed normalization is required to reduce the systematic error due to the temperature dependent quench. This quench also limits the sensitivity of melting curve analyses. At present, there is only one software package available that can remove the quench effects from the data (Wilhelm, Pingoud et al. 2003).

With SYBR Green I, the amplification of the correct target sequence can be confirmed. In most cases, nonspecific products have different lengths and therefore deviating melting temperatures (Ririe, Rasmussen et al. 1997).Hybridisation probes, molecular beacons and TaqMan probes are used for mutation detection (Lay and Wittwer 1997; Bernard, Ajioka et al. 1998; Bernard and Wittwer 2000), genotyping (Whitcombe, Brownie et al. 1998; Ulvik and Ueland 2001; Grant, Steinlicht et al. 2002; Randen, Sørensen et al. 2003) and SNP screening (Sasvari-Szekely, Gerstner et al. 2000; Mhlanga and Malmberg 2001).

11. Applications

Real-time PCR is used for absolute and relative quantifications of DNA and RNA template molecules and for genotyping in a variety of applications (Wilhelm and Pingoud 2003).

Quantitative real-time PCR is used to determine viral loads (Mackay, Arden et al. 2002),gene expression (Bustin 2000; Goerke, Bayer et al. 2001), titers of germs and contaminations (infood, blood, other body fluids and tissues) (Locatelli, Urso et al. 2000; Hernandez, Rio et al. 2001; Norton 2002), allele imbalances (Ruiz-Ponte, Loidi et al. 2000) and the degrees of amplification and deletion ofgenes (Chiang, Wei et al. 1999; Nigro, Takahashi et al. 2001).

Real-time PCR is also becoming increasingly important in thediagnosis of tumors, such as for the detection and monitoringof minimal residual diseases (Marcucci, Livak et al. 1998; Elmaagacli, Beelen et al. 2000; Amabile, Giannini et al. 2001; Krauter, Heil et al. 2001; Krauter, Hoellge et al. 2001), the identification of micrometastases in colorectal cancer (Bustin, Gyselman et al. 1999), neuroblastoma (Cheung and Cheung 2001) and prostate cancer (Gelmini, Tricarico et al. 2001). It has been used to quantify amplifications of oncogenes (Bieche, Laurendeau et al. 1999; Lehmann, Glöckner et al. 2000; Lyon, Millson et al. 2001; Konigshoff, Wilhelm et al. 2003) as well as deletions of tumor suppressor genes in tumor samples (Wilhelm and Pingoud 2003). Also, the response of human cancer to drugs has been studied (Au, Chim et al. 2002; Miyoshi, Ando et al. 2002;Reimer, Koczan et al. 2002). Other clinically relevant applications are cytokine mRNA profiling in immune response (Hempel, Smith et al. 2002; Stordeur, Poulin et al. 2002) and tissue-specific gene expression analysis (Bustin 2002; Poola 2003; Prieto-Alamo, Cabrera-Luque et al. 2003).

Also, the results of DNA chip experiments are validated by real-time PCR quantifications (Miyazato, Ueno et al. 2001; Rickman, Bobek et al. 2001;Crnogorac-Jurcevic, Efthimiou et al. 2002).

Chimerism analysis is possible when sequence-specific probes are utilized to differentiate and quantify alleles. High dynamic ranges can be achieved with allele-specific real-time PCR (Shively, Chang et al. 2003). Robust chimerism analyses with extremely large dynamic ranges based on insertion/deletion polymorphisms and on SNPs are also possible (Wilhelm, Reuter et al. 2002; Maas, Schaap et al. 2003). Genetic chimerisms have been monitored by Y-chromosome-specific real-time PCR for sex-mismatched transplantations (Fehse, Chukhlovin et al. 2001; Byrne, Huang et al. 2002; Elmaagacli 2002) and by allele-specific real-time PCR (Maas, Schaap et al. 2003; Shively, Chang et al. 2003). This combination of allele-specific amplification with real-time PCR has been shown to reveal detection limits of down to 0.01% for SNPs (Maas, Schaap et al. 2003). Real-time PCR is increasingly used in forensic analyses (Andreasson, Gyllensten et al. 2002; von Wurmb-Schwark, Higuchi et al. 2002; Ye, Parra et al. 2002), but also to monitor disease- or age-related accumulation of deletions in the mitochondrial genome (Mehmet, Ahmed et al. 2001; He, Chinnery et al. 2002).

Melting curve analyses are used for real-time competitive PCR (Al-Robaiy, Rupf et al. 2001; Lyon, Millson et al. 2001), gene dosage tests (Ruiz-Ponte, Loidi et al. 2000) and genotyping and SNP detection (Bullock, Bruns et al. 2002; Burian, Grosch et al. 2002; Randen, Sørensen et al. 2003). These applications will have a particularly strong impact on pharmacogenetics (Palladino, Kay et al. 2003). Profiling of DNA methylation is also possible by melting curve analysis (Worm, Aggerholm et al. 2001; Akey, Akey et al. 2002), which simplifies the analysis of epigenetic variations of the genome and developmental processes.

In brief, the advantages of real-time PCR are exploited in clinical diagnosis and the monitoring of infectious diseases and tumors. The technique is applied for the analysis of age dependent diseases, cytokine and tissue-specific expression, forensic samples, epigenetic factors like DNA methylation and for food monitoring. The field of applications is still growing rapidly, which suggests that real-time PCR will become one of the most important techniques in molecular life sciences and medicine (Wilhelm and Pingoud 2003).

12. Normalization

Gene expression analysis at the messenger RNA (mRNA) level has become increasingly important in biological research. Generally we detect RNAs to determine if differences protein expression could be explained at the transcriptional level. In particular, measurement of mRNA is needed in situations where quantification of the protein is difficult or cumbersome. Most recently, mRNA expression analysis is being used to provide insight into complex regulatory networks and to identify genes relevant to new biological processes or implicated in diseases (Hendriks-Balk, Michel et al. 2007).Common methods for RNA detection include: Northern blotting, in situ hybridization, qualitative RTPCR, RNase protection assay, competitive RT-PCR, microarray analysis, and quantitative real-time PCR. The specificity, wide dynamic range , ease-of-use , requiring a minimal amount of RNA, no post-PCR handling and avoiding the use of radioactivity, has made the real-time quantitative reverse transcription polymerase chain reaction (qRT-PCR) the method of choice for quantitating RNA levels (Radonic, Thulke et al. 2004). The technique has two main steps: CDNA synthesis by reverse transcription of mRNA and subsequent quantification of specific CDNAs by real-time PCR. It is in many cases the only method for measuring mRNA levels of vivo low copy number targets of interest for which alternative assays either do not exist or lack the required sensitivity so these specification has led to made it the "gold standard" for mRNA quantification (Huggett, Dheda et al. 2005). Most gene expression assays are based on the comparison of two or more samples and require uniform sampling conditions for this comparison to be valid. Unfortunately, many factors can contribute to variability in the analysis of samples, making the results difficult to reproduce between experiments. During the preparation of CDNA for real-time PCR analysis there is significant potential for small errors to accumulate. For example, differences in sample size, RNA extraction efficiency, pippetting accuracy and reverse transcription efficiency will all add variability to your samples (Huggett, Dheda et al. 2005). Not only can the quantity and quality of RNA extracted from multiple samples vary, but even replicates can vary dramatically due to factors such as sample degradation, extraction efficiency, and contamination. On the other hand, since many biological samples contain inhibitors of the RT and/or the PCR step, it is crucial to assess the presence of any inhibitors of polymerase activity in RT and PCR. so it is clear that we need to incorporate some normalization method to control for errors. The identification of a valid reference for data normalisation remains the most stubborn of problems and none of the solutions proposed are ideal. Normalization methods range from ensuring that a similar sample size is chosen to using an internal housekeeping or reference gene (Table 3) (Huggett, Dheda et al. 2005).

Normalisation strategy	*Pros*	*Cons*	*Note*
Similar sample size/tissue volume	Relatively easy	Sample size/tissue volume may be difficult to estimate and/or may not be biological representative	Simple first step to reduce experimental error
Total RNA	Ensures similar reverse transcriptase input. May provide information on the integrity (depending on technique used)	Does not control for error introduced at the reverse transcription or PCR stages. Assumes no variation in rRNA/mRNA ratio	Requires a good method of assessing quality and quantity
Genomic DNA	Give an idea of the cellular sample size.	May vary in copy number per cell. Difficult to extract with RNA	Rarely used. Can be measured optically or by real time PCR
Reference genes ribosomal RNAs (rRNA)	Internal control that is subject to the same conditions as the RNA of interest. Also measured by real time RT-PCR	Must be validated using the same experimental samples. Resolution of assay is defined by the error of the reference gene	Oligo dt priming of RNA for reverse transcription will not work well with rRNA as no polyA tail is present. Usually in high abundance
Reference genes messenger RNAs (mRNA)	Internal control that is subject to the same conditions as the mRNA of interest. Also measured by real time RT-PCR	Must be validated using the same experimental samples. Resolution of assay is defined by the error of the reference gene	Most, but not all, of mRNAs contain polyA tails and can be primed with oligo dt for reverse transcription
Alien molecules	Internal control that is subject to most of the conditions as the mRNA of interest. Is without the biological variability of a reference gene	Must be identified and cloned or synthesised. Unlike the RNA of interest, is not extracted from the within the cells	Requires more characterisation and to be made available commercially

There is good correlation between the RNA concentration used and the real time PCR estimation of the different amounts of HuPO cDNA (using omniscript reverse transcriptase).

(Huggett, Dheda et al. 2005)

Table 3. Comparison of the actual amount of RNA used in different reverse transcription reactions with the respective amount of HuPO
Comparison of the actual amount of RNA used in different reverse transcription reactions with the respective amount of HuPO
cDNA measured by real-time RT-PCR

12.1 Methods of normalization

1. Standardizing Sample size

The most basic method of normalization ensures that an experiment compares similar sample sizes and this is achieved by measuring tissue weight, volume or cell number. This method can reduce the experimental error of first stage of qRT-PCR. It seems to be straightforward, but we can't ensure that equal volume of different samples contain the same cellular material. Real-time RT-PCR experiments that rely on the extraction of RNA from complex tissue samples are averaging the data from numerous, variable subpopulations of cells of different lineage at different stages of differentiation (Bustin, Benes et al. 2005). This can be misleading, as is illustrated when sampling a similar volume of blood from HIV +ve patients. Patients with HIV that have less advanced immunosupression (CD4 counts X200 cells/ml) will yield a higher amount RNA than patients with CD4 counts p200 cells/ml. This is simply because there are fewer cells per milliliter of blood in the latter group. Even cellular subpopulations of the same pathological origin can be highly heterogeneous. Tumor biopsies, in particular, are made up not just of normal and cancer epithelial cells, but there may be several subclones of epithelial cancer cells together with stromal, immune and vascular components (Vandesompele, De Preter et al. 2002; Bustin, Benes et al. 2005). This variability can give us misleading or meaningless result to solve this we can use laser capture microdissection to normalize against the dissected area which can report the target mRNA levels conveniently as copies per area or cell dissected. In in vitro cell culture, due to different morphologies or clumping up of cells, it's hard to determine sample size (cell number) .we can treat them with buffers and/or enzymes till they could be counted, however these treatments surely could affect gene expression. This approach could not be applied for solid tumors for which the amount of cells cannot be determined accurately. To work around this problem, it was suggested to standardize the RT-qPCR data between samples using the amount of genomic DNA as an indicator reflecting the number of cells in each sample. However, these approaches do not account for the degradation of RNA or the efficiency of RT and PCR. So, while ensuring a similar sample size is important it clearly is not sufficient on its own (Huggett, Dheda et al. 2005).

2. Normalization with genomic DNA

Another method for normalization is measuring the amount of genomic DNA (gDNA). This appears to be an ideal method as it does not require reverse transcription for detection by real-time PCR (Bustin 2002). However, this approach do not account for the degradation of RNA or the efficiency of RT and PCR. .Moreover, in the case of normalization with genomic DNA, the fact of working with tumor cells can present additional problems because they tend to have abnormal karyotype. Therefore, the ratio between the amount of DNA and cell number is variable (Huggett, Dheda et al. 2005). Another major problem with using this strategy is that RNA extraction procedures are usually not designed to purify DNA, so the extraction rate may vary between different samples, with DNA yields often being low. In conclude Normalization against genomic DNA is rarely used since it is difficult to coextract with RNA and it may vary in copy number per cell (Huggett, Dheda et al. 2005).

3. Normalization with total RNA

The normalization of RT-qPCR results can be compared to the amount of total RNA used in the reverse transcription step. Not only does this facilitates normalization but circumvents problems associated with the linearity of the reverse transcriptase step. There are several methods for quantifying RNA; the most common is to measure the absorbance at 260 nm (A_{260}) with a UV spectrophotometer. The major advantage of this spectrometer, whose sensitivity is estimated at 5 ng/uL, is that it requires only 1 microL of sample, placed in direct contact with the optical system (Huggett, Dheda et al. 2005; Hendriks-Balk, Michel et al. 2007). However, contaminants absorbing at 260 nm, such as proteins, phenol or genomic DNA, can lead to overestimated results. Another optical system is flourimetry in which intercalating fluorescent nucleic acid is used, the kit RiboGreen ® Molecular Probes based on this principle. This is a more sensitive technique but does not discriminate RNA from DNA, and contaminants such as phenol can produce variable results (Huggett, Dheda et al. 2005). Since it is generally assumed that OD260 analysis is less accurate than the RiboGreen assay, we have compared RNA quantification data obtained using the RiboGreen assay with

OD260 analysis using a Genequant II (Pharmacia). The results (Fig. 2) suggest that both methods generate comparable results when the RNA concentration is not less than 100 ng/µl, with RiboGreen measurements lower than those obtained using the spectrophotometer. OD260 analysis becomes less reliable at lower RNA concentrations (Bustin 2002).

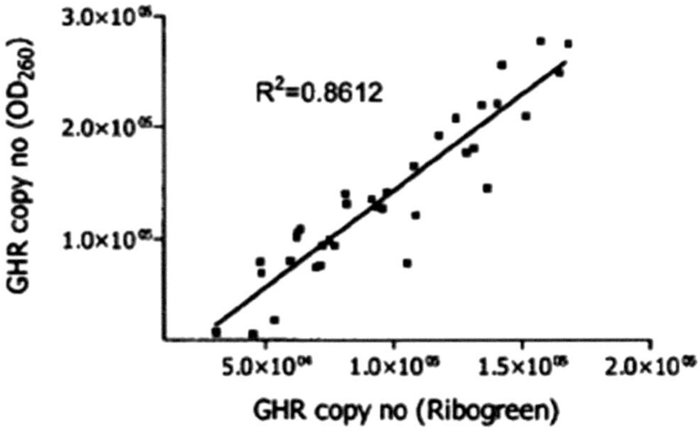

Fig. 2. Comparison of RNA quantification using the Genequant II and the RiboGreen fluorescent assay. RNA from 34 normal colon biopsies was quantitated using a standard Genequant II protocol which measures the absorbance at 260 nm. The same samples were then quantitated using a standard RiboGreen fluorescent assay. RT-PCR assays targeting the GHR were carried out and Ct normalised against the respective concentrations determined by the two methods. The scatterplot shows a good correlation between the two methods ($r2$=0·8612) (Bustin 2002).

What is also important, but often overlooked, is the need to assess the quality of RNA because degraded RNAs may adversely affect results. The opportune development of Agilent's 2100 Bioanalyser and LabChip technology has provided a new standard of RNA quality control as well as permitting concomitant quantification of RNA. The analysis is not influenced by contamination of phenol or proteins, against the presence of genomic DNA requires a correction of the measurement of the concentration of RNA (Bustin, Benes et al. 2005). This is particularly important when the RNA has been extracted from 'dirty' tissue such as the colon (Vandesompele, De Preter et al. 2002). This technique allows characterizing the RNA in a concentration between 5 and 500 ng/uL. For each sample, the software determines the ratio 28S/18S and assigns a RIN (RNA Integrity Number) which takes into account the entire electropherogram. The value of RIN ranges from 1 to 10, with 1 being totally degraded RNA, and 10 to a high-quality RNA. In addition to being fast and allow high throughput, this technique requires only 1 microL of sample. It is the simplest method and objective qualitative analysis of RNA, its use is recommended (Schmittgen and Zakrajsek 2000; Bustin 2002). Similar in concept, but requiring an additional RT-PCR assay, is normalization against one of the rRNAs. rRNA levels may vary less under conditions that affect the expression of mRNAs and the use of rRNA has been claimed to be more reliable than that of several reference genes in rat livers and human skin fibroblasts (Bustin 2002; Huggett, Dheda et al. 2005). But, a drawback is that it primarily measures ribosomal RNA (rRNA) whereas real-time PCR aims to determine mRNA expression and normalization for total RNA assumes that the rRNA:mRNA ratio is the same in all groups, which might not always be the case. Moreover, rRNA is not present in purified mRNA and the high abundance of rRNA compared to mRNA makes it difficult to subtract the baseline value in realtime PCR analysis. Thus, markers of rRNA such as 18S or 28S rRNA might also be suboptimal as normalization factors in many settings (Hendriks-Balk, Michel et al. 2007). Also, it has been reported that rRNA transcription is affected by biological factors and drugs. An important parameter to consider when normalized relative to the RNA is the quality of it. Differences in the quality of the samples strongly depend on the extraction step and are the source of the most common variations in RT-qPCR. It is therefore important to use the same method of extraction for all samples analyzed (Vandesompele, De Preter et al. 2002). The quality of RNA is defined by both its purity (no contamination) and integrity (non-degraded RNA). Its purity was determined by measuring the absorbance at 230 (organic contaminants) and 280 nm (specific proteins), RNA is considered pure ratios A_{260}/A_{230} and $A_{260}/A_{280} > 1.8$. With regard to the integrity, the traditional method was to visualize the bands of 28S and 18S ribosomal RNA on a gel electrophoresis. Indeed, it is difficult to analyze directly the mRNA; they represent only 1% to 3% of total RNA. We must therefore consider that the degradation of ribosomal RNA, the majority, reflecting the degradation of mRNA. Thus, the 18S/28S ratio assesses the integrity of RNA; a ratio close to 2 is considered an indicator of RNA with little or no gradient. However, this method requires a large amount of RNA (0.5-2 mg), and is not sensitive enough to detect slight damage. Normalizing a sample against total RNA has the drawback of not controlling for variation inherent in the reverse transcription or PCR reactions and it ignores the efficiency of converting RNA into CDNA. Also rRNA cannot be used for normalization when quantifying targets from polyA-enriched samples (Huggett, Dheda et al. 2005). A final drawback when using total RNA for normalization is the lack of internal control for RT or PCR inhibitors. All quantitative methods assume that the RNA targets are reverse transcribed and subsequently amplified with similar efficiency but the reaction is extremely sensitive to the presence of inhibitors, which can be reagents used in the extraction step (salts, alcohols, phenols), or components copurification organic (urea, heme, heparin, immunoglobulin G). These compounds can also inhibit the PCR reaction. Thus, two reactions with an equal amount of RNA, but the efficiencies of RT and / or PCR are different, will yield results that cannot be compared. Different methods exist to assess the presence of inhibitors in biological samples. First, it is possible to compare the efficacy of PCR for different dilutions (1/20 and 1/80 for example) of a sample. An alternative is to add a defined amount of a synthetic single-stranded amplicon CDNA samples, and comparing its amplification compared to a control without CDNA. However, these methods are limited to verify the absence of PCR inhibitors, and do not evaluate the effectiveness of the reverse transcription step (Bustin 2002).

4. Normalization with an artificial molecule (spike)

An interesting solution to control the two enzymatic reactions (RT and PCR) is added to RNA extracted an exogenous RNA, which will compare the amplification between the different samples. This sequence control should show no similarity to the target RNA, we will use such a specific mRNA from a plant when studying gene expression in humans. The main criticism of using spikes is that, while they can be introduced prior to extraction, unlike the cellular RNAs they are not extracted from within the tissue. Consequently, there may be situations (e.g. if the samples differ histologically) when the spike may not be a good control for the extraction procedure. The stages required to generate the alien molecule may also not be feasible for small laboratories wanting to perform limited amounts of real-time RTPCR (Schmittgen and Zakrajsek 2000; Argyropoulos, Psallida et al. 2006).

5. Normalization with reference genes

Reference genes represent the by far most common method for normalizing qRT-PCR data. Reference genes are often referred to as housekeeping genes assuming that those genes are expressed at a constant level in various tissues at all stages of development and are unaffected by the experimental treatment (Hendriks-Balk, Michel et al. 2007; Balogh, Paragh et al. 2008). Use of this endogenous control theory allows controlling all stages of the experimental protocol; its expression reflects not only the quantity and quality of RNA used, but the efficiencies of the RT and PCR. An advantage of reference genes as compared to total or rRNA is that the reference gene is subject to the same conditions as the mRNA of interest (Hendriks-Balk, Michel et al. 2007). The most commonly used reference genes include β-actin (ACTB), (GAPDH), (HPRT) and 18S rRNA. The other commonly used reference genes would be PGK1, B2M, GAPD, HMBS, HPRT1, RPL13A, SDHA, TBP, UBC and YWHAZ (Vandesompele, De Preter et al. 2002). The initial concentration of a target is usually derived from the C_T (cycle threshold), which is the number of amplification cycles where the amplification curve crosses the threshold line. This line is placed at the exponential phase, so as to be clearly distinguishable from background noise. For each sample, the C_T obtained for the genes of interest and reference must be converted to normalized expression ratio. For this, various options are available, they are integrated in the software provided with the various qPCR instruments or described in the literature. The relative standard curve method requires the construction, for the target gene and reference gene, a range made from a series of dilutions of a reference sample. These ranges to obtain standard curves, obtained by expressing the C_T as a function of log of the initial concentration of cDNA. Concentration values for each point of the range can be set arbitrarily in accordance with the dilution factors. Therefore, the relative amount of a target is determined by the C_T by interpolation with the standard curve. The standard expression of a gene of interest is determined by the following formula:

$$R = \frac{Relative\ amount\ of\ the\ gene\ of\ intrest}{Relative\ amount\ of\ the\ refrence\ gene}$$

In addition, a calibrator is typically used. This is a sample used as a basis for comparing results. The normalized ratio of each sample is divided by the normalized ratio of the calibrator. Thus, the calibrator becomes the reference 1x, and all other samples are expressed as a ratio relative to the calibrator. The method of $\Delta\Delta C_T$ uses a mathematical formule to calculate the ratio of expression of a target gene between two samples, normalized with reference gene. First, the differences ΔC_T between the values of C_T target gene and reference gene were determined for the test sample and control.

$$\Delta C_{T\,(sample)} = C_T\ (target\ sample) - C_T\ (reference\ sample)$$

$$\Delta C_{t\,(control)} = C_T\ (target\ control) - C_T\ (reference\ control)$$

Next, the $\Delta\Delta C_T$ between control and the sample is calculated:

$$\Delta\Delta C_T \, \Delta C_{T\,(control)} - \Delta C_{T\,(sample)}$$

Finally, the normalized ratio of expression of a target gene is determined by the formula: $2^{-\Delta\Delta CT}$.

Unlike the relative standard curve method, where the amplification efficiency (E) target genes and reference is directly taken into account when building ranges, the method of $\Delta\Delta C_T$ is assumed that the efficiencies of the two genes are equal to 100% (E = 2, with each cycle of the exponential phase, the concentration of PCR products is doubled). However, a difference in PCR efficiency of 3% ($\Delta E = 0.03$) between the two genes results in an error of 47% for the ratio of expression if $E_{target} < E_{ref}$ and 209% if $E_{target} > E_{ref}$ after 25 cycles. In addition, the error increases exponentially with larger variations of efficiency and a greater number of cycles. New models have been developed taking into account the efficiency of PCR target gene and reference gene. The most common is the model of Pfaffl, where the relative expression ratio (R) of a target gene between a sample and control is determined by the following formula:

$$R = \frac{(E_{target})^{\Delta C\ Ttarget(control\ -sample)}}{(E_{reference})^{\Delta C\ Treference(control\ -sample)}}$$

In this model of Pfaffl, the efficiency of PCR for a given gene is calculated from the construction of a calibration curve using the following formula: $E = 10^{[-1/\ gradient]}$. This method gives a good estimate of effectiveness, although it is possible that it is overestimated. However, this approach assumes that the amplification efficiencies between the diluted samples are identical, creating a linear relationship between C_T and amount of CDNA in the beginning. Therefore, some authors such as Liu and Saint have developed models that take into account standards of efficiency for each sample, the latter being determined by the kinetics of the amplification curve. However, with this kind of approach, the slightest error in the measurement of effectiveness is amplified and passed exponentially on the expression ratio calculated. The different models of normalization with reference genes therefore have all the advantages and disadvantages. At present, there is no time-honored method for the treatment of the results of RT-qPCR. Normalization to a reference gene is a simple method and frequently used because it can control many variables. An advantage of reference genes as compared to total or rRNA is that the reference gene is subject to the same conditions as the mRNA of interest (Bustin, Benes et al. 2005; Hendriks-Balk, Michel et al. 2007). What has become apparent over recent years is that there is no single reference gene for all experimental systems. Quantified errors related to the use of a single reference gene as more than three-fold in 25% and more than six-fold in 10% of samples. Today it is clear that reference genes must be carefully validated for each experimental situation and those new experimental conditions or different tissue samples require re-validation of the chosen reference genes (Balogh, Paragh et al. 2008). If inappropriate reference genes are used for normalization, the experimental results obtained can differ greatly from those using a validated reference gene. Validation of a reference gene requires removal of any non-specific variation in expression. This can be done using a recently introduced program called geNorm (freely available at http://medgen.ugent.be/~jvdesomp/genorm/) that mathematically identifies the most suitable reference gene for a given experimental condition. Because of the inherent variation in the expression of reference genes the use of multiple reference genes rather than one reference gene is recommended to ensure reliable normalization of real-time PCR. Several statistical programs can help to determine the most appropriate reference gene or set of genes (Hendriks-Balk, Michel et al. 2007; Borges, Ferreira et al. 2010).

13. Abbreviation

ACTB: Beta actin
B2M: Beta-2-microglobulin
C_T: Threshold Cycle
FRET: Fluorescence Resonance Energy Transfer
GAPD: Glyceraldehyde-3- phosphate dehydrogenase
HMBS: Hydroxymethyl-bilane synthase
HPRT: hypoxanthine ribosyltransferase
HPRT1: Hypoxanthine phosphoribosyl-transferase 1
HRM: High Resolution Melting
KPCR: Kinetic Polymerase Chain Reaction
LNA: Locked Nucleic Acid
LUX: Light Upon Extension
PCR: Polymerase Chain Reaction
PGK1: phosphoglycerokinase 1
PNA: Peptide Nucleic Acid
Q-PCR (QRT-PCR): Quantitative Real-Time Polymerase Chain Reaction
RPL13A: Ribosomal protein L13a
RT PCR: Reverse transcription PCR
SDHA: Succinate dehydrogenase complex, subunit A
TBP: TATA box binding protein
T_m: Melting Point
UBC: Ubiquitin C
YWHAZ: Tyrosine 3-monooxygenase/ tryptophan 5-monooxygenase activation protein, zeta polypeptide

14. References

Ahmad, A. I. (2007). "BOXTO as a real-time thermal cycling reporter dye." *Journal of Biosciences*32(2): 229-239.

Akey, D. T., J. M. Akey, et al. (2002). "Assaying DNA methylation based on high-throughput melting curve approaches." *Genomics*80(4): 376-384.

Akey, J., D. Sosnoski, et al. (2001). "Research Report Melting Curve Analysis of SNPs (McSNP®): A Gel-Free and Inexpensive Approach for SNP Genotyping." *Biotechniques*30(2): 358-367.

Al-Robaiy, S., S. Rupf, et al. (2001). "Research Report Rapid Competitive PCR Using Melting Curve Analysis for DNA Quantification." *Biotechniques*31(6): 1382-1388.

Amabile, M., B. Giannini, et al. (2001). "Real-time quantification of different types of bcr-abl transcript in chronic myeloid leukemia." *Haematologica*86(3): 252-259.

Andreasson, H., U. Gyllensten, et al. (2002). "Real-time DNA quantification of nuclear and mitochondrial DNA in forensic analysis." *Biotechniques*33(2): 402-411.

Au, W. Y., C. S. Chim, et al. (2002). "Real-time quantification of multidrug resistance-1 gene expression in relapsed acute promyelocytic leukemia treated with arsenic trioxide." *Haematologica*87(10): 1109-1111.

Ausubel, F. M., R. Brent, et al. (2005). "Current Protocols in Molecular Biology." *Hoboken, NJ: Wiley.*

Bengtsson, M., H. J. Karlsson, et al. (2003). "A new minor groove binding asymmetric cyanine reporter dye for real-time PCR." *Nucleic Acids Res*31(8): e45.

Bernard, P. S., R. S. Ajioka, et al. (1998). "Homogeneous multiplex genotyping of hemochromatosis mutations with fluorescent hybridization probes." *The American journal of pathology*153(4): 1055-1061.

Bernard, P. S. and C. T. Wittwer (2000). "Homogeneous amplification and variant detection by fluorescent hybridization probes." *Clinical chemistry*46(2): 147-148.

Bieche, I., I. Laurendeau, et al. (1999). "Quantitation of MYC gene expression in sporadic breast tumors with a real-time reverse transcription-PCR assay." *Cancer Res*59(12): 2759-2765.

Bio-Rad Laboratories, I. n. c. (2006). "Real-Time PCR Applications Guide."

BioInformatics (2003). "The Market for Real-Time PCR Reagents & Instrumentation." *Arlington VA: BioInformatics.*

Bonnet, G., S. Tyagi, et al. (1999). "Thermodynamic basis of the enhanced specificity of structured DNA probes." *Proc Natl Acad Sci*USA 96(5): 6171-6176.

Braasch, D. A. and D. R. Corey (2001). "Locked nucleic acid (LNA): fine-tuning the recognition of DNA and RNA." *Chem Biol*8(1): 1-7.

Brown, P. O. and D. Botstein (1999). "Exploring the new world of the genome with DNA microarrays." *Nat Genet*21: 33-37.

Bullock, G. C., D. E. Bruns, et al. (2002). "Hepatitis C genotype determination by melting curve analysis with a single set of fluorescence resonance energy transfer probes." *Clinical chemistry*48(12): 2147-2154.

Burian, M., S. Grosch, et al. (2002). "Validation of a new fluorogenic real-time PCR assay for detection of CYP2C9 allelic variants and CYP2C9 allelic distribution in a German population." *Br J Clin Pharmacol*54(5): 518-521.

Bustin, S., V. Gyselman, et al. (1999). "Detection of cytokeratins 19/20 and guanylyl cyclase C in peripheral blood of colorectal cancer patients." *British Journal of Cancer*79(11/12): 1813-1820.

Bustin, S. A. (2000). "Absolute quantification of mRNA using real-time reverse transcription polymerase chain reaction assays." *J Mol Endocrinol* 25: 169-193.

Bustin, S. A. (2002). "Quantification of mRNA using real-time reverse transcription PCR (RT-PCR): trends and problems." *J Mol Endocrinol*29(1): 23-39.

Bustin, S. A. and T. Nolan (2004). "Pitfalls of quantitative real-time reverse-trasncription polymerase chain reaction." *J Biomol Tech*15: 155-166.

Byrne, P., W. Huang, et al. (2002). "Chimerism analysis in sex-mismatched murine transplantation using quantitative real-time PCR." *Biotechniques*32(2): 279-286.

Caplin, B. E., R. P. Rasmussen, et al. (1999). "LightCyclerTM hybridization probes- the most direct way to monitor PCR amplification and mutation detection." *Biochemica*1: 5-8.

Cheung, I. Y. and N. K. Cheung (2001). "Quantitation of marrow disease in neuroblastoma by real-time reverse transcription-PCR." *Clin Cancer Res*7(6): 1698-1705.

Chiang, P. W., W. L. Wei, et al. (1999). "A fluorescent quantitative PCR approach to map gene deletions in the Drosophila genome." *Genetics*153(3): 1313-1316.

Costa, J. M., P. Ernault, et al. (2004). "Chimeric LNA/DNA probes as a detection system for real-time PCR." *Clin Biochem*37(10): 930-932.

Crnogorac-Jurcevic, T., E. Efthimiou, et al. (2002). "Expression profiling of microdissected pancreatic adenocarcinomas." *Oncogene*21(29): 4587-4594.

Crockett, A. O. and C. T. Wittwer (2001). "Fluorescein-labeled oligonucleotides for real-time pcr: using the inherent quenching of deoxyguanosine nucleotides." *Anal Biochem*290(1): 89-97.

Egholm, M., O. Buchardt, et al. (1992). "Peptide nucleic acids (PNA). Oligonucleotide analogs with an achiral peptide backbone." *J Am Chem Soc*114: 1895-1897.

Elmaagacli, A., D. Beelen, et al. (2000). "The amount of BCR-ABL fusion transcripts detected by the real-time quantitative polymerase chain reaction method in patients with Philadelphia chromosome positive chronic myeloid leukemia correlates with the disease stage." *Annals of hematology*79(8): 424-431.

Elmaagacli, A. H. (2002). "Real-time PCR for monitoring minimal residual disease and chimerism in patients after allogeneic transplantation." *Int J Hematol*76 Suppl 2: 204-205.

Espy, M. J., J. R. Uhl, et al. (2006). "Real-time PCR in clinical microbiology: applications for routine laboratory testing." *Clin Microbiol Rev*19(1): 165-256.

Fehse, B., A. Chukhlovin, et al. (2001). "Real-time quantitative Y chromosome-specific PCR (QYCS-PCR) for monitoring hematopoietic chimerism after sex-mismatched allogeneic stem cell transplantation." *J Hematother Stem Cell Res*10(3): 419-425.

Fronhoffs, S., G. Totzke, et al. (2002). "A method for the rapid construction of cRNA standard curves in quantitative real-time reverse transcription polymerase chain reaction." *Molecular and cellular probes*16(2): 99-110.

Gelmini, S., C. Tricarico, et al. (2001). "Real-Time quantitative reverse transcriptase-polymerase chain reaction (RT-PCR) for the measurement of prostate-specific antigen mRNA in the peripheral blood of patients with prostate carcinoma using the taqman detection system." *Clin Chem Lab Med*39(5): 385-391.

Gerard, C. J., K. Olsson, et al. (1998). "Improved quantitation of minimal residual disease in multiple myeloma using real-time polymerase chain reaction and plasmid-DNA complementarity determining region III standards." *Cancer research*58(17): 3957-3964.

Gibson, U. E., C. A. Heid, et al. (1996). "A novel method for real time quantitative RT-PCR." *Genome Res*6(10): 995-1001.

Gilliland, G., S. Perrin, et al. (1990). "Analysis of cytokine mRNA and DNA: detection and quantitation by competitive polymerase chain reaction." *Proceedings of the National Academy of Sciences*87(7): 2725-2729.

Giulietti, A., L. Overbergh, et al. (2001). "An overview of real-time quantitative PCR: applications to quantify cytokine gene expression." *Methods*25(4): 386-401.

Goerke, C., M. G. Bayer, et al. (2001). "Quantification of bacterial transcripts during infection using competitive reverse transcription-PCR (RT-PCR) and LightCycler RT-PCR." *Clinical and Vaccine Immunology*8(2): 279-282.

Grant, S. F. A., S. Steinlicht, et al. (2002). "SNP genotyping on a genome-wide amplified DOP-PCR template." *Nucleic Acids Research*30(22): e125-e125.

He, L., P. F. Chinnery, et al. (2002). "Detection and quantification of mitochondrial DNA deletions in individual cells by real-time PCR." *Nucleic Acids Res*30(14): e68.

Heid, C. A., J. Stevens, et al. (1996). "Real time quantitative PCR." *Genome Res*6(10): 986-994.

Hempel, D. M., K. A. Smith, et al. (2002). "Analysis of cellular immune responses in the peripheral blood of mice using real-time RT-PCR." *J Immunol Methods*259(1-2): 129-138.

Hernandez, M., A. Rio, et al. (2001). "A rapeseed-specific gene, acetyl-CoA carboxylase, can be used as a reference for qualitative and real-time quantitative PCR detection of transgenes from mixed food samples." *J Agric Food Chem*49(8): 3622-3627.

Higuchi, R., G. Dollinger, et al. (1992). "Simultaneous amplification and detection of specific DNA sequences." *Biotechnology*10: 413-417.

Higuchi, R., C. Fockler, et al. (1993). "Kinetic PCR analysis: real-time monitoring of DNA amplification reactions." *Biotechnology (N Y)*11(9): 1026-1030.

Holland, P. M., R. D. Abramson, et al. (1991). "Detection of specific polymerase chain reaction product by utilizing the 5′ → 3′ exonuclease activity of Thermus aquaticus DNA polymerase." *Proc Natl Acad Sci*USA 88(16): 7276-7280.

Isacsson, J., H. Cao, et al. (2000). "Rapid and specific detection of PCR products using light-up probes." *Molecular and cellular probes*14(5): 321-328.

Jansen, K., B. Norde´n, et al. (1993). "Sequence dependence of 40,6-diamidino-2-phenylindole (DAPI)-DNA interactions." *J Am Chem Soc*115: 10527-10530.

Klein, D. (2002). "Quantification using real-time PCR technology: applications and limitations." *Trends Mol Med*8(6): 257-260.

Konigshoff, M., J. Wilhelm, et al. (2003). "HER-2/neu gene copy number quantified by real-time PCR: comparison of gene amplification, heterozygosity, and immunohistochemical status in breast cancer tissue." *Clin Chem*49(2): 219-229.

Krauter, J., G. Heil, et al. (2001). "The AML1/MTG8 Fusion Transcript in t (8; 21) Positive AML and its Implication for the Detection of Minimal Residual Disease; Malignancy." *Hematology (Amsterdam, Netherlands)*5(5): 369-381.

Krauter, J., W. Hoellge, et al. (2001). "Detection and quantification of CBFB/MYH11 fusion transcripts in patients with inv(16)-positive acute myeloblastic leukemia by real-time RT-PCR." *Genes Chromosomes Cancer*30(4): 342-348.

Kubista, M. (2004). "Nucleic acid-based technologies: application amplified." *Pharmacogenomics*5(6): 767-773.

Kubista, M., J. M. Andrade, et al. (2006). "The real-time polymerase chain reaction." *Molecular Aspects of Medicine*27 95-125.

Kurata, S., T. Kanagawa, et al. (2001). "Fluorescent quenching-based quantitative detection of specific DNA/RNA using a BODIPY® FL-labeled probe or primer." *Nucleic Acids Research*29(6): e34.

Kutyavin, I. V., I. A. Afonina, et al. (2000). "3′-Minor groove binder-DNA probes increase sequence specificity at PCR extension temperatures." *Nucleic Acids Res*28(2): 655-661.

Lay, M. J. and C. T. Wittwer (1997). "Real-time fluorescence genotyping of factor V Leiden during rapid-cycle PCR." *Clin Chem*43(12): 2262-2267.

Le Pecq, J. B. and C. Paoletti (1966). "A new fluorometric method for RNA and DNA determination." *Anal Biochem*17: 100-107.

Lehmann, U., S. Glöckner, et al. (2000). "Detection of gene amplification in archival breast cancer specimens by laser-assisted microdissection and quantitative real-time polymerase chain reaction." *Am J Pathol*156(6): 1855-1864.

Li, Qingge, et al. (2002). "A new class of homogeneous nucleic acid probes based on specific displacement hybridization." *Nucleic Acids Res*30: e5.

Lin, M. H., C. H. Tseng, et al. (2001). "Real-time PCR for rapid genotyping of angiotensin-converting enzyme insertion/deletion polymorphism1." *Clinical biochemistry*34(8): 661-666.

Lind, K., A. Stahlberg, et al. (2006). "Combining sequence-specific probes and DNA binding dyes in real-time PCR for specific nucleic acid quantification and melting curve analysis." *Biotechniques*40(3): 315-319.

Liss, B. (2002). "Improved quantitative real-time RT-PCR for expression profiling of individual cells." *Nucleic Acids Res*30(17): e89.

Liu, W. and D. A. Saint (2002). "A new quantitative method of real time reverse transcription polymerase chain reaction assay based on simulation of polymerase chain reaction kinetics." *Analytical Biochemistry*302(1): 52-59.

Liu, W. and D. A. Saint (2002). "Validation of a quantitative method for real time PCR kinetics." *Biochem Biophys Res Commun*294(2): 347-353.

Livak, K. J. and T. D. Schmittgen (2001). "Analysis of relative gene expression data using real-time quantitative PCR and the 2(-Delta Delta C(T)) Method." *Methods*25(4): 402-408.

Locatelli, G., V. Urso, et al. (2000). "Quantitative analysis of GMO food contaminations using real time PCR." *Ital J Biochem*49(3-4): 61-63.

Luu-The, V., N. Paquet, et al. (2005). "Improved real-time RT-PCR method for high-throughput measurements using second derivative calculation and double correction." *Biotechniques*38(2): 287-293.

Lyon, E., A. Millson, et al. (2001). "Quantification of HER2/neu gene amplification by competitive PCR using fluorescent melting curve analysis." *Clinical chemistry*47(5): 844-851.

Maas, F., N. Schaap, et al. (2003). "Quantification of donor and recipient hemopoietic cells by real-time PCR of single nucleotide polymorphisms." *Leukemia*17(3): 621-629.

Mackay, I. M. (2004). "Real-time PCR in the microbiology laboratory." *Clin Microbiol Infect*10(3): 190-212.

Mackay, I. M., K. E. Arden, et al. (2002). "Real-time PCR in virology." *Nucleic Acids Res*30(6): 1292-1305.

Mackya, I. M. (2004). "Real-time PCR in the microbiology laboratory." *Clin Microbiol Infect*10: 190-212.

Marcucci, G., K. Livak, et al. (1998). "Detection of minimal residual disease in patients with AML1/ETO-associated acute myeloid leukemia using a novel quantitative reverse transcription polymerase chain reaction assay." *Leukemia: official journal of the Leukemia Society of America, Leukemia Research Fund, UK*12(9): 1482-1489.

Mattarucchi, E., M. Marsoni, et al. (2005). "Different real time PCR approaches for the fine quantification of SNP's alleles in DNA pools: assays development, characterization and pre-validation." *J Biochem Mol Biol*38(5): 555-562.

Mehmet, D., F. Ahmed, et al. (2001). "Quantification of the common deletion in human testicular mitochondrial DNA by competitive PCR assay using a chimaeric competitor." *Molecular human reproduction*7(3): 301-306.

Mhlanga, M. M. and L. Malmberg (2001). "Using molecular beacons to detect single-nucleotide polymorphisms with real-time PCR." *Methods*25(4): 463-471.

Miyazato, A., S. Ueno, et al. (2001). "Identification of myelodysplastic syndrome–specific genes by DNA microarray analysis with purified hematopoietic stem cell fraction." *Blood*98(2): 422-427.

Miyoshi, Y., A. Ando, et al. (2002). "Prediction of response to docetaxel by CYP3A4 mRNA expression in breast cancer tissues." *International Journal of Cancer*97(1): 129-132.

Muller, P. Y., H. Janovjak, et al. (2002). "Processing of gene expression data generated by quantitative real-time RT-PCR." *Biotechniques*32(6): 1372-1379.

Mullis, K. B. (1990). "The unusual origin of the polymerase chain reaction." *Sci Am*262: 56-61.

Mullis, K. B. and F. A. Faloona (1987). "Specific synthesis of DNA in vitro via a polymerase-catalyzed chain reaction." *Methods Enzymol*155: 335-350.

Myi, Q. "single-color real-time PCR instruction manual Biorad Company."

Nazarenko, I., B. Lowe, et al. (2002). "Multiplex quantitative PCR using self-quenched primers labelled with a single fluorophore." *Nucleic Acids Res*30(9): e37.

Nazarenko, I. A., S. Bhatnagar, et al. (1997). "A closed tube format for amplification and detection of DNA based on energy transfer." *Nucleic Acids Research*25(12): 2516-2521.

Nigro, J. M., M. A. Takahashi, et al. (2001). "Detection of 1p and 19q loss in oligodendroglioma by quantitative microsatellite analysis, a real-time quantitative polymerase chain reaction assay." *The American journal of pathology*158(4): 1253-1262.

Norton, D. M. (2002). "Polymerase chain reaction-based methods for detection of Listeria monocytogenes: toward real-time screening for food and environmental samples." *Journal of AOAC International*85(2): 505-515.

Nygren, J., N. Svanvik, et al. (1998). "The interaction between the fluorescent dye thiazole orange and DNA." *Biopolymers*46: 39-51.

Olson, M., L. Hood, et al. (1989). "A common language for physical mapping of the human genome." *Science*245: 1434-1435.

Palladino, S., I. D. Kay, et al. (2003). "Real-time PCR for the rapid detection of vanA and vanB genes." *Diagn Microbiol Infect Dis*45(1): 81-84.

Pfaffl, M. W. (2001). "A new mathematical model for relative quantification in real-time RT-PCR." *Nucleic Acids Res*29(9): e45.

Pfaffl, M. W., A. Tichopad, et al. (2004). "Determination of stable housekeeping genes, differentially regulated target genes and sample integrity: BestKeeper–Excel-based tool using pair-wise correlations." *Biotechnology letters*26(6): 509-515.

Poola, I. (2003). "Molecular assay to generate expression profile of eight estrogen receptor alpha isoform mRNA copy numbers in picogram amounts of total RNA from breast cancer tissues." *Analytical Biochemistry*314(2): 217-226.

Prieto-Alamo, M. J., J. M. Cabrera-Luque, et al. (2003). "Absolute quantitation of normal and ROS-induced patterns of gene expression: an in vivo real-time PCR study in mice." *Gene Expr*11(1): 23-34.

Randen, I., K. Sørensen, et al. (2003). "Rapid and reliable genotyping of human platelet antigen (HPA)-1,-2,-3,-4, and-5 a/b and Gov a/b by melting curve analysis." *Transfusion*43(4): 445-450.

Reimer, T., D. Koczan, et al. (2002). "Tumour Fas ligand: Fas ratio greater than 1 is an independent marker of relative resistance to tamoxifen therapy in hormone receptor positive breast cancer." *Breast Cancer Research*4(5): R9.

Reischl, U., C. T. Wittwer, et al. (2002). "Rapid Cycle Real-time PCR: Methods and Applications; Microbiology and Food Analysis." *New York: Springer-Verlag*.

Rickman, D. S., M. P. Bobek, et al. (2001). "Distinctive molecular profiles of high-grade and low-grade gliomas based on oligonucleotide microarray analysis." *Cancer research*61(18): 6885-6891.

Ririe, K. M., R. P. Rasmussen, et al. (1997). "Product differentiation by analysis of DNA melting curves during the polymerase chain reaction." *Anal Biochem*245(2): 154-160.

Ruiz-Ponte, C., L. Loidi, et al. (2000). "Rapid real-time fluorescent PCR gene dosage test for the diagnosis of DNA duplications and deletions." *Clin Chem*46(10): 1574-1582.

Saiki, R. K., S. Scharf, et al. (1985). "Enzymatic amplification of beta-globin genomic sequences and restriction site analysis for diagnosis of sickle cell anemia." *Science*230: 1350-1354.

Sails, A. D., A. J. Fox, et al. (2003). "A real-time PCR assay for the detection of Campylobacter jejuni in foods after enrichment culture." *Appl Environ Microbiol*69(3): 1383-1390.

Sasvari-Szekely, M., A. Gerstner, et al. (2000). "Rapid genotyping of factor V Leiden mutation using single-tube bidirectional allele-specific amplification and automated ultrathin-layer agarose gel electrophoresis." *Electrophoresis*21(4): 816-821.

Shively, L., L. Chang, et al. (2003). "Real-time PCR assay for quantitative mismatch detection." *Biotechniques*34(3): 498-502, 504.

Sjöback, R., J. Nygren, et al. (1995). "Absorption and fluorescence properties of fluorescein Spetrochim." *Acta Part*A 51: L7-L21.

Soheili, Z. and S. Samiei (2005). "Real Time PCR: Principles and Application." *Hepatitis*5(3): 83-87.

Solinas, A., L. J. Brown, et al. (2001). "Duplex Scorpion primers in SNP analysis and FRET applications." *Nucleic Acids Research*29(20): e96-e96.

Souaze, F., A. Ntodou-Thome, et al. (1996). "Quantitative RT-PCR: limits and accuracy." *Biotechniques*21(2): 280-285.

Stordeur, P., L. F. Poulin, et al. (2002). "Cytokine mRNA quantification by real-time PCR." *J Immunol Methods*259(1-2): 55-64.

Svanvik, N., A. Stahlberg, et al. (2000). "Detection of PCR products in real time using light-up probes." *Anal Biochem*287(1): 179-182.

Svanvik, N., G. Westman, et al. (2000). "Light-up probes: thiazole orange-conjugated peptide nucleic acid for detection of target nucleic acid in homogenous solution." *Anal Biochem*281: 26-35.

Thelwell, N., S. Millington, et al. (2000). "Mode of action and application of Scorpion primers to mutation detection." *Nucleic Acids Research*28(19): 3752-3761.

Tyagi, S., D. P. Bratu, et al. (1998). "Multicolor molecular beacons for allele discrimination." *Nat Biotechnol*16(1): 49-53.

Tyagi, S. and F. R. Kramer (1996). "Molecular Beacons: probes that fluorescence upon hybridization." *Nat Biotechnol*14(3): 303-308.

Uehara, H., G. Nardone, et al. (1999). "Detection of telomerase activity utilizing energy transfer primers: comparison with gel- and ELISA-based detection." *Biotechniques*26(3): 552-558.

Ulvik, A. and P. M. Ueland (2001). "Single nucleotide polymorphism (SNP) genotyping in unprocessed whole blood and serum by real-time PCR: application to SNPs affecting homocysteine and folate metabolism." *Clin Chem*47(11): 2050-2053.

Valasek, M. A. and J. J. Repa (2005). "The power of real-time PCR." *Adv Physiol Educ*29(3): 151-159.

Verhagen, O. J., M. J. Willemse, et al. (2000). "Application of germline IGH probes in real-time quantitative PCR for the detection of minimal residual disease in acute lymphoblastic leukemia." *Leukemia*14(8): 1426-1435.

von Wurmb-Schwark, N., R. Higuchi, et al. (2002). "Quantification of human mitochondrial DNA in a real time PCR." *Forensic Sci Int*126(1): 34-39.

Whitcombe, D., J. Brownie, et al. (1998). "A homogeneous fluorescence assay for PCR amplicons: its application to real-time, single-tube genotyping." *Clin Chem*44(5): 918-923.

Whitcombe, D., J. Theaker, et al. (1999). "Detection of PCR products using self-probing amplicons and fluorescence." *Nat Biotechnol*17(8): 804-807.

Wilhelm, J. and A. Pingoud (2003). "Real-time polymerase chain reaction." *Chembiochem*4(11): 1120-1128.

Wilhelm, J., A. Pingoud, et al. (2003). "Real-time PCR-based method for the estimation of genome sizes." *Nucleic Acids Res*31(10): e56.

Wilhelm, J., A. Pingoud, et al. (2003). "SoFAR: software for fully automatic evaluation of real-time PCR data." *Biotechniques*34(2): 324-332.

Wilhelm, J., H. Reuter, et al. (2002). "Detection and quantification of insertion/deletion variations by allele-specific real-time PCR: application for genotyping and chimerism analysis." *Biol Chem*383(9): 1423-1433.

Wilson, I. G. (1997). "Inhibition and facilitation of nucleic acid amplification." *Appl Environ Microbiol*63: 3741-3751.

Wilson, R. and M. K. Johansson (2003). "Photoluminescence and electrochemiluminescence of a Ru(II) (bpy)3- quencher dual-labeled oligonucleotide probe." *Chem Commun*21: 2710-2711.

Wittwer, C. T., M. G. Herrmann, et al. (2001). "Real-time multiplex PCR assays." *Methods*25(4): 430-442.

Wittwer, C. T., M. G. Herrmann, et al. (1997). "Continuous fluorescence monitoring of rapid cycle DNA amplification." *Biotechniques*22: 130-138.

Wong, M. L. and J. F. Medrano (2005). "Real-time PCR for mRNA quantitation." *Biotechniques*39(1): 75-85.

Worm, J., A. Aggerholm, et al. (2001). "In-tube DNA methylation profiling by fluorescence melting curve analysis." *Clinical chemistry*47(7): 1183 -1189.

Ye, J., E. J. Parra, et al. (2002). "Melting Curve SNP (McSNP) Genotying: a Useful Approach for Diallelic Genotyping in Forensic Science." *Journal of forensic sciences*47(3): 593-600.

Zipper, H., H. Brunner, et al. (2004). "Investigations on DNA intercalation and surface binding by SYBR Green I, its structure determination and methodological implications." *Nucleic Acids Res*32(12): e103.

20

PCR Advances Towards the Identification of Individual and Mixed Populations of Biotechnology Microbes

P. S. Shwed

Biotechnology Laboratory, Environmental Health Sciences and Research Bureau, Environmental and Radiation Health Sciences Directorate, Healthy Environments and Consumer Safety Branch, Health Canada, Canada

1. Introduction

Public health and safety, diagnostics and surveillance are aided by knowledge of the identity and genetic content of biotechnology microbes and their close relatives. Both types of information allow recognition and prediction of virulence and pathogenicity of microbes. PCR has played an important role in enabling the identification of micro-organisms and the distinction of pathogenic from non-pathogenic species, since the technical descriptions in the mid-1980s (Mullis et al., 1986; Mullis and Faloona, 1987). This DNA amplification technology allows the generation of large template quantity, a pre-requisite for cloning and for dideoxy DNA "Sanger" sequencing (Sanger et al., 1977). As such, PCR has been integral in first generation and phylogenetic marker sequencing projects (Bottger, 1989).

During the last decade, PCR has remained a cornerstone in microbial genetic characterization. Marker sequencing remains a component of the polyphasic characterization of microbial genomes in which genetic, morphological and biochemical data are reconciled. At the same time, great progress has been made in single cell microbial genetics and PCR miniaturization has been implemented in second generation sequencing platforms (Metzker, 2010). Collectively, these developments have resulted in increased numbers of whole genome sequences from individual microbes of "unculturable" microorganisms and outbreak strains such as Shiga toxin-producing *E. coli* strain O104:H4 detected in Europe during 2011 (Mellmann et al., 2011). High throughput sequencing has allowed for insights into natural and human environments and their mixed bacterial populations (Hamady and Knight, 2009; Mardis, 2011; Sapkota et al., 2010).

This chapter serves to highlight PCR advances that have enabled microbial identification during the last decade. At the level of single species, identifications can involve phylogenetic marker sequencing, or whole genome sequencing from individual cells or cultures. Mixed microbial populations, may be sorted, individually identified by sequencing or collectively sequenced using high throughput platforms. The potential and challenges of these new platforms, as well as their applications towards novel microbial strains that will be produced by synthetic biology approaches, will be discussed.

2. Current challenges in microbial identification

Collectively, microbes occupy a vast range of ecological niches and feature intrinsic diverse metabolic potential. Microbial biotechnology has enabled the screening and enhancement of strains for commercial applications such as: preservation and harvest of natural resources (bio-pesticides and bio-mining of metals), environmental remediation (improved soil/air/water quality) and applications for sustainable development. Often, biotechnology microbes are used as single species, while other commercial products involve mixtures of a few or many different species and strains.

Bacterial strains, that feature desirable phenotypic traits, have been traditionally isolated by high-throughput screening, or strains have been improved by random mutagenesis and screening. Currently, consensus identification and classification of bacterial strains is carried out by a polyphasic approach. Phenotypic data (biochemical tests, fatty acid composition), genotypic data and phylogenetic information, that includes genetic information, derived from PCR amplification of marker genes, are reconciled (Vandamme et al., 1996).

Discrimination of beneficial and harmful species is challenging in a number of genera that contain closely related species: *Burkholderia, Bacillus, Acinetobacter and Pseudomonas.* For example, in the *Burkholderia* genus, *B. cepacia* is a non-pathogenic soil bacterium that is being developed for the application of phytoremediation (Barac et al., 2004) and clinically, *B. cepacia* bacteria have been associated with infections and cystic fibrosis, as reviewed in (Coenye and Vandamme, 2003). Another prime example concerns the *Bacillus cereus sensu lato* family of bacteria. This group comprises the *Bacillus cereus* species *sensu stricto, B. anthracis,* strains and subspecies of *B. thuringiensis*, *B. mycoides*, and *B. weihenstephanensis*. Most *Bacillus cereus* organisms are common soil bacteria that are pathogenic to insects and invertebrates. Some species may cause contamination problems in the dairy industry and paper mills and may also be a causative agent of food poisoning. Select strains of *Bacillus anthracis* and a few *Bacillus cereus sensu stricto* strains are the only ones reported to cause fatal pulmonary or intestinal infections (Dixon et al., 2000).

Bacterial mixtures, that have been created or isolated in order to carry out a function, pose technical challenges to polyphasic characterization. Phenotypic data is typically derived for individual microbial isolates. However, current culture techniques cannot support a substantial fraction of microbial species (Handelsman, 2004) and there is risk of bias towards culturable species. In these cases, molecular methods that directly acquire genetic information may remove this hindrance.

2.1 PCR towards microbial identification

2.1.1 Phylogenetic marker amplification and analysis

Of all global markers, small subunit ribosomal RNA (16S rRNA) encoding genes are the best characterized genes for microbial systematics. The 16S rRNA gene is ubiquitous, highly conserved, but possesses enough variability to allow taxa specific discrimination. The gene is composed of nine hypervariable regions separated by conserved regions and sequences are available for numerous organisms via public databases such as NCBI, the Ribosomal Database project (Cole et al., 2009) and Greengenes (Desantis et al., 2006). The nine different variable 16S rRNA regions are flanked by conserved nucleotide stretches in bacteria (Neefs et al., 1993) and these could be used as targets for PCR primers with near-universal specificity.

It was during the mid-1980's that PCR first enabled molecular microbial ecology studies involving the 16S rRNA gene. Pace and colleagues first amplified 16S rRNA from bulk nucleic acid extractions using nearly "universal" primers, in order to sequence, classify and compare these to phylogenetic trees (Pace, 1997) (Lane et al., 1985) (Woese, 1987). At this time it was observed that not all environmental microorganisms were capable of colony formation and that by sequencing cloned ribosomal DNA, new microbial species could be revealed (Stahl et al., 1984),(Stahl et al., 1985).

Over the last few decades, a large number of primer sequences have been designed for amplification and sequencing of 16S RNA genes, as reviewed in (Baker et al., 2003). There are a number of databases available for the primer sequences. Some of these primers have been designed as taxa specific, whereas others have been designed to amplify all prokaryotic rRNA genes and are referred to as "universal".

16S rRNA sequences may offer limited taxonomic resolution, particularly for genera that feature close phylogenetic relationships. *B. cepacia* complex reference strains feature high similarity values (above 98%) which reflects a close phylogenetic relationship (Coenye and Vandamme, 2003). Also, up to 2% intraspecies diversity has been observed in *B. cepacia* rRNA sequences and they cannot be identified at the species level by simple comparison of 16S rRNA sequences. Similarly, for the *B. cereus* group, there is also insufficient divergence in 16S rRNA to allow for resolution of strains and species (Bavykin et al., 2004). In these cases, other global markers have been explored for strain discrimination such as the genes that encode: RNA polymerase subunits, DNA gyrases, heat shock and recA proteins and hisA. The strong functional and structural constraints for these gene products, limits the number of mutations that can occur in the genes and renders them useful as markers for relatedness.

Identification of distinct strains of a prokaryotic species can take place by multi-locus sequence typing, in which sequence mismatches in a small number of house keeping genes are analyzed (as reviewed in (Maiden, 2006)). In the case of prokaryotic identification of closely related species, a similar strategy designated multi-locus sequence analysis has been used for several studies and involves a two step process: rRNA sequencing in order to assign an unknown strain to a group (either genus or family), that in turn defines the particular genes and primers to be used for analysis. This two-tiered approach has allowed discrimination of *Burkholderia* strains and those of the *Bacillus cereus* group (reviewed in (Gevers et al., 2005)).

2.1.2 PCR as a component of genomic methods

During the last decade, various applications of DNA microarrays have been used to assess the risk of a particular microbe by enabling detection and/or identification at the species, subspecies or strain level, or presence of virulence genes (reviewed in (Shwed et al., 2007)). However, DNA amplification is rarely a technical component of these studies. However, as will be described in section 3.1, novel PCR amplification strategies are a component of the workflow for high throughput sequencing platforms.

3. Miniaturization of PCR

Arguably, the major PCR advancement of the last decade has been the development of miniaturized and parallelized platforms. Whereas previously PCR reactions were typically carried out at the microlitre scale, new configurations have enabled femtolitre scale reactions. In turn, higher throughput and cost efficiencies have been achieved.

One miniaturization has been achieved by reaction entrapment in thermodynamically stable "water in oil" nanoreactor microemulsion systems, such as reverse micelles, as described for enzymatic reactions (Klyachko and Levashov, 2003). These emulsions are easily prepared and stable under a wide variety of temperatures, pH and salt concentrations. The smallest droplets rival the scale of bacteria with diameters of less than one micrometre with volumes in the femtolitre scale.

Emulsion PCR was first reported for the directed evolution of heat-stable, heparin insensitive variants of *Taq* DNA polymerase (Ghadessy et al., 2001). The concept of emulsion PCR was to disperse template DNA into a water in oil emulsion such that most droplets contain a single template and only a few droplets contain more than one template. Amplification was carried out within the drops by PCR, so that each droplet generated an amplified number of clonal copies.

3.1 Convergence of miniaturized PCR with other technologies

During the last decade, advancements have been made in the engineering of microfluidic scale devices that integrate multiple analytical steps into "laboratory on chip" systems (as reviewed in (Liu and Mathies, 2009)). These devices allow the generation and manipulation of aqueous microdroplets at high rates and with high fidelity manipulation in microfluidic channels. PCR-based genetic analysis and sequencing can now be carried out at the picolitre to nanolitre volume scale, with the advantages of decreased thermal cycling times and reagent consumption along with increased throughput.

Microfluidic droplet PCR has been reported to allow 1.5 million parallel amplifications for target enrichment of loci in the human genome (Tewhey et al., 2009). In this instance, microfluidic chips were designed to merge 20 picolitre droplets that contain about 3 picograms of biotinylated fragments of template DNA (2-4 kb) with droplets that contain a pair of PCR primers that amplify specific sequences. This platform allowed a yield of more than one million merged droplets that are subjected to PCR. At the end of the amplification reaction, the emulsion is broken. After centrifugation, the aqueous phase, that contains the PCR products from all the droplets, is subjected to a second generation sequencing strategy.

During the last decade, several commercial second-generation sequencing platforms have been developed and these feature cyclic array sequencing strategies, involving new variations of PCR. In both cases, amplification of densely arrayed amplicons is achieved, in order to serve as features for *in situ* sequencing and imaging-based sequence by synthesis data collection (more detailed descriptions of second generation sequencing platforms are reviewed in (Shendure et al., 2011)). Common to all strategies, the first step is the *in vitro* generation of a shot gun genomic library, by the random fragmentation of DNA and the ligation of universal adaptor sequences. Afterwards, *in vitro* clonal amplification is carried out by one of two principal types of PCR, which generate template for sequencing. Table 1 shows how various commercial platforms use PCR to derive features that are sequenced.

Emulsion PCR is carried out as described above (section 3.0 and shown in Fig.1 A, B), with the exception that paramagnetic beads that are bound to one of the PCR primers on their surface, are used (Dressman et al., 2003). These beads allow the solid-phase capture of clonally amplified PCR amplicons from each emulsion PCR compartment. For some commercial pyrosequencing platforms, beads are then deposited on microfabricated arrays of picolitre scale wells to allow immobilization and *in situ* pyrosequencing.

Bridge PCR (Adessi et al., 2000; Fedurco et al., 2006) involves the use of spatially distributed oligonucleotides that are covalently attached to a support (shown in Fig. 1 C,D). A DNA library is hybridized as single stranded DNA to the support. Immobilized copies of the library are synthesized by extension from the immobilized primers. After denaturation, the template copies are able to loop and hybridize to an adjacent oligonucleotide on the support. Additional copies of the template are synthesized and the process is repeated on each template so that clonal clusters, each with about 2000 molecules are generated.

Instrument	PCR type	Sequence Method	Reference
454	Emulsion	Pyrosequencing	(Margulies et al., 2005)
Illumina	Bridge	Polymerase	(Fedurco et. al. 2006)
SOLiD	Emulsion	Ligase	(Shendure et al., 2005)

Table 1. PCR clonal amplification by second-generation sequencing instrument

4. Emerging PCR applications

4.1 Second generation sequencing from microbial mixtures

In recent years, complex microbial communities, such as those of the human gut intestinal tract, or those associated with biofilm infections, have been analyzed by second generation sequencing of shot gun libraries derived from either metagenomic DNA, or PCR amplified variable 16S regions amplified from metagenomic DNA prepared from a microbial mixture (Arumugam et al., 2011; Dowd et al., 2008).

Second generation platforms allow economies of scale in sequencing. PCR amplified products can be characterized without cloning, which saves time and costs. Also, the estimated costs per megabase of derived sequence are lower for the new platforms compared to first generation sequencing (Shendure et. al. 2011). Lastly, multiplexed runs, derived from 16S rRNA coding sequences from several communities, are feasible by using unique sequence barcodes during amplification (Hamady et al., 2008).

It has been proposed that sequencing of individual variable regions is sufficient for taxonomic differentiation of bacterial mixtures (Liu et al., 2007). The sequence read lengths of second generation platforms are generally short, but several new models have shown greater read lengths (Liu et al., 2008). On the other hand, direct sequencing of metagenomic DNA has been proposed to be less biased than that of PCR amplified DNA, due to lack of 16S primer bias (von Mering et al., 2007).

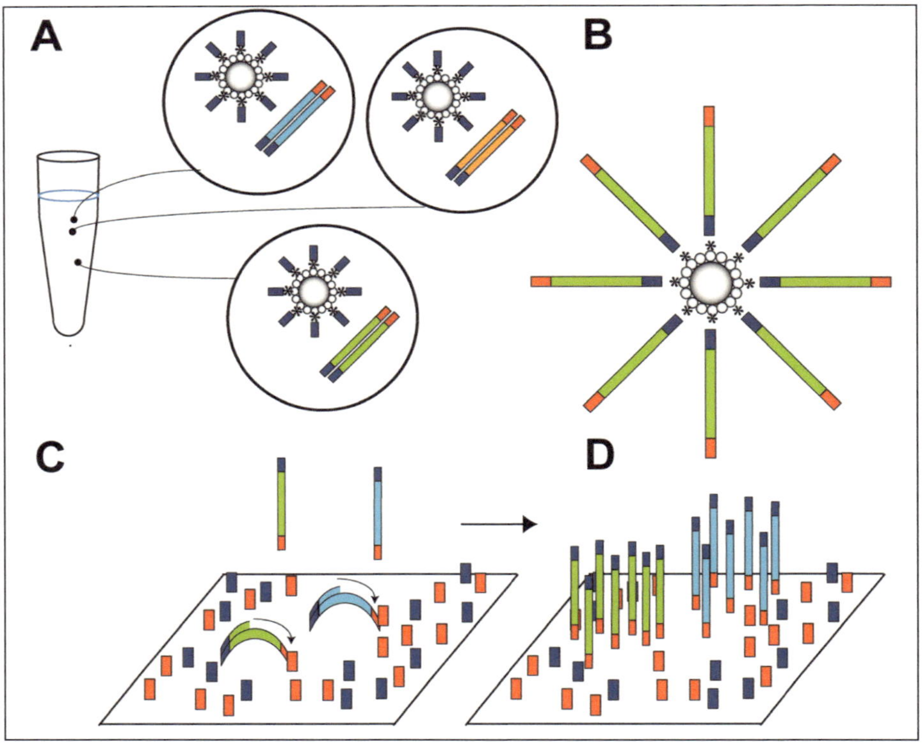

Fig. 1. PCR advancements towards second-generation sequencing
Panels A.B: Emulsion PCR
Panel A) A shot-gun DNA library is ligated to adaptors (blue and red bars), diluted, and PCR amplified in a water in oil emulsion, within aqueous microdroplets. The droplets contain streptavidin coated beads that carry one of the biotinylated PCR primers tethered to beads. Panel B) Where DNA is amplified in the presence of a bead, several thousand copies of the template will be captured.
Panels C,D: Bridge PCR
Panel C) A shot-gun DNA library is ligated to adaptors, made single stranded and hybridized to PCR primers that are immobilized with flexible linkers on a substrate. Bridge amplification occurs when primer extension occurs from immediately adjacent primers. Panel D) Immobilized clusters of about one thousand amplicons are formed after successive cycles of extension and denaturation.

The critical analytical step of taxonomic analyses of microbial diversity analysis is known as binning, where the sequences from a mixture of organisms are assigned phylogenetic groups. However, the outcome of binning results may range from kingdom level to genus level assignment, depending on the quality of data and the read length of data (Yang et al., 2010). One of the binning strategies in use is based on classification of DNA fragments based on sequence homology, using publically available reference databases such as Basic Local Alignment Search Tool (Huson et al., 2007; Meyer et al., 2008). The second strategy involves similarity to protein families and domains, such as in the phylogenetic algorithm CARMA (Krause et al., 2008).

Collectively, these identification approaches are limited by the use of reference databases of known species and genes from readily cultivated microbes. As a consequence, species within a microbial community that lack a reference sequence will remain unidentified.

4.2 PCR analysis of single cells

The analysis of complex mixtures of environmental bacteria will benefit from microfluidic digital PCR analysis that involves single cell sorting from mixtures of bacteria. Single bacterial cells can be isolated by various technologies, including: optical tweezers, micromanipulation, FACS, serial dilutions, or laser capture microdissection. In turn, experimentation that involves retrieving "needles in a haystack", such as searches for microbes featuring particular genes are facilitated by microfluidics technologies (Baker, 2010).

Characterization of environmental bacteria of the 1 microlitre volume termite hindgut model, exemplify the potential of cell sorting and PCR. This microenvironment contains about 10^6-10^8 microbial cells, comprised of unculturable species not detected in other environments (reviewed in (Hongoh, 2010)).

Otteson et al. (Ottesen et al., 2006), applied a microfluidic digital PCR characterization approach for the termite bacteria. In this study, individual cells were partitioned in a microfluidic array panel and served as templates for the simultaneous amplification of both rRNA and metabolic genes of interest. The digital PCR aspect involved ensuring that the partitioning was into reactions that contained an average of one template (bacterial cell) or less (Sykes et al., 1992). Retrieved PCR products from individual chambers allowed sequence analysis of both genes by standard methods and allowed the determination of new bacterial species that contribute to metabolism. More recently, microfluidic digital PCR was used to associate particular viruses that infect the bacteria of the termite gut, without culturing either the viruses or the hosts (Tadmor et al., 2011). Here, amplification of both rRNA gene and a viral marker gene was carried out from a PCR array panel containing individual microbes.

4.3 Whole genome sequencing from individual cells

Genomic sequences provide the most absolute indication of genetic variation and virulence potential for a bacterial strain. The documentation of the complete nucleic acid sequences of high priority beneficial and detrimental microorganisms in public databases are efforts that can greatly aid the identification of unknown strains. In studies involving closely related bacterial strains, shotgun library sequences can be assembled by mapping the reads to a reference genome.

Direct single bacterial cell genome sequencing can be carried out by multiple displacement amplification, using individually lysed bacteria and the few femtograms of DNA present in bacterial cells in order to generate template for shotgun sequencing. This reaction involves the use of ϕ29 DNA polymerase and random primers to amplify DNA templates under isothermal conditions (Dean et al., 2001).

Genomic sequencing from individual uncultured bacterial cells was first shown by Raghunathan et al., using *E. coli* cells that had been isolated by flow cytometry (Raghunathan et al., 2005). This report illustrated contamination as a technical challenge when working with individual microbial cells. The reaction involves random primers in order to initiate polymerization and this can result in amplification of contaminating DNA. In the case of poorly characterized or novel biotechnology microbes, the non-target DNA could confound conclusions about the target organism. In addition, there are biases introduced by multiple displacement amplification, particularly with the use of small input quantity of DNA. Segments of the chromosomes have been observed to be preferentially amplified. As well, chimeric rearrangements of DNA result from the linking of non-contiguous chromosomal regions (Zhang et al., 2006).

Despite these challenges, there have been recent reports that are more encouraging about the acquisition of finished genomic sequence derived from a single bacterial cell (Woyke et al., 2010). Multiple displacement amplification artifacts have been overcome with new computational algorithms, that can compensate for amplification bias and chimeric sequences, using short sequence reads (Chitsaz et al., 2011).

5. Conclusions and future challenges

The safe use of biotechnology microbes for public health and in the environment requires knowledge of the identity and genetic potential of these organisms. In the first decade of the 21st century, amongst the genetic tools available for genetic characterization, PCR remains a cornerstone. Advances in miniaturization and parallelism of PCR have enhanced throughput and enabled second generation sequencing platforms. These technological advancements have been linked to progress in single cell microbial genomics, whole genome sequencing and the characterization of microbial mixtures. Collectively, these developments have direct implications for the safety assessments that are carried out by industry and governments.

These recent technological advances will allow new human and environmental surveys. As an example, movements of genes amongst microbes by horizontal gene transfer mechanisms may be tracked. Environmental surveys of the movements of particular nucleotide sequences are now possible by metagenomic methods. Culture-independent methodology for genetic analysis will allow greater throughput. However, at present, computational hurdles remain for the wide-spread implementation of such technology.

Miniaturization has been a hallmark of progress in electronics and computing. By this measure, PCR miniaturization that has taken place to date is of relatively low order. At the same time, the complexity of biotechnology microbes developed for commercial applications is increasing. The advances in PCR and genomic technologies must be considered in parallel with the technical advancements that have been made towards the *de novo* construction of synthetic microbes. High throughput, high efficiency microfluidic devices can enable the encapsulation of novel genetic material in abiotic chassis (Szita et al., 2010). PCR and sequencing advancements will remain important for microbial genetic characterization.

6. Acknowledgements

Drs. Guillaume Pelletier and Azam Tayabali are thanked for constructive criticism of the manuscript. Open access charges are supported by the Canadian Regulatory System for Biotechnology Fund.

7. References

Adessi C, Matton G, Ayala G, Turcatti G, Mermod JJ, Mayer P, and Kawashima E. 2000. Solid phase DNA amplification: characterisation of primer attachment and amplification mechanisms. *Nucleic Acids Res.* 28 (20): E87.

Arumugam M, Raes J, Pelletier E, Le PD, Yamada T, Mende DR, Fernandes GR, Tap J, Bruls T, Batto JM, Bertalan M, Borruel N, Casellas F, Fernandez L, Gautier L, Hansen T, Hattori M, Hayashi T, Kleerebezem M, Kurokawa K, Leclerc M, Levenez F, Manichanh C, Nielsen HB, Nielsen T, Pons N, Poulain J, Qin J, Sicheritz-Ponten T, Tims S, Torrents D, Ugarte E, Zoetendal EG, Wang J, Guarner F, Pedersen O, de Vos WM, Brunak S, Dore J, Antolin M, Artiguenave F, Blottiere HM, Almeida M, Brechot C, Cara C, Chervaux C, Cultrone A, Delorme C, Denariaz G, Dervyn R, Foerstner KU, Friss C, van de GM, Guedon E, Haimet F, Huber W, van Hylckama-Vlieg J, Jamet A, Juste C, Kaci G, Knol J, Lakhdari O, Layec S, Le RK, Maguin E, Merieux A, Melo MR, M'rini C, Muller J, Oozeer R, Parkhill J, Renault P, Rescigno M, Sanchez N, Sunagawa S, Torrejon A, Turner K, Vandemeulebrouck G, Varela E, Winogradsky Y, Zeller G, Weissenbach J, Ehrlich SD, and Bork P. 2011. Enterotypes of the human gut microbiome. *Nature* 473 (7346): 174-180.

Baker GC, Smith JJ, and Cowan DA. 2003. Review and re-analysis of domain-specific 16S primers. *J Microbiol. Methods* 55 (3): 541-555.

Baker M. 2010. Clever PCR: more genotyping, smaller volumes. *Nat Meth* 7 (5): 351-356.

Barac T, Taghavi S, Borremans B, Provoost A, Oeyen L, Colpaert JV, Vangronsveld J, and van der LD. 2004. Engineered endophytic bacteria improve phytoremediation of water-soluble, volatile, organic pollutants. *Nat Biotechnol.* 22 (5): 583-588.

Bavykin SG, Lysov YP, Zakhariev V, Kelly JJ, Jackman J, Stahl DA, and Cherni A. 2004. Use of 16S rRNA, 23S rRNA, and *gyrB* gene sequence analysis to determine phylogenetic relationships of *Bacillus cereus* group microorganisms. *J Clin. Microbiol.* 42 (8): 3711-3730.

Bottger EC. 1989. Rapid determination of bacterial ribosomal RNA sequences by direct sequencing of enzymatically amplified DNA. *FEMS Microbiol. Lett.* 53 (1-2): 171-176.

Chitsaz H, Yee-Greenbaum JL, Tesler G, Lombardo MJ, Dupont CL, Badger JH, Novotny M, Rusch DB, Fraser LJ, Gormley NA, Schulz-Trieglaff O, Smith GP, Evers DJ, Pevzner PA, and Lasken RS. 2011. Efficient *de novo* assembly of single-cell bacterial genomes from short-read data sets. *Nat. Biotechnol.* 29 (10): 915-921

Coenye T, and Vandamme P. 2003. Diversity and significance of Burkholderia species occupying diverse ecological niches. *Environ. Microbiol.* 5 (9): 719-729.

Cole JR, Wang Q, Cardenas E, Fish J, Chai B, Farris RJ, Kulam-Syed-Mohideen AS, McGarrell DM, Marsh T, Garrity GM, and Tiedje JM. 2009. The Ribosomal Database Project: improved alignments and new tools for rRNA analysis. *Nucleic Acids Res.* 37 (Database issue): D141-D145.

Dean FB, Nelson JR, Giesler TL, and Lasken RS. 2001. Rapid amplification of plasmid and phage DNA using Phi 29 DNA polymerase and multiply-primed rolling circle amplification. *Genome Res.* 11 (6): 1095-1099.

Desantis TZ, Hugenholtz P, Larsen N, Rojas M, Brodie EL, Keller K, Huber T, Dalevi D, Hu P, and Andersen GL. 2006. Greengenes, a chimera-checked 16S rRNA gene database and workbench compatible with ARB. *Appl. Environ. Microbiol.* 72 (7): 5069-5072.

Dixon TC, Fadl AA, Koehler TM, Swanson JA, and Hanna PC. 2000. Early *Bacillus anthracis*-macrophage interactions: intracellular survival survival and escape. *Cell Microbiol.* 2 (6): 453-463.

Dowd SE, Wolcott RD, Sun Y, McKeehan T, Smith E, and Rhoads D. 2008. Polymicrobial nature of chronic diabetic foot ulcer biofilm infections determined using bacterial tag encoded FLX amplicon pyrosequencing (bTEFAP). *PLoS One.* 3 (10): e3326.

Dressman D, Yan H, Traverso G, Kinzler KW, and Vogelstein B. 2003. Transforming single DNA molecules into fluorescent magnetic particles for detection and enumeration of genetic variations. *Proc Natl Acad Sci U. S. A* 100 (15): 8817-8822.

Fedurco M, Romieu A, Williams S, Lawrence I, and Turcatti G. 2006. BTA, a novel reagent for DNA attachment on glass and efficient generation of solid-phase amplified DNA colonies. *Nucleic Acids Res.* 34 (3): e22.

Gevers D, Cohan FM, Lawrence JG, Spratt BG, Coenye T, Feil EJ, Stackebrandt E, Van de PY, Vandamme P, Thompson FL, and Swings J. 2005. Opinion: Re-evaluating prokaryotic species. *Nat Rev. Microbiol.* 3 (9): 733-739.

Ghadessy FJ, Ong JL, and Holliger P. 2001. Directed evolution of polymerase function by compartmentalized self-replication. *Proc Natl Acad Sci U. S. A* 98 (8): 4552-4557.

Hamady M, and Knight R. 2009. Microbial community profiling for human microbiome projects: Tools, techniques, and challenges. *Genome Res.* 19 (7): 1141-1152.

Hamady M, Walker JJ, Harris JK, Gold NJ, and Knight R. 2008. Error-correcting barcoded primers for pyrosequencing hundreds of samples in multiplex. *Nat Methods* 5 (3): 235-237.

Handelsman J. 2004. Metagenomics: application of genomics to uncultured microorganisms. *Microbiol. Mol Biol. Rev.* 68 (4): 669-685.

Hongoh Y. 2010. Diversity and genomes of uncultured microbial symbionts in the termite gut. *Biosci. Biotechnol. Biochem.* 74 (6): 1145-1151.

Huson DH, Auch AF, Qi J, and Schuster SC. 2007. MEGAN analysis of metagenomic data. *Genome Res.* 17 (3): 377-386.

Klyachko NL, and Levashov AV. 2003. Bioorganic synthesis in reverse micelles and related systems. *Current Opinion in Colloid and Interface Science* 8 (2): 179-186.

Krause L, Diaz NN, Goesmann A, Kelley S, Nattkemper TW, Rohwer F, Edwards RA, and Stoye J. 2008. Phylogenetic classification of short environmental DNA fragments. *Nucleic Acids Res.* 36 (7): 2230-2239.

Lane DJ, Pace B, Olsen GJ, Stahl DA, Sogin ML, and Pace NR. 1985. Rapid determination of 16S ribosomal RNA sequences for phylogenetic analyses. *Proc Natl Acad Sci U. S. A* 82 (20): 6955-6959.

Liu P, and Mathies RA. 2009. Integrated microfluidic systems for high-performance genetic analysis. *Trends in Biotechnology* 27 (10): 572-581.

Liu Z, Lozupone C, Hamady M, Bushman FD, and Knight R. 2007. Short pyrosequencing reads suffice for accurate microbial community analysis. *Nucleic Acids Res.* 35 (18): e120.

Liu Z, DeSantis TZ, Andersen GL, and Knight R. 2008. Accurate taxonomy assignments from 16S rRNA sequences produced by highly parallel pyrosequencers. *Nucleic Acids Research* 36 (18): e120.

Maiden MC. 2006. Multilocus sequence typing of bacteria. *Annu. Rev. Microbiol.* 60: 561-588.

Mardis ER. 2011. A decade's perspective on DNA sequencing technology. *Nature* 470 (7333): 198-203.

Margulies M, Egholm M, Altman WE, Attiya S, Bader JS, Bemben LA, Berka J, Braverman MS, Chen YJ, Chen Z, Dewell SB, Du L, Fierro JM, Gomes XV, Godwin BC, He W, Helgesen S, Ho CH, Irzyk GP, Jando SC, Alenquer ML, Jarvie TP, Jirage KB, Kim JB, Knight JR, Lanza JR, Leamon JH, Lefkowitz SM, Lei M, Li J, Lohman KL, Lu H, Makhijani VB, McDade KE, McKenna MP, Myers EW, Nickerson E, Nobile JR, Plant R, Puc BP, Ronan MT, Roth GT, Sarkis GJ, Simons JF, Simpson JW, Srinivasan M, Tartaro KR, Tomasz A, Vogt KA, Volkmer GA, Wang SH, Wang Y, Weiner MP, Yu P, Begley RF, and Rothberg JM. 2005. Genome sequencing in microfabricated high-density picolitre reactors. *Nature* 437 (7057): 376-380.

Mellmann A, Harmsen D, Cummings CA, Zentz EB, Leopold SR, Rico A, Prior K, Szczepanowski R, Ji Y, Zhang W, McLaughlin SF, Henkhaus JK, Leopold B, Bielaszewska M, Prager R, Brzoska PM, Moore RL, Guenther S, Rothberg JM, and Karch H. 2011. Prospective genomic characterization of the German enterohemorrhagic *Escherichia coli* O104:H4 outbreak by rapid next generation sequencing technology. *PLoS One.* 6 (7): e22751.

Metzker ML. 2010. Sequencing technologies - the next generation. *Nat. Rev. Genet.* 11 (1): 31-46.

Meyer F, Paarmann D, D'Souza M, Olson R, Glass EM, Kubal M, Paczian T, Rodriguez A, Stevens R, Wilke A, Wilkening J, and Edwards RA. 2008. The metagenomics RAST server - a public resource for the automatic phylogenetic and functional analysis of metagenomes. *BMC. Bioinformatics.* 9: 386.

Mullis K, Faloona F, Scharf S, Saiki R, Horn G, and Erlich H. 1986. Specific enzymatic amplification of DNA *in vitro*: the polymerase chain reaction. *Cold Spring Harb. Symp. Quant. Biol.* 51 Pt 1: 263-273.

Mullis KB, and Faloona FA. 1987. Specific synthesis of DNA *in vitro* via a polymerase-catalyzed chain reaction. *Methods Enzymol.* 155: 335-350.

Neefs JM, Van de PY, De RP, Chapelle S, and De WR. 1993. Compilation of small ribosomal subunit RNA structures. *Nucleic Acids Res.* 21 (13): 3025-3049.

Ottesen EA, Hong JW, Quake SR, and Leadbetter JR. 2006. Microfluidic digital PCR enables multigene analysis of individual environmental bacteria. *Science* 314 (5804): 1464-1467.

Pace NR. 1997. A molecular view of microbial diversity and the biosphere. *Science* 276 (5313): 734-740.

Raghunathan A, Ferguson HR, Jr., Bornarth CJ, Song W, Driscoll M, and Lasken RS. 2005. Genomic DNA amplification from a single bacterium. *Appl. Environ. Microbiol.* 71 (6): 3342-3347.

Sanger F, Nicklen S, and Coulson AR. 1977. DNA sequencing with chain-terminating inhibitors. *Proc Natl Acad Sci U. S. A* 74 (12): 5463-5467.

Sapkota AR, Berger S, and Vogel TM. 2010. Human pathogens abundant in the bacterial metagenome of cigarettes. *Environ. Health Perspect.* 118 (3): 351-356.

Shendure J, Porreca GJ, Reppas NB, Lin X, McCutcheon JP, Rosenbaum AM, Wang MD, Zhang K, Mitra RD, and Church GM. 2005. Accurate multiplex polony sequencing of an evolved bacterial genome. *Science* 309 (5741): 1728-1732.

Shendure JA, Porreca GJ, Church GM, Gardner AF, Hendrickson CL, Kieleczawa J, and Slatko BE. 2011. Overview of DNA Sequencing Strategies. In *Current Protocols in Molecular Biology*, John Wiley & Sons, Inc.

Shwed P.S., Crosthwait J, and Seligy VL. 2007. Comparative genomic and DNA microarray applications in risk assessment of biotechnology microorganisms. In *Recent Advancements in Analytical Biochemistry: Applications in Environmental Toxicology*, eds P Kumarathasan and R Vincent, 137-156. Research Signpost.

Stahl DA, Lane DJ, Olsen GJ, and Pace NR. 1984. Analysis of hydrothermal vent-associated symbionts by ribosomal RNA sequences. *Science* 224 (4647): 409-411.

Stahl DA, Lane DJ, Olsen GJ, and Pace NR. 1985. Characterization of a Yellowstone hot spring microbial community by 5S rRNA sequences. *Appl. Environ. Microbiol.* 49 (6): 1379-1384.

Sykes PJ, Neoh SH, Brisco MJ, Hughes E, Condon J, and Morley AA. 1992. Quantitation of targets for PCR by use of limiting dilution. *Biotechniques* 13 (3): 444-449.

Szita N, Polizzi K, Jaccard N, and Baganz F. 2010. Microfluidic approaches for systems and synthetic biology. *Curr. Opin. Biotechnol.* 21 (4): 517-523.

Tadmor AD, Ottesen EA, Leadbetter JR, and Phillips R. 2011. Probing individual environmental bacteria for viruses by using microfluidic digital PCR. *Science* 333 (6038): 58-62.

Tewhey R, Warner JB, Nakano M, Libby B, Medkova M, David PH, Kotsopoulos SK, Samuels ML, Hutchison JB, Larson JW, Topol EJ, Weiner MP, Harismendy O, Olson J, Link DR, and Frazer KA. 2009. Microdroplet-based PCR enrichment for large-scale targeted sequencing. *Nat Biotechnol.* 27 (11): 1025-1031.

Vandamme P, Pot B, Gillis M, De VP, Kersters K, and Swings J. 1996. Polyphasic taxonomy, a consensus approach to bacterial systematics. *Microbiol. Rev.* 60 (2): 407-438.

von Mering C, Hugenholtz P, Raes J, Tringe SG, Doerks T, Jensen LJ, Ward N, and Bork P. 2007. Quantitative phylogenetic assessment of microbial communities in diverse environments. *Science* 315 (5815): 1126-1130.

Woese CR. 1987. Bacterial evolution. *Microbiol. Rev.* 51 (2): 221-271.

Woyke T, Tighe D, Mavromatis K, Clum A, Copeland A, Schackwitz W, Lapidus A, Wu D, McCutcheon JP, McDonald BR, Moran NA, Bristow J, and Cheng JF. 2010. One bacterial cell, one complete genome. *PLoS One.* 5 (4): e10314.

Yang B, Peng Y, Leung HC, Yiu SM, Chen JC, and Chin FY. 2010. Unsupervised binning of environmental genomic fragments based on an error robust selection of l-mers. *BMC. Bioinformatics.* 11 Suppl 2: S5.

Zhang K, Martiny AC, Reppas NB, Barry KW, Malek J, Chisholm SW, and Church GM. 2006. Sequencing genomes from single cells by polymerase cloning. *Nat Biotechnol.* 24 (6): 680-686.

21

Lack of Evidence for Contribution of eNOS, ACE and AT1R Gene Polymorphisms with Development of Ischemic Stroke in Turkish Subjects in Trakya Region

Tammam Sipahi
Department of Biophysics, Medical Faculty, Trakya University, Edirne Turkey

1. Introduction

Nitric oxide (NO) is produced in the endothelial cells, neurons, glia, and macrophages by the nitric oxide synthase (NOS) isoenzymes. Endothelial nitric oxide synthase (eNOS) is a subgroup of this family of enzymes that catalyze the production of nitric oxide (NO) from L-arginine and oxygen, which causes vascular relaxation (1) by activates guanylate cyclase, which induces smooth muscle relaxation.

The reaction catalyzed by eNOS is:

$$\text{L-arginine} + 3/2\ \text{NADPH} + \text{H}^+ + 2\ \text{O}_2 = \text{citrulline} + \text{nitric oxide} + 3/2\ \text{NADP}^+$$

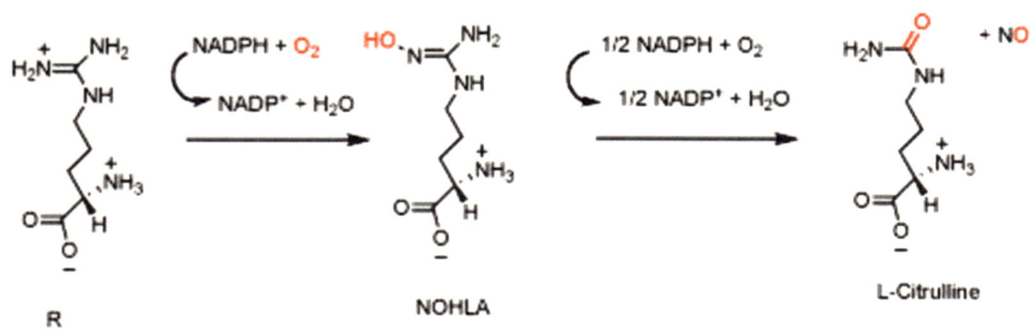

NO can also promote vasorelaxation indirectly by inhibiting the release of renin which converts angiotensinogen to angiotensin I. This is in turn cleaved to form active angiotensin II by Angiotensin-converting enzyme (ACE), the key component of the physiological control of blood pressure in human. Angiotensin II exerts its effects by binding to angiotensin II type 1, 2, 3, and 4 receptors (AT1R, AT2R, AT3R, AT4R). AT1R is the major mediator of physiological effects of angiotensin II. AT1R mediates its action by association with G proteins and followed by vasoconstriction. The activated receptor in turn couples to G proteins and thus activates phospholipases, increases the cytosolic

Ca^{2+} concentrations, which triggers cellular responses such as stimulation of protein kinases. Activated receptor also inhibits adenylyl cyclases and activates various tyrosine kinases (2).

Ischemic stroke, caused either by thrombosis or embolism, is the most frequent disease leading to disability and/or to death (3). The genetic differentiations varying with ethnic properties may be related to the arrangement of the classic and non-classic risk factors for ischemic stroke (4).

During the last two decades, there has been an increasing interest in the study of the different polymorphisms of genes of the renin-angiotensin system (RAS) and its association with the pathogenesis of stroke disease (5, 6). The RAS gene system comprises the angiotensinogen (AGT), renin, angiotensin I, angiotensin I-converting enzyme (ACE), angiotensin II, and angiotensin II receptors (7).

The ACE is a key component of both the RAS and the kinin-kallikrein system. ACE cleaves the carboxy-terminal dipeptide of angiotensin I, releasing the physiologically active octapeptide angiotensin II (8). Angiotensin II is a potent vasoconstrictive molecule that plays a key role in modulating vascular tone. Angiotensin II exerts its effects by binding to the major mediator AT1R. Human AT1R is present predominantly in vascular cells and in both kidney and adrenal gland mediating physiological actions of angiotensin II. AT1R mediates its action by association with G proteins that activate a phosphatidylinositol-calcium second messenger system, followed by vasoconstriction, hypertrophy, or catecholamine liberation at sympathetic nerve endings (9).

Our study aimed to assess the distribution of gene polymorphisms of ACE, AT1R and eNOS gene polymorphisms in ischemic stroke patients compared to healthy controls in the subjects from Trakya region.

The ACE gene maps to chromosome 17 (17q23.3), spans 21 kb, and comprises 26 exons and 25 introns, and is characterized by a polymorphism resulting from the presence (insertion) or absence (deletion) of a 287 base pairs fragment of a repeated Alu sequence at intron 16 hence, the corresponding designation of insertion (I) or deletion (D) of the two resulting alleles (10, 11).

The AT1R gene maps to chromosome 3 (3q21q25), spans 45.123 kb, and comprises 5 exons and 4 introns (12). AT1R entire coding region harbored only on exon 5, and is characterized by a polymorphism resulting from an A/C (adenine/cytosine) transversion located at position 1166 (A1166C polymorphism) in 3′ untranslated region (13).

The eNOS gene is located on chromosome 7q35-36 and comprises 26 exons spanning 21 kb (14). Three classes of genetic polymorphisms in eNOS have been identified: those in intron regions, those in the promoter, and those in exon regions (15).

The variable number of tandem repeat (27 VNTR) polymorphism in intron 4 of the eNOS gene (eNOS 4 a/b), and Guanine (G) to Thymine (T) conversion at nucleotide position 894 in exon 7 causing Glutamic acid (Glu) to Aspartic acid (Asp) change at 298 are two of the most encountered polymorphisms. This polymorphism was shown to affect the response of vascular endothelium and the NO levels of plasma (16, 17).

In view of the aging population stroke is becoming a major problem, it is the most frequent disease leading to disability (3) and estimates forecast a continuing increase in the incidence, prevalence, and mortality of stroke in the next decades.

2. Material and methods

The study included 341 subjects; 197 stroke patients and 144 controls **(Table 1)**. All participants gave informed consent that was approved by the local ethics committee. DNA was isolated from peripheral blood, collected into tubes containing ethylenediamine-tetraacetic acid (EDTA) by eZNA (EaZy Nucleic Acid Isolation) blood DNA kits (Omega Bio-tek, Doraville, USA). eNOS (4 a/b) and ACE (I/D) gene polymorphisms were identified using a polymerase chain reaction (PCR) technique (5, 18). The AT1R (A1166C) and eNOS (Glu298Asp) gene polymorphisms were identified using PCR technique and restriction fragment length polymorphism (RFLP) assay (5, 19).

	Control Group	Stroke Group	p
Hypertension (%)	61.3	83.1	<0.001
Current smoker (%)	3.6	28.3	<0.001
Diabetes mellitus (%)	17.6	33.8	0.001
Family history of stroke (%)	17.6	33.0	0.002
Age (years)	63.0 (17.0)	69.0 (14.0)	<0.001
SBP (mmHg)	120.0 (20.0)	140.0 (40.0)	<0.001
DBP (mmHg)	70.0 (10.0)	80.0 (20.0)	<0.001
FBG (mg/dl)	89.5 (18.3)	105.5 (41.0)	<0.001
TG (mg/dl)	117.5 (93.5)	145.0 (105.0)	0.008
TC (mg/dl)	189.0 (44.0)	190.0 (52.0)	NS
HDL-C (mg/dl)	39.0 (19.5)	38.5 (14.0)	NS
LDL-C (mg/dl)	120.5 (35.0)	124.0 (41.5)	NS

SBP/DBP; Systolic/Diastolic blood pressure, FBG; Fasting blood glucose, TG; Triglycerides, TC; Total cholesterol, HDL-C/LDL-C; High/Low density lipoprotein cholesterol, NS: Non-significant.

Table 1. Demographic and clinical characteristics of the control and stroke groups

PCR technique, developed in 1983 by Kary Mullis, is an in vitro indispensable scientific technique used in medical genetics and hereditary disorders researches to amplify a single (or a few copies) of a piece of DNA to generating millions of copies of a particular DNA sequence (20).

The method relies on thermal cycling, consisting of steps of thermal cycling which can be accomplished automatically with the DNA thermal cycler. First step is DNA denaturation. DNA denaturation is necessary first to physically separate the two strands in a DNA double helix at a high temperature in a process called DNA melting. The four bases found in DNA are adenine (A), cytosine (C), guanine (G) and thymine (T). A base on one strand normally binds only to T on the other strand, and C base on one strand normally binds only to G on the other strand. The two types of base pairs form different numbers of hydrogen bonds, AT forming two hydrogen bonds, and GC forming three hydrogen bonds. To separate the tow strands of DNA, typical strand separation temperatures (T_{ss}) are 95°C for 30 seconds, or 97°C for 15 seconds (21). For G and C rich region higher temperature may be appropriate (21). The second step is primer annealing. Primers contain sequences complementary to the target region of the DNA template. Primer annealing is required for initiation of DNA synthesis at a lower temperature. A temperature of 55°C is a starting degree for 20 base primers with equal GC/AT content (22). Annealing temperatures in the range of 55°C to 72°C generally yield the best results and occurs in a few seconds (21). The third step is primer extension. Primer extension depends upon the length of the target sequence. Extension at 72°C for fragments shorter than 500 base takes only 20 seconds, and fragments up to 1.2 kilo base 40 seconds is sufficient (23).

In the PCR a thermostable Taq DNA polymerase, an enzyme originally isolated from the bacterium *Thermus aquaticus,* are used. The half life of Taq DNA polymerase activity is larger than 2 hours at 92.5°C, 40 minutes at 95°C, and 5 minutes at 97.5°C (21). This DNA polymerase enzymatically assembles a new DNA strand from deoxynucleotide triphosphates (dNTPs), by using separated single-stranded DNA as a template and DNA primers. Because the primer extension products synthesized in one cycle can serve as a template in the next, the DNA template is exponentially amplified. Thus, 20 cycles of PCR yields about a million - fold (2^{20}) amplification (22). Since strand dissociation temperatures, primer annealing, product specificity, and Taq DNA polymerase activity affected by magnesium concentration, the magnesium ion concentration was optimized for all gene amplifications in the study. Also, a recommended buffer for PCR is 10 to 50 mM Tris-HCl pH 8.3, up to 50 mM KCl, and up to 0.1% detergents such as Tween 20 must be included. The PCR products of a particular segment of DNA in an ethidium bromide stained agarose gel visualized by UV transillumination. The minimum amount which can be detected by UV transillumination is larger than 10 ng DNA.

Restriction endonucleases are a set of enzymes expressed in bacteria against foreign DNA. Restriction enzymes cut or cleave double stranded DNA at specific recognition base sequences. In 1970 Smith H. et al identified the first restriction enzyme Hind II. Over 3000 of restriction enzymes have been isolated from different bacterial species (24, 25). Restriction enzymes can be used to distinguish single base changes in DNA (26). This method can be used to genotype a DNA sample without the need for expensive gene sequencing. The sample is first digested with the restriction enzyme to generate DNA fragments, and then the different sized fragments separated by gel electrophoresis. The choice of a restriction enzyme for PCR product is dictated by the product itself. All restriction enzymes require Mg^{2+} ions as a cofactor and 37°C is optimal for most of them to works. The recommended units and digestion buffer for 100% digestion with restriction enzymes is 10-20 units for 0.1

to 0.5 µg of PCR products and 10 mM Tris-HCl (pH 7.5 to 8.5), 10 mM $MgCl_2$, 100 mM KCl and 0.1 mg/mL BSA.

A genomic DNA were amplified by PCR technique in a total 25 µL PCR mixture containing 200 ng of DNA, deoxynucleotide triphosphates (0.2 mM of each), 0.5 nmol of sense and anti-sense oligonucleotide primers, 1X Taq buffer and 1.25 U of Taq DNA polymerase. eNOS (4a/b), ACE (I/D) and eNOS (Glu298Asp) gene polymorphism reactions were contained 2.5 mM $MgCl_2$ whereas AT1R (A1166C) gene polymorphism reaction were contained 1.5 mM $MgCl_2$. All reagents for PCR amplification and gel electrophoresis were purchased from Fermentas Life Sciences (ELİPS, Istanbul, Turkey). All other chemicals were from Sigma and Merck (BO&GA, Istanbul, Turkey) and of the highest purity available. DNA amplifications were performed with a Techne (TechGene) DNA Thermal Cycler.

3. ACE I/D gene polymorphism (rs 4646994)

The PCR primers with the sequences reported by Rigat B. et al. (27) were used. Sense and anti-sense primers were; 5′-CTGGAGACCACTCCCATCCTTTCT-3′ and 5′-GATGTGGCCATCACATTCGTCAGAT-3′, respectively. Normally the sense primer in Rigat et al. didn't contain (G), so our PCR products also didn't contain (G), and the anti-sense primer in Rigat et al. didn't contain (G); which is normally must be included, but it was instead of (G) contained (A). Figure 1 shows the sequencing of the region which contains ACE (D) polymorphism.

PCR Conditions

94°C	5 min	
94°C	**1 min**	
58°C	**1 min**	**30 Cycles**
72°C	**1 min**	
72°C	7 min	

The expected insertion (I) and deletion (D) alleles were visualized after electrophoresis on a 2% agarose gel and ethidium bromide staining under UV light transillumination **(Fig. 2)**. Preferential amplification of the D allele in the heterozygotes has led to their mistyping as DD homozygotes (28). To exclude this possibility, all DD homozygotes were retyped using I

CTGGAGA(G)CCACTCCCATCCTTTCTCCCATTTCTCTAGACCTGCTGCCTATACAG

TCACTTTTATGTGGTTTCGCCAATTTTATTCCAGCTCTGAAATTCTCTGAGCTCCCC

TTACAAGCAGAGGTGAGCTAAGGGCTGGAGCTCAAGGCATTCAAACCCCTACCA

G***ATCTGACGAATGTGATGGCCAC(G→A)TC***

Fig. 1. The sequencing of the region which contains ACE (D) polymorphism. Italic and bold letters were used for the primer sequences.

allele specific sense primer 5′-TTTGAGACGGAGTCTCGCTC-3′ and anti-sense primer, also reported by Rigat B et al. (27) were used. Amplification was performed with a DNA Thermal Cycler with 3 min of denaturation at 93°C, followed by 30 cycles with 1 min of denaturation at 93°C, annealing for 1.5 min at 68°C, and extension for 2 min at 72°C, followed by 3 min of extension at 72°C. When a DD sample amplified using the I-specific primer, it was retyped ID.

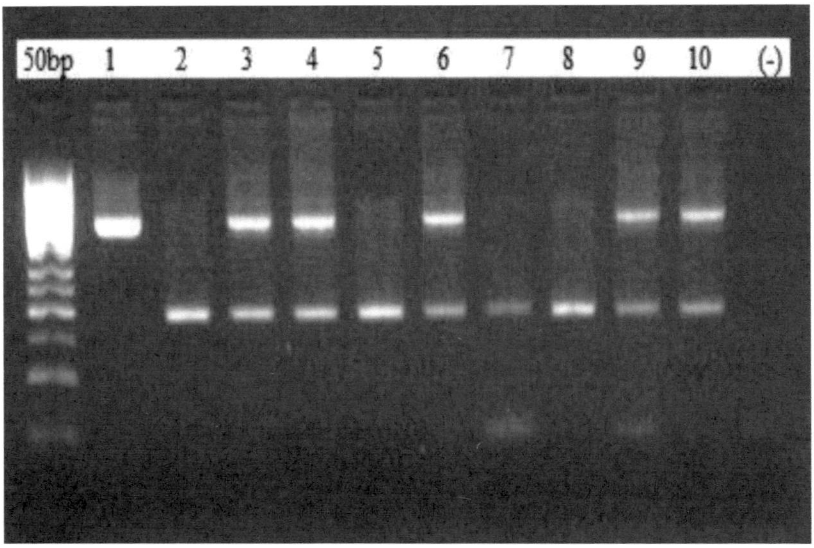

Fig. 2. PCR products of ACE gene I/D polymorphism. The DD (190 bp; lane 2, 5, 7, and 8), the ID (190 bp, and 490 bp, lane 3, 4, 6, 9, and 10) and the II (490 bp, lane 1), 50 bp is a size marker, (-) is a negatif control.

4. eNOS 4 a/b (27 VNTRs) gene polymorphism

The PCR primers with the sequences reported by Wang et al. (29) were used. Sense and anti-sense primers were; 5′-AGGCCCTATGGTAGTGCCTT-3′ and 5′-TCTCTTAGTGCTGTGGTCAC-3′, respectively. Figure 3 shows the sequencing of the region which contains eNOS 4 a/b (27 VNTRs) polymorphism.

PCR Conditions

94°C	1 min	
95°C	**25 sec**	**38 Cycles**
56°C	**35 sec**	
72°C	**40 sec**	
72°C	5 min	

The PCR products were electrophorized on 2.5% agarose gels, stained with ethidium bromide, and checked under UV light transillumination **(Fig. 4)**.

*AGGCCCTATGGTAGTGCCTT*GGCTGGAGGAGGGGAAA**GAAGTCTAGACCTGCTG**
CAGGGGTGAGGAAGTCTAGACCTGCTGCAGGGGTGAGGAAGTCTAGACCTGCTG
CAGGGGTGAGGAAGTCTAGACCTGCTGCGGGGGTGAGGAAGTCTAGACCTGCTG
CGGGGGTGAGGACAGCTGAGCGGAGCTTCCCTGGGCGGTGCTGTCAGTAGCAGG
AGCAGCCTCCTGGAAAAGCCCTGGCTGCTGCTTCTCCCCCAAGAGAGAAGGCTTC
TCCCGCCAGGCCAGTCCAGTGCAGCCCCTCACCCACACCCACTGCTACCCCAGTT
CCCCTGCTTCGGCCCGCACCCTCCCTCACACCCCAGCCCACAGACTCGGGGCTGG
CCTTAGTTACTGGAACGCCT*GTGACCACAGCACTAAGAGA*

Fig. 3. The sequencing of the region which contains eNOS 4 a/b (27 VNTRs) polymorphism. Italic letters were used for the primer sequences and bold letters were used for 27-bp repeats which are deleted in the VNTR 4a polymorphism.

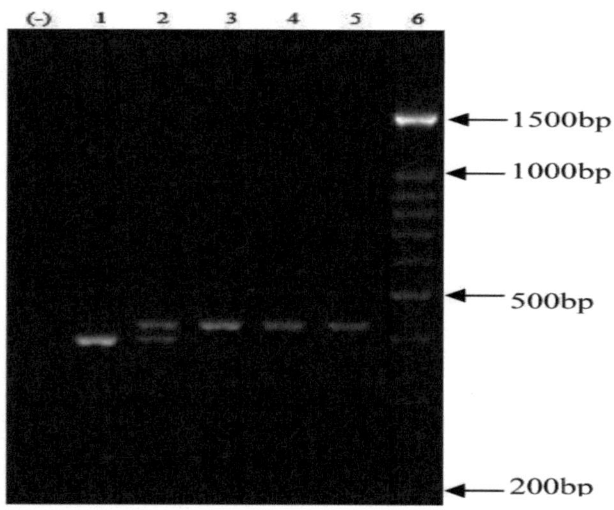

Fig. 4. PCR products of eNOS VNTR gene polymorphism. The aa genotype (394 bp; lane 1), the ab genotype (394 bp and 421 bp; lane 2), and the bb genotype (421 bp; samples 3, 4, and 5). Lane (-) is a negative control, and 6 is a size marker (O'RangeRuler 100bp DNA Ladder).

5. AT1R A1166C gene polymorphism (rs 5186)

AT1R A1166C gene polymorphism was identified with PCR technique followed by RFLP with the restriction enzyme HaeIII (30).

PCR primers were generated to amplify the 255 bp fragment encompassing the A1166C variant (sense and anti-sense primers were 5′-GCAGCACTTCACTACCAAATGGGC-3′ and 5′-CAGGACAAAAGCAGGCTAGGGAGA -3′, respectively) in a 25 μL PCR mixture. Figure 5 shows the sequencing of the region which contains AT1R A1166C gene polymorphism. The sense primer contains one mismatch (A→G) which was required for restriction site.

PCR Conditions

94°C	5 min	
94°C	**1 min**	
55°C	**1 min**	**35 Cycles**
72°C	**1 min**	
72°C	7 min	

The PCR products were electrophorized on 2% agarose gels, stained with ethidium bromide, and checked under UV light transillumination.

GCAGCACTTCACTACCAAATG(A→G)GCATTAGCTACTTTTCAGAATTGAAGGAGA
AAATGCATTATGTGGACTGAACCGACTTTTCTAAAGCTCTGAACAAAAGCTTTTC
TTTCCTTTTGCAACAAGACAAAGCAAAGCCACATTTTGCATTAGACAGATGACGG
CTGCTCGAAGAACAATGTCAGAAACTCGATGAATGTGTTGATTTGAGAAATTTTA
CTGACAGAAATGCAA***TCTCCCTAGCCTGCTTTTGTCCTG***

Fig. 5. The sequencing of the region which contains AT1R A1166C polymorphism. Italic and bold letters were used for the primer sequences. The underlined and bold letters represent the restriction site for HaeIII (5′-GG↓CC-3′).

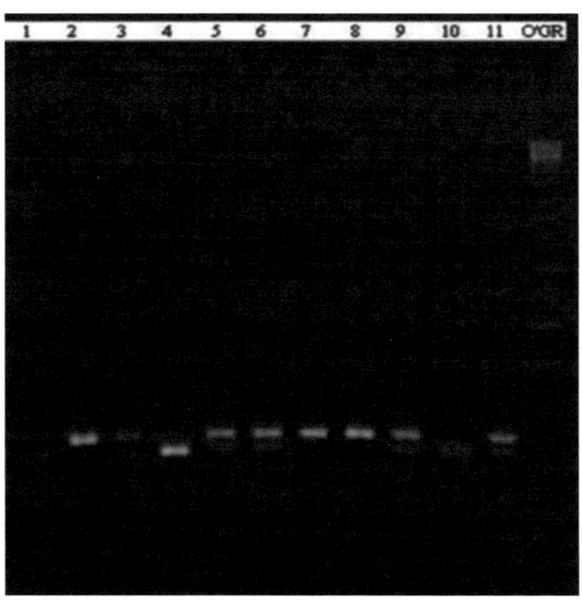

Fig. 6. EtBr stained gel of HaeIII digested PCR products of AT1R A1166C shows the AA genotype (255 bp; lane 2, 3, 7, and 8), the AC genotype (255 bp, 231 bp, and 24 bp; lane 5, 6, 9, and 11), the CC genotype (231 bp, and 24bp; lane 1, 4, and 10), lane O'GR is a size marker (100bp DNA Ladder).

Ten microliters of PCR product were digested with 5 unite of the restriction enzyme HaeIII (Takara Bio Inc, Japan) in 1 X M buffer (10 mM Tris-HCl, pH 7.5, 10 mM $MgCl_2$, 1 mM Dithiothreitol and 50 mM NaCl) for 2 hours at 37°C. When mutant allele (cytosine), digested with HaeIII that yield two fragments, whereas a wild allele (adenine) at nucleotide position 1166, had no cutting site for HaeIII, so that the PCR product was not cleaved into two fragments. The restriction digest products were visualized after electrophoresis on a 2.5% agarose gel and ethidium bromide staining (Fig. 6).

6. eNOS Glu298Asp (rs 1799983) gene polymorphism

Glu298Asp polymorphism of eNOS was identified with PCR technique followed by RFLP with the restriction enzyme BanII (19, 31).

PCR primers were generated to amplify the 248 bp fragment encompassing the eNOS Glu298Asp variant primers 5'-AAGGCAGGAGACAGTGGATGGA-3' (sense) and 5'-CCC AGTCAATCCCTTTGGTGCTCA-3' (anti-sense). Figure 7 shows the sequencing of the region which contains eNOS Glu298Asp gene polymorphism.

PCR conditions

eNOS Glu298Asp;

95°C	5 min	
94°C	**1 min**	**38 Cycles**
59°C	**1 min**	
72°C	**1 min**	
72°C	5 min	

The PCR products were electrophorized on 2% agarose gels, stained with ethidium bromide, and checked under UV light transillumination.

AAGGCAGGAGACAGTGGATGGAGGGGTCCCTGAGGAGGGCATGAGGCTCAGCCC

CAGAACCCCCTCTGGCCCACTCCCCACAGCTCTGCATTCAGCACGGCTGGACCCC

AGGAAACGGTCGCTTCGACGTGCTGCCCCTGCTGCTGCAGGCCCCAGAT**GATCCC**

CCAGAACTCTTCCTTCTGCCCCCCGAGCTGGTCCTTGAGGTGCCCCTGGAGCACCC

CACG***TGAGCACCAAAGGGATTGACTGGG***

Fig. 7. The sequencing of the region which contains eNOS Glu298Asp polymorphism. Italic and bold letters were used for the primer sequences. The underlined and bold letters represent the restriction site for Ban II (5'-G(A/G)GC(T/C)↓C-3').

Ten microliters of PCR product were digested with the restriction enzyme BanII to digest wild allele (guanine). When a guanine is at nucleotide position 894, resulting in a glutamic acid at amino acid position 298, BanII restriction enzyme produces two fragments of 163 and

85 bp. In contrast, when a thymine is at nucleotide position 894 (mutant allele), resulting in an aspartic acid in the amino acid sequence, the Asp298 variant had no cutting site for BanII, so that the 248 bp PCR product was not cleaved into 163 and 85 bp fragments. The restriction digest products were analyzed through electrophoresis on 2.5% agarose gel and ethidium bromide staining (Fig. 8).

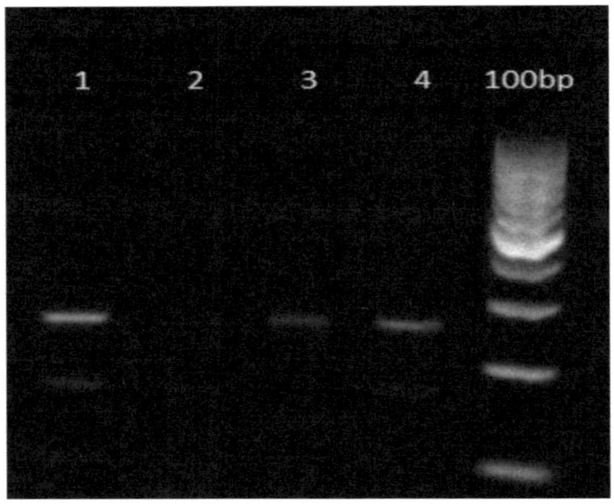

Fig. 8. EtBr stained gel of BanII digested products of eNOS gene Glu298Asp polymorphism. Line 1 and 4; GT alleles (85, 163, and 248 bp), line 2; GG alleles (85 and 163 bp), line 3; TT alleles (248 bp) and line 100bp; Gene Ruler 100 bp DNA Ladder.

7. Results and discussion

Table 2, 3, 4, and 5 shows the distributions of ACE I/D, eNOS (4 a/b), AT1R (A1166C), and eNOS Glu298Asp genotypes, respectively.

Statistical analyses were performed with the SPSS 15.0 software and STATA program. Genotypic distributions were in accordance with Hardy-Weinberg equilibrium in the stroke group as well as in the control group. Several studies have shown differences in the genotypic distributions of these genes while, others have shown no differences between the controls and patients. Our results didn't show any significant difference between the ischemic stroke patients and the controls ($p>0.05$) and suggested the lack of an association between the 4 gene polymorphisms and ischemic stroke (Table 2, 3, 4, and 5). So the 4 gene polymorphisms did not enhance the predictability of stroke.

	DD	ID	II
Controls (%)	34.3	49.7	16.1
Stroke Patients (%)	34.0	50.0	16.0
	Non-Significant	Non-Significant	Non-Significant

Table 2. Distribution of ACE (I/D) genotype frequency in the controls and stroke patients

	aa	ab	bb
Controls (%)	2.8	29.8	67.4
Stroke Patients (%)	2.0	35.0	63.0
	Non-Significant	Non-Significant	Non-Significant

Table 3. Distribution of eNOS (4 a/b) genotype frequency in the controls and stroke patients

	AA	AC	CC
Controls (%)	60.1	35.7	4.2
Stroke Patients (%)	58.0	34.6	7.4
	Non-Significant	Non-Significant	Non-Significant

Table 4. Distribution of AT1R (A1166C) genotype frequency in the controls and stroke patients

	GG	GT	TT
Controls (%)	49.3	45.8	4.9
Stroke Patients (%)	56.3	40.6	3.1
	Non-Significant	Non-Significant	Non-Significant

Table 5. Distribution of eNOS (Glu298Asp) genotype frequency in the controls and stroke patients

In our previous study about potential angiotensinogen (AGT) gene that predispose to hypertension, we failed to detect any relation between T174M and M235T gene polymorphisms of the AGT gene in the RAS and the development of hypertension (32).

Now, we are working on the AGT gene to clarify the role of T174M and M235T gene polymorphisms of the AGT gene in the stroke Turkish patients from Trakya region.

8. Conclusions

In addition to demographic and clinical characteristics, which are important in the developing of ischemic stroke, our data does not suggest that ACE (I/D), AT1R (A1166C), eNOS (4 a/b) and eNOS (Glu298Asp) gene polymorphisms, in contrast to other studies which shows a positive association between this gene polymorphisms and ischemic stroke, are a common cause of ischemic stroke in Turkish patients from Trakya region.

9. References

[1] Arnal JF, Dinh-Xuan AT, Pueyo M et al. Endothelium-derived nitric oxide and vascular physiology and pathology. Cell Mol Life Sci 55: 1078-1087. doi: 10.1007 / s000180050358, (1999).

[2] Matsusaka T, Ichikawa I. Biological functions of angiotensin and its receptors. Annu. Rev. Physiol. 59: 395–412. doi:10.1146/annurev.physiol.59.1.395. PMID 9074770, (1997).

[3] Saver J., Tamburi T. In: Neurogenetics (S.M.Pulst, Ed.), Oxford University Press, New York, 403-432, (2000).

[4] Hassan A, Markus H. S. Genetics and ischaemic stroke. Brain 123:1784-1812. doi:10.1093/brain/123.9.1784, (2000).

[5] Sipahi T., Guldiken B., Guldiken S., Ustundag S., Turgut N., Budak M., Cakina S., Ozkan H., Sener S. The Association of Gene Polymorphisms of the Angiotensin-Converting Enzyme and Angiotensin II Receptor Type 1 with Ischemic Stroke in Turkish subjects of Trakya Region. Trakya Univ. Tip Fak. Derg. (ISI), 1-8 pp., (2009).

[6] Hong S. H., Park H. M., Ahn J. Y., Kim O. J., Hwang T. S., Oh D., et al. ACE I/D polymorphism in Korean patients with ischemic stroke and silent brain infarction. Acta Neurol Scand. 117: 244-9, (2008).

[7] de Gasparo M., Catt K. J., Inagami T., Wright J. W., Unger T. International union of pharmacology. XXIII. The angiotensin II receptors. Pharmacol Rev. 52: 415-72, (2000).

[8] Soubrier F., Hubert C., Testut P., Nadaud S., Alhenc-Gelas F., Corvol P. Molecular biology of the angiotensin I converting enzyme: I. Biochemistry and structure of the gene. J Hypertens. 11: 471-6, (1993).

[9] Kim S., Iwao H. Molecular and cellular mechanisms of angiotensin II-mediated cardiovascular and renal diseases. Pharmacol Rev. 52: 11-34, (2000).

[10] Hubert C., Houot A. M., Corvol P., Soubrier F. Structure of the angiotensin I-converting enzyme gene. Two alternate promoters correspond to evolutionary steps of a duplicated gene. J Biol Chem, 266: 15377-83, (1991).

[11] Mattei M. G., Hubert C., Alhenc-Gelas F., Roeckel N., Corvol P., Soubrier F. Angiotensin-I converting enzyme gene is on chromosome 17. Cytogenet Cell Genet. 51: 1041-5 (1989).

[12] Szpirer C., Rivière M., Szpirer J., Levan G., Guo D. F., Iwai N., et al. Chromosomal assignment of human and rat hypertension candidate genes: type 1 angiotensin II receptor genes and the SA gene. J Hypertens. 11: 919-25, (1993).

[13] Bonnardeaux A., Davies E., Jeunemaitre X., Féry I., Charru A., Clauser E., et al. Angiotensin II type 1 receptor gene polymorphisms in human essential hypertension. Hypertension. 24: 63-9 (1994).

[14] Marsden P. A., Heng H. H. Q., Scherer S. W., Stewart R. J., Hall A. V., Shi X. M., Tsui L. C., Schappert K. T. The Journal of Biological chemistry, 268 (23): 17478-17488, (1993).

[15] Wang X. L., Wang J. Molecular Genetics and Metabolism, 70: 241-251, (2000).

[16] Leeson C. P., Hingorani A. D., Mullen M. J. et al. Glu298Asp endothelial nitric oxide synthase gene polymorphism interacts with environmental and dietary factors to

influence endothelial function. Circ Res 90:1153-1158. doi:10.1161/01.RES.0000020562.07492.D4, (2002).

[17] Metzger I. F., Serto´rio J. T., Tanus-Santos J. E. Modulation of nitric oxide formation by endothelial nitric oxide synthase gene haplotypes. Free Radic Biol Med 43: 987–992. doi:10.1016/j. freeradbiomed.2007.06.012, (2007).

[18] Sipahi, T. Basak A. A., Ozgen Z., Aksoy A., Omurlu I. K., Palabiyik O., Cakina S., Sener S. Lack of Evidence for Contribution of Endothelial Nitric Oxide Synthase Intron 4 VNTR Gene Polymorphisms To Development Of Ischemic Stroke In Turkish Subjects. Biotechnol. & Biotechnol. Eq. (ISI), 1372-1377 pp., (2009).

[19] Guldiken B., Sipahi T., Guldiken S., Ustundag S., Budak M., Turgut N., Ozkan H. Glu298Asp polymorphism of the endothelial nitric oxide synthase gene in Turkish patients with ischemic stroke. Mol. Biol. Rep. (ISI), 539-1543 pp., DOI: 10.1007/s11033-008-9348-7, (2009).

[20] Saiki R. K., Scharf S., Faloona F., Mullis K. B., Horn G. T., Erlich H. A., Arnheim N. Science. 37: 170-172, (1985).

[21] Innis M. A., Gelfand D. H. Sninsky J. J., White T. J. PCR Protocols A Guide to Methods and Applications. Academic Press, (1990).

[22] Henry A. Erlich H. A. PCR Technology Principles and Applications for DNA Amplification. Stockton Press, (1989).

[23] Rolfs A., Schuller I., Finckh U., Weber-Rolfs I. PCR: Clinical Diagnostics and Research. Springer-Verlag Berlin Heidelberg, (1992).

[24] Lorne T. Kirby. DNA Fingerprinting: An Introduction. W.H. Freeman and Company, New York, (1992).

[25] Roberts R. J., Vincze T., Posfai J., Macelis D. Rebase-enzymes and genes for DNA restriction and modification. Nucleic Acids Res 35 (Database issue): D269–70. doi:10.1093/nar/gkl891.PMC 1899104.PMID 17202163.http://www.pubmedcentral.nih.gov/articlerender.fcgi?tool=pmcentrez&artid=1899104, (2007).

[26] Zhang R., Zhu Z., Zhu H., Nguyen T., Yao F., Xia K., Liang D., Liu C. SNP Cutter: a comprehensive tool for SNP PCR–RFLP assay design. Nucleic Acids Res. 33 (Web Server issue): W489–92. doi:10.1093/nar/gki358. PMC 1160119. PMID 15980518. http://www.pubmedcentral.nih.gov/articlerender.fcgi?tool=pmcentrez&artid=1160119(2005).

[27] Rigat B., Hubert C., Corvol P., Soubrier F. PCR detection of the insertion/deletion polymorphism of the human angiotensin converting enzyme gene (DCP1) (dipeptidyl carboxypeptidase 1). Nucleic Acids Res. 20: 1433, (1992).

[28] Shanmugam V., Sell K. W., Saha B. K. Mistyping ACE heterozygotes. PCR Methods Appl. 3: 120-1 (1993).

[29] Wang X. L., Sim A. S., Badenhap R. F., McCredle R. M., Wilcken D. E. A smoking-dependent risk of coronary artery disease associated with a polymorphism of the endothelial nitric oxide synthase gene. Nat. Med. 2: 41-45, (1996).

[30] Berge K. E., Berg K. Polymorphisms at the angiotensinogen (AGT) and angiotensin II type 1 receptor (AT1R) loci and normal blood pressure. Clin Genet. 53: 214-9, (1998).

[31] Shimasaki Y., Hirofumi Y., Michihiro Y., Masafumi N., Kiyotaka K., Hisao O., Eisaku H., Takenobu M., Wasaku K., Yoshihiko S., Yoshihiro M., Yoshihiro O., Kazuwa N.

Association of the Missense Glu298Asp Variant of the Endothelial Nitric Oxide Synthase Gene With Myocardial Infarction. JACC. 31: 1506-10, (1998).

[32] Basak, A. A., Sipahi T., Ustundag S., Ozgen Z., Budak M., Sen S., Sener S. Association of Angiotensinogen T174M and M235T Gene Variants with Development of Hypertension in Turkish Subjects of Trakya Region. Biotechnol. & Biotechnol. Eq., 22, 984-989, (2008).

22

Analysis of Genomic Instability and Tumor-Specific Genetic Alterations by Arbitrarily Primed PCR

Nikola Tanic[1], Jasna Bankovic[1] and Nasta Tanic[2]
[1]*Institute for Biological Research "Sinisa Stankovic", University of Belgrade, Belgrade*
[2]*Institute of nuclear Sciences "Vinca", Belgrade*
Republic of Serbia

1. Introduction

It is now widely accepted that cancer development is a multistage process that results from an accumulation of mutations (Lengauer et al., 1998). Since spontaneous mutation rates in human cells are considerably lower then the large number of mutations observed in cancer cells, cancer cells must be a manifestation of the mutator phenotype. The mutator phenotype, also referred to as genomic instability, designates the increased mutation rate that occurs in neoplastic cells (Loeb, 1991). The induction of the genomic instability phenotype is emerging to be a crucial early event in carcinogenesis that enables an initiated cell to evolve into a cancer cell by achieving a greater proliferative capacity and genetic plasticity, which can overcome host immunological resistance, localized toxic environments and a suboptimal supply of micronutrients (Loeb, 1991; Cahill et al., 1999; Fenech 2002). Two distinct forms of genomic instability have been identified, microsatellite instability (MIN) and chromosomal instability (CIN). They probably encompass most characterized malignances (Lengauer et al., 1998; Breivik & Gaudernack, 1999). Genomic instability is present in all stages of cancer, from precancerous lesions to advanced cancers (Negrini et al., 2010; Markovic et al., 2008)

Measurements of instability have been performed by a variety of techniques, including flow cytometry, comparative genomic hybridization (CGH), allelotyping, and analysis of gene amplification rates (Vogelstein et al., 1989; Kallioniemi et al., 1994; Jass et al., 1994). These approaches, although informative, are generally cumbersome and somewhat impractical for widespread clinical use. Unlike these techniques, DNA fingerprinting methods, RAPD (Random Amplified Polymorphic DNA) and AP-PCR (Arbitrarily Primed Polymerase Chain Reaction) are rapid and simple procedures that examine the whole genome and detect the propensity of a tumor to undergo genomic rearrangements (Peinado et al., 1992; Perucho et al., 1996).

AP-PCR is a PCR-based DNA fingerprinting method that utilizes arbitrarily chosen primers to co-amplify multiple and independent sequences under low stringency conditions during the first cycles. It was first described by Welsh and McClelland (1990), who designed it to amplify multiple DNA fragments from anonymous regions of the genome. Initial cycles of the reaction are performed under low stringency conditions which are achieved with low temperatures during the annealing step of PCR and/or high magnesium concentration in the reaction. Under these conditions the arbitrary primer anneals to the best matches in the template. The priming events during the initial low stringency cycles are arbitrary since they depend on the nucleotide sequence of the PCR primer, which is arbitrarily chosen. Competition between these annealing events results in reproducible and quantitative amplification of many discrete bands. Further amplification of these sequences (discrete bands) under high stringency conditions produces a complex fingerprint which can be visualized by gel electrophoresis. The obtained band pattern is characteristic and representative of the genome used as template.

The large number of bands amplified with a single arbitrary primer generates a complex fingerprint that can be used to detect differences in the arbitrary amplified DNA sequences from two different but closely related genomes, like DNA from normal and cancer cells. Such differences correspond to somatic genetic alterations. In addition, AP-PCR method permits direct cloning and identification of altered variant bands i.e. altered DNA sequences. Therefore, this unbiased methodology allows for molecular karyotyping of somatically acquired genomic abnormalities, comparing related genomes, whereby one is a derivative of the other emerging via undefined and abnormal genomic events. Indeed, AP-PCR has been successfully used as a molecular alternative for cancer cytogenetics since it has proved to be capable of detecting chromosomal gains and losses as well as point mutations associated with carcinogenesis (Perucho et al., 1996; Chariyalertsak et al., 2005). This is based on the following favorable properties of the method: (i) the amplified bands usually originate from single copy sequences rather then from repetitive elements; (ii) there is no apparent bias for the chromosomal origins of the amplified bands, and therefore, fingerprints representative of the full chromosomal complement can be obtained by using a few arbitrary primers; (iii) the amplification is semi-quantitative, that is, the intensities of the amplified bands are almost proportional to the concentration of the corresponding template sequences.

Taking into account the potential and advantages of AP-PCR method, it seems as a reasonable approach to use this method to detect and quantify the level of genomic instability in various cancer samples. Therefore, we applied AP-PCR to measure genomic instability in samples of patients with Non Small Cell Lung Carcinoma (NSCLC) of various stages and grades, samples of patients with Malignant Gliomas of various grades (Anaplastic Astrocytomas and Glioblastomas) and samples of patients with Head and Neck Squamous Cell Carcinoma (HNSCC) and their premalignant lesions leukoplakias. Moreover, we aimed to identify some of these genomic alterations associated with the process of carcinogenesis in these types of tumors.

Here we describe the procedure for analyzing the level of genomic instability and identifying specific genetic alterations that occur during the tumorigenic process by Arbitrarily Primed PCR. This procedure involves the following steps: (i) comparative AP-PCR analysis of matching normal and tumor tissue and determination of the frequency of DNA alterations, a measurement of genomic instability; (ii) correlation between the level of genomic instability and histological grade and stage of each tumor; (iii) isolation and identification of altered amplified bands. Obtained results are presented and discussed in terms of the evolution of these types of tumors.

2. Materials and methods

2.1 Tissue samples and DNA extraction

Paired tumor and normal tissue samples (adjacent normal lung tissue and blood for malignant gliomas, HNSCC and leukoplakias) were analyzed. Specifically, 30 malignant glioma patients who underwent surgical resection at Clinic for Neurosurgery, Clinical Center of Serbia, 30 NSCLC patients who underwent surgery at the Institute for Lung Diseases and Tuberculosis, Clinical Centre of Serbia, 32 leukoplakia patients and 30 HNSCC patients who underwent surgery at the Clinic of Maxillofacial Surgery, School of Dentistry, University of Belgrade. Freshly excised tissue samples were partitioned for histopathology and DNA analyses. The specimens for DNA analyses were frozen in liquid nitrogen until DNA extraction. The samples were collected and used after obtaining informed consents and approval from the Ethics Committee, in accordance with the ethical standards laid down in the 1964 Declaration of Helsinki.

DNA was extracted using the phenol/chloroform/isoamyl alcohol method (Sambrook et al., 1989). The quality of the DNA was verified by electrophoresis on 0.8% agarose gel. The DNA concentration was assessed spectrophotometrically.

2.2 AP-PCR DNA fingerprinting

Genomic instability was determined by comparing the AP-PCR profiles of paired tumor and normal DNA samples of the same patient. Altogether, twenty primers were tested for the ability to generate informative fingerprints that distinguish tumor from normal tissue. Optimization of AP-PCR reactions was done for each primer according to Cobb (1997) and included the search for conditions that would yield profiles of moderate complexity in order to simplify the analysis (McClelland & Welsh, 1994). Normally, optimization of AP-PCR DNA fingerprinting would require each variable to be tested independently. An experiment investigating the effects and interactions of four critical reaction components (dNTPs, $MgCl_2$, primer and DNA), each at three concentrations, would require 81 (i.e., 3^4) separate reactions. However, using modified Taguchi method (Taguchi & Wu, 1980, as cited in Cobb, 1997) only nine reactions are required to perform the same optimization. Here an estimate of the effect of individual components is achieved by looking at the effects that component interactions have on the fingerprint. These interactions are determined by arranging those components that are likely to affect the reactions into an orthogonal array. The product yield for each reaction is used to estimate the effects of individual components on amplification. We varied the PCR components in the following final concentrations: dNTPs (0.2 mM, 0.4 mM, 0.6 mM), $MgCl_2$ (1.5 mM, 2.5 mM, 3.5 mM), primer (1.5 μM, 3.0 μM, 5.0 μM) and DNA (50 ng, 100 ng, 150 ng). DNA concentration did not affect AP-PCR fingerprints and it was used to validate the method. Namely, after optimal reaction conditions were established, each experiment included the analysis of two template concentrations (25 ng and 50 ng in a final volume of 25 μL) for each individual in order to exclude artifacts arising from impurities in the DNA preparations.

Twelve out of twenty primers produced informative profiles differentiating normal from tumor tissue. Primer sequences, AP-PCR conditions and reaction mixtures are given in Table 1.

Primer	Primer sequence	AP-PCR low-stringency conditions	AP-PCR high-stringency conditions	AP-PCR reaction mixture
CCNA1	5′-AAG AGG ACC AGG AGA ATA TCA-3′	95°C 30″ 45°C 2′ 72°C 1′	95°C 30″ 60°C 1′ 72°C 1′	0,2mM each dNTP; 3,5mM $MgCl_2$; 5µM primer; 1U Taq DNA
LRP-A	5′-GCT TCC GAG GTC TCA AAG C-3′	95°C 30″ 40°C 2′ 72°C 1′	95°C 30″ 58°C 1′ 72°C 1′	0,2mM each dNTP; 3,5mM $MgCl_2$; 5µM primer; 1U Taq DNA
MDR-A	5′-GTT CAA ACT TCT GCT CCT GA-3′	95°C 30″ 40°C 2′ 72°C 1′	95°C 30″ 58°C 1′ 72°C 1′	0,4mM each dNTP; 2,5mM $MgCl_2$; 5µM primer; 1U Taq DNA
E8S p53	5′-TAA ATG GGA CAG GTA GGA CC-3′	95°C 30″ 40°C 2′ 72°C 1′	95°C 30″ 58°C 1′ 72°C 1′	0,4mM each dNTP; 2,5mM $MgCl_2$; 5µM primer; 1U Taq DNA
GAPDH-S	5′– CGG AGT CAA CGG ATT TGG TCG TAT– 3′	95°C 30″; 50°C 2′; 72°C 1′	95°C 30″; 70°C 1′; 72°C 1′	0,4mM each dNTP; 2,5mM $MgCl_2$; 5µM primer; 1 U Taq DNA
GAPDH-A	5′–AGC CTT CTC CAT GGTGGT GAA GAC–3′	95°C 30″; 50°C 2′; 72°C 1′	95°C 30″; 72°C 1′; 72°C 1′	0,2 mM each dNTP; 2,5 mM $MgCl_2$; 3 µM primer; 1 U Taq DNA
E5A p53	5′–CAG CCC TGT CGT CTC TCC AG–3′	95°C 30″; 40°C 2′; 72°C 1′	95°C 30″; 55°C 1′; 72°C 1′	0,6 mM each dNTP; 3,5 mM $MgCl_2$; 3 µM primer; 1 U Taq DNA
p53 A	5′–TTG GGC AGT GCT CGC TTA GT–3′	95°C 30″; 40°C 2′; 72°C 1′	95°C 30″; 60°C 1′; 72°C 1′	0,2 mM each dNTP; 3,5 mM $MgCl_2$; 5 µM primer; 1 U Taq DNA
H61-5′	5′-AGG TGG TCA TTG ATG GGG AG-3′	94°C 1′; 45°C 2″, 72°C 2′	94°C 1′; 62°C 1′; 72°C 2′	0.4 mM each dNTPs, 2.5 mM $MgCl_2$ 5 µM primer; 1 U Taq DNA

Table 1. Primer sequences, AP-PCR conditions and reaction mixtures

The reactions consisted of an initial denaturation step (95°C for 5 min), 4 cycles at low-stringency conditions (specific for each primer), 35 cycles at high-stringency conditions (specific for each primer), and a final extension (72°C for 7 min) in a GeneAmp® PCR System 9700 (Applied Biosystems, Foster City, California, USA).

The AP-PCR products were separated on 6 – 8% non-denaturing polyacrylamide (PAA) gels and visualized by silver staining. Silver- staining procedure creates permanent record of the electrophoresis results and includes several steps: fixing, silver impregnation, development and stopping the reaction. In the fixing step, the gel is treated with 1% nitric acid solution to render the macromolecules in the gel insoluble and prevent diffusion during the subsequent staining steps. In the silver impregnation step, soluble silver ion (Ag+) derived from the silver nitrate, 12mM $AgNO_3$ solution, binds to nucleic acid bases fixed in gel. Generally, DNA bases promotes the reduction of silver ion to metallic silver (Ag0), which is insoluble and visible, allowing nucleic acid-containing bands to be seen. In order to prevent reduction of silver ion to metallic silver before the end of silver impregnation, this step is often performed in mildly acidic acid conditions. During the development step, formaldehyde reduces silver ions to metallic silver in process that only proceed at high pH, approximately 12. For that reason, sodium carbonate is included as one of the main component that render development solution alkaline. Stopping reaction step imply prevention of any further silver ion reduction by soaking the gels in the 10 % acetic acid solution. Finally, it should be emphasized that water washes are also included between some of the above mentioned steps in the silver staining procedure (detailed procedure is given in Table 2).

Step	*Solution*	*Time*
Fixation Pretreatment	10% Ethanol	10 minutes
Fixation	1% Nitric Acid Solution	3 minutes
Water Washing	Destiled H2O	2 x 1 minute
Silver Impregnation	12 mM Silver Nitrate Solution	30 minutes
Water Washing	Destiled H2O	3 x 1 minutes
Developing - Reduction	0.28 M Sodium Carbonat with 0.019 % Formaldehyde	Until desired images appear
Stopping Reduction	10 % Acetic Acid	5 - 10 minutes

Table 2. In-house procedure for silver- staining of PAA gels.

Gel images were acquired with the Multi-Analyst/PC Software Image Analysis System (Bio Rad Gel Doc 1000). Digitized images were loaded into the specialized public software Image J (National Institute of Health, USA, www.rsb.info.nih.gov/ij) and analyzed by the image enhancement function 'adapthisteq'. This function performs contrast-limited adaptive histogram equalization on small regions of the image, called tiles. Contrast of each tile is enhanced so that the histogram of the output region approximately matches a specified histogram. After equalization, adapthisteq combines neighboring tiles using bilinear interpolation to eliminate artificially induced boundaries.

2.2.1 Reproducibility

The problem of reproducibility of AP-PCR has been a matter of concern for quite some time (Meunier and Grimont, 1993; McClelland and Welsh, 1994). In our study, reproducibility was verified by at least three independent reactions and a reaction with a two-fold higher template concentration. Occasional irreproducibility was found to be due to template quality, where additional round of purification solved the problem. Template carry-over was routinely monitored by systematic incorporation of "no-template reaction" in each set of experiments. Day to day variation was found only in respect of band intensities. This variability was in the range of less than 10% (± 5%) as estimated by integration of densitometric scans. Interlab variation was not assessed but we presume that it does not affect the interpretation of data from this report.

2.3 Isolation, cloning and DNA sequencing of variant bands obtained by AP-PCR

Selected variant DNA bands, bands with altered mobility, were further characterized. The PCR amplicons resolved on the silver stained gels were gently removed with a hypodermic 22-gauge needle pre-wetted with the PCR master mix solution. The needle was dipped in the PCR master mix for 2 min and then discarded. The PCR products were reamplified with the same primers used for AP-PCR reactions at high-stringency conditions specific for each particular primer. The reamplified material was administrated on 1.5% agarose gels, purified using DNA Extraction Kit (Fermentas Life Sciences, Lithuania) and cloned with GeneJetTM PCR Cloning Kit (Fermentas Life Sciences, Lithuania) according to manufacturers' instructions. Plasmids were purified using GeneJetTM Plasmid Miniprep Kit (Fermentas Life Sciences, Lithuania).

Cloning process consisted of setting up the blunting and ligation reactions. Blunting reaction allows the conversion of PCR products generated with non-proofreading Taq DNA polymerase to DNA fragments with blunt ends using thermostable DNA Blunting Enzyme provided with the kit. The reaction consists of 10 µL of 2x Reaction Buffer, 2 µL of non-purified PCR product, 5 µL of nuclease free water and 1 µL of DNA Blunting Enzyme in 18 µL reaction mixture. The resulting blunt-ended DNA can be ligated efficiently into a vector, pJET1.2/blunt, using the included DNA Ligation Kit Solutions: 1 µL of pJET1.2/blunt Cloning Vector (50ng/ µl) and 1 µL of T4 DNA Ligase (5u/µl). The vector contains a lethal restriction enzyme gene that is disrupted by ligation of a DNA insert into the cloning site. As a result, only bacterial cells with recombinant plasmids are able to form colonies. Recircularized pJET1.2/blunt vector molecules lacking an insert express a lethal restriction enzyme which kills the host *E.coli* cell after transformation. This positive selection drastically accelerates the process of colony screening and eliminates additional costs required for blue/white selection. The reactions can be used directly for bacterial transformation and in vitro packaging procedures without further purification. All common laboratory *E.coli* strains can be directly transformed with the ligation product.

Before the transformation procedure, the preparation of competent bacteria of *E. coli* GM2163 strain was performed using TransformAid™ Bacterial Transformation Kit (Fermentas Life Sciences, Lithuania) according to the manufacturer instruction.

The next step was to recover plasmid DNA from recombinant *E.coli* cultures using GeneJET™ Plasmid Miniprep Kit (Fermentas Life Sciences, Lithuania). A single colony was picked from a freshly streaked selective plate for inoculation of 5 mL of LB liquid medium (Fermentas Life Sciences, Lithuania) supplemented with the ampicillin. A bacterial culture is harvested and lysed. The lysate is then cleared by centrifugation and applied on the silica column to selectively bind DNA molecules at a high salt concentration. The adsorbed DNA is washed to remove contaminants, and the pure plasmid DNA is eluted in a small volume of elution buffer or water. The purified DNA is ready for immediate use in all molecular biology procedures such as automated sequencing. Before sequencing, the ligation of DNA fragment into the plasmid was verified using restriction enzymes HindIII and EcoRI (Sigma-Aldrich Chemie GmbH, Germany). The fragments obtained after restriction were analyzed on 1% agarose gels. The sequencing was performed only after the presence of the DNA fragment in plasmid was confirmed by comparing the molecular weight of recombinant plasmid with DNA ladder.

Sequences were determined on ABI Prism 3130 Genetic Analyzer automated sequencer (Applied Biosystems, Foster City, CA, USA) using BigDye Terminator v3.1 Cycle Sequencing Kit (Applied Biosystems, Foster City, CA, USA). Sequencing was performed in both directions on several clones for each selected DNA band. The obtained sequences were analyzed using BLAST software in the NCBI GenBank and EBI (Sanger Institute) database.

The sequencing procedure itself involved: 1) two independent cycle sequencing PCRs, each with one primer only (5′ and 3′), for the sequencing in both directions; 2) precipitation of the amplicons; 3) their denaturation and 4) automatic electrophoresis. Cycle sequencing PCRs were performed on the GeneAmp® PCR System 9700 (Applied Biosystems, Foster City, CA, USA) using BigDye Terminator v3.1 Cycle Sequencing Kit (Applied Biosystems, Foster City, CA, USA) with the final concentration of 100-300 ng of the plasmid DNA and 4pmol of the primer under the following conditions: initial denaturation at 96°C for 1 min, 25 cycles at 96°C for 10 s, 50°C for 5 s, 60°C for 4 min and at 4°C indefinitely. The obtained PCR products were precipitated and EDTA (25 mM final) and EtOH (70-75% final) added.The mixture was incubated for 15 min at RT and then centrifuged 30-45 min at 6000 rpm and +4°C. The supernatant was removed, a new quantity of 70% EtOH added, followed by centrifugation for 25 min at 5000 rpm and +4°C. Supernatant was removed again and the obtained pellet dried at 90°C. Then, 15 µl of Hi-DI™ Formamide (Applied Biosystems, Foster City, CA, USA) was added for the denaturation at 95°C. 10 µl of the amplicons dissolved in formamide were subjected to the automatic electrophoresis and sequence reading on ABI Prism 3130 Genetic Analyzer automated sequencer (Applied Biosystems, Foster City, CA, USA). The obtained sequences were analyzed using BLAST software in the NCBI GenBank and EBI (Sanger Institute) database.

3. Results and discussion

Genomic instability was determined by comparing the AP-PCR profiles of DNA isolated from paired normal and tumor tissues of patients with non small cell lung cancer (NSCLC), malignant glioma, head and neck squamous cell carcinoma (HNSCC) and leukoplakia (L). Twelve out of twenty tested primers produced informative amplification profiles differentiating normal from tumor tissue or normal from leukoplakia (Table 1). Specifically, five primers produced informative sequence alterations that distinguish NSCLC from normal tissue, a set of four primers produced informative fingerprints differentiating malignant gliomas from normal tissue and another set of four primers produced informative sequence alterations that distinguish HNSCC and leukoplakias from their normal counterparts. The AP-PCR products were separated on 6-8% nondenaturing polyacrylamide (PAA) gels and visualized by silver staining. Typical fingerprints are shown in Figure 1.

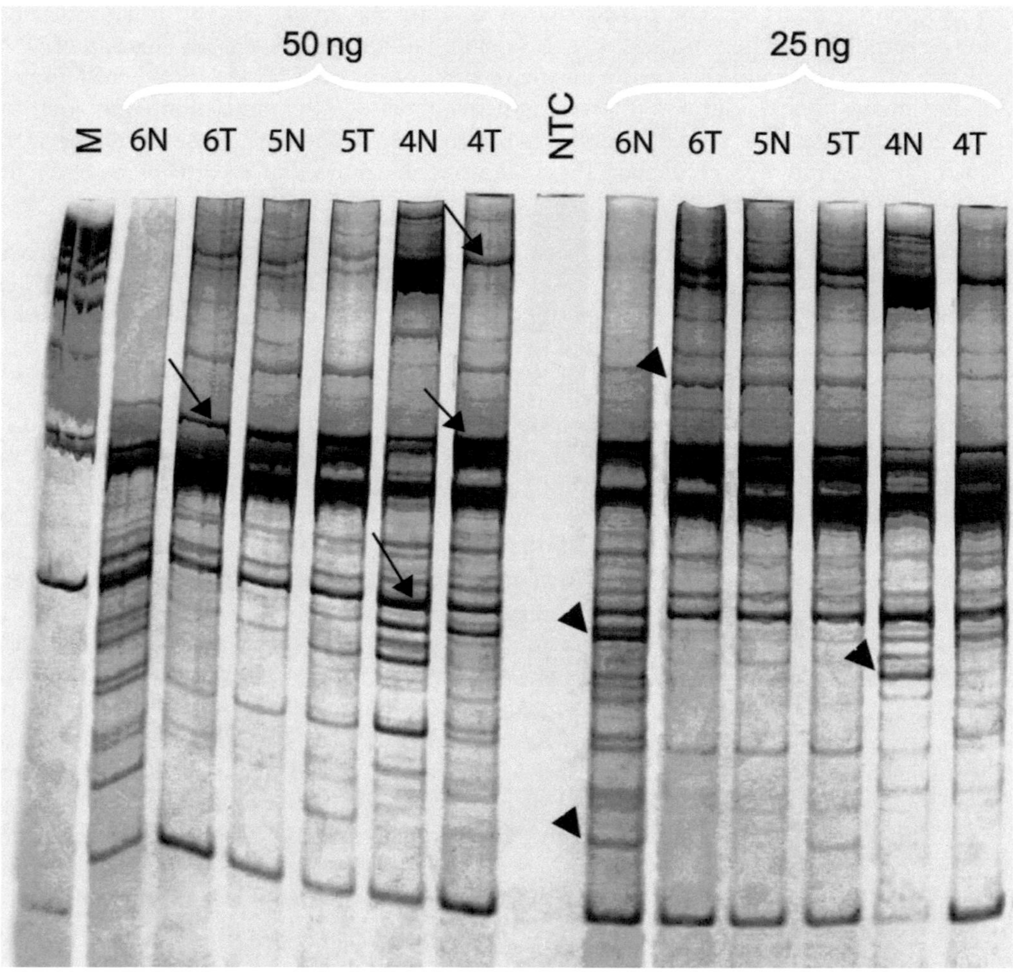

Fig. 1. AP-PCR fingerprint profiles of tumor (T) and normal (N) tissues from patients with NSCLC obtained with GAPDH AS primer. Reactions were performed in duplicate with 25 ng and 50 ng of DNA. Numbers 1–5 represent the patients; M–the DNA ladder; NTC–no template control. Arrows and arrowheads indicate examples of quantitative and qualitative changes, respectively.

This type of analysis differentiates individuals and, thus, displays the cardinal feature of the DNA profile analysis. Additionally, some bands are characteristic for the human genome, being common to all analyzed patients. Importantly, some electrophoretic bands were present in DNA profiles of tumor but not in normal tissue, and vice versa, indicating the mutational like events. The unbiased nature of AP-PCR profiling allows for the screening of anonymous regions of a genome without any prior knowledge of its structure (Welsh and McClelland, 1990; Williams et al., 1990) and provides information about two distinct types of DNA alterations: qualitative and quantitative. Qualitative differences, which represent microsatellite instability (MIN), are detected as mobility shifts in the banding pattern, i.e., the presence or absence of specific bands in tumor and control samples. Quantitative differences appear as altered band intensities and represent amplifications or deletions of existing chromosomal material as manifestations of chromosomal instability (CIN). Observed changes should be cautiously regarded as semi-quantitative and semi-qualitative due to the competitive nature of AP-PCR where sequence context may play unpredictable role. This situation may present a serious problem for simple to moderate patterns but not for complex patterns. Unfortunately, the former are preferred due to simplicity of interpretation. Since the profile is the result of a competition between many PCR products, the problem may appear with very simple profiles in analysis of similar but non-identical genomes. For this reason, it had been suggested to use profile pattern with more than 10 prominent PCR products of moderate complexity (McClelland and Welsh, 1994). We followed this reasoning and the necessary precautions for reproducibility and reliability of DNA profiling analysis in comparing DNA fingerprints of paired normal - tumor samples. We identified significant genomic instability in most cases as qualitative and quantitative electrophoretic changes. The qualitative alterations represented as a loss or a gain of a band are the result of mutations at the primer-template interaction sites leading to a mobility shift of a band. Quantitative changes were observed as bands of either decreased or increased intensity. Allelic losses, which may occur as a result of their linkage to suppressor genes, produce bands with decreased intensity. Gene amplification or chromosomal aneuploidy appears as bands with increased intensity.

For each type of DNA change, as well as for the total number of changes, the frequency of DNA alterations, a measurement of genomic instability, was calculated as the number of altered bands in the AP-PCR profile of tumor tissue divided by the total number of amplicons in the fingerprint of normal tissue from each patient. AP-PCR fingerprints were analyzed and qualitative and quantitative changes determined using image enhancement function 'adapthisteq' of the specialized public software Image J (Figure 2).

DNA alterations were detected in all analyzed samples with the frequency varying among different types of tumors (Table 3). The largest variation of the frequency of total DNA alterations was in NSCLC patients ranging from 8% to even 68%. The contribution of qualitative changes to overall genomic instability was significantly greater than the contribution of quantitative changes. This large range of instability raised the question of its distribution among samples of NSCLC patients. In other words we were interested to see if there was association between the level of genomic instability and any clinicopathological parameter.

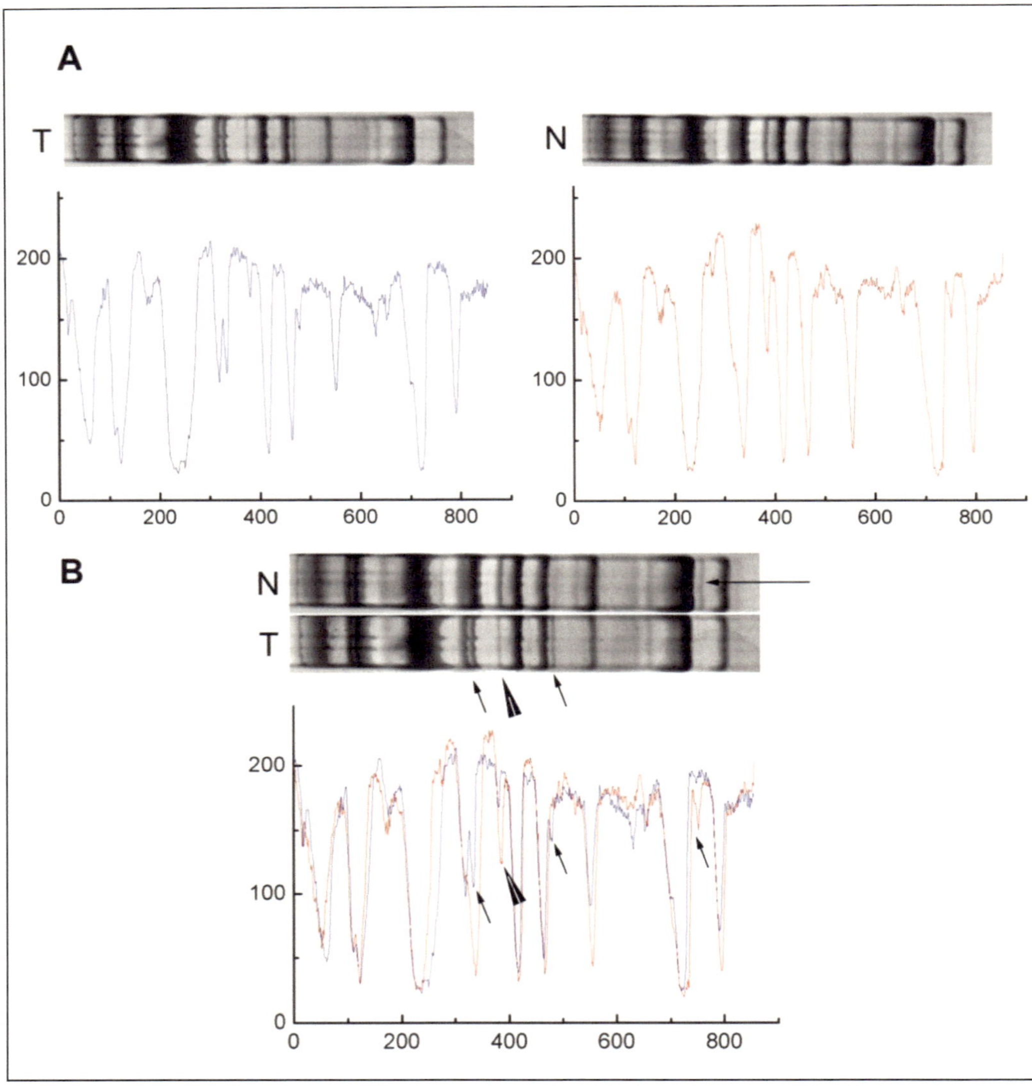

Fig. 2. AP-PCR fingerprinting analysis of genomic instability in glioma samples. AP-PCR profiles of tumor (T) and blood (N) tissues from the same patient obtained using MDRa primer, separated on 6% non-denaturing polyacrylamide (PAA) gel and corresponding contrast-limited adaptive histograms obtained using image enhancement function 'adapthisteq' of the specialized public software Image J (**A**). Arrows and arrowheads indicate examples of qualitative and quantitative electrophoretic changes respectively, clearly seen on the overlap of tumor and blood histograms (**B**).

The most noteworthy finding of this study was the association between the level of genomic instability and histological grades of NSCLC. Namely, we found the significant decrease of the total number of DNA alterations with increasing histological grade of the NSCLC. The same pattern was found for quantitative changes alone - the frequency of alterations decreased with the increase of the histological grade (Figure 3). These results support the idea that mutational alterations conferring genomic instability and the mutator phenotype occur early during tumor formation. The mutator phenotype hypothesis proposes that such phenotypes result from mutations in genes that maintain genomic stability in normal cells. Instability promotes mutations in other genes, oncogenes and tumor suppressor genes, providing the tumor cell with a selective growth advantage. These findings strongly support the increasingly popular explanation of neoplastic transformation in terms of Darwinian evolutionary mechanisms (Breivik, 2001; Breivik & Gaudernack, 1999; Cahill et al., 1999). Evolution through natural selection depends on two essential elements, the availability of

type of DNA alteration	frequency of DNA alterations			
	NSCLC	glioma	leukoplakia	HNSCC
qualitative	0.07 - 0.53	0.06 - 0.27	0.18 - 0.41	0.05 - 0.21
quantitative	0.01 - 0.16	0.05 - 0.27	0.07 - 0.16	0.07 - 0.21
TOTAL	0.08 - 0.68	0.14 - 0.49	0.30 - 0.48	0.12 - 0.31

Table 3. Measurement of genomic instability in various types of tumors.

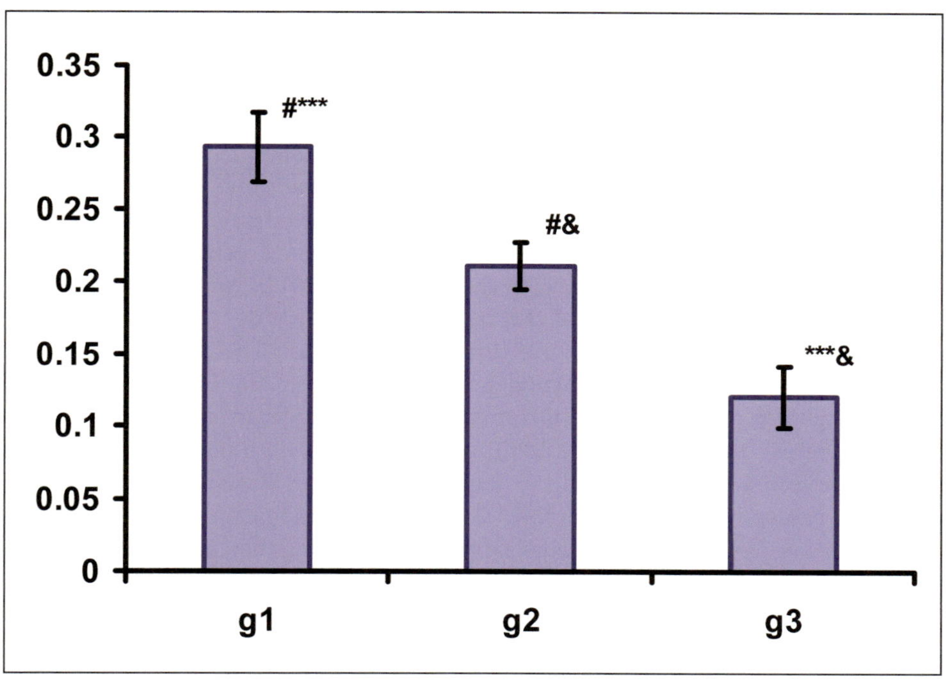

Fig. 3. The relationship between the total frequency of DNA alterations and the histological grades of the lung tumors. All values are presented as means ± SEM. # $p < 0.05$ when grade 1 was compared to grade 2; *** $p < 0.005$ when grade 1 was compared to grade 3; & $p < 0.05$ when grade 2 was compared to grade 3.

genetic variation and selection pressure (Dawkins, 1989.). In general evolutionary terms, it could be said that genomic instability accelerates the somatic evolutionary process by promoting genetic variation in an organism. Extensive genomic instability is thus expected in early phases of cancer progression (histological grade 1 in this study). At the same time, an increased mutation rate is expected to cause mutations that are deleterious or lethal at higher frequencies rather than mutations that have favorable effects on cellular proliferation. Consequently, elevated mutation rates must generally be regarded as disadvantageous to cellular growth (Tomlinson et al.,1996). Theoretical arguments suggest that the accumulation of large numbers of mutations can exceed the error threshold for cell replication and viability (Eigen, 1993). Only cells carrying reasonable number of mutations with favorable effects on cell growth would survive. Therefore, it seems probable that the expression of the mutator phenotype could be decreased and lost in the late phases of tumor progression. As a result, tumors may no longer exhibit a mutator phenotype but will nevertheless reveal its history, i.e. random mutations, throughout their genome (Loeb, 2001). In other words, the result showing the lower degree of genomic instability in advanced NSCLCs (grades 2 and 3) is not unexpected in the light of these arguments and could be considered as a marker of poor prognosis.

Following the study of genomic instability in NSCLC tissue samples, we made an attempt to identify some of detected DNA changes in order to identify genes that alter during NSCLC promotion and progression (Bankovic et al., 2010). Selected DNA bands with altered mobility were further characterized. Twenty one unique bands present only in tumor but not in normal tissue were retrieved from the gels and cloned. Variant bands that appeared in more than one sample (new bands with the same mobility), were chosen in order to identify DNA alterations common to as many NSCLC patients as possible. Bands (amplicons) with the same electrophoretic mobility were isolated and characterized from at least two patients in order to confirm that they represent the same DNA sequence. Three clones of each band were sequenced. Obtained sequences were submitted to homology or identity search in NCBI GenBank and EBI (Sanger Institute) databases. Following genes were identified: tetraspanin 14 (TSPAN14), cadherin 12 (CDH12), retinol dehydrogenase 10 (RDH10), cytochrome P450, family 4, subfamily Z, polypeptide 1 (CYP4Z1), killer cell immunoglobulin-like receptor (KIR), E2F transcription factor 4 (E2F4), phosphatase and actin regulator 3 (PHACTR3), PHD finger protein 20 (PHF20), PRAME (preferentially expressed antigen in melanoma) family member and solute carrier family 2 (facilitated glucose transporter), member 13 (SLC2A13). Moreover, we were able to identify types of mutations in revealed genes according to sequence data and BLAST search results and to examine their presence in relation to NSCLC subtype, histological grade and stage of the tumor, lymph node invasion and patients' survival. Examining their relation to the patients' clinicopathological parameters and survival we concluded that TSPAN14, SLC2A13 and PHF20 could have a role in NSCLC promotion, CYP4Z1, KIR and RDH10 would possibly play a role in NSCLC progression, while E2F4, PHACTR3, CDH12 and PRAME family member probably play important role in NSCLC geneses. Patients with altered E2F4 and PHACTR3 lived significantly shorter.

Unlike NSCLC samples, all leukoplakias demonstrated extensive instability in a relatively small range (Table 3). The frequency of total DNA alterations ranged from 0.30 to 0.48 and clearly distinguished two groups of leukoplakias: a group of six leukoplakias had a frequency of DNA alterations of 0.3–0.34 and was denoted as leukoplakias with a moderate degree of instability while the other group of 26 leukoplakias had a frequency of DNA alterations of > 0.4 and was denoted as leukoplakias with a high degree of instability (Tanic et al., 2009). However, such high levels of genomic instability in leukoplakia samples were a surprise mainly because they are defined as white patches or plaques of oral mucosa that cannot be rubbed off and cannot be diagnosed clinically or pathologically as other specific diseases and have been considered premalignant lesions only since recently (Neville & Day, 2002; Hunter et al., 2005). It is impossible to state, with precision, the proportion of leukoplakias that undergo malignant transformation. For oral mucosa, in general, up to 20% of leukoplakias exhibit dysplasia. Dysplastic leukoplakias have a greater probability of developing into cancer, although leukoplakias without evidence of dysplastic changes may also progress to highly aggressive squamous cell carcinoma. Still, the majority of leukoplakias fail to undergo malignant transformation. The frequency of malignant alterations in oral leukoplakia varies from study to study and ranges from 8.9 to 17.5% (summarized in Neville & Day, 2002). These facts and our finding were the reasons to include samples of Head and Neck Squamous Cell Carcinoma patients, identify and quantify genomic instability in these samples and compare obtained results with those of leukoplakia samples.

Obtained frequency of DNA alterations in HNSCC samples was significantly lower than that of leukoplakia samples, as shown in Table 3. When comparing mean frequencies of DNA alterations the result is even more convincing. Namely, mean frequency of total DNA changes was 0.42 for leukoplakia samples vs. 0.28 for HNSCC samples. Interestingly, contribution of quantitative changes to the total instability in HNSCC samples is significantly higher (0.21) than the contribution of qualitative changes (0.16) which is quite opposite in leukoplakia samples. In other words, the level of genomic instability decreased during HNSCC promotion from premalignant lesions but more serious alterations, quantitative changes as manifestations of chromosomal instability, were selected. These results fit nicely into Darwinian evolutionary theory of neoplastic transformation. High instability is present at the very beginning of HNSCC genesis, providing genetic variability in the population of premalignant cells, which is absolutely necessary for the evolution by natural selection. During tumor progression the level of instability decreases due to selection of genotypes that are better adapted to the micro-environment in which natural selection took place. However, the question remains: why the majority of leukoplakias with such a huge instability fail to undergo malignant transformation? The answer may be in exceeding the error threshold for cell replication and viability (Eigen, 1993) with so many mutations. In other words, it seems that leukoplakias with a high degree of genomic instability have less chance to develop into HNSCC, whereas leukoplakias with a lower (moderate) degree of genomic instability have a better chance of transforming, probably because they carry a certain number of mutations that have favorable effects on cell growth (Tanic et al., 2009).

Following the same reasoning as in the case of NSCLC we attempted to identify some of detected DNA changes in leukoplakias, with the aim of identifying tumor-specific alterations (Peinado et al., 1992) that could lead to the development of potential diagnostic markers involved in the genesis of HNSCC. To that end, nine variant bands present in leukoplakias but not in normal tissue, were selected. Unexpectedly, two different amplicons, originating from distinct leukoplakias, were identified as altered part of the TIMP-3 gene (tissue inhibitors of metalloproteinases 3), two were identified as mutated DNMT 3A gene (DNA (cytosine-5)-methyltransferase 3 alpha) and two represented copies of the Ty1-copia-like retrotransposon.

Further investigations of the detected genes in both, leukoplakia and NSCLC samples, on larger sample size, with special emphases on tumor promoting genes, are underway. We expect more detailed profile of their involvement in NSCLC and HNSCC after extensive analyses of their mutational status and detailed analyses of their expression profile at RNA and protein level in a larger sample. We expect that some of them might prove to be a good prognostic biomarkers for NSCLC or HNSCC patients.

Finally, we analyzed malignant gliomas, tumors that originate from glia, the most common and deadly brain tumors. All patients had histologically confirmed diagnosis of anaplastic astrocytoma (AA) or glioblastoma multiforme (GBM) according to the new World Health Organisation (WHO) classification. Anaplastic astrocytomas (WHO grade III) and glioblastomas (WHO grade IV) are two major groups of malignant gliomas. Glioblastomas are further classified as primary and secondary. Distinction between them is based on different genetic pathways leading to their development (Ohgaki & Kleihues, 2007; Van Meir et al., 2010). Primary glioblastoma develop rapidly *de novo*, without clinical or histological evidence of a less malignant precursor lesion. Secondary glioblastoma develop slowly progressing from low-grade diffuse astrocytoma (WHO grade II) or anaplastic astrocytoma (WHO grade III).

Examination of the extent of genomic instability revealed that samples of patients with anaplastic astrocytoma had similar level of total, microsatellite and chromosomal genomic instability as patients with glioblastoma multiforme, with very high values in both histological subtypes (Table 4). It was unexpected and, at first sight, looked like these results contradicted the expectation and results obtained from NSCLC and HNSCC samples. However, all analyzed grade IV glioblastomas were classified as primary glioblastomas (because glioblastoma diagnosis was made at the first biopsy, without clinical or histopathologic evidence of a less malignant precursor lesion), which are considered to be *de novo* tumors and not the progressive form of grade III astrocytomas. Therefore, obtained results are still consistent with the evolutionary theory of neoplastic transformation and the decrease of the level of genomic instability could be expected in secondary glioblastomas. In other words, extensive genomic instability might be used as diagnostic character where pathology cannot provide unambiguous distinction between primary and secondary GBM. Similar results were obtained by Nishizaki et al. (2002) who demonstrated that there was no significant difference in FISH heterogeneity between malignant gliomas of WHO grades III and IV. We expect that further research involving secondary glioblastomas will confirm our hypothesis and will provide additional confirmation for the evolutionary theory of tumor progression. Moreover, we hope that cloning and sequencing of amplified DNA bands showing genetic alterations specific for glioma genome, will allow the detection of new genes implicated in glioma pathogenesis and progression.

instability	Mean frequency		
	Anaplastic Astrocytoma	vs.	Glioblastoma Multiforme
microsatellite	0.15	vs.	0.16
chromosomal	0.19	vs.	0.16
TOTAL	0.34	vs.	0.33

Table 4. Mean frequency of DNA alterations in malignant glioma samples.

Finally, it is worth mentioning that measurements of genomic instability could be performed by another DNA fingerprinting technique, RAPD (Random Amplified Polymorphic DNA). Wang et al. (2002) measured genomic instability in various cancer types using RAPD and the instability they detected was in average higher than 40% for lung cancer tissues. In another study (Ong et al., 1998), DNAs from 20 lung cancer (18 non-small cell lung cancers and two small cell lung cancers) and their corresponding normal tissues were amplified individually by RAPD with seven different 10-base arbitrary primers. PCR products from RAPD were electrophoretically separated in agarose gels and banding profiles were visualized by ethidium bromide staining. The ability to detect genomic instability in 20 cancer tissues by each single primer ranged from 15 to 75%. DNA changes were detected by at least one primer in 19 (95%) cancer tissues. They concluded that these results seem to indicate that genomic rearrangement is associated with lung carcinogenesis and that RAPD analysis is useful for the detection of genomic instability in lung cancer tissues.

Misra A. et al. (2007) used RAPD to attempt to quantify the number of clonal mutations in primary human gliomas of astrocytic cell origin . They targeted genomic loci of a different nature and estimated that the number of overall alterations in tumor genome seemed to be greater than expected. They also observed a higher number of genetic changes in tumors of lower grade and suggested that it could be a consequence of an increased mutation rate in early tumorigenesis due to acquisition of a mutator phenotype. The increased extent of alterations occurring in tumors of a lower grade is consistent with our study. The results of Misra et al. showed the acquisition of a mutator phenotype early in tumorigenesis and support the mutator hypothesis proposed by Loeb (1991, 2001).

4. Conclusions

AP-PCR DNA fingerprinting is an efficient tool to quickly and easily screen a very large number of loci for possible DNA alterations in cancer cells. It has several advantages: first, minor amounts of template DNA are sufficient for analysis; second, it allows for the screening of anonymous regions of a genome without any prior knowledge of its structure; third, two types of DNA alterations could be detected in single reaction, chromosomal rearrangements and random mutations dispersed over the genome; and forth, possibility of reamplification, cloning and sequencing of variant bands enables the rapid identification of the genes probably linked to tumor progression. Here, we demonstrated the use of AP-PCR DNA fingerprinting in detection and quantification of genomic instability (microsatellite, chromosomal and total) in three types of tumors as well as in search for molecular biomarkers for cancer promotion and progression. Therefore, we conclude that AP-PCR

DNA fingerprinting is important and practically feasible technique for elucidating the genetic background of various tumors. Accordingly, we believe that this technique is rather neglected in contemporary research and should make a comeback because it still has a particularly promising future in experimental oncology.

5. Acknowledgments

This study was supported by Grant # III41031 from the Ministry of Education and Science, Republic of Serbia.

6. References

Bankovic, J.; Stojsic, J.; Jovanovic, D.; Andjelkovic, T.; Milinkovic, V.; Ruzdijic, S. & Tanic N. (2010). Identification of genes associated with non-small-cell lung cancer promotion and progression. *Lung Cancer*, Vol.67, No.2, pp. 151-159, ISSN 0169-5002.

Beivik, J. 2001. Don't stop for repairs in a war zone: Darwinian evolution unites genes and environment in cancer development. *PNAS*, Vol.98, No.10, pp. 5379–5381, ISSN 0027-8424.

Breivik, J. & Gaudernack G. (1999). Genomic instability, DNA methylation and natural selection in colorectal carcinogenesis. *Seminars in Cancer Biology*, Vol.9, No.3, pp. 245–254, ISSN 1044-579X

Cahill, D.P.; Kinzler, K.W.; Vogelstein, B, & Lengauer, C. (1999) Genetic instability and Darwinian selection in tumours. *Trends Cell Biology*, Vol.9, No.12, pp. M57–M60, ISSN 0962-8924.

Chariyalertsak, S.; Khuhaprema, T.; Bhudisawasdi, B.S.; Wongkham, S. & Petmitr S. (2005). Novel DNA amplification on chromosome 2p25.3 and 7q11.23 in cholangiocarcinoma identified by arbitrarily primed polymerase chain reaction. *Journal of Cancer Research & Clinical Oncology*, Vol.131, No.12, pp. 821–828, ISSN 1432-1335.

Cobb, B. (1997) Optimization of RAPD fingerprinting. In: *Fingerprinting Methods Based on Arbitrarily Primed PCR*, Micheli M.R. & Bova R., (ed), pp. 93-102, Springer Lab Manual, Springer-Verla, ISBN 3-540-61229-7, Berlin, Germany.

Dawkins, R. (1989). *The selfish gene* (edition 1). Oxford University Press, ISBN 0192860925, Oxford, UK.

Eigen, M. (1993). The origin of genetic information: viruses as models. *Gene*, Vol.135, No.1-2, pp. 37–47, ISSN 0378-1119.

Fenech, M. (2002). Chromosomal biomarkers of genomic instability relevant to cancer. *Drug Discovery Today*, Vol.7, No. 22, 1128–1137, ISSN 1359-6446.

Hunter, K.D.; Parkinson, E.K. & Harrison, P.R. (2005). Profiling early head and neck cancer. *Nature Reviews Cancer*, Vol.5, No pp. 127–135, ISSN: 0028-0836.

Jass, J.R.; Mukawa, K.; Goh, H.S. & Love, S.B. (1989) Clinical importance of DNA content in rectal cancer measured by flow cytometry. *Journal of Clinical Pathology*, Vol.42, No.3, pp. 254–259, ISSN 0021-9746.

Kallioniemi,O.P.; Kallioniemi, A.; Piper, J.; Isola, J.;Waldman, F.M.; Gray, J.W. & Pinkel, D. (1994). Optimizing comparative genomic hybridization for analysis of DNA sequence copy number changes in solid tumors. Genes Chromosomes Cancer, Vol.10, No.4, pp. 231–243, ISSN 1098-2264.

Lengauer, C.; Kinzler, K.W. & Vogelstein, B. (1998). Genetic instabilities in human cancers. *Nature*, Vol.396, pp. 643–649, ISSN 0028-0836.

Loeb, L.A. (2001). A mutator phenotype in cancer. *Cancer Research*, Vol.61, pp. 3230–3239, ISSN: 0008-5472.

Loeb, L.A. (1991). Mutator phenotype may be required for multi-stage carcinogenesis. *Cancer Research*, Vol.51, pp. 3075–3079, ISSN: 0008-5472.

Markovic, J.; Stojsic, J.; Zunic, S.; Ruzdijic, S. & Tanic, N. (2008) Genomic instability in patients with non-small cell lung cancer assessed by the arbitrarily primed polymerase chain reaction. *Cancer Invest*, Vol.26, No. pp. 262-8, ISSN: 0735-7907.

McClelland, M. & Welsh, J. (1994). DNA fingerprinting by arbitrarily primed PCR. *PCR Methods Application*, Vol.4, pp. S59-S65.

Misra, A.; Chattopadhyay, P.; Chosdol, K.; Sarkar, C.; Mahapatra, A.K.; & Sinha S. (2007). Clonal mutations in primary human glial tumors: evidence in support of the mutator hypothesis. *BMC Cancer*, Vol.7, pp. 190. ISSN 1471-2407.

Negrini, S.; Gorgoulis, V.G. & Halazonetis, T.D. (2010) Genomic instability-an evolving hallmark of cancer. *Nat Rev Mol Cell Biol*, Vol.11, pp. 220-228, ISSN 1471-0072.

Neville, B.W. & Day, T.A. (2002). Oral cancer and precancerous lesions. *CA Cancer J Clin*, Vol.52, No.4, pp. 195–215, ISSN 0007-9235.

Nishizaki, T.; Harada, K.; Kubota, H.; Harada, K.; Furuya, T.; Suzuki, M. & Sasaki K. (2002). Chromosome instability in malignant astrocytic tumors detected by fluorescence in situ hybridization. *Journal of Neurooncology*, Vol.56, No.2, pp. 159-165, ISSN 0167-594X.

Ohgaki, H. & Kleihues, P. (2007). Genetic pathways to primary and secondary glioblastoma. *Am J Pathol*, Vol.170, No.5, pp. 1445-53, ISSN 0002-9440.

Ong, T.M.; Song, B.; Qian, H.W.; Wu, Z.L. & Whong, W.Z. (1998). Detection of genomic instability in lung cancer tissues by random amplified polymorphic DNA analysis. *Carcinogenesis*, Vol.19, No.1, pp. 233-235, ISSN 0143-3334.

Peinado, M.A.; Malkhosyan, S.; Velazquez, A.; & Perucho, M. (1992). Isolation and characterization of allelic losses and gains in colorectal tumors by arbitrarily primed polymerase chain reaction. *Proc. Natl. Acad. Sci. USA*, Vol.89, No.21, pp. 10065–10069, ISSN 0027-8424.

Perucho, M. (1996). Microsattelite instability: the mutator that mutates the other mutator. *Nat. Med.*, Vol.2, No.6, pp. 630–631, ISSN 1078-8956.

Sambrook, J.; Fritch, E.F. & Maniatis, T. (1989). *Molecular cloning: a laboratory manual*. (2nd edition). Cold Spring Harbor Laboratory Press, Cold Spring Harbor, NY.

Tanic, N.; Tanic, N.; Milasin, J.; Vukadinovic, M. & Dimitrijevic, B. (2009). Genomic instability and tumor-specific DNA alterations in oral leukoplakias. *European Journal of Oral Sciences*, Vol.117, pp. 231 - 237, ISSN 0909-8836.

Tomlinson, I.P.; Novelli, M.R.; Bodmer, W.F. (1996). The mutation rate and caner. *Proc. Natl. Acad. Sci. USA*, Vol.93, No.25, pp. 14800–14803, ISSN 0027-8424.

Van Meir, E.G.; Hadjipanayis, C.G.; Norden, A.D.; Shu, H.K.; Wen, P.Y.; & Olson, J.J.(2010). Exciting new advances in neuro-oncology: the avenue to a cure for malignant glioma. CA Cancer J Clin, Vol.60, No.3, pp. 166-93, ISSN 0007-9235.

Vogelstein, B.; Fearon, E.R.; Kern, S.E.; Hamilton, S.R.; Preisinger, A.C.; Nakamura, Y. & White, R. (1989). Allelotype of colorectal carcinomas. *Science*, Vol.244, No.4901, pp. 207–211, ISSN: 0036-8075.

Wang, J.; Wang, Q. & Ye, F. (2002). Genetic instability in cancer tissues analyzed by random amplified polymorphic DNA PCR. *Chin Med J (Engl)*, Vol.115, No.3, pp. 430-432, ISSN 0366-6999.

Welsh, J. & McClelland, M. (1990). Fingerprinting genomes using PCR with arbitrary primers. *Nucleic Acid Research*, Vol.18, No.24, pp. 7213–7218, ISSN 0305-1048.

Williams, J.G.K.; Kubelik, A.R.; Livak, K.J.; Rafalski, J.A. & Tingey, S.V. (1990). DNA polymorphisms amplified by arbitrary primers are useful as genetic markers. *Nucleic Asids Research*, Vol.18, No.22, pp. 6531-6535, ISSN 0305-1048.

23

Analysis of Alternatively Spliced Domains in Multimodular Gene Products - The Extracellular Matrix Glycoprotein Tenascin C

Ursula Theocharidis and Andreas Faissner
Ruhr-University Bochum
Germany

1. Introduction

In 1977 it was discovered that the one-gene-one-enzyme hypothesis was not true (Chow et al.,1977; Berget at al., 1977). The primary transcription product can be spliced in different ways and give rise to several proteins depending on the exons being present in the final mRNA. This phenomenon is called alternative splicing and indeed is common to many genes. Several possible modes of alternative splicing are known and the most common one is the inclusion or exclusion of an exon, the exon skipping.

Based on polymerase chain reaction (PCR) techniques we developed a method to analyse combinations of alternatively spliced domains in multimodular gene products. This method was used to determine the combinatorial variability of tenascin C isoforms in the mouse central nervous system (Joester & Faissner, 1999) and in neural stem cells (von Holst et al., 2007).

Here, we present the method of amplifying different sized isoforms of a gene product with several alternatively spliced domains via PCR and the isolation and subcloning of the PCR products. Clones are analyzed for alternatively spliced domains contained therein by a dot blot *in vitro* hybridization method with domain-specific DNA probes which were generated using PCR.

2. Background information

Tenascin C is a multimodular glycoprotein of the extracellular matrix which is mainly expressed during central nervous system development and in pathological states such as brain tumours or lesions. We have studied the expression pattern of this molecule and its function *in vivo* and *in vitro* and collected evidence concerning its structural diversity. We and others determined its functions during neural development, in the adult neural stem cell niche and in lesions and tumours (Czopka et al., 2009, 2010; Dobbertin et al., 2010; Garcion et al., 2001, 2004; Garwood et al., 2011; Gates et al., 1995; Orend & Chiquet-Ehrismann, 2006; von Holst et al., 2007).

Tenascin C contains a constant part including eight contitutive fibronectin type III (fnIII) domains and a variable part of six alternatively spliced fnIII domains in the mouse which can be included independently into the gene product (figure 1).

The alternatively spliced fnIII domains of the tenascin C molecule have different functions, e.g. affecting the axon outgrowth of developing nerve cells or the migration potential of brain tumour cells (Rigato et al., 2002; Michele & Faissner, 2009; Broesicke & Faissner, personal communication). Therefore it is important to have a method to determine the isoform composition of the molecule in the tissue or cell cultures used.

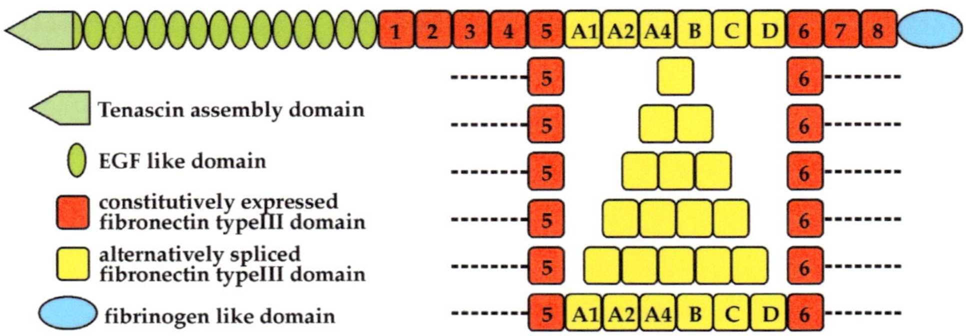

Fig. 1. Schematic representation of mouse tenascin C. The monomer consists of several distinct protein domains. At the N-terminal tenascin C assembly domain six monomers can be assembled to the so called hexabrachion (Erickson & Inglesias, 1984). 14,5 epidermal growth factor (EGF) like domains and eight constitutive fibronectin type III (fnIII) domains follow before the C-terminal globular lobe homologous to the beta- and gamma-chains of fibrinogen. Between the fifth and sixth constant fnIII domain up to six alternatively spliced domains can be inserted and an independent alternative splicing at each position could lead to the generation of 64 (=2^6) possible isoforms of the molecule. All possible numbers of domains can be inserted in the final splicing product, but the combination of cassettes is unclear in most cases. Only the largest variant necessarily contains all six alternatively spliced domains.

Gene products with different exons being alternatively spliced and inserted into the sequence can be distinguished by PCR when the sizes of the resulting mRNAs are different. A PCR analysis uses primers flanking the alternatively spliced region and results in amplicons with different sizes. These can be analysed by agarose gel electrophoresis and show bands in distinct positions. Tenascin C has six domains that can be alternatively spliced and independently inserted into the sequence. The analysis of these domains on an agarose gel shows the size of the resulting amplicons and therefore the number of inserted domains but leaves the question open which of the possible domains are included. A further analysis is therefore needed. We have shown that it can be performed using an *in vitro* dot blot hybridization technique to verify the exact domain combinations.

3. Analysis of isoform sizes in multimodular gene products by RT-PCR

When the sequence of the gene of interest is known primers can be generated which allow the amplification of the relevant region. The primers can either bind in the alternatively spliced region itself and therefore generate PCR products only when the target sequence is expressed. When the primers bind outside of the alternatively spliced part of the sequence the products can contain every possible insert additionally to the constant parts of the sequence which are defined by the primer binding sites. Additionally, isoforms without any insert can be amplified with these primer combinations.

Tenascin C has its variable region between the constantly expressed fnIII domains 5 and 6. We used two different primer combinations to determine the isoform pattern of the molecule in various tissues and cell cultures (figure 2). The primers 5s and 6as bind to the 5′end of the fifth and the 3′end of the sixth domain and result in PCR products with the smallest form containing only these two constant domains. Another primer pair we used was called 5for and 6rev and these bind to the 3′end of domain number 5 and the 5′end of domain number 6. The smallest amplicons are then represented by forms with one alternatively spliced fnIII domain. The further analysis was carried out with PCR products obtained with this primer pair.

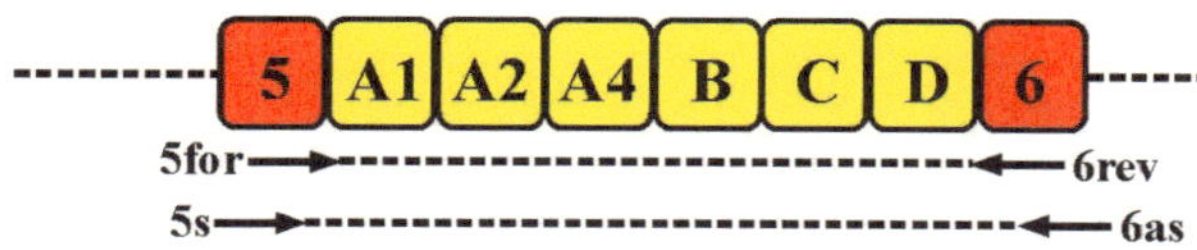

Fig. 2. Primer binding sites. Two different primer pairs were used to amplify the alternatively spliced region of tenascin C and analyse the expression profile of different isoforms. The primers 5s / 6as and 5for / 6rev bind to the constant fnIII domains 5 and 6 at their outer or inner tails, respectively. The primers 5s and 6as bind to the 5′end of the fifth and the 3′end of the sixth domain. The resulting amplicons therefore contain the minimum of two fnIII domains, namely 5 and 6. The further insertion of alternatively spliced cassettes increases the size of the PCR product. The primers 5for and 6rev bind to the 3′ end of fnIII domain 5 and the 5′end of fnIII domain 6. Only PCR products with the minimum of one alternatively spliced fnIII domain can be generated. Every additional domain increases the size of the amplicons by 273 bp, the size of the single domains.

3.1 Expression analysis by RT-PCR

The expression analysis can be performed on RNA isolated from tissue or cell culture material which was processed by reverse transcription. Several commercially available kits help to isolate total RNA or mRNA from tissue or cell cultures. The resulting RNA can then be used to generate cDNA by reverse transcription which can also be carried out using kits from different suppliers. If oligo-dT primers or random primers are used for the reverse transcription makes no difference in our experience. The generated cDNA is the template for the PCR which possibly needs some optimization steps to generate all bands of interest. According to our experience it is of outstanding importance to test the performance of different Taq polymerases in advance because not every enzyme from each supplier will work equally efficiently. Different polymerases in their respective buffer system show variable results and should be adapted to the reaction requirements.

The PCR conditions with regard to annealing temperature and time, elongation time as well as concentration of cDNA, primers and Magnesium must be worked out in advance. Addition of DMSO or betain may be needed and checked when the standard conditions don′t lead to the desired results. To be able to generate all the expected amplicons the longest product determines the elongation time. The rule of thumb to calculate 1 minute elongation time per 1000 base pairs gives a good estimation here.

The resulting PCR products can be processed on an agarose gel and the DNA bands made visible with ethidium bromide or a substitute. The concentration of the agarose must be high enough to discriminate between contiguous bands but sufficiently low that the longest products can enter the analysis area. A long gel chamber increases the migration way and a lower voltage over a longer time period narrows the single bands and makes the discrimination easier.

We used brain tissue from postnatal mice or cultures of neural stem cells to isolate total RNA and analysed the expression pattern of the alternatively spliced forms of Tenascin C in the respective system (Joester & Faissner, 1999; von Holst et al., 2007). In these cases we found isoforms of all possible sizes to be present and performed the further analysis for isoforms containing between one and six additional cassettes. The use of the primer pair 5s and 6as leads to DNA bands on the agarose gel where the smallest one is 546bp, representing only the two constant fnIII domains 5 and 6 of 273bp each. Every additional cassette increases the amplicon size by 273bp. Therefore, we can see a "ladder" structure of up to seven DNA bands on the agarose gel when using this primer pair (figure 3B). When the primers 5for and 6rev are present in the PCR mix instead we get products where the smallest isoform contains the minimum of one alternatively spliced fnIII domain. The larger bands represent the larger forms with up to six fnIII domains. Here, we get the maximum of 6 DNA bands on the gel (figure 3A).

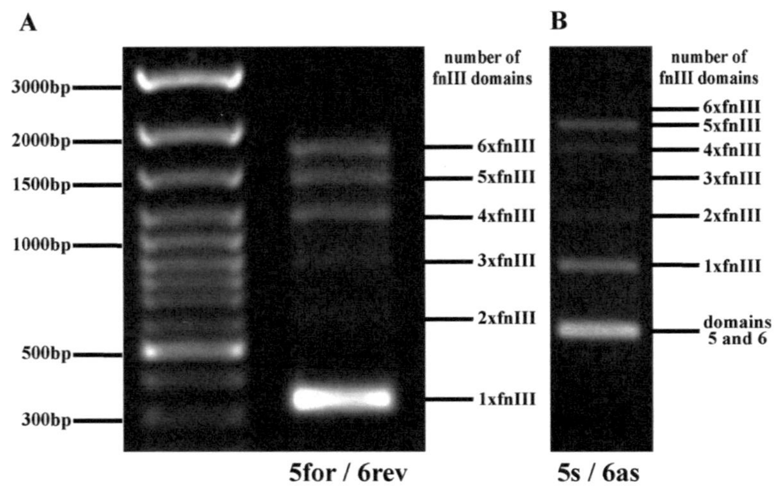

Fig. 3. Examples of tenascin C isoform PCRs. The primer pairs 5for / 6rev and 5s / 6as were used to amplify the alternatively spliced region of tenascin C. The PCR products were separated on an 1,2 % agarose gel. (A) The smallest amplicon generated using the primers 5for and 6rev contains only one of the alternatively spliced cassettes. The insertion of additional domains increases the product size by 273 bp. Up to six bands appear representing the different possible amplicon sizes. PCR products amplified with this primer pair were used for the further analysis of the domain expression profile after separation on an agarose gel. (B) The use of the alternative primer pair 5s and 6as leads to the generation of up to seven DNA bands on the agarose gel because the smallest band represents only the constant fnIII domains 5 and 6 without any insert. When alternatively spliced domains are included in the sequence the product size increases by 273 bp for each domain. Up to six domains can be added and therefore the largest DNA band on the gel represents the total of eight fnIII domains. This primer pair was mainly used for the analysis of expression profiles.

The agarose gel shows the expression profile of the alternatively spliced gene products in the analysed tissue or cells. The resulting amplicons answer the first questions in this respect: Are different forms expressed in parallel? Are all possible product sizes present? What is the ratio between different forms? Does the expression profile change with the conditions?

3.2 Cloning of resulting PCR products

The analysis of the PCR products on an agarose gel answers the question for size and ratio of the isoforms expressed but leaves open which of the possible domains are contained in the bands. Some further experimental steps are necessary to determine the domains being expressed. Because several domain combinations can migrate in the same position they must be separated from each other. This can be achieved by subcloning the different PCR amplicons and analysis of the resulting clones.

The PCR bands are cut out of the gel under visual control at an UV desk and the gel slices collected in separate tubes. It is important to use different knives for each band because otherwise DNA from other bands might be carried over and contaminate the samples. Isolation of the DNA from the gel can be performed using classic methods or commercially available kits. The elution should be done with the minimal amount of water to avoid problems with following reaction steps. For the subcloning of the PCR products we used the TOPO-TA cloning kit from invitrogen but any other similar kit will do. In our experience it is important to handle the bacteria quite carefully and leave them grow in antibiotic-free medium for 30 minutes after the transformation. Spread the bacteria to LB agar plates with appropriate antibiotics then and let them grow over night.

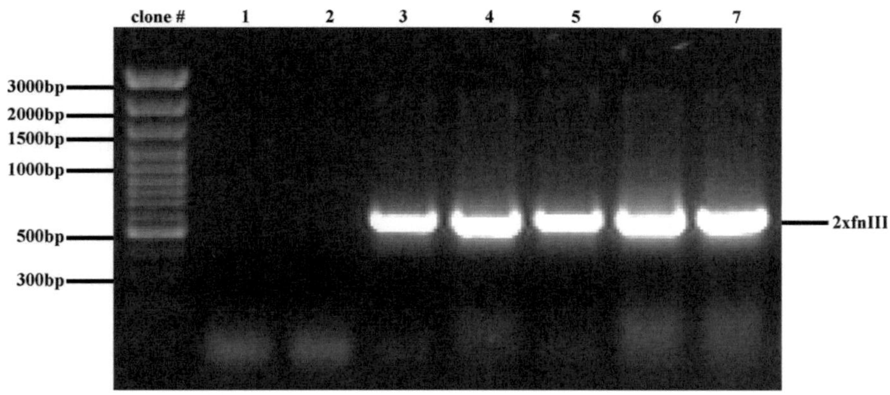

Fig. 4. Check for positive clones after direct colony lysis. The colonies grown after the cloning and transformation of the PCR products are checked for their content of fnIII domains. The primer pair 5for / 6rev was used to generate amplicons of the expected size when the clones have taken up the plasmids containing the fnIII domains. This example shows seven clones from a DNA band containing 2 fnIII domains. Two of the clones shown here do not contain any fnIII domains and are therefore not selected for the following screen. The other clones show PCR bands of the expected size and are analysed in the subsequent dot blot hybridizations.

The content of the resulting clones can quickly and easily be checked by direct lysis of the bacteria and a subsequent PCR with the primers used before. The colonies grown on the agar plate are picked with a pipette tip and transferred to another (the "master" plate) into numbered fields. The tip is then shaken in 25μL 70% ethanol in PCR tubes to lyse the bacteria. The master plate can be placed in the incubator while in the meantime the colonies are checked for their content. In an incubation step the ethanol is evaporated at 80°C for approx. 15

minutes before the PCR master mix containing buffer, primers and polymerase is added. The reaction conditions can be the same as before. The products can be analysed on an agarose gel and should show single bands in the expected position for each positive clone (figure 4). These can subsequently be picked from the master plate and propagated in miniprep scale. The plasmid DNA from the miniprep cultures can be isolated by alcaline lysis or with appropriate kits.

3.3 Analysis of clones - dot blot

Of course, these plasmids could be sequenced and their composition clarified by this method at this point. Because sequencing is not cheap when analysing hundreds of clones a method was developed that renders the identification of many samples in one step possible and is cheaper. The basis is a dot blot of the isolated plasmids to nylon membranes which subsequently can be used in hybridizations with domain specific probes.

The plasmid solutions should be adjusted to similar concentrations with water to have equal amounts of target DNA in the spots. The easiest way to apply the plasmids to the membranes is the use of a dot blot apparatus with a vacuum manifold, but it is also possible to spot the liquid using a master plate onto the membrane which is placed on filter paper. For our analysis we used Hybond N+ membranes from Amersham which were pre-wetted with 10x SSC (1,5 M NaCl; 150 mM Na_3Citrate, pH 7,0) buffer. Because there are six possible domains to detect (A1, A2, A4, B, C and D) and we used a negative control we prepared seven membranes with identical spot patterns. The plasmid solutions were diluted in 10x SSC in a volume of 100μL when we used the dot blot apparatus and 6μL when a pattern was used.

After application of the plasmid solutions the membranes are incubated for 10 minutes in denaturing buffer (500 mM NaOH; 1,5 M NaCl) and 10 minutes in renaturing buffer (500 mM Tris/HCl, pH 7,5; 1,5 M NaCl) to prepare the DNA for the hybridization. The nylon membranes are dried and baked at 80°C for two hours to have the DNA bound covalently to them. These dot blots can be stored for some time at room temperature.

3.4 Positive and negative controls

To determine the specificity of the method and to be sure that no false-positive or false-negative results appear the use of positive and negative controls is important. For every application appropriate controls must be defined. In our case we could exploit the fact that the fnIII domain number 6 is not included in the alternatively spliced region which we amplify with the primer pair 5for / 6rev in the initial PCRs. Therefore a probe detecting the fnIII domain 6 serves as negative control. Another control we use is a plasmid, called pJT1# which contains the constant part of tenascin C between the fnIII domains 2 and 8, but none of the alternatively spliced domains. The positive control is a plasmid containing all six alternatively spliced domains. On the dot blot it is applied in addition to the clones under investigation.

3.5 Generation of domain specific DNA probes by PCR

The hybridization of membrane-bound DNA with probes detecting defined DNA fragments identifies specific sequences in the bound nucleic acids. Probes detecting the desired target sequences are generated based on the cDNA of these fragments which are cloned into common plasmid vectors. We used the sequences of the tenascin C fnIII domains A1, A2, A4, B, C, D and 6 as negative control which were inserted in pBluescript II KS+ vectors.

These inserts are labelled to use them in expression studies. The labelling with fluorescein has several advantages. When using non-radioactive probes no special safety regulations must be obeyed. Additionally, the probes can be used for a longer time period. This is of special advantage when several probes are used in parallel. Manufacturer's data state that fluorescein-labelled probes are stable for 6 months without decreasing activity. Radioactively labelled probes would lose sensitivity after a few days because the isotopes disintegrate continuously. Indeed, the probes generated in our lab could be used for several years (Joester & Faissner, 1999; von Holst et al., 2007).

The labelling was performed with fluorescein-coupled dUTP (Amersham). The manufacturer's labelling kit could not be used because it is based on a random primer labelling method. This uses the hybridization of 8 to 10 bases long random primers to single DNA strands. In a polymerization mix with the labelled nucleotide the Klenow fragment of the DNA polymerase I generates the complementary probe. But the tenascin C domains are less than 300 base pairs long and therefore have only a few potential binding sites for random primers which may lead to only very short probes. The labelling efficiency is too low (1 labelled base in 50 bases, according to manufacturer's data) to achieve an appropriate labelling frequency. Therefore we developed a labelling protocol which uses a PCR method to generate dUTP-labelled DNA probes.

The Taq polymerase incorporates dUTP with less efficiency than unlabelled dTTP. Therefore the exclusive use of dUTP would show the optimum of labelled probe but only low yield of amplification product. A low amount of fluorescein-dUTP and higher amount of dTTP reverses this effect and leads to a higher product yield but low labelling efficiency. We adjusted the PCR conditions to the optimal yield of labelled amplification product.

The optimal reaction conditions required only 10 pg of plasmid DNA and a low amount of dNTPs (20 μM). The reaction mix contained 60 mM Tris/HCl, pH 8,5; 15 mM $(NH_4)_2SO_4$; 2 mM $MgCl_2$; 0,2 μM sense primer; 0,2 μM antisense primer and 1 Unit Taq polymerase in 25 μL volume. The cycling conditions are dependent on the hybridization temperature of the respective primers and the length of the expected product.

In the first labelling reactions with domain C different amounts of dTTP were replaced with FI-dUTP (3 to 50% equivalent to 0,6 to 10 μM FI-dUTP). The amount of the products increases with decreasing amounts of the labelled nucleotide. The labelling efficiency was also tested on dot blots with different concentrations of the plasmid containing the C domain. 1μL of the PCR products were used in the hybridization solution. After an over-night incubation the blots were developed with an alkaline phosphatase-coupled antibody detecting fluorescein. The detection sensitivity was proportional to the concentration of the FI-dUTP used in the labelling reaction. The subsequent labelling reactions were performed using 17,5 μM dTTP and 2,5 μM fluorescein-11-dUTP. All probes detecting the fnIII domains A1, A2, A4, B, C and D of tenascin C were labelled with this method and called FI-A1, FI-A2,... Figure 5 shows the resulting PCR amplicons.

The fluorescein-labelled probes were tested for their detection capability of different dilutions of the respective plasmids. The senstivity was different for the probes and therefore their concentration was adjusted in the hybridization solution. The hybridization results show that the sensitivity is equal between 3 pg up to 1 ng of the target sequence (figure 6A). This sensitivity is much higher than that seen for agarose gels stained with ethidium bromide which is in the range of 1 ng DNA. The sensitivity was tested regularly to adjust the stability or labelling efficiency of the different probes but none showed a significant reduction in detection efficiency over time.

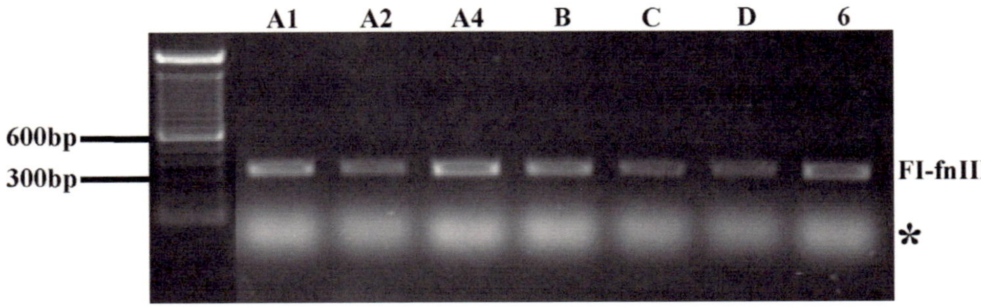

Fig. 5. Fluorescein-labelled probes on an agarose gel. The DNA probes labelled with FI-dUTP were applied to an agarose gel and show bands in the expected size of less than 300 bp. The asterisk designates the fluorescence signal of the non-incorporated nucleotides.

The specificity of the fluorescein-labelled probes was tested with seven dot blot stripes containing the plasmids pA1, pA2, pA4, pB, pC, pD, p6 and pJT1# in equal concentrations. The stripes were hybridized with the probes for the single fnIII domains (FI-A1, FI-A2,...) and the alkaline phosphatase reaction was developed (figure 6B). Highly stringent hybridization and washing conditions minimised the cross-reactivity of the probes with unspecific target sequences. These conditions included the hybridization and the first washing steps at 72°C

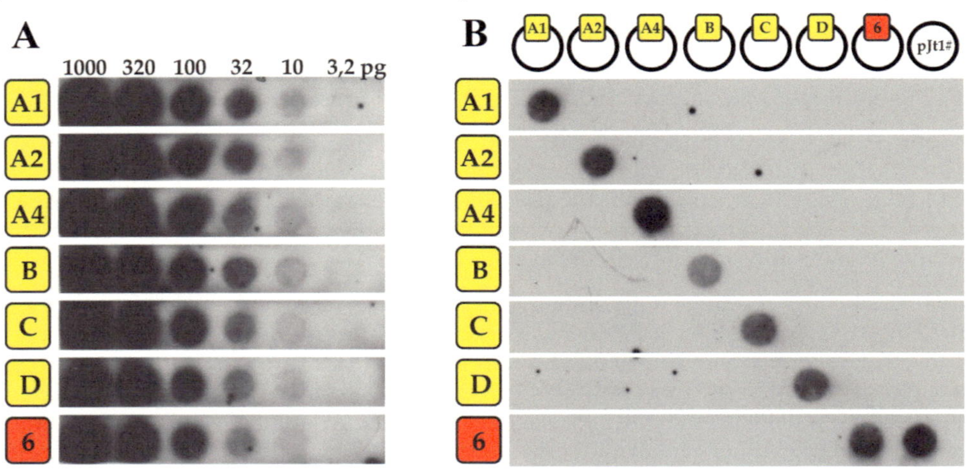

Fig. 6. (A) Sensitivity of fluorescein-labelled probes. The plasmids containing the single fnIII domains A1, A2, A4, B, C, D and 6 (designated pA1, pA2, ...) were diluted and applied to the nylon membranes in dots. The membranes were incubated with the respective probes and the reaction developed with an anti-fluorescein antibody coupled to alkaline phosphatase. A minimum of about 10 pg of target sequence was detected by each of the probes. The probes were diluted so that all of them detected their targets in a comparable way. (B) Specificity of fluorescein-labelled probes. Seven identical dot blots containing 10 ng of the plasmids pA1, pA2, pA4, pB, pC, pD, p6 and 19 ng pJT1# (corresponding to 1 ng target sequence) were hybridized with the different fluorescein-labelled probes. The probes detect their target sequences with high specificity. Although domains A1 and A4 are highly identical the probes do not show a significant cross-reactivity. The probe FI-6 detects the plasmid pJT1# which contains domain number 6.

and the use of 0,5% SDS in the washing buffer. The highest probability for a cross-reactivity exists between the domains A1 and A4 because their nucleotide sequence is 80% identical. Only in a few cases a light background signal could be detected when using these probes.

3.6 Hybridization

The generated probes are used in an *in vitro* hybridization protocol and applied to the nylon membranes containing the plasmid DNA from the clones which shall be analysed. The nylon membranes with the bound plasmids are washed in 5xSSC and pre-incubated in hybridization solution (5x SSC; 0,1% SDS; 5% dextrane sulfate; 5% liquid block (Amersham)) for 30 minutes at 72°C with gentle agitation. An appropriate amount of the probes which must be determined in preliminary experiments is added to 200μL of hybridization solution. The probes are denatured at 96°C for 5 minutes and applied to the membranes. The hybridization takes place over night at 72°C with constant agitation in a hybridization oven.

3.7 Signal detection

We used two different methods for the detection of the hybridized probes. A detection protocol to obtain chemoluminescence signals uses the fluorescein gene images CDP-*Star* detection system (Amersham). The other option was the development of a colour reaction using NBT and BCIP as alkaline phosphatase substrates.

When the DNA on the membranes was hybridized over night with the fluorescein-coupled probes the membranes can be washed for 2 x 15 minutes in wash buffer 1 (0,1x SSC; 0,5% SDS) at 72°C. To block unspecific binding sites they are incubated for one hour in blocking buffer (10% liquid block (Amersham) in detection buffer (100 mM Tris/HCl, pH 7,5; 300 mM NaCl)) before the alkaline phosphatase-coupled anti-fluorescein antibody (1:5000 in detection buffer with 0,5% BSA) is applied for an hour. Unbound antibody is washed away with wash buffer 2 (0,3 % Tween-20 in detection buffer) 4x 8 minutes. For the chemoluminescence detection the blot is moistened with a dioxetane-based substrate solution (CDP-*Star* detection reagent (Amersham). After 3 minutes the excess substrate solution is removed and the blot placed between two sheets of foil and laid on an autoradiographic film. Depending on the DNA concentration the optimal detection time was between 10 and 60 minutes.

For the alternative developing method the membranes are washed 3x 5 minutes in wash buffer A (100 mM Tris/HCl, pH 7,4; 150 mM NaCl; 0,3% Tween-20) after the antibody incubation, 2x 5 minutes in wash buffer B (100 mM Tris/HCl, pH 9,5; 100 mM NaCl) and 3x 10 minutes in TBS (50 mM Tris/HCl, pH 7,5; 150 mM NaCl). To develop the colour reaction the membranes are wetted with staining solution containing NBT and BCIP (Roche) and not shaken any more. The colour reaction will appear after five to 60 minutes. The reaction can be stopped with water and the membranes dried afterwards. Because the detection sensitivity is lower for the colour reaction a higher amount of plasmid DNA must be used in this case. 50 ng of target sequence in the spots lead to good results (data not shown).

3.8 Analysis of domain combinations

The read-out of the results is straight forward. The membranes show positive signals whenever the respective domain is present in the clone. Every plasmid DNA shows a specific pattern of positive and negative signals and therefore stands for the presence or absence of a given single domain.

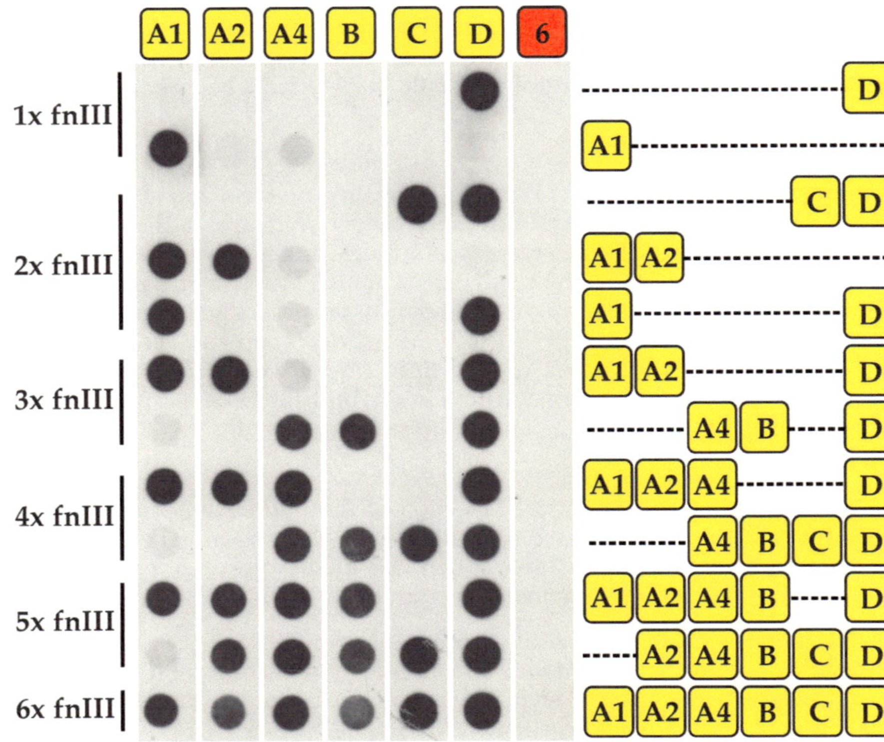

Fig. 7. Example of screening results. Plasmids containing different numbers of alternatively spliced fnIII domains were applied to nylon membranes in dots. Seven identical blots were generated and hybridized with the fluorescein-labelled probes FI-A1, FI-A2, FI-A4, FI-B, FI-C, FI-D and FI-6 as negative control. After the development of the alkaline phosphatase reaction the dot blots show positive signals whenever the respective fnIII domain is present in the plasmid. Therefore the domain combination of every single clone can be directly read out from the blots.

In intensive studies of the expression pattern of Tenascin C isoforms in the developing brain and in neural stem cells (Joester & Faissner, 1999 and von Holst et al., 2007) we detected 28 different isoforms of Tenascin C out of 64 possible ones which could theoretically be generated with six independently spliced domains ($=2^6$). We had several hundred clones to identify which contained between one and six alternatively spliced fnIII domains. The membranes we prepared were handled separately depending on the expected number of domains to be present in the plasmids. Figure 7 shows an example of the analysis of several clones with different numbers of fnIII domains. Plasmids from the distinct subcloning reactions were spotted onto seven nylon membranes and hybridized with the probes FI-A1, FI-A2, FI-A4, FI-B, FI-C, FI-D and FI-6 as negative control. The signals show that different combinations of fnIII domains can be contained in the plasmids. The clones with only one alternatively spliced domain displayed here for example contain the domains A1 or D, respectively. Indeed, these were the most common domains among single-domain clones when a complete screen was performed (Joester & Faissner, 1999; von Holst et al., 2007). The variability of domain combinations is higher in the middle-size clones with two, three or four alternatively spliced fnIII domains. The plasmids containing five additional cassettes on the other hand show usually the absence of fnIII domain C and only few miss A1. The extensive screens for the expression profiles of fnIII domains being expressed in postnatal mouse cerebellum or neural stem cells show that the possible variability among the clones is not utilised. 64 different isoforms of tenascin C would be theoretically possible but only 28 forms were found. Some combinations of domains were never seen like the direct link between the fnIII domains C and 6 or A4 and C.

3.9 PCR of single domains

To confirm the results of the hybridization and to clarify ambiguous signals we carried out PCRs for the single fnIII domains that were detected in the plasmid DNA. It is important to highly dilute the plasmid DNA and use only 10 to 20 pg plasmid DNA as template and to use highly specific PCR conditions. The specificity of the PCR conditions was confirmed before because the fnIII domains show high similarities and could therefore lead to false-positive signals when using standard PCR conditions on plasmid templates. We used a 2-step PCR with a high annealing temperature of the primers of 72°C and combined the annealing step with the elongation step to a 40-seconds 72°C incubation step. The cycling was therefore between 20 seconds 94°C and 40 seconds 72°C. We also skipped the final 5-minutes elongation step which we usually applied, especially for the addition of adenosines for cloning purposes.

Figure 8 shows the high specificity of these PCR conditions when the different fnIII domain-containing plasmids were used in these reactions. Only for those plasmids amplicons were generated when the respective primer pair was used. Therefore, we had an additional tool to confirm the dot blot results.

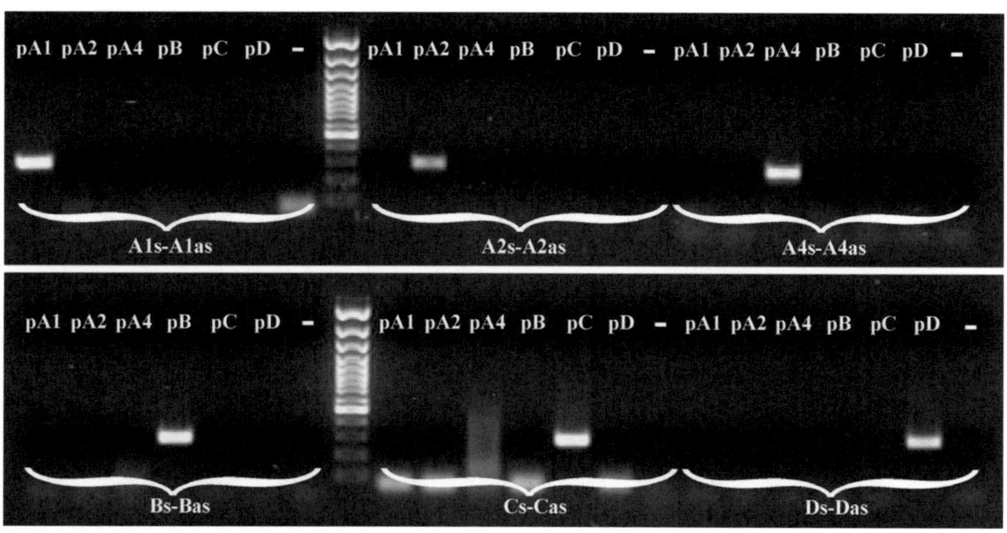

Fig. 8. PCRs for single fnIII domains. The primer pairs A1-s / A1-as, A2-s / A2-as,... were used in PCRs for the amplification of single fnIII domains. The domains have some similarities in their sequences. Therefore the PCR conditions must be highly specific. PCRs with stringent conditions amplify only products from plasmids containing the respective domain. Such conditions can be used to test plasmids with unclear domain composition.

4. Adaptation to general application

Many genes are subject to alternative splicing and most of them show an exon skipping mode which implies the inclusion or exclusion of single exons. When the possible exon structure leading to the appearance or absence of single domains is known the expression profile of

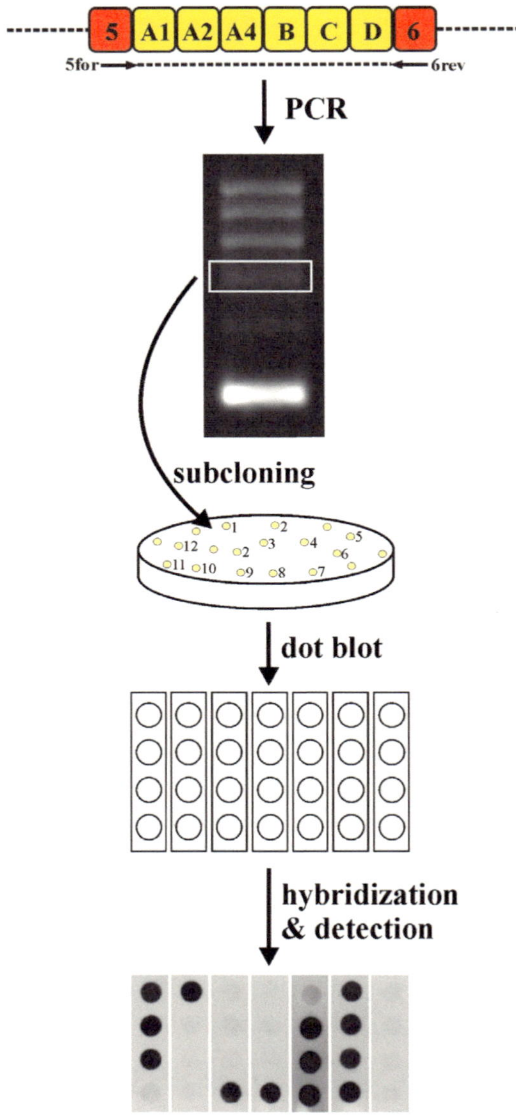

Fig. 9. Schematic presentation of the method. Primers flanking the alternatively spliced region of the molecule are used to generate PCR products of different sizes which are separated on an agarose gel. The single bands are cut out off the gel and subcloned separately. The resulting clones can be analysed with a dot blot hybridization procedure with non-radioactively labelled probes. Positive and negative signals display the domain composition of every single clone.

the domains can be analysed using the method presented here. Some preliminary steps are necessary before a screen for expressed domains can be started but when the system is set up once it can be used for the screening of many PCR products over a long time.

To start such a screen the following steps must be accomplished:
1. Clarify the domain sequence
2. Generate primers detecting the single domains
3. Clone the single domains into plasmid vectors
4. Use these vectors as templates for PCRs generating dUTP-labelled DNA probes

A screen includes the generation of the plasmids and the dot blot before the hybridization can start. Therefore conduct the following steps to start a screen:

1. Isolate RNA from the tissue or cell type under investigation
2. Prepare cDNA based on this RNA
3. Use this cDNA as template in PCRs for the alternatively spliced region of your gene
4. Separate the amplicons on an agarose gel and cut off the single bands
5. Clone the PCR products into plasmid vectors
6. Dilute the plasmid vectors to appropriate concentrations
7. Apply the plasmid solutions onto nylon membranes
8. Denature the DNA and bind it covalently to the membranes
9. Hybridize the DNA on the membranes with the probes
10. Wash under stringent conditions
11. Apply an antibody to the labelling marker
12. Develop the enzyme reaction
13. Read out your domain structure

Figure 9 shows the summary of the method:

5. Conclusion

With the method presented here we developed a possibility to unravel unknown structures of splice products for alternatively spliced transcripts. The example we analysed was the extracellular matrix molecule tenascin C but any other multimodular protein can be examined in a similar way. With some preliminary preparations an operational tool is at hand which makes the screening of many clones and therefore the generation of an expression profile possible.

6. Acknowledgements

The authors wish to thank very much Dr. Angret Joester for providing unpublished material concerning the dot blot assays. We also thank the German Research Foundation (DFG) (SPP 1048, Fa 159/11-1, 2, 3) and the GRK 736 for support.

7. References

Berget, S.M., Moore, C. & Sharp, P.A. (1977) Spliced segments at the 5′ terminus of adenovirus 2 late mRNA. Proc Natl Acad Sci U S A. 1977 Aug;74(8):3171-5.

Chow, L.T., Gelinas, R.E., Broker, T.R. & Roberts, R.J. (1977) An amazing sequence arrangement at the 5′ ends of adenovirus 2 messenger RNA. Cell. 1977 Sep;12(1):1-8.

Czopka, T., Von Holst, A., Schmidt, G., Ffrench-Constant, C. & Faissner, A. (2009) Tenascin C and tenascin R similarly prevent the formation of myelin membranes in a RhoA-dependent manner, but antagonistically regulate the expression of myelin basic protein via a separate pathway. Glia. 2009 Dec;57(16):1790-801.

Czopka, T., von Holst, A., ffrench-Constant, C. & Faissner, A. (2010) Regulatory mechanisms that mediate tenascin C-dependent inhibition of oligodendrocyte precursor differentiation. J Neurosci. 2010 Sep 15;30(37):12310-22.

Dobbertin, A., Czvitkovich, S., Theocharidis, U., Garwood, J., Andrews, M.R., Properzi, F., Lin, R., Fawcett, J.W. & Faissner, A. (2010) Analysis of combinatorial variability reveals selective accumulation of the fibronectin type III domains B and D of tenascin-C in injured brain. Exp Neurol. 2010 Sep;225(1):60-73. Epub 2010 May 5.

Erickson, H.P. & Inglesias, J.L. (1984) A six-armed oligomer isolated from cell surface fibronectin preparations. Nature. 1984 Sep 20-26;311(5983):267-9.

Garcion, E., Faissner, A. & ffrench-Constant, C. (2001) Knockout mice reveal a contribution of the extracellular matrix molecule tenascin-C to neural precursor proliferation and migration. Development. 2001 Jul;128(13):2485-96.

Garcion, E., Halilagic, A., Faissner, A. & ffrench-Constant, C. (2004) Generation of an environmental niche for neural stem cell development by the extracellular matrix molecule tenascin C. Development. 2004 Jul;131(14):3423-32.

Garwood, J., Theocharidis, U., Calco, V., Dobbertin, A. & Faissner, A. (2011) Existence of Tenascin-C Isoforms in Rat that Contain the Alternatively Spliced AD1 Domain are Developmentally Regulated During Hippocampal Development. Cell Mol Neurobiol. 2011 Oct 4. The paper has been published in (2012); add the correct designation: Cell. Mol. Neurobiol. 32(2), 279-287

Gates, M.A., Thomas, L.B., Howard, E.M., Laywell, E.D., Sajin, B., Faissner, A., Götz, B., Silver, J. & Steindler, D.A. (1995). Cell and molecular analysis of the developing and adult mouse subventricular zone of the cerebral hemispheres. J Comp Neurol, 361(2), 249-66.

Joester, A. & Faissner, A. (1999). Evidence for combinatorial variability of tenascin-C isoforms and developmental regulation in the mouse central nervous system. J Biol Chem, 274(24), 17144-51.

Michele, M. & Faissner, A. (2009) Tenascin-C stimulates contactin-dependent neurite outgrowth via activation of phospholipase C. Mol Cell Neurosci. 2009 Aug;41(4):397-408. Epub 2009 Apr 24.

Orend, G. & Chiquet-Ehrismann, R. (2006). Tenascin-C induced signaling in cancer. Cancer Lett, 244(2), 143-63.

Rigato, F., Garwood, J., Calco, V., Heck, N., Faivre-Sarrailh, C. & Faissner, A. (2002) Tenascin-C promotes neurite outgrowth of embryonic hippocampal neurons through the alternatively spliced fibronectin type III BD domains via activation of the cell adhesion molecule F3/contactin. J Neurosci. 2002 Aug 1;22(15):6596-609.

von Holst, A., Egbers, U., Prochiantz, A. & Faissner, A. (2007) Neural stem/progenitor cells express 20 tenascin C isoforms that are differentially regulated by Pax6. J Biol Chem, 282, 9172-9181.

24

Submicroscopic Human Parasitic Infections

Fousseyni S. Touré Ndouo
Medical Parasitology Unit,
Centre International de Recherches Médicales de Franceville (CIRMF), Franceville
Gabon

1. Introduction

Polymerase chain reaction (PCR) amplification provides a powerful tool for parasite detection. This chapter examines the use of PCR to diagnose malaria in patients with low parasite densities (submicroscopic infections, SMI) and also occult loaiosis (OL: *Loa loa* infection without detectable circulating microfilaria on standard microscopy). It provides therefore the issue of management of these kinds of infections with regard to the eradication policy of such pathogens.

1.1 Classification

i. **Malaria:** Malaria is caused by *Plasmodium* parasites, of which there are about 200 species (Levine ND 1980). These protozoans belong to the *Apicomplexa* phylum, *Sporozoa* class and *Haemosporidae* subclass (Levine ND 1970). They are obligatory intracellular parasites. Two successive hosts, humans and mosquitoes (Culicidea and Anophelinea), are necessary for their life cycle. Four main species infect humans, namely *Plasmodium falciparum*, *P. vivax*, *P. malariae* and *P. ovale*. A fifth species, *P. knowlesi*, is currently spreading in south-east Asia and Oceania. This species derived from chimpanzees has caused more than 250 human cases of malaria in Malaysia but is still considered to be zoonotic (Figtree et al. 2010). *P. falciparum* causes most life-threatening infections.
   Human is the intermediate host for malaria, wherein the asexual phase of the life cycle occurs. The sporozoïtes, inoculated by the infested female *Anopheles* mosquito, initiate this phase of the cycle from the liver, and continue within the red blood cells. From the mosquito bite, tens to few hundred invasive sporozoïtes are introduced into the skin. Following the intradermal drop, some sporozoïtes are destroyed locally by the immune cells, or enter into the lymphatic vessels, and some others can find blood circulation (Megumi L et al. 2007; Ashley M et al. 2008; Olivier S et al. 2008). The sporozoïtes that find peripheral blood circulation reach the liver within a few hours. It has been recently demonstrated that these sporozoïtes travel by a continuous sequence of stick-and-slip motility, using the thrombospondin-related anonymous protein (TRAP) family and an actin–myosin motor (Baum J et al. 2006; Megumi L et al. 2007; Münter S et al. 2009). The sporozoites migrate into hepatocytes and then grow within parasitophorous vacuoles and develop to the schizont stage which releases merozoites (Jones MK et al. 2006; Kebaier C et al. 2009). The entire pre-eryhrocytic phase lasts about 5–16 days depending

on the parasite species (5-6 days for *P. falciparum*, 8 days for *P. vivax*, 9 days for *P. ovale*, 13 days for *P. malariae* and 8-9 days for *P. knowlesi*. The pre-erythrocytic phase remains a "silent" phase, with little pathology and no symptoms, as only a few hepatocytes are affected (Ashley M et al. 2008). This phase is a single cycle, contrasting to the next, erythrocytic stage, which occurs repeatedly.

ii. **Loaiosis:** Filariasis are typically chronic tropical diseases caused by nematodes of the *Filariidae* family, transmitted by flies or mosquitoes. Eight species are currently known to infect humans, namely *Wuchereria bancrofti, Brugia malayi, B. timoni, Onchocerca volvulus, Loa loa, Mansonella perstans, M. ozzardi*, and *M. streptocerca*. Three groups of filariasis have been distinguished on the basis of their human target tissues: lymphatic filariasis (wuchereriasis and brugiasis); cutaneous dermal filariasis (loaiosis, onchocerciasis, and *streptocerca* mansonelliasis) and serous filariasis (*perstans* and *ozzardi* mansonelliasis) (Gentilini 1982). The vectors are blood-sucking flies and female mosquitoes. The microfilarial eggs or embryos are ingested by the vector when it bites an infected human. These microfilariae become infective stage L3 larvae after two successive mutes within the vector, and are transmitted to a new human host through a new blood meal or bite.

More than 3.3 billion people are exposed to filariosis, and an estimated 300 million people are infected. Loaiosis occurs in Africa, brugiaioses in South Asia, wuchereriasis in Africa and Asia, onchocerciasis in Africa, Central and South America and Asia (Yemen), *perstans* mansonelliasis in Africa and Central and South America, *ozzardi* mansonelliasis in Central and South America, and *streptocerca* mansonellaiosis in Africa.

L. loa infection (loaiosis) was initially described in 1770 by Mongin, in a female slave originating from West Africa and living on Saint Domingue island. Guyon et al. found the same worm in Gabon (Central Africa) in 1864. *L. loa* was first described in detail by Brumpt *et al.* in 1904, and then by Connors *et al.* in 1976. Although discovered in the Antilles, *L. loa* is restricted to Africa (Gentilini 1982). The adult worms live under the skin for about 15 years (Gentilini et al. 1982). The tabanides responsible for *L. loa* transmission are primarily *Chrysops dimidiata* and *silacea*, two forest species often present in the same hearth. Only the females are hematophagous, and they have diurnal activity.

1.2 Pathogenesis

i. **Malaria:** *P. falciparum* is responsible for most complicated forms of malaria and causes about 800 000 deaths a year, mostly among children in sub-Saharan African countries (WHO 2009). Malaria symptoms generally occur in three phases. After an incubation period of 7 to 10 days, symptoms begin with fever, aches and digestive disorders (febrile stomach upset). Then, when schizont rupture becomes synchronous, patients enter the feverish reviviscent schizogonic phase (periodic fever) of uncomplicated malaria. This phase is characterized by fevers typically appearing every 24 hours (third fever in infection by *P. vivax* or *ovalae*, every 48 hours, quartan fever in infection by *P. malariae* or *P. falciparum*), accompanied by a triad of symptoms: shivers, fever and sweating. Destruction of parasitized red blood cells leads to the release of malarial toxins and to TNF alpha production. The third phase, mainly seen with *P. falciparum*, corresponds to severe malaria (pernicious access), which sometimes occurs rapidly after infection. Clinical and biological signs are used to classify malaria (WHO 2000 gravity criteria, Imbert et al 2002). The reasons why some non immune individuals infected by *P. falciparum* develop severe malaria and die, while others have only uncomplicated malaria or remain asymptomatic, remain unclear (Marsh et al 1988). Severe anemia and cerebral malaria are responsible for most of the morbidity and mortality related to this disease in children. Despite abundant research, the pathophysiological mechanisms underlying severe forms are poorly understood. Several studies have implicated sequestration of *P. falciparum*-parasitized red blood cells (PRBC) in the lungs and brain (Taylor et al. 2001). This sequestration is characterized by PRBC adhesion (or cytoadherence), agglutination and rosetting. Cytoadherence of PRBC to host endothelial cells (EC) in brain and lung capillaries can obstruct the microvasculature, a phenomenon accompanied by changes in the T cell repertoire and by cytokine production (Mazier et al. 2000). This adherence is modulated by platelets (Brown et al. 2000) and is mediated by EC receptors such as CD36, intracellular adhesion molecule 1 (ICAM1), vascular cellular adhesion molecule 1 (VCAM1), CD31, integrins and hyaluronic acid (Hunt et al. 2003). PRBC adhesion can induce over-expression of inflammatory cytokines (Mazier et al. 2000) and EC apoptosis (Pino et al. 2003). Approximately 20% of *P. falciparum* isolates from Franceville, Gabon (Central Africa), were show to induce human lung endothelial cell (HLEC) apoptosis by cytoadherence (Touré et al. 2008). In addition, apoptogenic isolates were more frequent in children with neurological signs (prostration or coma), supporting the hypothesis that PRBC-mediated EC apoptosis could amplify blood-brain barrier disruption and dysfunction (Combes et al. 2005; Bisser et al. 2006). Whole transcriptome analysis revealed that 59 genes were more intensely transcribed in apoptogenic strains than in non apoptogenic strains (Siau et al. 2007). Knock-down of 8 of these genes by double-strand RNA interference significantly reduced the apoptogenic response in 5 genes (PF07_0032, PF10255, PFI0130c, PFD0875c, and MAL13P1.206). These five genes are known as *Plasmodium* apoptosis-linked pathogenicity factors (PALPF).

ii. Loaiosis: Loaiosis is characterized by calabar oedema (swelling) and conjunctivitis due to ocular passage of adult worms. Calabar oedema is transient and located on the face, limbs and back of the hands and fingers. *L. loa* is also called the "African eye worm". Meningoencephalic complications are an adverse effect of diethylcarbamazine treatment for hypermicrofilaremia. Other complications such as nephropathies, endocarditis, retinopathies, neuropathies and pneumonitis have been reported (Schofield et al. 1955; Hulin et al. 1994). Symptoms are more frequent in expatriates. Immunologically, loaiosis is characterized by hypergammaglobulinemia, hypereosinophilia and high IgE levels responsible for allergic symptoms. In endemic areas, loaiosis is the third reason for medical consultations in rural settings, although many microfilaremic subjects are asymptomatic. Occult loaiosis (amicrofilaremic infection) is defined as infection by the adult worm without peripheral microfilaremia on standard microscopy. Amicrofilaremic status is common among autochthonous residents and may be due to sequestration of microfilaria or to their massive destruction by the immune system, or to the presence of sterile adult worms. This form of infection is the most common in endemic areas. Other amicrofilaremic subjects are thought to be resistant. There is currently no way of discriminating between these two amicrofilaremic subgroups in the absence of (transient) ocular passage of adult worms.

1.3 Diagnostic challenges

i. **Malaria:** Light microscopy of blood smears remains the standard method for *Plasmodium* detection, both for clinical diagnosis and epidemiological surveys (Okell LC et al. 2009). However, sensitivity depends on parasite density in blood. In patients with low parasitemia, mixed infections, antimalarial treatment or chronic infection, microscopic diagnosis requires painstaking examination by an experienced technician. Low-density infections that cannot be detected by conventional microscopy are termed submicroscopic infections (SMI). *Plasmodium* species identification is mainly based on microscopic morphological characteristics but this is not entirely reliable (*Plasmodium vivax* resembles *P. ovale*). In addition, parasite morphology can be altered by drug treatment and/or sample storage conditions.

ii. **Loaiosis:** Human loaiosis differs from other filariasis by the fact that most patients have "occult" infection, with no circulating microfilaria. This peripheral amicrofilaremia can be due to microfilaria destruction by the immune system, and/or to their sequestration. These subjects cannot be diagnosed by microscopy and consequently go untreated, constituting a parasite reservoir. Before 1997, *L. loa* diagnosis was still based on microscopic examination and the prevalence was therefore underestimated. In contrast, because of cross-reactions, serological tests, and especially those based on total IgG detection, tend to overestimate prevalence. The existence of many cases of occult but symptomatic infection among residents in endemic areas implies the need for specific and sensitive detection.

2. PCR-based diagnosis of malaria

In 1993 a PCR method targeting the small subunit of the ribosomal RNA (SSUrRNA) gene was developed for use as an alternative to microscopy for detecting the four main *Plasmodium* species (Snounou et al. 1993, 1994, 1995). Nested PCR was used for its high sensitivity and specificity (Snounou et al. 1993). However, the nested reaction requires five separate PCR reactions and is therefore time-consuming, expensive and not always feasible in developing world laboratories. Several variants of this nested PCR method, such as semi-nested multiplex and one-tube multiplex have been developed (Mixon-Hayden T et al. 2010). In 1998 Jarra and Snounou showed that *Plasmodium* DNA is cleared very quickly from the bloodstream and that positive PCR amplification is usually associated with the presence of viable parasites. PCR positivity therefore indicates active *Plasmodium* infection. Since 1997, several PCR methods targeting other *P. falciparum* genes have been developed (Cheng et al. 1997; Filisetti et al. 2002). Their sensitivity has been estimated at 71%, 83% and 100% for the *MSP-2, SSUrRNA* and *STEVOR* genes, respectively (Oyedeji et al. 2007).

Real-time PCR has been reported to be able to improve parasite detection. Compared to *SSUrRNA* nested PCR, the real-time assay had a sensitivity of 99.5% and specificity of 100% for the diagnosis of malaria (Farcas GA et al. 2004). The real-time PCR method, specific for all *Plasmodium* species, avoids post-amplification sample handling and electrophoresis, and the result can be ready within 45 min (Farcas GA et al. 2004). This method would be useful for monitoring antimalarial drug efficacy, especially in areas of drug resistance (Lee MA et al. 2002).

More recently, it has been shown that dot18S (18SrRNA gene) and CYTB, two new molecular methods, are highly sensitive and allow high-throughput scaling up for many hundred samples (Steenkeste N et al. 2009). The CYTB is a nested PCR based on Plasmodium cytochrome b gene followed by species detection using SNP (single nucleotide polymorphism) analysis. The usefulness of these methods in detecting malaria has been demonstrated especially in low endemic areas.

2.1 Materials and methods

a. **Blood sampling**
   Samples must be collected in sterile tubes. For example, peripheral blood is collected in tubes containing an anticoagulant such as EDTA. However, some anticoagulants, such as heparin, inhibit the action of Taq DNA polymerase and should thus be avoided. Blood samples can also be collected in the form of drops on calibrated pre-punched paper disks (Serobuvard, LDA 22H, Zoopole, Ploufragan, France) (Ouwe-Missi-Oukem-Boyer et al. 2005).
b. **Microscopy**
   Thick and thin peripheral blood films were stained with Giemsa and examined by microscope. For microscope positive samples, the parasite load is expressed as the number of asexual forms of *P. falciparum*/µL of blood, assuming an average leukocyte count of 8000/µL.
c. **DNA template preparation**
   There are many useful techniques for DNA template processing. Plasmodial DNA extraction involves erythrocyte lysis and proteinase K digestion to prevent PCR inhibition.
   i. **DNeasy^R Blood & Tissue Kit:** Whole blood (200 µl) is used for DNA extraction with the DNeasy Blood & Tissue kit (QIAGEN, Hilden, Germany). Briefly, DNA extraction is carried out as follows. To a 1.5-ml tube containing 200 µl of while blood are added 20 µl of proteinase K solution and 200 µl of AL buffer (a detergent included in the kit). The mixture is pulse vortexed for 15 seconds and incubated for 15 minutes at 56°C. Two hundred microliters of cold ethanol is then added and the mixture is vortexed for 15 seconds. The mixture is transferred to a mini-column assembled on a 2-ml tube and centrifuged for 1 min at 8000 rpm. After centrifugation the 2-ml tube is discarded OK. The mini-column is recovered and placed on a new 2-ml tube. The mini-column is then washed with 500 µL of AW1 buffer (available in the kit) by centrifugation at 8000 rpm for 1 min. This washing step is repeated with another 500 µL of AW2 buffer, followed by centrifugation for 3 min at 14 000 rpm. The 2-ml tube is again discarded.
      The mini-column is placed on a 1.5-ml tube and 60 µl of AE elution buffer is added. This unit is left at room temperature for 10 min and then centrifuged for 1 min at 8000 rpm. The DNA is then recovered in the 1.5-ml tube and immediately used as a template or stored at 20°C.
   ii. **Dried blood-spot method** (DBS): DNA templates are extracted as described by Plowe CV et al in 1995. The dried blood spot is placed in 1 ml of phosphate buffered saline (PBS) containing 0.5% saponin and is incubated overnight at 4°C. The resulting brown solution is replaced with 1 ml of PBS and incubated for an additional 15-30 minutes at 4°C. Then, 200 µl of 5% Chelex 100 (Bio-Rad Laboratories, CA) is placed in clean tubes and heated to 100°C in a water bath. The disks are removed from the PBS and placed in the preheated 5% Chelex 100,

vortexed at high speed for 30 seconds and placed in a water bath at 100°C for 10 minutes with gentle agitation. The samples are then centrifuged at 10 000 *g* for 2 minutes, and the supernatant is removed and centrifuged as before. The supernatant is then collected in a clean tube and immediately used for PCR or stored at 20°C until use.

DNA can be also extracted from dried blood spots with several other methods, such as the QIAamp® DNA Mini Kit (QIAGEN, Hilden, Germany).

2.2 *P. falciparum* DNA amplification and detection

i. SSUrRNA gene amplification

Two microliters of DNA extract is amplified in a final volume of 25 µl containing 2.5 µl of 10X reaction buffer, 100 µM each dNTP (dATP, dGTP, dTTP, and dCTP), 0.5 pM each primer (rPLU5/rPLU6 (rPLU5 5'-CCT GTT GTT GCC TTA AAC TTC-3' and rPLU6 5'-TTA AAA TTG TTG CAG TTA AAA CG-3') for the primary reaction, and rFAL1/rFAL2 (rFAL 1 5'-TTA AAC TGG TTT GGG AAA ACC AAA TAT ATT-3' and rFAL 2 5'-ACA CAA TGA ACT CAA TCA TGA CTA CCC GTC-3') for the nested reaction) and 0.75 units of Taq DNA polymerase (QIAGEN, Hilden, Germany). The primer sequences (Table 1) are based on SSUrRNA sequences described elsewhere (Snounou et al. 1993). The PCR program is as follows: denaturation at 95°C for 5 min followed by 25 cycles (30 cycles in nested PCR) at 94°C for 1 min, 60°C for 2 min and 72°C for 2 min, with a final extension step of 5 min at 72°C.

ii. STEVOR gene amplification

The first round of amplification is performed with a reaction mix of 50 µl containing 5.0 µl of 10X reaction buffer, 200 µM each dNTP (dATP, dGTP, dTTP, and dCTP), 1.25 units of Taq

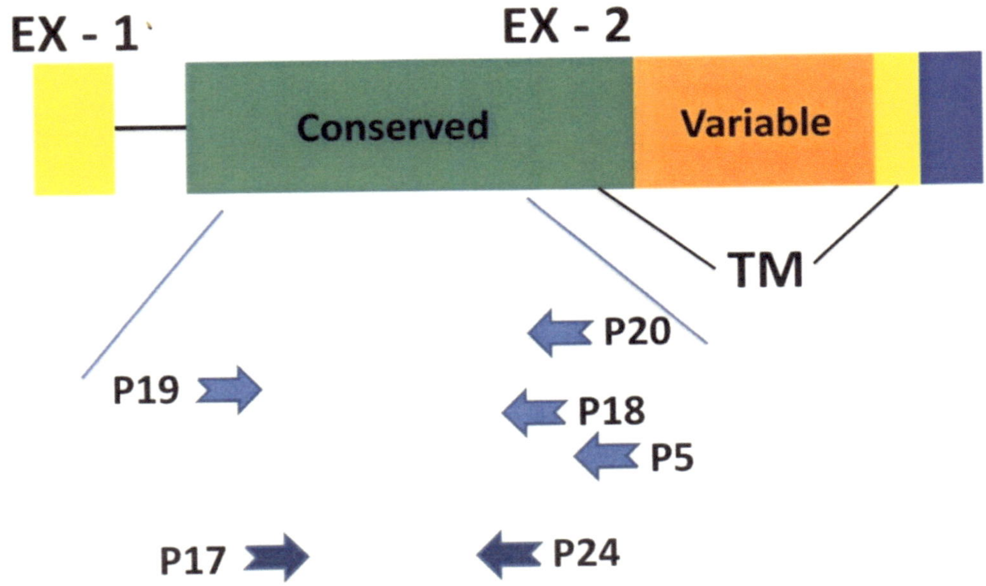

Schema 1. Schematic representation of the STEVOR PCR methodology (CHENG et al 1997).

DNA polymerase, 0.4 pM each primer (P5, P18, P19 and P20) (P5 5'-GGG AAT TCT TTA TTT GAT GAA GAT G-3', P18 5'-TTT CA(C/T) CAC CAA ACA TTT CTT-3', P19 5'-AAT CCA CAT TAT CAC AAT GA-3', P20 5'-CCG ATT TTA ACA TAA TAT GA-3') and 5 µl of DNA template. The PCR program is as follows: denaturation at 93°C for 3 min followed by 25 cycles of 30 s at 93°C, 50 s at 50°C and 30 s at 72°C, with a final extension step of 3 min at 72°C. Two microliters of the first-round PCR product is used for the second round of amplification, with a reaction mixture of 50 µl containing 5.0 µl of 10X reaction buffer, 200 µM each dNTP, 1.25 units of Taq DNA polymerase and 0.4 pM each primer (P17 and P24) (P17 5'-ACA TTA TCA TAA TGA (C/T) CC AGA ACT-3', P24 5'-GTT TGC AAT AAT TCT TTT TCT AGC-3'). The PCR conditions for the nested reaction are as follows: denaturation at 93°C for 3 s, followed by 25 cycles of 30 s at 93°C, 50 s at 55°C and 30 s at 72°C, with a final extension step of 3 min at 72°C.

iii. Detection procedures

Analysis of PCR products: After amplification, 10 µl of each PCR product is mixed with 1 µl of loading dye (0.25% bromophenol blue, 0.25% xylene cyanol and 40% w/v sucrose in water) and analyzed by electrophoresis on 1.5% agarose gel. The gel is stained with ethidium bromide or FluoProbes Gel Red (Interchim Montlucon, France) and the DNA is visualized and photographed under ultraviolet light.

***Plasmodium* SSUrRNA gene:**

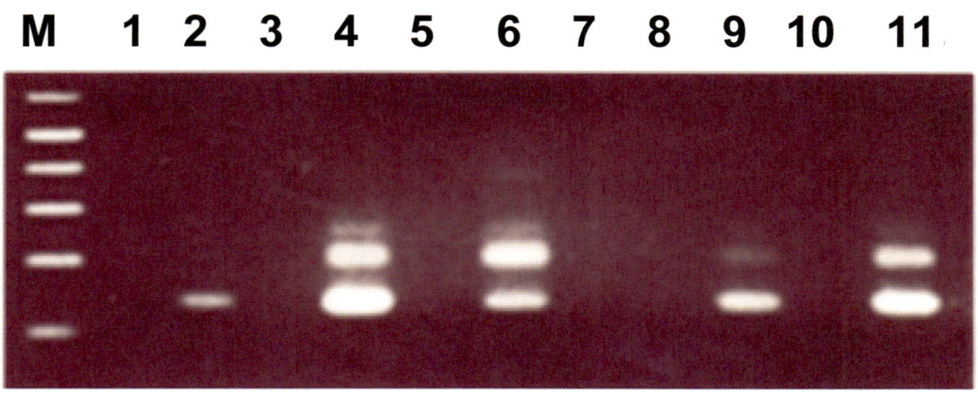

Fig. 1. Detection and speciation of *Plasmodium* by nested PCR using genus-specific primers and 1.5% agarose gel electrophoresis. Lanes 1 and 8: PCR-negative controls; lane 2: an individual with submicroscopic infection by *P. malariae* (size: 144 base pairs); lanes 3, 5, 7 and 10: PCR-negative individuals; lanes 4, 6, 9 and 11: individuals with submicroscopic co-infection with *P. falciparum* (size: 205 bp) and *P. malariae*. Lane M represents the DNA molecular weight marker (100 bp).

***P. falciparum* STEVOR gene**

Theoretically, three specific bands between 189-700 base pairs are generated using nested primers. We obtained a specific band of 250 bp for all *P. falciparum* isolates tested in Franceville, southeastern Gabon.

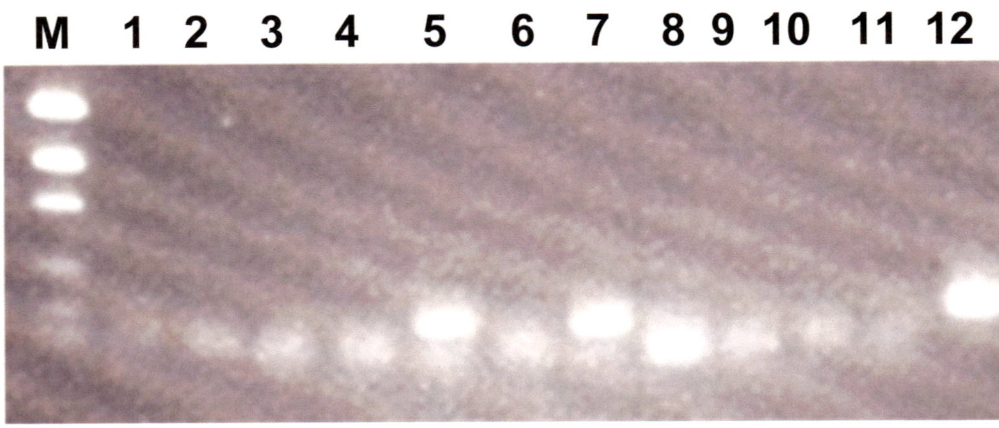

Fig. 2. Detection of the *Plasmodium* STEVOR gene by nested PCR using specific primers and 1.5% agarose gel electrophoresis. Lanes 1 and 10: PCR-negative controls; lanes 2, 3, 4, 6, 8, 9 and 11: PCR-negative samples; lanes 5 and 7: PCR-positive samples; lane 12: PCR-positive control; lane M: DNA molecular weight marker (123 bp).

3. PCR based diagnosis of *Loa loa*

In 1997, a PCR method (15r3-PCR) was developed to detect the repeat 3 region of the gene encoding the *L. loa* polyprotein in blood samples (Touré et al. 1997a, 1997b). Amicrofilaremic status is generally due to massive destruction of microfilaria, releasing parasite DNA into the bloodstream. These molecules may exist free in plasma, or be associated with cell-surface proteins, or even be contained in phagocytic cells. In addition, the adult worms can release DNA when they produce nonviable eggs or when they die after immune attack. The quantity of DNA released, whether from eggs, microfilaria or adult worms, is related to the parasite load of adult worms. The 15r3-PCR assay had a sensitivity of 95% with respect to detection of ocular passage of *L. loa* adult worms, and 100% compared to detection of microfilaremia.

3.1 Materials and methods

a. Blood sampling

   As previously mentioned blood samples must be collected by venipuncture into Vacutainer tubes containing an adequate anticoagulant such as EDTA.

b. Leukoconcentration

   Clinically, *L. loa* infection is diagnosed when migration of adult worms under the conjunctiva and/or skin is observed, or when a patient presents with classical symptoms. Diagnosis is classically based on standard microscopy. Microfilariae are the blood stage of *L. loa*. One milliliter of each blood sample is added to a 15-ml tube containing 9 ml of phosphate buffered saline (PBS), in duplicate. The mixture is treated with 600 µl of 2% saponin at room temperature for 15 min to lyse red blood cells, followed by centrifugation at 1000 *g* for 15 minutes at 4°C. The supernatant is discarded and the pellet is examined microscopically for microfilariae. The distinction between *L.*

loa and *M. perstans* microfilariae is based on size, motility, and by the presence of a sheath (*L. loa*).

Thick smears can be also prepared with venous blood and stained with Giemsa or hematoxylin-eosin to detect microfilariae. The QBC (Quantitative Coating Buffer) method is also used to detect *L. loa* microfilariae.

For microscopic detection of *L. loa* microfilariae, blood samples must be collected during the day, given the diurnal periodicity of human loaiosis.

c. Whole blood lysate processing

Whole blood (100 µl) is mixed with 500 µl of TE buffer (10 mM Tris pH 8; 0.1 mM EDTA pH 8) and spun at 10 000 *g* twice for 2 min, discarding the supernatant at each step. The pellet is resuspended in 500 µl of sucrose buffer (10 mM Tris pH 7.6, 5 mM $MgCl_2$ 1 M sucrose and 1% Triton X 100) and spun at 10 000 *g* twice for 2 min. After the final wash, the supernatant is discarded and the pellet is resuspended with 200 µl of prewarmed (56°C) proteinase K buffer (containing 20 mM Tris pH 8, 50 mM KCl, 2.5 mM $MgCl_2$, 100 µg/ml proteinase K and 0.5% Tween20), incubated at 56°C for two hours, then held at 90°C for 10 min. The DNA can be stored at 4°C for several days or at -20°C until required.

d. *L. loa* 15r3 gene amplification and detection

Primers corresponding to the 5' and 3' ends of the repeat 3 sequence of the gene coding for *L. loa* 15 kDa allergenic polyprotein are used. Primary amplification is done with a reaction mixture of 50 µl containing 2 µl of blood lysate, 1X PCR buffer (supplied by the manufacturer: 200 mM Tris-HCl pH 8.7, 100 mM KCl, 100 mM $(NH_4)_2SO_4$, 20 mM $MgSO_4$, 1% Triton x100, 1 mg/ml bovine serum albumin), 200 µM each dATP, dCTP, dGTP and dTTP, 1 µM each primer (15r3-1: 5'-AAT-CAG-GCA-AAT-AAT-GGC-ACA-AAA-3', 15r3-2: 5'-GCG-TTT-TCT-TCT-CAC-CAG-CTG-TCT-3') and 1 unit of DNA polymerase. Amplification is performed with a Perkin Elmer thermal cycler for 40 cycles: 94°C for 1 min (denaturation), 65°C for 1 min (annealing) and 72°C for 2 min

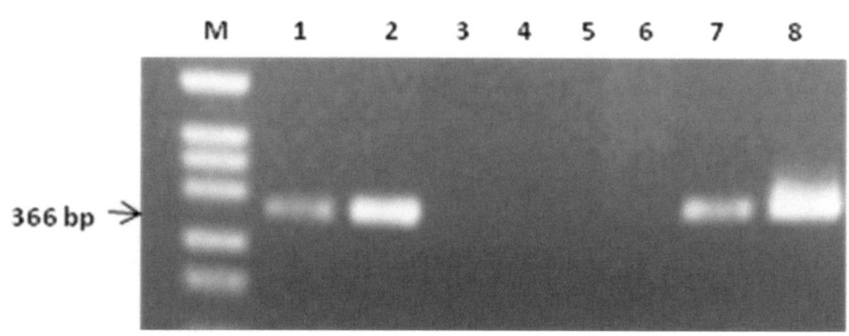

Fig. 3. Representative 1.5% agarose gel electrophoresis patterns of nested 15r3 PCR products. Lanes 1, 2 and 7: *L. loa* amicrofilaremic individuals (AMF) positive by PCR. Lanes 4 and 5: individuals negative by PCR. Lane 8: an individual with 100 *L. loa* microfilariae per ml and positive by PCR. Lanes 3 and 6: PCR-negative controls (no template); 5 µl of each nested PCR product was applied to each lane and revealed using UV transillumination after ethidium bromide staining. A fragment of 366 bp was observed with positive samples. Lane M: DNA molecular weight marker VI (Boehringer).

(extension), preceded by a "hot start" cycle at 96°C for 10 min, 80°C for 5 min and 94°C for 30 s. One microliter of product from the first-round amplification is used for a second round in the above conditions for 30 cycles. The following primers are used: 15r3-3: 5'GGC ACA AAA CAC TGC AGC AGT CCT3', and 15r3-4: 5'CAG CTG TCT CAA ATC GAA GAT TCT 3.'

4. Submicroscopic infection and disease management and control

i. Malaria:

The global strategy for malaria control is based on prevention, early diagnosis and prompt treatment. The detection limit of routine microscopy has been estimated to be about 100 parasites/milliliter, whereas PCR can detect as little as 0.01 parasite /micro liter (Mockenhaupt FP et al. 2002). Submicroscopic infection (SMI) including submicroscopic gametocytes is common in both symptomatic and asymptomatic individuals with malaria. A systematic review and analysis of field data carried out by Okell LC et al. in 2009 showed that the prevalence of *P. falciparum* was twice as high with PCR as with microscopy. In a village in Dienga, southeastern Gabon, PCR was performed on blood samples from asymptomatic individuals negative by microscopy: the prevalence of SMI (PCR positivity) was 13.7% by PCR and 7.2% by microscopy (Touré et al. 2006). A study carried out by Bouyou-Akotet et al. in 2010 in Libreville (capital of Gabon) showed an 18.2% prevalence of SMI in pregnant women. Recently, SMI was detected in 18% of symptomatic individuals in Franceville, southeastern Gabon, whereas the microscopic prevalence was 23% (author's personal data). It has been estimated that as many as 88% of infections remain undetectable by microscopy in low-transmission areas, where the PCR prevalence is generally under 10% (Okell LC et al. 2009). Thus, a high rate of SMI could undermine disease control programs. In endemic areas, it has been shown that *P. falciparum* SMI contributes to acute disease (Rogier C et al. 1996), and to malaria-associated anemia and inflammation (Mockenhaupt FP et al. 2002). It has also been shown that cerebral malaria is frequently associated with SMI in semi-immune individuals (Giha HA et al. 2005). Finally, Bouyou-Akotet et al. 2010 have demonstrated that SMI during pregnancy is associated with low birth weight, especially in primagravidae. As parasite resistance to antimalarial drugs is currently widespread and increasing, it is very important to identify resistant parasites in patients with SMI. Two major genes have been implicated in *P. falciparum* resistance to quinoline, namely *Pfcrt* (*P. falciparum* chloroquine resistance transporter) and *Pfmdr1* (*P. falciparum* multidrug resistance gene 1). Single-nucleotide polymorphisms (SNPs) in these genes are associated with resistance both *in vitro* and *in vivo* (Wongsrichanalai et al. 2002). Therefore, *P. falciparum* drug resistance is linked to particular parasite genotypes (Duraisingh et al. 1997). *P. falciparum* infection is generally polyclonal, and may thus involve both drug-sensitive and resistant genotypes. SMI detection can be used to evaluate the therapeutic effectiveness of anti-malarial drugs during mass treatments and preclinical trials.

SMI individuals are capable of infecting mosquitoes and contributing to human transmission (Coleman RE et al. 2004), mainly in areas of seasonal transmission (Nwakanma D et al. 2008). Microscopy fails to detect the parasite in 49.2% of all malaria cases and in 91.3% of gametocytemic individuals (Okell LC et al. 2009). Individuals whose blood smears are negative for gametocytes (submicroscopic gametocyte) are equally able to transmit the infection to mosquitoes as slide-positive individuals (Coleman RE et al. 2004). Thus, the SMI gametocyte reservoir may sustain malaria transmission despite efforts to fight malaria in endemic areas (Karl S et al. 2011). The prevalence of SMI, including submicroscopic gametocytes, must be assessed and taken into account in malaria control programs (Okell LC et al. 2009, Karl S et al. 2011).

Only patients with positive blood smears and/or rapid diagnostic tests (RDT) are routinely treated, while the treatment of patients negative by both methods depends on clinical signs and the physician's appreciation. These patients, including those with SMI, may represent more than 10% of infected individuals. In Gabon, SMI currently tends to be more frequent than microscopic infection, possibly due to better preventive policies and/or case management (Bouyou-Akotet et al. 2010). Treatment of all infected subjects, including those with SMI and submicroscopic gametocytes, would reduce the community parasite burden. Indeed, it has been shown that intermittent preventive treatment can reduce the prevalence and genetic diversity of *P. falciparum* malaria (Liljander A et al. 2010).

ii. *Loa loa*

Human loaiosis differs from other filariasis by the fact that most infected individuals do not have blood microfilariae detectable by standard microscopy. Since the first description of this filariasis, many epidemiologists have found a low prevalence of microfilaria despite local vector abundance. The notion that most patients clear their microfilaremia but continue to have (occult) infection is primarily based on the observation of adult worms during eye passage. The assumption that endemic resistant subjects also may exist (subjects able to completely eliminate *L. loa* infection) is still based on the same observations. Only a sensitive diagnostic test can confirm these assumptions. Our results have shown that 15r3-PCR is suitable for discriminating among endemic groups (microfilaremics, occult infected individuals (occults) and resistant subjects), as the results should be positive in the first two groups and negative in the last. Indeed, two-thirds of infected individuals in southeastern Gabon have occult loaiosis (OL) Touré et al. (1998, 1999a). This needs to be shown in a longitudinal study, however, as *L. loa* infection is characterized by its relative stability in humans and mandrills, the adult worm having a lifespan of about 15 years (Gentillini 1982, Pinder 1994).

This implies that the prevalence of loaiosis would be underestimated by microscopy. If *L. loa* DNA detection is a marker of active infection, all subjects positive by PCR should be treated. This would not have a major impact on health at the individual level but could reduce the parasite burden in the community, in turn reducing the intensity of transmission and resulting in public health benefits.

Studies of resistant individuals may provide interesting immunological information. Marked differences in humoral and cellular immune responses have already been noted between microfilaremic and amicrofilaremic patients (Pinder 1988, Akué 1997, Baize et al. 1997), as well as in the mandrill model (Leroy 1997). However, lacking a reliable method for diagnosing occult infection, it is not known if this difference is due to immunity directed against adult worms or against microfilaria. The identification of endemic groups ("microfilaremics", "occults" and "resistants") by 15r3-PCR method should allow immunological studies to be carried out with sera and cells from each endemic group, using antigens of each developmental stage of *L. loa*, and particularly infective larvae and adult worms. Such studies could help to identify possible cellular or humoral markers involved in resistance to infection, as well as the underlying mechanisms, including host genetic factors. These studies would open the way to investigations of the underlying molecular mechanisms.

In addition, the detection of OL by PCR will allow precise evaluation of filaricide effectiveness during mass treatment, and also that of new drugs in animal models. Pinder et al. showed in 1994 that experimental mandrill infection (*Mandrillus sphinx*) by human *L. loa* isolates led to the same parasitologic characteristics as the natural human infection. Thus, mandrills with occult infection (absence of microfilarae but presence of adult worms, as shown by 15r3-PCR positivity; Touré et al. 1998) can be used to evaluate macrofilaricidal drugs. It has been demonstrated that the 15-kDa polyprotein is conserved within human and simian *L. loa* (Touré et al. 1999b). L.15r3-PCR also detects simian occult *L. loa* and could be used to identify infected animals before their inclusion in preclinical trials.

Finally, serological tests using purified recombinant antigens or peptides offer much better specificity than those using crude antigens. When these antigens become available for loaiosis, immunoenzymatic methods like IgG4 ELISA will reach acceptable specificity. Comparison of ELISA and PCR results sould show whether or not specific IgG4 antibodies are markers of active *L. loa* infection.

5. Conclusion

The global strategy of eliminating the parasitic diseases especially malaria and filariasis is mainly based on prevention, early diagnosis and prompt treatment. However, most decisions still rely on microscopy diagnosis which is not always adapted in detecting all infections. Indeed, the success of any intervention depends of the effectiveness of tools and methods especially those allowing proper detection of parasites. PCR offers an exciting opportunity to diagnose submicroscopic malaria infections and occult loaiosis which may constitute a hidden reservoir of disease transmission. The detection of such infections would allow the accurate management of all cases necessary to progress from disease control to elimination.

6. References

[1] Levine ND. et al. A newly revised classification of the protozoa (1980). *Protozool.* 27, 37-58.

[2] Levine ND. Protozoan parasites of nonhuman primates as zoonotic agents (1970). *Lab Anim. Care* 20,377-82.

[3] Figtree M, Lee R, Bain L, Kennedy T, Mackertich S, Urban M, Cheng Q, Hudson BJ. Plasmodium knowlesi in human, Indonesian Borneo (2010). *Emerg Infect Dis.* Apr;16 (4):672-4.

[4] Megumi Y L, Alida Coppi, Georges Snounou, Photini Sinnis. Plasmodium sporozoites trickle out of the injection site. *Cell Microbiol.* 1 May 2007;9(5):1215–1222.

[5] Ashley M. Vaughan, Ahmed S. I. Aly, Stefan H. I. Kappe. Malaria parasite pre-erythrocytic stage infection: Gliding and Hiding. *Cell Host Microbe.* 11 September 2008;4(3):209–218.

[6] Olivier Silvie, Maria M Mota, Kai Matuschewski, Miguel Prudêncio. Interactions of the malaria parasite and its mammalian host. *Current Opinion in Microbiology* 2008;11:352–359.

[7] Jake Baum, Dave Richard, Julie Heale et al. A Conserved Molecular Motor Drives Cell Invasion and Gliding Motility across Malaria Life Cycle Stages and Other Apicomplexan Parasites. *The Journal of Biological Chemistry*. February 2006;281:5197-5208.

[8] Sylvia Münter, Benedikt Sabass, Christine Selhuber-Unke et al. Plasmodium Sporozoite Motility Is Modulated by the Turnover of Discrete Adhesion Sites *Cell Host & Microbe*. December 2009;6 (17):551-562.

[9] Malcolm K Jones, Michael F Good. Malaria parasites up close. *Nature Medicine* 2006; 12:170-171.

[10] Kebaier C, Voza T, Vanderberg J. Kinetics of Mosquito-Injected Plasmodium Sporozoites in Mice: Fewer Sporozoites Are Injected into Sporozoite-Immunized Mice. *PLoS Pathog* 2009; 5 (4):e1000399.

[11] Gentillini M., Duflo B. (1982). Filarioses. Médecine Tropicale- Paris: Flammarion. 173-195.

[12] Mogin Observations sur un ver trouvé dans la conjonctive à maribou. île Saint Domingue (1770). Jour. Med. Chirurgie Pharm. etc (Paris). 32 : 338-339. Guyon J (1964). Sur un nouveau cas de filaire, sous conjonctive, ou Filaria occuli des auteurs, observe au Gabon (Côte occidentale d'Afrique). *C. R. Séances Acad. Sci.* 59 : 743-748.

[13] Guyon J. (1864). Sur un nouveau cas de filaire, sous conjonctive, ou Filaria occuli des auteurs, observe au Gabon (Côte occidentale d''frique). *C. R. Séances Acad. Sci.* 59 : 743-748.

[14] Brumpt E. (1904). Les filarioses humaines en Afrique. *Compte Rendu de la Société de Biologie*, 56 : 758-760.

[15] Connor DH, Neafie RC, Meyers WM. (1976). Loiasis : Pathology of tropical and extra-ordinary diseases. Washington: Armed Forces Institute of Pathology. 356-359.

[16] World Health Organization (2009). WHO World Malaria report 2009.

[17] World Health Organization. Severe falciparum malaria (2000).*Trans R Soc Trop Med Hyg*; 94 (Suppl. 1): S1–S90.

[18] Imbert P and Gendrel D (2002). [Malaria treatment in children. 2. Severe malaria]. *Med Trop (Mars)* 62(6): 657-64.

[19] Marsh K, Marsh VM, Brown J, Whittle HC, Greenwood BM (1988). Plasmodium falciparum: the behavior of clinical isolates in an in vitro model of infected red blood cell sequestration.*Exp Parasitol* 65(2):202-8.

[20] Taylor HM, Triglia T, Thompson J, Sajid M, Fowler R, Wickham ME, Cowman AF, Holder AA. (2001). Plasmodium falciparum homologue of the genes for Plasmodium vivax and Plasmodium yoelii adhesrive proteins, which is transcribed but not translated. *Infect Immun*. Jun; 69 (6): 3635-45.

[21] Mazier D, Nitcheu J and Idrissa-Boubou M. (2000). Cerebral malaria and immunogenetics. *Parasite Immunol* 22(12): 613-23.

[22] Brown HC, Chau TT, Mai NT, Day NP, Sinh DX, White NJ, Hien TT, Farrar J, Turner GD. (2000). Blood-brain barrier function in cerebral malaria and CNS infections in Vietnam. *Neurology*. Jul 12;55(1):104-11.

[23] Hunt NH, Grau GE. (2003). Cytokines: accelerators and brakes in the pathogenesis of cerebral malaria. *Trends Immunol;* 24:491-499.

[24] Pino P, Vouldoukis I, Kolb JP, Mahmoudi N, Desportes-Livage I, Bricaire F, Danis M, Dugas B, Mazier D. (2003). Plasmodium falciparum--infected erythrocyte adhesion induces caspase activation and apoptosis in human endothelial cells. *J Infect Dis.* Apr 15;187 (8):1283-90. Epub 2003 Apr 2.

[25] Touré FS, Ouwe-Missi-Oukem-Boyer O, Bisvigou U, Moussa O, Rogier C, Pino P, Mazier D, Bisser S. (2008). Apoptosis: a potential triggering mechanism of neurological manifestation in Plasmodium falciparum malaria. *Parasite Immunol.* Jan;30 (1):47-51.

[26] Combes V, Coltel N, Faille D, Wassmer SC, Grau GE. (2006). Cerebral malaria: role of microparticles and platelets in alterations of the blood-brain barrier. *Int J Parasitol.* May 1;36(5):541-6.

[27] Bisser S, Ouwe-Missi-Oukem-Boyer ON, Toure FS, Taoufiq Z, Bouteille B, Buguet A, Mazier D. (2006). Harbouring in the brain: A focus on immune evasion mechanisms and their deleterious effects in malaria and human African trypanosomiasis. *Int J Parasitol.* May 1;36(5):529-40.

[28] Siau A, Toure FS, Ouwe-Missi-Oukem-Boyer O, Ciceron L, Mahmoudi N, Vaquero C, Froissard P, Bisvigou U, Bisser S, Coppee JY, Bischoff E, David PH, Mazier D. (2007). Whole-transcriptome analysis of Plasmodium falciparum field isolates: identification of new pathogenicity factors. *J Infect Dis.* Dec 1;196(11):1603-12.

[29] Schofield FD, (1955). Two cases of loiasis with peripheral nerve involvement. *Trans. Roy. Soc. Trop. Med. Hyg.* 49 : 588-589.

[30] Hulin C, et al., (1994). Atteinte pulmonaire d'évolution favorable au cours d'une filariose de type *Loa loa. Bull. Soc. Path. Ex.* 87: 248-250.

[31] Okell LC, Ghani AC, Lyons E, Drakeley CJ. (2009). Submicroscopic infection in Plasmodium falciparum-endemic populations: a systematic review and meta-analysis. *J Infect Dis.* Nov 15;200 (10):1509-17.

[32] Snounou, G. S., S. Viriyakosol, X. P. Zhu, W. Jarra, L. Pinheiro, V. E. Do Rosario, S. Thaithong, and K. N. Brown (1993). High sensitivity of detection of human malaria parasites by the use of nested polymerase chain reaction. *Mol. Biochem. Parasitol.* 61:315–320.

[33] Mixson-Hayden T, Lucchi NW, Udhayakumar V. (2010). Evaluation of three PCR-based diagnostic assays for detecting mixed Plasmodium infection. *BMC Res Notes.* Mar 31;3:88.

[34] Jarra W, Snounou G. (1998). Only viable parasites are detected by PCR following clearance of rodent malarial infections by drug treatment or immune responses. *Infect Immun.* Aug;66(8):3783-7.

[35] Cheng Q, Lawrence G, Reed C, Stowers A, Ranford-Cartwright L, Creasey A, Carter R, Saul A. Measurement of Plasmodium falciparum growth rates in vivo: a test of malaria vaccines. *Am J Trop Med Hyg.* 1997 Oct;57(4):495-500.

[36] Filisetti D, Bombard S, N'Guiri C, Dahan R, Molet B, Abou-Bacar A, Hansmann Y, Christmann D, Candolfi E. Prospective assessment of a new polymerase chain reaction target (STEVOR) for imported Plasmodium falciparum malaria. *Eur J Clin Microbiol Infect Dis.* 2002 Sep;21(9):679-81. Epub 2002 Sep 6.

[37] Oyedeji SI, Awobode HO, Monday GC, Kendjo E, Kremsner PG, Kun JF. (2007) Comparison of PCR-based detection of Plasmodium falciparum infections based on single and multicopy genes. *Malar J*. Aug 16;6:112.

[38] Farcas GA, Zhong KJ, Mazzulli T, Kain KC. (2004). Evaluation of the RealArt Malaria LC real-time PCR assay for malaria diagnosis. *J Clin Microbiol*. Feb;42(2):636-8.

[39] Lee MA, Tan CH, Aw LT, Tang CS, Singh M, (2002) Lee SH, Chia HP, Yap EP. Real-time fluorescence-based PCR for detection of malaria parasites. *J Clin Microbiol*. Nov;40(11):4343-5.

[40] Touré FS, et al., (1997a). Species-specific sequence in the repeat 3 region of the gene encoding a putative *Loa loa* allergen: a diagnostic tool for occult loiasis. *Am J Trop Med Hyg*. 56: 57-60.

[41] Touré FS, Odile Bain, Eric Nerrienet, et al. (1997b). Detection of *Loa loa*-specific DNA in blood from occult infected individuals. *Exp Parasitol*. 86: 163-70.

[42] Ouwe-Missi-Oukem-Boyer ON, Hamidou AA, Sidikou F, Garba A, Louboutin-Croc JP. (2005). The use of dried blood spots for HIV-antibody testing in Sahel]. *Bull Soc Pathol Exot*. Dec;98(5):343-6. French.

[43] Plowe CV, Djimde A, Bouare M, Doumbo O, Wellems TE (1995). Pyrimethamine and proguanil resistance-conferring mutations in Plasmodium falciparum dihydrofolate reductase: polymerase chain reaction methods for surveillance in Africa. *Am J Trop Med Hyg* 52: 565–568.

[44] Snounou G, Viriyakosol S, Zhu XP, Jarra W, Pinheiro L, do Rosario VE, Thaithong S, Brown KN. (1993). High sensitivity of detection of human malaria parasites by the use of nested polymerase chain reaction. *Mol Biochem Parasitol*. Oct;61(2):315-20.

[45] Mockenhaupt FP, Ulmen U, von Gaertner C, Bedu-Addo G, Bienzle U.(2002). Diagnosis of placental malaria. *J Clin Microbiol*. Jan;40(1):306-8.

[46] Touré FS, Mezui-Me-Ndong J, Ouwe-Missi-Oukem-Boyer O, Ollomo B, Mazier D, Bisser S. (2006). Submicroscopic Plasmodium falciparum infections before and after sulfadoxine-pyrimethamine and artesunate association treatment in Dienga, Southeastern Gabon. *Clin Med Res*. Sep;4(3):175-9.

[47] Rogier C, Commenges D, Trape JF (1996). Evidence for an age-dependent pyrogenic threshold of Plasmodium falciparum parasitemia in highly endemic populations. *Am J Trop Med Hyg*. Jun;54(6):613-9.

[48] Giha HA, A-Elbasit IE, A-Elgadir TM, Adam I, Berzins K, Elghazali G, Elbashir MI. (2005). Cerebral malaria is frequently associated with latent parasitemia among the semi-immune population of eastern Sudan. *Microbes Infect*. Aug-Sep;7(11-12):1196-203.

[49] Bouyou-Akotet MK, Nzenze-Afene S, Ngoungou EB, Kendjo E, Owono-Medang M, Lekana-Douki JB, Obono-Obiang G, Mounanga M, Kombila M. (2010). Burden of malaria during pregnancy at the time of IPTp/SP implementation in Gabon. *Am J Trop Med Hyg*. Feb;82(2):202-9.

[50] Wongsrichanalai C, Pickard AL, Wernsdorfer WH, Meshnick SR. (2002). Epidemiology of drug-resistant malaria. *Lancet Infect Dis*. Apr; 2 (4):209-18.

[51] Duraisingh MT, Drakeley CJ, Muller O, Bailey R, Snounou G, Targett GA, Greenwood BM, Warhurst DC (1997). Evidence for selection for the tyrosine-86 allele of the pfmdr 1 gene of Plasmodium falciparum by chloroquine and amodiaquine. *Parasitology*. Mar;114 (Pt 3):205-11.

[52] Coleman RE, Kumpitak C, Ponlawat A, Maneechai N, Phunkitchar V, Rachapaew N, Zollner G, Sattabongkot J. (2004). Infectivity of asymptomatic Plasmodium-infected human populations to Anopheles dirus mosquitoes in western Thailand. *J Med Entomol*. Mar;41(2):201-8.

[53] Nwakanma D, Kheir A, Sowa M, Dunyo S, Jawara M, Pinder M, Milligan P, Walliker D, Babiker HA (2008). High gametocyte complexity and mosquito infectivity of Plasmodium falciparum in the Gambia. *Int J Parasitol*. Feb;38(2):219-27. Epub 2007 Jul 18.

[54] Karl S, Gurarie D, Zimmerman PA, King CH, St Pierre TG, Davis TM. (2011). A sub-microscopic gametocyte reservoir can sustain malaria transmission *PLoS One*; 6 (6):e20805. Epub 2011 Jun 14.

[55] Liljander A, Chandramohan D, Kweku M, Olsson D, Montgomery SM, Greenwood B, Färnert A (2010). Influences of intermittent preventive treatment and persistent multiclonal Plasmodium falciparum infections on clinical malaria risk. *PLoS One*. Oct 27;5(10):e13649.

[56] Touré FS, Mavoungou E, Kassambara L, Williams T, Wahl G, Millet P, Egwang TG (1998). Human occult loiasis: field evaluation of a nested polymerase chain reaction assay for the detection of occult infection. *Trop Med Int Health*. 3: 505-511.

[57] Touré FS, Deloron P, Egwang TG, Wahl G (1999a). [Relationship between the intensity of *Loa loa* filariasis transmission and prevalence of infections] *Med Trop*. 59: 249-52. French.

[58] Touré FS, Ungeheuer MN, Egwang TG, Deloron P. (1999b). Use of polymerase chain reaction for accurate follow-up of *Loa loa* experimental infection in *Mandrillus sphinx*. *Am J Trop Med Hyg*. 61: 956-9.

[59] Pinder M, Everaere S, Roelants GE (1994). Loa loa: immunological responses during experimental infections in mandrills (Mandrillus sphinx). *Exp Parasitol*. Sep; 79(2):126-36.

[60] Pinder M Dupont A, Egwang TG. (1988). Identification of a surface antigen on *Loa loa* microfilariae the recognition of which correlates with the amicrofilaremic state in man. *Jour. Immunol*. 141 : 2480-2486.

[61] Akué JP, Hommel M., Devaney E. (1997). High levels of parasite- specific IgG1 correlate with the amicrofilaremic state in *Loa loa* infection. *Jour. Infect. Dis*. 175 : 158-163.

[62] Baize S, Wahl G, Soboslay PT, Egwang TG, Georges AJ. (1997). T helper responsiveness in human *Loa loa* infection: defective specific proliferation and cytokine production by CD4+ T cells from microfilaremic subjects compared with amicrofilaremics. *Clin. Exp. Immunol*. 108 : 272-278.

[63] Leroy E, Baize S, Wahl G, Egwang TG, Georges AJ. (1997). Experimental infection of a nonhuman primate with Loa loa induces transient strong immune activation followed by peripheral unresponsiveness of helper T cells. *Infection Immunity*. 65 (5) : 1876-1882.

25

Identification of Genetic Markers Using Polymerase Chain Reaction (PCR) in Graves' Hyperthyroidism

P. Veeramuthumari and W. Isabel
PG & Research Department of Zoology and Biotechnology, Lady Doak College, Madurai, Tamil Nadu India

1. Introduction

Thyroid is a butterfly shaped gland composed of two encapsulated lobes, located on either side of the trachea just below the cricoid cartilage. This is connected by thin isthmus and is composed of spherical thyroid follicles, which contain the hormone in colloidal form. T_3 and T_4 are active hormones secreted under the control of TSH from adenohypophysis of pituitary gland. T_3 is three to four fold more potent than T_4. It is involved in normal growth and development in children temperature regulation, metabolism, energy production and intelligence in both children adults. It ensures normal growth and development of nervous system [1].

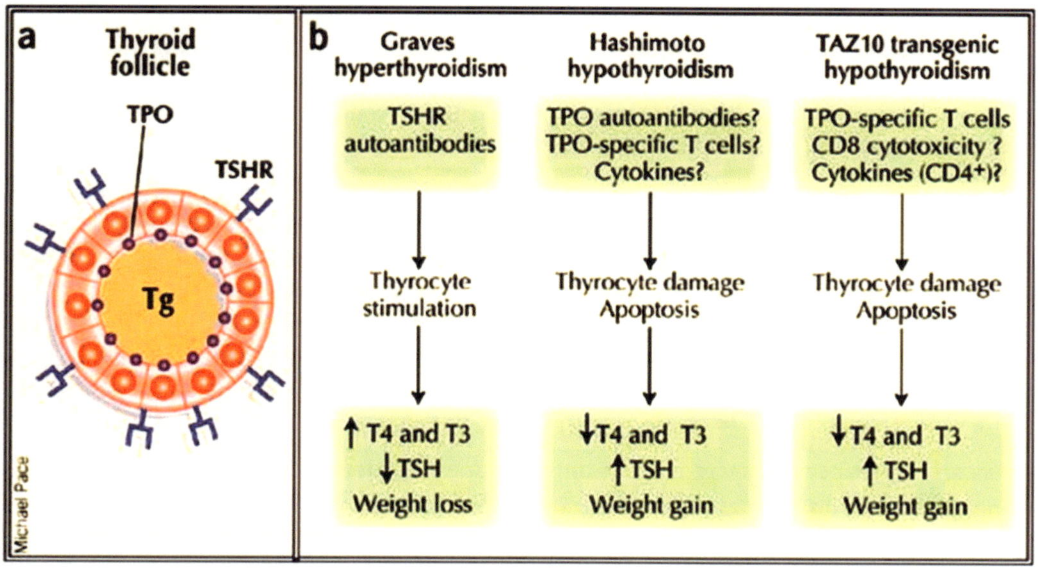

Fig. 1. Diagrammatic representation of variation of thyroid hormones in hypo and hyper thyroidism

The normal range of T_4 is suggested to be 77-155nmol/L, T3 to 1.2 -2.8nmol/L) and TSH to be 0.3-4 mU/L [2]. If the hormone levels are above or below the normal range, it leads to hyperthyroidism or hypothyroidism. The most common hypothyroid condition is Hashimoto's thyroiditis in adults and congenital hypothyroidism in children. Hyperthyroid conditions include Graves' disease, postpartum thyroiditis and thyrotoxicosis factitia.

Hyperthyroidism also leads to a number of complications like heart problems, brittle bones (Osteoporosis), eye problems (Graves' opthalmopathy) (Figure:2).

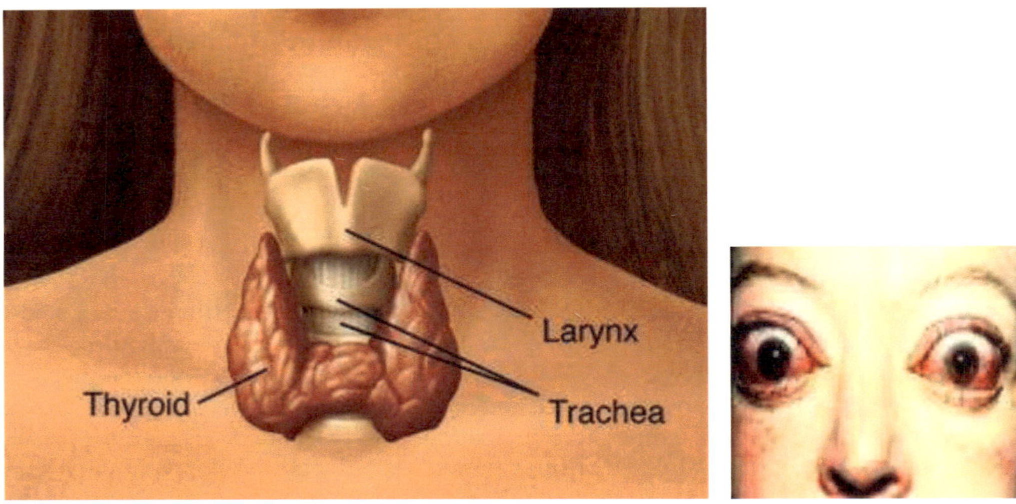

Fig. 2. Symptoms of Graves' disease

Hypothyroidism describes an under active thyroid gland that is producing low level of thyroid hormone. Hypothyroid patients experience a variety of symptoms, including weight gain, intolerance to cold, goiter (enlarged thyroid), dry coarse, skin, fatigue, constipation, decreased heart rate, poor memory and depression.

The most common form of hyperthyroidism is Graves' disease (GD), an autoimmune disorder accounting for 60-80 % of all cases, in which the antibodies produced by immune system stimulates thyroid gland to produce excess of thyroxine. Normally, the immune system uses antibodies to protect against viruses, bacteria and other foreign substances that enter the body system. In GD, the antibodies mistakenly attack the thyroid gland and occasionally the tissues behind the eyes and the skin of lower legs over the shins. Though the exact cause of GD is not known, several factors including a genetic predisposition are likely to be involved **(Figure:3).**

GD is an organ specific heterogeneous autoimmune disorder associated with T-lymphocyte abnormality affecting the thyroid eyes and skin. GD is also multifactorial disease that develops as a result of complex interaction between genetic susceptibility genes and environmental factors. Human leucocyte antigen (HLA) and cytotoxic T-lymphocyte associated molecule-4 (CTLA-4) are susceptibility candidates. CTLA_4 gene plays an important role in the development of GD, which is located on chromosome 2 q33.

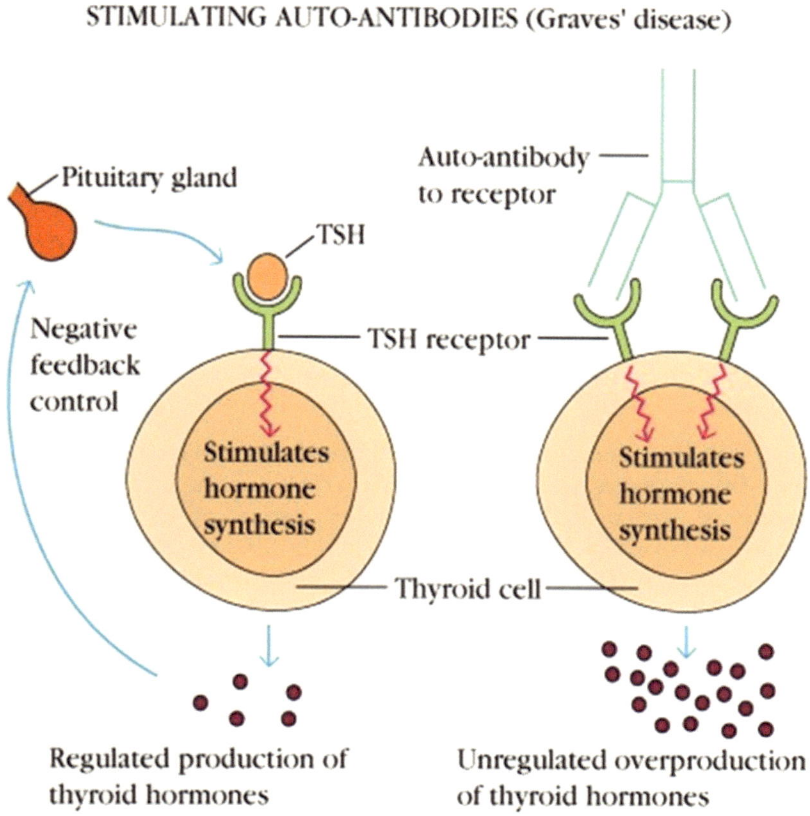

Fig. 3.

2. Cytogenetic location of CTLA-4 gene

Cytogenetic Location: 2q33

Molecular Location on chromosome 2: base pairs 204,732,510 to 204,738,682

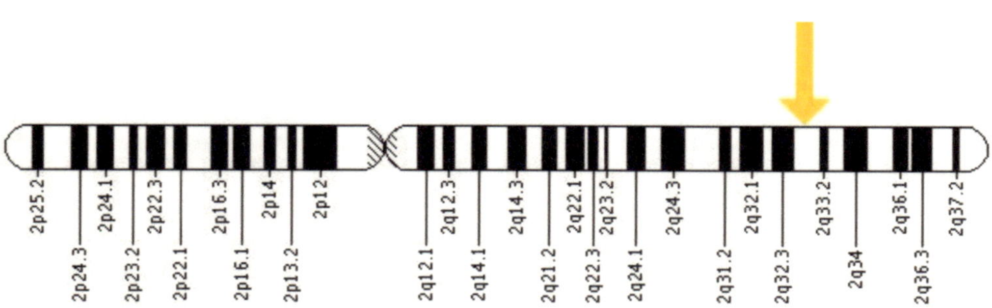

Fig. 4. The CTLA4 gene is located on the long (q) arm of chromosome 2 at position 33.

More precisely, the *CTLA4* gene is located from base pair 204,732,510 to base pair 204,738,682 on chromosome 2.

Activation of T cells requires 2 signals transduced by the antigen specific TCR and co stimulatory ligand such as CD28. CTLA-4, which is expressed on activated T cells, bind to B7 present on antigen presenting cells and functions as a negative regulator of T cell activation. CTLA-4 gene polymorphism confers susceptibility to several autoimmune diseases, such as Graves' disease (GD), Hashimoto's thyroiditis (HT), Addison's disease (AD), Insulin-dependent diabetes mellitus (IDDM), Rheumatoid arthritis (RA) and Multiple sclerosis.

The activity of T cells requires a co stimulatory signal mediated by CD28/B7 interaction. The CTLA-4 gene product delivers a negative signal to T cells and mediates apoptosis. This CTLA-4 gene product is a T cell surface molecule that binds to the B 7 molecule on the antigen presenting cells (APCs). The CTLA-4 gene expression on T cells may affect the course of ongoing immune process. TSH receptor antibody (TRAb) causes Graves' hyperthyroidism.

The GD will go into remission during antithyroid drug (ATD) treatment. Remission of GD is predicted by a smooth decrease in TRAb during (ATD) treatment. Treatment of GD may involve surgery or use of radioactive iodine or use of ATD like propylthiouracil, methimazole and carbimazole. The genetic susceptibility to GD is also conferred by genes in human leucocyte antigen (HLA) and several other genes that are not linked to HLA. The present paper describes the association of GD with the CTLA-4 gene.

The prevalence of hyperthyroidism has been reported to be 3.63% and hypothyroidism to be 2.97% especially the females being more affected by hyperthyroidism [3]. Hence the current study deals with A/G single nucleotide polymorphism (SNP) at position 49 (exon1, codon 17) of the CTLA-4 gene where in Thr/Ala substitution and can be a function related marker. It has been shown to be associated with GD in Caucasians, Japanese, Koreans, Tunisians, Hong Kong Chinese children [2,4,5,6,7,8,9,10] and South Indains [11,12].

The polymorphism cited (A/G polymorphism in exon 1, C/T polymorphism in the promoter, and micro satellite repeat in 3'-untranslated region of exon 4) in CTLA-4 gene have been reported to be associated with autoimmune endocrine disorder.

A/G polymorphism at position 49 in exon 1 of the CTLA-4 gene among South Indian population with Graves' hyperthyroidism has revealed the frequencies of the GG genotype and "G" allele to the significantly higher in GD patients. The study has also demonstrated that GD patients had higher frequencies of "G" allele (GG genotype) and lower frequencies of "A" allele (AA genotype) than control group.

Kinjo *et al.*, (2000) have also reported the relationship between the CTLA-4 gene type and severity of the thyroid dysfunction. At diagnosis, free T4 concentrations were shown to be more in patients with the GG genotype and low in patients with the AA genotype. GD patients were reported to have more "G" allele than the control, suggesting that the CTLA-4 GG genotype might induce down regulation of T-cell activation. If the function of CTLA-4 with "G" alleles at position 49 in exon 1 is impaired CTLA-4 function may have d\difficulty in achieving remission.

Identification of SNP

We can analyze and identify all types of gene SNPs by Polymerase Chain Reaction (PCR) thermal cycler.

Polymerase Chain Reaction

The polymerase chain reaction (PCR) is a laboratory (in vitro) technique for generating large quantities of a specified DNA. Obviously, PCR is a cell-free amplification technique for synthesizing multiple identical copies (billions) of any DNA of interest, which was developed in 1994 by Karry Mullis (Nobel Prize, 1993). PCR is now considered as a basic tool for any molecular biologist.

3. Primer designing [13]

As oligonucleotide primers are useful for polymerase chain reaction (PCR), oligo hybridization and DNA sequencing, proper primer designing is actually one of the most important factors/steps. Various bioinformatics programs are available for selection of primer pairs from a template sequence.

3.1 Guidelines for primer design

When choosing two PCR amplification primers, the following guidelines should be considered:

Primer length: It is accepted that optimal length of PCR primers is 18-22 bp (Wu et al., 1991)

Melting temperature (Tm): It can be calculated using the formula of Wallace et al., 1997, Tm (°C) = 2(A+T)+4(G+C). The optimal melting temperature for primers ranges between 52-58°C. Primers with melting temperature above 65°C should also be avoided because of potential for secondary annealing.

Primer annealing temperature: The two primers of a primer pair should have closely matched melting temperatures for maximizing PCR product yield. The difference of 5°C or more each can lead to no amplification.

$$T_a = 0.3 \times T_m(\text{primer}) + 0.7 \qquad (T_m \text{primer} = 14.9)$$

Where, Tm (primer) = Melting Temperature of the primers, T_m (product) = Melting temperature of the product.

GC Content: Primers should have GC content between 45 and 60 percent. GC content, melting temperature and annealing temperature are strictly dependent on one another

.Dimers and false priming because misleading results: Presence of the secondary structures such as hairpins, self dimer produced by intermolecular or intramolecular interactions in primers can lead to poor or no yield of the product.

Avoid Cross homology: To improve specificity of the primers it is necessary to avoid regions of homology

3.2 Software for primer design

NETPRIMER is software used to design and analyze the parameters of designed primer sequences using the following link http://premierbiosoft.com/netprimer/index.html

3.3 PCR standardization [13,14]

PCR is a revolutionary technique used in almost all molecular biology experiments. In PCR, the repeated three-step process of denaturation, primer annealing and DNA polymerase extension results in exponential amplification of target DNA. Initially PCR was reported with E.Coli DNA polymerase Klenow fragment in 1985. In 1988, the first report on PCR using thermostable Taq DNA polymerase was published. Since then PCR has been extensively modified and used for various applications such as cloning, sequencing, site-directed mutagenesis, diagnostics, genotyping, genome walking, amplification of RNA after reverse transcription for gene expression analysis amplification of a whole genome, etc.

The central components of a PCR reaction are oligonucleotide primers, thermostable DNA polymerase, target DNA, dNTPs and reaction buffer including $MgCl_2$. When a new PCR has to be developed, suitable primer pairs should be designed based on the target sequence,. Subsequently, the concentration of PCR components and the cycling conditions should be optimized.

Thermostable enzymes:

Thermostable enzymes should be selected based on the applications. High fidelity Taq DNA polymerase and proofreading recombinant enzymes are required for the amplification of more than 3 kb target sequence. For a standard PCR, 2 to 5 units of Taq DNA polymerase are recommended for a typical 100µl PCR.

Deoxynucleoside triphosphate (dNTPs):

For a standard PCR, 100 to 200 µM concentrations of dNTPs is used. The balanced solutions of all four nucleotides should be used to minimize the error frequency. The concentrations may be increased for Multiplex PCR and Repetitive PCR, where more than one PCR amplicons are expected.

Template DNA:

The purity and concentration of the template DNA are critical for a successful PCR amplification. For initial experiments, 0.1 to 200ng of the template DNA, based on the type can be used. For example, if it is a plasmid 0.1 to 1ng is sufficient. If the template is human genomic DNA, upto 200ng can be used.

Primer concentrations:

The primer concentration can affect the PCR. If the primer concentration is too low, amplifications will be failed; and if the concentration is too high, non-specific amplification will occur. Therfore, the primer concentration should be optimized empirically between 0.1 to 1µM final concentrations. The most straightforward way of optimizing a PCR with a given primer pair is to change the concentration of $MgCl_2$ or the annealing temperature or both.

Optimization of primer annealing temperature:

Optimization of the primer annealing temperature is the most critical step in PCR. The primer designing programs will suggest the Tm of the primers. In general, the annealing temperature should be set 2 to 5°C below the Tm of the primers. However, some oligonucleotides may not work optically at this temperature and hence the annealing temperature should be optimized using gradient PCR approach.

Optimization of $MgCl_2$ concentration:

Magnesium chloride is an essential component for PCR. It is a cofactor for Taq DNA polymerase. Mg^{++} promotes DNA/DNA interactions and forms complexes with dNTPs that are the actual substrates for Taq polymerase. When Mg++ is too low, primers fail to anneal to the target DNA. When Mg^{++} is too high, the base pairing becomes too strong and the amplicon fails to denature completely when you heat 94°C. $MgCl_2$ concentration should be optimized for every PCR reaction. All the components of the reaction mixture can bind to magnesium ion, including primers, template, PCR products and dNTPs. Therefore, the concentration of $MgCl_2$ has to be optimized for a new PCR. The most commonly used concentration of $MgCl_2$ is 1.5mM and it can be optimized empirically between 1.5 and 4.0mM.

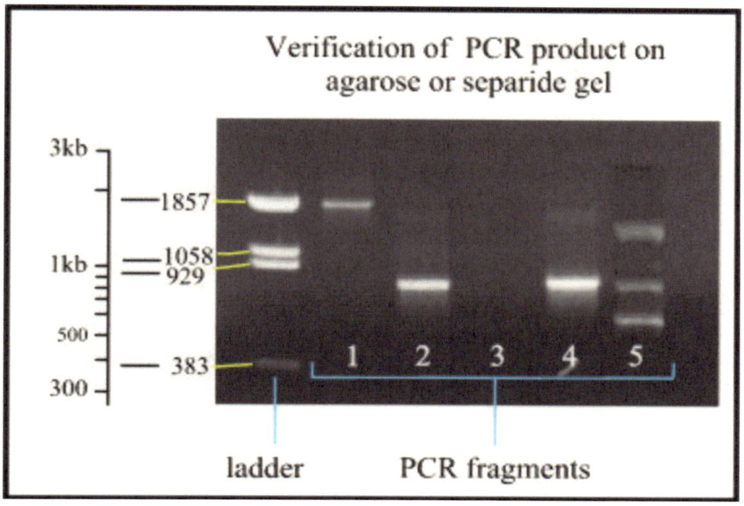

Fig. 5.

3.4 Genetic marker [15,16,17,18]

A **genetic marker** is a gene or DNA sequence with a known location on a chromosome that can be used to identify cells, individuals or species. It can be described and observed as a variation which may arise due to mutation or alteration in the genomic loci. A genetic marker may be a short DNA sequence, such as a sequence surrounding a single base-pair change (single nucleotide polymorphism, SNP), or a long one, like minisatellites. For many years, gene mapping was limited in most organisms by traditional genetic markers which include genes that encode easily observable characteristics such as blood types or seed shapes.

3.5 Some commonly used types of genetic markers

RFLP (or Restriction fragment length polymorphism)
SSLP (or Simple sequence length polymorphism)
AFLP (or Amplified fragment length polymorphism)
RAPD (or Random amplification of polymorphic DNA)
VNTR (or Variable number tandem repeat)

Microsatellite polymorphism, SSR (or Simple sequence repeat)
SNP (or Single nucleotide polymorphism)
STR (or Short tandem repeat)
SFP (or Single feature polymorphism)
DArT (or Diversity Arrays Technology)
RAD markers (or Restriction site associated DNA markers)
They can be further categorized as dominant or co-dominant.

Dominant markers allow for analyzing many loci at one time, e.g. RAPD. A primer amplifying a dominant marker could amplify at many loci in one sample of DNA with one PCR reaction. The dominant markers, as RAPDs and high-efficiency markers (like AFLPs and SMPLs), allow the analysis of many loci per experiment within requiring previous information about their sequence.

Co-dominant markers analyze one locus at a time. A primer amplifying a co-dominant marker would yield one targeted product. so they are more informative because the allelic variations of that locus can be distinguished. As a consequence, we can identify linkage groups between different genetic maps but, for their development it is necessary to know the sequence (which is still expensive and is considered one of their down sides). **Eg.** RFLPs, microsatellites, etc.,

3.6 Uses of genetic markers

- Genetic markers can be used to study the relationship between an inherited disease and its genetic cause (for example, a particular mutation of a gene that results in a defective protein). It is known that pieces of DNA that lie near each other on a chromosome tend to be inherited together. This property enables the use of a marker, which can then be used to determine the precise inheritance pattern of the gene that has not yet been exactly localized.
- Genetic markers have to be easily identifiable, associated with a specific locus and highly polymorphic, because homozygotes do not provide any information.
- Detection of the marker can be direct by RNA sequencing, or indirect using allozymes.
- Genetic Markers have also been used to measure the genomic response to selection in livestock.
- Natural and artificial selection leads to a change in the genetic makeup of the cell. The presence of different alleles due to a distorted segregation at the genetic markers is indicative of the difference between selected and non-selected livestock.

Hence, SNP (Single nucleotide polymorphism) in Graves' hyperthyroidism is used as marker to identify which mutation is responsible for causing GD and other hereditary diseases.

4. Analysis of CTLA-4 A/G polymorphism among South Indian population

4.1 Protocol used for A/G single nucleotide polymorphism (SNP) study in Graves' disease

Genomic DNA was prepared from peripheral white cells using standardized protocol. We have analysed CTLA -4 genotypes and allele with PCR. PCR was performed with oligonucleotide primers (Forward, 5′ - GCTCTACTTCCTGAAGACCT - 3′ and Revers, 5′ - AGTCTCACTCACCTTTGCAG - 5′)[2]. PCR was performed by initial denaturation 30 sec for 5 min. annealing for 45 sec at 57°C, extension for 30 sec at 72°C, denaturation 30 sec at 94°C (for 20 cycles) and final extension for 7 min at 72°C. The PCR product was confirmed by agarose (1.8%) gel electrophoresis. The presence of G alleles was determined in each subject by PCR amplification of CTLA-4, followed by diffusion with *Bbv1,* which acts on the G variation, but not on the A variation. It a G allele was at position 49, 88/74 bp fragments were obtained. This was confirmed by 2% agarose gel.

4.2 Restriction digestion

The amplified CTLA-4 gene should be digested with the restriction enzyme *Bbv1,*which is commercially available. A typical 30μl reaction mix was used . Modify the required volume proportionately.

PCR amplified product	- 20. 0μl
10x buffer	- 3.0μl
Bbv1(10units/ul)	- 1.0μl
Deionized water	- 6.0μl
Total	- 30.0μl

Incubated the reaction mixture at 37°C for 4 hrs and inactivated by heating at 70°C for 10 min. The product was confirmed using 2% agarose gel electrophoresis.

4.3 Results

The presence of genomic DNA confirmed by subjecting the agarose gel electrophoresis (0.7%) **(Figure 6).** The genomic DNA was then subjected to PCR and 162 bp fragments were obtained **(Figure 7).** The amplified PCR product digested with enzyme ***Bbv*1, t**he restriction enzyme acts on the G variation, but not on the A variation. If a G allele was at position 49, 88bp and 74bp fragments were obtained and the fragments were detected by 2% agarose gel electrophoresis **(Figure 8).**

In the present study, the G/G genotype was observed in 32 (40 %) GD patients and in 26 (32.50 %) individuals of the control group, A/G genotype was found in 37 (46.25 %) patients and in 25 (31.25 %) persons of the control group, A/A genotype was observed in 11 (13.75 %) patients and in 29 (36.25 %) persons of the control group and G allele was found in 50 (62.5%) GD patients and in 38 (47.5 %) persons of the control group, and A allele was found in 30 (37.5 %) GD patients and 42 persons (52.5%) of the control group **(Table 1).** There was significant difference ($p < 0.05$) in genotype and allelic frequency between the control group and GD patients. The present study also demonstrates an association between the CTLA-4 gene polymorphism in Graves' disease and with the remission rate of Graves' hyperthyroidism. Among the GD cases studied, only 2% had remission. The frequencies of GG genotype (40 %) and G allele (62.5%) were higher when compared to A/A genotype (13.75%) and A allele (37.5 %) (Table 1).

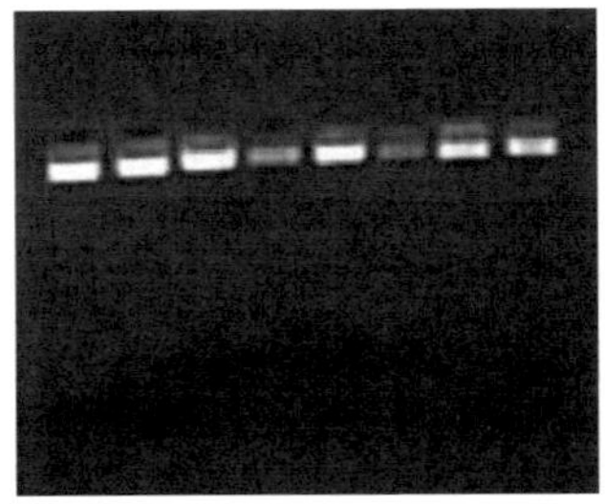

Fig. 6. Confirmation of human genomic DNA

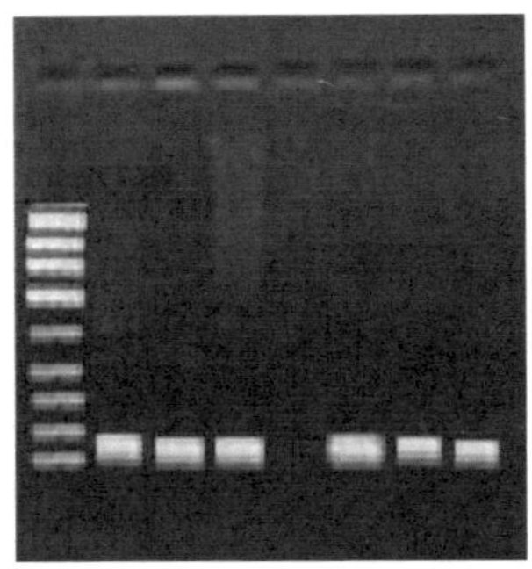

Fig. 7. CTLA-4 gene amplification

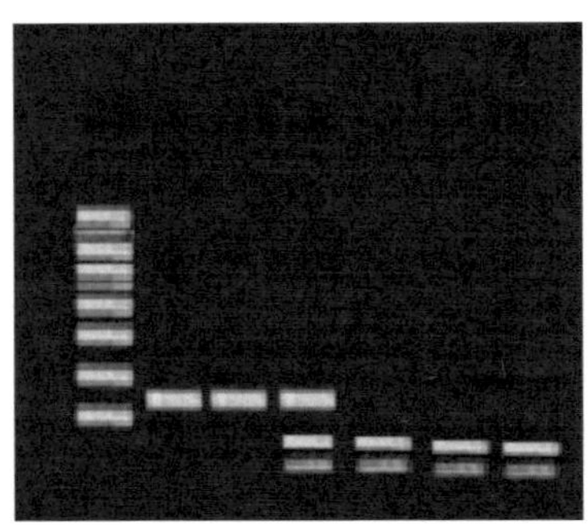

Fig. 8. Restriction analysis of CTLA-4 gene

GENOTYPE	GD patient (n=80)	*Control group* **(n=80)**
G/G	32 (40%)	26 (32.50%)
A/G	37 (46.25%)	25 (31.25%)
A/A	11 (13.75%)	29 (36.25%)
Allele		
G	50 (62.5%)	38 (47.5%)
A	30 (37.5%)	42 (52.5%)

Table 2. Prevalence of CTLA-4 gene genotype and allele frequency among South Indian

4.4 Discussion

In the present study genomic DNA was isolated from patients and control groups and was subjected to Agarose gel electrophoresis (0.7%). This enables easy visualization of DNA band patterns. After confirming the presence of genomic DNA, it was subjected to PCR and 162 bp fragments were obtained. The amplified PCR product was digested with enzyme ***Bbv*1.** The restriction enzyme acts on the G variation, but not on the A variation. If a G allele was at position 49, 88bp and 74bp two fragments were obtained. The PCR products were detected by 2% Agarose gel electrophoresis.

A/G polymorphism at position 49 in exon 1 of the CTLA-4 gene among Madurai population with Graves' hyperthyroidism revealed that the frequencies of the GG genotype and G allele were significantly higher in GD patients. This study has also revealed lower frequency (or absence) of A allele (AA genotype) than the control. CTLA-4 gene polymorphism has been reported to be associated with GD. CTLA-4 molecule is a member of the family of cell surface molecule CD28, which binds to B7. The CTLA-4/B7 complex competes with the CD28/B7 complex and delivers negative signals to the T-cells, which affects T-cell expansion, cytokine production, and immune responses as evidenced by Park *et.al.*[6] in Korean population, Yanagawa *et al.* [9] in Japanese population and Yanagawa *et al.* [8] in Caucasian population. However, we do not know how CTLA-4 gene polymorphisms may contribute to the development of Graves' hyperthyroidism.

Three polymorphism sites (A/G polymorphism in exon 1; C/T polymorphism in the promoter, and micro satellite repeat in the 3'-untranslated region of exon 4) in the CTLA-4 gene have been reported to be associated with autoimmune endocrine disorders. Kinjo *et. al.*, [2] have reported the relationship between the CTLA-4 gene type and severity of the thyroid dysfunction. At diagnosis, free T_4 concentrations were shown to be highest in patients with the GG genotype and lowest in patients with the AA genotype. GD patients have more G allele than control, suggesting that the CTLA-4 GG genotype might induce down regulation of T-cell activation. If the function of CTLA-4 with the G alleles at position 49 in exon 1 was impaired CTLA-4 function might have difficulty in achieving remission.

	Graves' Disease % (n = 144)	**Controls** % (n = 110)
Genotype		
G/G	50 (34.7)	26 (23.6)
A/G	62 (43.1)	46 (41.8)
A/A	32 (22.2)	38 (43.6)

Table 3. Frequency of the genotype and allele of A/G polymorphism at position 49 in exon 1 of CTLA-4 gene in GD patients and controls among Japanese -population. [2]

Bednarczuk *et. al.*, [4] analysed the association of CTLA-4 A49G polymorphism with Graves' disease in Caucasian and Japanese population. Their study also reveals that, CTLA -4 G allele and G/G genotype confer genetic susceptibility to GD in Caucasian and Japanese population.

The study of Kouki *et. al.*, [5] among patients with GD revealed there were more individuals with G/G (17.8 %GD vs 11.6% of controls) or A/G CTLA-4 exon 1 genotypes (64.4 % GD vs 53.5% control) and significantly fewer individuals with the A/A alleles (17.8 %GD vs 43.9

%control) when compared with controls. According to their findings, the frequency of the G allele was higher in GD patients (50%) than in controls (38.4%) in their population.

There was significant difference between the control group and GD patients both in genotype and allelic frequency. Therefore, in accordance with previously published results, the present study also demonstrates an association between the CTLA-4 gene polymorphism in Graves' disease and with the remission rate of Graves' hyperthyroidism. Among the GD cases studied, only 2% had remission and the frequencies of GG genotype and G allele were higher when compared to A/A genotype and A allele. GD patient with G allele in exon 1 of the CTLA-4 gene were required to continue Anti thyroid drug (ATD) treatment [19] for longer periods to achieve remission. Further studies will be required to determine a clear association of the CTLA-4 gene polymorphism with the remission of GD.

We have studied another gene polymorphism called PKD1 (C/T) at position 4058 in exon 45 which is responsible for causing autosomal polycystic kidney disease (ADPKD) among South Indian.

5. Short summary of C/T polymorphism in PKD1 gene

Polycystic kidney disease (PKD) is a group of monogenic disorders that result in renal cyst development in kidney leads to kidney failure. Autosomal dominant polycystic kidney disease (ADPKD) and autosomal recessive polycystic kidney disease (ARPKD) are two forms of PKD, which are largely limited to the kidney and liver, which extends from neonates to old age. ADPKD is a commonly inherited disorder in humans, with a frequency among the general population of 1 in 500. ADPKD caused by mutations in PKD1 gene (85%) located on human chromosome 16p13.3; the remaining 15% are caused by mutations in the PKD2 gene, located on human chromosome 4q21-23. A total of 60 ADPKD patients among South Indian (Madurai) population were analyzed. In genetic study, the genomic DNA was isolated, which would be subjects into PCR (Figure:9) and RFLP analysis (Figure:10). C/T polymorphism at position 4058 in exon 45 of the PKD1 gene among South Indian (Madurai) population with ADPKD revealed that the "TT" "CT' genotype and the frequency of "T" allele was found be significantly (at p=0.001) higher in the patients compared to control subjects. The study was demonstrated that ADPKD patients had higher frequencies of "T" allele and lower frequency of "C" allele than control subjects. The present study also has been supported by Constantinides *et al.*,[20]. Therefore, the study reveals that there was an association of C/T polymorphism in ADPKD and the prevalence of ADPKD among South Indian (Madurai) population.

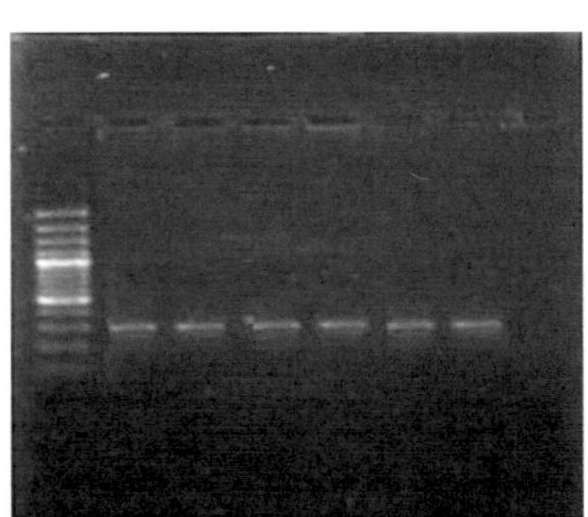

Fig. 9. PKD1 gene amplification

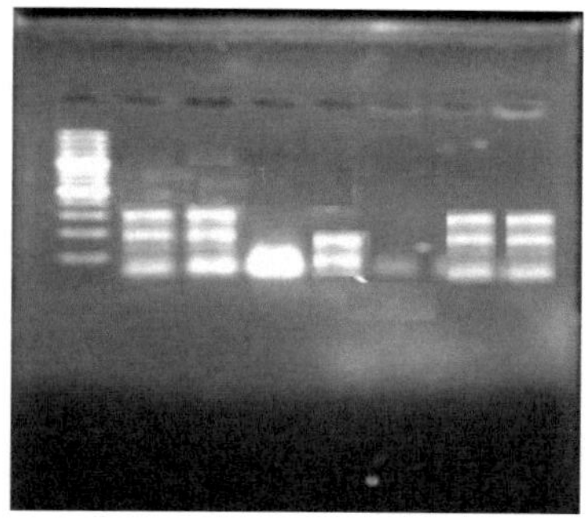

Fig. 10. Restriction digestion PKD1 gene

Hope this chapter will provide an insight on genetic screening of different disease and genetic disorders.

6. Acknowledgement

We would like to acknowledge Gunasekaran P, Dr.Sujatha K, Dr.Mahalakshmi A, UGC-NRCBS, School of Biological Science, Madurai Kamaraj University, Madurai, for valuable guidance in standardization of PCR and Primer designing. And also we like to acknowledge Dr.R. Shenbagarathai & faculty in PG & Research Department of Zoology and Biotechnology, Lady Doak College, Madurai, for their continuous support.

7. References

[1] Guyton. (1991) Text book of medical physiology, 1091 - 95.

[2] Kinjo Y., Takasu N., Komiya I., Tomoyose T., Takara M., Kouki T., Shimajiri Y., Yabiku K., and Yoshimura H. (2002) Remission of Graves' hyperthyroidism and A/G polymorphism at position 49 in exon 1 of Cytotoxic T- lymphocyte-associated molecule-4 gene. J. of clin. Endocrinol. and Metab. 87(6): 2593-2596.

[3] Velumani A., Kadival GV., Nirmala R. and Lele RD. (2005) Hyperthyroidism. Health screen. 17-21.

[4] Bednarczuk T., Hiromatsu Y., Fukutani T., Jazdzewski K., Miskiewicz P., Osikowska M. and Nauman J. (2003) Association of cytotoxic T-lymphocyte-associated antigen-4 (CTLA-4) gene polymorphism and non-genetic factors with Graves' ophthalmophathy in European and Japanese populations. European J. of Endocrin. 148:13-18.

[5] Kouki T., Sawai Y., Gardine C.A., Fisfalen M-E., Alegre M-L. and Degroot L.J. (2000) CTLA-4 gene polymorphism at Position 49 in Exon 1 reduces the inhibitory function of CTLA-4 and Contributes to the Pathogens of Graves, disease. The J. of Immunol. 165:6606-6611.

[6] Park Y.J., Chung H.K., Park D.J., Kim W.B., Kim S.W., Koh J.J. and Cho B.Y. (2000) Polymorphism in the promotor and exon 1 of the cytotoxic T lymphocyte antigen-4 gene associated with autoimmune thyroid disease in Koreans. Thyroid. 10:453 –459.

[7] Vaidya B., Imrie H. and Perros P. (1999) The cytotoxic T-lymphocyte antigen –4 is a major Graves disease locus. Hum. Mol. Genet. 8:1195-99.

[8] Yanagawa T., Hidaka Y., Guimaraes V., Soliman M. and DeGroot L.J. (1995) CTLA-4 gene polymorphism associated with Graves' disease in a Caucasian population. J. Clin. Endocrinol. Metab. 80:41-45.

[9] Yanagawa T., Taniyama M., Enomoto S., Gomi K., Maruyama H., Ban Y. and Saruta T. (1997) CTLA-4 gene polymorphism confers susceptibility to Graves' disease in Japanese. Thyroid. 7:843-846.

[10] Wang P-W., Liu R-T., Jou S-H.H., Wang S-T., Hu Y-H., Hsieh C-J., Chen M-C., Chen I-Y. and Wu C-L. (2004) Cytotoxic T lymphocyte associated molecule - 4 ploymophism and relapse of Graves' hypethyroidism after Antityroid withdrawal. J. Clin. Endocrinol. Metab. 89(1): 169-173.

[11] Veeramuthumari P, Isabel W, Kannan K. (2009) "A study on the level of serum T.Chol., TGL, HDL, LDL in patients with Graves' hyperthyroid in Madurai population" in Indian J of Endocrinol and Met.

[12] Veeramuthumari P, Isabel W, Kannan K. (2010) "A Study on the level of T3, T4,TSH and the association of A/G Polymorphism with CTLA-4 gene in Grave's Hyperthyroidism among South Indian Population" Indian J of Clin Biochem: 26, 66-69.

[13] Gunasekaran P, Sujatha K. (2010) UGC NRCBS Winter School on Gene Cloning and Expression in Bacteria, Lab Manual. 9-19.

[14] Satyanarayana U, Chakrapani U. ((2006) Biochemistry, Arunabha Sen Book and Allied (P) Ltd. 594-596.

[15] de Vicente, C., T. Fulton (2003). *Molecular Marker Learning Modules - Vol. 1.*. IPGRI, Rome, Italy and Institute for Genetic Diversity, Ithaca, New York, USA.

[16] de Vicente, C., T. Fulton (2004). *Molecular Marker Learning Modules - Vol. 2.*. IPGRI, Rome, Italy and Institute for Genetic Diversity, Ithaca, New York, USA..

[17] de Vicente, C., J-C. Glaszmann, editors (2006). *Molecular Markers for Allele Mining*. AMS (Bioversity's Regional Office for the Americas), CIRAD, GCP, IPGRI, M.S. Swaminathan Research Foundation.

[18] Spooner, S., R van Treuren and M.C. de Vicente (2005). *Molecular markers for genebank management*. CGN, IPGRI, USDA.

[19] Rodriguez S., Quinn F.B., Matthew W. and Ryan W. (2003) Benign thyroid disease. J. Pathol. 33-34.

[20] Constantinides R, Xenophoutos S, Neophyton P, Nomura S, Pierides A and Deltas CC: New amino acid polymorphism, Ala/Val4058, in exon 45 of the polycystic kidney disease 1 gene. Hum Genet 1997, 99:644-647.

26

Detection of Bacterial Pathogens in River Water Using Multiplex-PCR

C. N. Wose Kinge[1], M. Mbewe[2] and N. P. Sithebe[1]
[1]Department of Biological Sciences, School of Environmental and Health Sciences, North-West University, Mafikeng Campus, Mmabatho,
[2]Animal Health Programme, School of Agricultural Sciences, North-West University, Mafikeng Campus, Mmabatho
South Africa

1. Introduction

The aquatic environments receive a significant number of human microbial pathogens from point and non-point sources of pollution. Point-source pollution enters the environment at different locations, through a direct route of discharge of treated or untreated domestic sewage, industrial effluent and acid mine drainage (State of the Environment Report [SER], 2002). Non-point (or diffuse) sources of pollution comprises up to 80 % of the pollution entering major river systems thus are of significant concern with respect to the dissemination of pathogens and their indicators in water systems. They may be attributable to the run-off from urban and agricultural areas, leakage from sewers and septic systems, insecticides and herbicides from agricultural land, and sewer overflows (Stewart et al., 2008). Although majority of pathogenic microbes can be eliminated by sewage treatment, many end up in the effluent which is then discharged into receiving bodies of water. These pathogenic microbes have been implicated in human diseases linked with the use of contaminated water and food. Adequate sanitation and clean water, being two critical factors in ensuring human health, protects against a wide range of water-related diseases. These include diarrhoea, cholera, trachoma, dysentery, typhoid, hepatitis, polio, malaria, and filariasis (United Nations Department of Public Information [UNDPI], 2005).

Water is a vital natural resource because of its basic role to life, quality of life, the environment, food production, hygiene, industry, and power generation (Meays et al., 2004). With the rapid increase in world population and increased urbanisation, there is a massive strain on the existing water supply and sanitation facilities (UNDPI, 2005). In the developing world, poor access to safe water and inadequate sanitation continues to be a danger to human health (World Health Organisation [WHO], 2004). The water situation, in the African continent, has attracted a lot of concern from all sectors of government as it is estimated that more than 300 million out of the 800 million people who live on the continent are in water-scarce environments (United Nations Educational, Scientific and Cultural Organisation [UNESCO], 2004). In Northern Africa, the present water supply is unstable as population growth and economic development have surpassed the traditional water management practices, leading to water scarcity and pollution to a varying degree (UNESCO, 2004). According to Beukman and Uitenweerde (2002), Southern Africa faces very serious water challenges with an estimated half of the population lacking access to portable water and sanitation facilities. They further hinted that, by 2025, countries like Mozambique, Namibia, Tanzania and Zimbabwe will face more water pressures.

The scarcity of water does not only threaten food security, but also the production of energy and environmental integrity. This often results in water usage conflicts between different communities, and water contamination when humans and animals share the same source of water (Kusiluka et al., 2005). According to the Department of Water Affairs - DWA (2000), South Africa is a water scarce region, with 450mm rainfall per annum. This is lower than the world's 860mm average rainfall. Of the forty-four million people who live in South Africa, 12 million people were without access to portable water supply prior to 1994 (Momba et al., 2006). Although the South African government is making significant progress in ensuring the supply of potable water to all communities, 3.3 and 15.3 million inhabitants of South Africa are still identified to be living without access to potable water and adequate sanitation facilities (Council for Scientific and Industrial Research [CSIR], 2008). A total of 80% of the population live in the rural areas with the unavailability of potable basic water supplies and proper sanitation facilities (Kasrils, 2004; Reitveld et al., 2009).

Due to the scarcity of water in South Africa, extensive exploitation of water resources such as dams, pools, unprotected rivers and springs for domestic and other water uses, is common, particularly in the rural communities where access to potable water supply is limited (Younes and Bartram, 2001). In many developing countries with inadequate sanitation, faecal contaminations of environmental waters by enteric pathogens are very common and river water is major source of microbial pathogens (Sharma et al., 2010). In this study, we report the use of conventional identification, and multiplex PCR (m-PCR) method that permits the simultaneous detection of water-borne *Salmonella*, *Shigella*, *E. coli*, and *Klebsiella* bacteria spp. from rivers in the North West province of South Africa. The major rivers in the province include the Molopo, Groot Marico, Elands, Hex, and Crocodile Vaal, Skoonspruit, Harts and Mooi. These rivers are grouped into five catchment areas, which include the Crocodile and Elands, Marico and Hex, Marico and Molopo, Mooi and Vaal, and the Harts (SER, 2002; Department of Water Affairs [DWA], 2007). The water quality in these rivers has been impaired partly due to the frequent contamination of water sources with a number of pathogenic microorganisms from human as well as animal activities, which result in the spread of diarrhoeal diseases (Meays et al., 2004).

1.1 Bacterial pathogens in the aquatic environment

Microbial pathogens in water include viruses, bacteria, and protozoa (Girones et al., 2010). Currently, pathogenic bacteria have been identified as the major etiological agent in the majority of the waterborne outbreaks worldwide (WHO 2003; Liang et al., 2006). Bacillary dysentery caused by *Shigella* bacteria alone is responsible for approximately 165 million cases of bacterial diarrhoeal diseases annually. Of this, 163 million are in developing countries and 1.5 million in industrialized ones accounting for an estimated 1.1 million death cases each year (Sharma et al., 2010). Most members of the genus *Arcobacter* have been isolated from different environmental water sources including surface and ground water. Their presence has been correlated with that of faecal pollution indicators (Collado et al., 2008; Fong et al., 2007; Ho et al., 2006) as well as meat mainly from poultry, pork and beef (Collada et al., 2009; Houf, 2010; Wesley and Miller, 2010). Some members of the genus *Arcobacter*, like *A. butzleri*, *A. cryaerophilus*, and *A. skirrowii*, have been implicated in animal and human diarrhoeal cases, suggesting a faecal oral route of transmission to humans and animals (Gonzalez et al., 2007). *Helicobacter pylori* on the other hand, found to be present in surface water and wastewater has been implicated in gastritic, peptic, and duodenal ulcer diseases (Linke et al., 2010; Queralt et al., 2005).

Biofilms in drinking water distribution systems have been reported as possible reservoirs of H. *pylori* and attempts to culture these cells from water samples have proven unsuccessful (Linke et al., 2010; Percival and Thomas, 2009). Due to the fastidious nature of this bacterium, the lack of standard culture methods for environmental samples, and the controversy in its ability to survive in an infectious state in the environment, very few quantitative studies have been reported (Percival and Thomas, 2009). *Legionella pneumophila* is a ubiquitous bacterium in natural aquatic environments that can also persist in human-controlled systems containing water, such as air conditioning and plumbing infrastructures (Steinert et al., 2002). Furthermore, *Vibrio vulnificus*, an opportunistic human pathogen that cause gastroenteritis, severe necrotizing soft-tissue infections and primary septicaemia, has been found present in fish, shell fish, water, and wastewater. Infection generally, is associated with the ingestion of contaminated seafood and water (Harwood et al., 2004; Igbinosa et al., 2009). More so, the presence of enteric bacteria of the genera *Salmonella, Shigella, E.coli* and *Klebsiella* in water has been identified as a major threat to human health and causative agents for many diseases (Leclerc et al., 2001).

Salmonellae are the most frequent agents of bacterial gastroenteritis and typhoid in humans and a prime example of a water- and shell fish-transmitted human pathogen. It is frequently isolated from the marine environment where it can remain viable for several hours (Malorny et al., 2008; Westrell et al., 2009). Contamination with *Salmonella* has been reported in surface water used for recreational purposes, source of drinking water (Till et al., 2008) and irrigation (Gannon et al. 2004) underlining the possible risk associated to the use of such contaminated water. The typhoid caused by *Salmonella enterica* serotype Typhi remains an important public health problem in developing countries and the burden of typhoid fever worldwide is further compounded by the spread of multiple drug resistant *S. typhi* (Kim 2010; Lynch et al., 2009; Srikantiah et al., 2006). The runoff from fields with animal husbandry, and untreated sewage disposal contribute to the presence of *Salmonella* in natural water resources (Jenkins et al., 2008; Moganedi et al., 2007). Low numbers of *Salmonella* in food, recreational, surface and potable water sources may pose a public health risk given that their infective dose can be as low as 15–100 CFU (Cobbold et al., 2006; Seo et al., 2006).

Species of *Shigella* and enteroinvasive *Escherichia coli* (EIEC) are important human pathogens identified as the major cause of bacillary dysentery (Wanger et al., 1988; Szakál et al., 2003). The infective dose of *Shigella* cells is very low (10^1-10^4 organisms), whereas EIEC strains require a larger infectious dose (between 10^6 and 10^{10} organisms) (Rowe and Gross, 1984). Both *Shigella* spp. and EIEC carry a large invasion plasmid and express a similar set of proteins. Both of them are transmitted by direct contact from human to human or via contaminated food and water (Parsot, 1994; Rowe and Gross, 1984). Clinical features of bacillary dysentery caused by EIEC that resemble shigellosis include fever, severe abdominal cramps, malaise, toxemia, and watery diarrhea. The serotype *E. coli* O157:H7, an emergent pathogen of faecal origin frequently isolated from waters, has been implicated in food and water-borne disease outbreaks (Bavaro, 2009).

Bacteria of the genus *Klebsiella* are ubiquitous in nature and are a frequent cause of nosocomial infections (Horan et al., 1988). Their non-clinical habitats encompass the gastrointestinal tract of mammals as well as environmental sources such as soil, surface waters, and plants (Bagley, 1985). Environmental isolates have been described as being indistinguishable from human clinical isolates with respect to their biochemical reactions and virulence (Matsen et al., 1974). While the medical significance of *Klebsiella* obtained in the natural environment is far from clear, such habitats are thought to be potential reservoirs for the growth and spread of these bacteria which may colonize animals and humans (Knittel et al., 1977). Of the five identified *Klebsiella* species, *K. oxytoca* and *K. Pneumonia*, remain the most clinically important opportunistic pathogen, implicated in community-acquired pyogenic liver abscess and bacterial meningitis in adults (Casolari et al., 2005; Haryani et al., 2007; Keynan and Rubinstein, 2007), has been reported to be present in water (Syposs et al., 2005).

1.2 Methods used in detection of bacterial pathogens in water

Detection, differentiation, and identification of bacteria can be performed by numerous methods, including phenotypic, biochemical and immunological assays, and molecular techniques. These traditional methods for the detection and enumeration of bacterial pathogens have largely depended on the use of selective culture and standard biochemical methods. This classical microbiological methodology relies on the cultivation of specific bacteria, for example plate counts of coliforms. Drawbacks of these methods include firstly, pathogenic bacteria, which normally occur in low numbers, tend to incur large errors in sampling and enumeration (Fleischer, 1990). Secondly, culture-based methods are time-consuming, tedious; detect only one type of pathogen, and no valid identification of the pathogen (Szewzyk et al., 2000). Thirdly, many pathogenic microbes in the environment, although viable, are either difficult to culture or are non-culturable (Roszak and Colwell, 1987). Sometimes too, it is often difficult to achieve appropriate enrichment, which makes the work even more tedious.

Moreover, concentrations may be too low for cultural detection but still be high enough to cause infection. These limitations therefore make routine examination of water samples for pathogens like *Vibrio cholerae*, *Shigella dysenteriae*, *Aeromonas* spp. and *Campylobacter* spp., difficult. Instead, bacterial indicator species like *Escherichia* coli, which is a normal flora present in very high numbers in the gut of warm-blooded animals, is widely used as an indicator of faecal pollution, to estimate the risk of exposure to other pathogenic microbes present in animal or human wastes (Lund, 1994). However, *Escherichia coli* as well as some bacterial species like *Enterococcus faecalis*, once released into freshwater bodies, enter a viable but non-culturable (VBNC) state and express different set of activities, including virulence traits (Lleo et al., 2005). As a result, the current methodology is unsuitable for the detection of bacterial pathogens in water and the assessment of their virulence potential. Therefore, a molecular detection method is needed, since such methods are highly specific and sensitive.

Molecular methods used are typically based on the detection and quantification of specific segments of the pathogen's genome (DNA or RNA). To achieve this, the specific segments are subjected to *in vitro* amplification. These methods allow researchers to speedily and specifically detect microorganisms of public health concern (Girones et al., 2010). Recently, molecular techniques, specifically nucleic acid amplification procedures, immunocapturing, fluorescence *in-situ* hybridization (FISH), and polymerase chain reaction (PCR) have provided highly sensitive, rapid and quantitative analytical tools for detecting specific pathogens in environmental samples (Watson et al., 2004). These techniques are used to evaluate the microbiological quality of food and water, as well as microbial source-tracking (Albinana-Gimenez et al., 2009; Field et al., 2003; Hundesa et al., 2006). Most applied molecular techniques are based on protocols of nucleic acid amplification, of which the polymerase chain reaction (PCR) is the most commonly used.

PCR is a molecular tool that allows for the amplification of target DNA fragments using oligonucleotide primers in a chain of replication cycles catalysed by DNA polymerase (*Taq* polymerase) (Rompré et al., 2002). This tool is used for microbial identification and surveillance with high sensitivity and specificity (Watterworth et al., 2005). It has successfully been applied for the detection and identification of pathogenic bacteria in clinical and environmental samples, as well as for the investigation of food and water-borne disease outbreaks (Harakeh et al., 2006; Haryani et al., 2007; Hsu and Tsen, 2001; Riyaz-Ul-Hassan et al., 2004; Shabarinath et al., 2007). The use of quantitative PCR (qPCR) is rapidly becoming established in the environmental sector since it has been shown, in many cases, to be more sensitive than either the bacterial culture method or the viral plaque assay (He and Jiang, 2005). However, molecular protocols, unlike traditional culture-based methods, do not distinguish between viable and non-viable organisms hence the need for more information before replacing the current conventional methods by molecular ones.

Molecular techniques for the specific detection and quantification of bacterial pathogens also offer several advantages over conventional methods: high sensitivity and specificity, speed, ease of standardization and automation. As with the viruses, direct PCR amplification of some bacterial pathogens from water samples is difficult due to the presence of only low numbers of these bacteria in environmental sources. Therefore, an enrichment step is usually required prior to performing a PCR (Noble and Weisberg, 2005). Improved detection of pathogenic *E. coli* (Ogunjimi and Choudary, 1999) by immuno-capture PCR, and the sensitive detection of *Salmonella* (Hoorfar et al., 2000) and *Campylobacter* (Nogva et al., 2000) by real-time PCR have also been developed; but these procedures are all mono-specific and are either laborious or very expensive for routine use in water testing laboratories. More recent improvements have allowed simultaneous detection of several microorganisms in a single assay (Maynard et al., 2005; Straub et al., 2005; Marcelino et al., 2006). The use of such multiplex polymerase chain reaction (m-PCR) has provided rapid and highly sensitive methods for the specific detection of pathogenic microbes in the aquatic environment (Kong et al., 2002).

1.3 Multiplex polymerase chain reaction (m-PCR)

Following the application of PCR in the simultaneous amplification of multiple loci in the human dystrophin gene (Chamberlain et al., 1998), multiplex PCR has been firmly established as a general technique. To date, the application of multiplex PCR in pathogen identification, gender screening, linkage analysis, template quantitation, and genetic disease diagnosis is widely established (Chehab and Wall, 1992; Kong et al., 2002; Serre et al., 1991; Shuber et al., 1993). For pathogen identification, PCR analysis of bacteria is advantageous, as the culturing and typing of some pathogens has proven difficult or lengthy. Bacterial multiplexes indicate a particular pathogen among others, or distinguish species or strains of the same genus. An amplicon of sequence conserved among several groups is often included in the reaction to indicate the presence of phylogenetically or epidemiologically similar, or environmentally associated, bacteria and to signal a functioning PCR. Multiplex assays of this set-up distinguish species of *Legionella* (Bej et al., 1990), *Mycobacterium* (Wilton and Cousin, 1992), *Salmonella* (Chamberlain et al., 1998), *Escherichia coli,* and *Shigella* (Bej et al., 1991) and major groups of *Chlamydia* (Kaltenboek et al., 1992) from other genus members or associated bacteria. It has also been shown that multiplex PCR remains the ideal technique for DNA typing because the probability of identical alleles in two individuals decreases with the number of polymorphic loci examined. Reactions have been developed with potential applications in paternity testing, forensic identification, and population genetics (Edwards et al., 1991, 1992; Klimpton et al., 1993). Multiplex PCR can be a two-amplicon system or it can amplify 13 or more separate regions of DNA. It may be the end point of analysis, or preliminary to further analyses such as sequencing or hybridization. The steps for developing a multiplex PCR and the benefits of having multiple fragments amplified simultaneously, however, are similar in each system (Edwards and Gibbs, 1994).

1.4 Aim/Objectives of the study

To detect the presence of pathogenic *Escherichia coli, Klebsiella, Salmonella,* and *Shigella* species in water samples obtained from rivers in the North-West Province of South Africa, conventional typing and multiplex PCR methods were applied to enriched cultures. The objectives of the study were to use conventional methods to check for the presence and molecular tools to confirm the identity of *Escherichia, Klebsiella, Salmonella,* and *Shigella* species in river water. Our prognosis is that the results will emphasize the need for a rapid and accurate detection method for water-borne disease outbreaks and bacterial pathogens in water to protect human health.

2. Materials and methods

A total of 54 water samples were collected using sterile 500mL McCartney bottles, downstream, midstream, and upstream of the Crocodile, Elands, Hex, Mooi, Vaal, Molopo, Groot Marico, Harts and Skoonspruit rivers between November 2007 and March 2008 (Fig. 1). These rivers form the five major catchments in the province, which are the Crocodile and Eland, Marico and Hex, Marcio and Molopo, Mooi and Vaal, and Harts catchments. Samples collected were transported on ice to the laboratory for analysis.

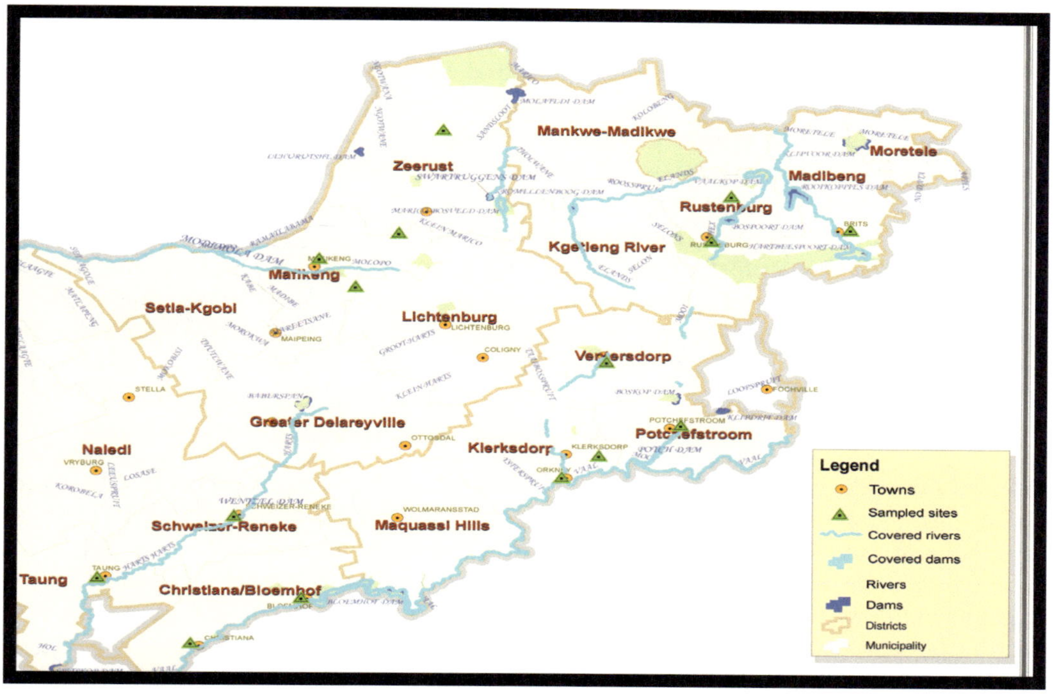

Fig. 1. A cross-section of the North West province map showing the rivers and dams sampled

2.1 Bacterial reference strains

Bacterial strains used for the experimental work (Table 1) were American Type Culture Collection (ATCC) cultures. The strains were grown on Nutrient Agar (Biolab, Merck, South Africa) under aerobic conditions at 37°C for 24 hours.

Bacterial Strains	Source	Reference	*Mdh*	*IpaH*	*IpaB*	*GapA*
Salmonella paratyphi	ATCC 9150	This study	–	–	+	–
Salmonella typhi	ATCC 14028	Hsu and Tsen, 2001	–	–	+	–
Shigella boydii	ATCC 9207	Wang et al., 1997	–	+	–	–
Shigella sonnei	ATCC 25931	Wang et al., 1997	–	+	–	–
Klebsiella pneumonia	ATCC 15611	Lu et al., 2000	–	–	–	+
K. oxytoca	ATCC 43086	This study	–	–	–	+
Escherichia coli	ATCC 25922	Lu et al., 2000	–	–	–	–

Table 1. Bacterial strains used in the study for evaluation of primer specificity

2.2 Selective isolation of *Salmonella*, *Shigella*, *E. coli* and *Klebsiella*

Water analysis for *Salmonella, Shigella, E. coli* and *Klebsiella* bacteria, was done using the spread plate method (American Public Health Association [APHA], 1998). In brief, 1mL of each water sample was enriched in 9mL of 2% buffered-peptone water (Biolab, Merck Diagnostics, South Africa) and serial dilutions performed. Aliquots of 0.1mL of each dilution were plated out on Eosin Methylene Blue (EMB) agar plates (Biolab, Merck Diagnostics, South Africa) for the presumptive isolation of *E. coli* and *Klebsiella*, and on Salmonella-Shigella agar for *Salmonella* and *Shigella* isolation. All plates were incubated at 37°C for 24 hours. Presumptive isolates were sub-cultured on fresh media plates incubated at 37°C for 24hours and then preserved on 2.3% w/v Nutrient agar plates for further analysis.

2.3 Bacterial Identification using triple sugar iron (TSI) agar test

All 2992 and 1180 presumptive isolates on EMBA and SSA plates, respectively were Gram stained using the method of Cruikshank et al., (1975) to confirm their morphology as Gram negative rod-shaped bacteria. All Gram negative isolates were subjected to the TSI test, a biochemical test, which distinguishes the *Enterobacteriaceae* from other intestinal Gram-negative bacilli by the ability of the organisms to catabolise the sugars glucose, lactose and sucrose present at different concentrations in the medium, and the production of acid and gas (Prescott, 2002). The test was performed as previously recommended (United States Pharmacopeia Convention; Inc., 2001). Briefly, isolates were streak-plated on TSI agar slopes and incubated at 37°C for 24hours. The results were interpreted as previously determined by Forbes and Weissfeld (1998).

2.4 Differentiation of *Salmonella*, *Shigella*, *E. coli* and *Klebsiella* using conventional serological assay

All *Salmonella, Shigella, E. coli* and *Klebsiella* candidate isolates obtained from culture plates and identified by Triple Sugar Iron [TSI] agar test, were differentiated by conventional serotyping (Ballmer et al., 2007). To test for surface antigens, *E. coli* Poly D1–D8; *Shigella boydii* Poly C, C1, C2 and C3, *Shigella dysenteriae* Poly A Types 1, 2, 3, 4, 5, 6, 7, *Shigella sonnie* Poly D Phase I and II, *Shigella flexneri* Poly B Types I, II, III, IV, V, VI; *Salmonella* O Poly O (Factors A–G, O2, O4, O7, O8, O9, O9, 46, O3, 10, O1, 3, 4) and O1 (Factors O11, O13, O6, 14, O16, O18, O21, O35), *Salmonella* H Poly Phase 1 and 2; and *Klebsiella* Capsular Types 1, 2, 3, 4, 5, 6 antisera (Inqaba Biotech, South Africa) were used.

2.5 DNA extraction

Genomic DNA was extracted from all positive bacteria cells inoculated in 5mL Luria Bertani (LB) broth (Merck, South Africa) following overnight incubation at 37°C in a shaker (Doyle and Doyle, 1990). The pellets obtained were re-suspended in 50µL of sterile distilled water. The concentration of the extracted DNA in solution was determined spectrophotometrically (UV Visible spectrophotometer model S-22, Boeco, Germany) by measuring the absorbance at 260 nm. The DNA in solution was used as a template for multiplex PCR.

2.6 Oligonucleotide primers and multiplex PCR method

Oligonucleotide primers used in the study were synthesized by Inqaba Biotech, South Africa. Sequences of the four PCR primer pairs for m-PCR, their corresponding gene targets and size of the expected amplifications are as shown (Table 2). The malate dehyrogenase gene (*Mdh*) of *E. coli* (Hsu and Tsen, 2001; Wose Kinge and Mbewe, 2011), the invasive pasmid antigen B gene (*IpaB*) of *Salmonella* spp. (Kong et al., 2002), the invasive plasmid antigen gene H (*IpaH*) of *Shigella* spp. (Kong et al., 2002; Wose Kinge and Mbewe, 2010), and the glyceralehye-3-phospahate dehydrogenase gene (*GapA*) genes for *Klebsiella* spp. (Diancourt et al., 2005; Wose Kinge and Mbewe, 2011) were simultaneously detected by multiplex polymerase chain reaction (m-PCR) assays. DNA from 50μL extract from enriched cultures was used for PCR amplification in a final volume of 25μL. The reaction mixture consisted of 2X PCR Master mix (0.05μL Taq DNA polymerase, 4mM $MgCl_2$, 0.4mM dNTPs) (Fermentas, Inqaba Biotechnical Industries (Pty) Ltd, South Africa), 0.3μM of IpaB, 0.2μM of IpaH and 1.0μM each of Mdh and GapA genes. PCR amplification was performed in a Peltier Thermal Cycler (model-PTC-220 DYAD™ DNA ENGINE; MJ Research Inc. USA) under the following conditions: heat denaturation at 94°C for 3 min, followed by 34 cycles of denaturation at 94°C for 30 s; annealing at 60°C for 60 s and extension at 72°C for 1 min. This was followed by a final extension step at 72°C for 7 min and 4°C hold. To create a negative control template DNA was excluded.

Organism	Target gene	Primer	Primer sequence (5′→3′)	Expected size (bp)
E. coli	*Mdh*	Mdh F	CGTTCTGTTCAAATGGCCTCAGG	392
		Mdh R	ACTGAAAGGCAAACAGCCAAG	
Salmonella	*IpaB*	IpaB F	GGACTTTTTAAAGCGGCGG	314
		IpaB R	GCCTCTCCCAGAGCCGTCTGG	
Shigella	*IpaH*	IpaH F	CCTTGACCGCCTTTCCGATA	606
		IpaH R	CAGCCACCCTCTGAGGTACT	
Klebsiella	*GapA*	GapA F	GTTTTCCCAGTCACGACGTTGTATGAA ATATGACTCCACTCACG	700
		GapA R	TTGTGAGCGGATAACAATTTCCTTCAG AAGCGGCTTTGATGGCT	

Table 2. Oligonucleotide primers used in this study

2.7 Electrophoresis and visualization of PCR products

Following amplification, 10μL of each sample was electrophoresed in a horizontal agarose (LONZA, South Africa) 1% w/v slab gel containing ethidium bromide (0.1μg/mL) in 1X TAE buffer (40 mM tris-acetate; 2 mM EDTA, pH8.3). The agarose gel was electrophoresed for six hours at 60 V. The gel was visualized with UV light (Gene Genius Bio Imaging System, SYNGENE model GBOX CHEMI HR). The relative molecular sizes of the PCR products were estimated by comparing their electrophoretic mobility with 100bp marker (Fermentas O' GeneRuler DNA ladder; Canada).

2.8 Specificity of primers

The specificity of the primers used for multiplex-PCR was confirmed against related enteric bacterial DNA. The DNA was extracted from 5mL of overnight bacterial suspensions cultured in Luria Bertani broth as described under section 2.5. The extracted DNA was then stored at -20°C for use in m-PCR.

3. Results and discussion

3.1 Differentiation of *Salmonella*, *Shigella*, *E. coli* and *Klebsiella* using conventional serotyping assay

In order to differentiate the bacterial isolates using surface antigens present, conventional serotyping by slide agglutination was performed using polyvalent antisera. The commercially available typing antisera are not sufficient to recognize all prevalent serotypes of *Salmonella*, *E. coli* and *Klebsiella* spp. In our study, the antisera assay was not used to identify these serotypes, but rather to determine if a given isolate was a member of the genera of interest or not. The percentages of *E. coli*, *Klebsiella*, *Shigella* and *Salmonella* isolates obtained, showing a positive agglutination to antisera, were calculated for each catchment area and results recorded as contained in Table 3. The results indicate a presence of *E. coli*, *Klebsiella*, *Shigella* and *Salmonella* spp. in all five catchments areas. According to Table 3, *E. coli* (the main indicator for faecal contamination) was present in all five catchment samples. The highest was 29% in the Crocodile and Elands catchment, followed by the Mooi and Vaal catchment with 24% agglutination with surface antigen specific antisera. The other three catchments were not free of *E. coli* although at lesser levels, comparably.

According to DWA and WHO standards, water meant for irrigation (DWA, 1996) and human consumption (WHO, 2001) should contain no *E. coli* bacteria. The use of such contaminated water for irrigation as well as direct consumption as it is before treatment would result in the transmission of potentially pathogenic bacteria to humans through contaminated vegetables and other crops eaten raw, as well as milk from grazing cattle. *Klebsiella* was highest in the Mooi and Vaal followed by Harts catchments with 19% and 11%, respectively. Podschun et al. (2001) also reported a high percentage (53%) distribution of *Klebsiella* spp. from surface water samples, the most common species being *K. pneumoniae*. Bacteria species of the genera *Escherichia* and *Klebsiella* are amongst the group of faecal coliforms. Generally, faecal coliform bacteria inhabit the gastrointestinal tract of all warm and some cold-blooded animals as normal commensals, hence their presence in any given water body is a clear indication of faecal contamination. Although their presence in water cannot be pinpointed to a specific source of faecal contamination, faecal material from human and animal sources can be regarded as high risk due to the possible presence of pathogenic bacteria (Harwood et al., 2000).

High levels of *Shigella* contamination were also seen in all catchments with 31% and 41% in the Crocodile and Elands catchment and Harts catchment, respectively. In general, there was lesser contamination with *Salmonella* compared to other faecal coliforms in all catchments with a maximum of 8% in Mooi and Vaal catchment. Water-borne pathogens often occur in reasonably low concentrations in environmental waters. Therefore, some form of filtration and proliferation are needed for pathogen detection (Hsu et al., 2010). Following filtration of the sample on membrane filters, bacteria retained on filters can then be detected by culturing in or on selective media. Additional steps, such as biochemical tests, serological assays, and molecular methods, are necessary for confirmation. The isolation and identification of *Shigella* spp. and *E. coli* are straightforward and well established (Echeverria et al., 1991, 1992). However, *Shigella* spp. and entero-invasive *E. coli* [EIEC] are genetically close and exhibit considerable antigenic cross-reactivity, thus differentiating between them using a single method can be difficult (Cheasty and Rowe, 1983; Lan et al., 2001; Kingombe et al., 2005; Yang et al., 2005).

The O and H antigen serotyping method provide important epidemiological information. However it is not appropriate for routine diagnostic use because of its high cost and the labour-intensive requirements (Ballmer et al., 2007). There is, therefore, an urgent need for an accurate and simple detection, identification, and differentiation technique for *Shigella* spp. and EIEC, especially for epidemiological studies. On the contrary, serotyping is currently the most widely used technique for typing *Klebsiella* species. It is based mainly on a division according to the K (capsule) antigens (Ørskov and Ørskov, 1984) and shows good reproducibility and capability in differentiating most clinical isolates (Ayling-Smith and Pitt, 1990).

River Catchments	*E. coli* %	*Klebsiella* %	*Shigella* %	*Salmonella* %
Crocodile and Elands	29	4	37	6
Marico and Hex	9	7	18	4
Marico and Molopo	9	4	12	1
Mooi and Vaal	24	19	15	8
Harts	7	11	41	6

Table 3. Prevalence of *E. coli, Klebsiella, Shigella* and *Salmonella* bacteria obtained by serotyping

3.2 Multiplex PCR

The m-PCR was designed to target genes specific to the four entero-pathogenic bacteria selected for this study. Results obtained showed the presence of *E. coli, Klebsiella, Shigella* and *Salmonella* contamination in the five catchment areas (Table 4). A total of 39% of *E. coli* was recorded for the Crocodile and Elands catchment and up to 45% of *Shigella* spp. was recovered from the Marico and Hex catchment. The presence of *Klebsiella* and *Salmonella* spp. was also observed with 10% and 11% in the Mooi and Vaal catchment, respectively. Of these bacteria species, contamination with *Shigella* was widespread in all catchments. Detection of the *IpaH* gene, which is present on both the chromosome and the *inv* plasmid of all *Shigella* spp., confirmed the presence of this bacterium in water (Hsu and Tsen, 2001). Understanding the ecology of *Shigella* had been limited mainly due to the lack of suitable techniques to detect the presence of *Shigella* in environment samples (Faruque et al., 2002).

In the present study, we used molecular techniques as well as conventional serotyping method to detect *Shigella* as well as *E. coli*, *Salmonella* and *Klebsiella* spp. in river waters with special reference to virulence genes. We standardized the assay by culturing the environmental water samples and simultaneously conducting m-PCR tests. In a similar study by Faruque et al. (2002) and Sharma et al. (2010), the *IpaH* gene was used as an indicator tool to detect the presence of *Shigella* in environmental waters. Fresh contamination of surface water by faecal material of dysentery patients is a possibility in developing countries where sanitation is poor resulting in the presence of *Shigella* in surface water. Several previous studies have also detected *Shigella* in surface waters or sewage samples and have indicated that *Shigella* strains can possibly be transported by surface waters (Alamanos et al., 2000; Faruque et al., 2002; Obi et al., 2004a; Pergram et al., 1998).

Similarly, amplification of the *Mdh* gene, which codes for malic acid dehydrogenase, a housekeeping enzyme of the citric acid cycle, and reportedly found in all *E. coli* strains (Hsu and Tsen, 2001), confirmed the presence of both commensal and pathogenic *E. coli* in the water samples. Although *E. coli* is usually present as harmless commensals of the human and animal intestinal tracts, pathogenic strains possess virulent factors that enable them to cause diseases and hence, constitute a potential risk to the health of consumers (Kuhnert et al., 2000). For the detection of *Salmonella* spp. the *IpaB* gene, which is a virulence gene found on the invasion plasmid of *Salmonella* spp., was selected for the PCR as it is reportedly present in most *Salmonell*a strains (Kong et al., 2002). *Salmonella* is isolated from water in lower numbers than indicator bacteria such as faecal coliforms, faecal streptococci and enterococci, which are several orders of magnitude higher (Sidhu and Toze, 2009).

However, low numbers (15-100 colony-forming units [CFU]) of *Salmonella* in water may pose a public health risk (Jyoti et al., 2009). In the aquatic environment this pathogen has been repeatedly detected in various types of natural waters such as rivers, lakes, coastal waters, estuarine as well as contaminated ground water (Haley et al., 2009; Levantesi et al., 2010; Lin and Biyela, 2005; Moganedi et al., 2007; Theron et al., 2001; Wilkes et al., 2009). Their presence has been attributable to runoff from fields with animal husbandry, addition of untreated sewage from nearby civilization contribute *Salmonella* in natural water resources (Moganedi et al., 2007; Jenkins et al., 2008). *Salmonella* contaminated waters might contribute through direct ingestion of the water or via indirect contamination of fresh food to the transmission of this microorganism. *Salmonella* prevalence in surface water and drinking water has not been uniformly investigated in different countries in recent papers.

Surveys of *Salmonella* in fresh surface water environment were mainly performed in industrialized nations, particularly in Canada and North America. Reports of *Salmonella* prevalence in drinking water were instead more frequent from developing nations reflecting the higher concern relating to the use of low quality drinking water in these countries. Overall, the scientific community has mainly recently focused on the prevalence of is microorganism in impacted and non-impacted watersheds (Haley et al., 2009; Jokinen et al., 2011; Patchanee et al., 2010), on the identification of the routes of salmonellae contamination (Gorski et al., 2011; Jokinen et al., 2010, 2011; Obi et al., 2004b; Patchanee et al., 2010), and on the influence of environmental factors on the spread of *Salmonella* in water (Haley et al., 2009; Jokinen et al., 2010; Meinersmann et al., 2008; Wilkes et al., 2009).

Although direct consumption of water by humans from these rivers was minimal throughout the study, indirect consumption through fishing was common. This was particularly evident in the Crocodile and Elands, Marico and Molopo, and the Mooi and Vaal catchment areas. This may be a cause for concern because fish in water bodies contaminated with human and animal waste, harbour a considerable number of bacteria such as *Salmonella, Clostridium botulinum, Vibrio cholerae, E. coli* and other coliforms, which could be transmitted to humans if eaten raw or under-cooked (Jayasinghe and Rajakaruna, 2005). Fish and shellfish accounts for 5% of individual cases and 10% of all food-borne illness outbreaks in the United States (Flick, 2008) and not only does fish constitute potential sources of bacteria, they also harbour antibiotic resistant bacteria that could be transmitted to humans resulting in the spread of a pool of antibiotic resistant genes into the environment (Miranda and Zemelman, 2001; Pathak and Gopal, 2005). This also might be compounded by the presence of opportunistic pathogens like *Klebsiella* species in water with serious health implications for consumers that utilize water directly or indirectly from the rivers, especially high risk patients with impaired immune systems such as the elderly or young, patients with burns or excessive wounds, those undergoing immunosuppressive therapy or those with HIV/AIDS infection. Colonization may lead to invasive infections and on very rare occasions, *Klebsiella* spp., notably *K. pneumoniae* and *K. oxytoca*, may cause serious infections, such as destructive pneumonia (Bartram et al., 2003; Genthe and Steyn, 2006).

River Catchments	***E. coli*** **%**	***Klebsiella*** **%**	***Shigella*** **%**	***Salmonella*** **%**
Crocodile and Elands	39	0	11	6
Marico and Hex	4	6	45	0
Marico and Molopo	0	6	5	1
Mooi and Vaal	15	10	5	11
Harts	0	0	23	9

Table 4. Prevalence of *E. coli, Klebsiella, Shigella* and *Salmonella* bacteria obtained by m-PCR

3.3 Specificity of primers

In order to evaluate and verify the specificity of the primers in this study, each primer pair was tested by PCR on DNA templates prepared from a panel of seven different bacterial control strains. The analysis indicated that all primer pairs showed specificities only for their corresponding target organisms (Table 1) and all four sets of PCR primers were targeted at a virulence-associated gene. The Mdh primers specifically amplified a 392bp malic acid dehydrogenase gene fragment from *E. coli* strain obtained from the American Type Culture Collection (Table 1) and 4-39% of isolates obtained from the different river catchments. The IpaH primers produced a specific 606bp amplimer in all *Shigella* spp. examined in this study (Table 1; Fig 2. lane 3), which included two species of the genus, viz., *S. sonnei* and *S. boydii,* which are known to be pathogenic to humans. In a previously reported study, Kong et al.

(2002) tested two virulence genes of *Shigella*, the *virA* gene and the *IpaH* gene and obtained more positive amplifications with the *IpaH* gene when compared with the *virA* gene.

Although the *virA* gene was previously reported by Villalobo and Torres (1998) to be specific for virulent *Shigella* spp., the *IpaH* gene was found to be more reliable in detecting *Shigella* spp. in environmental isolates (Kong et al., 2002; Wose Kinge and Mbewe, 2010). The IpaB primers were found to produce a specific 314bp amplimer, in all *Salmonella* spp. examined, which included *S. paratyphi*, and *S. typhimurium* (Table 1; Fig 2. Lanes 7 and 8) as well as 1-11% of the isolates tested. Similar results were obtained with the GapA primers which generated a 700bp amplimer specific to *Klebsiella*. The amplimers were confirmed by sequencing (Inqaba Biotech, South Africa) all showed a high percentage of sequence similarity (>90%) with published malic acid dehydrogenase, invasive plasmid antigen H and B, and glyceraldehydes-3-phosphate gene sequences in the GenBank database. Our results therefore, indicated that this particular set of primers were suitable for the specific detection of most general strains of *E. coli*, *Salmonella*, *Shigella* and *Klebsiella* from water samples.

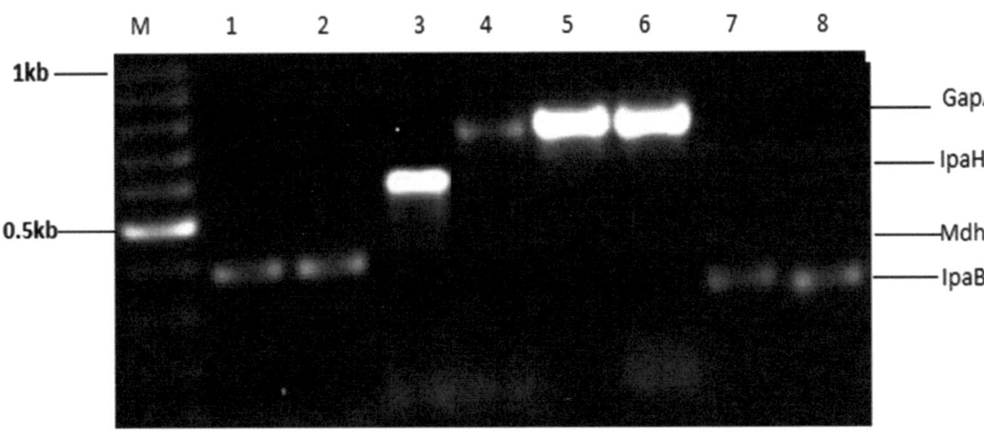

Fig. 2. Electrophoretic analysis of PCR-amplified target genes from six different bacterial pathogens. Mobilities of the different target gene amplicons are indicated on the right. Lane M, 100bp DNA ladder (size marker); lanes 1 and 2, Mdh amplicon of *Escherichia coli* ATCC 25922; lane 3, IpaH amplicon of *Shigella boydii* ATCC 9207; lane 4, GapA amplicon of *Klebsiella oxytoca* ATCC 43086; lanes 5 and 6, GapA amplicon of *K. pneumoniae* ATCC 15611; lane 7, IpaB amplicon of *Salmonella paratyphi* ATCC 9150, lane 8, IpaB amplicon of *S. typhimurium* ATCC 14028

4. Conclusion

Both conventional and molecular methods successfully identified bacteria of interest, however, the multiplex-PCR assays were sensitive and faster than conventional serotyping methods for detecting *E. coli*, *Salmonella*, *Shigella*, and *Klebsiella* spp. from river water samples. The 392bp *Mdh*, 314bp *IpaB*, 606bp *IpaH* and 700bp *GapA* genes were found to be specific and present in the control strains analyzed. Therefore, m-PCR screening of these strains for *Mdh*, *IpaB*, *IpaH* and *GapA* genes should provide a better indicator of possible presence of potentially pathogenic *E. coli, Salmonella, Shigella* and *Klebsiella* bacteria in river water. The water quality is affected by human activities around the areas, which include industrial processes, mining, agriculture and domestic usage. Thus, the main source of *E. coli, Salmonella, Shigella* and *Klebsiella* in these rivers may be discharge from wastewater effluent as well as domestic sewage around the catchment areas. Our results indicate that the water-borne and food-borne spread of these pathogens is possible due to drinking water contamination, recreational activities, and fisheries. Since the aquatic environment is implicated as the reservoir for these microorganisms, and consequently responsible for their transmission in humans, it is obvious that detailed studies on the pathogenic potential of the environmental strains will certainly contribute to understanding the virulence properties of these bacteria and to establish the importance of these significant pathogens of aquatic systems. The results thus emphasize the need for the implementation of a rapid and accurate detection method in cases of water-borne disease outbreaks and the need for more rapid detection of bacterial pathogens in water to protect human health. The ability to rapidly monitor for various types of microbial pathogens would be extremely useful not only for routine assessment of water quality to protect public health, but also allow effective assessments of water treatment processes to be made by permitting pre- and post-treatment waters to be rapidly analyzed.

5. References

Alamanos, Y., Maipa, V., Levidiotou, S., Gessouli, E. (2000). A community waterborne outbreak of gastro-enteritis attributed to *Shigella sonnei*. *Epidemiol. Infect.* 125, pp. (499–503)

Albinana-Gimenez, N., Miagostovich, M., Calgua, B., Huguet, J.M., Matia, L., & Girones, R. (2009). Analysis of adenoviruses and polyomaviruses quantified by qPCR as indicators of water quality in source and drinking water-treatment plants. *Water Res.* 43, pp. (2011-2019)

APHA. (1998). Standard methods for the examination of Water and Wastewater, 19th Ed. Amer. Public Health Assoc. Washington DC.

Ayling-Smith, B., & Pitt, TL. (1990). State of the art in typing: *Klebsiella* species. *J. Hosp. Infect.* 16, pp. (287–295)

Bagley, ST. (1985). Habitat association of *Klebsiella* species. *Infect. Control,* 6, pp. (52–58)

Ballmer, K., Korczak, BM., Kuhnert, P., Slicker, P., Ehricht, R., & Hächler, H. (2007). Fast DNA serotyping of *Escherichia coli* by use of an oligonucleotide microassay. *J. Clin. Microbiol.* 45, 2, pp. (370-379)

Bartram, J., Chartier, Y., Lee, JV., Pond, K., & Surman-Lee S (Eds) (2007): *Legionella and the Prevention of Legionellosis.* WHO Press, World Health Organization, Geneva, Switzerland.

Bartram, J., Cotruvo, J., Exner, M., Fricker, C., & Glasmacher, A. (2003). Heterotrophic plate counts in drinking water safety: the significance of HPCs for water quality and human health. WHO Emerging Issues in Water and Infectious Disease Series. London, IWA Publishing.

Bavaro, MF. (2009). *Escherichia coli* O157: what every internist and gastroenterologist should know. *Curr. Gastroenterol. Rep.* 11, 4, pp. (301-306)

Bej, AK., Mahbubani, MH., Miller, R., DiCesare, JL., Haft, L., & Atlas, RM. (1990). Multiplex PCR amplification and immobilized capture probes for detection of bacterial pathogens and indicators in water. *Mol. Cell. Prob.* 4, pp. (353-365).

Bej, AK., McCarty, SC., & Atlas, RM. (1991). Detection of coliform bacteria and *Escherichia coli* by multiplex polymerase chain reaction: Comparison with defined substrate and plating methods for water quality monitoring. *Appl. Environ. Microbiol.* 57, pp. (1473-1479)

Beukman, R. & Uitenweerde, T. (2002). Water Demand Management Phase II. International Conservation Union, Pretoria, pp. (12)

Casolari, C.., Pecorari, M., Fabio, G., Cattani, S., Venturelli, C., Piccinini, L., Tamassia, M.G., Gennari, W., Sabbatini, A.M., Leporati, G., Marchegiano, P., Rumpianesi, F., & Ferrari, F. (2005). A simultaneous outbreak of *Serratia marcescens* and *K. pneumoniae* in a neonatal intensive care unit. *The J. Hosp. Infec.* 61, pp. (312-320)

Chamberlain, JS., Gibbs, RA., Ranier, JE., Nguyen, PN., & Caskey, CT. (1988). Deletion screening of the Duchenne muscular dystrophy locus via multiplex DNA amplification. *Nucleic Acids Res.* 16, pp. (11141-11156).

Cheasty, T., & Rowe, B. (1983). Antigenic relationships between the enteroinvasive *Escherichia coli* O antigens O28ac, O112ac, O124, O136, O143, O144, O152, and O164 and *Shigella* O antigens. *J. Clin. Microbiol.* 17, pp. (681–684)

Chehab, FF., & Wall, J. (1992). Detection of multiple cystic fibrosis mutations by reverse dot blot hybridization: A technology for carrier screening. *Hum. Genet.* 89, pp. (163-168)

Cobbold, RN., Rice, DH., Davis, MA., Besser, TE., & Hancock, DD. (2006). Long-term persistence of multidrug-resistant *Salmonella enterica* serovar Newport in two dairy herds. *J. Amer. Vet. Med. Assoc.* 228, pp. (585–91)

Collado, L., Guarro, J., & Figueras, MJ. (2009) Prevalence of *Arcobacter* in meat and shellfish. *J. Food. Prot.* 72, pp. (1102– 1106)

Collado, L., Inza, I., Guarro, J., & Figueras, M.J. (2008) Presence of *Arcobacter* spp. in environmental waters correlates with high levels of fecal pollution. *Environ. Microbiol.* 10, pp. (1635–1640)

Collie, T.K., Strom, M.S., Sinigalliano, C.D., Moeller, P.D., & Holland, A.F. (2008). The coastal environment and human health: microbial indicators, pathogens, sentinels and reservoirs. *Environ. Health*, 7, 2, pp. (S3)

Cruikshank, R., Duguid, JP., Marmoin, BP.,& Swain, RH. (1975). Medical Microbiology, 12th ed. Longman, New York. 2, pp. (3-4)

CSIR. (2008). Water Quality and Pollution: Parliamentary Portfolio Committee on Water Affairs and Forestry.

Diancourt, L., Passet, V., Verhoef, J., Grimont, P.A., & Brisse, S. (2005). Multilocus sequence typing of *Klebsiella pneumoniae* nosocomial isolates. *J. Clin. Microbiol.* 42, 8, pp. (4178-4182)

Doyle, JJ., & Doyle, JL. (1990). A rapid total DNA preparation procedure for fresh plant tissue. *Focus*, 12, pp. (13-15).

DWA. (2000). Water use authorisation process for individual applications, DWAF, Edition 1 final draft for implementation and use, revision 3, December 2000.

DWA. (2007). Integrated Vaal River system WMR studies. Department of Water Affairs and Forestry, Pretoria.

Echeverria, P., Sethabutr, O., & Pitarangsi, C. (1991). Microbiology and diagnosis of infections with *Shigella* and enteroinvasive *Escherichia coli*. *Rev. Infect. Dis.* 13, Suppl. 4, pp. (S220–S225)

Echeverria, P., Sethabutr, O., Serichantalergs, O., Lexomboon, U., & Tamura, K. (1992). *Shigella* and enteroinvasive *Escherichia coli* infections in households of children with dysentery in Bangkok. *J. Infect. Dis.* 165, pp. (144–147)

Edwards, A., Civitello, A., Hammond, HA., & Caskey, CT. (1991). DNA typing and genetic mapping with trimeric and tetrameric tandem repeats. *Am. J. Hum. Genet.* 49, pp. (746 – 756)

Edwards, A., Hammond, HA., Jin, L., Caskey, CT., & Chakroborty, R. (1992). Genetic variation at five trimeric and tetrameric tandem repeat loci in four human population groups. *Genomics,* 12, pp. (241-253)

Edwards, MC., & Gibbs, RA. (1994). Multiplex PCR: advantages, development and applications. *Genome Res.* 3, pp. (S65- S75)

Faruque, SM., Khan, R., Kamruzzaman, M., Yamasaki, S., Ahmad, QS., Azim, T., Nair, GB., Takeda, Y., & Sack, DA. (2002). Isolation of *Shigella dysenteriae* type 1 and *S. flexneri* strains from surface waters in Bangladesh: comparative molecular analysis of environmental *Shigella* isolates versus clinical strains. *Appl. Environ. Microbiol.* 68, pp. (3908–3913)

Field, KG., Bernhard, AE., & Brodeur, TJ. (2003). Molecular approaches to microbiological monitoring: faecal source detection. *Environ. Monit. Assess.* 81, pp. (313-326)

Fleisher, JM. (1990). Conducting recreational water quality surveys: some problems and suggested remedies. *Mar Pollut Bull.* 21, 12, pp. (562–7)

Flick, GJ. (2008). Microbiological safety of farmed fish. Global Aquaculture Advocate, pp. (33-34)

Fong, TT., Mansfield, LS., Wilson, DL., Schwab, DJ., Molloy, SL., & Rose, JB. (2007). Massive microbiological groundwater contamination associated with a waterborne outbreak in Lake Erie, South Bass Island, Ohio. *Environ. Health Perspect.* 115, 6, pp. (856-864)

Forbes, S., & Weissfeld, F. (1998). Bailey and Scott's diagnostic microbiology, 10th Ed. Mosby, Inc., St. Louis, Mo. Harakeh, S., Yassine, H., El-Faidel, M. (2006). Antibiotic resistant patterns of *Escherichia coli* and *Salmonella* strains in the aquatic Lebanese environments. *Environ. Pollut.* 143, pp. (269-277)

Gannon, VP., Graham, T., Read, S., Ziebell, K., Muckle, A., & Mori, J., et al. (2004). Bacterial pathogens in rural water supplies in Southern Alberta, Canada. *J. Toxicol. Environ. Health,* 67, 20–22, pp. (1643–1653)

Genthe, B., & Steyn, M. (2006). Good intersectoral water governance- a Southern African decision-makers guide. Chapter on Health and Water. CSIR/NRE/WR/EXP/2006/0047/A.

Girones, R., Ferrus, MA., Alonso, JL., Rodriguez-Manzano, J., Correa AA., Calgua, B., Hundesa, A., Carratala, A., & Bofill- Mas, S. (2010). Molecular detection of pathogens in water- The pros and cons of molecular techniques. *Water Res.* 44, pp. (4325-4339).

González, A., Botella, S., Montes, RM., Moreno, Y., & Ferrús, M.A. (2007). Direct detection and identification of *Arcobacter* species by multiplex PCR in chicken and wastewater samples from Spain. *J. Food Prot.* 70, (341-347)

Gorski, L., Parker, CT.,Liang, A., Cooley, MB., Jay-Russell, MT., Gordus, AG. (2011). Prevalence, distribution and diversity of *Salmonella enterica* in a major produce region of California. *Appl. Environ. Microbiol.* doi: 10.1128/AEM.02321-10.

Haley, B.J., Cole, DJ., & Lipp, EK. (2009). Distribution, diversity and seasonality of water-borne *Salmonella* in a rural watershed. *Appl. Environ. Microbiol.* 75, pp. (1248–1255)

Haley, BJ., Cole, DJ., & Lipp, EK. (2009). Distribution, diversity and seasonality of water-borne *Salmonella* in a rural watershed. *Appl. Environ. Microbiol.* 75, pp. (1248–1255)

Harakeh, S., Yassine, H., & EL- Fadel, M. (2006). Antimicrobial resistant patterns of *Escherichia coli* and *Salmonella* strains in the aquatic Lebanese environments. *Environ. Pol.* 143, 2, pp. (269-277)

Harwood, VJ., Gandhi, JP., & Wright, AC. (2004). Methods for isolation and confirmation of *Vibrio vulnificus* from oysters and environmental sources: a review. *J. Microbiol. Methods*, 59, pp. (301-316)

Harwood, VJ., Whitlock, J., & Withington, V. (2000). Classification of antibiotic resistance patterns of indicator bacteria by discriminant analysis: use in predicting the sources of faecal contamination in subtropical waters. *Appl. Environ. Microbiol.* 66, 9, pp. (3698-3704)

Haryani, Y., Noorzaleha, A.S., Fatimah, A.B., Noorjahan, B.A., Patrick, G.B., Shamsinar, A.T., Laila, R.A.S., & Son, R. (2007). Incidence of *Klebsiella pneumoniae* in street foods sold in Malaysia and their characterization by antibiotic resistance, plasmid profiling, and RAPD-PCR analysis. *Food Cont.* 18, pp. (847-853)

He, J., & Jiang, S. (2005). Quantification of enterococci and human adenoviruses in environmental samples by real-time PCR. *Appl. Environ. Microbiol.* 71, 5, pp. (2250-2255)

Ho, HT., Lipman, LJ., & Gaastra, W. (2006) *Arcobacter*, what is known and unknown about a potential food-borne zoonotic agent!. *Vet. Microbiol.* 115, pp. (1–13)

Hoorfar, J., Ahrens, P., & Radstrom, P. (2000). Automated 5'Nuclease PCR Assay for Identification of *Salmonella enterica. J Clin Microbiol.* 38, 9, pp. (3429–35)

Horan, T., Culver, D., Jarvis, W., Emori, G., Banerjee, S., Martone, W., & Thornsberry, C. (1988). Pathogens causing nosocomial infections. *Antimicrob. Newsl.* 5, pp. (65–67)

Houf, K. (2010) *Arcobacter*. In: Dongyou, L. (Ed.), Molecular detection of foodborne pathogens, CRC Press, New York, USA, pp. (289–306)

Hsu, B-M., Wu, S-F., Huang, S-W., Tseng, Y-J., Chen, J-S., Shih, F-C., & Ji, D-D. (2010). Differentiation and identification of *Shigella* spp. and enteroinvasive *Escherichia coli* in environmental waters by a molecular method and biochemical test. *Water Res.* 44, pp. (949-955)

Hsu, S-C., & Tsen, H-Y. (2001). PCR primers designed from malic acid dehydrogenase gene and their use for detection of *Escherichia coli* in water and milk samples. *Int. J. Food Microbiol.* 64, pp. (1-11)

Hundesa, A., Maluquer de Motes, C., Bofill-Mas, S., AlbinanaGimenez, N., & Girones, R. (2006). Identification of human and animal adenoviruses and polyomaviruses for determination of sources of fecal contamination in the environment. *Appl. Environ. Microbiol.* 72, pp. (7886-7893)

Igbinosa, EO., Obi, LC., & Okoh, AI. (2009). Occurrence of potentially pathogenic vibrios infinal effluents of a wastewater treatment facility in a rural community of the Eastern Cape Province of South Africa. *Res. Microbiol.* 160, 8, pp. (531-537)

Jayasinghe, PS., & Rajakuruna, RMAGG. (2005). Short Communication: Bacterial contamination of fish sold in fish markets in the Central province of Sri Lanka. *J. Nat. Sci. Foundation Sri Lanka*, 33, 3, pp. (219-221)

Jenkins, MB., Endale, DM., & Fisher, DS. (2008) Most probable number methodology for quantifying dilute concentrations and fluxes of *Salmonella* in surface waters. *J Appl Microbiol*, 104, pp. (1562–8)

Jokinen, C., Edge, TA., Ho, S., Koning, W., Laing, C., Mauro, W., et al. (2011). Molecular subtypes of Campylobacter spp., *Salmonella enterica*, and *Escherichia coli* O157:H7 isolated from fecal and surface water samples in the Oldman River watershed, Alberta, Canada. Water Res. 45, pp. (1247–1257)

Jyoti, A., Ram, S., Vajpayee, P., Singh, G., Dwivedi, PD., Jain, SK., & Shanker, R. (2009). Contamination of surface and potable water in South Asia by Salmonellae: Culture-independent quantification with molecular beacon real-time PCR. *Sci. Total Environ.* 408, 6, pp. (1256-1263)

Kaltenboek, B., K.G. Kansoulas, and J. Storz. 1992. Two-step polymerase chain reactions and restriction endonuclease analyses detect and differentiate *ompA* DNA of the *Chlamydia* spp. *I. Clin. Microbiol.* 30, pp. (1098-1104)

Kasrils, R. (2004). A decade of delivery. Minister of Water Affairs and Forestry.

Keynan, Y., & Ruinstein, E., 2007. The emerging face of *K. pneumoniae* infections in the community. *Int. J. Antimicrob. Agents.* 30, pp. (385-389)

Kim, HM., Hwang, CY., & Cho, BC. (2010). *Arcobacter marinus* sp. nov. *Int. J. Syst. Evol. Microbiol.* 60, pp. (531–536)

Kingombe, CI., Cerqueira-Campos, M-L., & Farber, JM. (2005). Molecular strategies for the detection, identification and differentiation between enteroinvasive *Escherichia coli* and *Shigella* spp. *J. Food Prot.* 68, 2, pp. (239–245)

Klimpton, CP., Gill, P., Walton, A., Urquhart, A., Millican, E.S., & Adams, M. (1993). Automated DNA profiling employing multiplex amplification of short tandem repeat loci. *PCR Methods Applic.* 3, pp. (13-21)

Knittel, MD., Seidler, RJ., Eby, C., & Cabe, LM. (1977). Colonization of the botanical environment by *Klebsiella* isolates of pathogenic origin. *Appl. Environ. Microbiol.* 34, pp. (557–563)

Kong, RYC., Lee, SKY., Law, TWF., Law, SHW., & Wu, RSS. (2002). Rapid detection of six types of bacteria pathogens in marine waters by multiplex PCR. *Water Res.* 36, pp. (2802-2812)

Kuhnert, P., Boerlin, P., & Frey, J. (2000). Target genes of virulence assessment of *Escherichia coli* isolates from water, food and environment. *Fed. European Microbiol. Soc. Rev.* 24, pp. (107-117)

Kusiluka, LJM., Karimuribo, ED., Mdegela, RH., Luoga, EJ., Munishi, PKT., Mlozi, MRS., & Kambarage, DM. (2005). Prevalence and impact of water-borne zoonotic pathogens in water, cattle and humans in selected villages in Dodoma Rural and Bagamoyo districts, Tanzania. *Phy. Chem. Earth*, 30, pp. (818-825)

Lan, R., Lumb, B., Ryan, D., Reeves, PR. (2001). Molecular evolution of large virulence plasmid in *Shigella* clones and enteroinvasive *Escherichia coli*. *Infect. Immun.* 69, pp. (6303–6309)

LeClerc, H., Mossel, DAA., Edberg, SC., & Struijk, CB. (2001). Advances in the bacteriology of the coliform group: Their suitability as markers of microbial water safety. *Ann. Rev. Microbiol.* 55, (201-234)
Levantesi, C., LaMantia, R., Masciopinto, C., Böckelmann, U., Ayuso-Gabella, MN., & Salgot, M. (2010). Quantification of pathogenic microorganisms and microbial indicators in three wastewater reclamation and managed aquifer recharge facilities in Europe. *The Sci. Total Environ.* 408, 21, pp. (4923–4930)
Liang, JL., Dziuban, EJ., Craun, GF., Hill, V., Moore, MR., Gelting, RJ., Calderon, RL., Beach, MJ., & Roy, SL. 2006. Surveillance for waterborne disease and outbreaks associated with drinking water and water not intended for drinking — United States, 2003–2004. MMWR Surveillance Summit 2006, 55, pp. (5512-5561)
Lin, J., & Biyela, PT. (2005). Convergent acquisition of antibiotic resistance determinants amongst the *Enterobacteriaceae* isolates of the Mhlathuze River, KwaZulu-Natal (RSA). *Water SA.* 31, 2, pp. (0378-4738), ISSN
Linke, S., Lenz, J., Gemein, S., Exner, M., & Gebel, J. (2010). Detection of *Helicobacter pylori* in biofilms by real-time PCR. *Int. J. Hyg. Environ. Health,* 213, pp. (176-182)
Lleo, MM., Bonato, B., Benedetti, D., & Canepari, P. (2005). Survival of enterococcal species in aquatic environments. *FEMS Microbiol. Ecol.* 54, pp. (189-196)
Lund, V. (1994). Evaluation of *Escherichia coli* as an indicator for the presence of *Campylobacter jejuni* and *Yersinia enterocolitica*in chlorinated and untreated oligotrophic lake water. Abstract HRM37, Water Quality International 94. IAWQ 17 Biennial International Conference, Budapest, Hungary, 24–29 July 1994.
Lynch, MF., Blanton, EM., Bulens, S., Polyak, C., Vojdani, J., & Stevenson, J. (2009). Typhoid fever in the United States, 1999–2006. *J. Amer. Med. Assoc.* 302, 8, pp. (852–869)
Malorny, B., Löfström, C., Wagner, M., Krämer, N., & Hoorfar, J. (2008). Enumeration of *Salmonella* bacteria in food and feed samples by real-time PCR for quantitative microbial risk assessment. *Appl Environ Microbiol,* 74, pp. (1299- 304)
Marcelino, L.A., Backman, V., Donaldson, A., Steadman, C., Thompson, J.R., Preheim, S.P., Lien, C., Lim, E., & Veneziano, D. (2006). Accurately quantifying low-abundant targets amid similar sequences by revealing hidden correlations in oligonucleotide microarray data. *Polz MF. Proc. Natl. Acad. Sci. U. S. A.* 103, 37, pp. (13629-13634)
Matsen, JM., Spindler, JA., & Blosser, RO. (1974). Characterization of *Klebsiella* isolates from natural receiving waters and comparison with human isolates. *Appl. Microbiol.* 28, pp. (672–678)
Maynard, C., Berthiaume, F., Lemarchand, K., Harel, J., Payment, P., Bayardelle, P., Masson, L., & Brousseau, R. (2005). Waterborne pathogen detection by use of oligonucleotide-based microarrays. *Appl. Environ. Microbiol.* 71, 12, pp. (8548-8557)
Meays, C., Broersma, K., Nordin, R., & Mazumder, A. (2004). Source tracking faecal bacteria critical review of current methods. *J. Environ. Man.* 73, pp. (71-79)
Meinersmann, RJ., Berrang, ME., Jackson, CR., Fedorka-Cray, P., Ladely, S., & Little, E. (2008). *Salmonella, Campylobacter* and *Enterococcus* spp.: Their antimicrobial resistance profiles and their spatial relationships in a synoptic study of the Upper Oconee River basin. Microbial Ecology, 55, 3, pp. (444–452)
Miranda, CD., & Zemelman, R. (2001). Antibiotic resistant bacteria in fish from concepción bay, Chile. *Marine Pol. Bulletin,* 42, 11, pp. (1096-1102)

Moganedi, KLM., Goyvaerts, EMA., Venter, SN., & Sibara, MM. (2007). Optimisation of the PCR-*invA* primers for the detection of *Salmonella* in drinking and surface waters following a pre-cultivative step. *Water SA*. 33, 2, pp. (195-201)

Momba, MNB., Tyafa, Z., Makala, N., Brouekaert, BM., & Obi, CL. (2006). Safe drinking water is still a dream in the rural areas of South Africa. Case study: The Eastern Cape Province. *Water SA*. 32, 5, pp. (715-720)

Noble, RT., & Weisberg, SB. (2005). A review of technologies for rapid detection of bacteria in recreational waters. *J. Water Health*, 3, 4, pp. (381-392)

Nogva, HK., Bergh, A., Holck, A., & Rudi, K. (2000). Application of the 50-nuclease PCR assay in evaluation and development of methods for quantitative detection of *Campylobacter jejuni*. *Appl Environ Microbiol*. 6, 9, pp. (4029–36)

Obi, CL., Bessong, PO., Momba, MNB., Potgieter, N., Samie, A., & Igumbor, E.O. (2004a). Profiles of antibiotic susceptibilities of bacterial isolates and physico-chemical quality of water supply in rural Venda communities, South Africa. *Water SA*. 30, 4, pp. (515-520)

Obi, CL., Potgieter, N., Musie, EM., Igumbor, EO., Bessong, PO., Samie, A. (2004b). Human and environmental-associated non-typhoidal *Salmonella* isolates from different sources in the Venda region of South Africa. Proceedings of the 2004 Water Institute of Southern Africa (WISA), Biennial Conference, May 2004.

Ogunjimi, AA., & Choudary, PV. (1999). Adsorption of endogenous polyphenols relieves the inhibition by fruit juices and fresh produce of immuno-PCR detection of *Escherichia coli* O157: H7. *FEMS Immunol Med Microbiol*. 23, 3, pp. (213–20)

Ørskov, I., & Ørskov, F. (1984). Serotyping of *Klebsiella*. *Methods Microbiol*. 14, pp. (143–164)

Parsot, C. (1994). *Shigella flexneri*: genetics of entry and intercellular dissemination in epithelial cells. In: Dangl, J.L. (Ed.), Bacterial Pathogenesis of Plants and Animals. *Curr. Top. Microbiol. Immunol*. 192, pp. (217–241)

Pathak, SP., & Gopal, K. (2005). Occurrence of antibiotic and metal resistance in bacteria from organs of river fish. *Environ. Res*. 98, pp. (100-103)

Percival, SL., & Thomas, JG. (2009). Transmission of *Helicobacter pylori* and the role of water and biofilms. *J. Water Health*, 7, 3, pp. (469-477)

Pergram, GC., Rollins, N., & Esprey, Q. (1998). Estimating the costs of diarrhoea and epidemic dysentery in KwaZulu- Natal and South Africa. *Water SA*. 24, 1, pp. (11–20)

Podschun, R., Pietsch, S., Holler, C., & Ullmann, U. (2001). Incidence of *Klebsiella* species in surface waters and their expression of virulence factors. *Appl. Environ. Microbiol*. 67, 7, pp. (3325-3327)

Prescott, H. (2002). Laboratory exercise in Microbiology. Smith publishers, New York. 5th Ed. pp. (133)

Queralt, N., Bartolomé , R., & Araujo, R. (2005). Detection of *Helicobacter pylori* DNA in human faeces and water with different levels of faecal pollution in the north-east of Spain. *J. Appl. Microbiol*. 98, pp. (889-895)

Reitveld, LC., Haarhoff, J., & Jagals, P. (2009). A tool for technical assessment of rural water supply systems in South Africa. *Phy. Chem. Earth*, 34, pp. (43-49)

Riyaz-Ul-Hassan, S., Verma, V., & Qazi, G.N. (2004). Rapid detection of *Salmonella* by polymerase chain reaction. *Molecul. Cellul. Probes*, 18, pp. (333-339)

Rompré, A., Servais, P., Baudart, J., de-Roubin, M-R., & Laurent, P. (2002). Detection and enumeration of coliforms in drinking water: current methods and emerging approaches. *J. Microbiol. Methods* 49, pp. (31-34)

Roszak, DB., & Colwell, RR. (1987). Survival strategies of bacteria in the natural environment. *Microbiol. Rev.* 51, pp. (365- 79)

Rowe, B., & Gross, RJ. (1984). Facultatively anaerobic Gram negative rods. Genus II. *Shigella*. In: Krieg, NR., Holt, JG. (Eds.), Bergey's Manual of Systematic Bacteriology. Williams & Wilkins, Baltimore, MD, pp. (423–427)

Seo, KH., Valentin-Bon, IE., & Brackett, RE. (2002). Detection and enumeration of *Salmonella enteritidis* in homemade ice cream associated with an outbreak: comparison of conventional and real-time PCR methods. *J Food Prot.* 69, pp. (639–43)

SER. (2002). State of Emergency Report, North West Province, South Africa.

Serre, JL., Taillandier, A., Mornet, E., Simon-Bouy, B., Boue, J., & Boue, A. (1991). Nearly 80% of cystic fibrosis heterozygotes and 64% of couples at risk may be detected through a unique screening of four mutations by ASO reverse dot blot. *Genomics* 11, pp. (1149-1151)

Shabarinath, S., Sanath Kumar, H., Khushiramani, R., Karunasagar, I., & Karunasagar, I. (2007). Detection and characterization of *Salmonella* associated with tropical seafood. *Int. J. Food Microbiol.* 114, pp. (227-233).

Sharma, A., Singh, SK., & Bajpai, D. (2010). Phenotypic and genotypic characterization of *Shigella* spp. with reference to its virulence genes and antibiogram analysis from river Narmada. *Microbiol. Res.* 165, pp. (33-42)

Shuber, AP., Skoletsky, J., Stern, R., & Handelin, BL. (1993). Efficient 12-mutation testing in the CFTR gene: A general model for complex mutation analysis. *Hum. Mol. Genet.* 2, pp. (153-158)

Sidhu, JPS., & Toze, SG. (2009). Human pathogens and their indicators in biosolids: a literature review. *Environ. Int.* 35, pp. (187-201)

Srikantiah, P., Girgis, FY., Luby, SP., Jennings, G., Wasfy, MO., & Crump, JA. (2006). Population-based surveillance of typhoid fever in Egypt. *The Amer. J. Trop. Med. Hyg.* 74, pp. (114–119)

Steinert, M., Hentschel, U., & Hacker, J. (2002). *Legionella pneumophila*: an aquatic microbe goes astray. *FEMS Microbiol. Rev.* 26, pp. (149-162).

Stewart, J.R., Gast, R.J., Fujioka, R.S., Solo-Gabriele, H.M., Meschke, J.S., Amaral-Zettler, L.A., Del Castillo, E., Polz, M.F.,

Straub, TM., Dockendorff, BP., Quinonez-Diaz, MD., Valdez, CO., Shutthanandan, JI., Tarasevich, BJ., Grate, JW., & BrucknerLea, CJ. (2005). Automated methods for multiplexed pathogen detection. *J. Microbiol. Methods,* 62 (3), pp. (303-316)

Syposs, Z., Reichart, O., & Meszaros, L., 2005. Microbiological risk assessment in the beverage industry. *Food Control,* 16, 6, pp. (515-521)

Szákal, D., Schneider, G., & Pál, T. (2003). A colony blot immune assay to identify enteroinvasive *Escherichia coli* and *Shigella* in stool samples. *Diagn. Microbiol. Infect. Dis.* 45, 3, pp. (165–171)

Szewzyk, U., Szewzyk, R., Manz, W., & Schleifer, H. (2000). Microbial safety of drinking water. *Annu. Rev. Microbiol.* 54, (81-127)

Theron, J., Morar, D., Preez, MDU., Brozel, VS., & Venter, SN. (2001). A sensitive semi-nested PCR method for the detection of *Shigella* in spiked environmental water samples. *Water Res.* 35, pp. (869–874)

Till, D., McBride, G., Ball, A., Taylor, K., & Pyle, E. (2008). Large-scale freshwater microbiological study: Rationale, results and risks. *J. Water and Health,* 6, 4, pp. (443–460)

UNDPI. (2005). The International Decade for Action: "Water for Life" 2005-2015. United Nations Department of Public Information.

UNESCO. (2004). Water Programme for Africa, Arid and Water Scarce Zones 2004-2006.

United States Pharmacopeia Convention, Inc. (2001). The United States Pharmacopeia 25. Rockville, M.D.

Villalobo, E., & Torres A. (1998). PCR for detection of *Shigella* spp. in mayonnaise. *Appl. Environ. Microbiol.* 64, 4, pp. (1242–5)

Watson, CL., Owen, RJ., Said, B., Lai, S., Lee, JV., Surman-Lee, S., & Nichols, G. (2004). Detection of *Helicobacter pylori* by PCR but not culture in water and biofilm samples from drinking water distribution systems in England. *J Appl Microbiol.* 97, pp. (690-698)

Watterworth, L., Topp, E., Schraft, H., Leung, K.T., 2005. Multiplex PCR-DNA probe assay for the detection of pathogenic *Escherichia coli*. *J. Microbiol. Methods,* 60, pp. (93-105)

Wesley, IV., & Miller, GW. (2010). *Arcobacter*: an opportunistic human food-borne pathogen? In: Scheld, W.M., Grayson, M.L., Hughes, J.M. (Eds.), Emerging infections 9, ASM Press, Washington, DC, pp. (185–221)

Westrell, T., Ciampa, N., Boelaert, F., Helwigh, B., Korsgaard, H., Chríel, M., Ammon, A., & Mäkelä, P. (2009). Zoonotic infections in Europe in 2007: a summary of the EFSA-ECDC annual report. Euro Surveill. 14 (3) pii: 19100.

WHO. (2001). Guidelines for Drinking-Water Quality, 2nd Ed. Microbiological Methods, vol.1. World Health Organization, Geneva.

WHO. (2003). Emerging Issues in Water and Infectious Disease. Geneva, Switzerland: WHO Press, World Health Organization.

WHO. (2004). Evaluation of the costs and benefits of water and sanitation improvements at the global level. WHO/SDE/WSH/04.04, Geneva.

Wilkes, G., Edge, T., Gannon, V., Jokinen, C., Lyautey, E., & Medeiros, D., et al. (2009). Seasonal relationships among indicator bacteria, pathogenic bacteria, *Cryptosporidium* oocysts, *Giardia* cysts, and hydrological indices for surface waters within an agricultural landscape. *Water Res,* 43, pp. (2209–2223)

Wilton, S., & Cousins, D. (1992). Detection and identification of multiple mycobacterial pathogens by DNA amplification in a single tube. *PCR Methods Applic.* 1, pp. (269-273)

Wose Kinge, C., & Mbewe, M. (2010). Characterization of *Shigella* species isolated from river Catchments in the North West Province of South Africa. *S. Afr. J. Sci.* 106, 11/12, pp. (1-4)

Wose Kinge, CN., & Mbewe, M. (2011). PCR and sequencing assays targeting *mdh* and *gapA* genes for *Escherichia coli* and *Klebsiella* bacteria species identification in river water from the North West Province of South Africa. *Life Science J.* 8, S1, pp. (104-112)

Yang, F., Yang, J., Zhang, X., Chen, L., Jiang, Y., Yan, Y., Tang, X., Wang, J., Xiong, Z., Dong, J., Xue, Y., Zhu, Y., Xu, X., Sun, L., Chen, S., Nie, H., Peng, J., Xu, J., Wang, Y., Yuan, Z., Wen, Y., Yao, Z., Shen, Y., Qiang, B., Hou, Y., Yu, J., Jin, Q. (2005). Genome dynamics and diversity of *Shigella* species, the etiologic agents of bacillary dysentery. *Nucleic Acids Res.* 33, 19, pp. (6445–6458)

Younes, M., & Bartram, J. (2001). Waterborne health risks and the WHO perspectives. *Int. J. Hyg. Environ. Health*, 204, pp. (255-263)

27

PCR-RFLP and Real-Time PCR Techniques in Molecular Cancer Investigations

Uzay Gormus[1], Nur Selvi[2] and Ilhan Yaylim-Eraltan[3]
[1]Istanbul Bilim University, Department of Medical Biochemistry,
[2]Ege University, Department of Medical Biology
[3]Istanbul University, Department of Molecular Medicine
Turkey

1. Introduction

1.1 Polymerase Chain Reaction (PCR)

The polymerase chain reaction (PCR) is a rapid scientific method for generating a 10^6-10^7-fold increase in the number of copies of discrete DNA or RNA sequences (Boehm,1989; Imboden et al,1993). The use of PCR technology has greatly increased the ability to study on genetic material. PCR is a rapid and reliable molecular biology technique that allows quick replication of mainly DNA, the starting material can be a single molecule of rRNA or mRNA. It was developed by Kary Mullis in 1983, and he was awarded the Nobel Prize in 1993. PCR method is useful in situations of limited amount of DNA sample as in forensics, prenatal testing, because it amplifies a single or a few copies of DNA creating millions of copies of the region(1). The ability to quickly produce large quantities of genetic material has enabled significant scientific advances including DNA fingerprinting and sequencing of the human genome. As PCR technology allows taking specimen of genetic material even from just one cell, copy its genetic material several times, this facilitates genetic studies. Currently, besides research purposes, PCR technology is heavily used in diagnosis and patient management especially for viral diseases such as AIDS and hepatitis. Other than detection of infectious organisms, this technology is also useful for determination of genetic polymorphisms or mutations of individuals (Stahlberg,2011).

The method relies on thermal cycles of repeated heating and cooling of the reaction for DNA melting. Double stranded DNA can be disrupted by heat or high pH, giving rise to single stranded DNA. The single stranded DNA serves as a template for synthesis of a complementary strand by replicating enzymes, DNA polymerases. In order to imitate the accelerated form of DNA replication for a gene region, a special form of DNA polymerase is used. This DNA polymerase should be resistant to the thermal denaturation. Most of the PCR applications employ Taq polymerase, an enzyme isolated from the bacterium Thermus aquaticus, but there are some other heat-stable DNA polymerases used by the same purpose. Most polymerases require short regions of double stranded nucleic acid to initiate synthesis. For in vitro PCR reactions, this can be provided by synthetic oligonucleotides of about 21-25 bp that are complementary to the negatice strand of main DNA molecule. Those oligonucleotide sequences are known as 'primer' and chosen due to the DNA region that we want to amplify. In PCR, two synthetic primers that flank the region of interest are used; one primer is complementary to the negative strand of DNA and second primer to the positive strand. The primers must be oriented that DNA synthesis proceeds across the regions defined by the primers. By this way, only a single region of giant DNA molecule can be amplified. As only one amplification is not enough, PCR is a cyclic process to generate 10^6-10^7-fold increase in a gene region; each PCR cycle contains three steps. Those thermal cycling steps are necessary separate two strands in the DNA double helix at a high temperature by a process called DNA melting. There are three main sequentially repeating steps of PCR:

- *Denaturation* of DNA duplex (94-98°C),
- *Annealing* of primers (37-60°C),
- *Extention* (elongation) of primers by polymerase reaction (~72°C)

In the *denaturation* step, the purpose is to separate strand to be ready for replication, denaturation temperature is higher than the other steps. In the *annealing* step, at a lower temperature, each strand is used as templates for DNA synthesis. The selectivity of PCR results from this step by the usage of primers complementary to the targeted DNA region under specific thermal cycling conditions. After this, there is *extention* step continuing by the heat-stable DNA polymerase to amplify the target DNA region (Boehm,1989).

After 20 cycles of amplification, a million copies of DNA can be generated from a single copy. After several rounds of amplification (about 40 times), the PCR product is analysed on an agarose gel an sis abundant enough to be detected with an ethidium bromide stain.

After this stage, to detect the changes on the DNA sequence, the classical PCR-RFLP method (the next heading) can be used. But also specific DNA sequences can be detected without opening the reaction tube (Higuchi, 1992). Recently, after first preliminary studies the technique developed to get both structural and quantitative informations about the amplified DNA region by real-time PCR devices using flourescent dyes, as we will mention in following headings.

1.2 PCR-Restriction Fragment Length Polymorphism (PCR-RFLP) method

In contemporary, there are several forms of PCR that are extensively modified to perform a wide array of genetic manipulations. PCR-RFLP (PCR-restriction fragment length polymorphism) is one of those that was preliminary to most of classified PCR methods. RFLP is a technique referring to a difference between restriction enzyme sites on DNA samples, broken into pieces (digested) by those restriction enzymes and the resulting fragments are separated according to their lengths by gel electrophoresis.

Restriction endonucleases are specific enzymes that can cleave specific nucleotide sequences; because of that property, it is possible for them to discriminate nucleotide changes in DNA. Sometimes they can effect the loci other than the target one, but the important part of the procedure is the possible polymorphism or mutation loci to be detected whether the cleavage site is intact or not. If there is a change in the cleavage site of restriction endonuclease, it will not cleave the site, or by addition of the mutation, there may ocur a previously not existing cleavage site.

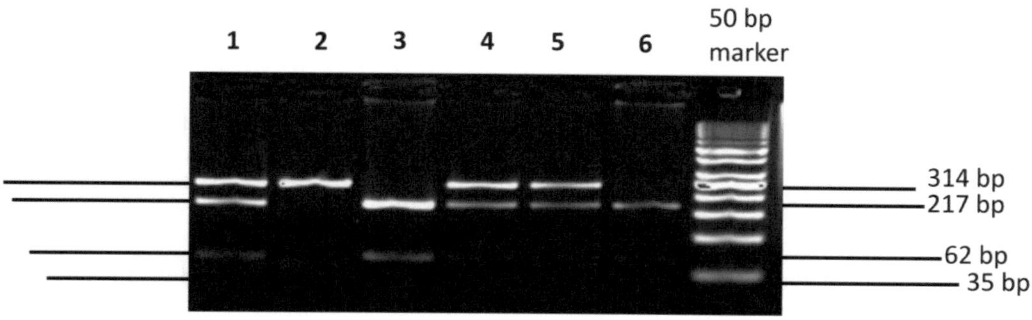

Fig. 1. An example of RFLP results from our laboratories. As shown in the figure, there is a 50 bp marker to compare with our own results and detect the basepair (bp) length. After treatment and incubation with specific restriction enzyme:

- The uncut homozygote cases (having the same alleles) were expected to be having only one 314 bp band (as in number 2).
- The cut homozygote cases (having the same alleles) were expected to be having three bands of 217, 62, 35 bp (as in number 3 and 6)
- The heterozygote cases (having two separate alleles) were expected to be having four bands of 252, 217, 62, 35 bp (as in number 1,4 and 5)

1.3 Types of PCR

- **Conventional PCR:** This is the DNA-based PCR, primers target specific sequences on DNA and amplification follows the usual steps of denaturation, annealing and elongation.
- **Reverse transcription-PCR:** mRNA or rRNA can be the main material to be amplified. The first step is the enzymatic 'reverse transcriptase' reaction to transcribe RNA to cDNA. Subsequent steps are similar to conventional PCR steps (Tania et al, 2006).
- **Asymmetric PCR:** It can be used for generation of single strand for sequencing studies. This can be done by adjusting primer concentrations to favor one strand; by this way after first cycles, only the strand complementary to the first strand continues to be copied.
- **Nested PCR:** In this type of PCR, there are two stages of the procedure; in the first part, by using a set of primers, a fragment is amplified. After this, by using another primer set, a sub-region of the previously amplified region is re-amplified. Main aim is to increase sensitivity and specificity.
- **Real-time PCR.**

2. Real-time PCR

Real-time PCR (PCR with real time) is also known as kinetic PCR, QPCR, QRT- PCR. Automated thermal cycling devices have been improved by using Taq DNA polymerase which is thermostable and continued to be developed by fluorescence luminescence techniques(Higuchi et al,1992; Logan j et al, 2009.). Real-time PCR is easy to perform, providing reliable results with high accuracy as well as rapid quantification. Quantification of polymorphic DNA regions and genotyping single nucleotide polymorphisms are detected by using the real-time PCR reaction. For gene expression analyses, the mRNA levels can be done quantitatively by reverse transcriptase–PCR (RT-PCR) reaction (Tanie Eet al, 2006). By this way, it is possible to monitor gene outputs numerically in many different fields, from the drug-resistant tumor cells to the chemotherapy scanning and also to the molecular determination of tumor stages. The use of gene expression analysis is getting increased in many notable fields of biological research. Gene profiling opens new possibilities to classify the disease into subtypes and guide a differentiated treatment.

This method has been preferred especially in the samples, the analysis of which cannot be possible, or in the samples, the cytogenetic analysis of which are turned out as auxiliary techniques to the molecular analysis. Therefore, it has became one of the indispensable methods. The introduction of real-time PCR technology has significantly improved and simplified the quantification of nucleic acids, and this technology has become a valuable tool for many scientist working in different disciplines. Especially in the field of molecular diagnosis, real-time PCR-based assays took their advantage (Pfaffl, 2004).

2.1 Real-time PCR protocols

Real-time PCR has been preferred as one of the favored methods in molecular studies and in routine analyses, since the process takes short time as 20-30 minutes,it provides fast heating and cooling cycles of 30-40 times, in addition to these, it benefits the control of PCR reaction on a computer monitor (Wittwer, 1997). High sensitivity of real-time PCR makes the technique applicable to very small samples, such as fine needle aspirates. Real-time PCR instruments can simultaneously amplify and detect, eliminating the need to open tubes containing PCR products and therefore reducing the risk of future contamination (Lyon, 2009). Additionally nested PCR and touchdown PCR can be performed using real-time PCR Machine. There are various real-time PCR machines that are used mostly in laboratory experiments:

- ABI Prism 7700
- LightCycler2/Lightcycler 480 Probes (Roche, Mannheim, Germany)
- i-cycler (BioRad)

2.2 Probing techniques

Today, fluorescence is exclusively used as the detection method in real-time PCR. The fluorescent reporters can be divided into two categories: nonspecific and sequence-specific labels (Wilhelm, 2003).

Nonspecific labels: These are DNA-binding dyes such as SYBR Green I (Wittwer et al,1997; Zipper et al, 2004.) and BEBO (Bengtsson et al, 2003), which become strongly fluorescent when they are bound to double-stranded DNA. SYBR Green I binds all double-stranded DNA molecules regardless of their sequence. The Double-stranded DNA bindind dye SYBR Green I is proven to be effective. Maximum excitation of SYBR Green I dye occurs at 497 nm. Maximal emission of DNA stained with SYBR Green I occurs at 521 nm. The specifity and sensitivity of SYBR Green I detection can be monitored by performing a melting curve analysis after using the amplification reaction with external standard.

Differentiation of single point mutant alleles from wild type allele is not possible with SYBR Green I but it is possible to detect small deletions/insertions (10 to 20 bp).

2.2.1 Hybridization probes (pair of sequence-specific, single-labeled probes)

Sequence-specific probes are based on oligonucleotides or their analogs that one or two fluorescent dyes are coupled.

There are some types of probes with two dyes (Holland et al 1991; Tyagi et al, 1996; Tyagi et al, 1998; Caplin et al 1999): a) hydrolysis probes (TaqMan® probes), b) molecular beacons, c) hybridization probes.

a. **Hydrolysis probes:** This probe is a single oligonucleotide labeled with two different fluorophores. The fluorophore near the 3' end(acceptor) acts as a fluorescence emission "quencher" of the other one near the 5' end(donor) (Holland et al 1991). As soon as Taq DNA polymerase hydrolyzes the probe via its 5' exonuclease activity during a combined annealing/extension step, the 5' fluorophore (donor) is liberated. Therefore, its emission can no longer be suppressed by the quencher and can be measured in the fluorimeter. TaqMan real-time PCR is one of the two types of quantitative PCR methods, and uses a fluorogenic probe which is a single stranded oligonucleotide of 20-26 nucleotides and is designed to bind only the DNA sequence between two PCR primers. In this case, two primers with a preferred product size of 50-150 bp, a probe with a fluorescent reporter or fluorophore such as 6-carboxyfluorescein (FAM) and tetrachlorofluorescin (TET) and quencher such as tetramethylrhodamine (TAMRA) covalently attached to its 5' and 3' ends are required, respectively (Giller et al, 2011).
b. **Hybridization probes**: In this case, there are two oligonucleotides that hybridize to adjacent internal sequences of the same amplicon (Witther et al, 2011). For instance, the 5' oligonucleotide (donor) has a fluorescence in label at its 3' end. The 3' oligonucleotide(acceptor) has either LightCycler-Red 640 or LightCycler-Red 705 at its 5' end. Only after hybridization to the template DNA, two probes come in close proximity, resulting in fluorescence resonance energy transfer (FRET) between the two fluorophores. During FRET, fluorescein, the donor fluorophore, is excited especially by the light source of the LightCycler Instrument, and part of the excitation energy is transferred to either LightCycler-Red 640 or LightCycler-Red 705, the acceptor fluorophores. Emitted fluorescence of these acceptor fluorophores are then measured by the LightCycler Instrument. Specific detections are performed with these probes. For example, the mutation detections are analysed via the external and internal standards .
c. **Molecular Beacon Probes:** A molecular beacon is one oligonucleotide labeled with two different fluorophores, an acceptor and a donor. Due to the specific secondary structure formed by the oligonucleotide (beacon), acceptor (quencher) and donor dyes are in close proximity. A molecular beacon unfolds while binding to the growing PCR product, thereby separating the dyes and enhancing the fluorescence of the donor dye. Four different fluorophores can be designed to detect different point mutations simultaneously (Vincent et al, 2005).

2.3 Melting curve analysis

At the beginning of a melting curve analysis, the reaction temperature is low and the fluorescence signal is high. As the temperature steadily increases, the fluorescence will suddenly drop as the reaction reaches the melting point (T_m) of each DNA fragment. More specific analysis of PCR reactions can be performed with SYBR Green I because of its specific melting behaviour, identification/differentiation of multiple specific PCR products (multiplex PCR) with SYBR Green I, genotyping and mutation analyses with hybridization probes. Melting Curve Analysis has many advantages (Wittwer et al, 2009). Just like gel

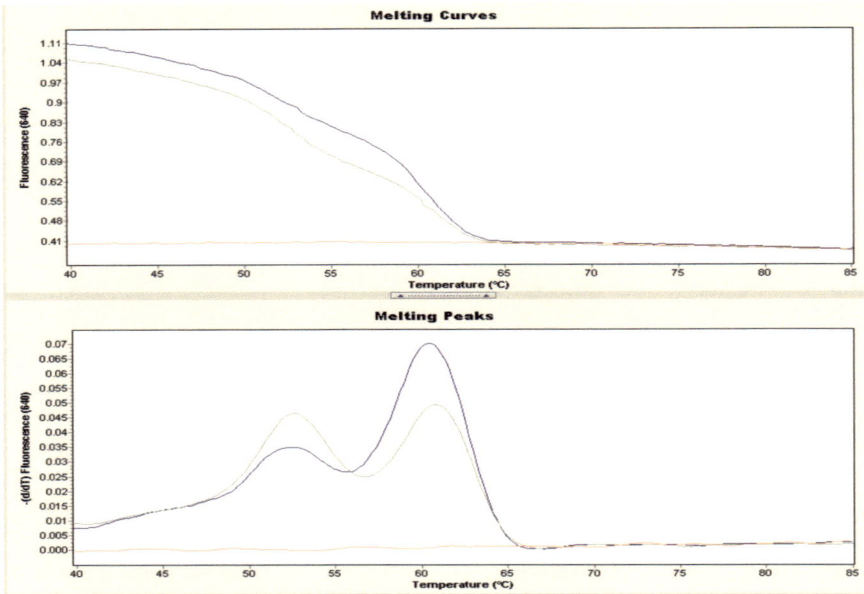

Fig. 2. Heterozygote result indicating two melting curves (53.0 °C and 62.0 °C)

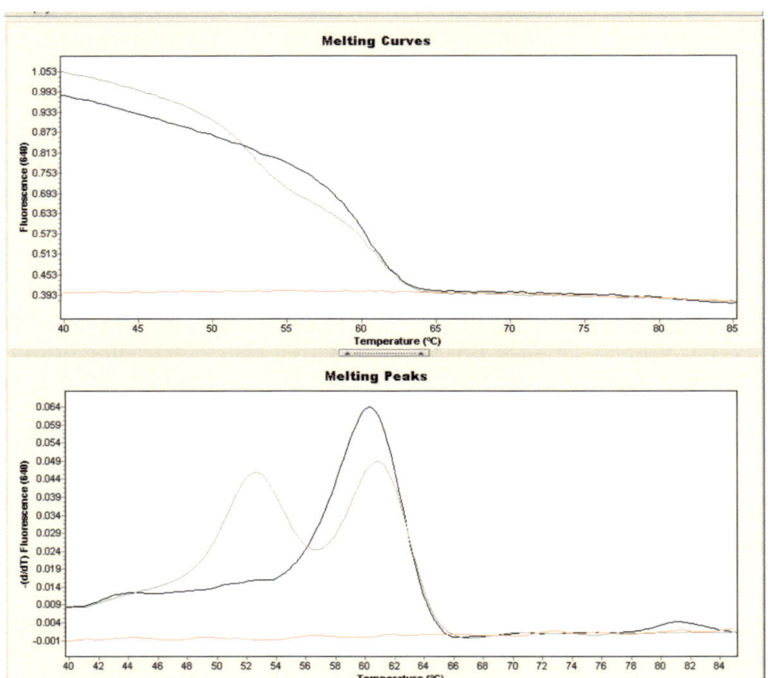

Fig. 3. Wild type result which indicate one melting curve (62.0 °C)

electrophoresis, melting point analysis permits clear identification of the amplicon, since each PCR product possesses a characteristic melting point. Moreover, nonspecific products (primer dimers) can also be identified by this method. If performed with hybridization probes, melting point analysis can also detect point mutations. For instance, the acquired Janus Kinase 2(JAK2) V617F point mutation can be found in more than 90% patient with polycytaemia, and in 50% of patients with other chronic myeloproliferative diseases. For instance in the figures 2-4, our own laboratory results are given. Myeloproliferative neoplasms JAK2V617F-mutation analysis results are shown as melting curve analyses. The genotype is identified by running a melting curve with specific melting points (Tm).

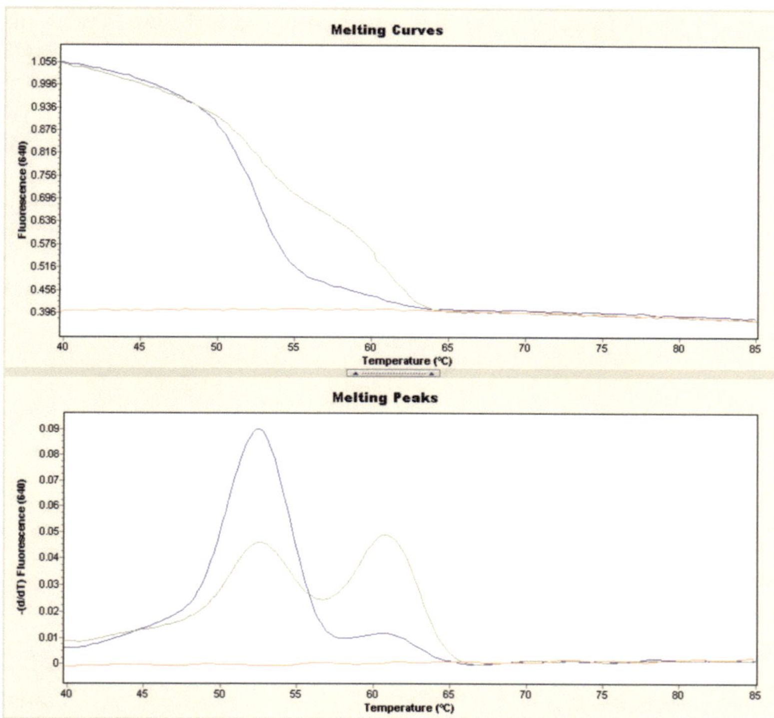

Fig. 4. Mutant result which indicate one melting curve (53.0 °C)

2.4 High-Resolution Melting Analysis (HRMA)

High resolution melting is a post-PCR-based method for detecting DNA sequence variation by measuring changes in the melting of a DNA duplex (Martin-Nunez et al, 2011). Melting analysis using new instruments have been designated for high-resolution melting curve analysis (HRM or HRMA) based on its ease of use, simplicity, flexibility, cost-effectivity, nondestructive nature, superb sensitivity, and specificity (Vossen et al, 2009). It enables researchers to rapidly detect and categorize genetic mutations and single nucleotide polymorphisms(SNPs), identify new genetic variants without sequencing (gene scanning) or determine the genetic variation in a population (e.g. viral diversity) prior to sequencing. SYBR®Green I is introduced into a sensitive conventional dye for PCR product melting analysis. High-resolution melting analysis have been used clinically to detect somatic changes in select exons of oncogenes such as *EGFR*,53 *KRAS*,54 *PDGFRA*,55 *KIT*,56 *BRAF*,57 and*TP53* (Bastien et al, 2008).

2.5 Gene expression analysis

Conventional microarrays have limitations in flexibility, speed, cost, and sensitivity. Gene expression analysis by microarray techniques and real-time PCR offers new possibilities to classify malignant tumors, such as lymphomas, into more distinct subtypes for diagnosis and treatment (Schmit et al, 2010; Bagg et al, 1999; Stahlberg et al, 2005.). The study of biological regulation usually involves gene expression assays and requires quantification of RNA frequently. In the past, conventional gel- or blot-based techniques were used for these assays. However, these techniques often have limitations in speed, sensitivity, dynamic range, and reproducibility required by current experimental systems. In contrast, real-time PCR methods, can easily meet these requirements. Reverse transcription PCR (RT-PCR) is a common and powerful tool for highly sensitive RNA expression profiling. Quantification by real-time PCR may be performed as either absolute measurements using an external standard, or as relative measurements, comparing the expression of a reporter gene with that of a presumed constantly expressed reference gene (Stahlberg et al, 2005).

A flow- chart, represents the steps of Real-time PCR and its applications, is given in Figure 5.

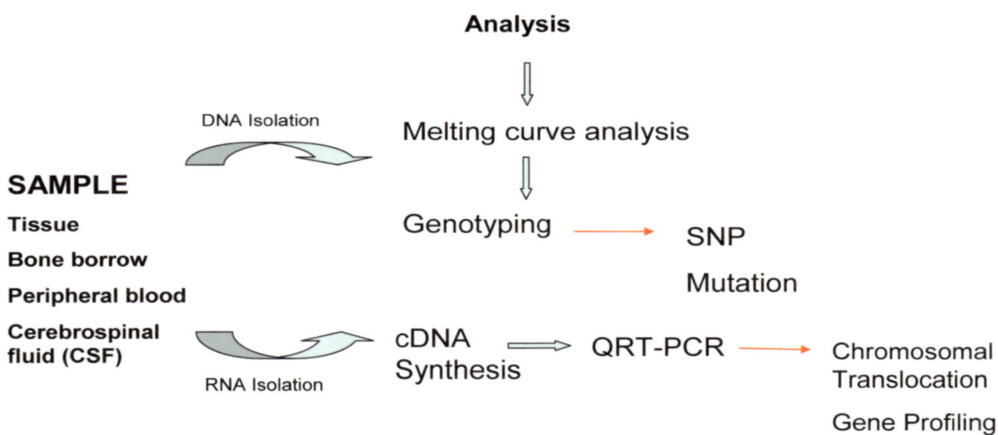

Fig. 5. Real-time PCR Flow-chart

2.6 Epigenetic studies with PCR

Epigenetic information is usually lost during the PCR because of the insensitivity of DNA polymerase, it cannot distinguish between methylated and unmethylated cytosines. After PCR, any methylated allele will be extremely diluted. Therefore, something must be done to preserve methylated form of DNA. Treatment with sodium bisulfite will deaminate cytosine to uracil, the rate of deamination of 5-methylcytosine to thymine is slower than the conversion of cytosine to uracil, thus it is assumed that the only cytosines remaining after sodium bisulfite treatment are derived from 5-methylcytosines. By this way, during subsequent PCR cycles, uracil residues are replicated as thymine residues, and 5-methylcytosine residues are replicated as cytosines. The efficiency of the method is about

99% in appropriate conditions, but this method needs intense attention while chosing primers and arranging study conditions(Gulley et al.).

3. PCR-based studies in cancer research

The advances in molecular techniques provide new molecular targets for diagnosis and therapy of cancer. These advances can provide both researchers and clinicians with precious information concerning the behavior of tumors. Therefore, these tumors can detect at earlier stages when the tumor burden is smaller and be potentially more curable currently. After the human genome project has completed, the application of highthroughput technologies for polymorphism detection for explaining molecular mechanism for complex disease has created very important opportunities (Khoury,1997).

Single nucleotide polymorphisms (SNPs) offers widespread use in gene mapping of genetic disorders, in the delineation of genetic influences in multifactorial diseases such as cancer, cardiovascular disease, in haplotype mapping, and as genetic markers to predict responses to drugs (Riddick et al, 2005). However, for example, there are some inconsistent results regarding the relationship between the presence of polymorphic forms of genes encoded detoxifying enzymes and chemotherapeutic response. It has been reported that the genetic polymorphism analysis in peripheral blood may not be enough representative for the status in tumour tissue. For instance, Uchida et al reported that individuals heterozygous for the 28-bp polymorphism in thymidylate synthase (TS) gene may have increased risk for cancer that are homozygous for this polymorphism due to loss of one allele during carcinogenesis(Uchida et al, 2004). They also showed that the response to 5-FU-based chemotherapy in these cases was comparable to cases where the individual was homozygous. Therefore, it may be excellent to determine the genotype of polymorphisms in tumour cells than in peripheral blood.

Some data obtained from combined genotype studies have demonstrated that these data may have significance for models of cancer prognosis or treatment. But, many researchers suggest that larger studies will be needed also to investigate the effect of specific treatment modalities in cancer. While investigating the post-initiation stages of cancer, four basic parts can be dedicated to gene polymorphisms affecting: (a)growth control of cell (cell proliferation, differentiation and death); (b)factors involved in tumour invasion and metastasis (immune and inflammatory responses, extracellular matrix remodelling, angiogenesis and cell adhesion); (c)effects of hormones and vitamins on growing tumours; (d)outcome of cancer therapy (cancer pharmacogenetics) (Loktionov, 2004). Quantitation of gene expression in tumor or host cells has another an enormous importance for investigating the gene patterns responsible for cancer development, progression and response or resistance to therapy.

Analysis of transcriptional activity of tumor cells or detection of possible new tumor markers by polymerase chain reaction (PCR), reverse transcriptase PCR (RT-PCR) and quantitative reverse transcriptase polymerase chain reaction (qRT-PCR) techniques have the potential to change cancer diagnosis and treatment (Mocellin, 2003). New molecular tecniques for diagnosis offers the promise of accurately matching patient with treatment. It has been shown that there is a resultant significant effect on improved disease outcome. Currently, the real-time reverse transcription polymerase chain reaction (qRT-PCR), has a potential to become an important analytical tecnique for the mRNA detection in tissue biopsies or body fluids. qRT-PCR is especially promising in prognostic assays and monitoring response to treatment for cancer patients. It is known that histopathological staging in cancer defines patient prognosis. However, there are some limitations in the prognostic heterogeneity of patients within a given tumour stage. According to this view, not all patients with lymph node-negative are treated and not all patients with lymph node-positive tumours die from their cancer. So, more accurate staging protocols are needed for detection clinical tumour staging by using molecular techniques.

Gene expression analysis is one of the most important parameter that utilises the qRT-PCR assay's potential for generating quantitative data (Skrzypski, 2008; Schuster et al, 2004). It is reported that the detection of disseminated tumor cells in peripheral blood obtained from colorectal cancer patients by RT-PCR could be an effective method for identifying patients for adjuvant therapy. It is known that the mRNA for prostate specific antigen (PSA) is expressed only by prostatic cells. RT-PCR are suggested as a potentially more sensitive assay for the detection of cells expressing PSA mRNA in peripheral blood or in extraprostatic tissues. Some studies suggest that the molecular detection of circulating tumor cells (CTC) and micrometastases may help develop new prognostic markers in patients with solid tumors (Ghossein et al, 2000). It has been reported that prostatic tissue specific markers and melanoma related transcripts were detected by RT PCR in the peripheral blood, bone marrow and lymph nodes of patients with localized and advanced tumors. Currently, many reliable methods emerged with fast and efficient mechanisms for screening and monitorizing large populations for genetically linked traits and for cancer-related genes discovery.

In addition to gene expression profiling, real-time PCR is also useful to detect chromosomal aberrations. Non-random chromosomal translocations are frequently associated with a variety of cancers, particularly hematologic malignancies and childhood sarkomas (Peter et al, 2006). For example t(15,17) translocation is found only in the leukemic cells. Only in patients with acute promyelocytic leukemia (APL) and the other forms of leukemia, t(1;19) translocation is found with B-cell precursor acute lympoblastic leukemia (ALL). Quantitative analysis provide small number of remaining malignant cells (minimal residual disease, MRD) in patients to be revealed whose disease is in a clinical remission. Therefore, quantitative results are very important in terms of detection in malignancies and MRD. For example, *BCR-ABL* quantification monitors MRD and therapy of chronic myelogenous leukemia (Lyon et al, 2009). Using the real-time PCR Instrument as a closed tube, rapid amplification and real-time fluorescence detection system, for example quantitative measurement of the BCR-ABL expression level can be performed with a minimum risk of cross contamination. Relative expression levels of different samples may be calculated by standardizing the amount of BCR-ABL transcripts in a sample to the amount of an endogenous expressed housekeeping gene. The values for BCR-ABL and housekeeping gene for each sample are calculated by the real-time PCR software by the comparing the crossing points to the standard curve. A normalized target value (the ratio of BCR-ABL/housekeeping) is then derived by dividing the amount of BCR-ABL by the amount of housekeeping gene. The chromosomal aberration examples in various leukemia types can be detected by RNA quantification, shown in Table 1. On the other hand, melting analysis of the PCR product or the probe is used to confirm detection of the correct product.

ACUTE LYMPHOBLASTIC LEUKEMIA (ALL)			ACUTE NON-LYMPHOBLASTIC LEUKEMIA (ANLL)	
t(9;22)	BCR-ABL	Translocation	t(15;17)	PML-RARα Translocation
t(1;19)	E2A-PRL	Translocation	t(8,21)	AML1-ETO Translocation
t(12;21)	TEL-AML1	Translocation	inv (16)	CBFβ- MYH11 Inversion
t(4;11)	MLL-AF4	Translocation	**CRONIC MYELOID LEUKEMIA (KML)**	
Multidrug resistance 1 (MDR1)			t(9;22)	BCR-ABL Translocation

Table 1. The chromosomal aberrations that can be detected by RNA quantification.

4. References

Adam Bagg, Bhaskar Kallakury. Molecular pathology of Leukemia and Lymphpma. *Am.J.Clin.Pathol.* 112, 76-92, 1999.

Alexandre Loktionov. Common gene polymorphisms, cancer progression and prognosis *Cancer Letters* Volume 208, Issue 1, 10 May 2004, Pages 1-33

Anders Stahlberg†, Neven Zoric, Pierre Åman and Mikael Kubista Quantitative real-time PCR for cancer detection: the lymphoma case: *Expert Rev. Mol. Diagn.* 5(2), 2005.

Bastien R, Lewis TB, Hawkes JE, Quackenbush JF, Robbins TC, Palazzo J, Perou CM, Bernard PS: High-throughput amplicon scanning of the TP53 gene in breast cancer using high-resolution fluorescent melting curve analyses and automatic mutation calling. *Hum Mutat*, 29:757–764, 2008.

Bengtsson M, Karlsson JH, Westman G, Kubista M. A new minor groove binding asymmetric cyanine reporter dye for realtime PCR. *Nucleic Acids Res.* 31(8), E45, 2003.

Caplin BE, Rasmussen RP, Bernard PS, Wittwer CT. LightCyclerTM hybridization probes - the most direct way to monitor PCR amplification and mutation detection. *Biochemica* 1, 5–8, 1999.

Corlnne D. Boehm. Use of Polymerase Chain Reaction for Diagnosis of Inherited Disorders. *CLIN. CHEM.* 35/9, 1843-1848 (1989).

Devita VT, Hellman S, Rosenberg SA. (2005). Cancer princibles and Practise of Oncology.7th Edition, 2005.

Elaine Lyon and Carl T. Wittwer. LightCycler Technology in Molecular DiagnosticsJournal of Molecular Diagnostics, Vol. 11, No. 2, 2009.

Ghossein RA, Carusone L, Bhattacharya S. Molecular detection of micrometastases and circulating tumor cells in melanoma prostatic and breast carcinomas. *In Vivo.* 2000 Jan-Feb;14(1):237-50.

Gillet JP, Gottesman MM. Advances in the molecular detection of ABC transporters involved in multidrug resistance in cancer. *Review.Curr Pharm Biotechnol.* 12(4):686-92, 2011

Gulley ML, Shea TC, Fedoriw Y. Genetic tests to evaluate prognosis and predict therapeutic response in acute myeloid leukemia. *J Mol Diagn.* 2010 Jan;12(1):3-16. Epub 2009 Dec 3.

Higuchi, R., Dollinger, G., Walsh, P. S., and Griffith, R. "Simultaneous amplification and detection of specific DNA sequences." *Biotechnology* 10:413–417,1992.

Holland PM, Abramson RD, Watson R, Gelfand DH. Detection of specific polymerase chain reaction product by utilizing the 5′–3′ exonuclease activity of Thermus aquaticus DNA polymerase. *Proc. Natl Acad. Sci.* 88(16), 7276–7280, 1991.

Imboden P, Burkart T, Schopfer K. Simultaneous detection of DNA and RNA by differential polymerase chain reaction (DIFF-PCR). *PCR Methods Appl.* 1993 Aug;3(1):23-7.

Khoury, M. J. (1997) Genetic epidemiology and the future of disease prevention and public health. *Epidemiol. Rev.* 19, 175–180.

Logan J, Edwards K, Saunders N. Real-Time PCR: Current Technology and Applications, 2009.

Martín-Nunez GM, Gómez-Zumaquero JM, Soriguer F, Morcillo S. High resolution melting curve analysis of DNA samples isolated by different DNA extraction methods. *Clin Chim.Acta*, 2011.

Michael W. Pfaffl, Quantification strategies in real-time PCR, Chapter 3 pages 87 - 112, 2004.

Mocellin S, Rossi CR, Pilati P, Nitti D & Marincola FM. (2003). Quantitative real-time PCR: a powerful ally in cancer research. *Trends in Molecular Medicine.* 9, 5, 189-195.

Peter D. Aplan, Causes of oncogenic chromosomal translocation,Trends in Genetics, 22(1), 2006.

Riddick DS, Lee C, Ramji S, Chinje EC, et al. (2005). Cancer chemotherapy and drug metabolism. *Drug Metab. Dispos.* 33: 1083-1096.

Schmit A, Molecular markers in hematology and oncology, 22;99(19):1143-52, 2010.

Schuster R, Max N, Mann B, Heufelder K, Thilo F, Gröne J, Rokos F, Buhr HJ, Thiel E & Keilholz U. (2004). Quantitative real-time RT-PCR for detection of disseminated tumor cells in peripheral blood of patients with colorectal cancer using different mRNA markers. *Int J Cancer.*10, 108, 2, 219-27.

Skrzypski M. (2008). Quantitative reverse transcriptase real-time polymerase chain reaction (qRT-PCR) in translational oncology: lung cancer perspective. *Lung Cancer.* 59, 2, 147-54.

Ståhlberg A, Kubista M & Aman P. (2011). Single-cell gene-expression profiling and its potential diagnostic applications. *Expert Rev Mol Diagn.*11, 7, 735-40.

Tania N, Hands RE & Bustin SA. (2006). Quantification of mRNA using real-time RT-PCR. *Nature Protocols.* 3, 1559-1581.

Tyagi S & Kramer FR. (1996). Molecular beacons: probes that fluorescence upon hybridization. *Nature Biotechnol.* 14, 3, 303–308.

Tyagi S, Bratu DP & Kramer FR. (1998). Multicolor molecular beacons for allele discrimination. *Nature Biotechnol.* 16, 1, 49–53.

Uchida K, Hayashi K, Kawakami K, Schneider S, Yochim JM, Kuramochi H, Takasaki K, Danenberg KD & Danenberg PV. (2004). Loss of heterozygosity at the thymidylate synthase (TS) locus on chromosome 18 affects tumor response and survival in individuals heterozygous for a 28-bp polymorphism in the TS gene. *Clin Cancer Res.* 10, 433–439.

Vossen RH, Aten E, Roos A & den Dunnen JT. (2009). High-resolution melting analysis (HRMA): more than just sequence variant screening. *Huma mutant.* 30, 860-6.

Wilhelm J & Pingoud A. (2003). Real-timepolymerase chain reaction. *Chembiochem.* 4, 11, 1120–1128.

Wittwer CT & Farrar JS. PCR Troubleshooting and Optimization, *Hybridization Probes in PCR*, 2011.

Wittwer CT & Kusukawa N. Real-time PCR and melting analysis. (2009). *Molecular Microbiology: Diagnostic Principles and Practice* (ed 2).

Wittwer CT, Herrmann MG, Moss AA & Rasmussen RP. (1997). Continuous fluorescence monitoring of rapid cycle DNA amplification. *Biotechniques.* 22, 130–138.

Zipper H, Brunner H, Bernhagen J & Vitzthum F. (2004). Investigations on DNA intercalation and surface binding by SYBR Green I, its structure determination and methodological implications. *Nucleic Acids Res.* 32, 12, 103.